THE PUBLICATIONS OF
THE CHAMPLAIN
SOCIETY
LXIX

GENERAL EDITOR
ROGER HALL

Eight hundred copies of this volume have been printed. Seven hundred and fifty have been reserved for members of the Society and subscribing libraries. The balance are reserved for editorial purposes and for sale to the public.

This copy is No.

THE
PUBLICATIONS OF
THE CHAMPLAIN SOCIETY

THE MEANING OF LIFE

THE SCIENTIFIC AND SOCIAL EXPERIENCES
OF EVERITT AND ROBERT MURRAY

1930–1964

TORONTO
THE CHAMPLAIN SOCIETY

Joburg congratulates Robert for winning the first CSM scientific achievement award at the 1963 meetings of the Canadian Society of Microbiology in Guelph, Ontario
LAC, Photographic Division, E.G.D. Murray Collection, E/008406496

THE MEANING OF LIFE

THE SCIENTIFIC AND SOCIAL EXPERIENCES
OF EVERITT AND ROBERT MURRAY
1930–1964

DONALD H. AVERY AND MARK EATON

TORONTO
THE CHAMPLAIN SOCIETY
2008

© The Champlain Society 2008
ISBN 978-0-9810506-0-7
No. 69 in the General Series
Legal deposit third quarter 2008

No part of this publication may be reproduced, stored in a retrieval system or transmitted, in any form or by any means, without the prior written permission of the publisher except by a reviewer who wishes to quote brief passages in connection with a review written for inclusion in a magazine or newspaper or, in case of photocopying or other reprographic copying, a license from ACCESS (Canadian Copyright Licensing Agency), 1 Yonge St., Ste. 1900, Toronto, ON M5E 1E5, fax 416-868-1621.

LIBRARY AND ARCHIVES CANADA CATALOGUING IN PUBLICATION

Murray, E. G. D. (Everitt George Dunne), 1890-1964.
The meaning of life : the scientific and social experiences of Everitt and Robert Murray, 1930-1964 / [edited by] Donald H. Avery and Mark Eaton.

(The publications of the Champlain Society ; 69)
Correspondence of Everitt and Robert Murray, 1930-1964.
Includes bibliographical references and index.

1. Murray, E. G. D. (Everitt George Dunne), 1890-1964 – Correspondence.
2. Murray, Robert, 1919- – Correspondence.
3. Murray, Freda, 1895-1990- – Correspondence.
4. Murray, Doris, 1916-1984 – Correspondence.
5. Bacteriology – Research – Canada – History.
6. Bacteria – Classification.
7. Bacteriology – Canada – History.
8. Bacteriology – History.
9. Bacteriologists – Canada – Correspondence.
10. Microbiologists – Canada – Correspondence.
I. Avery, Donald, 1938-
II. Eaton, Mark, 1974-
III. Murray, Robert, 1919-
IV. Champlain Society
V. Title.
VI. Series: Publications of the Champlain Society ; 69

QR31.M87A3 2009   579.3092'271   C2008-907319-3

Typesetting by ParaGraphics, Toronto
Productions services by Becker Associates

CONTENTS

Illustrations	....	xi
Introduction	....	xiii
Note to the Reader	....	cxvii
Acknowledgements	....	cxxv

CHAPTER ONE	Family and Professional Correspondence: 1930-1939	1
CHAPTER TWO	Scientific Warfare and the Threat of Biological Warfare	87
CHAPTER THREE	Adjusting to Wartime Challenges: 1939-1945	155
CHAPTER FOUR	Family Correspondence: 1946	247
CHAPTER FIVE	Family Correspondence: 1947	347
CHAPTER SIX	Family Correspondence: 1948-1949	451
CHAPTER SEVEN	Family Correspondence: 1950-1951	543
CHAPTER EIGHT	Family Correspondence: 1952-1955	625
CHAPTER NINE	Family Correspondence: 1956-1964	721

Select Bibliography	....	765

APPENDIX A	The Murray Family Profile	773
APPENDIX B	Professional Achievements of Everitt and Robert Murray	776
APPENDIX C	Selection of Freda Murray's Poetry	782
APPENDIX D	Profile of Doris Murray's Painting Career	786

Index	....	789
Champlain Society	....	809

APPENDIX E*	Selected Documents on Scientific, Medical, and Social Issues, 1945-1964	819
APPENDIX F*	Tributes following Joburg's death in July 1964	957
APPENDIX G*	Praise for Robert Murray, 1984-2008	973
APPENDIX H*	Publications of EDGM	979
APPENDIX I*	Publications of RGEM	984

* Available for download from www.champlainsociety.ca/meaning/

ILLUSTRATIONS

FRONTISPIECE vi
Joburg congratulates Robert for winning the first CSM scientific achievement award at the 1963 meetings of the Canadian Society of Microbiology in Guelph, Ontario.

NEWSCLIPPING lxiii
Saving Richard Osnos: Joburg as Medical Celebrity (1935)

SOUTH AFRICA AND THE UNITED KINGDOM . . . . Facing 450
1. The First Medical Murray: George Alfred Everitt Murray in Johannesburg (1892)
2. The Murray residence in Johannesburg in the 1890's
3. Kate Murray and her four sons during the Anglo-Boer War
4. Captain Everitt Murray on "Old Soldier" during his 1916 visit to South Africa
5. Bacteriologist at work: Everitt in the Cambridge Pathology Laboratory during the late 1920's
6. Personalities of the Cambridge Pathology Department (1930)
7. Joburg in his McGill Laboratory (1939)
8. Joburg and Fred Smith: Scientific cooperation from Cambridge to McGill
9. Bob and Joburg share their love of fishing at the Columbus Club, Quebec
10. "Soup's on." Freda and Joburg in their Bark Lake cottage
11. Robert and Susan enjoy the sun at Bark Lake
12. Scientists meet on the eve of war: The Third International Congress for Microbiology, New York City, September 1939

THE MURRAYS IN WAR AND PEACE, 1939-1964 . . . . Facing 642
13. Joburg leaves Biological Warfare Meeting (1944)
14. Joburg captivates his audience at the Suffield BW Station (1947)
15. "Extended honeymoon:" Doris and Robert visit Toronto's Royal Ontario Museum (1944)
16. Robert Murray in his UWO Medical Laboratory, Victoria Hospital (1954)
17. Two outstanding microbiologists: Robert Murray and Carl Robinow (1954)

18 Bob Murray's lecture at the 1970 Congress of Microbiology in Mexico City
19 Joburg receives an honorary degree (D.Sc.) from McGill University (1955)
20 Joburg greets Soviet Virologist Zhdanov at the IAMS 1962 Montreal Congress
21 Former prime minister Louis St. Laurent and Joburg exchange stories at a Quebec City reception (1962)
22 The Murray family and friends enjoy the 1963 CSM Conference in Guelph, Ontario
23 The Murray children pose for a picture in the early 1950's
24 Tom, Peter and Alice celebrate their father's D.Sc. (Hon) award at the May 2007 McGill Convocation

APPENDIX D: OVERVIEW OF DORIS MURRAY'S PAINTING CAREER . . .   786
Cover of the 1985 Exhibit
Doris Marchand Murray, 1916-1984
Doris Murray and her paintings

INTRODUCTION

AN OVERVIEW

Research in the life sciences has had an enormous impact on so many aspects of our contemporary society. Nor is this a new phenomenon. As one commentator observed in 1968, "We are now, though we only dimly realize the fact, in the opening stages of the Biological Revolution – a twentieth century revolution, which will affect human life far more profoundly than the great Mechanical Revolution of the nineteenth century or the Technological Revolution through which we are now passing."[1] The ramifications of these dramatic changes in scientific medical research are evident in many different ways. Above all, the development of antibiotics and antivirals has greatly improved the ability of health care workers to cope with infectious and chronic diseases, while the dramatic 1953 revelations about the structure and functions of DNA and other genetic material have revolutionized our understanding of pathogens and the human genome.[2] Indeed, during the past fifty-five years, there has been a virtual explosion in biotechnology research as scientists moved "to the frontier where the outer edges of genetics, biochemistry, and microbiology were merging alongside a flood of new technologies, such as electron microscopy, crystallography, cell culture, and virology ... and steeply rising capabilities for information storage and analysis."[3]

This volume documents the careers of two remarkable microbiologists who were involved in this new frontier of scientific discovery – Everitt George Dunne Murray (1890-1964)[4] and his son, Robert George Everitt Murray (b. 1919). Their collective experiences cover over eighty years of Canadian microbiology and immunology.

1 G. Rattray Taylor, *The Biological Time Bomb* (New York: New American Library, 1968), 13.
2 The DNA saga was followed by two other major developments: the 1973 discovery of a recombinant DNA technique whereby genetic material was transferred from one organism to another, and the culmination in 2000 of projects to sequence the entire human genome. Gary Zweiger, *Transducing the Genome: Information, Anarchy, and Revolution in the Biomedical Sciences* (New York: McGraw-Hill, 2001), 201-15.
3 The fact that the development of the transistor and the double helix emerged within a decade of each other was fortuitous since "without massive computer power, assembling the fragments of code into a complete human genome sequence would have been beyond the capacities of even an army of clerks." Matt Ridley, "Foreword," in *Inspiring Science: Jim Watson and the Age of DNA*, ed. John Inglis et al. (Cold Spring Harbor, NY: Cold Spring Harbor Laboratory Press, 2003), xviii-xix; Donald Fredrickson, *The Recombinant DNA Controversy, A Memoir: Science, Politics, and the Public Interest 1974-1981* (Washington, DC: National Academies of Sciences, 2001), 11.
4 By the late 1920s most of Everitt's close friends were using his nickname "Joburg," after his birthplace, Johannesburg. Despite the occasional spelling of "Jo'burg," we have consistently used "Joburg." Family members in South Africa tended to call him either "Ever" or "Biff." His grandchildren used the term "Opa" ("grandfather" in German and Dutch).

The senior Murray, for example, was involved primarily with the challenging task of locating, isolating, identifying, and treating bacterial pathogens: he was one of those famous microbe hunters of the early twentieth century.[5] In contrast, while Robert shared his father's background in clinical bacteriology, he increasingly gravitated towards fundamental scientific research specializing in bacteriophage studies, bacterial cytology, and electron microscopy. In a recent interview, Robert Murray reflected on how microbiology has changed since those days when the Murrays were two of Canada's foremost microbiologists:

> 1953 brought the major realization that DNA had a structure that was clearly built for duplication and that the sequence of bases could provide for a code ... It took until the 90's for sequencing to become a reasonably rapid procedure for both RNA and DNA to the extent that we have the complete genome sequence for more than 300 bacteria and more weekly ... the old microbiologists like EGDM would be a bit puzzled and struggle with the concepts but would be glad for us and the solutions to age old problems.[6]

Canada, like many other countries, has recently honoured many of its outstanding medical scientists through the 1994 establishment of the Canadian Medical Hall of Fame (CMHF). Under its mandate, the CMHF is dedicated "to celebrating the accomplishments of Canada's medical and health sciences heroes," through an annual system of nomination and election. At present, there are seventy-one CMHF laureates, chosen on the basis of their contributions in basic research, in applied medical research, or as builders in medicine. After examining the professional backgrounds of the seventy-one laureates, we noted that medical researchers, focusing on infectious diseases, are unrepresented, despite the 2007 election of Félix d'Herelle for his work on bacteriophage.[7] Perhaps other gifted microbiologists will soon follow.

Scientific Dimensions

This biographical study, as portrayed through personal and professional correspondence, starts in 1930 when Everitt Murray, after a distinguished scientific career with the British Medical Research Council and the Pathology Department of Cambridge

---

[5] This term was made famous by Paul de Kruif's famous book, *Microbe Hunters* (London: J. Cape, 1930). There have also been similar contemporary studies such as Hilary Koprowski and Michael Oldstone, eds., *Microbe Hunters – Then and Now* (Bloomington, IL: Medi-Ed Press, 1996). Everitt Murray was an authority on the pathogens causing meningitis and dysentery, and the discoverer of *Listeria*.

[6] University of Western Ontario Archives (UWO Archives), R.G.E. Murray Collection, Robert Murray, "Developing streams affecting biomedical sciences, esp. Microbiology" (hereafter cited as Robert Murray, "Developing streams affecting biomedical sciences, esp. Microbiology"). DNA is an abbreviation for deoxyribonucleic acid, while RNA is ribonucleic acid.

[7] The CMHF is located in London, Ontario. See http://www.cdnmedhall.org.

University, accepted an offer to become chairman of the Department of Bacteriology and Immunology at McGill University.[8] It ends in London, Ontario, after Everitt, having retired from McGill, had joined his son at the University of Western Ontario, as guest research professor in the Faculty of Medicine, a position he held until his death in July 1964.

Everitt and Robert Murray shared a lifelong interest in studying bacteria and placing their various discoveries within an effective classification system. In the case of Everitt, this scientific focus was institutionalized during the late 1930s when he became actively involved with the International Nomenclature Committee of the International Association of Microbiological Societies (IAMS) and the important publication *Bergey's Manual*, which had become the bible for the world's bacteriologists after 1923.[9] Robert shared his father's passion for bacterial taxonomy and made his own unique contribution to this important scientific field. As he describes in his Recollections, "Both Murrays started bacterial practice in clinical microbiology, having to contend with the isolation and identification of bacteria and so had to have a basic interest in classification and, eventually, in taxonomy."[10]

Throughout his Canadian career, Everitt was a leading advocate of the importance of combining clinical observation and laboratory analysis. From his perspective, this was a complementary rather than a divisive relationship, and he thoroughly enjoyed the time he spent examining patients at the Royal Victoria Hospital, observing their symptoms, and taking biological samples back to his laboratory for analysis. Given his clinical interests, it was not surprising that E.G.D. was an active member of the Canadian Public Health Association and a frequent speaker at its conferences

---

8 The medical faculty at McGill dates back to 1824 with the establishment of the Montreal Medical Institution. Its development was strongly influenced by Sir William Osler (1874-84), the generous support of rich patrons such as Lord Strathcona, external assistance from the Rockefeller and Carnegie foundations, and the activities of many eminent medical scientists. Unlike the University of Toronto, McGill received relatively little support from its provincial government. In 1911, for example, the former received a grant of $750,000, compared to the $3,000 McGill obtained. Marianne Fedunkiw, *Rockefeller Foundation Funding and Medical Education in Toronto, Montreal and Halifax* (Montreal and Kingston: McGill-Queen's University Press, 2005), 12, 88.

9 The first edition of *Bergey's Manual of Determinative Bacteriology* appeared in 1923, through the efforts of a number of prominent scientists associated with the Society of American Bacteriologists. Under the editorship of David H. Bergey, three editions of the *Manual* were published (in 1925, 1930, and 1934). In 1936, it was decided to create a separate educational trust for the purpose of publishing subsequent editions and assisting research in taxonomy. The original trustees were David Bergey, Robert S. Breed, and Everitt G.D. Murray. By 2006, nine editions of the *Manual* had been published. See the Web site of the Bergey's Manual Trust, http://www.bergeys.org.

10 UWO Archives, R.G.E. Murray Collection, Robert Murray, "Then and Now" (hereafter cited as Robert Murray, "Then and Now").

and workshops.[11] Closely related was his involvement with veterinary medicine and its various Canadian organizations, along with a complementary interest in Arctic bacteriology.

Another important dimension of their respective careers was the Murrays' strong commitment to improving the quality of research in their discipline, to organizing their fellow microbiologists nationally, and to fostering international cooperation between scientists. As a result, Everitt and Robert were extensively involved with a number of important scientific organizations such as the Canadian Society of Microbiologists (which they founded), the Royal Society of Canada, and the influential Society of American Bacteriologists (SAB), one of the world's largest single life-science membership organizations.[12] In 1954, Everitt came within 147 votes of being elected president of this organization, which he regarded as "a real compliment" since he was running against a prestigious American microbiologist.[13] Robert was more fortunate, being elected SAB president in 1973, while also serving as editor of the Society's journal from 1969 to 1979. At the international level, the Murrays regularly attended meetings of the IAMS, an organization which cut across the ideological divide created by the Cold War. Significantly, when the IAMS met in Montreal in August 1962, Everitt was named president of the Congress, while Robert served as program organizer.

The Murrays also assumed important roles in shaping their respective academic institutions. Between 1930 and his retirement in 1955, Everitt was a force to be reckoned with at McGill University, a situation that was aptly described by J.B. Collip, a friend and long-term colleague:

> The staff of his Department grew to nearly 60 persons, and his Department of Bacteriology and Immunology became recognized as one of the leading centres in Canadian Microbiology ... His dictum that teaching in Bacteriology should be provided wherever and whenever the need arose has led to a close liaison between his Department and many Faculties of the University during his tenure of office.[14]

---

11 Until 1951, the CPHA Laboratory section was a major meeting place for Canadian microbiologists. See R.G.E. Murray, "Memories of the Laboratory Section Meetings: 1940's, '50's and '60's," in *Sixty Years of the Conjoint-Christmas Laboratory Meetings & Its Role in the Development of Microbiology in Canada, 1932-1992*, ed. J. Michael Dixon (Ottawa: Canadian Association for Clinical Microbiology and Infectious Diseases, 1992), 27-29.

12 The SAB, now called the American Society of Microbiology, was founded in 1899 with a membership of fifty-nine scientists. In 2008, it has more than 43,000 members, with over one-third located outside the United States.

13 Joburg to Bob Murray, 7 December 1954. See chapter 8.

14 J.B. Collip, "Professor E.G.D. Murray: An Appreciation," *Canadian Medical Association Journal* 92 (9 January 1965): 95-97.

Less well known was Everitt's long stint on the McGill Senate, where he often defended the principles of free speech and secular education.[15]

In turn, his Senate career led to an extended appointment as one of McGill's three representatives on Montreal City Council between 1947 and 1955. These eight years as City Councillor were enhanced by Everitt's ability to function in the French language and his range of contacts with eminent francophone doctors and academics. Being able to overcome the "two solitudes" provided Everitt with a unique vantage point to comment on political and social developments in Quebec during the postwar years when Mayor Camillien Houde, Premier Maurice Duplessis, and Prime Minister Louis St. Laurent were in their heydays. In 1955, with his retirement from McGill imminent, Everitt reluctantly left civic politics, an experience he had thoroughly enjoyed despite his periodic grumbling about idle talk and wasted time.

Robert Murray continued the family tradition for academic excellence by establishing his own impressive reputation as a teacher, clinician, and researcher. These talents were recognized in 1949 when he became full professor and chairman of the UWO Department of Bacteriology and Immunology at the tender age of thirty. Six years later, Robert was faced with the difficult choice of whether to remain at the University of Western Ontario (UWO) or accept McGill's generous offer to succeed his father as head of the Department of Bacteriology and Immunology. He chose the former for a variety of reasons. High on the list was the fact that he had built his own impressive research operation at UWO; also, his family fully enjoyed the quality of life in southwestern Ontario. An additional factor was the strong possibility that Joburg would soon be appointed visiting research professor at the UWO Collip Research Medical Centre. This took place on 1 July 1955.

Since Everitt and Robert were now UWO colleagues and London neighbours, this virtually ended their decade-long correspondence. Instead, detailed discussions about scientific trends, medical issues, and academic developments were conducted in personal conversations. As a result, this volume has only intermittent Murray family correspondence between July 1955 and Everitt's death on 6 July 1964.

Methodology and Major Themes

The Meaning of Life has many dimensions. Above all, it is a dual scientific biography of two distinguished microbiologists who, during their respective careers, were among the leading practitioners of their discipline, not only within Canada but also on the international stage. In attempting to appreciate the Murrays' scientific status, we addressed a number of important questions. What were their respective

---

15 In 1946 Joburg was one of the major opponents of efforts to create a Faculty of Divinity Studies. In his opinion, such a faculty would further polarize the McGill campus.

individual goals and achievements? How did they influence their Canadian scientific peers, graduate students, and the general public? And what is their individual and collective legacy? In his Recollections, Robert provides an interesting overview of the Murrays' place in Canadian scientific medical history:

> Are we remembered? ... Each of us held positions of record, received awards relevant to a scientific career, had long-term membership in Societies and International Committees. ... Are we still quoted in current papers – not often and, disappointingly, not in recent histories of biology. ... But we helped to put some approaches to structure and function on the map. ... Bergey's Manual will be with us for a while as one marker of the assembly of biological understanding ... We figure in various ways in the fields of bacterial classification even unto the naming of taxa and some bacteria. ... The conclusion has to be that each of us served our times to the best of our abilities, and were amply awarded by a happy engagement with science and appreciative colleagues.
> It has been a great life for both of us.[16]

This project has greatly benefited from Robert Murray's many insights, past and present. In addition, we have drawn on the approaches adopted by other medical and scientific biographies, including those of a number of famous British, American, and European microbiologists.[17] Above all, we were impressed by Michael Bliss's outstanding profiles of Sir Frederick Banting, Sir William Osler, and Harvey Cushing and we profited greatly from his work. Within the Canadian context, we consulted the various volumes of the Canadian Medical Lives series, which examines the life and times of twenty-four medical practitioners.[18] Finally, we were intrigued by the recent biography of Félix d'Herelle (1873-1949), a self-educated medical researcher from Montreal, whose discovery and promotion of bacteriophage therapy made him an international celebrity during the interwar years. The fact that his career intersects with the Murray experience was an added bonus.[19]

16 UWO Archives, R.G.E. Murray Collection, Robert Murray, "Thoughts about where we stand as scientists and significance of our work" (hereafter cited as Robert Murray, "Thoughts about where we stand as scientists").
17 As subsequent footnote references will demonstrate, we examined the career patterns of the following important scientists: Oswald Avery, J.B. Collip, Max Delbrück, René Dubos, Alexander Fleming, Howard Florey, Salvador E. Luria, Linus Pauling, Wilder Penfield, Guilford Reed, Tom Rivers, Ernest Rutherford, Leo Szilard, and James Watson.
18 Thomas Morley was the general editor of this series, and it was funded by Associated Medical Services.
19 The high point in d'Herelle's career was his discovery in 1915 of phage (possibly along with the British scientist F.W. Twort). During the next twenty-four years, he was associated with many of the world's leading centres for medical research: the Pasteur Institute in Paris, the Yale University Medical School, the University of Leiden, and research institutes in Moscow and Georgia in the USSR. In addition, there was a sojourn in India during the 1920s, when he studied the possibility of using bacteriophage to combat cholera, along with Igor Asheshov, a

Our book is, however, quite different from these other studies since it is a documentary collection, not a biographical narrative. Moreover, it deals with two medical scientists, a father and a son, over an extended period of time. And it discusses a wide range of scientific and social topics. Indeed, throughout this volume there is a running commentary about exciting new biological discoveries and scientific controversies, both in the correspondence between the Murrays and in other relevant contemporary documents. Among the subjects under review were the implications of the post-1945 antibiotic revolution, the merits of socialized medicine, Canada's role in the World Health Organization, and the emerging field of bacteriophage research, which contributed to the discovery of the structure and functions of DNA. This latter field also demonstrated the important convergence of different scientific research fields after 1945 – namely, biological cell structure, molecular biology, molecular genetics, biochemistry, and information technology.

On a related front, we explored the ability of the Murrays to shape the microbiology curriculum in the teaching of medical and science students at three universities: the University of Cambridge (1925-30); McGill University (1930-1955); and the University of Western Ontario (1945 to the present).[20] Moreover, during their respective careers, both Murrays were involved in the supervision of many graduate students who subsequently obtained prestigious jobs, hospital appointments, and government positions. In turn, these former students provided glowing accounts of their mentors during their respective academic rites of passage, such as Everitt's retirement from McGill in 1955 and Robert's from the University of Western Ontario in 1984. We have also noted the many tributes and obituaries that followed Everitt's death in July 1964.[21]

Another important theme discussed in this volume was how three generations of Murrays served their country in time of war. This commitment began with George Alfred Everitt Murray (1862-1941), a third-generation resident of Transvaal, South Africa, who provided essential medical assistance to the British military forces during the Anglo-Boer War (1899-1902). Everitt continued the family tradition during the First World War, accumulating an impressive list of medical achievements while serving at home and overseas with the British Royal Army Medical Corps. In World War Two, Everitt, now based at McGill, was extensively involved with defence science,

---

White Russian scientist who would later obtain a position at the University of Western Ontario. William Summers, *Félix d'Herelle and the Origins of Molecular Biology* (New Haven: Yale University Press, 1999).

20 "Microbiology" is used throughout the text, even though "bacteriology" was the more common term prior to the 1950s.

21 These statements are included in the Appendices, along with a detailed list of the scientific publications of Everitt and Robert.

notably his coordination of Canada's biological warfare program throughout most of the war years. Robert Murray's experiences were quite different since he was a medical student until the last year of the war, and his service was limited to homeland defence.[22] During the Cold War, Robert was also fortunate that the Canadian government did not conscript medical researchers, despite periodic discussions of this possibility.

Nevertheless, Robert's professional career development, like that of his father, was disrupted by wartime service. Indeed, there are interesting parallels between Everitt's 1918 decision not to obtain a higher medical degree and Robert's 1945 viewpoint that an MD, rather than a PhD in microbiology, would provide the basis for his scientific career. The fact that both men, at war's end, had family responsibilities was an additional incentive to seek immediate employment.[23]

Family and Social Trends

This book also examines the life and times of three generations of Murray women: Kathleen (Kate) Dunne Murray (1862-1936), Harriet Winifred (Freda) Hardwick Woods Murray (1895-1990), and Doris Marchand Murray (1916-84). Given the chronological scope of this study, however, only passing reference is made to the South African experiences of Kathleen, notably her role during the Anglo-Boer War (1899-1902) and her influence in shaping the careers of her four sons – Everitt, Thorkill, Roger, and Allan.[24] In contrast, there are extensive references throughout the text to the activities of Freda and Doris, both in terms of family issues and in their influence on the professional careers of the principals of this volume – Everitt and Robert Murray.[25]

Freda and her "mate Ever" had many overlapping interests which included scientific debates, academic freedom, social reform, good theatre, and, above all, a love of nature and wilderness adventures. Indeed, shortly after arriving in Montreal in 1930, they built their cottage at Bark Lake in the Laurentian Mountains of Quebec, which became the focus of family life during the summer months. For the Murrays, their version of Canadian "cottaging" was quite unique, consisting of long and arduous canoe trips throughout the Laurentian wilderness that were vividly described in long family letters and personal chronicles.[26] Over the years, Freda celebrated these moments in a series of creative poetry and short stories. Two of these accounts,

---

22 Joburg did not want his son to be involved with Canada's biological warfare operation, since most of the research was classified and therefore would not advance Bob's academic career.
23 Everitt's first child, Bob, was born in 1919; while Alice, Bob's first child, was born in 1945.
24 Information from Marge Shearing, "My Ain Folk: Family Book" (unpublished, 1989), in the possession of Robert Murray.
25 Ibid.
26 Two of the more substantial accounts were "Lakes and White Water (Parts 1-4), and Camp Journal (1948)," and "Canoe Trip 1951."

"Daybreak" and "Reflections," were published in the *Federation of Ontario Naturalists* in 1959 and 1960.[27] In addition, Freda was a committed environmentalist, at a time when that movement had few supporters in Quebec or elsewhere in Canada.[28]

Doris Marchand (1916-84) shared the Murrays' love of science and art, drawing on her own unique family background. Her father was Dr. Richard Werner Marchand, a well-known writer on evolution and a founding member of the Rockefeller Institute of Medical Research branch at Princeton University; while her mother, Dr. Grace Blair Wilkinson Marchand, taught zoology at the college and high-school levels.[29] During the summer of 1938, Doris met Robert Murray when they both attended a summer course in invertebrate zoology at Mount Desert Biological Laboratory in Maine. They were married six years later in Montreal. Doris's interest in and knowledge of scientific matters was a tremendous asset for her husband as he embarked upon his outstanding career in microbiology.[30] With three small children – Alice was born in 1944, Peter in 1946, and Tom in 1949 – Doris's artistic career remained on hold until the mid-1950s. During the next three decades, she gradually established herself as a productive and gifted Canadian painter of watercolours, having eight individual exhibitions and four group showings between 1963 and 1978.[31]

It is fortunate that we could include letters written by Freda and Doris Murray since they provide valuable counter-perspective in our understanding of the scientific careers of the male Murrays. They also give fresh insights into the context of the late 1930s, given the enormous social dislocation of the Great Depression and the imminent threat of another global conflict. During the Second World War these letters discuss issues associated with personal safety, family separation, economic deprivation, and arrested career development. The most extensive correspondence relates to

---

27 Fortunately, these and other writings have been preserved in a special album prepared by Sarah Schmidt, youngest daughter of Susan Murray Robinson, in 1996. This volume also contains a series of fascinating interviews with Freda Murray about her life and times. There is also Robert Murray's thoughtful and sensitive recollection "A 'Mum for All Seasons': The 20's to the 40's," as well as his sister Sue's account, "My Mother Had a Profound Influence on Me." See UWO Archives, R.G.E. Murray Collection.

28 Her firm grip of the subject was evident in 1946 when she prepared a lengthy report, for the Bark Lake Protective Association, on the problems of fish stocks in the various Laurentian lakes. See chapter 4.

29 "Dr. R.W. Marchand, Scientist, Is Dead: Writer on Evolution ... Was Editor and Lecturer ... Made Study of Mosquitoes," *New York Times*, 25 March 1936; Obituary of Mrs. R.W. Marchand, ibid., 14 March 1959. Doris's two brothers were also professionals: Erich Marchand, a physicist at the Eastman Kodak Company, and John F. Marchand, a lawyer in New York City. See also UWO Archives, R.G.E. Murray Collection, "Grace Blair Watkinson Marchand (her own notes)."

30 UWO Archives, R.G.E. Murray Collection, Robert Murray, "Notes on Doris Murray."

31 See Appendix D for more information on the artistic work of Doris Murray.

the many economic and social challenges of the postwar years. Indeed, if space had permitted, this section of the manuscript would have been greatly expanded.

Conclusion

The Meaning of Life explores various dimensions of Canadian medical and scientific history while also revealing many important trends in Canadian social history during the period 1930-1964. The extensive and detailed correspondence between Everitt and Robert sheds new light on major research trends in the life sciences, on major issues affecting Canadian public health, and on the relationship between university scientists and the state, in peace and war.

Within their respective academic environments, Everitt and Robert were able to enjoy the benefits of Canada's scientific reward system. They both chaired major departments of microbiology and immunology and were important power brokers at their respective universities. They also assumed important roles in the Canadian Society of Microbiologists, the Canadian Public Health Association, and the Royal Society of Canada. In addition, they were the recipients of honorary degrees, dedicated scientific Festschriften, and prestigious awards. The most important of these was the Flavelle Medal of the Royal Society for scientific and medical excellence, awarded to Everitt in 1953 and to Robert in 1984.[32]

While this is a historical study, periodic references are made to questions of how the research activities of the Murrays relate to current medical and scientific problems. In the case of Everitt, for instance, there was his important work on the pathogens that were responsible for three serious diseases – meningitis, dysentery, and listeriosis (caused by the bacterium *Listeria monocytogenes*). Significantly, during the summer of 2008 there was a serious listeriosis outbreak in Canada, primarily associated with tainted meat, which resulted in over fifty cases across the country, including thirteen confirmed deaths as of 9 September 2008.[33] Everitt was also in the forefront during the 1950s in examining the incidence of antibiotic-resistant staphylococcal infections,[34] which are among the "superbugs" that have increasingly been found in many Canadian hospitals.[35] While Robert shared many of his father's

---

32 "RSC: The Academies of Arts, Humanities and Sciences of Canada," http://www.rsc.ca. Several of Everitt's friends also received the Flavelle Medal: J.B. Collip (1936), James Hubert Craigie (1942), Guilford Reed (1947), and George Lyman Duff (1956); and also his son, Bob (1984).
33 Public Health Agency of Canada, "Listeria Monocytogenes," Update: September 8, 2008, http://www.phac-aspc.gc.ca/alert-alerte/listeria/listeria_2008-eng.php; *Toronto Star*, 27 August 2008.
34 See Appendix E for an account of Joburg's involvement with the National Research Council's special study of staphylococcal infections in Canadian hospitals.
35 In the United States it was estimated that over 90,000 people died annually from drug-resistant staphylococcus infections. Michelle Shephard, "'Superbug' tied to 90,000 U.S. deaths,"

clinical interests, and was also concerned about antibiotic-resistant pathogens, his research interests gravitated more towards scientific issues. And, during his long and distinguished career he made many important contributions in the research fields of bacteriophage, cytology, and taxonomy. The annual Robert Murray Lectureship at the University of Western Ontario, featuring many of the world's outstanding microbiologists, serves to remind us of Bob's valuable scientific legacy.[36]

In closing, we remind our readers that this book is not a systematic analysis of the intricacies of microbiology or a thorough discussion of the discipline since 1930. Nor do we attempt a detailed examination of the major trends in Canadian scientific medicine or the research role of Canadian universities on public health issues. These challenging tasks require the expertise of those who have the necessary scientific and medical expertise. We are social historians, not microbiologists. On the other hand, through our extensive use of relevant secondary sources, we have attempted to place the Murray correspondence within the appropriate scientific and social context. In terms of organization, the book has three distinct sections: Everitt's career in the United Kingdom; the Murrays' adjustment to Canadian society in the 1930s and to the impact of the Second World War; and the post-1945 period, which witnessed so many changes in Canadian medicine and scientific research in the life sciences.

THE MURRAY FAMILY BEFORE WORLD WAR TWO
The South African Background

Everitt was born into the third generation of Murrays to call South Africa home. His grandfather Walter (1837-1924) had been a successful farmer in the Graaff Reinet district of the Cape Colony, acquiring a major estate, Roodebloem, in 1873, where he lived the rest of his life. His extensive livestock operation was characterized by the use of scientific techniques in developing irrigation systems, in eliminating sheep scab disease from his herds, and in organizing the Zwart Ruggens Farming Association.[37] Walter's wife, Anna Elizabeth Southey (1836-1914), also had strong roots in this emerging country, being closely related to Robert Southey, Poet Laureate of South Africa, and niece of Sir Richard Southey, governor of Griqualand West.[38]

Toronto Star, 17 October 2007.

36 The R.G.E. Murray Lectureship was established in 1999, with a grant of over 1 million dollars. The first invited speaker was Julian Davies, professor emeritus from the University of British Columbia. Other distinguished scientists have followed during the past eight years. See Web site of the Department of Microbiology and Immunology, UWO, http://www.uwo.ca/mni.

37 In 1815 the United Kingdom gained control over the Cape Colony and its many Dutch (Boer) settlers. By 1899, there were four distinct political units: the Cape Colony, Natal, the Orange Free State, and the Transvaal. In 1910, the Union of South Africa was created as a self-governing dominion in the British Empire. In 1961 it became an independent republic.

38 Peter Everitt Murray, "Thurtell/Murray Family History" [unpublished; latest entry, April 10, 1998].

As the eldest son of this achieving family, George Murray (1862-1941) was under pressure to develop a successful professional career. And these expectations were quickly realized. After graduating from Graaff Reinet College with many prizes, he went to England in the early 1880s for his medical education.[39] This took the form of a bachelor of medicine (MB) program at Durham University (1882-84), where he passed his first two L.R.C.P. examinations. This was followed by three years of clinical work at St. Bartholomew's Hospital, where he won the Abernethy Prize in Surgery while also serving as house surgeon. In June 1887 he wrote the final examinations for his MB, earning top marks in the honours category. In November 1887 he passed both examinations of Royal College of Surgeons (R.C.S.), "only the tenth instance in the history of the College [Barts] of the two examinations being passed in one week."[40]

It was at this stage in his career that George met Kathleen (Kate) Dunne, a nursing sister at the hospital. After a short courtship, they decided to get married. This happy event occurred in 1889, in South Africa.

George Murray's decision to return home, despite attractive career options in the United Kingdom, were the result of a combination of strong family ties, professional considerations, and favourable economic conditions. Given the somewhat glutted British medical market, there were many incentives for a young medical graduate to establish himself in the frontier community of Johannesburg, with its booming gold-based economy and the prospect of becoming one of the community's leading medical experts.[41] And this soon materialized, despite the challenging economic, social, and political circumstances.[42]

During the 1890s, the medical situation in the South African colonies, notably the Transvaal, was quite diverse. First, there were many different groups of white medical practitioners: those with university training from the United Kingdom or Europe; those with only clinical experience, often obtained through military service in the British or Colonial armed forces; and those lacking any formal training, who relied on folk remedies or quack medicine.[43] Second, many medical doctors faced severe

---

39 In 1880 George was sent to England for further education, spending his first few years at the home of his great-uncle, the Reverend Alexander Thurtell, Rector of Oxborough. He then began his studies at St. Bartholomew's Hospital. Shearing, "Family Book."
40 Ibid.; Anne Digby, *Diversity and Division: Health Care in South Africa from the 1800s* (New York: Lang, 2006), 122-38.
41 In 1885 there were only about thirty-five doctors in all of Transvaal; five years later there were that many just in Johannesburg, which quickly emerged as a major city. Edmund Burrows, *A History of Medicine in South Africa, Up to the End of the Nineteenth Century* (Cape Town: A.A. Balkema, 1958), 270.
42 According to one historical account, George Murray's "name will always stand in the medical history of Johannesburg." Ibid., 278.
43 The Colonial Medical Council registered and regulated five types of health care professionals:

economic problems as a result of high travel expenses and difficulties in collecting their fees. For the fortunate and well-connected, these problems could, however, be overcome by full-time or part-time public appointments as district surgeons or hospital consultants.[44] A third factor was the impact of the diamond and gold rushes of the late nineteenth century in attracting a number of medical adventurers who were motivated by visions of instant wealth, not adherence to the Hippocractic oath. This group included Lander Starr Jameson (MD, London, 1877), a close associate of Cecil Rhodes, the controversial business magnate and British imperialist.[45] Finally, there were the divisions between English-speaking doctors, with their links to British universities and medical organizations, and doctors of Dutch (Afrikaner) background, generally located in rural areas, who "resented the growing medical and cultural imperialism of the British."[46]

George Murray's medical career in South Africa reflected many of these trends. Among his many assets was the achievement of great distinction in his British medical courses and clinical training, which facilitated his appointment in 1890 as one of five honorary medical practitioners at the newly created Johannesburg General Hospital.[47] He soon became chief of surgery, a position he held until the 1930s.[48] Another major achievement was George's ability to work closely with Afrikaner doctors, such as his colleague John van Niekerk, within both the hospital setting and the broader field of medical activism. In 1898, for example, van Niekerk, as president of the Transvaal Medical Society, hosted the Fifth Congress of the emerging South African Medical Association, which elected as its new president "the much-respected gold fields surgeon, Dr. G.A.E. Murray."[49] On this occasion, many delegates expressed

medical practitioners, dentists, apothecaries, nurses and midwives, and chemists. Digby, *Diversity and Division*, 145.

44 By 1910 as many as one in three doctors in the Cape region had state positions. Ibid., 158.
45 By 1890 Rhodes controlled most of the diamond industry through his De Beers Mining Company, and he was prime minister of the Cape Colony. He was subsequently implicated in the unsuccessful Jameson Raid of December 1895, which attempted to overthrow the Transvaal government of Paul Kruger. Burrows, *A History of Medicine in South Africa*, 257, 274.
46 It was estimated that almost nine out of ten white doctors licensed in the Cape region between 1810 and 1910 obtained their medical training in the United Kingdom, with the remainder coming from Holland and Germany. The ratio was, of course, not as high in the Transvaal and Orange Free State. Digby, *Diversity and Division*, 146, 192.
47 The hospital was officially opened by President Kruger of the South African Republic, with much pomp and ceremony. John van Niekerk, who was of Afrikaner background, was appointed Resident Surgeon. The hospital had 130 beds, but expanded to 300 by 1910. Burrows, *A History of Medicine in South Africa*, 271.
48 George Murray was licensed to practice in 1890. However, even before this occurred, he was asked to appear as an expert witness in the so-called Hospital Scandal in Pretoria. Ibid., 278.
49 There had been a series of congresses throughout the 1890s in an attempt to create a national medical body. In 1896, the Fourth Congress had divided the country into fourteen districts,

confidence that South Africa's medical professionals should be independent from the British Medical Association and instead work together in advancing further medical reforms, so critical "in the shaping of the destinies of a new country."[50]

George Murray's identification with South Africa was reinforced through his strong family loyalties. Above all, there was his close interaction with his parents and his seven brothers and sisters, who were located near Walter's estate, Roodebloem, in Graaff Reinet. For professional and social reasons, George was particularly close to his younger brother Frank (1873-1907), who had followed a similar career pattern in the United Kingdom,[51] completing his university medical studies and then becoming hospital house surgeon at Barts in 1902. Frank had subsequently returned to South Africa, where he became a popular general physician prior to his untimely death in 1907.[52] On her side of the family, Kathleen Murray maintained close contact with her father, Captain John J. Dunne, who was "a larger than life character, a swashbuckler Irishman, soldier of fortune, a famous fisherman, and a bit of a rolling stone."[53] His occasional visits were times of merriment and adventure. In addition, Kathleen communicated with her older sister, Mary Chavelita Dunne, a feminist writer whose books and lifestyle during the 1890s scandalized British middle-class sensibilities.[54]

---

with one delegate coming from each region. Plans were also made to establish the *South African Medical Journal*, which included George Murray on its editorial board. Ibid., 362.

50 Advocates of a separate national medical organization pointed out that only one-third of doctors in the country belonged to the British Medical Association, that distance made the relationship tenuous, and that South Africa had "a rather different spectrum of diseases." Reference was also made to major public health initiatives in the Cape Colony, such as the Medical and Pharmacy Act (1891), the creation of the Colonial Bacteriological Institute (1891), and the establishment of a Department of Public Health. Ibid., 280, 356-65; Digby, *Diversity and Division*, 216.

51 Percy W. Laidler and Michael Gelfand, *South Africa: Its Medical History 1652-1898; a Medical and Social Study* (Cape Town: C. Struik, 1971); Burrows, *A History of Medicine in South Africa*. Despite the difference in their ages, the two brothers shared many activities, including a common love of hunting and horse racing.

52 Frank died after a serious riding accident. He was only thirty-four. Shearing, "Family Book."

53 Prior to the 1890s Captain Dunne and his family spent time in a number of countries – Australia, New Zealand, Chile, the United Kingdom, and the United States – before returning to Dublin, Ireland. In 1875, his wife, Isabel, died, which meant that he and his eldest daughter Chavelita were left to raise the younger children. Among his many interests was a great love of fishing, and his book *How and Where to Fish in Ireland* (date unknown) was a best-seller. UWO Archives, R.G.E. Murray Collection, Robert Murray, "A Personal Narrative" (hereafter Robert Murray, "A Personal Narrative"). This is one of a series of recollections prepared by Robert Murray for this volume. They are all deposited at the UWO Archives. See Note to the Reader, Sources.

54 Chavelita Dunne (1859-1945) was educated in Ireland and Germany, demonstrating talent in languages (she later spoke six) and creative writing. Her father was, however, insistent that she train as a nurse, which resulted in a series of "rebellious" incidents. In 1887, for example, she eloped to Norway with one of her father's friends, Henry Higginson, a married clergyman.

Throughout these years, Kathleen was also busy raising her four sons – Everitt (born in 1890), Thorkill (1892), Roger (1895), and Allan (1901).

Everitt, usually known as "Ever" (or "Biff"), spent his early childhood at the family home, 24 Plein Street in Johannesburg, with occasional sojourns to his father's hunting lodge at Godwan River Estate. In addition, there were visits to Grandfather Murray's estate near Graaff Reinet, where he was exposed "to life in the open veldt, to the magic of knowing about nature, to then abundant wild life … in company of native children who knew more about dangers and snakes etc. and from them they learned at least one native language."[55] He also enjoyed his periodic encounters with his maternal grandfather, Captain J.J. Dunne, whose love of fishing greatly inspired the young Everitt. And, of course, there were his three brothers, whose company he clearly enjoyed, although his family nickname, "Biff," suggests some degree of older-brother pugilism.[56]

The outbreak of the Anglo-Boer War in October 1899 had an enormous impact on the Murray family. During the early stages of the conflict Johannesburg was under Boer control, which meant that the Murrays were regarded with suspicion, along with the rest of the English (*outlander*) population. This situation changed dramatically with the "liberation" of the city in May 1900, which the English community celebrated by raising a monster Union Jack flag, which Kathleen Murray and her friends had secretly constructed. Victory was, however, a distant goal, and the next two years were characterized by vicious guerrilla warfare, with both sides guilty of terrible atrocities. Throughout this bloody conflict George and Kathleen Murray were heavily involved in treating British casualties at the Johannesburg General Hospital and in the temporary military field hospitals. Later, they assumed the duties of Medical Superintendent and Matron on the hospital ship *Norman* as it transported the seriously wounded back to the United Kingdom. For their dedicated

---

During her three years in Norway Chavelita became deeply involved with Henrik Ibsen, August Strindberg, and Knut Hamsun, whose work she translated into English while carrying on a brief romance. In 1891 she married Canadian novelist George Egerton Clairmonte, and they moved to Ireland, where Chavelita completed two books of short stories, *Keynotes* (1893) and *Discords* (1894), which examined women's sexuality, psychology, and wild nature. At this time she adopted her literary name, George Egerton. These publications created a major literary controversy on both sides of the Atlantic: for her admirers she was a crusader for the "New Woman," while her detractors portrayed her as a strange sexual hybrid. In 1895 she was divorced (after many affairs), and six years later she married Golding Bright, a London literary agent. Her only son, George, was killed in World War One. Kimberly Van Hoosier-Carey, "George Egerton," in *Nineteen-Century British Women Writers: A Bio-Bibliographical Critical Sourcebook*, ed. Abigail Burnham Bloom (Westport: Greenwood Press, 2000).

55 Robert Murray, "Then and Now."
56 Within the Murray family there was a juxtaposition between formal names and nicknames: Everitt (Ever, Biff), Thorkill (Thor), Roger (Rory), and Allan (Lal).

public service, Queen Victoria bestowed the medal of Knight of Grace of St. John of Jerusalem on George, while Kathleen received the Lady of Grace of St. John of Jerusalem award.[57]

During the war years, Everitt and his three brothers resided with their grandfather Walter Murray at his Roodebloem estate. These uncertain times, living away from direct parental control and near the veldt, had a definite effect on the four Murray boys. This was particularly true for Everitt, whose love of nature and fascination with zoology were greatly reinforced. After the cessation of hostilities in May 1902, with the family reunited in Johannesburg, another form of education would prevail. Both Kathleen and George were determined that their sons should immediately improve their academic standards, in terms of both scholarly achievement and religious awareness. As a result, in 1906, Everitt and Rory were sent to England to continue their schooling at the Downside Jesuit boarding school. It was an experience that neither enjoyed.

Significantly, after 1906 Everitt would return only once to South Africa, a visit that occurred during the First World War while he was en route from India to England. In contrast, Rory immediately gravitated back home after his graduation from the Woolwich Military College, eventually taking over his father's farm holdings at Godwan River Estate (Ngodwana region) in the Eastern Transvaal.[58] Thor and Lal would likewise spend their entire lives in South Africa, once they had completed their British education.[59]

To what extent did Everitt maintain contact with his family after 1906? And to what degree did he continue to identify with his South African heritage? Superficially, at least, it appears that after 1906 Everitt was rather detached from his parents, siblings, and other Murray relatives. Although there was regular correspondence between himself and Rory, there is no evidence that he maintained contact with his other brothers, in part because of a bitter family feud.[60] Nor did Everitt seem that attached to his mother Kathleen, perhaps because of her insistence that all her sons attend Jesuit boarding schools in South Africa and England, "which probably taught him a lot but he did not enjoy the experience." In fact, as an adult Everitt was a circumspect agnostic.

---

57 Library and Archives Canada (LAC), E.G.D. Murray Papers, vol. 17, E.G.D. Murray to W.R. Le Fanu, Librarian of the Royal College of Surgeons, London, England, 12 February 1942.
58 Lt. Colonel Rory Murray pursued a military career, serving with great distinction during the First and Second World Wars. See chapter 3, note 21.
59 Thor was sent to school in England, and later to Trinity College, Cambridge, to study law. During the First World War he served in Europe with the British armed forces, and then returned to South Africa. He died in 1968, a lonely bachelor. His brother Lal's educational and career background is sketchy, other than his marriage to and divorce from Unita (Nita) Thorburn, who later married his brother Rory.
60 This serious family split occurred when Rory married Lal's former wife, Nita.

Everitt's relationship with his father was more complex. On one hand, he admired George's medical achievements and social prestige. Certainly, according to contemporary accounts, George Murray "was a very handsome man of six foot four, and was respected and beloved throughout South Africa for his kindness and dignity."[61] Nor was it accidental that Everitt chose a medical career, emulating his father's close association with St. Bartholomew's Hospital. But, unlike George, he did not become a surgeon; instead, his chosen profession was bacteriology. There were a number of reasons for this career choice: his reluctance to compete with his father's surgical reputation; his great interest in zoology and natural history; and, above all, exciting new opportunities in the field of bacteriology.[62] Indeed, in 1909, when Everitt began his studies at Christ's College, Cambridge University, many promising young medical scientists gravitated towards this discipline. As one of his contemporaries wrote, "bacteriology had recently made great strides on the continent and was at that time dominating medical thought ... and we younger men welcomed with enthusiasm the opportunity of learning about it."[63]

Unfortunately, there are few surviving letters between George and his son, either during Everitt's period of medical training or during the 1920s when he became a fully qualified bacteriologist. What correspondence does exist relates to the early 1930s, a period when his father reluctantly retired as chief surgeon of the Johannesburg General Hospital, because of growing problems of dementia. But at this stage in his career, Everitt was at McGill University, trying to create a university department, supporting a family on a limited university salary, and, after 1939, becoming heavily involved with wartime research. Indeed, it was during the war, on 5 July 1941, that George Murray passed away, with many accolades. One of the most sensitive came from a medical colleague at the Johannesburg General Hospital: "He was the perfect general practitioner, and a surgeon of great judgement and skill. He was beloved by his patients and, what is perhaps more rare, he was much beloved and respected by all his medical colleagues. In the medical history of Johannesburg, his name will always stand for all that is best and most worthy in the medical profession."[64]

His father's death, it seems, further reduced Everitt's interest in revisiting South Africa, particularly since his former homeland was moving towards extreme institutionalized racism.[65] Yet in many ways Everitt's emotional connection with South

---

61 Excerpt from *Plarr's Lives of the Fellows of the Royal College of Surgeons of England* (London: Simpkin, Marshall Ltd., 1930).
62 Since Everitt never stated his motivation clearly, this line of argument is drawn from various impressions from the documents and from interviews with Robert Murray.
63 Recollections of Dr. Louis Cobbett, cited in Mark Weatherall, *Gentlemen, Scientists and Doctors: Medicine at Cambridge 1800-1940* (Cambridge: Boydell Press, 2000), 182.
64 E.B. Fuller, *South African Medical Journal*, 9,8,1941, cited in Shearing, "Family Book."
65 Other deterrents for visiting South Africa were the fact that Kate Murray died in April 1936,

Africa remained a factor throughout his life.[66] Through his regular correspondence with Rory, he was kept informed about important social and political developments in the country, along with selected family gossip. In terms of his own profile, during the 1920s he earned the nickname "Joburg" on the basis of his Johannesburg background, and it became an important part of his persona. Moreover, he maintained a lifelong interest in South African geography, zoology, and history,[67] and he would sometimes participate in popular celebrations of the 1902 British victory.[68] In addition, there were periodic ceremonial occasions when he represented the South African connection.[69] In October 1944, for example, Everitt requested permission to meet the Earl of Athlone, the new Governor General of Canada, at McGill's fall Convocation, for Athlone had known his father when he was Governor General of South Africa during the 1920s.[70]

On occasion, Everitt was also contacted by South African expatriates who either remembered his father or wanted to praise E.G.D.'s own scientific accomplishments. One such letter was sent in April 1961: "Coming from the South African Institute for Medical Research it was particularly interesting to meet the first white baby born in Johannesburg and to see how well he had done in life."[71] Significantly, in June 1955, Joburg's identification with South Africa was a major theme in the *McGill News* article about his forthcoming retirement:

> His knowledge and love of outdoor life goes back to his boyhood in South Africa, from which he will occasionally tell wonderful red-blooded tales of adventures in hunting,

and that the Graaff Reinet region was a hotbed of Afrikaner right-wing racist sentiment. See Graaff Reinet, "Step Back in Time," http://www.graaffreinet.co.za.

66 His interest in South African medical scientists was evident when he named his 1926 discovery of *Listeria*, in honour of Sir Spencer Lister, a specialist in tropical medicine. There is, however, no record of his response to the awarding of the first Nobel Prize in Physiology or Medicine (1951) to Max Theiler for his work on yellow fever. Robert Murray would become closely associated with another major South African medical scientist, Sydney Brenner, who won this same award in 2002. Robert Murray, interview by Donald Avery, 25 April 2008.

67 On 19 March 1959, historian James J. Talman, chief librarian at the University of Western Ontario, thanked Everitt for his donation of a book on the South-Eastern Transvaal Lowveld, which he claimed was "a real addition to our collection ... (since) we are trying to acquire everything we can bearing on the economic and physical geography of Africa. It is a continent much in the public eye." LAC, E.G.D. Murray Papers, vol. 19, file 1959 Correspondence.

68 In February 1954 the Montreal Branch of Officers of the South African War Veterans Association invited Joburg to speak at their annual dinner "in commemoration of the 54th Anniversary of the Battle of Paardeberg and the Relief of Ladysmith, to be held in our Club room, 1419 Drummond Street." LAC, E.G.D. Murray Papers, vol. 19, file 1954 Correspondence.

69 During the spring of 1944, while he was in England, Joburg had an opportunity to meet South African Premier Jan Smuts. He was quite impressed.

70 See chapter 3.

71 Victor Bokhenheuser, Children's Hospital-Buffalo-NY to Everitt, 30 April 1961. LAC, E.G.D. Murray Papers, vol. 19, file 1961 Correspondence.

in scaring off thieving natives, or in coping with drunken miners – tales which gain a certain mythological grandeur from his insistence that he was by far the smallest, weakest, meekest, and mildest of the males of his large family. These memories are very precious to him and he rather enjoys being addressed by his nickname of 'Jo'burg', which he acquired in his Cambridge days ...[72]

Yet despite periodic reminders of their South African identity, the only member of the Canadian Murray family to visit Roodebloem and the Godwan River Estate (Johannesburg region) was Freda, who embarked upon a four-month excursion during the fall of 1968, when she was in her seventies.[73] Overall, her impressions were very favourable, although she did have one major concern: "I really like this marvelous country and the good people I meet. I must admit that I would not like to be mixed up in any of its politics though: they seem to be the root of much bitterness of feeling & I carefully avoid the subject."[74]

Everitt's Formative Years in the United Kingdom, 1906-1930
The Changing Character of Scientific Medical Research

In 1906 Everitt Murray began his long association with British educational institutions. His initial experience was with Downside School, a Jesuit boarding facility, along with his brother Rory. This was followed by his "little-go" examination for entry to Cambridge University, where he was accepted as a student at Christ's College in 1909. During the next three years Everitt took Natural Sciences Tripos subjects for his medical studies, which included chemistry, physiology, anatomy, and pathology, as well as a zoology elective. After having completed his BA degree, "five years thereafter he applied ... for an MA, which (was) granted." He then entered his father's old hospital, Saint Bartholomew's, for his clinical training in 1911-1912.[75]

St. Bartholomew's (Barts) was one of Britain's foremost teaching hospitals, with a medical legacy that dated back to the twelfth century. But it was during the

---

72 Cited in Shearing, "Family Book."
73 Peter Murray has information about this trip, which we hope will be deposited in the R.G.E. Murray Collection, UWO Archives.
74 UWO Archives, R.G.E. Murray Collection, Freda Murray to Robert Murray, 13 October 1968. Robert Murray has recently commented that, while he regrets not having visited South Africa, he was deterred from going "by the obviously rightist embellishment of observations on the part of Roger Murray (my age and Rory's son) with ... whom I would have to spend some time with if I went to Knysna, a beautiful area where he lived and farmed." UWO Archives, R.G.E. Murray Collection, Robert Murray, "Response to Avery, 19/04/08."
75 During his college days, Joburg spent summer holidays in Normandy learning French and enjoying good wines. UWO Archives, R.G.E. Murray Collection, Robert Murray, "A Personal Narrative: Background, environmental and educational influences for the careers of EGDM (Joburg) and RGEM (Bob)."

nineteenth century, with the gradual development of scientific medicine and pervasive public health standards, that Barts and the other major British teaching hospitals became essential training grounds for a new class of professional physicians and surgeons. This transition from the more traditional forms of medical education produced many intense debates and controversies throughout the period 1880-1920.[76] Above all, there was the question of whether the established medical elites, as represented by the royal colleges of physicians and surgeons, would continue to dominate, or whether they would be supplanted by science-based medical specialists.[77]

In the 1880s, when George Murray had emerged as one of Barts' leading young surgeons, medical education stressed a holistic approach to patient care. Students were expected to have acquired a solid humanities educational background, with a good knowledge of classical studies, before they began their hospital training. Once they were in the medical system, emphasis was placed on the bedside "art" of patient observation and treatment, which, it was argued, was vastly superior to the work of "narrow" specialists in remote laboratories.[78] Yet despite this defensive stance, by the turn of the century even the old-guard traditionalists were forced to recognize the value of medical theories and standardization, and the advantages of using the microscope and other emergent medical technology. This shift towards a more science-based and standardized medical education was evident by the end of the 1890s, when all eleven teaching hospitals in greater London taught the same core medical subjects, with similar modes of instruction.[79] Above all, the focus was on the teaching laboratory. In turn, medical certification boards increasingly demanded greater scientific knowledge, as did the Central Medical Council established in 1892 (later the Medical Research Council).[80]

The distinguished career of Sir William Osler personified this transition towards the new scientific medicine.[81] Osler's experience was unique since he had the advantage of participating in medical education at four universities in three different

---

76 Keir Waddington, *Medical Education at St. Bartholomew's Hospital 1123-1995* (Suffolk: Bury St. Edmonds, 2003).
77 Hubert Lechevalier, *Three Centuries of Microbiology* (New York: Dover Publications, 1974).
78 In this period Barts was often criticized for being too conservative and a resolute defender of the medical status quo. Yet it was the largest teaching hospital, particularly after the 1879 construction of its new buildings. In 1886, for example, of the 439 new entrants to medicine in England, 77 were at Barts, or roughly 17 per cent of the total.
79 There was, however, a certain tension between hospital administrators, who viewed their public ward patients as objects for medical charity, and the lecturers, who increasingly saw them as "vehicles for teaching and research."
80 For a detailed analysis of the role of preliminary and preclinical scientific education, see J.H. Warner, *The Therapeutic Perspective: Medical Practice, Knowledge and Identity in America, 1820-1885* (Princeton, NJ: Princeton University Press, 1986).
81 Michael Bliss, *William Osler: A Life in Medicine* (Toronto: University of Toronto Press, 1999).

countries: McGill University (1874-84), University of Pennsylvania (1884-88), Johns Hopkins University (1888-1905), and Oxford University (1905-19). During the earlier part of his career, Osler advanced the ideal image of the family doctor and the benefits of bedside medicine, which were reinforced by his deeply ingrained religious values.[82] Gradually, however, he embraced the tenets of scientific medicine, particularly after his appointment in 1905 as Regis Professor of Medicine at Oxford University. In his concerted attempts to modernize Oxford's medical educational system, which he described as "resolutely classical and unscientific," Osler championed creating chairs of pharmacology and physiology and streamlining the medical examination system.[83] He also found time to participate in the Haldane Commission on University Reform in 1913, which cited German advances in laboratory medicine as models that British universities should emulate.[84]

The emergence of bacteriology as a separate and important discipline was an integral part of the triumph of scientific medicine. There were a number of reasons for this phenomenon. First, there were the major discoveries by that outstanding scientific trinity: Louis Pasteur in France, Robert Koch in Germany, and Joseph Lister in the United Kingdom. Through their individual and collective achievements, the isolation, diagnosis, and treatment of many bacterial diseases was now possible. Second, there was the development of a more demanding scientific methodology that stressed the value of focusing on specific bacterial pathogens, while at the same time recognizing the complex relationship between the "germ" and the immune system of the human or animal host. Third, under the so-called Koch postulates, a rigorous procedure was initiated to determine the identity of pathogens: detection and isolation from an infected subject; growth in a controlled laboratory environment; and replication of the disease symptoms by using inoculated animal trials. This methodology was increasingly standardized. Fourth, there was the emergence of an expanded international medical research system of cooperation that dated back to the work of the early microbe hunters. Although Pasteur, Koch, and Lister had enormous impact in their respective countries, they insisted that the implications of their research should surmount national boundaries and be shared with life scientists throughout the world. Finally, throughout the early twentieth century there was a strong correlation between the growing status of

---

82 In 1905 Osler published the fifth edition his influential textbook, *The Principles and Practice of Medicine* (the original version had been published in 1872).

83 Osler's choice for the chair of physiology was Charles Sherrington, who won the Nobel Prize for that discipline in 1932.

84 Ibid., 310-16. Osler was quite impressed with the work of German bacteriologists and chemists such as August von Wassermann and Paul Ehrlich for their discovery of the syphilis spirochete and for their Salvarsan chemical treatment.

bacteriology and the development of the public health movement, particularly in the United Kingdom.

Throughout the western world, acceptance of the "germ theories" proceeded at different stages, reflecting the dynamics of specific national medical cultures. By 1900, however, there was a general consensus that most major diseases were caused by bacteria, or by unidentified other pathogens, a form of diagnosis that reinforced the "popular military analogy of invading germs in conflict with the body's defences."[85] In addition, there was general agreement that "laboratory-based ideas and practices were important resources in the reshaping of prevention, diagnosis, treatment and patient management ... [and] ongoing movement of medicine to an expert, science based profession."[86] According to some medical historians, this shift in medical paradigms was much more pronounced in the United States than in the United Kingdom because of "the relative backwardness of British bacteriology and science generally."[87] On the other hand, other scholars have argued that British acceptance of scientific medicine assumed a more incremental pattern, and that earlier initiatives "such as disinfection, isolation, antisepsis, anti-inflamatory remedies and vaccination, were redefined as germ practices after 1865."[88]

Whatever the explanation, by the turn of the century British universities, hospitals, and government health agencies increasingly embraced scientific medicine.[89] Of special importance was the recognition that laboratory investigations provided the basis for standardized analysis, avoiding "the messy clinical reality of an infectious disease, with its variety of symptoms and individual manifestations, its regional, seasonal and environmental variations, and its often unpredictable outcomes." This meant that it was now possible to study the properties of different pathogens, their impact on the host, and even pathological changes in the bacteria themselves through analysis and manipulation within a laboratory setting.[90] According to microbiologist Robert Murray, this focus on pathogenesis meant a greater appreciation of the "interaction between the pathogenic bacteria with the tissues and function of the host ...

---

85 Michael Woroboys, *Spreading Germs: Disease Theories and Medical Practices in Britain 1865-1900* (Cambridge: Cambridge University Press, 2000), 6.
86 Ibid., 7, 14.
87 Ibid., 235; also, see Warner, *The Therapeutic Perspective*, 160-198
88 Woroboys, *Spreading Germs*, 3, 234-38.
89 W.D. Foster, *A History of Bacteriology and Immunology* (London: William Derck, 1970); A.M. Silverstein, *A History of Immunology* (San Diego, CA: Academic Press, 1989); Charles Rosenberg, *Explaining Epidemics and Other Studies in the History of Medicine* (New York: Cambridge University Press, 1992).
90 Olga Amsterdamska, "Standardizing Epidemics: Infection, Inheritance, and Environment in Prewar Experimental Epidemiology," in *Heredity and Infection: The History of Disease Transmission*, ed. Jean-Paul Gaudillière and Ilana Löwy (London: Routledge, 2001), 139.

[because it] concentrated on the recognition and isolation of new or variant pathogenic microbes, on the recognition of toxins."[91] But it was a slow process. Indeed, by the First World War, the level of laboratory experimental technology remained relatively primitive, with all viruses and many groups of bacteria still "beyond the capture of microscopy and the new laboratory techniques."[92]

Given these technical and conceptual limitations, it is not surprising that there was a series of controversies within the evolving fields of bacteriology and immunology. One of these involved the cause and prevention of epidemics during the late nineteenth century, notably those associated with the outbreaks of cholera (1881-82), smallpox (1884-85), and bubonic plague (1896).[93] In each of these epidemics, there were common problems in isolating, identifying, and categorizing pathogens and in determining "their physical, chemical and biological effects on the body." As well, there were questions about the apparent dichotomy between infection and disease among individuals and groups, a paradox that baffled Robert Koch and other great microbe hunters of the nineteenth century.[94] A related set of questions involved the relative importance of pathogens, heredity, and environment in the cause of disease, a debate that extended into the 1920s when there were competing theories proposed by the gifted W.W.C. Topley of the London School of Hygiene and Tropical Medicine and Leslie Webster of the Rockefeller Institute.[95]

Closely related was the emergence of different theories about immunology. Although by 1900 there was a general consensus that the body had a comprehensive defence system against infection, based on the production of antibodies, little was known about the specific process of molecular interaction.[96] In addition, only limited progress had been made in developing effective chemotherapy for dealing with pneumonia, diphtheria, scarlet fever, tuberculosis, or the bacterial pathogens that

---

91 Robert Murray, "Developing streams affecting biomedical sciences, esp. "Microbiology."
92 Woroboys, *Spreading Germs*, 3. In 1914 it was not possible to detect and isolate viruses because of their small size.
93 During the 1881 cholera epidemic the Robert Koch research group in Germany worked closely with the Royal Medical and Chirurgical Society in London. Ibid., 235. In 1896 the British government established a Plague Commission to investigate the India outbreak; the group included many of the leading bacteriologists of that era. Ibid., 233, 235.
94 Andrew Mendelsohn, "Medicine and the Making of Bodily Inequality in Twentieth Century Europe," in *Heredity and Infection*, ed. Gaudillière and Löwy, 21-22.
95 Topley's theories emphasized the importance of studying the effect of epidemics on collectivities rather than individuals, seeing these outbreaks as a herd phenomenon of susceptibility. Amsterdamska, "Standardizing Epidemics," 142-44; Lise Wilkinson and Anne Hardy, *Prevention and Cure: The London School of Hygiene & Tropical Medicine: A 20th Century Quest for Global Public Health* (London: Kegan Paul, 2001), 125-29.
96 Peter Nara, "Humoral Immunity," in *Immunotherapy for Infectious Diseases*, ed. Jeffrey Jacobson (Totowa, NJ: Humana Press, 2002), 3-22.

caused streptococcal and staphylococcal infections. Indeed, despite the activities of German scientists such as Paul Ehrlich with his "silver bullet" chemical treatments, it was not until the late 1930s that the sulpha drugs (sulphanilamide) became available, ushering in a new era in the use of antibiotics.[97]

The early twentieth century also witnessed close interaction between specialists in scientific medicine and public health, at least in Britain, Europe, and North America. In part, this represented a shared appreciation of the value of preventive medicine, which was enhanced by the discovery of disease-causing pathogens and by the development of various immunological strategies. In Britain, for example, the important adoption of the diphtheria antitoxin in 1884 provided "an entirely new weapon in public health medicine, a specific remedy that had the potential to make isolation hospitals curative institutions."[98] These developments were reinforced by a series of legislative initiatives such as the Public Health Act (1875), the Notification of Diseases Act (1889), the London Hospitals Act (1893), and the National Insurance Act (1911).[99] Collectively, these measures institutionalized the process of specialization in medicine and the public's right for adequate health care protection. For public health professionals, they also meant greater recognition and status, while providing increased "opportunities for conflict with general practitioners over the control of patients and confidentiality."[100]

Medical Education at the University of Cambridge, 1900-1914

Almost all of the major developments in scientific medicine found expression at the University of Cambridge prior to the outbreak of the First World War.[101] This is important for our story since Everitt Murray was involved with the university as a medical preclinical student during the years 1909-12, concentrating most of his studies on the

---

97 The mercury-based Salvarsan was perhaps the most notable chemical treatment, given its extensive use against syphilis. Carol Moberg and Zanvil Cohn, *Launching the Antibiotic Era: Personal Accounts of the Discovery and Use of the First Antibiotics* (New York: Rockefeller University Press, 1990).

98 Woroboys, *Spreading Germs*, 237.

99 The National Insurance Act (NIA) provided for the establishment of the Medical Research Committee (Council). Joan Austoker and Linda Bryder, eds., *Historical Perspectives on the Role of the MRC: Essays in the History of the Medical Research Council of the United Kingdom and Its Predecessor the Medical Research Committee, 1913-1953* (New York: Oxford University Press, 1989).

100 By 1911 there was a marked improvement in the pay, tenure, and respect for public health doctors. This was reflected in the creation of the British Institute of Preventive Medicine, the expansion of programs for diplomas of public health, and the establishment of the public health journal, *Journal of State Medicine*. Woroboys, *Spreading Germs*, 238.

101 George Murray's involvement with the bachelor of medicine program (MB) during the 1880s reflected many of the changes that affected medical students at Cambridge during that same time period.

Natural Sciences Tripos. Because of the First World War, however, he did not actually complete his Cambridge bachelor of medicine (MB), but became a qualified medical doctor through another avenue. During the 1920s, Everitt returned to Cambridge, becoming a fellow at Christ's College and lecturer in the Department of Pathology, until 1930 when he joined McGill University.[102] Therefore, an understanding of how the University of Cambridge changed throughout the period 1900-1930 provides useful historical context for this study.

At the turn of the century, Cambridge was still an elitist university, with approximately 3,500 students, many drawn from the prestigious public schools, with about 10 per cent of the total enrolled in the medical program.[103] During the previous three decades the university had experienced a range of administrative reforms and curriculum changes, although the nineteen constituent colleges retained a degree of independence, including an ability to award fellowships.[104] On the other hand, there was an important growth in the influence of the natural science departments – physiology, biochemistry, and pathology – usually in alliance with the influential Regius Professor Physic.[105] This trend was greatly facilitated by the growing status of the Natural Sciences Tripos (NST), which had been created in 1851 with a relatively narrow academic constituency.[106] By 1910, for example, over two hundred male and ten female students, many of whom were "medicals," were enrolled in the NST.[107] As a result, there was a steady influx of students who thronged to Cambridge "for basic science, for medical science, for medical degrees and for rowing[108] ... and then set off,

102 Between 1938 and 1941, Robert Murray was enrolled in the Cambridge preclinical medical program, until the war situation forced his return to Canada.
103 During the 1880s there was much debate about whether increased numbers of medical students were a liability or an asset. On the negative side, there were references to the "medicals" being obsessed with "the narrow necessities of professional training." In contrast, in 1885 the *British Medical Journal* claimed that the arrival of "the hard-working purposeful medical student had added a needed element of steadiness and energy to the undergraduates." Weatherall, *Gentlemen, Scientists and Doctors*, 248-49.
104 Trinity and St. John were two of the largest colleges, having over one-half of the total population in 1900. In addition, there were two women's colleges, Girton and Newnham, which had only limited academic standing until after 1945. Christopher Brooke, *A History of the University of Cambridge*, vol. 4 (Cambridge: Cambridge University Press, 1993), 66-68; Laurence Fowler and Helen Fowler, eds., *Cambridge Commemorated: An Anthology of University Life* (Cambridge: Cambridge University Press, 1984).
105 Between 1870 and 1910 there were Sir George Paget, Sir Michael Foster, and Sir Clifford Allbutt.
106 Founded in 1870, the Cambridge medical school attracted ninety students by 1883, second only to St. Bartholomew's Hospital. Weatherall, *Gentlemen, Scientists and Doctors*, 259.
107 In 1910 the NST represented 18 per cent of the total number of students taking Tripos. By this stage, all were divided into Part I and Part II. Ibid., 157.
108 Ibid., 211.

mostly to London ... [whose] teaching hospitals gathered many of the most successful and most ambitious of the great physicians and surgeons of the century."[109] While this trend benefited the preclinical science departments, it had a disastrous impact on the teaching of clinical medicine at the university and at the affiliated Addenbrooke's Hospital, further intensifying tensions between the two institutions.[110]

During Everitt Murray's student days there were several ways of becoming a university-trained doctor. First, it was necessary to register with the General Medical Council as a medical student, after having completed an examination (usually the "little-go") which required basic knowledge of the classics and mathematics.[111] Second, there was the option of acquiring a bachelor of medicine (MB) at Cambridge by taking the various mandatory and option courses.[112] Under this system, the first MB examinations were in biology, chemistry, and physics, with the second MB examinations focusing on anatomy, physiology, and pathology. An alternative academic route was to take all the necessary preclinical medical courses through the NST, since, it was argued, "exposure to the university's high class scientific minds would make them scientific and successful doctors."[113] Satisfactory clinical work at a certified hospital was the next stage in becoming a licensed medical doctor.[114] This could be an arduous undertaking since the obligations were often numerous and onerous: "to act as in-patient clerks, in- and out-patient dressers, post-mortem clerks, gynaecological clerks; to attend medical out-patients, demonstrations on insanity, a fever hospital, ophthalmic and other special departments; instruction in vaccination; and to deliver at least twenty women in labour, not to mention attending lectures on medicine, midwifery and gynaecology, surgery and surgical anatomy, the use of drugs, medical

---

109 The London hospitals also had large numbers of patients, many suffering from infectious diseases, given the city's unhealthy environment during the nineteenth century. Ibid., 166-68.
110 The selection in 1892 of Sir Clifford Allbutt as Regius Professor Physic was bitterly opposed by many of the leading physicians at the hospital, notably P.W. Latham, the Downing Professor of Medicine. Indeed, it was not until 1900, after Latham's death, that Allbutt was given full access to the hospital facilities for teaching and clinical purposes. Ibid., 166-68.
111 Other options were the London University matriculation examination or the examinations of the Apothecaries' Society or the Royal College of Surgeons.
112 By 1907, students could take the first MB in October of their first year. Later, it was decided that all students would complete six months of ward work before taking the second MB. Weatherall, *Gentlemen, Scientists and Doctors*, 223, 235.
113 Ibid., 252.
114 The term MD within the British medical educational system means advanced standing, based on an original thesis, and is equivalent to a medical PhD in North America. According to Robert Murray this meant being attached "to a particular Professor/Researcher to gain experience in a particular field, and after a number of years (10 for one I know) submit the research papers published and a review of the work in supplication for a doctorate." Robert Murray, "Then and Now."

jurisprudence, mental diseases, and clinical pathology."[115] Once this was completed, there was the option of taking the third MB examinations, or, if sufficiently senior and talented, "to take the Cambridge MD."[116] In addition, there were other less challenging options for becoming a qualified medical practitioner provided through the London Colleges Conjoint Examinations or through the Society of Apothecaries.[117]

The development of bacteriology at Cambridge occurred under the direction of Charles Roy, chair of the Pathology Department, and his able assistant John George Adami, who had previously worked at the Pasteur Institute investigating "whether it was possible to confer immunity against infectious diseases."[118] This line of enquiry was subsequently extended to clinical work at Addenbrooke's Hospital, under Roy's supervision, where "bacteriological investigations were usually done to confirm diagnosis rather than to establish it, but in a few cases bacteriological findings caused the clinicians to change their minds."[119] Unfortunately, support for bacteriology among hospital clinicians remained limited during the nineteenth century, in part because they were "reluctant to take cases of diseases such as diphtheria that were believed to be infectious."[120] The situation improved under Roy's two successors, Alfredo Kantrack (1897-99) and German Woodhead (1899-1921), since both had impressive credentials in bacteriology and public health.[121] Indeed Woodhead was directly involved with the Royal Commission on Tuberculosis (1901) and the 1911 National Insurance Plan, which resulted in a number of important initiatives by the British government in dealing with infectious diseases. This role was expanded during the First World War when Cambridge's Pathological Laboratory was used for testing antitoxins and for investigation "into cerebro-spinal fever ... supported by the Medical Research Committee, work subsequently continued after the war by E.G.D. Murray."[122]

Everitt Murray's Scientific and Medical Career

Did the important changes in bacteriology and scientific medicine prior to 1914 influence Everitt Murray's professional development? Did he believe in the value of a close

---

115 Weatherall, *Gentlemen, Scientists and Doctors*, 161.
116 Ibid., 238.
117 Ibid., 213-15.
118 This partnership ended in 1892 when Adami left to become the chair of pathology at McGill University, a position he maintained for the next thirty years. Marie Adami, ed., *J. George Adami: A Memoir, Together with Contributions from Others* (London: Constable, 1930).
119 Weatherall, *Gentlemen, Scientists and Doctors*, 165.
120 Ibid., 154.
121 Kantrack had previously been director of bacteriology at St. Bartholomew's Hospital, while Woodhead had been director of the Conjoint Laboratories of the Royal College of Physicians and Surgeons. He was also interested in tropical medicine.
122 Weatherall, *Gentlemen, Scientists and Doctors*, 167.

relationship between laboratory analysis and clinical activities, both in the hospitals and the broader public health movement? Did his career as a preclinical medical student at Cambridge demonstrate his potential as a talented bacteriologist?

Overall, the answer to all these questions is affirmative. Certainly there is ample evidence that Everitt shared the views of his scientific contemporaries "that science and medicine must be close allies."[123] In his preclinical program, he followed the more elitist and demanding NST route rather than merely passing the MB examinations, which were often regarded as "regrettable but necessary obstacles in the way of qualification ... (whereas) the training ground of the elite was to be, as always in Cambridge, the tripos."[124] He also appears to have been strongly influenced by those, such as John Ryle, Regius Professor, who emphasized the symbiotic relationship between the university laboratory and the hospital clinic. Under this system, "A stimulus of an idea originating in the ward may well promote fresh enquiry in the laboratory. A method elaborated in the laboratory, may well have essential value in the particular investigation of a disease-process in the ward."[125] In the end, Everitt's academic performance at Cambridge was highly competent, but not outstanding. As his transcript showed: "1909 matriculated, 1912 Tripos in the Natural Sciences, 2nd class N.S.T. ... 1st class marks in Zoo [Zoology]."[126]

This pre-medical educational experience was followed by clinical studies at St. Bartholomew's Hospital, where he took a range of medical courses and worked closely with Mervyn Gordon, a key member of Frederick W. Andrewes's Department of Bacteriology.[127] Everitt's clinical duties at Barts appear to have gone well, suggesting that his lifelong appreciation of the importance of hospital ward work was already well established.[128] But the most important aspect of his Barts experience was the opportunity to focus on bacteriology, an experience that was inspired and facilitated by his mentor Mervyn Gordon, "a most able experimental bacteriologist ... (whose) work

---

123 Brooke, *A History of the University of Cambridge*, 166.
124 Ibid., 260.
125 Weatherall, *Gentlemen, Scientists and Doctors*, 283.
126 LAC, E.G.D. Murray Papers, vol. 17, file Cambridge (1929-30), Christ's College, Cambridge, Memorandum (no date).
127 After studying physiology at Oxford and St. Bartholomew's Hospital, Gordon obtained the degrees of BM (1898), BSc (1901), and DM (1903). Despite his many scientific achievements, he never held a university chair or readership, being content to pursue his various research projects at Barts and develop the clinical bacteriological skills of promising medical students. Lawrence Garrod, "Mervyn Henry Gordon, 1872-1953," *Obituary Notices of Fellows of the Royal Society* 9, no. 1 (November 1954): 153-63.
128 In contrast, the famous physiologist Sir Henry Dale, who attended Barts a decade earlier, regarded his clinical experience as having seriously "delayed my scientific development." W.S. Feldberg, "Henry Hallett Dale 1875-1968," *Biographical Memoirs of Fellows of the Royal Society* 16 (1970): 89.

on the meningococci, the cause of cerebrospinal meningitis, and upon streptococci, was of great practical importance."[129] In addition, Gordon had already established a reputation as a pioneer in virus research for his work on vaccinia and variola (causative agent of smallpox), as well as the virus causing poliomyelitis. Gordon's personal influence on Everitt is evident in a number of ways: by providing a model for dedicated and exacting scientific research; by demonstrating the importance of teamwork within the laboratory; and by his warmth and humour in interpersonal relations.[130] As his biographer writes, "In the years in which I knew him he drew on an apparently inexhaustible supply of anecdotes, told with immense gusto, often Rabelaisian but never offensive."[131] The Gordon-Murray working relationship would become even more intense after the outbreak of the First World War in August 1914.[132]

British Medical Research during the First World War

There is considerable historical debate about whether wartime conditions advance or delay fundamental scientific research.[133] In the case of military medicine, the situation is equally murky. Some scholars have argued, however, that during both world wars there were important medical advances that sharply reduced the number of battlefield deaths from infection and disease. Indeed, this trend was already obvious during the Anglo-Boer War (1899-1902), when there was a sharp decline in the mortality rate from battlefield wounds because of greater use of antiseptic methods and more rapid treatment of casualties. As well, British military officials enjoyed some success in dealing with enteric fever, typhoid, and dysentery.[134] Part of the credit for these medical

---

129 Geoffrey Bourne, *We Met at Bart's: The Autobiography of a Physician* (London: F. Muller, 1963), 25, 174.

130 Everitt also seems to have enjoyed good relations with Sir F.W. Andrewes, an eminent pathologist, as well as with his son, Christopher H. Andrewes, who became one of Britain's leading virologists during the 1930s. The young Andrewes was also actively involved with *Bergey's Manual*. D.A.J. Tyrrell, "Christopher Howard Andrewes, 7 June 1896-31 December 1987," *Biographical Memoirs of Fellows of the Royal Society*, 37 (1991), 38-52.

131 Garrod, "Mervyn Henry Gordon," 159.

132 Everitt would become a licensed medical doctor in 1916. The official citation read: "qualified 1916 (M.D.)" LAC, E.G.D. Murray Papers, vol. 17, file Cambridge (1929-30), Christ's College, Cambridge, Memorandum (no date).

133 See Everett Mendelsohn et al., eds., *Science, Technology and the Military*, 2 vols. (Boston: Kluwer Academic Publishers, 1988); Stuart Leslie, *The Cold War and American Science* (New York: Columbia University Press, 1993); Guy Hartcup, *The Challenge of War: Britain's Scientific and Engineering Contributions to World War Two* (New York: Palgrave, 1970); Guy Hartcup, *The Effects of Science on the Second World War* (New York: Palgrave, 2000).

134 The British Empire mobilized about 250,000 troops for service in South Africa, with some 30,000 coming from the self-governing dominions. In total there were approximately 22,000 deaths, with a ratio of 50 per 1,000. Bill Rawling, *Death Their Enemy: Canadian Medical Practitioners and War* (Ottawa: AGMU Marquis, 2001), 63-101.

improvements was attributed to the actions of the Royal Army Medical Corps (RAMC), created in 1899, which received valuable assistance in South Africa from civilian doctors, such as George Murray.[135] On the other hand, the subsequent British Royal Commission on the South African War produced a long list of criticisms about lack of medical supplies, inadequate treatment facilities, and faulty diagnosis.[136]

The challenge of reducing battlefield deaths was much greater during the First World War.[137] After a few months, the possibility of a short, low-casualty conflict had been dramatically replaced by a grim war of attrition, where machine guns, artillery, and poison gas turned the Western Front into a massive killing ground.[138] Nor were any of the belligerents initially prepared for this scale of battlefield carnage, although remedial measures were soon adopted. In the case of the United Kingdom, for instance, the Royal Army Medical Corps was given a mandate for treating the wounded and preventing disease outbreaks, at home and abroad.[139] This latter role included the development of new vaccines and toxins, as well as trying to understand the pathogenesis of specific disease agents, notably those causing typhoid, meningitis, and dysentery.

While there is a rich body of personal accounts of wartime doctors during the First World War, two case studies reveal some important dimensions about the impact of war on medical professionals. One of these was Frederick Banting, whose involvement with trench warfare included his September 1918 assignment to the No. 13 Canadian Field Ambulance casualty clearing station. In the words of his biographer, this experience had an enormous long-term impact on Banting: "He proved the worth of his training as a doctor, and found it exhilarating to know that he was literally saving men's lives as he tended to their wounds."[140] While not as well known, the military medical experiences of RAMC bacteriologist Captain Everitt Murray also provide an interesting example of how the war affected medical careers and personal development.

In Britain, the task of recruiting bacteriologists was carried out by the RAMC, assisted by a number of well-established medical researchers such as Sir Frederick

---

135 In order to improve British military medicine, in 1902 the government established the RAMC College. John S.C. Blair, *In Arduis Fidelis: Centenary History of the Royal Army Medical Corps*, 2nd ed. (Edinburgh: Lynx Pub, 2001), 18-93. Two years later Canada created the Royal Canadian Army Medical Corps.
136 Deaths from gunshot wounds to the bowel or cranium declined substantially over levels in previous British wars. The incidence of disease was also reduced to about 126 per 1,000 troops between 1899 and 1902. Ibid., 72.
137 Military medicine had also been important during the Spanish-American War of 1898-99. See Vincent Cirillo, *Bullets and Bacilli: The Spanish-American War and Military Medicine* (New Brunswick, N.J.: Rutgers University Press, 2004).
138 Richard Gabriel, *A History of Military Medicine* (Boulder, CO: Greenwood Press, 1992).
139 Blair, *In Arduis Fidelis*, 94-120.
140 Michael Bliss, *Banting: A Biography* (Toronto: McClelland & Stewart, 1984), 41.

Andrewes, Sir Archibald Garrod, Sir David Bruce, and Sir Almroth Edward Wright. The latter, a lieutenant colonel in the RAMC, was often described as "Britain's first academic immunologist," because of his pioneering work on immunity to infectious diseases and his role in the development of an antityphoid vaccine.[141] Indeed, one of Wright's first wartime triumphs was his insistence that all British troops be vaccinated against typhoid, thereby saving countless lives. In 1914, he was also determined to shift his St. Mary's Hospital immunological team to France in order to deal with the high incidence of wound infections, notably those caused by gas gangrene (*Clostridium welchii*) and tetanus. In addition, the Wright group, along with other RAMC units, carried out a range of other important military medical studies, including the effects of chemical antiseptics on wound healing, the problems of shifting the recently wounded back to England, and the challenge of dealing with venereal disease. Wright also proved rather innovative by encouraging close cooperation with French bacteriologists at the Pasteur Institute, despite some opposition from his military superiors.[142]

Overall, the performance of the RAMC and other allied medical initiatives were reasonably successful in reducing the ratio of deaths from battlefield injuries and from war-related infectious diseases. Both ratios declined dramatically in comparison with the earlier Anglo-Boer War.[143] On the Western Front, for example, typhoid cases were fewer than 10 per 1,000; tetanus, 1.47 per 1,000; and dysentery, 0.79 per 1,000. Unfortunately, quite a different situation existed on the Eastern Front, where the British Military Mission to Serbia was overwhelmed by a serious outbreak of typhus in 1915-16.[144] Even worse were conditions in the Middle East, where virulent strains of dysentery caused thousands of casualties among British forces.[145]

Mervyn Gordon and his protegé Everitt Murray were directly involved with the RAMC campaign against infectious diseases. In 1915, at the request of Sir Frederick Andrewes of St. Bartholomew's Hospital, they carried out an extensive investigation of the outbreak of meningitis among recruits gathered in the numerous army camps across the country. After careful analysis they concluded that the disease was

---

141 Frank Diggins, "The Life and Times of Almroth Wright," *Biomedical Scientist* (March 2002); *Oxford Dictionary of National Biography*.
142 One of Wright's staff, Alexander Fleming, provided an important study of the interaction of gas gangrene and streptococcus bacteria, which was published in a 1915 issue of *Lancet*. Fleming would later become famous for his discovery of penicillin in 1928. Kevin Brown, *Penicillin Man: Alexander Fleming and the Antibiotic Revolution* (Sparkford, UK: Sutton, 2004).
143 G.W.L. Nicholson, *Seventy Years of Service: A History of the Royal Canadian Army Medical Corps* (Ottawa: Borealis Press, 1977), 68-113.
144 The Royal Canadian Army Medical Corps also maintained field hospitals in Salonika. Rawling, *Death Their Enemy*, 159-205.
145 Blair, *In Arduis Fidelis*, 204-235.

meningococcal meningitis. Their laboratory studies also revealed that there were three distinct types of the bacterium, which they effectively typed and described in their report, "The Identification of the Meningococcus," published in the Journal of the RAMC in 1915. It was widely praised for its sophisticated methodology and important observations.[146]

In 1916, Everitt's wartime career took another direction when he officially joined the RAMC. By this stage he was a qualified medical doctor after taking his "qualifying examination set by the Society of Apothecaries."[147] Because of his expertise and availability, Everitt was soon transferred to the British army medical laboratory in Basra, Iraq, with a mandate to determine the causes of Shiga dysentery, through an isolation of the bacillus. It was a formidable undertaking. Indeed, during the course of this dangerous research he became ill with the disease, and was sent to Kasauli, India, for recuperation.[148] Once healthy, Everitt was ordered back to the United Kingdom via South Africa, which provided an opportunity to visit his family in Johannesburg. A photograph taken during this joyous occasion reveals a smartly dressed and confident young officer astride a beautiful horse – in sharp contrast with the realities of trench warfare![149]

After his return to England in 1917, Everitt was appointed to the Royal Army Medical Corps College in Millbank "sorting through the collection of cultures of dysentery bacilli (*Shigella* species) to classify them using techniques he had learned on meningococci."[150] As a member of the Vaccine Department at the College, he worked closely with Lt. Colonel David Harvey and the Commandant, the redoubtable Sir David Bruce, whose ability to incite medical controversies was legendary.[151] Never-

---

146 This was the first time there had been a systematic analysis of the different types of meningococcus, which was an important research achievement, not only in understanding this pathogen but also in facilitating the typing of other dangerous germs, notably staphylococcus and pneumococcus. Robert Murray, interview by Donald Avery, 2 February 2008.

147 Robert Murray, "A Personal Narrative." It is not clear why Everitt did not return to Cambridge to complete his MB; perhaps it was wartime conditions which "prevented him from wasting his time, as he thought, on further examinations." G.S. Wilson, "Obituary: Everitt George Dunne Murray 21 July 1890-6 July 1964," *Journal of Pathology and Bacteriology* 91, no.2 (1966): 641.

148 E. Dolev, *Allenby's Military Medicine: Life and Death in World I Palestine* (London: I.B. Tauris, 2007).

149 See photograph in the text relating to the First World War.

150 Robert Murray, "A Personal Narrative."

151 Sir David Bruce (1855-1931) received his medical education at the University of Edinburgh. In 1885 he joined the British Army Medical Service (later the RAMC), and was credited for discovering two deadly pathogens: *Brucella melitensis*, the causative organism for undulant (Malta) fever; and *Trypanosoma brucei*, the cause of sleeping sickness. Although scheduled to retire in 1914, he was appointed Commandant of the RAMC during the war years, although he wanted active service.

theless, Everitt appears to have got along very well with Bruce, and he appreciated his research assignments and medical liaison responsibilities, which included periodic visits to the Pasteur Institute, where he developed a number of lifelong friendships.[152] During the fall of 1918 he also published a series of reports in the *Journal* of the RAMC about his ongoing research on the dysentery bacilli, based on fifty-three separate strains, including the ones he brought back from Mesopotamia. Microbiologist G.S. Wilson described Murray's research methodology and conclusions as follows:

> The methods he used were essentially those taught him by Gordon, namely fermentation of the sugars and agglutination and absorption of agglutinins. The classification he reached corresponds closely with that which we now recognize ... His class 1 included the Shiga strains, his class 2 the Schmitz strains, his class 3 the Flexner strains, and his class 4 what were later referred to as Sonne strains. He paid great attention to technique, used the macroscopic method of agglutination, worked with living organisms, standardized his suspensions by opacity, incubated his tubes for 24 hr at 55 (degree) C in a moist atmosphere, and expressed the degree of agglutination by a simple system of pluses.[153]

Being based at the RAMC College in the greater London region provided an opportunity of resuming his courtship of Freda Woods, now a nurse in the Volunteer Aid Detachment (VAD) program.[154] In March 1918 they were married, and a year later Robert was born. Not surprisingly, these were challenging times given the uncertainty of the war, the challenges of finding adequate housing, and the difficulties of reconciling family pressures and priorities. Because of Everitt's surviving letters, it is possible to understand some of the problems this newly married couple had to overcome.

On the positive side, there was George Murray's March 1918 letter of congratulations about their wedding, along with his descriptions of the activities of Thor and Rory, who were temporarily in Johannesburg, after spending most of the war period in England and the Western Front.[155] In contrast, there was the grim news that Captain Thomas Cecil Hardwick Woods, Freda's only brother, had been killed on the Western Front. Consumed by grief, Freda temporarily sought sanctuary in the Spalding region of southeastern England, far away from the war effort.[156] This left Everitt

---

152 An extensive account of Everitt's recollections about his activities at the RAMC College, his impressions of Sir David Bruce, and his visits to Paris is included in his 9 March 1962 speech to the Osler Society, University of Western Ontario. See chapter 1.
153 G.S. Wilson, "Everitt George Dunne Murray," 642-43.
154 Freda's location in London was based, in part, on her attempts to become independent of her family, who had sold the family home without any consultation. This was also the focal point of medical assistance. Robert Murray, interview by Donald Avery, 25 April 2008.
155 UWO Archives, R.G.E. Murray Collection, "Letters from Ever, 1918," compiled by Sarah Schmidt, George Murray to Joburg, 28 March 1918.
156 In early May, Everitt received a letter from Freda's mother about her son's death, which expressed satisfaction that Everitt's work "keeps you in England."

in London, corresponding daily with his beloved "mate,"[157] while also trying to deal with his mother, Kathleen Murray, who had come to England in order to ensure the safety of her three sons.[158] Predictably, her relations with Everitt were stormy!

Another set of themes deals with Everitt's various research projects at the Royal Army Medical College. These included consultations with bacteriology experts at St. Bartholomew's Hospital and Cambridge University, where he spent several enjoyable days in early May 1918. The results were mixed. On one hand, he was grateful that "the Cambridge atmosphere has stimulated my mind to think of 6 experiments to do that may help me out of my difficulty."[159] On the other hand, despite his meetings with Professors Nuttall, Marshall, Shipley, and Corbett, there were no easy answers to his research problems: "Anyway, there must be a way of persuading the little microbes to give up their secret. The trouble is I don't know anything like enough & I must set about learning some more." This awareness of his own scientific limitations did not deter Everitt. Instead, he was convinced that, after the war, he should pursue an advanced degree, the doctor of science (DSc), which required five years of study after an MA. "It is only given on first class work ... If I had taken my M.A. in 1915 when I became eligible, I should now have 3 of those years behind me; but that is a chance missed. Now it resolves itself into a question of money & consequently must wait. Though I have not done enough yet to get the Dr.S. I hope I can do so during the next 5 yrs. In any case we will have a good try."[160]

In 1919, Everitt was discharged from the RAMC, a process that included his appointment as an Officer of the Order of the British Empire (Military Division). It was a great honour! Shortly afterwards, he obtained a temporary position as a senior demonstrator of pathology at St. Bartholomew's Hospital, before accepting a position with the British Medical Research Council, first in London and then in the Cambridge area.[161] By the mid-1920s Everitt was finally in a position to pursue his great ambition of becoming a university professor of bacteriology.

In many ways Everitt's wartime experiences were similar to those of other British bacteriologists of his generation. First, they were given the responsibility and

---

157 The term "mate" was a constant form of endearment in the correspondence between Freda and Everitt throughout their lives.
158 There were several incidents involving Kathleen's complaints that she was being ignored, and Everitt's curt response. He did, however, take her to dinner on a number of occasions. Joburg to Freda, 8 May 1918.
159 Joburg to Freda, 4 May 1918. While en route, Everitt had met with Professor Andrewes at Barts, having "a long & arduous discussion." Joburg to Freda, 3 May 1918.
160 Everitt also enjoyed revisiting Christ's College, and was delighted that most of the remaining staff recognized him. He was, however, appalled that his old rooms were full of army cadets and were in "a frightful state." Joburg to Freda, 5 May 1918.
161 Ibid.

resources to analyze a number of infectious diseases and to propose innovative forms of treatment. For the gifted W.W.C. Topley, for example, his RAMC assignment included fieldwork in Serbia analyzing the etiology of typhus;[162] while W.J. Tulloch's RAMC duties meant specializing in tetanus wound infections and the treatment of cerebrospinal infections while a member of the RAMC College.[163] The involvement of these young bacteriologists in disease and infection control was often accompanied by a strong sense of personal commitment and frustration; as Lt. Colonel Almroth Wright explained, "it makes me unhappy to see those wounded boys lose their limbs or their lives through infection, which could if we had the knowledge be cured."[164] A second trend was the strong bond between those bacteriologists who served with the RAMC, which continued into the interwar years. A third factor was the continuity between their wartime activities and the types of research many of them carried out during the 1920s. In the case of Everitt Murray, for instance, there was "a life interest in Neisseria meningitis, in enteric diseases particularly Shingella dysentery, in the identification of bacteria ... and in bacterial classification and taxonomy."[165]

Another traumatic event affecting British bacteriologists was the outbreak of Spanish influenza (1918-19).[166] This devastating global pandemic killed upwards of 50 million people, with about 240,000 of these deaths occurring in the United Kingdom.[167] The situation was further aggravated by the ineffectiveness of treatment, the

---

[162] When war broke out, W.W.C. Topley (1896-1944) was working at the Institute of Pathology at Charing Cross Hospital in London. In 1915, as a member of the RAMC, he was attached to the British Military Sanitary Expedition to the Serbian Forces, and collected many specimens during a severe outbreak of typhus. After returning to his London laboratory, he carried out a range of experiments that convinced him of the importance of "the bacteriological aspects of preventive medicine." His paper "The Spread of Bacterial Infection," published in *Lancet* in 1919, was based on his wartime research with the spread of typhus. Wilkinson and Hardy, *Prevention and Cure*, 220-26.

[163] W.J. Tulloch (1887-1966) began teaching at University College Dundee in 1914 as a lecturer in bacteriology, later entering the Royal Army Medical Corps College, where he remained until the end of the war. In 1921 he was appointed professor and chair of bacteriology at Dundee, as well as bacteriologist at the Dundee Royal Infirm, responsible for public health in the region. Between 1945 and 1956 he was dean of the Faculty of Medicine.

[164] Cited in Brown, *Penicillin Man*, 63.

[165] Robert Murray, "A Personal Narrative."

[166] Studies based on careers of bacteriologists Alexander Fleming and W.W.C. Topley demonstrate the strong personal anguish and sense of helplessness in dealing with the 1918 pandemic. See Brown, *Penicillin Man*, 64-65; Wilkinson and Hardy, *Prevention and Cure*, 125-29.

[167] Estimates of the number of deaths range between 40 and 100 million. In the United States, about 450,000 died, compared with about 16 million in India. Recently, samples of the 1918 virus strain (H5N1) have been subjected to genome sequencing, allowing researchers to understand why it was so lethal. See Jeffrey Taubenberger and David Morens, "1918 Influenza: The Mother of All Pandemics," *Emerging Infectious Diseases* 12, no. 1 (January 2006), http://www.cdc.gov/ncidod/EID/vol12n001.

high ratio of young adult fatalities, and above all the lack of knowledge about the viral pathogen.[168] Everitt's duties at the RAMC College in London meant an obligation to treat scores of serious influenza cases, along with the effects of a variety of "opportunistic" bacterial pathogens – pneumococci, streptococci, and staphylococci."[169]

Scientific Medical Research during the Twenties

After the First World War, public health issues became a high priority for successive British governments and their medical agencies. Of particular importance was the role assumed by the state-funded Medical Research Council (MRC), which claimed a mandate to supervise all experimental, clinical, and laboratory research – more specifically, "of supporting fundamental scientific research, maintaining that progress in physiology, biochemistry and pathology was essential to the development of clinical practices."[170] During the 1920s these MRC guidelines helped shape the pattern of British medical research in a variety of ways. First, they drew a sharp line between academic specialists, on one hand, and general practitioners and hospital clinicians on the other. Indeed, MRC officials consistently asserted that medical researchers were solely under their jurisdiction and were not subject to the mandate of either of the royal colleges (Surgery, Physicians). Second, the MRC insisted that within the academic cohort of medical research there should be a hierarchy based on scientific achievements, a bias that determined the composition of MRC expert committees. This emphasis on medical elitism was reinforced by the ability of the MRC to obtain large-scale grants from the US-based Rockefeller Foundation, which provided over 2.5 million pounds for teaching and medical research in British universities during the interwar years.[171] In addition, one-third of the government funds allocated to the MRC were directed to the operation of the National Institute for Medical Research in London, initially consisting of departments of bacteriology, applied physiology, biochemistry, pharmacology, and medical statistics.[172] Finally, the MRC maintained its own group of researchers such as Everitt Murray, who worked at the Milton Road

---

168 It was only in 1919 that the real cause of the influenza was identified as a virus, and even longer before the viral nature of the infection was generally accepted. Brown, *Penicillin Man*, 64-65.
169 Robert Murray, telephone interview by Donald Avery, 20 January 2008; Brown, *Penicillin Man*, 65.
170 Joan Austoker, "Walter Morley Fletcher and the Origins of a Basic Biomedical Policy," in *Historical Perspectives on the Role of the MRC*, ed. Austoker and Bryder, 26.
171 Ibid.
172 Joan Austoker and Linda Bryder, "The National Institute for Medical Research and Related Activities of the MRC," in *Historical Perspectives on the Role of the MRC*, ed. Austoker and Bryder, 35-57. The NIMR was established in 1914 at Hampstead in Greater London, but it enjoyed its greatest success during the interwar years under the chairmanship of Sir Henry Dale.

Field Laboratories in Cambridge, where he continued his wartime study on "the biology of the Meningococcus."[173]

During the interwar years Cambridge University was one of the world's major centres for scientific activity. In physics, Sir Ernest Rutherford's Cavendish Laboratory produced a number of Nobel laureates,[174] with Rutherford himself gaining this honour in 1908, shortly after his return from McGill University.[175] The life sciences at Cambridge also experienced impressive growth during this decade, particularly the biochemistry laboratory of Gowland Hopkins[176] and the Pathology Institute of Henry Roy Dean, newly formed in 1922. Part of the reason why these two scientists were so successful was their ability to obtain extensive funding from the MRC, from the Rockefeller Foundation, and from Sir William Dunn, a noted philanthropist.[177] In turn, this facilitated the expansion of their respective laboratory facilities, the recruitment of new faculty, and an ability to remain independent of university funding and control. Hopkins and Dean were also influential in shaping the character of preclinical medical education through their promotion of the Natural Sciences Tripos.[178]

---

173 Joan Austoker, "Walter Morley Fletcher," ibid., 23-33. Fletcher was secretary of the MRC between 1914 and 1933. He was a Trinity College, Cambridge, graduate, specializing in physiology. Robert Murray, "A Personal Narrative."

174 James Chadwick, the first of Rutherford's "boys" to be honoured, won the prize in 1937 for his discovery of the neutron. He was followed by Otto Hahn, John Cockcroft, and Patrick Blackett. Jeffrey Hughes, "'Brains in Their Fingertips': Physics in the Cavendish Laboratory, 1880-1940," in *Cambridge Minds*, ed. Richard Mason (Cambridge: Cambridge University Press, 1994), 160-76.

175 Ernest Rutherford (1871-1937) is recognized as one of the outstanding scientists of the twentieth century for his work on radioactivity and his discovery of the atomic nucleus. Between 1898 and 1907 Rutherford was Macdonald Professor of Physics at McGill, where he carried out important research on the transformation theory of radioactivity, as well as influencing the careers of many outstanding young Canadian and foreign scientists. Mario Bunge and William Shea, eds., *Rutherford and Physics at the Turn of the Century* (New York: Cambridge University Press, 1979); A.S. Eve, *Rutherford: Being the Life and Letters of the Rt. Hon. Lord Rutherford, O.M.* (Cambridge: Cambridge University Press, 1939).

176 Hopkins would be awarded the Nobel Prize in 1929 for his discovery of vitamins. In 1922, physiologist A.V. Hill was similarly honoured, joined ten years later by his colleague Lord E.D. Adrian (along with Charles Sherrington).

177 Between 1900 and 1930, the various Rockefeller philanthropic ventures had an enormous impact on university education in general, and medical research in particular, throughout the western world. Administratively, these funds were channeled through the Rockefeller Institute for Medical Research (1901), the General Education Board (1903), and the Rockefeller Foundation (1913). In addition, the Rockefeller Hospital in New York City opened in 1910, being "the first American institution to apply the full-time clinical system." Fedunkiw, *Rockefeller Foundation Funding*, 24.

178 Henry Roy Dean (1879-1961) was professor of pathology at the University of Cambridge for

Within Cambridge, Dean emerged as an important academic power-broker by advancing the goals of his Institute, by insisting on natural sciences courses for pre-clinical medical education, and by his broader concern for preserving Cambridge traditions.[179] In terms of his discipline, he was determined to achieve a transformation in the status of pathology at Cambridge "from something between an ancillary in clinical diagnosis ... to a lively science with its roots in physiology, bacteriology and biochemistry, and its branches in medicine and surgery."[180] Another aspect of Dean's operation was his ability "to pick men who were potential scientific winners; he then backed them heavily."[181] As a result, by the late 1920s Dean had personally recruited an outstanding group of researchers: Howard Florey,[182] Robert Webb, Meredith Swann, G.S. Wilson, Alan Drury, and Everitt Murray.[183] Nor did this connection end after they left Dean's Cambridge laboratory: "It was said that it was his ambition to fill every chair of pathology in the British Commonwealth with his own men, and he did not fall short of his mark."[184]

During the 1920s, Murray's renewed association with Cambridge University

---

thirty-nine years from 1922, and also served as Master of Trinity Hall (1929-1954). His scientific background was impressive. After graduating from Oxford and St. Mary's Hospital, Dean carried out research in Germany with the famous bacteriologist August von Wassermann, before returning to England as assistant bacteriologist at the Lister Institute, London. In 1912 he was appointed professor of pathology at Sheffield University, which he left in 1915 in order to assume a similar chair at the University of Manchester (1917-21). "Obituary: H.R. Dean, M.D., LL.D., D.Sc., F.R.C.P.," *British Medical Journal* 1 (5225) (25 February 1951): 595-596.

179 Dean was determined that pathology have equal standing with anatomy and physiology within the second MB examinations, while, at the same time, he discouraged the creation of a separate Medical Science Tripos. Brooke, *A History of the University of Cambridge*, 170-71.

180 Tribute in *The Lancet* (1961), cited in ibid., 168. Dean was one of the last faculty members hired under the old statutes, which meant that his appointment was for life.

181 Gwyn Macfarlane, *Howard Florey: The Making of a Great Scientist* (Oxford: Oxford University Press, 1979), 154.

182 Born in Australia, Howard W. Florey began his British educational experience at Oxford University in 1922, studying physiology with the legendary Sir Charles Sherrington. During the period 1926-1931 he was connected with Dean's laboratory. After obtaining his PhD in 1927, Florey was appointed to the Chair of Pathology at Sheffield in 1934 and then to the coveted Chair in Pathology at Oxford in 1935. Florey is, however, best known for his Nobel Prize in Physiology or Medicine in 1945 for the discovery of penicillin and its curative effects, along with Sir Alexander Fleming and Ernest Chain. E.P. Abraham, "Howard Walter Florey: Baron Florey of Adelaide and Marston, 1898-1968," *Biographical Memoirs of the Royal Society of London* 17 (1971), 255-301.

183 Cambridge obtained a block grant of 700,000 pounds for its scientific departments and for the University Library. Brooke, *A History of the University of Cambridge*, 170-71. During the mid-thirties Dean became vice-chancellor of the university. Ibid., 262-82.

184 During the 1920s many of the more ambitious medical students took the NST, "specializing in those subjects which gave exemption from Parts I and II of the MB." Brooke, *A History of the University of Cambridge*, 172.

was determined by two factors: his MRC research project at the neighbouring Field Laboratories, and the fact that Christ's College needed a supervisor of studies for its Natural Sciences Tripos. As a result, Everitt was appointed a Fellow of the College and subsequently became associated with Dean's Pathology Institute, at a time when it was developing a comprehensive program in pathology, bacteriology, and immunology for the Natural Sciences Tripos.[185] By 1926 Joburg, as he was now commonly called, became lecturer in pathology, responsible for organizing and teaching the new Tripos course in pathology, part II. From all accounts, he carried out his duties admirably, demanding high standards of performance from himself and his students. As one of Joburg's colleagues remembers, "Perfection in technique he insisted on ... and woe to the man who was clumsy or careless. His chin would jut forward, his moustache bristle and with a baleful eye he would fly at the offender." A.A. Miles and E.T.C. Spooner, who were two of Joburg's students during that period, provided their own impressions:[186]

> Murray appeared to us as a beardless, anti-clerical, slow-speaking perfectionist with a strong taste for technique and a stronger one for argument. ... He would not have fitted into Lord Snow's ideas of a Fellow of Christ's; for example, at high table he was wont to mutter audibly when the chaplain ... sat down to dinner – 'bloody rainmakers'; and when he found a (Fellow) ... who had just lost the leather-bound volume of lectures he had used for the last 20 years, said, 'Serve him damn well right, he will have to bring them up to date now.'[187]

Among his departmental colleagues Joburg was also well respected for his research activities, both his individual work on meningococcus and his 1926 joint project, with Webb and Swann,[188] on *Bacterium monocytogenes* and *Listeria monocytogenes*.[189] Another dimension of his Cambridge experience was his involvement with

---

185 George Nuttall (1862-1937) was a senior member of the Department of Pathology, although most of his interests appeared to have been associated with the Institute of Parasitological Research. In 1929 he served as one of Everitt's referees for the renewal of his Christ's College Fellowship.
186 Both Miles and Spooner would have impressive microbiology careers in the United Kingdom, with Spooner becoming a member of the Cambridge Pathology Department during the 1930s, while Miles found a position at the University of London. Another of these Tripos stars was Frederick Smith, who joined Joburg in his McGill laboratory in 1932. Robert Murray, "A Personal Narrative."
187 Ibid.
188 In 1926 Swann died from an infection contracted while doing a post-mortem pathological examination.
189 The term *Listeria* was apparently in honour of the South African bacteriologist Sir Spencer Lister, who was widely respected for his expertise. He was a good friend of George Murray, who often consulted with him on problems of infectious diseases. UWO Archives, R.G.E.

academic affairs, particularly on debates over salary levels, and ensuring that Pathology obtained its fair share of "the total allotment of remuneration."[190] Throughout the late 1920s Joburg also helped Dean in supervising the construction of the new pathology building. According to his boss, "every detail in fixed and moveable equipment, in water, heat, electric supply, in fitting up rooms for special purposes and in the purchase of special apparatus has been done by Murray."[191] By 1930, when Joburg was considering leaving Cambridge for McGill, Dean summarized his colleague's many achievements: "E.G.D. Murray is a very close personal friend of mine, a most loyal and devoted colleague and my right hand in the department. ... Murray is a very good bacteriologist and has published research work of distinction and he has proved himself an excellent teacher, lecturer and organizer."[192]

Nor was Dean the only eminent British scientist to praise Murray's research achievements. Although his eight publications might be considered meagre by today's microbiology standards, during the interwar years considerable emphasis was placed on high-quality scientific papers that developed new paradigms. And by this criterion, Murray's work was highly regarded by the bacteriological elite in the United Kingdom and North America, in part because his investigations covered a number of important fields. Four of Murray's studies dealt with the bacterium meningococcus, ranging from the joint 1915 report with Melvyn Gordon to his more sophisticated report of 1929, which was published by the MRC.[193] Everitt also continued his wartime research on the categorization of the different strains of dysentery, along with some preliminary work on the etiology of tuberculosis.[194] In addition, he

Murray Collection, "Letters from Ever, 1918," compiled by Sarah Schmidt, George Murray to Joburg 28 March 1918.
190 G.S. Wilson, "Everitt George Dunne Murray," 648.
191 According to one of his colleagues, Joburg's involvement with the new laboratory reinforced his low estimation of architects "and his opinion of ours in particular was quite blistering." Ibid., 649.
192 McGill University Archives, RG 2, C67 – Medicine 1930-1933, Department of Bacteriology and Immunology, Professor Murray file 1256 (hereafter cited as McGill Archives, B & I), Dean to Dean Charles Martin, 22 July 1930.
193 E.G.D. Murray (with M.H. Gordon, M.D., C.M.G., C.B.E., F.R.S.), "The Identification of the Meningococcus," *Journal of the Royal Army Medical Corps* (1915); E.G.D. Murray and J.A. Glover, "Cerebro Spinal Fever," in *The Practise of Medicine in the Tropics*, vol. 2, ed. W. Byam and R.G. Archibald (London: Henry Froude and Hodder and Stoughton, 1923); E.G.D. Murray, "Some Aspects of Meningococcal Virulence: A Report to the Medical Research Council on Work Carried Out at the University of Cambridge Field Laboratories," *Journal of Hygiene* 22, no. 2 (January 1924): 175-207; E.G.D. Murray, *The Meningococcus*, Special Report Series No. 4 (Medical Research Council, 1929), 124; E.G.D. Murray, "The Meningococcus," in *A System of Bacteriology in Relation to Medicine*, vol. 2 (London: Privy Council Medical Research Council, 1930), 291-325.
194 E.G.D. Murray, "Attempt at Classification of *B. dysenteriae*, based upon the Agglutinating

moved into new areas of enquiry, such as his innovative analysis of the pathogen associated with rabbit infections, which he called *Bacterium monocytogenes*. In turn, this work contributed to the discovery of a similar organism, *Listeria monocytogenes*, the causal agent of listeriosis, a rare but lethal food-borne infection. This fascination with listeria would remain throughout his professional career.[195]

During the 1920s, however, it was Murray's research on meningococcal meningitis that attracted most of the scientific accolades. According to Sir Graham Wilson, one of Britain's most accomplished bacteriologists, "Murray's work on the meningococcus was carried out primarily to study the properties of the organisms. Most previous work had been inspired by the needs of diagnosis or the wish to cure ... Murray realized the importance of studying the biology of the coccus itself. Failure to do so was the reason why, in his opinion, hospital pathologists had contributed so little to the treatment of the disease."[196]

At the end of the twenties Everitt and Freda Murray, with their two children, Bob (age eleven) and Susan (age four), were comfortably located in Cambridge society. They had many professional and personal friends, including the Deans, who periodically entertained them at their big house in Land's End, Cornwall. There were also connections with Freda's family in Suffolk,[197] and Joburg's more distant connection with his Murray and Thurtell relatives.[198] Above all, the Murrays enjoyed

---

Properties of 53 strains," *Journal of the Royal Army Medical Corps* (October – November 1918); E.G.D. Murray, "Staining Tubercle Bacilli in Formol-Fixed Tissues," *Journal of Pathology and Bacteriology*, 27 (1924): 118-119.

[195] E.G.D. Murray, R.A. Webb, and M.B.R. Swann, "A Disease of Rabbits Characterized by a Large Mononuclear Leucocytosis Caused by Hitherto Undescribed Bacillus Bacterium Monocytogenes," *Journal of Pathology and Bacteriology* (1926); Robert Murray, "A Personal Narrative."

[196] G.S. Wilson, "Everitt George Dunne Murray," 641. Graham Selby Wilson (1895-1987) was a student at King's College, London, in 1912, prior to obtaining his medical degree at Charing Cross Hospital, before joining W.W.C. Topley at the London School of Hygiene and Tropical Medicine during the 1920s. Their textbook, *The Principles of Bacteriology and Immunity* (London: E. Arnold, 1931) was widely cited. In 1990, it was published in its eighth edition as *Topley & Wilson's Principles of Bacteriology and Immunity* (London: E. Arnold, 1990).

[197] Freda was born in Riverside, California, where her parents attempted an ill-fated fruit farming venture. Upon their return to England, they settled in the family house at Blundeston, Suffolk. Her brother Thomas was killed during the First World War. Reflections of H.W.H. Woods, interview by Sarah Schmidt (unpublished manuscript in possession of Robert Murray), 1996.

[198] Peter Murray has prepared an extensive genealogical survey of the two families in England and in South Africa. Apparently, in 1829 Alfred Thurtell took the additional name Murray upon his marriage to Mary Everitt, because of the notoriety surrounding the trial and execution of his first cousin John Thurtell. Six years later, Alfred migrated to the Cape region of South Africa. While most of the Thurtell side of the family remained in England, some members did come to South Africa and to North America. See Appendix A.

their Cambridge home, with its roomy interior, lovely garden, and proximity to local hiking trails. "Then as later," Robert Murray has written, "she (Freda) did much in observing nature around her and what she was not familiar with was clarified by Opa (Joburg) who had included zoology at Cambridge."[199] Vigorous walking also became more prevalent after Everitt eventually abandoned his motorcycle, previously used for going back and forth to the MRC Field Station, and after 1925, "the family never had a car or vehicle more complex than a bicycle."[200] This lack of personal transportation did not deter the Murrays from going on vacations, although these excursions were determined by school vacations. As his Cambridge colleagues noted, even during the 1920s Joburg "had great plans and high hopes for his son Bobby; perhaps to become one day a scientist more dedicated even than his father."[201]

In 1927, young Robert began his academic studies at the Summerfields boarding school, "which has a high reputation for producing scholarship students at the major Public Schools, like Eton, Winchester, Rugby."[202] As he remembers, "Entry in 1927 was a bit traumatic, but survivable, because the first year was considerate of change involved for boys who had led a protected life. Subsequent years were a different matter: classes were small, there were demands for performance, and not doing your best or better was met with punishment, often corporal, and students moved up in particular subjects as soon as a new level was attained."[203] All this would change in July 1930 when McGill University offered Everitt a job as professor and chairman of the Department of Bacteriology and Immunology, and the position of chief bacteriologist at the Royal Victoria Hospital.

Coming to Canada

There are three major questions associated with the appointment of E.G.D. Murray at McGill University. First, why did McGill choose Joburg when there were so many other qualified scientists? Second, why did he accept a job in distant Canada when there were promising opportunities in the United Kingdom? And, finally, how did he change the status of bacteriology at McGill and influence Canadian medical science?

During the summer of 1930, dean of medicine Charles Martin and principal Arthur Currie of McGill weighed a number of factors before they decided that Murray was their man. Of special importance was Martin's determination to improve the

---

199 UWO Archives, R.G.E. Murray Collection, Robert Murray, "A 'Mum for All Season': The 20's to the 40's."
200 Robert Murray, "A Personal Narrative." Joburg did not drive a car in Canada, and Bob only obtained his driver's licence in 1963.
201 Comment by Robert Webb, cited in G.S. Wilson, "Everitt George Dunne Murray," 648.
202 Robert Murray, "A Personal Narrative."
203 Ibid.

quality of his faculty through aggressive recruitment and creative funding overtures to the Rockefeller Foundation.[204] These policies dated back to December 1919 when the McGill medical faculty had qualified, as one of five Canadian recipients, for a Rockfeller grant of a million dollars, which was calculated to bring about an immediate upgrade in the quality of its North American medical education.[205] On one hand, "it offered medical schools a greater opportunity to revaluate their curriculum, infrastructure, and faculty requirements and their applicant pool," while at the same time emphasizing "that modern scientific medicine required laboratories, equipment, and clinical teaching facilities."[206] In the case of McGill,[207] the Rockefeller assessors observed that the medical faculty had "been doing excellent work on a budget of only $114,000 and should with Toronto be more liberally aided than any other of the Canadian schools."[208] Not surprising, news of the Rockefeller grant was well received by members of McGill's medical faculty, who predicted a new era of research excellence. These sentiments were aptly expressed by pathologist Horst Oertel, who had been hired in 1919 to replace the eminent scientist J. George Adami: "during the last ten years the changes in all biological sciences including pathology have been so profound and have brought about such fundamental alterations in our knowledge, that the older systems and methods of research and teaching are antiquated ..."[209]

204 In February 1918 McGill received a 1 million dollar grant from the Carnegie Corporation, which was matched by an aggressive fundraising campaign by Montreal's wealthy English-speaking community, and by the Quebec government. Fedunkiw, *Rockefeller Foundation Funding*, 24, 90-95.
205 In 1909-10, Alexander Flexner of the Rockefeller Institute for Medical Research had examined 155 North American medical schools. His subsequent report had praised a number of "winners," including the medical facilities at the University of Toronto and McGill, while acknowledging the potential of emerging faculties at the University of Manitoba and the University of Alberta. In contrast, the medical programs at the University of Western Ontario, Queen's University, and Laval University were deemed inadequate. Ibid., 5-7.
206 A number of prominent Canadians lobbied the Rockefeller Foundation on behalf of Canadian medical schools, notably William Lyon Mackenzie King, Vincent Massey, and Sir William Osler. The latter sent a special appeal on behalf of McGill, arguing that this financial support was crucial "to see Montreal keep in first rank as a medical centre." Ibid. 5, 78.
207 In 1918-19, McGill had 463 men and 8 women enrolled in its five-year medical program, compared to Toronto's 619 men and 79 women. Ibid., 47.
208 In 1920 McGill and Toronto each received 1 million dollars, while Dalhousie and the University of Manitoba received $500,000. A sum of $25,000 was provided to both the University of Alberta and the University of Montreal. Rockefeller officials were concerned about the situation at Laval and the University of Montreal, since they had the responsibility of "serving a population of two million isolated from practically all English influence with poorly trained physicians and inadequate public health service." In the end, Montreal did receive a grant of $25,000 for development purposes. Ibid., 48, 95.
209 In February 1919 the Cambridge-trained Adami had announced, after being at McGill since 1892, that he had accepted a position as vice chancellor at the University of Liverpool. His

Dean Martin was strongly influenced by this new era of generous funding and the search for medical excellence. This was reflected in his supervision of two major building projects – the biology facility, and the Pathology Institute – which were financed through the Rockefeller grant.[210] At the same time, Martin was busy recruiting a number of outstanding medical scientists. In 1924, for instance, he was able to lure Jonathan Meakins, a highly regarded surgeon, back from the University of Edinburgh, by offering him a position as chairman of the faculty of medicine and physician-in-chief at the Royal Victoria Hospital. Four years later he added two other research stars: the eminent Canadian biochemist J.B. Collip, and Wilder Penfield, an outstanding US-born neurosurgeon.[211] In all three of these appointments there was an assumption that McGill should create world-class centres of scientific excellence,[212] the most famous of these being Penfield's Montreal Neurological Institute (MNI), which was established in 1934.[213] Everitt's hiring was therefore an integral part of Martin's determination to upgrade McGill's medical faculty in general, and its bacteriology component in particular.

But why was Murray selected? One of the key factors was the strong endorsement he received from prestigious scientists such as W.W.C. Topley of the London School of Hygiene and Tropical Medicine, and H.R. Dean. Another was McGill's determination to hire a bacteriologist who would combine effective university teaching, demonstrate appropriate clinical skills, and, above all, possess impressive research credentials.[214] Unlike other possible candidates, Murray met this criterion. Another

support for the German-trained Oertel as his successor created a bitter debate, since many argued that such an appointment was unpatriotic. Ibid., 64, 91.
210 During his twenty-seven years at McGill, Adami had elevated the status of pathology at McGill, holding the endowed chair that was funded by Lord Strathcona in 1893. Adami's many publications included his two-volume study *The Principles of Pathology* (1908-09), and he was president of the Association of American Physicians in 1912. Locally, he was active in the Canadian Association for the Prevention of Tuberculosis and the Montreal City Improvement League. Marie Adami, ed., *J. George Adami: A Memoir*, 33-90.
211 Joseph Hanaway, *McGill Medicine*, vol.2, *1885-1936* (Montreal and Kingston: McGill-Queen's University Press, 2006), 110-135, 222, 235, 238.
212 In biochemistry, there were two additional quality appointments: David Thomson and Hans Selye, both of whom worked closely with Collip in producing over two hundred scientific papers during the next ten years. Ibid., 223; Alison Li, *J.B. Collip and the Development of Medical Research in Canada* (Montreal and Kingston: McGill-Queen's University Press, 2003), 95-130.
213 In 1928 Penfield came to McGill with his close associate, surgeon William Cone, and they established the neurosurgical unit at the Royal Victoria Hospital. The Rockefeller Foundation grant of $1,250,000 was crucial for the establishment of the Montreal Neurological Institute (MNI). Ibid., 239; Wilder Penfield, *No Man Alone: A Neurosurgeon's Life* (Boston: Little Brown, 1977). See chapter 3, note 130.
214 The exchange of letters between McGill officials in 1930 provides a short list of possible

advantage was that he did not face competition from American bacteriologists, since McGill officials were determined to hire British, and preferably someone from Cambridge or Oxford.[215] A final asset was Murray's willingness to make an immediate commitment and move his family to Canada during the fall of 1930.[216]

Is it surprising that Joburg took the McGill job, rather than remain in Britain? In our opinion, no! While some scientists, such as Howard Florey, might have experienced a rapid career track during the early 1930s, this was the exception, reserved for the truly gifted and well-connected.[217] In contrast, Murray's scientific profile was good, but not outstanding, and he did not have a PhD. Nor did Joburg relish the prospect of being a perpetual lieutenant for H.R. Dean in an institute where bacteriology played second fiddle to pathology.[218] Moreover, he was not deterred by the prospect of leaving the United Kingdom since he was already a professional migrant, having parted ways with his South African homeland in 1906. Coming to Canada, therefore, represented an exciting adventure with great job prospects, rather than a fearful venture.[219]

Montreal Society and McGill University during the Thirties

On 4 November 1930 the Murrays arrived in Canada.[220] Despite their well-planned trip, the first two months were a challenge for Joburg and Freda as they sought suitable lodging, enrolled the children in school, and adjusted to Montreal society.[221] Nor

candidates. It is clear they did not want to hire a specialist in public health medicine. See chapter 1.

215 It is not clear how important the Rutherford connection was in influencing Martin and Currie in their decision. But it obviously reinforced the institutional linkages between Cambridge and McGill.
216 See chapter 1; also, Stanley B. Frost, *McGill University for the Advancement of Learning*, vol. 2, *1985-1971* (Kingston and Montreal: McGill-Queen's University Press, 1980).
217 In 1927, his first year at Dean's laboratory, Florey had been awarded a Rockefeller Foundation Fellowship by the MRC. He was also closely connected with Sherrington at Oxford and with A.N. Richards, one of the rising stars of American scientific medicine. Abraham, *Howard Walter Florey*.
218 It has been suggested that the opportunity to create a department of bacteriology independent of pathology was another incentive, along with the fact that "much of his work was closely related to clinical problems ... (favouring) his attachment to the staff of a large hospital." Robert Murray, "A Personal Narrative"; Sclater Lewis, *Royal Victoria Hospital, 1887-1947* (Montreal: McGill University Press, 1969), 221-23.
219 Robert Murray, "A Personal Narrative." It is instructive that all of the 1930-31 "well-wish" letters applauded his decision, stressing the advantages of the promising new venture. See chapter 1.
220 Robert Murray has provided a vivid account of the move from Cambridge to Montreal. See chapter 1.
221 In his 29 December 1930 letters to his brother Rory, Joburg described the Montreal situation

was this an easy undertaking, since their new urban environment had a number of challenging features.[222] First, Montreal was a divided city, separated by language, ethnicity, and class. For the English-speaking community there were distinct neighbourhoods and urban landscapes that had gradually evolved in response to large-scale immigration from the United Kingdom and the influx of Anglo-Quebecers from rural areas. The fact that English was the language of work in Montreal's booming manufacturing and financial sectors provided additional incentives for anglophone migrants to gravitate towards Canada's leading metropolitan centre.[223] Secondly, in Montreal wealth was concentrated in the hands of a powerful Anglo-Canadian elite, whose mansions "were concentrated within an area referred to as the Golden Square Mile, whose residents were estimated to have controlled 70 % of all Canadian wealth at the turn of the century."[224] Thirdly, after the 1880s the size of metropolitan Montreal gradually increased through the absorption of twenty-four separate municipalities, a process that coincided with the eventual triumph of two populist French Canadian mayors, Médéric Martin and Camillien Houde.[225] Significantly, this annexation movement did not affect affluent anglophone communities such as the towns of Westmount and Mount Royal, which continued to be self-governing and independent communities.[226] Fourthly, many of the English-speaking elite, in keeping with their belief in the stewardship of wealth, provided philanthropic assistance to less fortunate members of the Anglo-Protestant community through their support of churches, schools, hospitals, and charitable organizations.[227] This assistance became

in very positive terms, despite the cold and snowy weather, and made special reference to "a game on the ice here called Hockey."

[222] According to the 1931 census, Montreal had a population of 1,020,018, of whom over 60 per cent were French-speaking.

[223] The 1931 census showed that 95 per cent of all Quebecers whose mother tongue was English were of British origin. There were, however, 130,000 residents of the province who were neither English nor French, with Jews and Italians representing over half of this figure. Ronald Rudin, *The Forgotten Quebecers: A History of English-Speaking Quebec, 1759-1980* (Montreal: Institut québécois de recherche sur la culture, 1985), 153-73.

[224] Rudin, *Forgotten Quebecers*, 205.

[225] The last anglophone mayor of Montreal was John Guerin (1910-21). He was defeated by Martin, who dominated civic politics until the late 1920s when he was displaced by Houde. Alan Gordon, *Making Public Pasts: The Contested Terrain of Montreal's Public Memories, 1891-1930* (Montreal: McGill-Queen's University Press, 2001), 40-43.

[226] In 1941 Westmount had a population of about 25,000, of whom 88 per cent were English-speaking and 70 per cent were of British origin. Margaret Westley, *Remembrance of Grandeur: The Anglo-Protestant Elite of Montreal, 1900-1950* (Montreal: Libre expression, 1990). 203.

[227] By 1900, the condition of the English-speaking working class, particularly those of Irish background, was a matter of some concern, as reflected by the controversial study by Sir Herbert Ames, *The City below the Hill* (Montreal, 1897; repr. Toronto: University of Toronto Press, 1972); see also Terry Copp, *Anatomy of Poverty: The Condition of the Working Class in Montreal*,

particularly important during the Great Depression of the 1930s, when Montreal had to borrow vast sums of money to provide relief assistance.[228] As a result of widespread poverty, there was considerable social unrest in Montreal during the 1930s, as reflected by growing numbers of industrial strikes, demonstrations by the Communist Party of Canada, and the activities of the newly founded Commonwealth Commonwealth Federation (CCF), including a small core group from McGill.[229]

The presence of these CCF "radicals" was not well received by Montreal's wealthy elite, who regarded McGill as their educational chattel and "a linchpin in the whole English Montreal community."[230] Traditionally, the university's board of governors included the corporate leaders of the Canadian Pacific Railway Company, the Bank of Montreal, the Royal Bank, Macdonald Tobacco, and the Molson brewing company, who expected the university to reflect their values. And indeed, a number of McGill's 5,000 students came from affluent backgrounds, a characteristic that helped shape McGill's cultural character. This was evident in the proliferation of fraternities and sororities and in the glorification of McGill's athletic teams, which in 1929 won twelve out of fifteen intercollegiate championships. Yet despite these social amenities, McGill's corporate and academic leaders regarded their institution as "the Harvard of the North," with many outstanding scholars and scientists. These included a medical faculty that dated back to the nineteenth century, along with its partner, the Royal Victoria Hospital.[231] By the late 1920s Dean Martin and Principal Arthur Currie had implemented a number of ambitious policies designed to make McGill's medical school one of the best in North America.[232]

The hiring of Murray was an integral part of this grand design.[233] And Joburg was

1897-1929 (Toronto: McClelland & Stewart, 1974).
228 Quebec municipalities were required to assume 28 per cent of the costs of relief, compared with about 15 per cent in the rest of the country. In Montreal the municipal debt rose from $252 million in 1930 to $345 million in 1937. By the 1940s the city was bankrupt. Ibid., 244.
229 The most active members of the CCF group at McGill were lawyer Frank Scott, economist Eugene Forsey, and social worker Leonard Marsh. The Spanish Civil War (1936-39) further polarized the university and the city, particularly since the Spanish military junta under General Franco was strongly supported by the French Canadian religious and political elites. Frost, *McGill University*, 170-78; Wendell MacLeod et al., *Bethune: The Montreal Years* (Toronto: James Lorimer, 1978).
230 In 1930, only about 4 per cent of Quebec's population attended university. At McGill, approximately 10 per cent of the university's students were women. Frost, *McGill University*, 168, 209, 268.
231 The Royal Victoria Hospital was founded in 1887 after George Stephen and Donald Smith of the Canadian Pacific Railway jointly donated 1 million dollars. Its board of governors included many prominent businessmen. Ibid., 168, 209.
232 The hierarchical character of the Royal Victoria Hospital was described by a former nurse: "The chief doctors wore their cutaway coats and striped trousers on their rounds. They were lords of creation. When they spoke, everybody was supposed to jump." Ibid., 211.
233 Earlier (in 1915) Queen's University hired bacteriologist Guilford Reed (1886-1955), one of

eager to make his unique contribution. This was evident in his assessment of the merits and disabilities of the Department of Bacteriology and Immunology, which was submitted to the dean of medicine on 23 February 1931. Overall, it was a devastating critique of the "general disorganization in both the R.V. [Royal Victoria] Hospital and the Medical Faculty, so far as Bacteriology is concerned."[234] Specifically, Joburg criticized the serious manpower shortage of his department, which had only one professional bacteriologist and one technician; the poor performance of the hospital's clinical laboratory; the faulty training in bacteriology for medical students; and the lack of an acceptable bacteriology graduate program.[235] Although Dean Martin accepted all of Murray's comments, he did suggest that perhaps his new colleague didn't fully appreciate the broader structural problems affecting McGill during the depression years.

Throughout the thirties there were serious cutbacks in university revenues and also in faculty salaries. This unfortunate situation was further complicated by the interference of Chancellor Edward Beatty in university governance after the death of Principal Currie in 1933. In addition, there were poorly coordinated efforts to obtain additional grants from the Rockefeller Foundation, making even the highly promoted Montreal Neurological Institute a questionable undertaking. Within the medical faculty there were also serious tensions between clinicians at the teaching hospitals and McGill's medical scientists.[236]

Despite this litany of problems, Joburg soon transformed McGill's Department of Bacteriology into one of the best in the country.[237] This was accomplished by his aggressive lobbying for additional faculty and staff, once he had demonstrated an ability to deliver quality bacteriology courses for science and medical students. One of his most important hires was Fred Smith, one of the original Part II Tripos students from Cambridge, who brought great scientific skills and administrative ability to the department.[238] Another positive development was the enhanced relationship

---

Joburg's closest friends, to upgrade its medical faculty. Originally from Nova Scotia, Reed had obtained a PhD in biochemistry from Harvard University in 1912. His early work on the influenza virus during the 1918 pandemic established his reputation as an outstanding microbiologist. A.A. Travill, *Just a Few: Queen's Medical Profiles* (Kingston: Faculty of Medicine, Queen's University, 1992), 148-154.

234 See chapter 1, letter from Dr. Carleton W. Stanley, May 1931.
235 Until 1944 the Department of Bacteriology was a joint hospital and university operation, which created an uneasy financial situation.
236 Hanaway, *McGill Medicine*, 110-134, 150; Frost, *McGill University*, 187-209.
237 Joburg was fortunate that Willy Clark, his technician in Dean's laboratory, was willing to accompany him to McGill. This fine partnership lasted until the mid-1930s when Clark took a job with the pharmaceutical company Ayerst, McKenna and Harrison, where he had a rewarding career.
238 Fred Smith joined the department in 1931 after a short stint with the Rockefeller Foundation in New York City. See chapter 1, note 87.

between his department and clinicians at the Royal Victoria Hospital, the result of careful negotiations and mutual understanding.[239] These talents were also evident in Joburg's ability to establish effective communication with many of McGill's other high profile academics – J.B. Collip (endocrinology), Otto Maass (chemistry), Jonathan Meakins (surgery), Horst Oertel (pathology), and Wilder Penfield (MNI).[240]

Murray's own reputation within McGill was substantially boosted through a series of internal and external controversies.[241] One of these occurred in the spring of 1932 when Joburg carefully distanced his department from exaggerated claims, made by some scientific colleagues, about a revolutionary methods of treating the disease brucellosis.[242] His identification with good science was further enhanced by his insistence, despite pressure from McGill's administration, that a graduate program in bacteriology was premature, since his department lacked the necessary faculty and laboratory resources. There would be no shortcuts. Another asset was his role as a consultant with Ayerst, McKenna and Harrison, part of Montreal's emerging pharmaceutical industry, which provided Joburg with additional income and expanded his scientific contacts.[243] His awareness of innovative vaccine research was also evident when he introduced a new pneumoccocus serum into the Royal Victoria Hospital, along with an improved technique for typing the deadly Type III pneumococcus. As well, his own staphylococcus toxoid was well regarded because of its "greatly improved potency."[244]

Murray's immunological reputation was given a further boost in March 1935,

---

239 By 1939 the Department of Bacteriology and Immunology maintained small laboratories at the Alexandra and Children's Memorial Hospital, along with a four-person laboratory team at the Royal Vic, headed by Murray. Lewis, *Royal Victoria Hospital*, 221-23.

240 Many of the leading surgeons at Royal Vic – notably Drs. Archibald, Jonathan Meakins, and Wilder Penfield – were committed to fundamental research, and therefore appreciated Murray's laboratory. Lewis, *Royal Victoria Hospital*, 222. For more information on Meakins see chapter 1, note 26.

241 There are some intriguing parallels between Joburg's experiences at McGill and those of pathologist George Adami (1892-1919). First, they both came from the University of Cambridge, with strong connections with the leading pathologists and bacteriologists of their respective eras. Second, they were both associated with Christ's College. Third, upon their arrival at McGill, both Murray and Adami called for major reforms in the medical program. Fourth, both became involved in the public health movement, locally and nationally. See Marie Adami, ed., *J. George Adami: A Memoir*.

242 See chapter 1.

243 Joburg was particularly impressed by the development of a pneumococcus serum by Oswald Avery's laboratory at Rockefeller Hospital, New York City, which he quickly introduced into the Royal Victoria Hospital. His connection with Avery dated back to his Cambridge days. See chapter 1.

244 Some types of pneumococcus, notably Type III, had a very high death rate. Lewis, *Royal Victoria Hospital*, 221-23.

when he was dramatically airlifted from Montreal to Detroit with his "Magic Serum," required for the emergency treatment of young boy's mysterious blood disease. As the *Detroit Free Press* described, "The boy's father, Max Osnos (a wealthy industrialist) making a desperate attempt to save the child ... chartered ... the 250 mile-an-hour plane ... to carry Dr. E.G.D. Murray of Montreal, to Detroit with a serum, which, it was hoped, would turn the trend of the disease threatening the life of the boy, Ronald."[245] And so it did!

During the 1930s Joburg continued his research activities on meningococcus and *Listeria monocytogenes*, as well as some novel work on Arctic bacteriology. He was also actively involved with a number of medical and scientific organizations such as the Canadian Medical Association, the Canadian Public Health Association (Laboratory Section), the Royal Society of Canada, and the International Association of Microbiological Societies (IAMS).[246]

This latter organization, originally founded in 1927, tried to address concerns that the international classification of bacteria was being controlled by botanists rather than by experts in bacteriology. Once the IAMS had established a separate identity, it sought to organize and educate the world's bacteriologists through periodic congresses and through the operation of specialized bodies such as the Nomenclature Committee on Determinative Bacteriology. Like many international scientific organizations of that era, the IAMS was dominated by a small group of British and North American experts. These included Ralph Terence St. John-Brooks of the National Collection of Type Cultures at the Lister Institute and H.R. Dean at Cambridge, along with three American microbiologists, R.E. Buchanan, Iowa State University; Robert Breed, Cornell University; and Thomas Rivers, an eminent virologist at the Rockefeller Institute of Medical Research. Since Joburg had previously participated in the founding of IAMS, being secretary of the British section, he quickly became Canada's leading advocate of this international organization.

Murray also believed that there should be a separate Canadian society of bacteriologists which "would secure Canadian representation in international activities."[247] As a result, in 1938 he launched a campaign to create the Canadian Society of Microbiologists, contacting many of the country's leading life scientists. Among this group, Frederick Banting (University of Toronto), R.D. Defries (Connaught Laboratories), and Guilford Reed (Queen's University) were particularly active. Joburg's contacts with these and other medical luminaries were reinforced by his involvement with a

---

245 *Detroit Free Press*, 4 March 1935.
246 In 1937 the name was changed to the International Association of Microbiologists.
247 Murray met considerable opposition from other Canadian scientists who called for a Canadian branch of the large and powerful Society of American Bacteriologists. Murray to Dr. C.K. Johns, 13 March 1939; see chapter 1.

Saving Richard Osnos: Joburg as Medical Celebrity (1935)
Peter Murray Collection, Associated Press Newswire

survey of Canada's medical research facilities, carried out in 1938 by the newly formed National Research Council's Associate Committee on Medical Research (ACMR).[248]

Social Adjustment

Another set of questions concerns the adjustment of the Murray family to Canadian society. Joburg's situation was, of course, the most complex, since there were three distinct stages in his life: South Africa, 1890-1906; England, 1906-30; and Canada, 1930-64. Did this mean that he had divided loyalties? In terms of his South African identity, the answer is clear: Joburg had no intention of returning to his former homeland.

The British equation is more complicated since Joburg treasured his many professional friends and remained keenly interested in British academic appointments and gossip about Cambridge University. He and Freda also enjoyed entertaining visiting British scientists, in their residence at 3590 University Avenue or at the prestigious University Club. On the other hand, their return visits to the United Kingdom were few and far between.[249] In fact, by the late 1930s the major links with Cambridge were Joburg's negotiations to ensure that Robert be admitted to Christ's College for his pre-medical education.[250] This goal was achieved in late 1938 through personal diplomacy and by Bob's strong academic performance at Lower Canada College and McGill University.[251]

On the eve of the Second World War, the Murrays were firmly entrenched in Montreal's Anglo-Canadian society, located in the famous one square mile. As such, they had access to its many social amenities. Thanks to wealthy patrons, Montreal supported a range of cultural organizations that included the Montreal Museum of Fine Arts, the Montreal Symphony Orchestra, and the Montreal Repertory Theatre. On the literary front, there were the venerable St. James Literary Society and the rapidly

---

248 Banting, as the first chairman of the ACMR, took this survey very seriously, visiting most of the country's medical schools, major hospitals, and provincial public health laboratories as well as interviewing some three hundred medical researchers. LAC, National Research Council Papers (NRC) 88/89, vol. 5-4-M-4-7, ACMR organizational report, 19 April 1938.

249 When Joburg came to Montreal he grew a beard, which substantially changed his public image. Indeed, in 1944 while he was in England on war matters, Freda speculated that his visits with family and friends must have been eventful: "I suppose the beard caused no little amazement (or should I say amusement!!)..." Freda to Everitt, 5 June 1944; see chapter 3.

250 Christ's was one of the seventeen men's colleges, described by one eminent scientist as "a smallish college with a decent reputation." Brooke, *A History of the University of Cambridge*, 267.

251 According to Hugh MacLennan, who taught at Lower Canada College during the late 1930s, the masters were primarily young British academics "who had little regard for Canadian history ... while the students lived in the English section of Montreal." As a result, there was a tendency for LCC graduates to assume that "their country was not Canada but the British Empire." Westley, *Remembrance of Grandeur*, 68-69.

expanding Canadian Club of Montreal. The former was composed largely of "the wealthy Old Guard members, who were primarily interested in British and European arts, letters, history and institutions ... while the rising new rich and the middle class were interested in Canadian issues and joined the Canadian Club."[252]

Social exclusiveness was even more pronounced among the city's men's clubs, notably the St. James (1857), the Mount Royal (1899), the University (1907), and the Mount Stephen (1923). Each had its own unique clubhouse, "usually an imposing and luxurious building, including a well-stocked library, rooms for playing cards and billiards, private rooms, and a respectable dining room and modern kitchens and with as many employees as a luxury hotel."[253] Many members of these clubs also belonged to elite sporting organizations such as the Montreal Hunt Club, the Montreal Racquet Club, the Royal Montreal Golf Club, and the Royal St. Lawrence Yacht Club.[254] As well, there were large numbers of impressive country homes and resorts on the St. Lawrence and Ottawa rivers and in the Laurentians, which affluent Montrealers used during the "unhealthy" summer months, when the threat of typhoid or other gastrointestinal disease was prevalent.[255] In addition, for the avid outdoorsmen, there were a number of male-only fishing lodges, whose members included many rich and influential American businessmen and politicians. By the 1960s, these private groups controlled 2,200 reserves of crown lands, covering almost 30,000 square miles.[256]

252 Both organizations were founded in the 1890s. The 1929 creation of the *McGill Fortnightly Review* provided an important venue for the promotion of Canadian culture. Ibid., 221, 233.
253 The membership of the St. James Club, founded in 1857, was dominated "by well-heeled business people," who controlled Montreal's economic, social, and political life and who were "almost exclusively Anglophone." Today, the club membership is about 75 per cent Francophone. François Hudon, *History of the Saint-James Club of Montreal*, limited ed. (Montreal: Saint-James Club of Montreal, 2001), 11, 45.
254 A 1928 survey listed 105 clubs, most being Anglo-Protestant. Many of Montreal's elite had multiple memberships. For instance, Colonel Herbert Molson was a member of twenty-four different clubs. Ibid., 158.
255 These included Como on the Ottawa River, the Hermitage Club on Lake Memphremagog, Knowlton on Brome Lake, and Metis Beach on the lower St. Lawrence. Ibid., 95-101. In 1927 Montreal experienced a serious outbreak of typhoid, largely due to an unsanitary sewage disposal system and the inadequate pasteurization of milk.
256 During the nineteenth century successive Quebec governments leased much of the best brook trout and salmon streams to private clubs. While these private enclaves were often criticized, there were counter-arguments that the clubs provided badly needed funds, maintained stewardship over the resource, and, above all, provided badly needed jobs in remote areas for managers, guides, wardens, cooks, maids, and maintenance personnel. *New York Times*, 30 July 1989; Sylvain Gingras, *A Century of Sport: Hunting and Fishing in Quebec* (St. Raymond, Quebec: Editions Rapides Blancs, 1994).

Overall, the Murrays were an integral part of the affluent Anglo-Montreal way of life. But they had their own set of social priorities. First, the university was their most important professional and social focus, since Joburg enjoyed his McGill colleagues, whether it was beer and meatballs with the "Greenhouse Follies" group or more formal dining at the University Club.[257] Second, during the academic term, he and Freda maintained an active social life in Montreal, enjoying English-language music theatre, good movies, fine dining, and their mutual participation in the North American Arctic Institute.[258] While most of his friends were English-speaking, Joburg's fluency in French facilitated contact with a number of French Canadian scientists, among them Armand Frappier of the University of Montreal, although their relationship was professional rather than personal, given their different social circles.[259] Third, given their residential location, educational experiences, and social networks, the Murray children were even more insulated within Montreal's Anglo-Protestant community than their parents. In recent years, Robert Murray has expressed regret that his English-speaking schooling at Westmount High School and Lower Canada College was not offset by being "farmed out to some French Canadian family during the summer holidays."[260] Nor were Robert's experiences unique. Murray Ballantyne, the son of an influential businessmen, has described at length how, during the 1930s, "one could live and die in the Square Mile without ever pronouncing a word of French, nearly without hearing one."[261]

With the arrival of summer, the focus of Murray family life shifted to their beloved cottage on Bark Lake, a twisting six-mile body of water in the Laurentians, north of Montreal. As Robert Murray remembers, "I think of those times in the context of the house at 3590 University Street and the Cottage at Bark Lake, which seemed their natural environments, from which time they escaped to 'the bush' for rest and recreation ... and in going on fishing and canoe trips – a truly Canadian form of energetic relaxation and returning to country ways."[262] As a youth, Robert spent his

---

257 One aspect of English customs that prevailed was afternoon tea when "people weren't necessarily invited, they just dropped in! Granny Woods' 'Indian Tree' china tea service was always used." UWO Archives, R.G.E. Murray Collection, Susan Murray Robinson, "My Mother Had a Profound Influence on Me."
258 As Bob Murray remembers, "Social pretensions were not for her [Freda]. She thoroughly enjoyed company, but once friendships were made she expected them to 'drop in' or be brought by one of us (Bob & Sue) or Opa, when they were always welcome." Robert Murray, "A 'Mum for All Seasons.'"
259 Armand Frappier, *Un Rêve, une lutte: autobiographie* (Sillery: Presses de l'Université du Québec, 1992).
260 Robert Murray, interview by Donald Avery, 10 October 2007.
261 Westley, *Remembrance of Grandeur*, 93.
262 During one of these trips, the Murrays bought a brown springer spaniel puppy which they named Uncle Fudge. He became their constant companion and was a favourite of all family

own summers at Bark Lake in the company of other English-speaking youth, enjoying "boats and water fun ... (and) my father started me on fishing for bass and pike."[263] In addition, there was the possibility of a special treat, fishing with his father at the exclusive Columbus Club, whose property was located between Nominingue and Kiamika in the Laurentians.[264]

By 1939, the Murrays were fully adjusted to Canadian society, and were grateful that their adopted country provided so many marvellous opportunities, both professional and personal. This sense of gratitude and public service would be evident during the Second World War when Joburg became deeply involved in vital aspects of Canada's defence medicine, and Freda pursued her own important wartime projects.

THE MURRAY FAMILY IN WAR AND PEACE, 1939-1964
Impact of the Second World War
Everitt, McGill, and Canadian Defence Science, 1939-1945

During the Second World War Canada was involved in an epic struggle against Nazi Germany, Fascist Italy, and Imperial Japan, and for the second time in the twentieth century, the country was forced to commit itself to total war. An important aspect of Canada's war effort was the contribution of the scientific and medical communities, many of whom were university faculty members. The complex relationship between the country's universities and the federal government was outlined on 16 September 1939 by General A.G. McNaughton, President of the National Research Council (NRC), the organization mandated to coordinate wartime defence science: "Owing to the possibility of the present war extending over a very long period and the need for ... large number of well trained men in all branches of pure and applied science, including medicine, dentistry and agriculture. ... Students now pursuing successfully these fields will serve their country in the most valuable way by continuing their university training until graduation."[265]

McGill and the University of Toronto, given their size and diversity of academic disciplines, were most involved with war projects directed by the NRC. McGill responded to McNaughton's appeal by creating a War Advisory Board under the direction of C.F. Martin, dean emeritus of the Medical Faculty,[266] and physicist David

---

members, who often indulged his insatiable appetite for asparagus. Robert Murray, "A 'Mum for All Seasons.'"
263 Robert Murray, interview by Donald Avery, 4 December 2007.
264 Half of the private fishing camp licenses were cancelled during the late 1960s, and the newly elected Parti Quebecois government abolished the entire system after 1976. *New York Times*, 30 July 1989.
265 LAC, National Research Council Collection (NRC), vol. 46, 17-15-9-2.
266 For the development of military medicine by Canada's allies, see Mark Harrison, *Medicine*

Keys.²⁶⁷ One of the Board's first programs was to prepare a comprehensive inventory of the university's manpower resources by circulating a detailed questionnaire about defence science potential.²⁶⁸ In a related initiative, Grant Fleming, McGill's dean of medicine, requested information about how each department in his faculty "could assist in training technical workers for war service, and ... what research work your Department could conduct." One of the most intriguing proposals came from E.G.D. Murray, who set forth an elaborate outline of how his colleagues "can and should contribute importantly in the military and civil emergency: because of the danger of epidemics of bacterial diseases in massed troops, because practically all war wounds are infected wounds."²⁶⁹ He also stressed the importance of preparing for a possible enemy use of biological weapons.²⁷⁰ Significantly, Joburg would soon coordinate Canada's biological warfare (BW) defensive programs.

Throughout the war years, the NRC earned high marks by effectively mobilizing the country's scientists and by creating a range of sophisticated weapon systems for the Canadian and British Armed Forces. This included radar and proximity fuses, explosives and ballistics, chemical and biological warfare, medical research and aviation medicine, and atomic research and development. Canada's scientific war was carried out in conjunction with the country's primary wartime allies, Great Britain and the United States.²⁷¹ Within the context of alliance warfare Canadian scientists often played a pivotal role, particularly in the shadowy and secret field of biological warfare. Although the 1925 Geneva Protocol had outlawed the use of both chemical and biological weapons, by the late thirties it was feared that these weapons of mass destruction would be used by the Axis powers, despite international conventions and ethical strictures.

In Canada, the call for biological warfare preparedness was initiated in 1937 by Sir Frederick Banting, Canada's sole Nobel laureate, on the grounds that the danger of a German germ warfare attack was grave and immediate. Between 1937 and his death in

and Victory: British Military Medicine in the Second World War (Oxford: Oxford University Press, 2004), and Albert Cowdrey, *Fighting for Life: American Military Medicine in World War II* (New York: Free Press, 1994).

267 Ibid.; McNaughton to Dean Martin, 24 October 1939.

268 Of particular importance was the impressive wartime role assumed by Otto Maass as coordinator of Canada's explosives program; director of the chemical and biological warfare operation; and special assistant to the NRC president, C.J. Mackenzie.

269 During the early stages of the war, Murray, along with Guilford Reed of Queen's University, was heavily involved with the NRC Committee on Gas Gangrene, which included close liaison with medical researchers in the United States. See chapter 2.

270 LAC, E.G.D. Murray Papers (MG 30-B-91), vol. 1, file War Service: Dean Fleming to Heads of Departments, 11 September 1939; ibid., Murray to Fleming, 14 September 1939.

271 Donald Avery, *The Science of War: Canadian Scientists and Allied Military Technology during the Second World War* (Toronto: University of Toronto Press, 1998), 14-46.

February 1941, Banting carried on a determined campaign to convince skeptical military officials in Canada and the United Kingdom of the importance of developing an effective BW civil defence system, carrying out an extensive research program on BW pathogens, and developing an offensive retaliatory capability. As part of his strategy for BW defensive planning, Banting assembled a talented research team, based at the University of Toronto, but including other experts such as Charles Mitchell, animal pathologist with the Department of Agriculture, and E.G.D. Murray of McGill.[272]

After the fall of France in June 1940, Banting and Murray became regular correspondents since both believed that an "unscrupulous Germany would undoubtedly use biological weapons," and that "the only safe defensive position against any weapon is afforded by a thorough understanding which can only be gained by a complete preparation for the offensive use of that weapon."[273] The Banting-Murray pattern of cooperation helps, in part, to explain why Murray succeeded Banting as Canada's foremost BW expert, responsible for coordinating Canada's biological warfare policies at home and with its wartime allies. The other reasons were Murray's impressive reputation as a bacteriologist and his range of scientific contacts at home and abroad.

During the period 1942-45, Canada's BW planners at the National Research Council and the Army's Directorate of Chemical Warfare and Smoke developed a number of important facilities and programs. First, there was a secret organization, the M-1000 (later C-1) Committee, which served as the central coordinating agency both in directing Canadian research programs and also in establishing close relations with British and American scientists. Second, BW research and testing facilities were established at the Experimental Station Suffield; at Grosse Île, Quebec;[274] and at Guilford Reed's biological weapons laboratory, Queen's University, under a secret arrangement negotiated with Principal Wallace in April 1942.[275] Reed's laboratory soon became the nerve centre for Canada's fundamental and applied research on possible BW agents and protective measures (vaccines and antitoxins). This challenging task of coordinating research, development, and testing of agents and munitions was greatly enhanced by the ability of the small C-1 Committee to work together in a constructive way. The close friendship between Reed and Murray was a key ingredient.

As chairman of the C-1 Committee, Joburg assumed a crucial role in shaping and directing Canada's wartime BW program. This became a full-time job after June

272 Ibid.
273 NRC Papers, RG 77/87-88, vol. 69, file 36-5-0-1 E.G.D. Murray to Frederick Banting 25 June 1940. Cited in chapter 2.
274 Avery, *Science of War*, 151-75.
275 The Queen's BW laboratory was located in the top floor of the Pathology Building until 1946 when it was moved off-campus as a distinct government laboratory. LAC, Defence Research Board Kingston (Headquarters Correspondence) HQS 4354-33-1, Major General J.V. Young, Master General of Ordnance, Memorandum for Colonel R.L. Ralston, Minister of Defence.

1943, when President Cyril James of McGill reluctantly agreed that Murray be seconded to the Directorate of Chemical Warfare and Smoke for the duration of the war. As a result, Murray spent most of the next three years either in Ottawa planning BW strategy, visiting C-1 research and testing facilities across the country, or attending policy meetings in Washington, DC.[276]

By the end of the Second World War, Canadian bacteriologists, biochemists, and veterinarians had made a number of significant contributions to the allied BW effort.[277] On the defensive side, there was a unique version of the botulinum toxoid (A and B), developed at Queen's, that could have been used to immunize Canadian troops, if intelligence reports about possible German use of botulinum toxin munitions during the D-Day landings had materialized. In addition, there was the joint US-Canadian project at Grosse Île, which produced an effective vaccine against rinderpest, a deadly animal virus. Grosse Île was also the site of another top secret program, the production of over 380 litres of virulent anthrax spores for a possible BW weapon. According to British and American strategists, the only way to deter a German BW attack was to develop a massive retaliatory capability. As a result, by the spring of 1944 plans were underway to produce two million anthrax bombs, sufficient to "drench" many of Germany's major cities. Fortunately, this policy was never implemented.[278]

The spectre of biological warfare was reinforced during the fall of 1944 when hundreds of large Japanese balloons began drifting eastward across the North American continent, with possible BW payloads. In responding to this threat, the Canadian government implemented a BW civil defence emergency program, coordinated by E.G.D. Murray, working closely with American scientific and military authorities. While no germ warfare attacks occurred, this incident established an important precedent for future North American cooperation against weapons of mass destruction. Indeed, this initiative was greatly expanded during the Cold War with the creation of an integrated Canadian-American aerospace defence system against Soviet bombers and missiles.[279]

In September 1947 Joburg's important wartime achievements were publicly acknowledged when he was awarded the United States Medal of Freedom "as a member of the Joint United States-Canadian Commission and later as Chairman of the

---

276 See documents in chapter 2.
277 On 25 July 1945 Murray submitted his resignation as chairman of the C-1 Committee to Otto Maass. Predictably, he recommended Guilford Reed as his successor. See HQS, 4354-33-17, Murray to Maass, 25 July 1945.
278 Avery, *Science of War*, 151-75; see also Erhard Geissler and John Ellis van Courtland Moon, eds., *Biological and Toxin Weapons: Research, Development and Use from the Middle Ages to 1945* (Oxford: Oxford University Press, 1999).
279 Avery, *Science of War*, 168-75.

Committee for Direction of Biological Warfare Research in Canada."[280] But why was he such a successful wartime administrator? One important asset was his high reputation among his scientific colleagues on the C-1 Committee and at the two BW testing sites, who included many of Canada's leading bacteriologists and veterinarians.[281] This bond was reinforced by their shared concerns about the threat and about the importance of working quickly, along with their British and American scientific counterparts, in developing defensive measures and retaliatory/deterrent weapons. Murray also played an important part in the gradual evolution of the Anglo-American-Canadian BW alliance system because of his ability to facilitate tripartite cooperation. In these difficult negotiations, Joburg had an advantage since he knew most of the key British defence science administrators because of his previous career at Cambridge. This was a special asset in dealing with the often arrogant Paul Fildes, who directed the British BW program at Porton Down.[282] On the other side of the equation, his prewar involvement with the Society of American Bacteriologists was an asset in working with key American BW planners such as Dean E.B. Fred, General Kelser, George Merck, Richard Shope, and Ira Baldwin. Significantly, Joburg's wartime friendships continued into the postwar years, particularly with his American colleagues.[283]

In recent times, as awareness about Canada's secret biological warfare program has increased, there have been serious questions whether the activities of E.G.D. Murray and his C-1 Committee colleagues violated contemporary scientific and medical values.[284] While this is a complex issue, the historical record demonstrates that biological warfare research was viewed, by most scientists working in this field, as crucial for national security, since the country faced powerful and ruthless adversaries. Frederick Banting captured the essence of this challenge in his diary entry of 20 October 1939: "This war is one of brains and scientific ingenuity rather than

---

280 UWO Archives, E.G.D. Murray Collection, Office of the Military Attache, United States Embassy to Murray, 9 September 1947; Murray to Colonel Williamson, 11 September 1947.
281 For Everitt it was certainly an advantage that two of the key defence science administrators – Otto Maass (Department of National Defence) and J.B. Collip (Associate Committee on Medical Research) – were his McGill colleagues and close friends.
282 Sir Henry Dale, W.W.C. Topley, and Lord Edgar Stamp were three other important British BW administrators.
283 See the 1946-47 correspondence in chapter 2.
284 In Canada, the publication of John Bryden's *Deadly Allies: Canada's Secret War, 1937-1947* (Toronto: McClelland & Stewart, 1989) created considerable controversy, particularly because of Bryden's inaccurate claim that large-scale anthrax field trials had taken place at Suffield during the Second World War, and that the region was still seriously contaminated. In contrast, Sheldon Harris's *Factories of Death: Japanese Biological Warfare 1932-45 and the American Cover-up* (London: Routledge, 1994), based on solid research, demonstrates how large numbers of human subjects had been used by the Japanese BW program.

one of manpower, mass infantry and massacre. Our job is to lick the Germans under Hitler."[285]

War and the Murray Family

The impact of the Second World War created a number of challenging situations for the Murray family. Most of these involved Joburg and his onerous wartime commitments as chairman of the C-1 Committee. As a result of his hectic administrative schedule, Murray had virtually no time to supervise the operation of McGill's Department of Bacteriology and Immunology or to pursue his own research projects. Moreover, given the tight secrecy surrounding this weapon system, Joburg could not discuss his activities with either friends or family.[286] Fortunately, there were occasions when he and Freda could arrange special visits to their Bark Lake retreat, times which Joburg jealously guarded. As Bob Murray remembers, "He would arrange a holiday and tell 'them' that 'It will cost you $100 to find me, and more to bring me back.'"[287]

Joburg's return to McGill in September 1945 was accompanied by a flurry of activity, as he attempted to rebuild his department, buffeted by wartime budget restrictions and diversion of faculty resources. While he was pleased with Principal Cyril James's public declaration that McGill was "proud of the accomplishments of its scientists in the last six years," he wanted promises of additional academic positions, not platitudes. Fortunately, his department managed to retain its core faculty group, notably Fred Smith, Gertrude Kalz, and J.W. Stevenson, but the challenge of regaining its 1939 scientific status was formidable.[288] Moreover, in terms of his own career development, he saw the war years as a period of sacrifice and frustration. As he wrote Bob, "the war set me back 5 years & I shall have difficulty in catching up, if ever I do so."[289]

Freda Murray had her own set of wartime challenges. The most serious were the long periods of separation from her beloved "Ever" because of his biological warfare

---

285 University of Toronto Archives, Diary of Sir Frederick Banting.
286 Bob Murray remembers that he and his father would talk generally about Canada's BW program while fishing at Bark Lake, but without any of the logistical details or policy concerns. At times, there were references to Joburg's various confrontations with Paul Fildes and Lord Stamp, while his views towards E.B. Fred and the other BW scientists was most positive. He did, however, have some reservations about US dominance in the field of biological warfare at the end of the war. Robert Murray, interview by Donald Avery, 10 October 2007.
287 Robert Murray, "A 'Mum for All Seasons.'"
288 Ibid. Joburg admired Dean Jonathan Meakins, but felt that it was his duty to push hard for the sake of the department. His relationship with James was more complex, and he felt rather aggrieved about the circumstances surrounding his return to the university in the fall of 1945. See chapter 2.
289 UWO Archives, E.G.D. Murray Collection, E.G.D. Murray to Robert Murray, 17 November 1946.

responsibilities.[290] She also had concerns about Bob's safety while he was studying in England and during his return to Montreal in 1941, given the ferocity of the German aerial blitz on British urban centres and the threat of U-boat attacks during the ocean crossing. On the domestic front, there were educational, career, and personal issues involving her daughter, Susan, a somewhat rebellious teenager.[291] These tensions continued after Susan's graduation from high school during her subsequent employment as "a draughtsman & cartographer" at Dorval Airport, under Ferry Command, a job that meant irregular hours and uncertain transportation.[292]

Outside the home, Freda had a number of important wartime commitments that taxed her energy and ingenuity. The most important of these was her involvement with the Montreal chapter of the Imperial Order of the Daughters of the Empire [IODE] food conservation campaign to assist the common-law wives and families of Canadian servicemen who were not getting their pay and allowances, at a time when Canada's welfare state was still underdeveloped.[293] Conditions in Montreal were particular grim for these female dependents because of limited housing, inadequate child-care services, and censorious moral standards that branded "sexual disloyalty" as grounds for disqualification for any social assistance. Moreover, "as soldiers and veterans transferred their primary allegiances and allowances from mothers to wives, mothers not only felt displaced, but suffered tangible consequences."[294] Freda's attempts, through the IODE, to deal with these social problems, were described by her daughter Susan:

> During the war, on certain days of the week, there would be a group of ladies from the I.O.D.E., all wearing blue smocks over their clothing, volunteers who were there to separate & bundle those recyclable items ... *junk* to most people but usable for the War Effort. The sale of this 'junk' brought money to run the efficient 'Emergency Food Supply' that Mum had organized & which brought relief to the families of soldiers who were desperately poor.[295]

---

290 Freda's correspondence in the spring and summer of 1944 demonstrates the difficulties of wartime separation. See chapter 3; also, Peter Murray Personal Collection, Markham, Ontario.

291 Susan Murray Robinson relates that she greatly missed Bob, particularly since "there was about a four year period during my formative years when I did not have my brother to champion my efforts of being of the 40's and not the Edwardian Era! I suppose I was also somewhat rebellious or resentful in my early teens, for I was not allowed to 'date' or have boyfriends until the age of sixteen." Susan Murray Robinson, "My Mother Had a Profound Influence on Me."

292 Ibid.

293 Susan claims that Freda's sense of British "class distinction" was substantially reduced by this wartime commitment, and particularly by her involvement in the Montreal campaign against slums. Ibid; also see chapter 3.

294 The administrative procedures of the federal Dependents' Allowance Board were moralistic and harsh. Magdalena Fahrni, *Household Politics: Montreal Families and Postwar Reconstruction* (Toronto: University of Toronto Press, 2005), 74.

295 Susan Murray Robinson, "My Mother Had a Profound Influence on Me."

Robert Murray's wartime experiences also had a number of interesting dimensions. First, there was his November 1938 academic sojourn to Cambridge, during the Munich Crisis, as a medical student. During the initial nine months, this homecoming provided a wonderful opportunity of renewing family contacts and previous friendships, while adjusting to the heavy demands of his Pathology NST program.

In many ways, Cambridge had not substantially changed since the Murrays' departure in 1930. It was still a relatively small, elitist academic institution, with impressive scientific and scholarly achievements, that excluded women students from full academic status.[296] While there had been some important changes in the governing structure and academic procedures, the "split" institutional distinction between university membership and college loyalties remained intact. The medical education program also remained fragmented, despite demands for a more independent faculty and a separate Medical Tripos.[297] On the other hand, during the previous eight years, Cambridge medical students had become increasingly aware of problems associated with poverty, limited health services, and the challenge of fascism at home and abroad. These trends were reflected by the formation of the Medical Peace Campaign and the growing popularity of social medicine, which was strongly endorsed by Regius Professor John Ryle as "clinical medicine activated ... by social conscience as well as scientific interest."[298] Closely related were the activities of a small group of leftist students, drawn from the humanities and the sciences, who believed that British society was morally bankrupt.[299] In contrast, they viewed the Soviet Union as a model of how science and Marxism could converge and create a dynamic and caring society, while also providing leadership in containing fascism.[300]

Robert's Cambridge agenda did not include involvement with political protest or consultations with the university's scientific left.[301] Throughout 1939, he was too busy

---

296 Women students, and their colleges, gained full status at Cambridge after the Second World War.
297 The Cambridge Medical Students Society was formed in 1921. Editorial, *Medical Society Magazine* (1936), cited in Weatherall, *Gentlemen, Scientists and Doctors*, 269.
298 Ryle was Regius Professor of Physic between 1936 and his resignation in 1941. Many of his ideas about social medicine were outlined in his 1942 book, *Social Pathology and the New Age of Medicine*." Ibid., 299.
299 Many of the leading Cambridge leftists were members of the secret Apostles society. These included John Cornford, James Klugman, Michael Straight, and the famous group of spies – Guy Burgess, Kim Philby, Donald Maclean, John Cairncross, and Anthony Blunt. Andrew Sinclair, *The Red and the Blue: Intelligence, Treason and the Universities* (London: Weidenfeld and Nicolson, 1986); Yuri Modin, *My Five Cambridge Friends* (Paris: Headline, 1994).
300 Cambridge's left-wing scientists included J.B. Bernal, Hyman Levy, J.B.S. Haldane, Lancelot Hogben, Patrick Blackett, and Joseph Needham. Gary Werskey, *The Visible College: A Collective Biography of British Scientists and Socialists of the 1930's* (London: A. Lane, 1988).
301 On occasion, both Everitt and Robert did encounter the biochemist Joseph Needham on

arranging his academic schedule, with the assistance of his College tutor, the physicist C.P. Snow,[302] who "carefully outlined the course of study that led toward medicine ... and we agreed which courses would be essential (i.e. all of them) in the first two years ... You had to take a least two major subjects during Part I and they were not fooling around about the degree of concentration offered."[303] Ironically, a number of Bob's professors had previously been Joburg's colleagues; these included H.R. Dean, Tibby Marshall, E.T.C. Spooner, and E.D. Adrian (later Lord Adrian). In addition, his pathology program, as Part II of the Natural Science Tripos, had been devised by his father back in 1925! On the social side, there were mellow hours spent with fellow pre-medical students at their favourite Cambridge pubs.[304] This pleasant scenario changed dramatically in September 1939 when Nazi Germany invaded Poland. As Bob remembers, "all this work and fun took place in a country fearing war in 1938-39 and at war from Sept. 1939 on."[305]

Like the rest of Britain, the University of Cambridge was transformed by the impact of total war. First, there was the threat of an imminent German invasion, which persuaded some faculty members to send their families to Canada or the United States for safekeeping. Second, there was the possibility that the intense German bombing of London would shift towards neighbouring centres. Third, the demands of the British scientific military effort drew numerous Cambridge faculty, particularly those from the medical and natural sciences, into the war effort.[306] Although Bob Murray, as a visiting Canadian student, was protected from the British draft, there were no guarantees that this exemption would continue if the war situation deteriorated further. As a result, by the end of 1940 Joburg was convinced that his son should leave Cambridge and continue his medical studies at McGill. But returning home was not an easy undertaking, as Robert Murray describes:

purely scientific issues. UWO Archives, R.G.E. Murray Collection, Robert Murray, "The Cambridge Experience" (hereafter cited as Robert Murray, "The Cambridge Experience").

302 Charles Percy Snow (1905-1980), renowned British novelist and scientific administrator, was a Christ's College tutor from 1934 to 1945. During the Second World War he worked with the British Cabinet's scientific advisory committee, particularly in the recruitment of scientists to work on radar, the atomic bomb, and other military technologies. Despite Snow's subsequent literary fame for his critique of Cambridge's "two cultures," Bob found his tutor rather humourless and severe. Ibid; also see C.P. Snow, *The Two Cultures, and a Second Look* (Cambridge: Cambridge University Press, 1964).

303 Robert Murray, "The Cambridge Experience." In addition, he was issued a "grey book" for signed records of medical subjects taken "in case I did in fact go on with medicine in the UK."

304 Bob's personal choice of pubs was The Eagle, where Jim Watson and Francis Crick often met in 1953 prior to their dramatic announcement about the structure of DNA.

305 Robert Murray, "The Cambridge Experience."

306 By 1941 most colleges were only one-half full, and many retired faculty were rehired to replace those recruited into wartime service. Brooke, *A History of the University of Cambridge*, 505-510.

The conscription Board required formal evidence of acceptance as a medical student at McGill and Canada House in London had to start the longer process of arranging for an Exit Permit and a passage by ship to Canada. ... There was time to fill preferably to some purpose to learn physical diagnosis. ... A visit to Addenbrooke's Hospital introduced me to two of the Honorary Physicians who were willing to take a student on Ward. ... Also had a night job three days a week as Fire Watcher ...[307]

Romantic reasons provided another incentive for Robert's return. During the previous three years he had been involved in an intense relationship with Doris Marchand, who he had met during the summer of 1938 at the Mount Desert Biological Laboratory in Maine. After Bob's departure for England, this became a long-distance romance, sustained by their many "love letters." Although only Robert's side of the correspondence has survived, it provides some marvellous insights into a range of personal and political events that occurred during these tumultuous years.[308] These included the challenges of Robert's preclinical medical studies, the difficulty of being a student in wartime Britain, the impact of the Blitz on English civilian morale, and the failure of the United States to become involved in the conflict against Nazi Germany. The most important subject was, of course, how he and Doris could share their lives together.[309]

In wartime, this was not an easy goal to achieve. Even after Robert's return to Montreal in 1941, there was the harsh reality that they lived in different countries, and that Bob's exemption from the Canadian draft was dependent on the completion of his McGill medical program. The situation, however, greatly improved during the fall of 1943 when Doris left her job at the Rockefeller Institute of Medicine in New York City and came to Montreal. On 15 January 1944 they were married in a small ceremony at Divinity Hall Chapel on the McGill campus. After a brief honeymoon, Robert began his duties with the Royal Canadian Army Medical Corps in Barrie, Ontario. In 1945 Alice, their first child, was born.

As a wartime medical student, Robert Murray's training was accelerated, in part because of McGill's special crash course and also because he had already obtained a degree in pathology from Cambridge. He also benefited from the fact that his studies in bacteriology were carried out in his father's laboratory, now under the direction of acting chairman Fred Smith, while his clinical work at the Royal Victoria Hospital

---

307 Robert Murray, "A Personal Narrative."
308 Bob Murray did have an opportunity to return home during the summer of 1939, and was in New York City visiting Doris Marchand when war was declared. His father was also in the city, attending the meeting of the International Association of Microbiological Societies. Robert Murray, interview by Donald Avery, 13 June 2007.
309 Bob's descriptions of crossing the U-boat–infested North Atlantic in November 1941 are quite remarkable, in terms of both detail and colourful prose. See chapter 3.

was closely supervised by Dr. Gertrude Kalz. Although both were family friends, there was no favouritism or lowering of professional standards.[310] With graduation in the spring of 1944 came another set of choices. Should he consider taking a PhD after his war service was completed? And if so, should he follow some of his friends, such as Roger Stanier, and seek admission to bacteriological programs at Harvard, Berkeley, or other prestigious US universities?

There were a number of reasons why Robert decided against taking a doctorate in microbiology.[311] Above all, there was the example of his father's scientific career: Everitt, as a medical doctor, had achieved an impressive record as a university teacher, hospital clinician, and academic researcher in the UK and in Canada. From Bob's perspective in 1945, there was also the feeling that as a medical doctor he would have access to all the necessary medical information, and could master the essential scientific challenges "without wasting two years pursuing a Ph.D."[312] This incentive to normalize his life and obtain a secure job was reinforced by his responsibility for his wife and their recently born daughter. And finally, there was an attractive job offer from G. Edward Hall, the newly appointed dean of medicine at the University of Western Ontario.[313]

The Western position had many attractive features.[314] First, there was the opportunity of working closely with the departmental chair, Igor N. Asheshov, a well-travelled White Russian bacteriologist who was a specialist in the exciting field of bacteriophage research.[315] Secondly, the UWO medical faculty was gradually being transformed through the imaginative and aggressive recruitment campaign of Dean G. Edward Hall. Among the new recruits were a group of gifted researchers

---

310 See Bob Murray's description of his McGill experience in chapter 3.
311 A study of sixty-six American medical schools during the 1930s discovered that forty-nine of the heads of the bacteriology departments had an MD degree rather than a PhD. In contrast, only seventeen of the heads in biochemistry were MDs. William Rothstein, *American Medical Schools and the Practice of Medicine: A History* (New York: Oxford University Press, 1987), 155-58.
312 Robert Murray, interview by Donald Avery, 13 June 2007.
313 The detailed negotiations leading to Bob's appointment at UWO are documented in chapter 3. G. Edward Hall (1907-1972) was one of Canada's foremost medical scientists, university administrators, and public figures during a career that spanned four decades. Between 1944 and 1947 he was dean of medicine at Western, and then became university president from 1947 to 1967. See chapter 3, note 73.
314 The UWO Faculty of Medicine was established in 1881, four years after the founding of the university. Murray L. Barr, *A Century of Medicine at Western* (London, ON: University of Western Ontario, 1977), 61-88.
315 Asheshov joined UWO in 1937. In 1942 the Department of Bacteriology and Immunology became a separate academic unit, distinct from Pathology. At that time, Asheshov was named chair. Ibid., 376, 430.

that included Roger Rossiter (biochemistry), Otto Edholm (physiology), Alan Burton (biophysics), Edgar Hobbs (clinical preventive medicine), Murray Barr (anatomy), James Stevenson (physiology), and Bob Noble (endocrinology).[316] Moreover, when Hall became president in 1947, he chose the eminent endocrinologist Bert Collip as his successor, further enhancing the image of Western's medical school as a place of innovation and creativity.[317] Collip described these favourable developments, in part, as an explanation of why he was abandoning McGill after so many years:[318]

> no endocrinology work of any significance has been done in this Institute [McGill] ... [and] I cannot have many regrets at leaving the top floor of the west wing of the Medical building now so hopelessly inadequate for modern laboratory space. ... At London on the other hand the situation is almost ideal. It's a small school strictly limited to 60 students per year. The departments for the most part are staffed by young men in their thirties with already proven abilities, with fine personalities and a cooperative spirit. ... The Med Library is one of the best in Canada, a brand new 600 bed hospital is just across the street from the teaching and research labs and there is the closest cooperation between the surgical and medical dept. ... It really gives me a great thrill to see all of this ...[319]

Did Joburg influence Robert's appointment? There is no evidence that this occurred, or that Robert asked his father to intervene.[320] On the other hand, it was obviously an asset having a father who was one of Canada's leading bacteriologists and who shared many wartime experiences with Dean Hall. The most important factors, however, were Bob's own achievements: his exemplary academic record at Cambridge and McGill, and his ability to convince Igor Asheshov that they could develop a creative partnership.

---

316 Other notable additions to the faculty were Theodore Coffey (physical medicine and rehabilitation) and D. Cecil McFarlane (ophthalmology). Bob Noble came with Collip in 1947, while James Stevenson arrived in 1950. Ibid.

317 James Bertram Collip (1892-1965) was a pioneer in endocrine research, especially in its biochemical aspects. In the early 1920s at the University of Toronto, Collip worked on the insulin project with Frederick Banting and Charles Best, and he shared in J.J.R. Macleod's portion of the 1923 Nobel Prize in Physiology or Medicine. In 1928 he assumed the chair of the Department of Biochemistry at McGill University. He was dean of medicine at the University of Western Ontario from 1947 to 1961. He was elected a Fellow of the Royal Society of Canada in 1933. M.L. Barr and R.J. Rossiter, "James Bertram Collip, 1892-1965," *Biographical Memoirs of Fellows of the Royal Society* 19 (December 1973); Li, *J.B. Collip*.

318 Collip's motivation for leaving McGill after nineteen years was also related to his disappointment in not being appointed dean of medicine (Fred Smith got the job); his uneasy working relationship with some of his colleagues in the Institute of Endocrinology; and Western's generous offer.

319 Collip to Charles Martin (former dean of medicine), April (undated) 1947, cited in Li, *J.B. Collip*, 142-43.

320 See the relevant sections in chapter 3.

And this quickly took place. After a four-year apprenticeship, in 1949 Bob Murray was appointed full professor and chairman of the Department of Bacteriology and Immunology, a position he held for the next thirty-five years. A rare achievement indeed!

Postwar Developments

Between 1945 and 1964 there was amazing growth in the size and influence of Canadian universities. As historian Paul Axelrod has described, "The Second World War, which disrupted and transformed so many facets of Canadian life, was instrumental in altering public perceptions about the respectability and value of higher education."[321] In terms of students, the surge of veterans, assisted by federal assistance programs, created a unique educational environment since these young men and women were highly motivated to succeed.[322] Unfortunately for Canada's university system, the dramatic increase in enrolment was not matched by sufficient government funding, although the situation varied from province to province. Ontario universities, for example, were more fortunate than their counterparts in Quebec, where the Duplessis government (1944-1960) showed little interest in expanding its support of higher learning.[323] This situation became most acute after the recommendations of the 1949 Massey Commission, which called for increased university funding through federal-provincial cost-sharing arrangements.[324] Quebec refused to participate.[325]

These developments had a powerful influence on the twelve Canadian medical faculties that were operative in 1950.[326] In particular, there was the ongoing challenge

---

321 Paul Axelrod, *Scholars and Dollars: Politics, Economics and the Universities of Ontario 1945-1980* (Toronto: University of Toronto Press, 1982), 14.
322 Because of the veterans, university enrolment soared from 38,000 in 1944-45 to over 80,000 in 1947-48. The National Conference on Canadian Universities predicted that these numbers would increase to over 125,000 by 1965. Ibid., 19, 23.
323 Until the 1960s the salaries of university professors remained "humiliating low." Ibid., 33.
324 The full title was the Royal Commission on National Development in the Arts, Letters and Sciences.
325 Under Ottawa's 1951 funding formula McGill was to receive $615,270 in federal funds. Threats from the Duplessis government, however, forced McGill to decline this financial support. Frost, *McGill University*, 251-52.
326 After the war, the University of Ottawa (1945) and the University of British Columbia (1950) joined the ten existing schools: McGill (1822), University of Montreal (1843), University of Toronto (1843), Laval University (1852), Queen's University (1854), Dalhousie University (1868), University of Western Ontario (1882), University of Manitoba (1883), University of Alberta (1913), University of Saskatchewan (1926); see N. Tait McPhedran, *Canadian Medical Schools: Two Centuries of Medical History, 1822 to 1992* (Montreal: Harvest House, 1993). For general background see Yves Gingras, "Financial Support for Post-Graduate Students and the Development of Scientific Research in Canada," in *Youth, University and Canadian Society: Essays in the Social History of Higher Education*, ed. Paul Axelrod and John Reid (Kingston: McGill-Queen's University Press, 1989), 300-341.

of allocating funds to the various subspecialties, which became more pronounced, often in response to technological advances such as the heart-lung machine, the dialysis machine, the automatic respirator, and the electron microscope. "Hospitals, traditionally organized by departments, had to adjust to subspecialties, and a new breed of cat, the full-time clinician-scientist."[327] Within Canada's medical community, a 1948 survey showed that there were 955 persons involved in research at universities, along with 278 technicians, 315 graduate students, and 311 part time clinicians.[328] The trend towards more research-based Canadian medicine was enhanced by the formation in 1958 of the autonomous Medical Research Council, even though its funding resources were far short of those achieved by the US National Institutes of Health or the British MRC.[329]

The gradual movement towards universal and comprehensive medical insurance was another important development in these years. Of particular importance was the establishment of the joint federal-provincial Hospital Insurance and Diagnostics Services Act of 1958, and Saskatchewan's bold decision in 1962 to implement its own medicare system.[330] Four years later the Pearson government would launch its own federal medical insurance scheme, drawing on the recommendations of the Royal Commission on Health Care (the Emmett Hall Commission).[331]

The Murrays and Scientific Medicine, 1945-55

The postwar expansion in funding for medical research had a powerful impact on the postwar careers of Everitt and Robert Murray. Indeed, this issue is extensively discussed in correspondence between the two men throughout the period 1945-1955, from the perspectives of their respective academic institutions. These were also years when microbiology came of age, with exciting developments in the fields of antibiotics, bacterial physiology, cell biology, taxonomy, genetics, molecular biology, and microscopy. Much of this research was dependent on complex and expensive

---

327 McPhedran, *Canadian Medical Schools*, 21.
328 Cited in Li, *J.B. Collip*, 164. Only fifty-one of the 1948 group of Canadian medical researchers were full-time.
329 Previously medical research had been coordinated at the federal level by the Medical Subcommittee of the National Research Council.
330 In 1958 five provinces joined the federal hospital insurance system – Alberta, Saskatchewan, Manitoba, Newfoundland, and British Columbia. The following year, four other provinces endorsed the scheme, although Quebec was not affiliated until 1961. C. David Naylor, *Private Practice, Public Payment: Canadian Medicine and the Politics of Health Insurance 1911-1966* (Kingston: McGill-Queen's University Press, 1986).
331 In 1969 Ontario passed its own Ontario Health Insurance Plan (OHIP). In 1984 Canada's medicare system was substantially revised by the Canada Health Act, particularly by its strong stance against extra billing by physicians. Ibid.

technology, a far cry from the meagre resources of the prewar years. "We used to joke we were 'card-carrying loop and needle bacteriologists'," writes Robert Murray, "and that was not far from the truth if you add to it a microscope, stains, an incubator, suitable media for cultivation and an infinity of patience and observational skill."[332]

For the senior Murray, operating his department after 1945 brought familiar challenges.[333] These included hiring new faculty, coping with the high turnover of technicians, adding new programs in virology and mycology, supervising the construction of a new laboratory, and, above all, trying to operate an expensive department within a cash-strapped university.[334] A major part of the problem was McGill's inability either to accept federal funding or to secure alternative private sources. But, as his colleagues have documented, "E.G.D. was really in his element during these years and his influence as a man and a teacher reached its highest level. Many microbiologists in senior positions in Canada and to some extent in the U.S.A. are products of those years."[335]

One of the unfortunate aspects of McGill's financial problems was a serious "brain drain," whereby many promising young faculty members and graduate students accepted positions in other provinces or gravitated towards well-endowed American universities.[336] This latter trend greatly disturbed Joburg, who deplored Canada's loss of scientific talent. That is not to say that Joburg was a narrow nationalist. On the contrary, since arriving in Canada in 1930, he regularly participated in the annual meetings of the Society of American Bacteriologists and had a wide range of contacts among American scientists. Many of these were involved with *Bergey's Manual*, engaged in *Listeria* research, or were members of the International Association of Microbiological Societies (IAMS).[337] His stature in the United States was enhanced in 1947 when

---

332 Robert Murray, "Developing streams affecting biomedical sciences, esp. Microbiology."
333 After 1945 there were several new empire builders at McGill, notably R.F. Vivian, chairman of the Department of Health and Social Medicine, and psychiatrist Ewen Cameron, who transformed the Allan Institute into a major North American centre for the treatment of mental illness before his sudden resignation in 1964. For the latter subject see Theodore Sorkes and Gilbert Pinard, eds., *Building on a Proud Past: 50 Years of Psychiatry at McGill* (Outremont, Quebec: Productions Immeda, 1995).
334 Joburg's 1946-47 letters abound with references to various confrontations with Principal James and Dean Meakins, and to aggressive departments encroaching on B & I turf. Nor did the situation improve when his colleague Fred Smith became dean in 1947, although he greatly lamented Smith's sudden death in September 1949.
335 Gertrude Kalz and Hugh Starkey, cited in G.S. Wilson, "Everitt George Dunne Murray," 650.
336 The elite US universities (Harvard, Yale, Berkeley, etc.) were able to obtain substantial grants for medical research through the National Institutes of Health and the National Science Foundation. See Toby Appel, *Shaping Biology: The National Science Foundation and American Biological Research, 1945-1975* (Baltimore: Johns Hopkins University Press, 2000).
337 In 1954 Joburg was surprised that he had been nominated for the position of SAB vice-president

he agreed to submit several articles for an important scientific study, *Bacterial and Mycotic Infections of Man*, edited by René Dubos of the Rockefeller Institute of Medical Research. One of these submissions was a joint paper with his colleague Gertrude Kalz, "The Cultivation and Identification of Pathogenic Bacteria."[338] As he explained, "Our purpose has been to help the practitioner and the medical student to understand the requirements and limitations of the laboratory in contributing to reach a correct diagnosis." Both of these scholarly contributions were favourably reviewed.[339]

Yet, despite his many positive experiences in dealing with American-based scientists, during the 1950s Joburg became increasingly shocked by the destructive impact of the Cold War on the US scientific community. In his opinion, young Canadian researchers should be aware of the pitfalls of pursuing graduate studies in the US, since "at present they have to sign that they are prepared to be called up for military service & they are called up at once!"[340] In 1953, Joburg extended this advice to his son: "I think it probable that you will have to consider offers from time to time. ... It is sometimes hard to be sure of the security & political influences [which] have an awkward prominence in some places in the USA."[341]

While E.G.D. rarely visited the United Kingdom after 1945, he was quite interested in maintaining contact with his many scientific and medical friends. Indeed, his correspondence with Robert contains numerous references to the death of prominent British scientists, controversial academic appointments, and a good dose of Cambridge University gossip.[342] He was also quite intrigued by the fact that many younger British medical scientists were dissatisfied with the Labour government's evolving state medical system[343] and sought academic jobs in Canada.[344] While Joburg shared

---

(future president). Although he lost, it was only by 149 votes – a good showing "for a Canadian against an American." Joburg to Bob, 6 November, 7 December 1954.

338 René Dubos was regarded as one of the foremost experts on bacterial diseases and antibiotics. This volume – *Bacterial and Mycotic Infections of Man* (Philadelphia: J.B. Lippincott, 1948) – was accompanied by a related study edited by virologist Tom Rivers, director of the Rockefeller Institute Hospital, *Viral and Rickettsial Infections of Man* (New York: J.B. Lippincott, 1948). Carol Moberg, *René Dubos: Friend of the Good Earth* (Washington, DC: ASM Press, 2005); *Tom Rivers: Reflections on a Life in Medicine and Science; an Oral History Memoir*, prepared by Saul Benison (Cambridge, MA: MIT University Press, 1967), 460-61.

339 Joburg to Dubos, 8 October 1947; Dubos to Joburg, 28 October 1948. See Appendix E.

340 Joburg to Bob, 17 February 1951.

341 Joburg to Bob, 27 January 1953. Bob shared his father's concerns about the anti-Communist hysteria.

342 In February 1947, for example, Bob informed his father that bacteriologist E.T.C. Spooner, a mutual friend from Cambridge University, had been appointed to a chair at the University of London.

343 In 1947 Western hired two British scientists: Roger Rossiter, a biochemist from Oxford, and Otto Edholm, a physiologist from King's College, London. Bob to Joburg, 26 February 1947.

344 Three obituaries were of particular interest: Almroth Wright (1947), Joseph Barcroft (1947), and Mervyn Gordon (1954).

many of the prevailing biases against government-controlled medicine, he often complained that in Canada "the cost of sickness has reached such absurdity that the hospitals are forcing state medicine to the forefront."[345] It is also instructive that, during the late fifties, some of his most flattering comments about innovative medical research focused on developments in Saskatchewan, where the CCF government of Tommy Douglas was poised to introduce a comprehensive medical insurance system.[346]

Throughout the postwar years, Joburg's involvement with Canada's bacteriological community was far-reaching and diverse. First, there was his interaction with most of Canada's elite microbiologists, some of whom had shared his wartime BW experiences, either as members of the C-1 Committee or through involvement with specific research projects.[347] For this reason, his relationship with Guilford Reed was particularly close.[348] Moreover, throughout the postwar years Reed and Murray arranged a number of cooperative activities; for example, the placement of younger faculty and graduate students, and the exchange of biological specimens.[349] A related trend was Joburg's close association with many of the leading figures of the Canadian Public Health Association (CPHA), including R.D. Defries of the Connaught Laboratories and his colleague, G.D.W. Cameron, who in 1946 became deputy minister of the Department of Health and Welfare.[350] Joburg's reputation as an innovative scientist was evident in April 1956 when Deputy Minister Cameron suggested he would like to meet his old friend in Ottawa for an "old-time chat" about mutual interests: "We have been doing a lot of thinking on questions of policy in aiding research and the views of those who have long experience in the field would certainly represent a valuable contribution."[351]

Joburg also had an impressive profile within Quebec medical circles. This was evident in November 1949 when, as part of a government-appointed subcommittee,

---

345 Joburg was aghast at the costs associated with Bob's extended stay at the Ottawa Civic Hospital following his 1951 attack of meningitis. He also claimed that "it would have been a much heavier bill had it been the R.V.H." Everitt to Bob 21 October 1951.

346 In September 1954 Joburg visited Wendell MacLeod, the controversial dean of medicine at the University of Saskatchewan, and was impressed with his creativity and energy.

347 For example, Guilford Reed (Queen's), Philip Greey (Toronto), James Craigie (Connaught Laboratories), and Charles Mitchell (Department of Agriculture) had been members of the C-1 Committee.

348 For example, on 4 April 1946 Joburg informed Bob that J.W. Stevenson would be joining his department as an assistant professor, replacing Ted Roy: "He has been working with G.B. Reed at Queens all the war. A very good fellow & I hope I can make him happy here."

349 An even more elaborate system was established between McGill and Western after 1948 when Robert became acting chairman of the Department of Bacteriology and Immunology.

350 In January 1951 Joburg was asked, on the recommendation of G.D.W. Cameron, whether he would give a paper at the annual meeting of the CPHA on the topic "The Emergency for Research in Unexplored Fields of Public Health." See Appendix E.

351 Don Cameron to Joburg, 12 April 1956. See Appendix E.

he assessed the capabilities of all Quebec City laboratories "which could possibly do research on Public Health problems."[352] Four years later, in 1953, he was a member of the subcommittee on research of the Health Sciences Inquiry of Quebec, along with his close friend, bacteriologist Armand Frappier of the University of Montreal.[353] After providing an extensive analysis of many of the more outstanding health problems, the Frappier Report called for a number of crucial measures, including the creation of a provincial office of medical research, with a separate budget.[354]

This commitment to improve public health standards in Canada did not end with Joburg's retirement from McGill. For example, between 1957 and 1960 he was chairman of the NRC Associate Committee on Control of Hospital Infections, which sought solutions for the growing problems of antibiotic-resistant staphylococcal bacteria.[355] Nor was it surprising that Joburg should have been asked to direct this pioneering investigation, since within Canada, he was one of the leading skeptics of the viewpoint that antibiotics could fully conquer pathogens.[356] Such attitudes were, in his opinion, merely faulty boosterism that did not recognize the adaptability of bacteria, the complex relationship between host and pathogen, and the fact that antibiotics themselves created new disease environments. In fact, he noted that recent US studies had shown "that pneumonia is just as common as in pre-sulpha-antibiotic days, but the age incidence is shifted & the mortality reduced." Robert shared his father's concerns, commenting in January 1951 that the antibiotic business "was getting infernally complicated and hard to rationalize."[357]

Joburg's insistence on the need to adopt more critical and holistic approaches to medical problems was revealed in a number of ways. First, his interest in the linkages between animal and human pathogens led to his study of various zoonotic diseases,

---

352 Joburg was not impressed with the quality of these labs, which included those associated with Laval University. In his opinion, only four met the necessary standards, a problem he attributed to "bad organizational design." Joburg to Bob, 27 November 1949.
353 During the 1930s, Armand Frappier founded the Institut de microbiologie et d'hygiène at the University of Montreal. It is now known as Institut Armand-Frappier, and is one of Canada's major research centres for the life sciences. In 1953 Frappier was chairman of the Health Sciences Inquiry of Quebec (Technical Committee), a group that included E.G.D. Murray. He was also president of the Canadian Society of Microbiologists during the early 1950s, and a major authority on Quebec health issues.
354 Health Sciences Inquiry of Quebec (Technical Committee), 1953: LAC, E.G.D. Murray Papers, vol. 5.
355 See Appendix E.
356 In part, Murray was influenced by the views of René Dubos, one of the world's foremost experts on drug-resistant bacteria. As early as 1944 he had shown that pathogens could evade sulpha drugs "by producing resistant strains." He later extended this model to the use of penicillin. Moberg, *René Dubos*, 62–80.
357 Joburg to Bob, 4 February 1951; Bob to Joburg, 23 January 1951.

and was integral to his discovery in 1926 of the bacterium *Listeria monocytogenes*.[358] His work in this area was extended in the late 1940s when he became interested, as a member of the Arctic Institute, in tracing disease patterns among Arctic animals and subsequent incidents of gastrointestinal outbreaks among the Inuit.[359] Second, there was his long-time association with the Canadian Veterinarian Association and his support of various attempts to upgrade veterinary medicine and animal pathology.[360] These initiatives were reinforced by his friendship with Charles Mitchell, one of Canada's leading veterinarians, and his former colleague H.R. Dean, who in 1947 established a Chair of Animal Pathology at Cambridge.[361]

Despite his heavy administrative duties, Joburg managed to publish a number of impressive scientific papers after 1945.[362] Many of these involved the classification of bacteria through his involvement with *Bergey's Manual* and the International Association of Microbiology Societies.[363] In addition, there was a group of detailed scientific papers based on rigorous laboratory analysis. Most of these dealt with listeria, tuberculosis, and problems of penicillin resistance.[364] Joburg also delighted in preparing

---

[358] E.G.D. Murray, R.A. Webb, and M.B.R. Swann, "A Disease of Rabbits Characterized by a Large Mononuclear Leucocytosis Caused by Hitherto Undescribed Bacillus Bacterium Monocytogenes," *Journal of Pathology and Bacteriology* (1926).

[359] Joburg and Freda both supported the Montreal branch of the Arctic Institute, founded in 1944. He was made a Charter Associate in 1948, in part because of his work in analyzing the incidence of Shigella dysentery, and his view that there were "important bacteriological possibilities to be investigated in the Arctic." Joburg to Max Dunbar, 22 February 1951 (see Appendix E).

[360] E.G.D. Murray, "Reflections on the Future of Veterinary Medicine," *Canadian Journal of Comparative Medicine* 12 (October 1948): 283.

[361] Mitchell had worked closely with Joburg on BW during the war; after 1945, he was chairman of the Animal Diseases Research Institute, Department of Agriculture. See Appendix E.

[362] E.G.D. Murray, "A Synopsis of the History of Medical Bacteriology," in René J. Dubos, ed., *Bacterial and Mycotic Infections of Man* (Philadelphia: J.B. Lippincott, 1948); E.G.D. Murray and Gertrude G. Kalz, "The Cultivation and Identification of Pathogenic Bacteria," ibid.; E.G.D. Murray (with G.D. Denton), "Plaster of Paris as a Source of Infection in Tetanus and Gas Gangrene," *Canadian Medical Association Journal* (April 1, 1949).

[363] E.G.D. Murray was co-editor (with R.S. Breed and A.P. Hitchens) of *Bergey's Manual of Determinative Bacteriology*, 5th ed., and contributor of sections: Neisseriaceae, Parvobacteriaceae, Diplococcus, Streptococcus, Listerella, and Corynebacterium; Robert S. Breed, E.G.D. Murray, and A. Parker Hitchens, "The Outline Classification Used in the Bergey Manual of Determinative Bacteriology," *Bacteriological Reviews* 8, no. 4 (December 1944): 255-260; E.G.D. Murray, R.S. Breed, and A. Parker Hitchens, "Notes on the New Edition of *Bergey's Manual of Determinative Bacteriology*," *Chronica Botanica* 9 (1945): 226-227.

[364] E.G.D. Murray, "The Value and Function of Scientific Societies," *Canadian Journal of Comparative Medicine* 11, no. 2 (February 1947): 47-52; E.G.D. Murray, "Research in Public Health," *Canadian Medical Association Journal* 66 (March 1952): 275-277, 308; E.G.D. Murray, "Nature in Man's World," *American Scientist* 42, no. 1 (January 1954): 130-135; E.G.D. Murray, "Destiny and Determinism in Infectious Disease" (the "Foundation Lecture" of the Academy of

satiric articles that lampooned poorly informed government officials, sensationalistic journalists, and posturing medical "experts." This latter group was the target of his scorn in a November 1954 missive. "The CBC started on Thursday weekly interviews," he wrote Robert, "with pairs of members of the Montreal Medico-Chirurgical Society; this first one was a physician & head of Dermatology of the MGH (Montreal General Hospital) and I could not believe such a display of ignorance was possible. There was no indication of any understanding of the principles of etiology nor of treatment."[365] Equally devastating was his controversial address to the 1952 meeting of the Canadian Society of Microbiologists, "Why Be a Microbiologist?"

> There is a good story of the brilliant Wyatt G. Johnston, who it is claimed was the first bacteriologist in Canada and was afterwards Professor of Hygiene at McGill University. After he examined a stained film of pus, Johnson made the diagnosis of gonorrhoea and the physician, protested that it was not possible as his patient was a man of standing and reputation. "Well" rejoined Wyatt Johnston "it must be a case of Immaculate Infection".[366]

Although Joburg belonged to over twenty-five professional associations, his role in creating the Canadian Society of Microbiology (CSM) in 1951 was one of his most rewarding experiences. First, it fulfilled his previous efforts to create a national microbiology organization; and secondly, this goal was achieved largely through the efforts of his son, who directed an imaginative and effective recruitment campaign.[367] Throughout the postwar years, Joburg was also active in the Canadian Medical Association, the Montreal Medico-Chirurgical Society, the Canadian Public Health Association, the Society of American Bacteriologists, the International Association of Microbiological Societies, and, above all, the Royal Society of Canada.[368] This latter affiliation was reinforced in 1953 when he was awarded the Royal Society's Flavelle Medal for scientific and medical excellence.[369]

Another scientific priority was his long association with *Bergey's Manual*; he was one of the original members of the Board of Trustees in 1936, and continued to exert

Medicine, Ottawa, read March 4, 1955), *Canadian Medical Association Journal* 72 (1955); E.G.D. Murray, "In Quest of Obscurities in Bacterial Infections," *Bulletin of the Academy of Medicine* 30, no. 9 (June 1957); E.G.D. Murray, "Reflections on Clinical Bacteriology" (presented as the Twelfth Augustus B. Wadsworth Lecture at the Meeting of the New York State Association of Public Health Laboratories, Albany, New York, October 20, 1961), *New York State Journal of Medicine* 62, no. 4 (February 1962): 501-507.

365 Joburg to Bob, 6 November 1954.
366 Excerpts from this controversial paper are cited in Appendix E.
367 References to the problems of establishing the CSM abound in the Murray correspondence of 1951-52.
368 Joburg was, however, often critical of some of the policies of his organizations, and he often expressed his criticisms forcefully. See Appendix E.
369 Many of Everitt's close friends, as well as his son, Bob, were also Flavelle medalists; see note 32.

considerable influence on its editorial policies throughout his career.[370] In 1956-57, for example, he and Bob worked very hard on the revised edition of the *Manual*, which was a considerable improvement over the earlier 1948 edition because of its new emphasis on viruses.[371] In 1964, Bob succeeded his father as a trustee, holding this position until 1990, as well as being vice-chairman of the Board in 1966-67. In recognition of his exemplary service to the organization, in 1994 he was awarded the Bergey Medal, "given in recognition of life-long contribution to the field of systematic bacteriology."[372] Some of the major debates about bacteria classification have been summarized in Robert's Recollections:

> The troubling problem was that in Bergey's Manual the bacteria were classified as a sub-group of the Plant Kingdom and named in botanical terms as Schizomycetes. Joburg argued strenuously in Bergey meetings that a separate description and independent name was needed. It was clear to us and to others like C.B. van Niel and R.Y. Stanier that they were not "fission fungi" and the only organisms anything like them were the blue-green algae, which in their turn were not really like algae proper. Bob was by the mid-50's interested in the structure of bacterial cells and aware of biochemical/genetic/physiological peculiarities then in active research, so was glad to be present and contribute to discussions with the Bergey's Manual Trustees. It was unsuccessful but a beginning and an introduction.[373]

Robert Murray's Scientific Career, 1945-55

Robert Murray's initial years at Western's medical school were both challenging and productive. Unlike McGill and the University of Toronto, the UWO program was constrained by its small size (eighteen full-time faculty) and by its separation from the main campus. Since the early 1930s Victoria Hospital had served as the focal point for academic medical research, which meant that Robert Murray and his colleagues were strongly affected by the hospital's attempts to become a modern and efficient institution. During the postwar years the most serious problems in achieving these goals were lack of hospital beds, inadequate operating facilities, and restricted space for the various medical scientific disciplines. The situation was further complicated by the 1948 decision of the UWO Board of Governors to transfer the medical school

---

370 By 1994 there had been nine editions of the *Manual*: 1923, 1925, 1930, 1934, 1939, 1948, 1957, 1974, and 1994. In 1977 there was an abbreviated version called *The Shorter Bergey's Manual of Determinative Bacteriology*.
371 In May 1952 Bob attended the Bergey dinner in Boston since it coincided with the SAB conference. He was, however, rather critical of the proceedings, which he found "both undigested and indigestible." Bob to Joburg, 6 May 1952.
372 "History of Bergey's Manual and the Trust," http://bergeys.org.
373 Robert Murray, "A Personal Narrative: Background, environmental and educational influences for the careers of EGDM (Joburg) and RGEM (Bob)."

to the north-end campus, and the building of the University Hospital. It took seventeen years to complete the process![374]

Meanwhile, Robert Murray and other members of the medical school were struggling to teach record numbers of students, a high percentage of whom were veterans.[375] In 1945, Dean Hall had introduced a new two-stage program that consisted of two pre-medical years followed by four years of medicine. Hall also decreed "that the medical class of September 1945 would be composed of only service veterans and that because of their maturity they would only need one year of pre-meds and thus forty-eight men were started on this medical school program."[376] Overall, Robert Murray enjoyed teaching these highly motivated and mature students, although, as a novice lecturer, he greatly appreciated Igor Asheshov's assistance in refining "what I had learned about teaching bacteriology in many and continued conversations."[377] He

---

374 In 1946 the Hospital Trust asked London City Council for permission to build a 1-million-dollar patient wing for two hundred beds, along with a modern cancer clinic, including the installation of a cobalt "bomb" radiation facility. This building expansion continued throughout the next two decades. John Sullivan and Norman Ball, *Growing to Serve: A History of Victoria Hospital, London, Ontario* (London, ON: Victoria Hospital Corporation, 1985), 140-53.

375 Some 1,100,000 men and women served in the Canadian armed forces during the Second World War. Those who qualified academically could obtain free university education plus a living allowance. Some 30,000 veterans took advantage of this program. In 1947 there were 2,200 full-time students registered at the University of Western Ontario.

376 Dr. Howie Cameron, "Recollections," in *Dr. G. Edward Hall: Scientist, Medical Administrator and UWO President: A Profile of the Man and His Times*, ed. Donald Avery (University of Western Ontario, Symposium, 11-12 April 2003).

377 UWO Archives, R.G.E. Murray Collection, Robert Murray, "Conversations and Science." Igor N. Asheshov, or "Ash" as friends called him, was born in 1891 in Nishni-Novgorod, Russia. He obtained his medical education at Saratov Imperial University from 1911 to 1916, graduating with a Diploma of Physician (equivalent to the MD in Canada). During the First World War he served in the medical services of the Imperial Army, and then with the Czarist forces during the Revolution. In 1919-20 he managed to get attached to the British Military Mission in South Russia, and later worked at the Bacteriological Laboratory of the British Military Hospital in Salonika, Greece. From there he went to Ragusa, Yugoslavia, where from 1921 to 1928 he was director of the State Bacteriological Laboratory, specializing in cholera and bacteriophage. From 1935 until 1937 he was a research officer of the MRC (UK) working with W.W.C. Topley at the London School of Hygiene and Tropical Medicine on phage acting on the typhoid bacillus. This connection proved important in 1937 when Asheshov applied for and was accepted as an associate professor of bacteriology in the Faculty of Medicine, UWO. In 1942 he was appointed associate professor and head of the Department of Bacteriology and Immunology. By 1945 he was promoted to professor and head, positions he held until 1948 when he went to the New York Botanical Garden to head up research on the inhibition of viruses for the National Foundation for Infantile Paralysis. This latter project ended in 1953 (see Appendix E) and Asheshov went to the UK, where he worked in the Lister Institute for a few years before a stroke brought inactivity. He died in 1961. Murray Barr, *A Century of Medicine at Western* (London, ON: University of Western Ontario, 1977), 476-77.

also benefited from the technical help he received in the laboratory from Dr. Josephine Bittner and Miss Frieda Strelitz, as well as his interaction with several younger scientists such as the biochemist R.H. Pearce. Murray's academic learning curve further accelerated in July 1948 when Asheshov abruptly left UWO for a scientific position in New York City, leaving him responsible for running the Department of Bacteriology and Immunology. Fortunately, during this challenging transition, Robert benefited from the support of his medical school colleagues, as well as receiving sage advice from his father.[378]

In many ways, Joburg's greatest source of pride was not his own scientific exploits, but rather Robert's academic success. And he assumed an important role in grooming his son for academic distinction, by regularly reminding him about the need for realistic priorities and strategies. First on the list was quality and timely research: "Don't worry about finding large & seemingly important projects, work at whatever presents itself & squeeze what you can out of it & publish anything informative." In contrast, Joburg claimed that administrative duties were "sterile fields full of worries & dissentions & jealousies & whatever you may do is destroyed by a single stroke at any moment leaving nothing at all ... [and] once you are trapped in it, it is hard to escape to freedom & solid work."[379] A second dictum was the importance of becoming "one of the workers in the research team of the University, making your own essential contribution part of your own scientific world ... [with] a definite agreement of time, opportunity & facility for your own research."[380] In this regard, he often reminded Robert that he should not repeat his father's mistakes: "I have wished several times I had written up some of my work at the time it was done. Several quite interesting things in my note books of 1917 on are cropping up in papers just coming out. ... No work is ever complete but it can make a useful contribution."[381] A third theme was the need to place individual career development over institutional loyalty. By the spring of 1948 Joburg had reluctantly concluded that, despite a promising beginning, the hierarchy at the University of Western Ontario did not appear sufficiently appreciative of Bob's scientific talents: "There are some jobs going which you are very well qualified. ... I think Western is a good place with good potentialities but I don't feel you are getting

---

378 There is some confusion about the circumstances of Asheshov's departure, along with his two technicians (Elizabeth, his new wife, and Miss Strelitz, a long-time associate). A rather lurid account of this event is provided by Ross Baxter Willis, former university comptroller, in his recollections, *Western 1939-1970: Odds and Ends* (London, ON: University of Western Ontario, 1980), 54.
379 Jogurg to Bob, 3 March 1946.
380 Joburg to Bob, 6 March 1947.
381 In another note, Everitt reiterated one of his constant themes: "Lab people, & it seems bacteriologists more than others, have neglected insisting on their rights. It is time they took a firm stand." Joburg to Bob, 29 February 1948; Joburg to Bob, 7 March 1948.

quite the share of it. ... I have kept quiet ... but I was disappointed in the distribution of available space & facilities when I was in your department last."[382] Fortunately for all concerned the situation changed dramatically during the next twelve months,[383] particularly after Bob accepted a summer research offer from the University of Washington.[384] In July 1949, at the age of thirty, Bob was appointed full professor and chairman of the UWO Department of Bacteriology and Immunology.[385]

And there was an added bonus. Acting on Robert's strong recommendation, Western hired Carl Robinow,[386] a European-trained bacteriologist.[387] This was an impressive addition to the department, since Robinow's many achievements included collaboration with René Dubos in the publication of his outstanding book *The Bacterial Cell in Its Relation to Problems of Virulence, Immunity and Chemotherapy*.[388] From Bob's perspective, Robinow's presence was indispensable, since he was "such a good friend and colleague ... [and] a real cytologist with considerable experience."[389] In short order, the Murray-Robinow partnership produced a number of publications

382 At the end of the year Joburg informed Bob that his friend Guilford Reed "would be glad to get you" if Western could not make a long-term commitment. Joburg to Bob, 4 April 1948; Joburg to Bob, 19 December 1949.
383 During the spring of 1948 Bob had a job interview with the Laboratory of Hygiene of the Department of Health and Welfare where, despite the lure of good salaries and job security, Bob doubted whether "very much fundamental research, if any, would occur." Bob to Joburg, 28 April 1948.
384 Bob thoroughly enjoyed his time in Seattle, working with bacteriologist Charles A. Evans and his colleagues. Professionally, there were a number of joint publications, and socially the family revelled in the West Coast environment. See chapter 6.
385 Joburg was delighted with this development, although he continued to caution Bob not to become over-burdened with administrative responsibilities: "In any case the greatest happiness & satisfaction is in research. I often look back to the days when I worked for the M.R.C. & wish I could return to something like it again." Bob to Joburg, 5 July 1949; Joburg to Bob, 27 November 1949; see also Robert Murray, "Thoughts about where we stand as scientists."
386 Carl Robinow (1909-2006) had graduated in medicine at the University of Hamburg, but left Germany in 1935. During the next fourteen years he carried out research at the Carlsberg Institute in Copenhagen, St. Bartholomew's Hospital in London, the Strangeways Research Laboratory at Cambridge University, and several research centres in the United States. Barr, *A Century of Medicine at Western*, 562.
387 Bob had originally met Robinow at the May 1948 meetings of the SAB in Minneapolis, and then had invited him to give a series of lectures at Western in April 1949. Both encounters were a tremendous success. Bob to Joburg, 23 May 1948; Bob to Joburg, 4 April 1949.
388 René Dubos, *The Bacterial Cell in Its Relation to Problems of Virulence, Immunity and Chemotherapy*, with an addendum by C.F. Robinow (Cambridge, MA: Harvard University Press, 1945). This book had an immense influence on scientists of the post-1945 period. According to Nobel laureate Joshua Lederberg, it was "the work from which I can say I learned most of the microbiology I know ... Dubos was far ahead of the times in so many ways." Cited in Moberg and Cohn, eds., *Launching the Antibotic Era*, ix-x.
389 Robert Murray, "Thoughts about where we stand as scientists."

dealing with phage and cytology, all generously funded by the NRC and Canada's MRC.[390] According to Robert Murray, these experiments "involved trying to solve problems of how to do good bacterial cytology and improve description of structure by improving the methods for staining preparations for light microscopy and for fixing, embedding, sectioning and giving differential contrast to the structure of bacterial cells."[391] Their work became even more sophisticated after 1954 when they were able to purchase an electron microscope, thanks to a private grant from the Bickell Foundation. In turn, the operation of this new technology was greatly assisted "by a long visit from Aksel Birch-Andersen from Copenhagen ... a co-researcher of distinction. Eventually we ... solved most of the problems so that there was high quality work and remarkably good resolution for some 14 years."[392]

In summary, between 1945 and 1955 Robert Murray's research activities went through a number of stages. First, there were a series of experiments using phages, encouraged by his colleague and mentor Igor Asheshov (Ash), an expert in this exciting field.[393] As Bob reported, "It seems to me aside from the interest of the field that I would be wasting my time if I did not learn all that I could of the business from him ..."[394] Moreover, despite Ash's abrupt departure from Western in 1948, Bob continued his phage experiments with a number of collaborators, both within his department and in the United States.[395] Second, there were a series of experiments[396] in the detection and assay of hyaluronidase, the swarming of some species of bacillus bacteria (spore structure), and some imaginative work on the impact of antibiotics on streptococcus.[397] Third, there was his involvement, along with Carl Robinow, with bacteria

390 R.G.E. Murray and C.F. Robinow, "A Demonstration of the Disposition of the Cell Wall of *Bacillus cereus*," *Journal of Bacteriology* 63 (1952): 298-300; C.F. Robinow and R.G.E. Murray, "The Differentiation of Cell Wall, Cytoplasmic Membrane and Cytoplasm of Gram-positive Bacteria by Selective Staining," *Experimental Cell Research* 4 (1953): 390-407.
391 Robert Murray, "Thoughts about where we stand as scientists."
392 As early as 1952, Bob had invited electron microscopy experts from the United States to visit Western and discuss the potential of this technology. UWO Archives, R.G.E. Murray Collection, Robert Murray, "Bob Murray and bacterial cytology" (hereafter cited as Robert Murray, "Bob Murray and bacterial cytology.")
393 R.G.E. Murray, D.H. Gillen, and F.C. Heagy, "Cytological Changes in *Escherichia coli* Produced by Infection with Phage T2," *Journal of Bacteriology* 59 (1950): 603-615.
394 Bob to Joburg, 22 April 1947.
395 J.F. Whitfield and R.G.E. Murray, "The Effects of the Ionic Environment on the Chromatic Structure of Bacteria," *Canadian Journal of Microbiology* 2 (1956): 245-26; J.F. Whitfield and R.G.E. Murray, "Observations on the Initial Cytological Effects of Bacteriophage Infection," *Canadian Journal of Microbiology* 3 (1957): 493-449.
396 Bob had a succession of impressive graduate research assistants: Bob Elder (1949), Fred Heagy (1950), Phil Fitz-James (1950), and J.F. Whitfield (1957).
397 R.G.E. Murray and R.H. Pearce, "The Detection of Assay of Hyaluronidase by Means of Mucoid Streptococci," *Canadian Journal of Research* E. 27 (1949): 254-264; R.G.E. Murray and R.H. Elder,

cytology research (understanding cell structures and membranes), a research undertaking that was strongly supported by Dean J.B. Collip and President Edward Hall.[398] "Dean Hall certainly can get things done, when he goes after them," Bob told his father. "He has obtained a grant to set up a unit for fundamental research on cell physiology. ... His idea is to get various people together with different lines of training and get them to attack cellular problems from a fresh angle. Very praiseworthy."[399]

In his Recollections, Robert Murray has summarized all of the above scientific developments: "Bob chose to work with Phage T2 ... and soon found that there [were] obvious cytoplasmic and nuclear changes starting almost immediately after infection... Between us [him and Robinow] ... in the next five years or so, we described several distinct patterns of cytological events among on the T phages acting on *E. coli*, some other phage systems, and lysogenesis (incorporation of phage in the genome) with J.F. Whitfield and G. Bertani."[400] At the time, Robert was particularly enthusiastic when Bertani visited Western in 1952: "He is a very pleasant, quiet and capable fellow doing excellent work in Luria's lab on the less dramatic but most interesting lysogenic (or 'symbiotic') phages. We got together to try and find out details of what happens when you establish a symbiotic infection. We got some fascinating morphological data that sets up excellently for the next year's work along that line."[401]

Bacteriophage Research and the Discovery of DNA

At this stage in the discussion, it is appropriate to ask the question: Why was phage research so important in the development of the fields of molecular genetics and molecular biology during the postwar years?[402] There are, of course, scores of books

"The Predominance of Counter-clockwise Rotation during Swarming of Bacillus Species," *Journal of Bacteriology* 58 (1949): 351-359; R.G.E. Murray and L.J. Loeb, "Antibiotics Produced by Micrococci and Streptococci that Show Selective Inhibition within the Genus *Streptococcus*," *Canadian Journal of Research* E. 28 (1950): 177-185; L.J. Loeb, A. Moyer, and R.G.E. Murray, "An Antibiotic Produced by *Micrococcus epidermitis*," *Canadian Journal of Research* E. 28 (1950): 212-216.

398 Bob to Joburg, 22 April 1947.
399 Joburg maintained "that physiologists have neglected intracellular functions & so have the pathogists ... & the more important cytoplasmic functions have been neglected." Bob to Joburg, 27 October 1946; Joburg to Bob, 3 November 1946.
400 Robert Murray, "Bob Murray and bacterial cytology."
401 By 1952 Salvador Luria had moved from the University of Indiana to the Massachusetts Institute of Technology, where he remained until retirement. Bob also had a successful collaborative arrangement with W.G. Wyckoff, an electron microscopy specialist at the National Institutes of Health, "on the phage infection series of T5 and its relatives." Bob to Joburg, 22 February 1952; Bob to Joburg, 6 May 1952.
402 Working with phage had a number of advantages. First, it was possible to generate suitable quantitative data overnight, given the rapid rate of phage to complete its life cycle; second, only simple equipment was required to prepare a phage assay; third, there was an opportunity to standardize phage experimental procedures; fourth, phage research could lead in many

that address this issue; but one of the most useful summaries comes from a recent biography of Max Delbrück: "Phage intended as the atom of genetics, opened up the field of molecular biology ... due to the fact that it is a stage in the life cycle of a bacterial virus in which it consists solely of DNA."[403] Significantly, Delbrück himself viewed the phage research endeavour as gradual and evolutionary: "The history of research on bacterial viruses or bacteriophages, now 30 years old, is fraught with controversy. ... today many lines of biological research are converging on a central problem, the organization of the cell. A feeling similar to that which inspired the physicists of 50 years ago when the structure of matter and the constitution of atoms became the focus on which all efforts converged."[404] The actual planning and development of modern phage research occurred during the Second World War when Delbrück and Salvador Luria, another émigré scientist, began their profitable collaboration. This was further enriched by the 1945 establishment of an annual series of phage seminars at Cold Spring Harbor in Long Island, New York.[405] An interesting aspect of these seminars was the intensified debate about the compatibility of laboratory and clinical-focused biological research, since the Delbrück group claimed their goal was "fundamental biology, quite uncontaminated by the MD [medical doctor] spirit."[406]

After 1940, a remarkable series of discoveries revolutionized the biological and life sciences.[407] While not all of this research can be attributed to the phage group, it was an important catalyst, along with other independent scientific researchers.[408] Of

important directions, including understanding the gene and DNA. Ernest Peter Fischer and Carol Lipson, *Thinking about Science: Max Delbrück and the Origins of Molecular Biology* (New York: Norton, 1988).

403 Ibid., 256. Delbrück (1906-1981) came from a prominent Berlin academic family. He initially pursued a career in theoretical physics, studying with Max Born and Niels Bohr. After his departure for the United States in 1937, he gravitated towards biology and phage research. Almost all of his academic career was in the United States, primarily at the University of California, Pasadena (1947-81).

404 During the 1930s, there was still uncertainty whether phages were viruses, despite the earlier work of Félix d'Herelle. Max Delbrück, Harvey Lecture (1945-46 Series), cited ibid., 132, 115.

405 Luria and Delbrück began their collaboration in 1938, and were later joined by biochemist Alfred Hershey. S.E. Luria, *A Slot Machine, a Broken Test Tube: An Autobiography* (New York: Harper & Row, 1984).

406 Max Delbrück in a 7 January 1947 letter to biochemist Mark Adams, who he was trying to persuade to come to the California Insitute of Technology. Cited in Fischer and Lipson, *Thinking about Science*, 177.

407 The chronological construct does not imply that major discoveries in molecular genetics and molecular biology ended after 1964; in fact, they accelerated during the 1970s. John Cairns, *Phage and the Origins of Molecular Biology*, centennial ed. (Cold Spring Harbor, NY: Cold Spring Harbor Laboratory Press, 2007).

408 The activities of the phage group had already been strongly endorsed by Erwin Schrödinger in his influential book, *What Is Life* (1945). Schrödinger, a prominent German-born physicist,

particular importance was an experiment on the transformation of pneumococcus in 1944 by the Oswald Avery team, which demonstrated "that *some* genetic information is stored in macromolecules that were identified as nucleic acids (DNA)."[409] Although the Delbrück/Luria group did not fully explore the consequences of Avery's work,[410] they did develop other important experiments and theories.[411] These included an analysis of the biological phenomenon of replication and the importance of concentrating the research focus on a set of seven phages, active on the same host – "the E. coli strain B and its mutants."[412] Another set of priorities was to determine the nature of genes: what do they consist of, how do they reproduce, and how do they act?[413] At the same time, there were the influential courses given by C.B. van Niel at the Hopkins Marine Station, Monterey, California, "which in a unique way nurtured respect for general microbiology as a discipline and recognition of microbial cells as research resources for biology."[414] And throughout this period, the technological potential of the electron microscope continued to be utilized by phage researchers, providing micrographs of the effects "of the phages on the host cells."[415]

The second sequence of exciting events occurred in 1946 when Joshua Lederberg and Edward Tatum announced that they had found bacteria which combined certain

was one of the founders of the quantum theory. His book, based on lectures presented at Trinity College, Dublin, gained great popularity among scientists. Ibid., 162-63.

[409] Although the 1944 Avery team experiments (MacLeod and McCarty) are now regarded as crucial in the discovery of DNA, in 1944 the full importance was not recognized. As a result, Avery was never awarded a Nobel Prize. Ibid., 150-52; Olga Amsterdamska, "From Pneumonia to DNA: The Research Career of Oswald T. Avery," *Historical Studies in the Physical and Biological Sciences* 24, part 1 (1993), 1-40.

[410] Dubos, one of Avery's long-term colleagues and friends, consistently insisted that the 1944 Rockefeller Institute experiments were crucial in the discovery of DNA, and downplayed the contributions of the phage group because of their deficiency in structural chemistry. René Dubos, *The Professor, the Institute, and DNA* (New York: Rockefeller University Press, 1976), 154-160.

[411] Among Luria's contributions was his 1946 discovery of reactivation of irradiated phage, and phage mutants, that could get around the bacterial resistance. He also mentored James Watson, his student at the University of Indiana, who was later one of the co-discoverers of the structure of DNA. Luria, *A Slot Machine*, 88-96.

[412] This set was called Type (T1, T2...T7) which "all infect and ultimately lyse their host cell, and are regarded as virulent phages." Ibid., 153-55

[413] Delbrück participated in two important meetings at Cold Spring Harbor that dealt with this topic: the 1941 symposium, Genes and Chromosomes: Structure and Organization, and the 1946 symposium, Heredity and Variation in Microorganisms. Ibid., 159.

[414] Max Delbrück was one of those who profited from the van Niel course, as did Roger Stanier, a Canadian-born microbiologist who had a powerful influence on American science. UWO Archives, R.G.E. Murray Collection, "Recollections."

[415] The electron microscope, developed in Germany during the late 1930s, soon became an essential research tool in well-funded laboratories in the United States. Thomas Anderson, "Electron Microscopy of Phages," in Cairns, *Phage and the Origins of Molecular Biology*, 65.

traits of two parental strains – the basis of genetic recombination.[416] Six years later, at the first International Phage Symposium in Paris, there was another major announcement: the so-called "blender experiment" of Alfred Hershey and Martha Chase, which "demonstrated that the DNA of phage contained *all* genetic information."[417] And, of course, the penultimate event occurred in April 1953 when James Watson and Francis Crick shocked the scientific community with their double helix theory for the structure of DNA, which revealed "with one blow the mystery of gene replication."[418] This monumental achievement was, in many ways, a merger of the two dominant schools of molecular biology: "the one structural and three-dimensional, in effect the British X-ray crystallographers plus Linus Pauling, the other school genetical and one-dimensional, in large part the American phage group."[419]

Not surprisingly, Crick[420] and Watson[421] became instant celebrities, as the DNA saga moved from scientific enquiry to public celebration.[422] Indeed, it was generally

---

[416] In 1958 Lederberg would share in the Nobel Prize in Physiology or Medicine "for his discoveries concerning genetic recombination and the organization of the genetic material of bacteria," along with Edward Tatum and George Beadle "for their discovery that genes act by regulating definite chemical events."

[417] According to René Dubos, this ingenious experiment established "that in bacteriophage infection, most of the viral DNA penetrates the infected bacterium, whereas the viral protein remains outside." Dubos, *The Professor*, 157; Fischer and Lipson, *Thinking about Science*, 194-95.

[418] Max Delbrück's statement, April 1953, cited in Fischer and Lipson, *Thinking about Science*, 200-202. There is a vast literature dealing with the DNA story.

[419] H.F. Judson, *The Eighth Day of Creation* (New York: Simon & Schuster, 1979), 209. Linus Pauling (1901-1994) was regarded as one of the world's outstanding physical chemists, publishing more than 500 scientific papers and books, most focusing on the structures of molecules and the nature of the chemical bond. He was awarded the Nobel Prize in Chemistry in 1954. The reason he almost discovered DNA in 1952-53 is astutely discussed in Thomas Hager, *Force of Nature: The Life of Linus Pauling* (New York: Simon & Schuster, 1995), 391-431.

[420] Francis Crick (1916-2004) was trained as a physicist, but moved into biological research after the Second World War at the Cavendish Laboratory at Cambridge University, working on protein crystallography under the supervision of Sir Lawrence Bragg, Max Perutz, and John Kendrew. In October 1951 he began his collaboration with the visiting James Watson on the structure of DNA: Crick was thirty-five and Watson twenty-three. He is remembered as having remarkable intellectual power and intuition "in all matters structural and biological." Judson, *Eighth Day*, 108-111; Leslie Orgen, "Francis Crick (1916-2004)," *Science* 305 (20 August 2004): 118.

[421] James Watson (b. 1928) remains a scientific celebrity not only for his DNA work but also for his role as a scientific administrator, his involvement in a number of scientific debates, and his controversial book *The Double Helix* (New York: Atheneum, 1968). This work has been revised on a number of occasions, most recently in 2001. See also his book *DNA: The Secret of Life* (New York: Alfred Knopf, 2004).

[422] The work in biophysics by Max Perutz, John Kendrew, and Francis Crick on the protein crystallography of hemoglobin was an important stage in the development of molecular

acknowledged that "the publication of the structure of DNA will rank in science history with Newton's theory of gravitation or Darwin's theory of an evolutionary origin for species."[423] Canadian life scientists also shared this enthusiasm about the DNA discoveries and related research. Robert Murray, in particular, was acutely aware of the various phage experiments and theories because of his close association with Asheshov and Robinow.[424] In addition, in 1950 he and Luria had published similar papers "on cytological studies of bacteria cells infected with several of the T-phages in 1950, both using the techniques of C.F. Robinow, which showed that the first and almost immediate effect of infection was disruption of the nucleoid."[425] Robert Murray's research also attracted the attention of another important phage researcher, Nobel laureate Joshua Lederberg, who "was interested in my work on the cytology of phage infections because he had recently shown that phages could transfer genes from one bacterium to another."[426]

Reconciling Two Research Careers, 1955-64

By 1955 Robert Murray's professional career was well advanced. At Western he was an effective chair of department, heeding his father's advice about the importance of building an effective teaching and research academic unit while, at the same time, lobbying for scarce university resources.[427] Moreover, his major concern, that Robinow might accept a more lucrative offer from Selman Waksman's laboratory at Rutgers University,[428] was effectively countered by President Hall's successful campaign

---

microbiology at Cambridge. Soraya de Chadarevian, *Designs for Life: Molecular Biology after World War II* (New York: Cambridge University Press, 2002), 61-117.

423 Judson, *Eighth Day*, 200. Leo Szilard, the famous anti-war scientist, was another physicist turned molecular biologist, who carried out important research in the United States. William Lanouette, *Genius in the Shadows: A Biography of Leo Szilard, the Man behind the Bomb* (Chicago: University of Chicago Press, 1992).

424 Robert did not attend any of the phage seminars at Cold Spring Harbor, a situation that he now regrets. At this stage in his career, however, he wanted to preserve the summer months for family activities. Robert Murray, interview by Donald Avery, 26 April 2008.

425 UWO Archives, R.G.E. Murray Collection, Robert Murray, "Recollections."

426 On two occasions Robert visited Lederberg's laboratory at the University of Wisconsin, and their correspondence continued when Lederberg moved to Stanford University.

427 Until 1952 Western was somewhat strapped for funds, but Bob appeared to have gained a fair share for his department. With the advent of federal support, things improved immensely, which meant a substantial increase in salaries. At the same time, Bob commented on the appearance of negative "medical politics," which was associated with "a long smoldering resentment of the way Ed Hall cleaned things up around here 8-10 years ago." Bob to Joburg, 4 May 1953.

428 Waksman was awarded the Nobel Prize in Physiology or Medicine in 1952 for his discovery of streptomycin, the first antibiotic effective against tuberculosis. During his career, Waksman published over 197 papers on antibiotics. Rollin Hotchkiss, "From Microbes to Medicine:

to acquire an electron microscope.[429] It was a fortuitous acquisition. During the next fourteen years, this technology would greatly assist the Murray-Robinow scientific partnership.

Robert's growing reputation for innovative research was reinforced by his performances at meetings of the Society of American Bacteriologists,[430] the Public Health Association (Laboratory Section),[431] and the newly formed Canadian Society of Microbiologists.[432] Given the small size of the North American microbiological community, Joburg was quickly informed about his son's many scholarly triumphs. In December 1947, for example, Charles Mitchell sent the following message: "Your presence at the recent meetings of the Laboratory Section (PHA) was missed and also your leadership in discussion. However, the father was well represented in the son, who gave an extraordinarily interesting talk, supplemented by a home-made film on bacterial swarming."[433]

Other honours soon followed. One of these was an invitation in February 1951 to join the editorial board of the prestigious *Journal of Bacteriology*, published by the Society of American Bacteriologists.[434] It was the beginning of an impressive career as a scientific editor.[435] Robert also received two important awards during this period: the Harrison Prize, awarded by the Royal Society of Canada in 1957, and the

Gramicidin, René Dubos and the Rockefeller," in *Launching the Antibiotic Era*, ed. Moberg and Cohn, 12-15.

[429] In November 1952 Bob, Robinow, and Chris Hannay had attended a stimulating meeting of the Electron Microscope Society which demonstrated that "the detail and clarity and resolution of cytoplasmic structure was magnificent." Bob to Joburg, 18 November 1952; Bob to Joburg, 29 November 1952.

[430] At the May 1953 SAB meetings, Robert's paper "got lots of favourable comments and a number of the really interested people got hold of me afterwards." Bob to Joburg, 6 May 1953.

[431] Bob helped organize the 1948 PHA meetings, including some papers from his laboratory. He was, however, quite frustrated in his dealing with R.D. Defries and others at the Connaught Laboratories. Bob to Joburg, 7 November 1948.

[432] The CSM held its first meeting in Ottawa in July 1951.

[433] On another occasion, Joburg related how "four people who were at French Lick (Indiana) have told me of your triumphant success there. It is a great compliment to you that you were chosen the spokesman for the bacteriologists and the way you did it is said to have impressed and captured everyone .... (Lyman) Duff (Dean of Medicine at McGill), was one of the people full of praise." Joburg to Bob, 6 November 1954.

[434] The *Journal of Bacteriology* was the original publication of the SAB (now American Society of Microbiologists). In 1937, another journal, *Bacteriological Reviews*, was established to provide another forum for SAB members. In 1997 the journal was renamed *Microbiology and Molecular Biology Reviews*. See John Ingraham and Robert G.E. Murray, "Minireviews: A Bit of History," *Microbiological and Molecular Biology Reviews* (June 1999), 263-64.

[435] Bob was a member of the editorial board of the *Journal of Bacteriology*, 1951-54 and 1980-86. His status with *Bacteriological Reviews* was more prestigious since he was general editor between 1969 and 1979.

Canadian Society of Microbiologists Award in 1963.[436] On occasion, Bob followed his father's example by popularizing science through public lectures, including a short radio talk in April 1947 on the value of immunizing children against certain infectious diseases.[437]

The question of joint research was the subject of Joburg's letter on 24 January 1948: "I have often wished we could work together; though I have had to give so much time to other things that I often feel I am out of it now."[438] These feelings of research opportunities lost were a recurring theme in Joburg's correspondence, reinforcing the fact that he "would have liked more recognition of his scientific contributions to bacteriology."[439] Yet, as some of his contemporaries observed, he was neither an effective self-promoter nor a widely published scientist. "Looking back," remarked Ashley Miles, "it seems to me that Joburg was one of the few effective people of their generation, who, like Bill Topley, developed skepticism almost into a cult. Forty years ago it was a valuable cult because the technique of evaluating published work was not very far advanced ... And yet as though to counterbalance this, he was wont to express himself in terms of absolute assurance. He was, in fact, a strange mixture of skepticism and dogmatism."[440]

Yet, in reality, during the postwar years the senior Murray made significant contributions to Canadian biological research within a broader scientific context. In a reflective assessment of his father's scientific contributions during the postwar years, Robert Murray made these comments:

> All along and into retirement he spent special effort in the giving of special and general lectures and discussions in varied aspects of medical bacteriology, infectious diseases and the importance of science in the practice of medicine. HIS MAJOR CONTRIBUTION ... was his involvement with Bergey's Manual of Determinative Bacteriology as an Editor/Trustee a major lexicographic task which took enormous dedication, application and thought applied to the classification and description of all known bacteria – in essence providing a sort of 'bible' for the bench-worker in diagnostic bacteriology, the researcher and the teacher.[441]

---

436 Bob to Joburg, 7 February 1951. In 1952 Bob became editor of the newly formed *Canadian Journal of Microbiology*, a position he held until 1960.
437 Bob enjoyed this brief experience of being a media scientist, particularly since he received several compliments "on the suitability of the Murray voice for Radio!" The transcript, which he sent to Montreal, was well received by his father and sister. Bob to Joburg, 26 February 1947; Joburg to Bob, 13 April 1947.
438 Joburg to Bob, 24 January 1948.
439 Views of microbiologist Sir Ashley Miles, cited in G.S. Wilson, "Everitt George Dunne Murray," 652.
440 Ibid.
441 Robert Murray, "Response to Avery, 19/04/08."

In addition, Joburg assisted his son in a number of important projects. One of these was the establishment of the Canadian Society of Microbiologists (CSM) in July 1951, a cause that Joburg had originally championed during the late thirties. And, indeed, it is questionable whether Robert would have started his organizational effort in 1949 without his father's support. As Joburg wrote, "I am encouraged by you interesting yourself in getting such an organization moving and I think you might well succeed. The attitude of various opponents has changed and I think you will get more support than I did."[442] Equally important for Robert's CSM campaign was his father's range of contacts among academic and government microbiologists, particularly his strong links with prominent French Canadian medical scientists such as Armand Frappier.[443]

In June 1951, 173 microbiologists met in Ottawa and formed the Canadian Society of Microbiologists, with Robert Murray as president.[444] During the next four years, the CSM substantially increased its membership, published a journal, and explored a number of contentious ethical issues. One of these was whether newly discovered micro-organisms should be patented, since "workers in the field of microbiology and related subjects can "tailor-make" micro-organisms to do specific jobs." In contrast, other microbiologists pointed out the dangerous precedent whereby "a new micro-organism would have to have been "invented" or caused to occur and not merely discovered."[445] This important debate, which involved academic researchers, government agencies, and the pharmaceutical industry, would gain further intensity during the 1970s, with the development of recombinant DNA manipulation.[446] Another CSM initiative was the presentation to the Massey Commission of a position paper which emphasized the unique role of the CSM in "obtaining representative opinions of Canadian microbiologists," and in providing "an official body to present these opinions to government, international congresses, etc."[447] In 1953, for example, the CSM began its long association with International Association of Microbiological

---

442 Joburg to Bob, 20 March 1949.
443 In February 1952 Bob finally had an opportunity to meet Frappier, and was most impressed. Joburg was pleased: "Armand F. is a nice fellow & very competent. I get on with him extremely well." Bob to Joburg, 10 February 1952; Joburg to Bob, 17 February 1952.
444 Although 700 scientists were approached, in 1951 only 173 were prepared to make a commitment. Bob to Joburg, 9 April 1951.
445 By June 1954, membership had increased to 380, drawn from all over the country. In August the CSM issued the first edition of its official publication, *Canadian Journal of Microbiology*, with Bob Murray as editor. Meanwhile, Joburg was president of the affiliated Quebec Society of Bacteriology, 1954-55. See Appendix E.
446 Robert Bud, *The Uses of Life: A History of Biotechnology* (Cambridge: Cambridge University Press, 1993); Gary Zweiger, *Transducing the Genome: Information, Anarchy, and Revolution in the Biomedical Sciences* (New York: McGraw-Hill, 2001).
447 Resumé of the Organizational Meeting, Ottawa University, June 7-8, 1951.

Societies, choosing to send Everitt and Robert Murray as its official delegates to its Seventh International Congress.[448]

These meetings, held in Rome, provided Joburg and Robert with a marvellous opportunity to advance their mutual interests in fostering international scientific cooperation. This included their joint participation in general discussions about the International Code of Nomenclature of Bacteria and Viruses, followed by Bob's paper on his cytology research at one of the scientific sessions. Another priority was their support for the creation of an international culture collection system, which would reinforce IAMS policies in terms of nomenclature, classification, and taxonomy, and provide badly needed support for national initiatives, such as the formation of the Canadian Committee on Culture Collections.[449] Even more important was their campaign to gain sufficient support for the CMS bid to host the 1962 IAMS Congress in Montreal.[450] After the Congress, the Murrays decided to reinforce some of their research connections at major microbiology laboratories. For Joburg, this meant a quick trip to the Pasteur Institute in Paris, followed by an enjoyable visit to the University of Cambridge, where he and Freda socialized with many of their old academic friends.[451] Meanwhile, Robert was busy investigating shared interests in the Stockholm laboratory of C.-G. Hedén, establishing a long and successful scientific liaison.[452]

To what extent did the father and son have similar research interests by the 1950s? This question is best answered by Robert: "Both Murrays started bacterial practice in clinical microbiology ... However, the father looked at the reasons for disease production and environmental modification, while the son concentrated on cellular aspects with a biochemical slant on physiology and worked towards how things were put together. ... By the 1960's I was no longer doing anything much with clinical material for research purposes. ... I was gradually moving into purer sciences."[453]

Many of these differences became increasingly evident in their discussion of new scientific trends. For instance, Joburg saw little merit in applying statistical methods to

---

448 Minutes of the Fourth Annual Meeting, June 24-26, Queen's University, in *CSM Newsletter* 3, no. 1 (August 1954).
449 N.E. Gibbons of the NRC was one of the founders of the Canadian Committee on Culture Collections, while S.T. Cowan was primarily responsible for the British National Collection of Type Cultures. By 1959, the IAMS developed an overall strategy for establishing an international culture collection system. See Appendix E.
450 Everitt would later become president of the 1962 Congress, while Bob would be program coordinator.
451 UWO Archives, R.G.E. Murray Collection, Robert Murray, interview by Donald Avery, 11 March 2007.
452 C.-G. Hedén, an old friend of Robinow, had visited UWO in September 1951. Bob to Joburg, 19 September 1953. See chapter 7, note 5 for information about Carl-Göran Hedén. See also Appendix E.
453 Robert Murray, "Thought about where we stand as scientists and significance of our work."

microbiological problems. Nor was he interested in methods for using the phase or electron microscope.[454] He also found some of the debates about bacteriophage research confusing: "I don't know the characters & destructions of phages & find papers difficult to read because of the terminology used."[455] In addition, there were occasions when Bob found it necessary to reassure his father that the leading genetic scientists were doing exciting and important work: "You are being quite severe about genetical work on bugs," he wrote in December 1950. "In some cases this is warranted. In the case of Lederberg's work I think it has to be accepted and I am pretty sure that his work is very carefully controlled."[456] On the other hand, there were many other examples where Joburg demonstrated an ability to anticipate and advance crucial new areas of scientific enquiry, notably mycology and virus research.[457] Above all, one should not overlook his innovative and dedicated research on listeria,[458] involving a network of experts around the world, which eventually became "one of the few important modal organisms for those researchers working on pathogenesis."[459] Some of this work was completed by Bob after his father's death in July 1964, and resulted in a number of important publications.[460]

Did Joburg have any problems working with women microbiologists? Or was he primarily motivated by considerations of good science, not gender? There are several reasons for raising these questions. First, there now exists a rich body of literature about the importance of gender in determining career patterns in the life sciences, past and present.[461] Second, since Joburg's professional activities covered approximately

454 Joburg to Bob, 1 December 1946; Joburg to Bob, 1 February 1948.
455 Joburg to Bob, 8 March 1952.
456 Bob to Joburg, 1 December 1950.
457 Even more frustrating was that the rival University of Toronto's medical school was heavily involved with virus research, notably through the work of A.J. Rhodes and C.E. van Rooyen. See Appendix E for a discussion of the 1955-56 controversy over the status of virology.
458 In 1955 Joburg was appointed adjunct research professor in the faculty of medicine at the University of Western Ontario, a position he held until his death in July 1964.
459 E.G.D. Murray, "The Story of Listeria," Flavelle Medallist's Address, *Transactions of the Royal Society of Canada* 57, series 3 (June 1953); E.G.D. Murray, "A Characterization of Listeriosis in Man and Other Animals," *Canadian Medical Association Journal* 72 (1955): 99-103; Robert Murray, "Response to Avery, 19/04/08."
460 B.K. Ghosh and R.G.E. Murray, "Fine Structures of *Listeria monocytogenes* in Relation to Protoplast Formation," *Journal of Bacteriology* 93 (1967): 411-26; R.A. Tadayon, K.K. Carroll, and R.G.E. Murray, "Factors Affecting the Yield and Biological Activity of Lipid Extracts of *Listeria monocytogenes*," *Canadian Journal of Microbiology* 15 (1969): 421-28; B.K. Ghosh and R.G.E. Murray, "Fractionation and Characterization of the Plasma and Mesosome Membrane of *Listeria monocytogenes*," *Journal of Bacteriology* 97 (1969): 426-40; R.A. Tadayon, K.K. Carroll, and R.G.E. Murray, "Purification and Properties of Biologically Active Factors in Lipid Extracts of *Listeria monocytogenes*," *Canadian Journal of Microbiology* 16 (1970): 535-544.
461 See Louise S. Grinstein, Carol A. Biermann, and Rose K. Rose, *Women in the Biological Sciences: A Biobibliographic Source Book* (Westport: Greenwood Press, 1997); Robert Fisher,

sixty years in two countries, his responses to issues of gender and sexism provides an interesting case study.[462] After examining the historical record, it appears that Joburg was often involved in successful cooperative ventures with women scientists, particularly during his long career at McGill University. For example, his long association with microbiologist Gertrude Kalz was based on mutual respect and an effective division of labour. These traits were quite evident during the preparation of their important 1948 joint article on pathogenic bacteria, as well as in subsequent collaborative ventures such as Kalz's useful article on bacteriology for surgeons.[463] An even more impressive pattern of cooperation was evident in Joburg's long working arrangement with Sarah Branham, MD, of the National Institute of Health, Bethesda.[464] Indeed, they directed the activities of the IAMS Neisseria Sub-Committee for almost fifteen years, until Sarah's death in 1962.[465] And they were well matched, with Branham's expertise on different strains of influenza and meningococcal viruses complementing Joburg's bacterial approach. Significantly, both of these gifted microbiologists were associated with major discoveries of pathogens, the importance of which was fully appreciated only after their deaths.[466]

Social and Political Issues

Fairness on scientific matters was a common virtue for both Murrays. Closely related was their liberal and pro-civil rights response to major political and social issues of

---

Making Science Fair: How Can We Achieve Equal Opportunity for Men and Women in Science (Lanham, MD: University of America Press, 2007); Ruth Watts, *Women in Science: A Social and Cultural History* (New York: Routledge, 2007).

462 The controversy over Rosalind Franklin's contribution to the discovery of DNA is an intriguing case study of a brilliant experimentalist, who "faced many challenges as a woman in science." Rena Selya, "Essay Review: Defined by DNA: The Intertwined Lives of James Watson and Rosalind Franklin," *Journal of the History of Biology* 36 (2003), 591-597.

463 See E.G.D. Murray and Gertrude Kalz, "The Cultivation and Identification of Pathogenic Bacteria," in Dubos, ed., *Bacterial and Mycotic Infections*; Joburg to Bob, 9 April 1950. While Joburg assisted Gertrude in this latter work, he made it clear that it was her project.

464 Sarah Branham (1888-1962) received her PhD degree as well her MD from the University of Chicago. During her long career in public health, she is credited with the promotion of microbiology and the growth of molecular immunology. Her papers are located at the US National Library of Medicine.

465 The Murray-Branham correspondence reflects their shared research interests, and they often exchanged live meningococcus strains. See Appendix E.

466 For Joburg, it was of course his identification with *Listeria*; while Branham's work on *Neisseria catarrhalis*, a serious pulmonary pathogen, in 1970 was transferred to a new genus, named in her honour. *Branhamella catarrhalis* continues to attract serious medical attention. Richard J. Wallace, "In Honor of Dr. Sarah Branham, a Star Is Born: The Realization of *Branhamella catarrhalis* as a Respiratory Pathogen," *Chest* 90, no.3 (1986), http://www.chestjournal.org (accessed 20 April 2008).

this period.[467] In particular, they deplored the negative impact of anticommunist witch hunts on the US and Canadian scientific communities.[468] In March 1952, for example, Bob was shocked to learn of the persecution of his friend David Bonner, an American microbiologist, who had been refused security clearance at the Oak Ridge research station "on the basis of much malicious hearsay ... [indicating] it seems to be a 'crime' to be at all independent in thought even scientifically. ... [We] can only hope that some good sense, fair play and integrity will prevail in Canada."[469] In his response to the Bonner case, Joburg was critical of what he regarded as Cold War science: "The authority is in the hands of the wrong kind of people. The implications are bad & unjust without any possibility of redress for malicious actions. Science is being used as a political tool, both nationally & internationally, to the great disadvantage of science & to the danger of scientists. ... It is best not to have anything to do with any of the classified work."[470] Indeed, despite his extensive involvement with Canada's biological warfare program during Second World War and the early Cold War,[471] in February 1952 he informed Robert that he had "resigned from everything to do with DRB ... It surprises me that self respecting bacteriologists tolerate the situation."[472]

For the Murrays, the challenge of protecting civil liberties in Canada was of ongoing concern. This was evident in February 1952, when Hewlett Johnson, the so-called Red Dean of Canterbury, was prevented from speaking at the University of Western Ontario by a rowdy audience who, in the words of Bob Murray, "shouted him down

---

467 There is an extensive body of literature on the topic of Cold War repressive policies in Canada and the United States. See Reginald Whitaker and Gary Marcuse, *Cold War Canada: The Making of a National Insecurity State, 1945-1957* (Toronto: University of Toronto Press, 1994); Jessica Wang, *American Science in an Age of Anxiety: Scientists, Anticommunism and the Cold War* (Chapel Hill: University of North Carolina Press, 1999).
468 In February 1946 Bob tried to understand the implications of the Royal Commission on Espionage. Bob to Joburg, 18 February 1946. The extent to which Canadian scientists were affected by the Cold War is discussed in Avery, *Science of War*, and Whitaker and Marcuse, *Cold War Canada*.
469 Bob to Joburg, 2 March 1952.
470 Joburg also rejected Soviet bloc accusations that the United States had used biological warfare in Korea and northern China. He was, however, critical that "the statements reported in the press by the US 'authorities' are inaccurate & misleading. They might at least take care that what they say is true." Joburg to Bob, 6 April 1952.
471 During the early stages of the Cold War, Joburg had chaired the Bacteriological Warfare Research Panel of the Defence Research Board (DRB), which included many of his wartime scientific associates: Guilford Reed, Charles Mitchell, Otto Maass, and P.H. Greey. He resigned from this body during the fall of 1949.
472 Joburg was, however, aware that it would be difficult for Bob to avoid involvement with civil defence programs, given the volatile political situation. Joburg to Bob, 17 February 1952.

... instead of listening and raising hell afterwards."[473] While Joburg had little sympathy for this particular British "fellow-traveler," he felt that Western's administrators, including President Hall, had over-reacted: "it would help a lot to have some sense & judgement exercised by those in high positions."[474] A second incident occurred in September 1960 when Joburg felt it necessary to defend the reputation of a former student, A.J. Romeyn, who had been dismissed from the University of Alberta for alleged communist proclivities. In his opinion, these allegations were baseless and unfair, and he urged University of Manitoba officials to hire Romeyn since "he was well trained ... [and] determinedly research minded."[475] This emphasis on scientific skill, not political correctness or negative stereotyping, was also evident in Joburg's willingness to accept European displaced persons with scientific training into his McGill laboratory, as long as they demonstrated professional competence and hard work.[476] Unfortunately, this policy was rarely followed by other Canadian universities, who usually relegated DP doctors to menial positions.[477]

A final example of Joburg's commitment to public service was his willingness to represent McGill University on Montreal City Council.[478] Over an eight-year period from 1947 to 1955, he faithfully attended Council meetings, despite tedious and litigious proceedings along with nasty political machinations.[479] In February 1955, he provided Robert with a graphic account of one of these sessions: "There was a caucus this afternoon from 3:30 to 6:45 on the threat of the Mayor [Camillien Houde] & the Chairman of the Executive Committee to cut off all contributions to charity. Many speeches, nearly all in the same direction (me too) & eventually the Executive was

---

[473] President Hall had made things worse afterwards by publicly defending the hecklers, and thereby appearing "as a supporter of rowdyism." Bob to Joburg, 22 February 1952.

[474] At the same time, he was supportive of administrative responsibility and of efforts to prevent a situation whereby "panics & witchhunts are developed ... for political purposes." Joburg to Bob, 8 March 1952.

[475] Joburg to Professor J.C. Wilt, Department of Bacteriology and Immunology, 15 September 1960; see Appendix E.

[476] Everitt's account of hiring a young Latvian doctor for his lab provides one such example. Joburg to Bob, 28 November 1949.

[477] Another incident, the application of George Ling to do an MSc in bacteriology at UWO, reveals the same emphasis on scientific abilities rather than racial or ethnic background. In this case, the applicant was of mixed Chinese and Afro-Canadian background. Bob to Joburg, 3 November 1946; Joburg to Bob, 17 November 1946.

[478] In 1940 a new system for Montreal elections was introduced that allocated one-third of the seats for property owners, one-third for general election, and Category C for members of public bodies, including two seats for McGill University. John Cooper, *Montreal: A Brief History* (Montreal: McGill-Queen's University Press, 1969), 165-180.

[479] Joburg was particularly concerned about heath issues and the problems of Montreal's slums, as well as the various power plays by Camillien Houde, who served seven terms as mayor between 1928 and 1954. Ibid.

requested to restore the donations, with only two dissenting votes. The Mayor made a stupid legalistic speech; he has a lot to learn."[480]

Family Matters

With the end of the Second World War, there was a concerted attempt on the part of Canadians to secure stable jobs, find suitable housing, and build happy lives. For the senior Murrays this challenge had already been met, although they were naturally concerned about the future life chances of their son and daughter. Fortunately, Bob had an attractive position at UWO, although it was not until 1950 that he and Doris could afford to buy a home. By that time they had three children: Alice, who was five years old; Peter, four; and Tom, one. A similar pattern emerged for Sue and her husband, Blake Robinson, who were able, through hard work and financial sacrifice, to secure their own home in Pointe Claire, a suburb of greater Montreal, on the eve of the prosperous fifties.[481]

Throughout the postwar years, correspondence between the Murrays of Montreal and London focused on problems of housing, low university salaries, rising taxes, cost of living, and family health problems, which included the cycle of children's infections such as measles, chicken pox, and mumps. When these illnesses occurred, Joburg and Robert usually would discuss the symptoms, identify the pathogen, and suggest a remedy. There were, however, several serious medical crises that tested family resolve. One of these was Blake's ongoing problems with dysentery; the other was Robert's serious attack of meningitis in June 1951 while attending the founding conference of the Canadian Society of Microbiologists.[482]

But for the most part, the 1950s were good times for the Murray family. In Montreal, Joburg and Freda had a wide array of interesting social options, including interesting plays and films, intellectual discussions at the Arctic Institute, fine dining at the University Club, and exciting hockey games at the Forum, where Joburg could cheer on his beloved Canadiens. They also seemed to win the Stanley Cup every year![483] In addition, there was a constant flow of scientific visitors, including many

---

480 Despite his dislike of the corruption and chicanery associated with the Houde administration, Joburg personally got along with Montreal's colourful mayor. Joburg to Bob, 14 February 1955; Robert Murray, interview by Donald Avery, 2 February 2008.
481 Sue and Blake were married in June 1945. Despite coming from a medical family, with his father being a physician in Montreal, Blake pursued a career as an insurance underwriter. He and Sue had four children: Doug, Joan, Christopher, and Sarah.
482 Despite Joburg's careful culturing of Blake's stools, no disease pattern was detected. Bob's medical expenses at the Ottawa Civic Hospital, for his fifteen-day stay, amounted to about $340.00. Both he and Joburg were outraged by these charges. Bob to Joburg, 8 October 1951.
483 Joburg's fascination with hockey was revealed in 1930, when he sent his first letter to his brother Rory. See chapter 1. Shortly afterwards he purchased season's tickets which he retained until the move to London, Ontario.

old friends from the United Kingdom, who were entertained appropriately *chez* Murray, 3590 University Avenue.[484] And, on occasion, there was news from South Africa, when brother Rory provided grim briefings about the outrages committed by the right-wing and racist National Party.[485] But for Joburg, these were remote problems, and his letters to Bob *never* mentioned South Africa.[486]

With the arrival of spring, all thoughts turned to the Bark Lake cottage, where the Murray clan gathered for extended holidays under the watchful eye of Freda, nicknamed "Agie" and "Gago" by her seven grandchildren.[487] These family excursions at Bark Lake also provided opportunities for Joburg and Robert to renew their mutual passion for trout fly fishing and making the perfect fishing lure. In addition, every year the senior Murrays would take an extended canoe and fishing trip into neighbouring lakes – reinforcing their love of the Canadian wilderness.[488] During these arduous excursions, Joburg was constantly amazed by his wife's bushcraft abilities. As he reported to Bob: "She does slap through the bush with an extraordinary certainty of where she is ... It always surprises me how she recognizes every rise or mountain & says exactly where it is & its relation to lakes & other contours."[489] Nor did these wilderness exploits escape the attention of Joburg's McGill colleagues, as indicated by the following passage: "E.G.D. is an angler par excellence, he would rather fish than eat, drink, or sleep ... his first and main love is fly fishing and he is an expert ... he ties his own flies with a skill envied by the Montreal professionals. ... Mrs. Murray shares his angling enthusiasm ..."[490]

Elsewhere, the Murrays of London were thriving. As a junior faculty member, Robert associated socially with his university colleagues, notably biochemist Roger Rossiter, who shared his love of canoeing and good science. They were also charter

---

[484] During the early fifties, the Murrays moved into a small apartment on Pine Avenue, also near the McGill campus.

[485] Given the level of harassment, Nita predicted the emergence of a police state, and suggested that she and Rory might actually leave their beloved Godwan River estate and flee to Southern Rhodesia. Nita to Joburg, 1 May 1953.

[486] In a recent interview Robert Murray expressed some regret that he did not visit South Africa, but was "aghast" at the racist environment, including even views expressed by his cousin Roger Murray (Rory's son). UWO Archives, R.G.E. Murray Collection, Robert Murray, "Response to Avery 19/04/08."

[487] Bob's three children were Alice, Peter, and Tom; while Sue's four children were Doug, Joan, Chris, and Sarah. Sue's husband, Blake, often referred to Freda as "The Admiral." Susan Murray Robinson, "My Mother Had a Profound Influence on Me."

[488] Spending time at Bark Lake during the fall hunting season was risky since, in the words of Joburg, there "were too many 'hunters' in the bush for comfort or safety." Joburg to Bob, 6 November 1954.

[489] Joburg to Bob, 15 October 1946.

[490] *McGill News*, Spring 1954, cited in Shearing, "Family Book."

members of the Chiron Club, which brought like-minded faculty together for intellectual discussion and good times. On occasion, there were also formal university receptions to meet with visiting dignitaries, such as Sir Henry Dale, who received an honorary degree at the 1951 UWO Convocation. On the domestic front, one of the most important developments was Doris's decision to resume her watercolour painting and her subsequent involvement with the London artistic community.

An event of some importance for the Murray family was the November 1951 Canadian visit of Princess Elizabeth and her consort, Prince Philip. While the "royals" visited both of their cities, the Montreal celebrations were more extensive and grandiose. Moreover, as a member of City Council, Joburg was invited to both the special welcoming ceremony at McGill University and the gala reception at the Hotel Queen Elizabeth. It was a grand event! The Murrays' identification with their British background was reinforced the following year when King George VI suddenly died.[491] Predictably, Joburg had his own analysis of the political consequences of this sad event: "The King's death was a shock ... Queen Elizabeth will be splendid too & ... it is a good fortune that the Old War Horse, Winston, is there to advise and encourage her. It would have been a ... sorry situation if She [sic] had been forced to rely on Atlee & Co."[492]

Despite this harsh indictment of the British Labour Party, it is interesting that Joburg and Robert rarely discussed Canadian political personalities or the merits of specific political parties. The only exceptions were their mutual distaste for Montreal's mayor, Camillien Houde,[493] and Premier Maurice Duplessis of Quebec. In the latter case, their negative comments usually focused on the Union Nationale's disastrous educational policies. As Joburg wrote in March 1953: "The report that Quebec will not allow Universities to get the Massey Report subsidy offered by Ottawa ... means the loss of something like $600,000 for McGill. The upshot is that projected improvements are completely out."[494]

In June 1955, the senior Murrays' life in Montreal came to an end when Joburg retired from McGill after twenty-five years of service. His retirement provided the occasion for an impressive farewell party, accompanied by a long list of tributes from scientific associates, wartime BW partners, academic colleagues, and personal friends.[495] Shortly afterwards, Joburg accepted an offer to become adjunct research

---

491 In June 1953 Joburg received a special medal in commemoration of Her Majesty's Coronation. See LAC, E.G.D. Murray Papers, vol. 19, file 32.
492 Joburg to Bob, 17 February 1952.
493 In 1947 both Murrays were outraged that the federal government had made Houde an honorary citizen of Canada. Bob wondered how "Mackenzie King or St. Laurent allowed that." Bob to Joburg, 12 January 1947.
494 Joburg to Bob, 8 March 1953.
495 See Appendix E.

professor at the Collip Laboratory at the University of Western Ontario. Ironically, while Joburg's new position was being negotiated, Bob was formally asked whether he was interested in succeeding his father as chairman of McGill's Department of Bacteriology and Immunology.[496] He declined the offer. As a result, after many years of talking about the prospect of joint work, the two Murrays at last became colleagues and neighbours.

But for Joburg, "snoozy" retirement was not an attractive prospect, after so many years of being an extremely active scientist and administrator. As he explained:

> The end of the road with an uncertain path beyond it is not very encouraging. We are sure we can make something of it in time, but the queasy shadows have to be looked at and we have to test the strength of our own resourcefulness. It will be rather like packing an unfamiliar trail in the dark, feeling for the hard ground with your feet, but we will find the lake.[497]

This sense of personal vulnerability was intensified by the sudden death on 21 February 1955 of Guilford Reed,[498] with whom he had shared so many experiences in peace and war.[499]

Retirement Years and Joburg's Last Hurrah!

Joburg's fears about enforced inactivity certainly did not materialize. During the next nine years, he received a number of job offers from Canadian and American institutions. The most attractive of these was an invitation in 1959 from the National Institutes of Health, Bethesda, to assume responsibility for some of its infectious disease research.[500] While he was flattered by the offer, Joburg made it clear that he was

---

496 See the correspondence on McGill's generous job offer in June 1955, and Bob's reasons for declining the position.

497 Joburg to Bob, 6 January 1955. Joburg had been pleased to receive a special letter of congratulations from Principal Cyril James: "It has a personal touch and says things won't be the same without me & praises what I have done; it has a wish for future happiness." He was also touched that during his last class the medical students honoured his contribution: "They applauded for a long time before I could thank them." Joburg to Bob, 6 January 1955; Joburg to Bob, 7 December 1954; Joburg to Bob, 14 January 1955.

498 In 1949 Joburg had been severely shaken by the sudden death of his longtime colleague Fred Smith, who had joined his department during the early thirties. At the time Smith was dean of medicine at McGill.

499 Reed had a stroke on 7 February and died two weeks later. Joburg visited Reed in his Kingston hospital, and attended his funeral, which attracted many of Canada's leading microbiologists. Joburg to Bob, 14, 21, and 27 February 1955.

500 Joburg was flattered by all these offers, particularly the NIH position. In this instance, his refusal was based on two concerns: "my limited interests confined to bacterial infections and my wish to avoid being too encumbered (with administration)." Joburg to Dr. George Williams, Clinical Center, NIH, 24 February 1959. See Appendix E.

fully committed to his new role as adjunct scientist at the UWO Collip Laboratory, "which I now have in working order and I am doing work that interests me."[501] Nor was this an idle comment. Joburg took his duties as a senior medical adviser seriously, and he was involved with a number of UWO committees. The most important of these was his 1960 role as consultant "for the new Basic medical Sciences Building to be erected on the University Campus." Indeed, his detailed and comprehensive 1962 report formed the basis of the medical faculty's overall strategy for the future allocation of resources and space.[502]

Robert Murray was immensely pleased and proud that his father had adjusted so well to his new academic environment.[503] Despite the fact that he was based at Victoria Hospital, many miles from the university campus, Bob tried to meet with his father daily, either at his Collip Institute laboratory or at the UWO faculty club.[504] They also had many opportunities to discuss mutual scientific and medical interests at their respective homes, along with many recreational outings with the Bob and Doris's three children.[505] Joburg and Freda did, however, miss the cosmopolitan qualities of Montreal, with its range of excellent restaurants, its many interesting cultural activities, and its proximity to Bark Lake.[506]

In many ways, Robert's personality was quite different from that of his father. But the two many shared many values. One of these was a sense of honesty and fair play; another was a commitment to personal loyalty. Both these traits were in evidence in 1953, when Robert was involved in a complex situation involving his former mentor, Igor Asheshov. After enjoying five years of research funding at the National Foundation for Infantile Paralysis based in New York, Asheshov's contract had been abruptly

---

501 Ibid.
502 According to the dean of medicine, if sufficient funds had been available, "I can assure you that almost all of the recommendations would have been put into effect." G. Edward Hall to Joburg, 4 March 1960; Dean O.H. Warwick to Joburg, 31 January 1962.
503 There was, however, concern about Joburg's health since he had his first severe heart attack shortly after the move to London, Ontario, in 1955. According to Sue Murray, her mother "nursed him fiercely, (and fearfully) through his long illness & saw to it that nothing would get in the way of his recovery. His illness, of course, put an end to their wonderful canoe trips in the wilderness they loved, a great sadness to them both." Susan Murray Robinson, "My Mother Had a Profound Influence on Me."
504 Western's medical school had been located at Victoria Hospital, in the south-central part of London, since 1930. In contrast, the university was located in the northern part of the city.
505 The only exception occurred in 1961, when Robert Murray spent six months of his sabbatical in Geneva and the other six months at the University of Copenhagen laboratory of his friend Professor Aksel Birch-Andersen. Robert Murray, conversation with Donald Avery, 11 March 2008. See chapter 9.
506 The senior Murrays did, however, enjoy their home on Regent Street, which was close to UWO. Most years they also spent May to October at Bark Lake, while Robert's family normally vacationed with friends in the Muskoka region. Robert Murray, interview by Donald Avery, 2 February 2008.

terminated. In his desperate search for an institutional affiliation, he hoped that his friend Robert Murray might "use the grapevine" to explore job possibilities in Ontario, hopefully at "the citadel of the Connaught Lab."

Not surprisingly, Robert discussed this situation with his father: "We had a very pleasant visit from Ash and Elizabeth a week ago. ... He sends his best regards. The work he has and is doing is most interesting and worthwhile; some activity against animal viruses has been shown for three of his antiphage agents. ... Despite his good work I hear that the Nat. Found. Infantile paralysis is not likely to maintain his grants – they are more interested in the possible quick rewards of the tissue culture and globulin work. So the fundamentals perish."[507] Yet, despite the endorsement of the two Murrays, the Connaught virologists were not interested in either Asheshov or his phage/viral research project, rejecting his application in a rather insulting manner.[508] In the end, Asheshov moved to London, England, upon receiving an attractive offer from the Lister Institute, where he remained until his death during the early sixties.[509]

Another example of Bob's ability to combine friendship and scientific cooperation was his long association with Roger Stanier, an outstanding microbiologist at the University of California, Berkeley.[510] The two men shared common interests in cytology, use of the electron microscope, taxonomy, and improving the teaching of microbiology at the university level. All these subjects were explored in their extensive correspondence and their periodic encounters at scientific conferences and seminars. Stanier's reputation soared during the late 1950s after he and two Berkeley colleagues published their classic textbook, *The Microbial World*, in 1956.[511] Four years later, Stanier completed another important study, *The Bacteria*, which provided an excellent survey of major research trends in microbiology, including an important article by

---

507 Bob to Joburg, 19 October 1952.
508 Asheshov was particularly offended by the letter from Connaught that suggested that his research career, based on long-term grants, was really "only suited to younger men who have few responsibilities to bear, and can continue to live the life of the grasshopper and the cricket." Asheshov to Bob Murray, 13 July 1953.
509 Asheshov to Bob Murray, 2 September 1953.
510 Roger Stanier (1916-1982) was born in Victoria, British Columbia, and educated at the University of British Columbia and the University of California, Berkeley. During the Second World War he was appointed Junior Research Bacteriologist at the NRC, before joining Merck & Company (Montreal) in the production of penicillin. In 1947 he returned to Berkeley as a member of the Microbiology Department, where he was to remain until 1971. Patricia H. Clarke, "Roger Yate Stanier 22 October 1916–29 January 1982 (Elected F.R.S. 1979)," *Biographical Memoirs of Fellows of the Royal Society of London* (December 1986): 543-568.
511 *The Microbial World* (Englewood Cliffs, N.J: Prentice-Hall, 1959) was the result of combined effort by Roger Stanier, Ed Adelberg, and Mike Doudoroff, all excellent microbiologists and well-respected in the profession. His co-editors gave Stanier credit for being the inspiration behind their classic study. Subsequent editions were written solely by Roger Stanier.

Bob Murray, "The Internal Structure of the Cell." The Murray-Stanier scientific linkages were also sustained through reciprocal seminar invitations, although Berkeley was more generous towards visiting scholars than Western.[512] Robert also valued Stanier's opinions about how to organize an international scientific conference, a subject he explored after being appointed program coordinator for the 1962 Montreal IAMS Congress. Stanier's recommendations were, however, not really appropriate for the occasion: "the most scientifically productive type of meeting is the small conference, comprising not more than 50 participants, all of whom are capable of making significant contributions to the discussion of a limited, but important scientific subject ... attendance should be by invitation only."[513]

The Montreal IAMS Congress was an enormous undertaking; the Murrays' efforts in organizing it involved considerable work, organization, and creativity. Fortunately, they had a number of advantages in dealing with these challenges. First, Joburg's role as president of the Congress reflected his long involvement with the IAMS, including its famous September 1939 Congress in New York City, on the eve of war.[514] He also had a good institutional memory and a good understanding of why some international congresses were successful while others languished. Second, in terms of logistics, there was the daunting task of vetting the many proposals, arranging the appropriate sessions, and dealing with precious scientific egos. A third consideration was the need for the Congress to obtain official endorsement and financial support from Canada's political leaders. Since the Congress was based in Montreal, Joburg made a special appeal to Premier Jean Lesage, whose government had launched Quebec's "Quiet Revolution": "Microorganisms are ... important to everyone and knowledge of microbiology is increasingly essential," he wrote. "The opportunities for exchange of knowledge and ideas at a Congress such as this are of great importance ... We are confident that the Government of Quebec will assist."[515] The Murray team also attempted to persuade Prime Minister John Diefenbaker and Governor General Major-General Georges P. Vanier to attend, but to no avail.[516]

---

512 Bob visited Berkeley in March 1963, with Stanier coming to Western in September 1964.
513 Stanier to Bob Murray, 6 February 1959. See Appendix E.
514 Joburg attended the IAMS meetings in Rome (1953) and Stockholm (1958). See Appendix E.
515 The task of negotiating with the Lesage government was entrusted to Leo Gauvreau, director of the CMS organizing team, and the influential Armand Frappier. E.G.D. Murray to the Honourable Jean Lesage, 10 January 1961. See Appendix E.
516 Among the premiers, Tommy Douglas of Saskatchewan demonstrated the greatest enthusiasm for the Montreal congress. T.C. Douglas to E.D.G. Murray, 3 February 1961. At the federal level, the Diefenbaker government was represented by the Minister of Agriculture, Alvin Hamilton, and by Governor General Vanier, who eventually sent an official greeting, but did not attend. See Appendix E.

Overall, the IAMS Congress from August 19 to 25 was a great scientific success.[517] The Murrays were particularly delighted by the large and diverse international contingent of microbiologists, including about forty representatives from various Soviet scientific organizations: the Academy of Sciences of the USSR, the All-Union Microbiological Society, and the I.I. Menshikov All-Union Society of "Epidemiologists, mikrobiologists [sic] and infectionists."[518] Joburg and Bob thoroughly enjoyed the many scientific sessions and informal conversations with many of the world's outstanding microbiologists, while Freda and Doris had the opportunity to attend the formal banquet and appreciate the significance of the event.[519] Afterwards, of course, there was still much work to be done: editing the fifty-five symposium papers and the twelve panel discussion papers, and arranging for their publication. This was eventually completed, with publication in August 1963.[520]

A vivid account of Joburg's role at the Congress has been provided by G.S. Wilson, a former Cambridge colleague:

> I said goodbye to him, not knowing it would be the last time, at the Montreal Congress of Microbiology in 1962. As President of the Congress, he reigned supreme. His generosity was unbounded. He had a large suite of rooms at the top of the Queen Elizabeth Hotel where he and Freda dispensed lavish hospitality to their friends – and they were numerous. He was still elementary and rugged, but he had mellowed with age, and he communicated to all around him that spirit of bonhomie and good fellowship for which we shall long remember him.[521]

On 6 July 1964, Everitt Murray died of a sudden heart attack. During the next few months, many letters of condolence were sent to Freda and Bob, followed by a series of effusive obituaries. Collectively, the comments about Joburg's personality and achievements emphasized similar themes: dedicated scientist, committed teacher, creative internationalist, loyal friend, and joyful personality. In some ways, Claude

---

517 In addition to his role as chairman of the Organizing Committee and president of the Congress, Joburg remained Canada's representative on the International Committee for Bacterial Nomenclature and chairman of the Subcommittee on Neisseriaceae.
518 N.D. Ierusalimsky (Vice-Chairman, USSR Academy of Sciences) to Dr. N.E. Gibbons, Secretary-Treasurer CSM (Ottawa), 23 April 1962. Joburg also received special thanks from Professor Viktor Zhdanov, Director of the Institute of Virology, USSR Academy of Sciences, 17 September 1963. The fact that such a large contingent of Soviet scientists should attend this event, two months before the October 1962 Cuban Missile Crisis, was quite remarkable.
519 Given the large size of the Congress (fifty sessions and over 2,000 delegates), there were the inevitable logistical problems. One of these was the last-minute rush to obtain a Soviet flag for the opening ceremony, which was sent by express from the Department of External Affairs in Ottawa. Robert Murray, interview by Donald Avery, 2 February 2008.
520 See Appendix E.
521 G.S. Wilson, "Everitt George Dunne Murray," 653.

Dolman of the University of British Columbia said it best:

> I need not tell you how much the fellow-attendants at the various occasions EGD patronized so faithfully will miss the stimulus and reassurance of his rare personality – his critical queries, discriminating comments, remarkable anecdotes, affectionate leg-pulling, and his pungent ... wit."[522]

POSTSCRIPT: THE MURRAY FAMILY AFTER 1964

Members of the Murray family responded to Joburg's death in various ways. Freda had the greatest adjustment since "her world had suddenly disintegrated ... (but) she soon pulled herself together & made a most valiant effort to live her life as fully as she knew how." During the next twenty-one years, her annual cycle of activity was divided between her London home, with Bob and his family nearby, followed by long sessions with Susan in the Montreal region, and then summers at the Bark Lake cottage. She also revisited England and Scotland in the mid-sixties to see friends and relatives there; and survived a life-threatening heart ailment in 1966, recovering well enough to enjoy Expo '67. The following year she embarked upon a four-month tour of South Africa, visiting members of the Murray family in the Cape region, the family enclave at Graaff Reinet, and then northward to Godwan River Estate and Johannesburg. She was of course saddened that Unita and Rory had recently passed away, although she did spend considerable time with their son Roger who, she remarked, "has a remarkable likeness to Rory as a young man as I knew him in 1918."[523]

Not surprisingly, given her naturalist inclinations and experiences, Freda's descriptions of her extensive South African travels were vivid and compelling.[524] "As the sun rose," she wrote, "my first glimpse of the veldt! The game was a greater thrill than all the magnificent scenery I have lately seen & how I wished Opa could have been with me – he loved it so much!"[525] There was another emotional moment when Freda toured George Murray's old house at Godwan River: "It is a lovely old place ... rooms full of treasures ... many portraits & caricatures by Grandpa Dunne, cabinets of beautiful silver & glass that he personally bought when Grandpa's [George's] estate

---

522 Excerpts from many of these letters and obituaries are located in Appendix D. Norman Gibbons, Secretary General of the Executive Committee of the IAMS, in his letter to Freda, made special reference to Joburg's role as chairman of the Seventh Congress: "he did a magnificent job, and you and he, as charming hosts, made it a great success."
523 Nita Murray died suddenly on 21 October 1967, and Rory shortly afterward on 21 April 1968. UWO Archives, R.G. E. Murray Collection, Freda Murray to Robert Murray, 19 November 1968.
524 At the end of this tour, Freda travelled to Copenhagen for a visit with Robert and Doris, who were spending the year at the laboratory of Professor Aksel Birch-Andersen. She wrote Bob sixteen letters and three postcards about her South African sojourn.
525 UWO Archives, R.G.E. Murray Collection, Freda Murray to Robert Murray, September 26 and December 3, 1968.

was sold ... everything just as it was when Rory died."[526]

In 1985, at the age of ninety, Freda reluctantly gave up her home in London and came to live with Susan and her family in Pointe Claire, Quebec. She died in 1990.

Susan Murray Robinson greatly missed her beloved Opa, who had been a source of constant personal support throughout her life. Her problems were compounded when, on 29 May 1969, her husband Blake died after a short illness. Fortunately, she was able to find solace from her four children and from Freda, who provided crucial "comforting strength." And she gradually began to rebuild her life. In 1972, during a visit to the United Kingdom, Susan met Don King, a widower with four children. They married in 1972 and established their home in Pointe Claire, Quebec. Throughout these years, Sue maintained close contact with Bob and Doris in London, although extended family times at Bark Lake were less frequent.

For Robert, his father's death meant the loss of a devoted parent, a trusted friend, and an inspiring scientific confidant. Throughout different stages of their respective lives, the two men had shared many wonderful personal and professional experiences, even when overseas schooling, wartime service, and separate scientific careers kept them apart. Evidence of their mutual affection and respect is vividly revealed in their extensive correspondence. In March 1948, for example, Joburg sent the following message after hearing about Bob's appointment as acting head of his department: "I am immensely proud of you, my boy, for what you have done & the way you have carried yourself to gain this tribute ..."[527] Robert's praise of his father's influence is no less complimentary:

> "There were many influences once I had got started; the research projects were my own devising, which burgeoned thanks to my father's continuing encouragement and support ... He NEVER told me what to do but was open to discussion about anything and everything letting me choose my own way. ... His advice was respected by me and vice versa."[528]

After 1964 Bob's professional career continued to soar, and he was soon firmly established as one of Canada's most outstanding microbiologists. Doris's artistic career also gained momentum during these years. In 1963, for instance, she hosted her first individual exhibition at the London Public Library and Art Gallery. By 1984, she had hosted eight individual and four group exhibitions of her watercolour

---

526 In contrast to her favourable comments about the rural regions and the cities of Cape Town, Port Elizabeth, and Durban, Freda found Johannesburg crowded, dirty, and racist: "Here I am, in this sprawling collection of skyscrapers, shabby back streets seething with crowds of people, masses of cars ... it is just too much of everything – a mixture of the hideously expensive & the desperately poor." Ibid., Freda to Bob, 3 December 1968.
527 Joburg to Bob, 6 March 1948.
528 Robert Murray, "Response to Avery 19/04/08."

paintings, in London and elsewhere. Tragically, in 1984, Doris died after a lengthy struggle with cancer. After a period of mourning, Bob married Marion Luney, an old family friend.[529]

Robert Murray's Scientific Career, 1964-2008

This portrayal of the scientific contributions of Everitt and Robert Murray ends in July 1964. It would, however, be unfortunate if the reader did not appreciate the full extent of Robert Murray's illustrious scientific career during the next twenty years, when he was one of Western's scientific stars, and then after his retirement in 1984. As one recent commentator described, "Professor Murray's scientific discoveries asserted Canadian microbiology on the international stage. ... He was one of the pioneers of bacterial cytology and in the use of thin-section electron microscopy for the analysis of microbial structure; this technique was further elaborated to include the freeze fracture of bacterial membranes."[530]

These achievements can be assessed in various ways. First, there are his many awards and honours. In recognition of his scientific exploits, Robert received the Flavelle Medal of the Royal Society in 1984; the J.R. Porter Award of the American Society of Microbiologists in 1987; and the Bergey Medal of Bergey's Manual Trust in 1994. These were accompanied by a series of civil awards, including the Centennial Medal in 1967 and the Queen's Jubilee Medal in 1978 and again in 2002. He was made an officer of the Order of Canada in 1998. Robert has also been awarded honorary degrees of Doctor of Science by the University of Western Ontario (1985), the University of Guelph (1988), the University of Victoria (2002), and McGill University (2007).

A second measure of Robert's achievements is his highly regarded body of publications (see Appendix I). In addition, there have been four major occasions when Robert's scientific peers have had an opportunity of expressing their admiration of his contribution to microbiology in Canada and internationally. The first of these, a special symposium entitled "Horizons in Microbiology," was held in 1984 to acknowledge Robert's retirement and appointment as professor emeritus by the University of Western Ontario. On this occasion, the many complimentary statements were aptly summarized by M.R.J. Salton, Department of Microbiology, New York University of

---

[529] Marian was born in 1919 in London, Ontario. Her father, Frederick Luney, was a pathologist associated with St. Joseph's Hospital. After two years of pre-medical studies, Marion became a laboratory technician at St. Joseph's. They were married on 18 May 1985. By this stage, two of Bob's three children, Alice and Peter, were also married. Robert Murray, conversation, 11 March 2008.

[530] Official Tribute to Robert Murray: Honorary Degree of Doctor of Science, McGill University, 21 June 2007.

Medicine: "It is a great pleasure that we come here together to honour Bob. ... Over the years we've admired tremendously the elegant and the beautiful work that he has done, and as Professor Robinow has said, he is truly an operational active scientist ..."[531] A second set of testimonials was included in the special April 1988 issue of the *Canadian Journal of Microbiology*, a journal that Robert and his father had created during the early 1950s. As the editors noted, "In the 32 years since that issue, R.G.E. Murray has become a scientist of the highest international reputation, and the Dean of Canadian microbiologists."[532]

In 1993 another colloquium, titled "Murray-Microbiology-McGill," celebrated Robert's graduation from McGill fifty years earlier. This event featured scientific presentations by four past presidents of the Canadian Society of Microbiologists, and was attended by thirty-five of Bob Murray's classmates, who gathered "to celebrate a colleague whose distinctions are international." And more recently, McGill again acknowledged one of its most famous graduates at the spring convocation of May 2007, when Robert was awarded the degree of Doctor of Science *honoris causa*.[533] The picture of this happy occasion, with Robert surrounded by his three children, provides a suitable transition from these introductory comments.

---

[531] In his introductory comments, Carl Robinow praised Robert as a brilliant scientist who "kept himself well informed ... about important events on the frontiers ... of his subject." UWO Archives, R.G.E. Murray Collection, tape recording of the symposium.

[532] *Canadian Journal of Microbiology* 34, no. 4 (April 1988): 359. This volume was divided into seven different sections, reflecting Robert's scientific interests and activities: Journal editor; Shape and form; Growth of cells; Motility and taxis; Growth "in the wild"; Inheritance; Classification in a diverse kingdom.

[533] There were fourteen honorary degrees presented at the spring convocation. See http://www.mcgill.ca/newsroom.

NOTE TO THE READER

There are a number of reasons for the title of this book – *The Meaning of Life: The Scientific and Social Experiences of Everitt and Robert Murray, 1930–1964*. First, it reflects the revolution in the life sciences that occurred during the period covered by the Murray correspondence, when scientific researchers steadily unraveled the mysteries of biological organisms, notably by the 1953 discovery of the structure of DNA and other related experiments.[1] Second, the title relates to the experiences of the Murray family, over three generations, and within three different countries—South Africa, the United Kingdom, and Canada. Within this section of the book, the impact of three major wars on the Murray family is discussed in some detail.[2] A third dimension of the meaning of life theme relates to the important contributions Everitt and Robert Murray made in the fields of medical and scientific research, and their correspondence provides many fascinating insights into the development of microbiology as a scientific discipline, the evolution of scientific medical research, and revolutionary changes in our understanding of pathogens and the human genome. While the Murrays' research interests often overlapped, there were also fundamental differences, reflecting the dominant scientific structures and social ambience of their respective eras. Everitt was a product of the early twentieth century, when microbiology was gradually emerging as a distinct research field, trying to overcome the limitations of inadequate technology, primitive laboratories, and insufficient funding. In contrast, Robert rode the wave of the post-1945 explosion in biological research, characterized by interdisciplinary cooperation, particularly in DNA-related studies, and substantial financial support from governments and foundations.

This project began during the spring of 2002 when Robert Murray, our colleague at the University of Western Ontario, indicated his willingness to consider the possibilities of publishing a select body of family correspondence. More specifically, we

---

1 There are a number of studies that use a variation of this title, linking science and life. See Mark Vernon, *Science, Religion and the Meaning of Life* (New York: Palgrave Macmillan, 2007); Soraya de Chadarevian, *Designs for Life: Molecular Biology after World War II* (New York: Cambridge University Press, 2002); Christian De Duve, *Life Evolving: Molecules, Mind, and Meaning* (Oxford: Oxford University Press, 2002); Evan Harris Walker, *The Physics of Consciousness: Quantum Minds and the Meaning of Life* (Cambridge, MA: Perseus, 2000); Connie Barlow, *Evolution Extended: Biological Debates on the Meaning of Life* (Cambridge, MA: MIT Press, 1994); Erwin Schrödinger, *What Is Life? The Physical Aspects of the Living Cell; With, Mind and Matter & Autobiographical Sketches* (New York: Cambridge University Press, 1992); *Tom Rivers: Reflections on a Life in Medicine and Science; an Oral History Memoir*, prepared by Saul Benison (Cambridge, MA: MIT University Press, 1967).
2 These military conflicts are the Anglo-Boer War (1899–1902), the First World War (1914–18), and the Second World War (1939–45).

were interested in the scholarly value of approximately four hundred letters Robert exchanged with his father, E.G.D. (Joburg) Murray, between 1945 and 1955, when Everitt was winding up his illustrious career in Montreal and Robert was beginning his own remarkable scientific journey in London, Ontario. In time, this initial cohort of documentation was expanded by additional family correspondence, along with over two thousand letters, reports, and articles from the personal collections of Robert and Everitt Murray at Library and Archives Canada. Other family information and photographs were supplied by Peter Murray, Robert's eldest son, who resides in Markham, Ontario. We also profited from our ongoing consultations with Robert Murray, now eighty-nine years of age. With his remarkable memory and formidable knowledge of the major trends in the development of modern microbiology, Bob has been an invaluable resource in the preparation of this volume. The R.G.E. Murray Collection, University of Western Ontario Archives, contains Robert's Recollections, a range of personal interviews, and other family material.

As the Table of Contents demonstrates, the Murray correspondence and related information have been organized into nine separate chapters, nine appendices, and two sections for relevant photographs. Unfortunately, because of space constraints, it was not possible to include all of this material in the published volume. Information from appendices E, F, G, H, and I can, however, be easily accessed through the Champlain Society web-site, www.champlainsociety.ca/meaning/, by using the special on-line index. In addition, readers can consult the one page summary of the important themes discussed in Appendix E, "Selected Documents on Scientific, Medical and Social Issues," located on page cxx of this section.

Throughout the text, there are many references to scientific terms, specific disease causative agents, and laboratory techniques, which are briefly explained in the footnotes. The Murray Glossary also provides a quick reference to the many scientific organizations, academic institutions, and government agencies that appear in the text; while the Select Bibliography identifies most of the secondary sources used in this study, as well as suggesting books for further reading. The reader should also be aware of the extensive cross-reference system that identifies most of the host of individuals the Murrays encountered during their respective scientific careers. This is a prominent feature of the Index for the published volume.

PRIMARY SOURCES

Library and Archives Canada (Ottawa)
- E.G.D. Murray Papers
- Robert G.E. Murray Papers
- Roger Stanier Papers
- The Department of National Defence: Biological Warfare Records, Directorate of Chemical Warfare and Smoke (Headquarters Papers, HQS)

University of Western Ontario Archives
- R.G.E. Murray Collection (Robert Murray Recollections)
  1. A Personal Narrative: Background, environmental and educational influences for the careers of EGDM (Joburg) and RGEM (Bob)
  2. Developing streams affecting biomedical sciences, esp. Microbiology
  3. Then and Now
  4. Conversations and Science
  5. Bob Murray and bacterial cytology
  6. The McGill Experience
  7. The Cambridge Experience
  8. Thoughts about where we stand as scientists and the significance of our work
- R.G.E. Murray Collection (Correspondence)
  - Freda Murray Papers
  - Correspondence Robert Murray and Doris Marchand
- R.G.E. Murray Collection (Related Documents)
  - Marge Shearing, "My Ain Folk: Family Book" (unpublished, 1989) A 'Mum for All Seasons': The 20's to the 40's
  - Notes on Doris Murray
  - "My Mother Had a Profound Influence on Me" (Susan Murray [Robinson] King)
- R.G.E. Murray Collection (Interviews)
  - Mark Eaton: Transcripts of Interviews with Robert Murray, 2005-08
  - Donald Avery: Transcripts of Interviews with Robert Murray, 2006-08

Peter Murray Collection (Markham, Ontario)
- Thurtell/Murray Family History (Work in Progress).
- Freda Murray Papers.
- Murray Family Photographs.

McGill University Archives
- RG2, C67, Faculty of Medicine (1930-1933), Department of Bacteriology and Immunology, E.G.D. Murray, file 1256
- RG 2, C-100, Principal Cyril James Correspondence, Faculty of Medicine, Bacteriology and Immunology, file 2724

GUIDE TO APPENDIX E
Selected Documents on Scientific, Medical, and Social Issues
Champlain Society Web-site

1. Everitt and McGill University, 1945-55 — Pages 819-54
    Major Trends
    Representing McGill on Montreal City Council
    Joburg's Retirement from McGill
    Post-1955 McGill Connection
2. Robert Murray and the University of Western Ontario, 1945-55 — 854-71
    Conversations and Science
    Bob Murray's Correspondence with Igor Asheshov
    Bob Murray's Correspondence with Roger Stanier
3. Joburg's Correspondence: The Western Connection, 1955-64 — 872-90
    General and Personal
    UWO Medical Science Building
    The Royal Society of Canada and UWO
    UWO and the Canadian Association of Microbiology
    Professional Advice
    Staphylococcal Infection in Canadian Hospitals
4. Canadian Medical and Public Health Issues, 1945-64 — 891-911
    Canadian Medical Association
    College of Physicians & Surgeons
    Canadian Public Health Issues and Hospitals
    Arctic Institute of North America and Bacteriology
    Veterinary Medicine and Microbiology
5. Defining Microbiology: An Evolving Discipline, 1945-64 — 912-28
    Everitt and René Dubos
    International Association of Microbiologists:
    Nomenclature Committee (Sara Branham)
    Canadian Committee on Culture Collections
6. National and International Microbiology Organizations — 929-56
    Canadian Society of Microbiologists
    Société de Microbiologie de la Province du Québec
    American Scientific Associations and Journals
    International Association of Microbiologists:
        The 1962 Montreal Congress

GLOSSARY: SCIENTIFIC ORGANIZATIONS AND THE MURRAYS[3]

Arctic Institute of North America (Montreal)	Joburg & Freda, members, 1935-64
Ayerst, McKenna & Harrison Ltd (Montreal)	Joburg, drug consultant, 1934-55
Bergey's Manual Trust (1923- 2008) bacterial/virus classification	Joburg & Bob, key members
Canadian Association of Bacteriologists	Joburg's 1938 venture
Canadian Society of Microbiologists (1951) *Canadian Journal of Microbiology*	Bob, founding president and first editor of the journal
Canadian Culture Collections (1950s)	Joburg helped create
Canadian Medical Association Founded in 1867, it now has 65,000 members National & Quebec branch	Joburg active in both
Canadian Medical Hall of Fame Established in 1994, London, ON.	
Canadian Public Health Association Founded in 1912	Joburg & Bob active in Laboratory Committee
Christ's College University of Cambridge	Joburg, 1908-12; Bob, 1939-41
Collège des Médecins et Chirurgiens du Québec Founded in 1847	Joburg, member

---

[3] See also Appendix B: Professional Achievements of Everitt and Robert Murray.

Graaff Reinet, east-central South Africa	Focus of activity for Murray family
Health Service Inquiry of Quebec (1953)	Joburg, an important consultant
International Association of Microbiological Societies (IAMS), Established in 1927 8th Congress, Montreal (July 1962)	Joburg, Congress President
McGill University, Office of the Principal General Sir Arthur Currie, 1920-33 F. Cyril James, 1939-62	
McGill University, Department of Bacteriology & Immunology	Joburg, head, 1930-55
Medical Research Council (UK) Established 1913	Joburg, researcher, 1919-30
Medical Research Council (Canada) Established 1969 Previously NRC Medical Council Canadian Institutes of Health Research (2000)	Joburg & Bob, many grants
Montreal Chapter of the Imperial Order of the Daughters of the Empire: wartime activities	Freda Murray, organizer
National Academy of Sciences (US) Established in 1863	Joburg & Bob, connections
National Department of Defence, Directorate of Chemical Warfare & Smoke, established in 1941; M-1000 & C-1 Committees (BW)	Joburg, chair, 1942-45

National Research Council of Canada Established in 1916	
NRC: Associate Committee of Medical Research Established in 1938	Joburg and Bob, members
Associate Committee on Control of Hospital Infections (1951-64)	Joburg, chair.
Royal Army Medical Corps (RAMC) Established in 1899	Joburg served, 1916-19
Royal Army Medical College (UK)	Joburg member, 1917-19
Royal Canadian Army Medical Corps Established in 1904	Bob served, 1944-45
Royal Society of Canada Established in 1882	Joburg elected 1938, Bob 1958
Flavelle Medal for scientific/medical excellence	Joburg, 1953; Bob, 1984
Royal College of Physicians & Surgeons (Canada) Established in 1929	Joburg, member
Microbiology certification committee	Joburg, member
Royal Victoria Hospital, Montreal (1887)	Joburg apppointment, 1930-55
St. Bartholomew's Hospital (Barts), London	George M. trained, 1884-87 Joburg trained, 1912-16
Society of American Bacteriologists, Established 1899 (Now American Society of Microbiology)	Joburg, candidate for president, 1954 Bob, president, 1972-73
University of Cambridge Founded in 13th century Pathology Institute, established 1922	Joburg, researcher, 1926-30

University of Western Ontario,
Faculty of Medicine
    G. Edward Hall, dean, 1944-47   Bob appointed, 1945
    J.B. Collip, dean, 1947-61   Joburg, Visiting Professor, 1955-64

University of Western Ontario,
Department of Bacteriology &
Immunology
    I. Asheshov, head, 1944-47   Bob Murray, head, 1949-74;
                                                   Professor Emeritus, 1984 to present

ACKNOWLEDGEMENTS

Many people have contributed to the genesis, development, and completion of this study. Our greatest debt is owed to Professor Robert G.E. Murray, who provided us with a wide range of correspondence in his possession, along with full access to the E.G.D. Murray and R.G.E. Murray collections at Library and Archives Canada (LAC). As a result, we were afforded a unique opportunity of charting the scientific careers of two of Canada's most eminent microbiologists, as well as portraying the dynamics of Murray family life within three different national contexts – South Africa, the United Kingdom, and Canada. Robert greatly assisted our editorial efforts by explaining important scientific concepts, clarifying medical terminology, and identifying a host of professional and personal contacts that he and his father established throughout their long careers. Donald Avery would also like to thank Marion [Luney] Murray for her generous hospitality during a number of editorial sessions with her husband at the Murray home in London, Ontario. Robert's eldest son, Peter, assisted our undertaking through his comprehensive analysis of the Murray genealogy and his willingness to share his own collection of family correspondence and photographs.

In terms of our archival explorations, we received considerable help from Archivist Gordon Burr and Mary Houde at the McGill University Archives in locating and copying relevant material about Everitt Murray's twenty-five years as a teacher and researcher of microbiology at McGill. In another location, our extensive use of the E.G.D. Murray and R.G.E. Murray collections at LAC was facilitated through the cooperation of a number of archivists and staff, including those working in the photograph division. We are also grateful for the willingness of Archivist Robin Keirstead, Anne Daniel, and other members of the University of Western Ontario Archives and Research Collection Centre to accept the challenging task of organizing and cataloguing the R.G.E. Murray Collection, the "motherlode" for this project and a valuable source for other researchers.

A special word of appreciation is directed to Heather Carson who, as a work-student assistant, carried out the laborious and time-consuming task of transcribing the original cohort of 380 Murray letters during the academic year 2003-04. Sara Ølholm Hansen also rendered valuable assistance in dealing with the different chapters, in the preparation of the index, and in providing general support for Mark Eaton's editorial efforts. For his part, Donald Avery would like to thank Dr. Irmgard Steinisch, also a historian, for her willingness to critique numerous drafts of her husband's introductory essay, despite her own busy professional schedule.

Of course, this project would not have evolved without the endorsement and guidance of Roger Hall, general editor of the Champlain Society. Although *The Meaning of Life*, with its emphasis on twentieth century medical/scientific themes, is quite different from the Society's many previous publications, Roger believed that our undertaking could make an important scholarly contribution, and therefore should be supported. For this we are grateful. In turn, Barbara Tryfos, our excellent copy-editor, has enhanced all aspects of our manuscript, and we greatly appreciate her professionalism and dedication. We are also appreciative of the work of Gwenne Becker (Becker Associates), and her team of technical and editorial experts, in printing such a quality product, in terms of both the text and accompanying photographs.

Finally, we would like to celebrate the results of our scholarly partnership. Throughout this venture the two editors have worked together in a creative and constructive manner, reinforced by mutual tolerance and good humour. Significantly, the professional relationship between Donald Avery and Mark Eaton was transformed during the completion of this volume, notably in 2007 when Eaton obtained his PhD in History at the University of Western Ontario and embarked upon his career as a professional historian. He is presently employed at the departments of Political Science and English, as well as the Canadian Studies Centre, at the University of Aarhus (Denmark), and his book on Canadian ideas about nuclear weapons is being evaluated for publication by the University of British Columbia Press. For his part, Avery has continued his scholarly activities, and will be completing his monograph *Biological Weapons, Terrorism and Pandemics: The Canadian Experience since 1939*, which will be published in the near future by the University of Toronto Press.

CHAPTER ONE

FAMILY AND PROFESSIONAL CORRESPONDENCE: 1930-1939

BACKGROUND
Everitt (Joburg) Murray
Cambridge Days and Wartime Service[1]

... The time at Bart's [St. Bartholomew Hospital] as a medical student was truly crucial to Joburg's development as a scientist and as a major contributor to bacteriology. As is true for all the students some time had to be spent in touch with laboratory medicine and at Bart's there were several on the medical staff that were important figures at the time. There was J.H. "Dropsy" Drysdale, an incomparable teacher at the bedside (a man fabled in stories of the time), Sir Frederick Andrewes who was a scientist and lecturer of note as a Professor of Pathology, Sir Archibald Garrod the sometime Professor of Medicine who was the first to recognize that there were "inborn errors of metabolism", and, most important to Joburg, among the pathologists was the all-round capable and interestingly erudite character embodied in Mervyn Gordon. Those were stirring times among people of substance for a young student for War was declared in 1914 when he was half way in his clinical studies. Sir Frederick Andrewes was consulted by the government and the War Office about emergent problems, among them what could be done about the large number of cases of meningitis in recruits then being gathered in army camps. Andrewes asked Mervyn Gordon to look into this which he did with great energy showing, with the active help of his student E.G.D. Murray, that it was meningococcal meningitis and isolating a number of strains of that organism.

... Mervyn Gordon's good opinion of this young medical bacteriologist in the making ... paralleled pressure from the War Office, for there were more than enough other problems and people to work on them were in short supply (e.g. wound infections, trench fever, trench foot, enteric fevers, dysentery, pneumonia, etc.). Normally Joburg would have waited for the qualifying M.B. examination from Cambridge ... But in war, time and tide wait for no man, so he took the immediately available qualifying examination set by the Society of Apothecaries – the LMSSA – and joined

---

1 University of Western Ontario Archives (UWO Archives), R.G.E. Murray Collection, Robert Murray, "A Personal Narrative: Background, environmental and educational influences for the careers of EGDM (Joburg) and RGEM (Bob)." This is one of a series of recollections prepared by Robert Murray for this volume. They are all deposited at the UWO Archives. See Note to the Reader, Sources.

the RAMC (Royal Army Medical Corps). He was wanted alright but, true to service experience from time immemorial, sent to investigate dysentery in Mesopotamia traveling as a medical officer on a troopship going there via Dar es Salaam, an east African port. Finally in camp near Basra he had to improvise means to isolate dysentery bacilli. The likely outcome was that he became very ill with Shiga dysentery and was invalided to Kasauli in India, where there was some interesting bacteriology going on.

By late 1917 he was sent back to England via South Africa and he broke the journey with his only visit "home". He was appointed to the Royal Army Medical College in London sorting through the collection of cultures of dysentery bacilli (*Shigella* species) to classify them using techniques he had learned on meningococci. He attached to the Vaccine Department of the RAM College, headed by Sir David Bruce, and this involved him in going to France to consult with authorities in the Institut Pasteur, making lifelong friends there in the doing. The result of his wartime experience was a life interest in *Neisseria meningitidis*, in enteric diseases particularly *Shigella* spp., in the identification of bacteria (the practicalities of medical bacteriology), and in bacterial classification and taxonomy.

Despite all the excitements of his life between 1914 and 1918 Joburg did not lose touch with the beautiful Freda Woods [Harriet Winifred Hardwick Woods] he had met as a teenaged young lady through his distant cousin Nicholas Everitt of Oulton Broad, Suffolk. She had spent much of the first half of the war nursing in the VAD [volunteer nurse] program. As soon as he was back and appointed to the RAM College they were married and when their first sub-culture was on the way they moved to a house in the western suburb of Ruislip where Bob was born in May 1919.

The army was certainly appreciative of what Joburg had accomplished during the war both in and out of the service and some senior colleagues must have made a great pitch on his behalf because he was made an Officer of the Order of the British Empire (Military Division). This is no light honour.

Joburg at the St. Bartholomew Hospital
Extract from *We Met at Bart's* by Geoffrey Bourne

Another incident in which Eccles figured involved Professor Murray, who was at that time still a student, but who was bedecked with long impressive moustachios, the ends of which he could insert into his ears. He also wore steel-rimmed spectacles. A rain storm drove him from the square to take refuge in Theatre B, in which at the time, Eccles was operating, Murray, the sole audience, was sitting in the front row of the gallery and, after the operation, Eccles walked over and asked where he came from. "South Africa" said Murray. Eccles then made lots of inquiries about the staff, the organisation, and the work of the Johannesburg General Hospital, which

Murray answered pretty well since his father was senior surgeon there. Eccles, taking his audience of one for a distinguished visitor, asked him down into the theatre for the next operation, which he demonstrated in detail and with some egotistical pride. This invitation astonished Wilson, then Eccles's Assistant Surgeon, and also the students who were acting as dressers, but they lay low like Brer Rabbit and said nothing. Then Murray had tea in the Surgeon's Room and, on shaking hands while parting at the door, Eccles asked him, "By the way, what was your medical school?" "I'm a student here," replied Murray and fled.[2]

How Joburg Met Freda [Harriet Winifred Hardwick]: England, 1911-17
Well, to say how I met Opa [Joburg] ... at Blunderston [Suffolk]. I was only, after all sixteen. ... I heard the front door ... and there was a large car which I recognized as belonging to Nick Everitt, [Joburg's distant cousin] who lived with his father at Oulton, Broad, which was about two and a half miles away. And Nick was sitting there with a strange young man ... And Nick said, "Oh good morning ... I thought I'd come along and, um, bring young Everitt with me. ... He's just here from South Africa you know, arrived about three days ago. He's come to stay with me. I thought I would take him around the country." ... After a while the young man said something in a very cultured, pleasant tone, in an accent which I didn't recognize at all, and then Nick said ... "We must be going" ... and the car bumbled off. ... The next day ... a bicycle arrived in the afternoon, pedaled by Opa ... And from that day on Opa was a constant visitor! He used to come and play tennis, used to come and dig gardens. ... We had a wonderful time. ...

Opa was twenty one ... He came to Cambridge to study medicine ... and at St. Barts Hospital in London. ... We had a very good time ... this went on for a long time, until 1915 when I had to make up my mind ... & made some applications, and was taken on at a hospital that was 300 miles away; as a junior nurse ... and there I was for three years, with one leave. ... I didn't see Opa again until 1917. He quite unexpectedly turned up, being driven in a huge van ... which had been fitted as a medical ambulance ... he was doing research all over England ... during the terrible epidemic of Cerebral Spinal Fever (meningitis). ... [We] were married in 1917.[3]

---

2 W. McAdam Eccles, one of the senior surgeons at St. Bartholomew Hospital, and a member of the Royal College of Surgeons, was described as "bombastic and opinionated." Geoffrey Bourne, *We Met At Barts: The Autobiography of a Physician* (London: Frederic Muller Ltd., 1963), 122, 154. Joburg had a copy of the book, which was presented to him by the author, and he commented extensively on various sections of the text.

3 "Reflections HWH Murray," (1895-1990), recorded by Sarah Schmidt; presented to R.G.E. Murray, Christmas 1996.

Joburg Relects on His Wartime Experiences: Osler Society Banquet[4]
Speech by Professor E.G.D. Murray on March 9th 1962, when he was Honorary President of the Osler Society, University of Western Ontario

Reflections

... I have no recipe for Instant History, and, besides, it is possible that you have no more taste than I have for being catechized in an after-dinner speech. So, at the risk of being accused of wallowing in a froth of anecdote, I shall tell you some little stories of people who have in a varying degree built something in the structure of medicine. ...

So I would have you reverently accept the few stories I shall tell you, for in a way we must pass amongst the shades of the great without disparagement. Reminiscence naturally turns to those whose example and teaching and encouragement awakened interests which shaped a career to be followed or which gave purpose and principles to modulate work through the years.

My lasting interest in zoology came from a diversity of men who maintained and secured the continuance of the fame of the department at Cambridge. Cecil Warburton,[5] who I last saw when he was 102 and still quick of memory, compiled ...

At Barts, though still a student, Professor Sir Frederick Andrewes[6] gave me a bench in the Path Lab, on the condition that I did not neglect my clinical studies. There I made friends with Dr. M.H. Gordon[7] whose example and encouragement directed me into a permanent interest in bacteriology. Gordon was one of the first to classify Streptococci by their fermentation of sugars and to elucidate the retting of flax to make linen by bacterial activity. ...

In 1915 Gordon invited me to join him in the War Office Central Cerebro-spinal Fever Laboratory, and, working with the meningococcus, this resulted in the first serological typing of bacteria. This was my introduction into the Royal Army Medical College, Millbank, and there I met many interesting people in the laboratories and in the Officers Mess. In 1917 I returned there from Mesopotamia, India and other places to work in the Vaccine Department under Colonel, later General David Harvey.

The Commandant of the College was General Sir David Bruce,[8] who had discovered *Trypanosoma brucei* and proved it the cause of *'Ngana* in African cattle. He also

---

4 Library and Archives Canada (LAC), E.G.D. Murray Papers, MG 30, B91, vol. 19, file Osler Society.
5 Cecil Warburton (1854-1958) was a member of Christ's College, Cambridge, and an expert in the field of medical entomology.
6 Sir Frederick Andrewes (1859-1932) was a highly regarded pathologist at St. Bartholomew's Hospital and an early exponent of medical bacteriology. Waddington, *Medical Education at St. Bartholomew's*, 131-32.
7 Mervyn Gordon (1872-1953) was Everitt Murray's mentor and co-author in their work on the meningococcus in 1914-15. He spent his research career at St. Bartholomew's Hospital focusing on meningococcus and Hodgkin's disease. See Introduction, note 127.
8 Sir David Bruce (1855-1931) was an expert in pathology and bacteriology, and spent most of his professional life as a military physician. See Introduction, note 151.

discovered *Brucella melitensis* the cause of Malta fever, and this led to its association with goats, work associated with Zamut and with Horrocks both of whom I met. Sir David was an upstanding broad shouldered big man, well over six feet, with smooth black hair brushed with a quiff, and a big black moustache. When roused his words were angry and well larded with expletive oaths; he was not difficult to rouse and I used this when he often came to my lab, which was the only one that had a small fireplace, and he told me wonderful tales.

Lady Bruce, who had assisted Sir David in many investigations, had a bench and microscope in the corner of his office; she was a small woman with bobbed gray hair and very quietly went on working while the business of the College came and went over Sir David's desk. But she was not diffident …

Harvey said Bruce ran the African Sleeping Sickness Commission like a martinet. I always found him resolute to get what you needed for your work but you had to have good reasons to convince him. He gave me much wise advice on many things …

At the RAM College we made the vaccines (Typhoid-Paratyphoid TAB, Cholera, Dysentery etcetera) in batches of many litres comprising thousands of doses. The cultures washed off hundreds of Roux bottles were pumped from flask to flask in 6 to 10 litre volumes … You learned to carefully wipe the bench clean of tiny bits of grit before putting down a flask of uncertain strength holding 10 or more pounds weight of hard gained living and dangerous culture … So one day a burst flask soaked my clothes from the chest down with a dense suspension of thousands of millions of living cholera vibrios per cc. I stripped out of my clothes and sent them to be sterilized and with a towel around my waist I proceeded with cleaning up the lab. Suddenly a high-ranking War Office Staff Officer opened the door and came in. He stared and spluttered "Is this the garb that His Majesties Officers wear in the College now? What is The Corps coming to with these Temporary Gentlemen----------". When Colonel Harvey followed in, he laughed and explained that this was a kind of accident we could not avoid, and said "You had better mind where you walk – this is living cholera," and the visitor vanished in rude haste without making an apology. …

I had many friends at l'Institute Pasteur and the dearest of them was Professor Maurice Nicolle,[9] who used to bite the ends off his words as he spoke with an enviable row of magnificent teeth. This mannerism gave an impression of indomitable determination. I treasure a letter from him about a paper I wrote on dysentery; he was then extensively paralyzed by a stroke and that one page letter cost him over four hours struggle with a typewriter. Some, like Roux, had worked with Pasteur, others later had added to the fame of the great Institute. Besredka pointed to his microscope

---

9 Maurice Nicolle (1862-1932) was later professor of bacteriology at the Pasteur Institute in Tunis. His younger brother Charles (1866-1936) won the 1928 Nobel Prize for Physiology or Medicine for his work on disease transmission for typhus and typhoid.

and said, "Look at that". I looked and said *Tretonema pallidum*. What of it", and was astonished to hear him say "Mais Oui, but it is the most virulent spirochaete in the world" "How do you know that?" I asked and he replied, "It must be. It is a Parisian spirochaete".

University of Cambridge
Memorandum [No Date][10]

1. We desire to bring to the notice of the College certain matters in connection with Mr. E.G.D. Murray's election by the College Council to a Fellowship in Class II. We feel that an injustice has been done to Mr. Murray which should be acknowledged and explained to the College; and that the College should consider whether steps should not be taken to restore Mr. Murray to the seniority which he has previously enjoyed and which it was the intention of the Council should continue.
2. In January 1929 enquiries were made by the Department of Pathology (in connection with the filling of a senior lectureship) as to the probability of Mr. Murray being transferred to a Class II Fellowship at the expiration of this Class IV Fellowship on December 1st, 1929. Notice that the matter would be discussed was given at a meeting of the Council on February 5th and the discussion took place on March 12th. It was agreed ... to continue Mr. Murray in his Fellowship; and a member of the Council was authorized to acquaint Mr. Murray with this opinion.
3. Eight of the nine members of the old Council are members of the Council appointed last October. There has never been any suggestion that the opinion to which they were parties in March has been changed, nor any suggestion that Mr. Murray's Fellowship should be other than a continuous prolongation with a change from Class IV to Class II. Nevertheless, *per incuriam*, Mr. Murray's Fellowship was allowed to lapse on December 1st, 1929, and re-election then became necessary. Mr. Murray was re-elected on December 11th.
4. The Council have carefully considered the Statutes and are clear that under them no action is possible to restore to Mr. Murray the privileges of precedence which he has lost. After December 1st he ceased to be a Fellow and all Fellows junior to him became automatically senior. ...
5. So far as financial considerations are involved the Council has passed Orders to secure Mr. Murray from any financial loss. We feel, however, that every effort should be made to restore to Mr. Murray the full privileges which he would have enjoyed but for the neglect which resulted in the lapsing of his Fellowship. It is

---

10 LAC, E.G.D. Murray Papers, vol. 17, file Cambridge (1929-30): memorandum [n.d.]. It appears that this document was prepared between October 1929 and July 1930.

impossible to forecast what value may attach to seniority of this kind in the course of years. It is easy to say that it is a matter of no very great importance; it is not difficult to envisage situations in which seniority might be the determining factor for the decision of important matters in our College life. ...

7. We see no way of restoring his precedence to Mr. Murray except by a specific Statute. A Statute must be a matter for consideration by the College. ...11

1. *Letters from* Sir Frederick Andrewes, Sir Leonard Rogers, Colonel Harvey, Dr. Gordon, Professor Nuttall.[12]
2. *Publications.* 1. On classification of Meningococci; 2. On classification of Dysentery bacilli; 3) Chapter in Bram and Archibald's "Practice of Medicine in the Tropics & on Meningo; 4) Some aspects of Meningococcal Virulence. Journ of Hygiene 1923.
3. *Posts held.* 1. On the staff of the War Office Central Cerebro-spinal Fever Laboratory. 2. Detailed for special work on Bacillary Dysentery in Mesopotamia. 3. On the Staff of the Vaccine Department of War Office. 4. Member of the War Office Committee on Bacillary Dysentery. 5. Senior demonstrator of Pathology at Barts. 6. Member of the Medical Research Council sub-committee on certain anti-bacterial sera.
4. *Present Position.* Research bacteriologist to M.R.C. (Medical Research Council). Annual grant of £600, liable to reduction or cessation. Pathological School-Tripos Part II.
5. *Degrees.* ... matriculated 1909, qualified 1916 (M.D.); Tripos. 2nd class N.S.T. 1912, 1st class marks in Zoo (Zoology). War record. O.B.E., Capt. R.A.M.C. (Royal Army Medical Corps)
6. *Research ability.* Proven and not promised. Posts held and referees. Election would give a very definite standing.

   [Joburg notation on margin] Not strictly correct. I was already appointed a lecturer until the age of 65. I made the enquiry because I wanted to know early on whether or not the College would make me a Class II Fellow for, if not, I must miss no opportunity to get an appointment elsewhere. In answer to my enquiry I received a letter from the Master and a verbal communication from (J.T) Saunders.[13]

11 It is not clear whether this uncertainty about his status at Cambridge influenced Murray's decision to accept the job offer from McGill University.
12 The memorandum listed these five referees. Three of them – Andrewes, Gordon, and Colonel Harvey – have already been identified. Sir Leonard Rogers (1868-1962) was an expert in veterinary medicine, leprosy, and amoebic dysentery; during the 1920s he was a lecturer at the London School of Hygiene and Tropical Medicine. George Nuttall created the Institute of Parasitological Research at Cambridge; see Introduction, note 185.
13 J.T. Saunders was professor of zoology at Cambridge. Joburg had studied zoology in his first two years at Christ's College and remained interested in this subject his entire life.

Family Life in Cambridge, 1920-1930[14]

Cambridge was the happy result and home for the first eleven years of his life ... There are only mental snapshots of the first years in Cambridge and many more for the three years spent until 1925 in a house with a big long and interesting garden in the village of Waterbeach, near the river Cam and fen country. Freda was an active mother and an energetic and knowledgeable gardener. Bob was constant company in that activity and the walks about the village, along the path along the river, picking blackberries in the hedgerows and, in company with father fishing local waters or in the workshop/garage fixing his motorbike or making things. There was lots to do and some time was given to playing with the son of the local doctor of similar age, but this was a relatively minor part of the apprenticeship to life. The garden and its "holes and corners" and its wild life from beetles to birds were entertainment enough until at five there was a bicycle to ride, in company, and soon on the many local paths. Joburg's involvement with the University and with Christ's College became more engaging in 1925 and inevitably a move into town and a house on Milton Road (and abandonment of the motorbike).

There was a motorcycle for Joburg to get to his work at the Field laboratories from Waterbeach. After a couple of years there was a newer large AJS with sidecar. The latter would hold Freda and Bob with a minimum of luggage strapped on it was transport to fishing places or to a summer's holiday. Two summers were spent with the Deans at The Lizard, near Lands End. The trips there by motorcycle and sidecar were memorable to the extent that Bob remembered features of The Bull Inn at Devizes when he revisited it some 70 years later. Strangely, after 1925 the family never had a car or vehicle more complex than a bicycle.

The move brought new life to Bob because next door lived the family of Bernard Jones (later Sir Bernard), the Professor of Aeronautical Engineering (later Director of the Royal Aircraft Establishment) which included a girl and two boys. These were real playmates for the next two or three years and a year or so of experience brought the special freedom of independent bicycle riding about town and country. The town was not yet saturated with cars, although the massiveness of buses gave moments of excitement, and there were places to visit with and without parental accompaniment. There were special lectures to be taken to where there were fascinating lecture demonstrations on anthropological adventures, string figures, or complexity of machines, etc., which were put on by the Cambridge Philosophical Society. Even more interesting were visits to Joburg's laboratory and the rabbits and guinea pigs in the "animal house".

On occasions at age 7 or 8 there were times with old "Bones", as the family referred to our professorial neighbour, that were sort of magical – with him playing a penny

---

14 UWO Archives, R.G.E. Murray Collection, Robert Murray, "A Personal Narrative."

whistle sitting on a stump with the kids dancing round him or with him designing and making tents of extraordinary shapes using a treadle sewing machine on the back lawn or with him telling us about flying a Moth back and forth between two points while filming the way strands of wool attached to surfaces behaved in the slipstream. (He came home from the latter experiences looking a bit green!). (The son nearest Bob's age was Melville Jones who later became a physiologist) ... There were a few family friends of long standing and particularly J.T. (Fanny) and Margaret Saunders and daughter Frances, and David and Joan Stockdale and son Tom (both became important to Bob when he returned as a Cambridge student), and only a few other families seen fairly regularly. (Fanny Saunders was a fellow student of Joburg's and also a Fellow at Christ's College, becoming registrar of the University as well as a distinguished Zoologist).

Childhood diseases and the arrival of a sister in 1926 were markers in the time for attendance at a nearby "dame school" and getting ready for the next stage in the life of an upper-middle-class English boy – attendance at a "prep school". This was Summerfields, a boarding school on the north edge of Oxford, which has a high reputation for producing scholarship students at the major Public Schools, like Eton, Winchester, Rugby, etc. Entry in September 1927 was a bit traumatic but survivable because the first year was considerate of the change involved for boys who had led a protected life. Subsequent years were a different matter; classes were small, there were demands for performance, and not doing your best or better was met with punishment, often corporal, and students moved up in particular subjects as soon as a new level was attained. It was a Church of England community with the Headmaster a minister and Chapel compulsory every day and twice on Sundays. Games were also compulsory. There were some excellent masters including a novelist (L.A.G. Strong), major poet (C. Day Lewis) and some who helped to make a demanding life bearable.

Holiday (vacation) times were most important. ... Again there was an adult component to them for we were often sharing with friends of the family and there were other children in the pack in rentals at Lyme Regis and in the Hope and Anchor Inn at Hope Cove. There may have been more than a little truth in what the Summerfields Headmaster wrote in a letter to the parents: "He seems to be used to the company of adults". This phase of school life ended early in October 1930 with the exciting move and introduction to Canada in the Port of Montreal with the Cross on the mountain brightly that late evening. ...

When I was at Summerfields school my father wrote to me faithfully every week and my mother wrote maybe every two weeks ... Receiving the letters was important for me not just because they were touch with home but because they were interesting and informative of what was going on. My father wrote in a very fine, small and controlled script while my mother's was a more energetic, flowing and ample line as befitted her descriptions of nature around her and homey things. My father wrote real letters about what was going on in life, about people met and seen. ... Sometimes his letters were illustrated with a diagram or a cartoon ...[15]

COMING TO CANADA[16]
Getting the McGill Position

Dr. F.C. Harrison[17]
Bank of Montreal
Waterloo Place, London, United Kingdom

July 9th, 1930

My dear Harrison,

Sir Arthur[18] has just shown me your letter about (Professor K.B.) Maitland ... Robert Muir of Glasgow suggested Dunlop, and Sir Arthur is cabling you this morning asking you to find out more about him.

Of course, you know better than any of us what kind of man we most need, and whatever we do, we do not want to be satisfied with mediocrity. It seems to me far better to wait for the right man than to get an appointee who would land us for many years in an unenviable position.

We hope that Maitland is not too much of a Public Health man, and that he has a wide general knowledge of the subject which would make him of use in other departments of the University. Dale [Sir Henry], of the National Research Council, recommended Maitland to me, and, I dare say, he would be the best choice in England. ...

Oertel[19] will be in London next week. ... I asked him to get in touch with you, and Sir Arthur cabled you this morning to that effect. Of course, it is important to get a man, if possible, in the fall in order that we can carry on satisfactorily. ...

Very sincerely yours,
C.F. Martin[20]
Dean of Medicine, McGill

---

15 UWO Archives, R.G.E. Murray Collection, Robert Murray, "About letter writing by parents."
16 McGill University Archives, RG2, C67, Faculty of Medicine (1930-1933), Department of Bacteriology and Immunity (hereafter cited as McGill Archives, B&I), Professor Murray File 1256.
17 F.C. Harrison had been responsible for teaching bacteriology at McGill during the late 1920s. He abruptly left his position during the spring of 1930, moving to London, England.
18 Sir Arthur Currie was one of Canada's most famous war heroes during the First World War. He was principal of McGill between 1919 and 1933, years that witnessed considerable growth in the size of the university's student body and faculty. Frost, *McGill University*, 130-96.
19 Horst Oertel (1873-1956) came to McGill in 1921 after an impressive career in German and American universities. He quickly established himself as an outstanding researcher and teacher, and in 1924 he was appointed director of the Pathological Institute. He retired in 1938 after a somewhat tumultuous career at McGill. Hanaway et al., *McGill Medicine*, vol. 2: 237-38.
20 Ibid., 110-135, 231-32. Charles Martin (1868-1953) began his long career in the McGill Faculty

July 7, 1930

Dear Sir Arthur,

I promised you that I would look at likely individuals well versed in Bacteriology so that you might have some information in case you decide to make an appointment.

I have from time to time since arriving in London visited a number of places & made enquiries, which, with my previous information, I beg to put before you.

You told me of the respective difficulties you have had with English & American selections & my first suggestion obviates these, as he is a Canadian by birth – Prof. K.B. Maitland of the Public Health Dept., University of Manchester. ...[21] He is, I believe, a Toronto graduate, but has been in England for some time & was working at the Lister Institute where he accomplished some valuable research along medical lines. From the Lister he went to the teaching post at Manchester, where he has been giving extensive courses in Bacteriology. ... I met him here three years ago & personally he seems keen, active, enthusiastic & young. I believe he would be a good appointment.

The other, Prof. W.W.C. Topley of London Sch. of Hygiene ... would not be so easy to transplant,[22] but his chief assistant Dr. Wilson would be my second choice.[23] He has recently with Topley published a very good two volume Bacteriology textbook.[24]

Both these men [Maitland & Wilson] are medical men & would fit in well with

---

of Medicine in 1893, specializing in clinical medical autopsies. He also gained an international reputation for his chapter on organic diseases of the stomach in Sir William Osler's seven-volume series *Modern Medicine*. As dean of medicine at McGill between 1923 and 1936, Martin instituted a number of reforms and aggressively recruited excellent faculty, including Jonathan Meakins, Wilder Penfield, and E.G.D. Murray.

21 Maitland remained at Manchester but continued to show interest in a job with McGill. In 1944 he wrote Principal Cyril James enquiring about post-war opportunities. James was not encouraging. McGill Archives, B & I, RG 2, C-100, Cyril James Correspondence, file 2724.

22 W.W.C. Topley (1896-1944) was one of Britain's most outstanding bacteriologists prior to the end of the Second World War. In 1922 he was appointed to the chair of bacteriology at the University of Manchester, before moving in 1927 to the newly established chair of bacteriology and immunology at the London School of Hygiene and Tropical Medicine (LSHTM). During the next twenty years he published many important papers and books with two colleagues, epidemiologist Major M. Greenwood and bacteriologist G.S. Wilson, also of the LSHTM. Topley's paper "The Spread of Bacterial Infection" (*Lancet*, 1919), based on his wartime research with the spread of typhus, was of major importance, as was his joint article with G.S. Wilson, "The Spread of Bacterial Infection, the Problem of Herd Immunity," *Journal of Hygiene* 21 (1923): 243-249. See Wilkinson and Hardy, *Prevention and Cure*, 120-45.

23 See Introduction, note 196 for information on Graham Selby Wilson. As an associate of W.W.C. Topley at the LSHTM, he quickly developed into one of Britain's leading experts in the bacteriology of public health and preventive medicine. Ibid.

24 The first edition of *The Principles of Bacteriology and Immunity*, by Topley and Wilson, was published in 1929. It was regarded as the "bible for bacteriologists." Ibid., 129-32.

the best of the work at McGill & would command the respect of both medical men & research workers & further I believe from inquiries may have teaching ability. ...

... I hope these inquiries will be of some assistance to you. ...

 Sincerely,

  F.C. Harrison

Devonshire Club
St. James's SW.1
London, 17 July 1930

Mr. dear Martin,

 I arrived last night and this morning (and afternoon) had a session with Boycott and Topley, who is head of the new, magnificent School of Public Health and has probably the most intimate knowledge of all available bacteriologists in Great Britain ... we took up Dunlop first. Both Boycott and Topley were unanimous that while he was a good man with equally good personal qualifications, he was too limited and not generally broad enough to represent Bacteriology as an academic and scientific subject, and that he could not go further than a 'hospital pathologist'. For that reason he did not get the appointment at University College.

 We then took up a list of possible candidates. One of the difficulties is that the British universities have latterly raised their salaries considerably so that $6000 is itself not a strong inducement unless certain other matters may contribute to induce a candidate to transfer his British chances overseas. We finally got down to the following names, in order of their desirability:-

 J.W. McLEOD of Leeds. He has recently expressed a desire to leave Leeds; is 44 years old; the head of the list, and, if agreeable, the most desirable. Whether he will go to Canada is uncertain. ...

 Then comes E.G.D. MURRAY of Cambridge. I am told also very excellent, perhaps a little brusk and sharp, but sympathetic to students and an excellent organizer. His candidature is also uncertain and I am to write Dean (H.R.) to sound him out. He is between 40 and 41 years old.

 Younger, but also very highly thought of are WILSON SMITH and C.H. Andrews, at present in the Mount Vernon Institute of the M.R.C. (Medical Research Council). Topley thinks particularly highly of Wilson Smith. Both are about 30 years ...

 All then agree that timbre from the U.S. is not to be recommended as their men are not sufficiently broadly trained and original for new paths in bacteriology.

 Please impress the seriousness of this appointment upon Currie and the necessity of very careful selection. ...

 Sincerely,

  Horst Oertel

Devonshire Club
21st July 1930

My dear Martin,

... We considered all the men which he (Harrison) mentioned in his letter to Sir Arthur. Wilson has now such a good post with the new School of Hygiene in London, that he is, of course, disinclined to consider another billet, especially since he is financially well fixed.

Tulloch, I hear, is very good, but, as they tell me, is a very difficult Scot, who would not transplant well, especially not to a foreign soil. Our Department of Bacteriology is in such need of complete reorganization and readjustment to University and Hospital demands, that a sympathetic and strong character is rather essential. In my last letter I wrote you about Murray at Cambridge, and reports about him as bacteriologist, teacher and colleague are exceedingly good. Dean (H.R.) has invited me to go down to Cambridge, and I think I shall take a look at, and have a talk with, Murray, at the same time.

McLeod in Leeds, of whom I also wrote you, has an excellent reputation, but is, so I am told confidentially, not a good teacher for beginners. Everyone here counsels caution, especially in a department which needs so much building up.

As regards Dunlop (they) ... doubt he will ever distinguish himself in his department.

I shall do my best and keep you advised.
    Sincerely,
      Horst Oertel

Department of Pathology
University of Cambridge,
July 22nd, 1930

My dear Martin,[25]

Yesterday I had a letter from Oertel asking whether E.G.D. Murray would consider the Chair of Bacteriology at McGill. I have written to Oertel and asked him to come down and see Murray and me in Cambridge.

E.G.D. Murray is a very close personal friend of mine, a most loyal and devoted colleague and my right hand in the department. I state deliberately and with the intention that you shall believe me when I say that I shall suffer a very serious loss if Murray goes to Montreal.

---

25 Henry Roy Dean (1879-1961) was professor of pathology at the University of Cambridge for thirty-nine years from 1922, and also served as Master of Trinity Hall from 1929 to 1954. See Introduction, note 178.

Murray is a very good bacteriologist and has published research work of distinction and he has proved himself an excellent teacher, lecturer and organizer. The courses which he has invented for part II of the Natural Sciences Tripos are admirable. He took a very large share indeed in the planning and equipment of our new laboratory. If the general layout and equipment of the rooms is my own work, it is true to say that every detail in fixed and moveable equipment, in water, heat, electric supply, in fitting up rooms for special purposes and in the purchase of special apparatus has been done by Murray.

Frankly I am devoted to him and can ill afford to lose him. I am sure that you will be very wise to secure him if you can for McGill and in view of Murray's future and bearing in mind the opportunities which a Chair at a great school like McGill offers, I shall and indeed have already, advised Murray to go if he has the chance.

Murray has a very good job here; he is our senior lecturer and under me has charge of most of the running of the lab. His position as my chief of staff is in fact recognized by a special payment. He has charge of the bacteriology teaching in the department other than the course given by Dr. Graham Smith to the D.P.H. candidates. He really has a good job here in every way and as far as I can estimate the situation and taking into account as far as I can the cost of things on both sides of the sea, I am disposed to think he will lose rather than gain by leaving us and going to you.

Murray has not come to a decision. He is fond of the work here and the laboratory which he has so largely made and there is a strong bond of personal friendship between him and me. On the other hand, he is quite rightly ambitious to hold an independent Post and quite naturally he would regard it as a great honour if he were asked to go to McGill. He believes that he would have great opportunities for work and for the development of bacteriology if he went to you, and personally I believe that if he was satisfied as to conditions and if you asked him he would go. That you would do well to ask him I am perfectly sure.

Murray has a nice wife, a boy of about twelve and a girl of about four.

My regards and kindest thoughts to you and Mrs. Martin. Meakins[26] has sent his son to us at Trinity Hall and I am very proud.

    Yours ever,

    H.R. Dean

---

26 Jonathan Campbell Meakins (1882-1959) was a highly regarded Canadian surgeon and medical administrator. An early protégé of Charles F. Martin at the Royal Victoria Hospital, Meakins was appointed to the McGill medical faculty in 1911. After outstanding war service, when he developed a reputation for pulmonary research, he accepted a position as professor of medicine and therapeutics at Edinburgh University. In 1924 he returned to McGill to become chairman of the faculty of medicine, physician-in-chief at the Royal Victoria, and director of the University Clinic. Between 1941 and 1947, Meakins was dean of medicine. Two other members of his family, son Jonathan Fayette and grandson Jonathan Larmonth, became leading surgeons at the Royal Victoria Hospital. Neville Terry, *The Royal Vic: The Story of Montreal's Royal Victoria Hospital, 1894-1994* (Kingston and Montreal: McGill-Queen's University Press, 1994), 86-97.

TELEGRAM
L.C.O. Cable, phoned CPR
August 13th, 1930

E. Murray
Pathological Department
Cambridge, England

I take pleasure in offering you the Chair of Bacteriology McGill University salary Six thousand dollars beginning September first next with two hundred pounds travelling expenses. Writing.
    Currie

August 13, 1930

Dr. E.G. Murray
Department of Pathology
University of Cambridge
Cambridge, England

... I cabled you in order that you might have time to think the matter over prior to receiving my letter.

    The Professorship of Bacteriology in this University is a general university chair, i.e., the general lectures are open to members of the University irrespective of their faculty. The laboratory is housed in the Pathological Institute where the demonstrations and all lectures are given.

    In addition to having supervision over this laboratory with respect to University teaching, you will also be given the opportunity of supervising and taking charge of the Bacteriology in the Royal Victoria Hospital, with which the Pathological Institute is connected. The two buildings are connected by means of a tunnel, so access to the Hospital is easy from the Laboratory of Bacteriology.

    It is my belief that there is great opportunity here at McGill to organize a Bacteriology Department which will do credit to the School and to the community, and I certainly hope that if you come you find sufficient opportunity for both teaching and research to your liking.

    I can assure you that it will be our pleasure to endeavour to make your stay here both pleasant and profitable, and I think you will find among the members of the Medical Faculty men of a high standard of scientific attainment and very congenial from a personal standpoint.

    Should you decide to accept this invitation, will you cable me at once, in order that

we may make our arrangements at as early date as possible?
>Faithfully yours,
>>Sir Arthur Currie
>>Principal, McGill University

>>>>TELEGRAM
>>>Canadian Pacific Railway Company's Telegraph
>>>>August 15, 1930: 12:05 Am
>>>>>Cambridge

Sir Arthur Currie
McGill University Montreal

I accept Chair of Bacteriology McGill University with great pleasure. Writing when letter received.
>Murray

>>>>>TELEGRAM
>>>>October 16 1930 6:50 pm

Sir Arthur Currie
McGill University Montreal

Matters here arranged enabling me to sail November 7 Duchess of Atholl. Please cable day which my engagement will date. Writing.
>Murray

>>>>>CPR CABLE
>>>>October 16th, 1930

Murray, University Cambridge, England

Engagement dates from November first.
>Currie

Christ's College, Cambridge
19 October 1930

Sir Arthur Currie, G.C.M.G., K.C.B,
Vice-Chancellor, McGill University

Dear Mr. Vice-Chancellor,

I had the advantage of a long conversation with Professor Oertel and time to make enquiries before receiving your cable, so I had no difficulty in coming to a decision. I believe, with you, that there is a great opportunity at McGill and it will be my endeavour to maintain and develop the Department of Bacteriology in a manner worthy of a renowned foundation.

The rapid development of the offer of the Chair at McGill University, with the time of year at which it occurred, made it impossible for Professor Dean to make immediate arrangements ... to replace me. I feel sure you will understand my desire to minimize inconveniences for him and to his department at the beginning of the teaching year. In addition I had a number of obligations, to the University, to my College, to the Medical Research Council and to the Vth International Botanical Congress. I also required a little time in which to order my private affairs. Bearing in mind my duty at McGill University and prompted by my desire to take up my appointment, I have succeeded in doing so much that I am able to sail for Canada in the 'Duchess of Atholl' from Liverpool on November 7th. In achieving this I feel I have done my duties to both Universities; this belief is shared by others here and I hope is reflected in your dating my stipend from the 1st of November.

Your letter and the assurances of my many friends convince me that my association with McGill can only be happy from both the official and the personal standpoint.

I look forward to taking up my duties with considerable pleasure.

    Yours sincerely,
    E.G.D. Murray

R.E. Priestly
Secretary of the General Board
The Registry of the University Cambridge
31 October 1930

Dear Murray,[27]

I have to acknowledge the receipt of your resignation as University Lecturer in the Department of Pathology as from 31 December 1930, and to inform you that the

---

27 LAC, E.G.D. Murray Papers, vol. 17, file Cambridge University.

General Board yesterday granted you leave of absence with stipend from November 3 until the end of the Michaelmas Term.

    Yours sincerely,

        R.E. Priestly

McGill University
Principal's Office
November 3, 1930

Memo (Sir Arthur Currie)[28]

I saw Professor Oertel to-day and confirmed with him the understanding I had with Dean Martin with regard to Professor Murray, the new professor of Bacteriology. In answer to questions from Dean Martin some days ago, I told him that Bacteriology would be regarded as a University subject, that Professor Murray would be a member of the Faculty of Science, in the Faculty of Arts and Sciences and that he would submit [a] separate budget each year. Professor Oertel approved of this understanding and promised to work in most cordial agreement with Professor Murray.

We also discussed Professor Bruere and I [told] him that he could assure Bruere that we would keep him employed as long as we fairly could.[29]

    Sir Arthur Currie

Dr. F.M.G. Johnson[30]
Acting Dean
Faculty of Science
McGill University

4 November, 1930

Dear Dr. Johnson,

The other day I discussed with Dean Martin the status of Professor Murray when he comes to us to succeed Dr. Harrison as Professor of Bacteriology. (By the way, Professor Murray sails from England next Friday and will be here, of course, a week later).

---

28 McGill Archives, B & I, 1930-33.
29 Bruere was a bacteriologist at the Royal Victoria Hospital between 1915 and 1930, while also doing some of the teaching in the Department of Bacteriology and Immunology. Much of his clinical work involved doing serological tests for syphilis and the typing of pneumococcus. See Sclater Lewis, *Royal Victoria Hospital, 1887-1947* (Montreal: McGill University Press, 1969), 220.
30 Ibid. One of Murray's responsibilities was teaching bacteriology to science students.

I told Dr. Martin that Bacteriology was a university subject, that the Department would submit its own budget. Hitherto, as you know, it had a very close association with the Department of Pathology. Professor Murray will have to serve the Department of Medicine and also your Department of Pure Sciences, but the point I wish to make clear to you is that Professor Murray is *not* under the direct and immediate control of the Department of Pathology. I confirmed this understanding with Professor Oertel, who has promised the greatest degree of cooperation.

 Ever your faithfully,
 Principal

6 November 1930

Dear Sir Arthur Currie,

I have your letter of 4th November in reference to Professor Murray and I have noted the contents,

 Yours respectfully,
 Fred Johnson
 Acting Dean, Science division
 Faculty of Arts and Science

Secretary Bursar's Office
McGill University
3 Dec 1930

Dear Professor Murray,

I have pleasure in informing you that the Board of Governors, at a meeting held on the 1st instant, appointed you Professor of Bacteriology.[31]

 Superintendent
 Royal Victoria Hospital
 Montreal, Quebec

---

31 LAC, E.G.D. Murray Papers, vol. 17, file 14-18. Murray also inherited W.W. Beattie, who had been appointed lecturer in bacteriology in 1926, after spending time at the Rockefeller Institute in New York City. In 1933 Murray would send Beattie to the Queen Charlotte's Hospital in London for advanced training on streptococcus. Unfortunately, he died in a traffic accident the following year. See Lewis, *Royal Victoria Hospital*, 218-30.

5 March 1931

Dear Professor Murray,[32]

I have the honour to inform you that at the Annual Meeting of the Board of Governor, held on the 26th February, you were appointed as Bacteriologist to this Hospital.

Reaction to Murray's Move to McGill

Norman McLean[33]
Christ's College Lodge,
Cambridge University
14 August 1930

Dear Murray,

Thanks for your thoughtful kindness in letting me know so soon of the offer which has come to you from Canada & of your intention to accept it. It is a calamity for us to lose you, but I rejoice with you in this recognition of the great work you have done in research & teaching & I trust thereby your usefulness will be increased.

Sincerely,
Norman McLean

Cork University Biological Station.
Loughine, Skibbereen' Co. Cork.
August 29th 1930.

My dear Ever,

Your great news has filled us with a very mixed lot of feelings. Primarily with joy at this recognition of yourself, not only by McGill but also by people at home: then with dismay at the thought that instead of seeing more of you all in the future, as we had hoped, we shall see even less than we have during the last few years. Our hope is that Cambridge may call you back to something even bigger before very many years have passed. ...

With our very best wishes and love to you all,
Lul[34]

---

32 Ibid.
33 Ibid. McLean was Master of Christ's College throughout the 1920s.
34 Ibid. This is presumably Louis Renouf, founder of the Cork University Biological Station. See Louis P.W. Renouf, "Preliminary Work of a New Biological Station," *Journal of Ecology* 19, no. 2 (August 1931): 410-38.

British Museum (Nat. Hist.)
Cromwell Road, London, S.W.7
Aug. 23 1930

Your letter arrived this morning giving the important news that you are off to Canada. I hasten to congratulate you upon your appointment to the Chair of Bacteriology at McGill and to wish every success in the venture and happiness for all of you in the country of pine and torrent. You will regret leaving Cambridge which has been your home so long, I know, but I am sure you will find good friends in Montreal. ...

How does Freda like the idea of Canada? The long, cold winter is in my mind the greatest drawback to the dominion. The Summer however recompenses for the months of snow.

With my love to all of you and once more wishing you every happiness and the best of good fortune out in Canada.

    Yours,
      Stanley[35]

30/9/30

My Beloved,[36]

... I think, by dint of some very hard work on my part, I can so fix things that we can count on going (to Canada) during the first week in November. I am aiming for that date & await any serious arguments against it. ...

... As we will have to live for a while somewhere before we find permanent quarters in Montreal, we shall need one (perhaps 2) trunks which can travel in the hold & contain clothes & necessities not wanted on the voyage but needed after landing. Our other effects will be stored meanwhile ...

Should you want me between now & Friday morning you can ring up College (894) between 8 pm onwards, or Lab (2232) between 12 & 1 pm when I shall endeavour to be in for certain. ...

    All my love to you,
      Your mate,
        Ever

---

35 Ibid. Stanley Smyth Flower (1871-1946) was the son of Sir William Flower, director of the British Museum (Natural History). One of Britain's foremost zoologists, Stanley Flower spent considerable time in Asia and the Middle East, and was director of the Giza Zoo in Cairo during the 1920s.

36 Murray Material, Personal Collection of Peter Murray, Markham, Ontario.

> Bacteriological Department
> University College
> Dundee
> 1st October, 1930

E.G.D. Murray, Esq., O.B.E.
Christ's College,
Cambridge

My Dear Murray,

I don't know what you will think of me for not writing earlier, but the fact is that I was unaware of your appointment to the Chair in McGill until Monday last when Stewart in Leeds told me about it. I cannot say how glad I am for your sake and I cannot say how sorry I am that you are leaving this country, for believe me, one of the very pleasant recollections I have is the happy time I spent with you in July 1929. I hope that everything will prosper with you in Canada and I consider McGill very lucky in acquiring your services.

I need hardly say that the silent Craigie joins me in extending to you these very warm good wishes in which I hope Mrs. Murray will share.[37]

Yours sincerely,
William John Tulloch[38]

The African World, October 11 1930[39]

Dr and Mrs GAE Murray, of Johannesburg, have received news that their eldest

---

[37] LAC, E.G.D. Murray Papers, vol. 17, file 14-18. Reference to the "silent Craigie" refers to microbiologist James Craigie (1899-1978) who worked with Tulloch at Dundee until 1931 when he became research assistant at the Connaught Laboratories in the University of Toronto. His work on viruses, notably vaccinia, the causal agent of smallpox, and poliomyelitis, greatly enhanced his scientific status. As a result, in 1940 he was appointed Professor of Virus Infections at Toronto, and in 1946 he was elected president of the American Society of Bacteriologists. The following year he returned to the United Kingdom to work in the laboratories of the Imperial Cancer Research Fund in London. Sir Christopher Andrewes, "James Craigie, 25 June 1899 - 26 August 1978," *Biographical Memoirs of the Royal Society of London* 25 (1979), 233-40.

[38] Ibid. W.J. Tulloch (1887-1966) had an impressive scientific and medical career. He began teaching at University College, Dundee, in 1914 as a lecturer in bacteriology, before entering the Royal Army Medical Corps College, where he remained until the end of the war. Working with Sir David Bruce, he specialized in wound infections, particularly tetanus, and the treatment of cerebrospinal infections. In 1921 he was appointed professor and chair of bacteriology at Dundee, as well as bacteriologist at the Dundee Royal Infirm, responsible for public health in the region. Between 1945 and 1956 he was dean of the Faculty of Medicine.

[39] Ibid.

son, Dr. Everitt Murray, M.A., OBE, Fellow of Christ Church, Cambridge, has been offered and has accepted, the Chair of Bacteriology at the McGill University, Canada.

Cambridge University Reporter, 11 November 1930[40]
"Report of the General Board on the replacement of a University Lectureship in the Department of Pathology by two University Demonstratorships"
    The GENERAL BOARD beg leave to report to the University as follows:
    The Board have received the following report from the Faculty Board of Biology 'B':

24 October 1930

    The BOARD OF THE FACULTY OF BIOLOGY 'B' have considered the appointments on the staff of the Department of Pathology rendered necessary by the election of Mr E.G.D. Murray to the Chair of Bacteriology in the McGill University at Montreal.

    Mr Murray has been responsible for the greater part of the lectures and classes in bacteriology for Part II of the Natural Sciences Tripos and since the retirement of Dr Cobbett has undertaken in addition to his Tripos work the lectures and practical classes in bacteriology which are given in both the Lent Term and the Long Vacation to medical students in preparation for the Third Part of the Second M.B. Examination. Mr Murray has in fact given about 75 lectures a year and devoted at least 220 hours in the year to demonstrations.

    The Board propose that the teaching work hitherto undertaken by Mr Murray shall be carried out by two University Demonstrators. ...

H. R. DEAN, *Chairman.*	F. J. W. ROUGHTON.
HUMPERY ROLLESTON.	W. B. HARDY.
T. S. HELE.	F. F. BLACKMAN.
E. D. ADRIAN.	S. W. COLE.
J. T. WILSON.	F. C. BARTLETT.
JOSEPH BARCROFT.	

The General Board and the Faculty Board both expressed the opinion that Mr Murray had too great a burden of teaching last year, but it was at that time impossible to re-arrange the lecture programme. The General Board welcome this opportunity to divide the teaching he has carried out between two junior University Teaching Officers who will be in a position to give a reasonable proportion of their time to research.

    The Board therefore recommend
    That, on the resignation of Mr E.G.D. Murray, the University Lectureship held by

---

40 Ibid.

him be suppressed, and that two additional University Demonstrator-ships be established in the Department of Pathology.

    Christ's College, Cambridge
    30/9/1930

Oswald T. Avery
The Hospital of the Rockefeller Institute for Medical Research
NYC, USA
22 January 1931

Dear Dr. Murray,[41]

    I enjoyed very much our visit together and appreciate greatly your sending me the reprints. I look forward with pleasure to reviewing again this interesting work. With the assurances of my heartiest good wishes for success in your new venture, which looks most promising.

    Yours sincerely,
        Oswald Avery

Department of Pathology,
University of Cambridge
January 9th, 1931

My dear Murray,

    Many thanks for your letter. I shall look forward with great interest to a letter with information about the other half of the world and how you find it. I believe that you have got a big job but I am sure that you will make a good job of it. The committee approved. Dr. Beattie and his name will appear on the pink paper next July. The committee decided to accept your resignation as from the date of appointment of a successor. That is to say, there will be no election to the committee until after the next meeting of the committee and we shall be all very pleased if you are able to attend the next meeting of the committee, which will he held in Glasgow on Thursday, July 2nd.

    Every good wish to you for 1931 and please give my kindest regards to Martin and Oertel and to the assistant Dean, whose name I never can remember, and the very nice man who does pathology at the other hospital.

    Yours ever,
        H.R. Dean[42]

---

41 LAC, E.G.D. Murray Papers, vol. 17, file 14-12. Oswald Avery, who made key contributions in the discovery of genes and the structure of DNA, is generally recognized as one of the most important American microbiologists of the period 1920-1960. See René Dubos, *The Professor, the Institute, and DNA* (New York: Rockefeller University Press, 1976).

42 Ibid. Dean and Joburg maintained their friendship and professional contacts during the next

Prof. W.W.C. Topley,
Division of Bacteriology and Immunology,
London School of Hygiene and Tropical Medicine,
University of London
12th January, 1931

Professor E.G.D. Murray[43]
Department of Bacteriology,
McGill University,
Montreal,
Canada.

Dear Murray,

Very many thanks for your letter. It is good to have news of you. What you say about conditions at McGill sounds very cheering. I should think that the opportunities are tremendous, and I have no doubt that you will make the best of them. I quite agree with you that nothing is to be gained by hurry when one is starting a new job. There is nothing for it but to put a couple of years or so aside, and then start one's own work again when things are running smoothly.

    With kindest regards and very best wishes,
        Yours ever,
           W.W.C. Topley

Professor E.G.D. Murray
McGill University
Department of Bacteriology
December 29, 1930

My dear old Rory,[44]

This is your birthday & we wish you the very best of everything. We hope you are very happy now & that the New Year will be a prosperous one.

We are settling down pretty well. Everyone here is exceedingly kind and thoughtful of our welfare & happiness. Nothing seems to be too much trouble for them to do if it is going to help us. We were met at the boat; pushed through the customs; taken off to comfortable lodgings where we lived for the first ten days & then we moved to a convenient little flat (apartment it is called here), furnished & let at what is a

---

three decades.
43 Ibid. See note 22.
44 Ibid. This letter was the start of extended correspondence between the two brothers, lasting until Everitt's death in 1964. Rory would call him "Biff."

reasonable rent here, where we shall live until May. By that time we shall have found a place of our own. This flat was found for us by friends too.

We have been entertained on all sides with the greatest friendliness. We like the people very much indeed & we like the place too. We can find nothing but good to say of Canada from our point of view.

Many things here are more expensive than in England, but many others are absolutely or relatively cheaper. Rents are very high indeed & clothes are very dear; fruit & some other things like telephone, gas, electricity etc. are much cheaper; vegetables are dearer but you get more for your money as every bit you buy is useable (no dirt, no useless leaves or roots etc – splendidly cleaned); there are many conveniences in the houses which are unknown in England.

The winter weather is cold but dry. The ground has been covered in snow for weeks past & the temperature has not been above freezing (32°F) & at times it has been zero & even 8°F below zero. So cold then that your breath freezes to icicles in your nostrils but it is not unpleasant when you are correctly dressed. We are told that colder weather will come. The houses are kept rather too warm, 70° to 75°F, but it is grateful to go into the warm when the weather is in the region of zero outside, or even 10° to 20° of frost. The streets are kept remarkably clean of snow by various methods & cars (automobiles here – cars are trains) run all the time. In the snowy weather the dominant noise is that of the chains on the wheels of the cars.

Bob[45] is at school & is quite happy & is making friends. His greatest joy is skating, at which he spends as much time as he can. He is quite good at it now & is starting to try to cut figures. He has had some fun toboganning & is looking forward to skiing. Freda[46] is busy housekeeping & discovering the shopping difficulties & taking care of Susan.[47] Servants are out of the question for us as wages are too high. Sue rather misses her garden & the freedom it gave her, but with luck we may find a house in which she will be less cramped than in a flat.

They play a game on the ice here called Hockey. There are both Professional & Amateur teams: both are good but the Pros are marvelous. It is the fastest game I have ever seen & the skill of the individual players is marvelous & must be seen to be believed. There are, of course, outstanding players who are miracles of speed, coordination & quick reaction. We greatly enjoy seeing it.[48]

---

45 Bob Murray, who was born in 1919, was eleven years old.
46 Freda Murray (Harriet Winifred Hardwick Woods) married Everitt in 1917.
47 Susan (Sue) was the second Murray child. She was born in 1926, seven years after her brother Robert.
48 Everitt became an avid fan of the Montreal Canadiens, owning season tickets for many years. Bob Murray has warm memories of going with his father to the Forum, Montreal's hockey arena.

I have a great deal to do as the department was rather disorganized & it will take me at least two years to get it straight. I am beginning to see daylight now & I think I shall be able to make a good show of it & I can see great possibilities. It is most encouraging to find that those in authority are ready to back me up in what I want to do, so I do not feel discouraged. If I can make good the effort will not be wasted as the higher authorities are well aware of the task.

Well, old chap, here's our love & good wishes to you & Nita & may the years to come make up for the past.

  Your loving brother,
    Ever

McGill University
Montreal
Department of Bacteriology
Professor E.G.D. Murray
July 24, 1931

Dear old Rory,

Here is the article on the Pollination of Avocado which you wished to have. I have not yet seen abut the moccasins but there will be no difficulty in getting them. I have not done so as the shoe sizes here are different to the English ones & it will be safer if you send me the measure in inches of your foot and a tracing on a piece of paper. Moccasins are made in various forms: shoe form & boot form. I have a pair of the boots for the woods & for fishing & I find them very comfortable. They are the best form of footwear for that purpose I have met.

The family are all at Bark Lake, about 80 miles north of here, having a splendid time. We have rented a cottage there set in a clearing in the woods on the edge of the lake.[49] The woods are almost impenetrable & our place can only be approached by water. I sent a description of it to Dad & I asked him to pass it on to you. We have an outboard motor (Johnson) & so can get about easily. I spent a fortnight with them & I shall go up again soon. The fishing is fairly good, mainly bass, though there are some Grey Trout & some other fish (catfish, whitefish, pickerel). Unfortunately there are no Speckled Trout.

The bass are the gamest fighters I have ever met & I had great sport with them – the largest I got was 3-1/2 lbs. They take a wet fly & in the evening a dry fly & fight like blazes. Our 1-1/2 pounder I got on a fly broke water 9 times before I netted him. They are good eating too.

---

49   The Murrays rented the cottage of McGill friends during the summer of 1931, their first in Canada. UWO Archives, R.G.E. Murray Collection, Robert Murray, "A 'Mum for All Seasons': The 20's to the 40's."

I am glad to have the photographs. I am sorry you did not include one of yourself. Do so next time. I would like a view of the farm taken from the top of the *Live Kelm* to compare with the one I took in 1917. I would love to see the farm now.

I wish Dad would go to you at the farm & abandon practice altogether.[50] He is getting old & the people have mostly gone & the young ones want younger men. He is having a hell of a miserable time & simply dare not do or say anything. Every week in his letter he tells me something & then he says "do not mention it in your letters". I cannot say anything or enquire about friends & relations as it makes trouble for him. In any case I only write to him & I have no wish that he should take my letters to the house. I do not write to them nor wish even to see them nor hear from them. I wish he would stand firm & break away. He has no call to support *Thor & Lall* & he could make an allowance to Mother.

I gather they are still trying to make hell for Nita & you.[51] If my curses came true they would not trouble anyone further.

Things promise very well here but it will take 2 or 3 years of very hard work to get the department on its feet. However I am not letting things stand still & I have convinced the authorities who are now prepared to give me everything they can. All will be well in time.

With love from us all to Nita, John & your old self.

Your loving brother,

Ever

24 Plein St.
Johannesburg
June 7th, 1933

My dear son,

Thus far no letter from you. You must be back again after the holiday ... Saturday will be the last days of the winter meeting at the Turf Club. Things in much the same condition. Domestic affairs unchanged and my existence is a misery. I got nothing but abuse:[52] it cannot be avoided; and as one is practically penniless and there is precious little work one does not know what to do. The only thing is to hope for change in the state of affairs. The weather is cold, as it is winter it must be expected. ... I wish

---

50 At one time George Murray had a flourishing medical practice in Johannesburg, as well as his position as chief of surgery in the Johannesburg General Hospital.
51 Nita's previous marriage was to Everitt's younger brother Allan (Lal). This complicated relationship deeply divided the Murray family, with Everitt strongly defending Rory.
52 This could relate to George Murray's dementia, which was becoming progressively worse; also, it might be a reference to the temper and sharp tongue of his wife, Kate Murray, "who might well have been very difficult and sharply critical of everything and anything." R.G.E. Murray, personal communication, 2 September 2004.

there was something interesting to tell you.
    Much love to you all.
        Your loving father,
            GAE Murray.

                                          24 Plein St.
                                          Johannesburg
                                          June 21st, 1933

My dear Son,

    Yours of 21st of May just delivered. I am pleased to hear that he is so well developed and that he is doing so well at school and I am glad to hear that Freda is improving in health and hope she will soon be quite well. Susan ought soon to recover from the effects of the flu!

    I hope your experiments will produce good results. You are having summer, and it is our winter, it has been very cold thus far, and there was rain during the night and early morning ...

    ... things are still bad, and there is practically no work to do. Plenty of fellows come in and want to borrow a few pounds, but I am practically penniless, so just have to clear them out.

    Let us hope for improvements.
    Much love to you all.
        Your loving father,
            GAE Murray.

                                          24 Plein St.
                                          Johannesburg
                                          July 12th, 1933

My dear Son,

    No letter from you today. I hope you are all in good health. The sun is quite bright this afternoon but the wind is cold. Winter is nearly half over, I shall be pleased when it is quite finished, one gets terribly cold hands especially when driving a car. There is nothing of interest to tell you from here. There is precious little work to do, and practically nothing coming in. I have not heard from Graaff Reinet[53] for some time. They must all be well otherwise I would have heard about it. Domestic affairs unchanged ...

---

53 Graaff Reinet was the family home, established by Walter Murray during the 1880s. Most of George Murray's immediate and extended family remained in this region, located in the eastern and southern area of South Africa. See Introduction.

I am well but practically idle.
    Much love to you all.
        Your loving father,
           GAE Murray

Relocation: Robert Murray remembers the family's move.[54]

My recollections are hazy about when I knew for sure that we were going to Canada. ... I remember that I did know about there being only a month or two of the Michaelmas term at Summer Fields School (just outside Oxford) to be "endured" before the exciting voyage. I remember leaving my last class and going outside for a last gesture of throwing my notebook in the air as a sort of statement of release. ... The furniture had already been sent on its way and we stayed a night or two at The Blue Boar before going to London to stay at my mother's uncle's house on Westbourne Terrace. During the few days there we went to a Theatre or two; one was a Jack Hulbert and Bea Lillie show and the other was Follow a Star, a top musical of the day. Then the train ride to Liverpool for embarking on the Duchess (of Atholl) which seemed enormous. I remember the rough steep seas on the Irish Sea and then the big rollers of the Atlantic which brought on the sea-sickness. We were not in the First Class and the cabin was down a bit in a ship noted for a rolling movement and near to a gangway to the engine room so there was a good smell of hot oil much of the time. After a sickening few days we came to Belle Isle and a rather peaceful Gulf of St Lawrence.

Not much detail in the memory about anything but a growing excitement with Canada in sight. Some clear snapshots in the mind of looking up at Quebec City and the Chateau, of going under the Quebec bridge and watching the mast not hit it, of crossing Lake St. Peter and realizing it was a mighty river full of shipping, and nightfall before getting to the port of Montreal. The sight of the illuminated cross on the Mountain was dominating. We went to a boarding house on Sherbrooke Street (two houses from Peel) belonging to the Foster family (he was an out-of-work architect and there were two grown children and a couple of other boarders; kindly and helpful Mrs. Foster ran the show) and they made us very welcome and were especially kind to me by seeing that I was introduced to Canada in a way that meant something. ... The son, Cary Foster, walked me all over town and taught me to be independent and a bit savvy; he also had been to Lower Canada College and put me in mind of a more suitable school after a rather unhappy half year at Westmount High School.

We stayed there [at Foster's] about a month while an apartment was sought and rented on Dorchester St.; we were on the third floor of an apartment house about half

---

54 R.G.E. Murray, letter to Donald Avery, 4 December 2007.

way between Atwater and Green Avenues in Westmount. ... Most important, half a block away on our side street there was a playground with an open air ice rink and a warm up shack with a Quebec heater looked after by a kind French Montrealer – one of the many jobless being given some chance by the city. We got some ancient skates to fit and I quickly learned skating among a lot of boys and girls of all ages. A few came from below the tracks, which were only a few houses away. That rink was an important learning place.

Altogether too soon there was school and I suppose some contact told my father that I should go to one of the Provincial schools – how the grade level was decided, goodness knows, but I ended up in Westmount High, about 6 blocks away, in Grade 9. I was way ahead in math, Latin and English and I was certainly out of my element for that 6 months. The boys were all older than I was and so there was a fair amount of friction, but a few were helpful and protective. One was a fellow rink rat and was from a poor but nice family across the tracks and the other lived a block away. They made for company at a difficult time and there were many kindnesses. Fairly soon my father was taken to a hockey game in the Forum and he was enchanted by the fast and clever play. Soon he got tickets for the rest of the season and for many years had a pair of seats a few rows back from the visitors bench and we went together.

Some time close to when that first school year ended my father and I went out to Lower Canada College in Notre-Dame-de-Grâce for an interview with C.S. Fosbery, the Headmaster. He was encouraging and gave me his own sort of simple tests and I was accepted for the Fall Term. With summer came our introduction to Bark Lake thanks to Hugh Starkey whose father, Professor of Public Health at McGill, was a long-time summer resident with a house on the main bay. It was yet another introduction to Canadian life and just as influential as was the City. There were new friends and experiences. I got to know boys who were not far off, especially those of the Moore family who started me on boats and water fun just as my father started me on fishing for bass and pike as a different sort of water fun.

MURRAY AND MCGILL, 1931-38
Building a Respectable Department
Robert Murray, Recollections[55]

The McGill situation presented some tough challenges. The department was essentially defunct. What existed was a laboratory providing service in clinical bacteriology and serology to The Royal Victoria Hospital with a professional staff of one and a senior technician. There was little else but space available for laboratory teaching

---

[55] Robert Murray, "A Personal Narrative."

and for non-existent teaching faculty. What existed was part-time from Pathology or other hospitals. Everything had to be built up from ground zero. By 1937 he had the staff to teach both medical and science students with, aside from the basic course, an honours full final year science course in bacteriology and related skills. The forces were three teachers with medical degrees and research backgrounds, and two medical clinical bacteriologists providing hospital service with 2-3 technicians, and a Cambridge trained Chief Lab Man to see that everything worked. A year and a half later there was an added general bacteriologist as a teacher and two research associates. All this accomplished during the hard part of the Great Depression. He was trying to go further and after the war accomplished the addition of a teacher/researcher concentrating on dental bacteriology and another on mycology.

J.B. Collip Recalls: Professor of Endocrinology, McGill University (1926-1948)[56]
In 1930 Dr. Murray came to Canada to join the Faculty of Medicine of McGill University and was appointed Professor and Head of the Department of Bacteriology and Immunology, where he remained for 25 years. ... One can perhaps appreciate some of his feelings when upon his arrival at McGill, he discovered that he was the sole member of his Department and that he was to have the assistance of one trained technician. He surveyed his new inheritance – a large bare laboratory with chemical benches and a half-dozen rooms with little or no equipment. The situation would have been most discouraging to any but a born pioneer. Before winter set in Professor Murray had made available adequate facilities for the teaching of medical students. Before winter yielded to spring, the 'Chief' and his new, full, bristling, and unusually well-trimmed beard, was marked as a man to be reckoned with in the years ahead at McGill.

Professor C.F. Martin[57]
Dean of the Faculty of Medicine
McGill University
Montreal, Que.

Feb. 23, 1931

Dear Mr. Dean:
... I have had charge of the Department only three months and during that time I have had to assimilate a considerable amount of information derived from a variety

---

56 "Professor E.G.D. Murray: An Appreciation," *Canadian Medical Association Journal* 92 (January 8, 1965): 95-97. James Bertram Collip, (1892-1965) a pioneer in endocrine research, taught at McGill until 1947 when he accepted an invitation to become dean of medicine at the University of Western Ontario, a position he held until 1961. See Introduction, note 317.
57 McGill Archives, B & I.

of sources. Information of a general nature, relating to procedures and inter-relationships within the University, and information of a particular nature, relating to the structure and activities of my own Department. As I am still in a phase of adjustment to new conditions and the matter on which I am reporting has many aspects, I cannot pretend to be in a position to view them all in their true perspective. Therefore, I crave some indulgences.

I have no doubt you will discover, here and there, reasons to differ from my opinions, and that other interpretations than those I have made will occur to you. With your greater experience, and your knowledge of the circumstances, the aims and ideals of the University, you have a vantage point from which to see better ways of attaining the end in view than the ways I have suggested, and you may well see a loftier and more desirable objective than I have realized possible. Recognizing this, I would ask you to allow me to have the benefit of your criticism and guidance.

I started to prepare this report because I felt the present position of the Department should be reviewed, its immediate needs formulated and its possibilities indicated. Since then I received your letters of December 15th, 1930, and February 6th, 1931 and I hope I have embodied the answers to your questions in my report.

    I have the honour to be, Sir,
        Yours obediently,
          E.G.D. Murray

                              Faculty of Medicine
                              Office of the Dean
                              May 14, 1931

Sir Arthur Currie[58]
Principal-McGill University
Montreal

Dear Sir Arthur,

Prior to any discussion on the budget for Bacteriology, I thought you would like to glance over the report of Professor Murray with respect to the Department of Bacteriology, past, present and future. It is very well put together and the list of the contents will enable you to select such things as are of more special interest. His tentative budget will be in our hands this week.

    Faithfully yours,
        C.F. Martin
           Dean

58 Ibid.

Report by C.W.S. (Carleton W. Stanley)[59]
Bacteriology-Professor Murray's Report to the Dean of the Faculty

May 1931

This is now the third report I had read from Dr. Murray of the Department of Bacteriology.

The first one gave the impression that Dr. Murray found absolutely nothing in his new surroundings which justified his new milieu being called a "Department of Bacteriology" at all. Here he gives chapter and verse for his complaint.

Allowing for the fact that a newcomer is pulled up sharply by contrasts and can see the defects more quickly than the qualities, this is still a pretty serious criticism. It indicates a general disorganization in both the R.V. Hospital and the Medical Faculty, so far as Bacteriology is concerned. It is fairly notorious, in certain circles, that the R.V. Hospital needs overhauling and modernization. Personally, I have had only one small brush with the place, but I saw clearly that the place is badly organized, and that in ward the conditions were *unsanitary* to the highest degree. Pasteur and Lister might as well never have been born. As for the hospital side, therefore, I am prepared to believe fully in Dr. Murray's criticism.

We are concerned, for the moment, with the University side.

Now one must not be too deeply shocked or disappointed. Call a thing a 'department' in any institution, and you will soon find how blind, conservative and mentally lazy human beings are. (e.g., after protesting for five years in my own department against the loose use of the word Honours, in no way advanced or special ...)

I do not think, therefore, that one should look for a target to blame in the case of Bacteriology. But at least it might be worth while to ask *ourselves* how such a state of affairs has come about, and whether there are more dank corners of the kind.

I recommend:-

1. That we satisfy ourselves, first of all, concerning Dr. Murray's good judgment, and his disinterestedness.
2. That we then concede to him as much as possible his requests so as to allow him to institute reforms.
3. That Dr. Murray's cooperation be secured in refraining from a hue and cry.
4. That notwithstanding (3), this case may be made an object lesson to the Faculty and other Departments.

---

59 Ibid. Carleton Stanley was associate professor of Greek. It is not clear why he would prepare this report, since he does not appear in any of the McGill committees for this period. *McGill University Calendar for the Session 1928-1929* (Montreal, 1928).

Report to the Dean of the Faculty of Medicine of the Present State of the Department of Bacteriology with Tentative Recommendations for the Development of the Activities of the Department in Teaching, Research and Applied Bacteriology in Relation to Medicine[60]

By

Professor E.G.D. Murray

CONTENTS

PART I: REVIEW OF THE PRESENT POSITION OF THE DEPARTMENT
1. The Personnel of the Department
2. Research Workers
3. The Housing of the Department and the Distribution of Workers
4. The Equipment of the Department
5. Finances of the Department
6. The Undergraduate Teaching
7. Graduate Studies
8. Clinical Bacteriology
9. The Relationship of the Department of Bacteriology to other Bacteriological Laboratories in the University

PART II: RECOMMENDATIONS FOR THE RE-ORGANIZATION OF THE DEPARTMENT
1. The Personnel of the Department
   a. The Minimum Personnel
   b. The Distribution of Duties
   c. The Qualifications Desirable, the Method of Selecting and the Promotion of the Staff of the Department
2. Undergraduate Teaching
3. Examinations & Examiners
4. Graduate Studies & research for higher Degrees
5. Provision for Persons outside the Department to Work in the Laboratory
6. Accommodation
7. Equipment of the Department
8. Finances of the Department
   a. The Comparison between the Contributions of the University and the Hospitals to the Department
   b. The Budget for the Reconstituted Department
   c. The Establishment of a Research Fund ...

---

60 Ibid.

PART III: THE PROSPECTS & PURPOSE OF THE DEPARTMENT
There are two ends which the Department may justifiably look to achieve:
1. To become of consequence within the University and the Hospital.
2. To become of consequences in the work of Bacteriology at large ...

These considerations immediately raise the question of the demand for bacteriologists. ... I gather that the General Hospital has no whole time bacteriologists of standing on its staff ... that other Hospitals in Montreal content themselves with students or Interns ... [and] that these conditions prevail elsewhere in Canada. It seems possible these deficiencies will be realized in the course of a few years. ...

I am not aware that McGill has trained bacteriologists who on emerging from its laboratories would be fitted to take up these appointments ... So we need to consider how the men should be trained who would put McGill in the strong position ...: a combination of Medicine and the B.Sc. Honours course in Bacteriology and Immunology and a Research Studentship at McGill, followed by two or three years in England or elsewhere. ...

It is improbable that consistently progressive development of the Department will be achieved without the stimulus of an ideal ... [which] does not depend entirely on the zeal and capability of its leader, it requires the qualities of loyalty, efficiency, initiative and spirit de corps in every member ... [which] they have given me during the three months I have had charge of the Department.

 I respectfully submit this report for consideration.
  E.G.D. Murray

Responsible Research

Dean G.S.H. Barton[61]
Macdonald College
P. Que.

September 22, 1932

Dear Dean Barton:
This morning I was present at the oral examination of Mr. R.R. Thompson on the very interesting questions of *Brucella Abortus* and undulant fever. After the examination there was some discussion by the examiners and we felt that it was desirable to express thanks to you for the generous support which Macdonald College had given to this experimental work.

It was further agreed that it was desirable to call your attention to the importance of further assistance for the investigation of a subject of great importance to

---

61 McGill Archives, B & I.

the health of the community, as well as those who are engaged in the production of milk, cream, butter, cheese and ice cream. It is desirable, if possible, to obtain further financial aid in such investigations. In this case it is possible to make a triple claim – high scientific interest, public health, and industrial development.

I am sending a copy of this letter to Sir Arthur Currie.

    Yours very truly,
        Dr. A.S. Eve[62]
        Dean, Graduate Faculty

                                              Macdonald College
                                                McGill University
                                                        1 Nov. 1932

Professor E.G.D. Murray[63]
Bacteriology Department
McGill University
Montreal, Que.
Dear Murray:

I enclose a brief note about [R.R.] Thompson's work on *Br. Abortus* brucellosis/undulant fever), which is drawn up for the Gazette (Montreal) to serve as a bait for solid encouragement to continuation of the work. I know that Dean Barton will do nothing until something is in print, but once it is in he will direct some funds towards us ... The work cost the Department more than it could afford last year. ... You will note that there is no reference to department or individuals, though the obvious inference will be that it is your department. Well, I am just as good at kicking out reporters as you are, if they come here. ...

Let me know as soon as you can, as the Dean is leaving shortly, next week I think.

    Yours sincerely,
        P.H.H. Gray
        Assistant Professor and Head of the Department

---

62 Ibid. A.S. Eve (1862-1948) was head of the Department of Physics before becoming dean of graduate studies at McGill. His long association with Nobel Prize winner Ernest Rutherford was reflected in his extensive biography of this outstanding nuclear physicist.

63 Ibid.

Department of Bacteriology
Prof. E.G.D. Murray
Nov. 3, 1932

Professor P.H.H. Gray[64]
Department of Bacteriology
MacDonald College
St. Anne de Bellevue, Que.

Dear Gray:

It is no affair of mine whether or not you publish the article in the "Gazette", but you cannot expect me to agree to it and I wish to make it perfectly clear that this Department is not concerned with the article in any way. Any article contributed should have the stamp of the Department responsible for it and if you are not prepared to acknowledge it you had better not publish it at all. I object very strongly to there being even a remote possibility of this Department being involved and in view of your statement "though the obvious inference will be that it is your Department", I must insist that you make it perfectly clear that I have nothing to do with it.

The article itself contains statements that I cannot subscribe to and, on reading it over, I cannot see that it is going to help your Department or obtain for you any support. ... With regard to the errors, in the first place Robert Bruce was King of Scotland and the work you referred to was done by David Bruce; secondly, Brucella infections in man hardly resemble tuberculosis; thirdly, Brucella infections in man hardly resemble typhoid fever; fourthly, Bacillus typhosus is not commonly excreted in the milk ... of cattle, as might be implied in this article. There are other minor misstatements and unfortunate constructions. On the whole, I think newspaper publication of this article would do more harm than good.

    Yours sincerely,
    E.G.D. Murray

Nov. 3, 1932

Sir Arthur W. Currie, G.G.M.G., K.C.B.[65]
The Principal and Vice-Chancellor,
McGill University,
Montreal, Que.

64 Ibid.
65 Ibid.

Dear Sir Arthur,

In view of the conversation we had a little while ago, I enclose a copy of the letter I have received from Professor Gray, of MacDonald College, and a copy of my reply. I hope very sincerely that no newspaper article will be published which might be interpreted wrongly and which might be attributed to my Department.

I am more than ever convinced that a procedure such as this indicates is fundamentally unsound and will do more harm than good.

Yours sincerely,
E.G.D. Murray

Nov.4, 1932

Dear Sir Arthur:

I certainly think it would be highly undesirable to publish the memorandum Professor Gray submitted to me for my opinion, but I also think that unrestricted newspaper publicity is in itself undesirable.

I do not attach very much importance to the statement about Dean Barton. ... I can say, however without any fear of contradiction, that Professor Gray's Department is very seriously handicapped at MacDonald College, as I explained to you in my letter of July 27th, 1931, and I must admit that were I in his shoes I would feel very strongly about the situation.

Yours sincerely,
E.G.D. Murray

Search for Excellence

Copy for Sir Arthur Currie[66]

Professor E.G.D. Murray
Department of Bacteriology
McGill University

January 18, 1933

Dear Prof. Murray:

We are having a meeting of the Science Division Committee ... and the question of the future status of Bacteriology must be considered with a view to presentation to the Faculty of February 1st. ...

Will you please let me once again beg you to do your utmost to receive graduate students in Bacteriology in your Department in the Pathological Building, in which you are placed. I should regard it as a profound misfortune if we were to lose this

---

[66] Ibid.

subject from our Graduate Faculty, and I will do all in my power to assist you to retain it.

>Yours very truly,
>A.S. Eve
>Dean, Graduate Studies

Copy for Sir Arthur Currie[67]

January 21, 1933

Dear Professor Eve:

I fully appreciate your attitude and am very grateful for the interest you take. ...

I still maintain that it is not possible at the present time for me to take Graduate Students in Bacteriology, since the conditions essential to such a step have not been met by the University. The position outlined in my letters of December 16th, 1931, January 30th, March 12th and October 19th, 1932 remains unaltered and I have nothing to add to what I have said in those letters. ...

I am sorry to have to take this stand but there is quite definitely no accommodation in this Department for Graduate students nor will my present appropriations meet the expense they would involve.

>Yours sincerely,
>E.G.D. Murray

Copy for Sir Arthur Currie[68]

Dean A.S. Eve
Faculty of Graduate Studies
McGill University

January 24, 1933

Dear Dean Eve:

I have just had a prolonged interview with Professor Murray apropos of the question of graduate students and thought it well to place before you the facts as I see them to-day.

Let me say, in the first place, that Professor Murray is only too eager to take on graduate students, not only for the purpose of carrying out research, but equally with a view to training them to take positions here or elsewhere in Bacteriology. He

---

67 Ibid.
68 Ibid.

feels – and I think with justice – that there would be no object in taking on graduate students to do research unless he could be assured of results that are well worthwhile, and which would do credit to the Department and the University. Rather than embark on this enterprise under unsatisfactory conditions, he believes that it is far better to wait until adequate facilities are available.

As you know, he and his department not only carry out on the course in Bacteriology (and, I may say, to the great satisfaction of the students), but they have a great deal of routine hospital work to do, which occupies fully all Professor Murray's time and that of his staff. This routine involves not only bacteriological examinations, but consultations on the wards, and also, with at least a few doctors, every day.

In addition to this, each member of his staff is engaged in special research, which adds still further to their burden. Accommodation in the bacteriological laboratory is, unfortunately, so restricted that there really is not adequate room for those that are there now.

Secondly, the budget for the Department is so limited that only with the greatest thrift is he able to make ends meet. I may add that the income from private patients, which ordinarily goes to the head of the department as private perquisites, is pooled for the benefit of the department in various ways. The expense entailed in graduate student work involves the use of animals (a considerable expense), as well as materials and supplies of all kinds.

Thirdly, the presence of graduate students in this department would, of necessity, if the work is to be worth while, occupy a considerable portion of the time of the existing staff, and, as already stated, this is now more than filled with the routine of the laboratory.

To summarize, if graduate work in Bacteriology is to be done with credit to the University, it really necessitates more space, more budget and an increase in the staff. Personally, I feel, after talking with Professor Murray, that his contention with respect to worth while work is justified, and neither he nor the Medical Faculty would be interested in graduate work of mediocre kind.

In conclusion, I merely want to repeat that it was a matter of very sincere regret to Professor Murray to have to turn down the application of more than a dozen graduates seeking the benefits of his training and facilities.

    Sincerely yours,
        C.F. Martin
        Dean, Faculty of Medicine

P.S. Don't forget that Murray is one of the outstanding bacteriologists on this continent. He came here under adverse conditions and has already achieved splendid results in organizing the Department, in arranging an excellent course and satisfying the Hospital staff with his consultative work. Those were the essentials for which he was brought here, and I question the wisdom of hurrying him into an endeavour for which he does not feel he has the inclination under existing conditions.

Celebrity Status

The Gazette, Montreal March 4, 1935
"Montreal Bacteriologist Flies with Serum to Save Child's Life"
Carrying several tiny tubes of a little known serum, Prof. E.G.D. Murray, O.B.E., Chairman of the Department of Bacteriology at McGill University, sped last night to Detroit in a Vultee low-wing monoplane, the fastest transport ship in the United States in answer to a frantic call to save the life of an eight-year old boy who lay near death with a rare blood disorder. He left St. Hubert airport at 6:47 p.m., piloted by Roscoe Kent, and was the only passenger in the giant air-liner.

In Detroit Prof. Murray was met at the city airport by a detail of police and rushed to the hospital in a police car. Plans for the immediate administration of the serum were made at the hospital where eight-year old Ronald Osnos is stricken with a baffling disease.

From McGill sources it was learned last night that Prof. Murray took with him the serum for the treatment of an obscure blood poison. The serum is said to have been discovered in Australia in 1930 and has been used successfully here during the past year. Prof. Murray had to leave so hurriedly that little was known by his laboratory assistants of the case.

The type of blood poison treated by the serum with which Prof. Murray armed himself was regarded as 100 per cent fatal before the serum was discovered, a McGill physician told the Gazette last night. This particular kind of blood infection usually develops in certain individuals after the recurrence of boils or carbuncles. To most persons these minor ailments are not serious but for some unknown reason certain individuals develop blood poisons. If the serum is not used, death usually occurs within three or four days.

The serum is produced in horses by the infection of certain poison. Soon after it was developed in Australia doctors connected with the Connaught Laboratory, Toronto, a research centre associated with the University of Toronto, experimented with the serum and the results proved to be fairly successful. Prof. Murray learned of it through that course and introduced it to Montreal medical circles.

The plane used on the trip was similar to that used by Col. James H. Doolittle when he established his transcontinental record. American and Canadian airways officials co-operated in bringing the serum to Detroit in the fastest time possible. ... Prof. Murray had been notified by wire to prepare for the flight and he was picked up at St. Hubert airport. ... The flight to Detroit was made via Buffalo and was accomplished in three and a half hours.

Detroit Free Press, March 4, 1935
While an eight-year old Detroit boy lay in Harper Hospital seriously ill from a baffling blood disease that threatened his life for the last nine days, one of the world's fastest transport planes made a special flight to Montreal to bring to the boy's aid a doctor who might save his life.

The boy's father, Max Osnos, of 18612 Muirland Ave., making a desperate attempt to save the child, Sunday afternoon chartered an eight-passenger Vultee plane ... The 260 mile-an-hour plane was to carry Dr. E.G.D. Murray, of Montreal, to Detroit with a serum, which, it was hoped, would turn the trend of the disease threatening the life of the boy, Ronald. ...

Montreal Daily Star, March 17, 1935
... It is practically two weeks since Prof. Murray gave the serum treatment by a method that had been worked out in Department of Bacteriology and Immunology at McGill University and the Royal Victoria Hospital, and even as he left Detroit on March 6 the boy was showing signs of holding his own at least, although nothing definite could be stated as to his ultimate recovery.

But the days have come by and steadily young Ronald has been regaining his grip on life. Many enquiries have been made of Professor Murray as to the progress in the case, and last evening, in order to satisfy the public interest, Dr. Murray made enquiries at Detroit and issued a statement which read, "The conditions of Ronald Osnos in the Harper Hospital, Detroit, continues to be favourable. Although the boy is not yet out of danger the doctors in the case now feel justified that he is likely to recover."

McGill Consultant

Whooping cough vaccine (1937)

Ayerst, McKenna & Harrison Limited
Biological and Pharmaceutical Chemists
781 William Street, Montreal, Canada
February 16th, 1937

Professor E.G.D. Murray,[69]
Department of Bacteriology,
McGill University, Montreal, Quebec.

---

69 LAC, E.G.D. Murray Papers, vol. 4, file Ayerst, McKenna and Harrison. This Montreal-based pharmaceutical company was closely involved with a number of McGill faculty members, notably endocrinologist J.B. Collip and members of the biochemistry department, in the production of Emmenin (estrogen), the first female sex hormone manufactured in Canada. In 1943 the company merged with Wyeth, based in Madison, New Jersey, which was one of the largest pharmaceutical companies in the world.

Dear Dr. Murray:[70]

I thought you might be interested in the enclosed booklet, published by the Glaxo Laboratories, inasmuch as you will note on page 84 some remarks on their whooping cough vaccine, also a note on page 86 regarding Royal Air Force T.A.B. vaccine trials, and on page 92 some material under the heading of "The Intradermal Use of Vaccines in Gonorrhea."

This publication is issued by the Glaxo Company to physicians on their mailing list, and we thought it might be well for you to have the opportunity of seeing how they forward their claims with the members of the medical profession.

    Yours very truly,
    W.J. McKenna
    Ayerst, McKenna & Harrison Limited

Memo re Pneumococcus Serum (1937)

                                                              Prof. E.G.D. Murray
                                                                  April 2, 1937

To: Messrs Ayerst, McKenna & Harrison
Memorandum on Pneumococcus Serum[71] *by Professor E.G.D. Murray*
The recent work by Goodner and Horsfall, in Dr. Avery's laboratory at the Rockefeller Hospital, New York, shows a very remarkable advantage of Rabbit serum over Horse serum in the treatment of pneumonia. The evidence is given in their papers and they discuss its significance. The chief reason for it is considered by them to be due to the smaller size of the formed antibody and its consequent more rapid and more abundant penetration.

At the recent meeting of the American Association of Immunologists, held at Chicago, Dr. Horsfall gave a detailed account of the treatment of a number of cases of pneumonia with Rabbit anti-pneumococcus serum, with very striking results and decided advantages over Horse serum. I believe this work is going to create a demand for Rabbit serum of this type; partly because of the results reported and importantly because of the great authority of Dr. Avery.

I am told that Mulfords, in the U.S.A., are preparing to make Rabbit anti-pneumococcus serum on a large scale. I know the Rockefeller Hospital is making it on

---

70 Ibid.
71 Ibid. Oswald Avery's research team at the Rockefeller Institute Hospital spent many years analyzing the bacterium pneumococcus and was successful in developing a serum for Type I of the pathogen. Olga Amsterdamska, "From Pneumonia to DNA: The Research Career of Oswald T. Avery," *Historical Studies in the Physical and Biological Sciences* 24, part 1 (1993): 1-40. Everitt was very interested in this project.

a sufficient scale for extensive therapeutic trial in New York and their results will determine the situation.

It seems to me that you might with advantage work to be in a position to meet this situation in Canada as soon as it arises.

    Prof. E.G.D. Murray

Dr. L.P. Strean[72]
Ayerst, McKenna & Harrison Limited,
645 Wellington St.,
Montreal, Que.

February 4th, 1939

Dear Dr. Strean,

Thank you for your progress report of February 3rd, 1938.

I am quite satisfied with the changes introduced relating to the Pneumococcus broth, the centrifuging of the Staphylococcus toxin, the filtration of the Anti-Pneumococcus serum, and the changes in the cages, diet etcetera for the rabbits.

I would like to know more about the rate of de-toxification of the Staphylococcus toxin by the reduced amount of formalin, taking into account haemolysis, necrosis and lethal dose. Up to now the Staphylococcus toxoid has given satisfactory clinical results and good antitoxin titrations in patients and I wish to be sure that the longer time factor with reduced formalin is not deleterious.

The titres reported (supplemented by telephone) of the Anti-Pneumococcus serum Types VIII, VII and V are very satisfactory and it is important to secure the necessary clinical trials.

I look forward to studying your observations on the peculiarities of various strains of pneumococcus. I feel sure it is important to take such things into account

I wish to congratulate you on this excellent work you have done.

    Yours sincerely,
    EGDM

---

72 Ibid.

RESEARCH ACTIVITIES, 1935-39
Correspondence on Listeria Research

November 21, 1935

Doctor Caspar G. Burn,[73]
Department of Pathology,
Yale University School of Medicine,
New Haven, Conn.

Dear Doctor Burn:

Could you let me have sub-cultures of the organism you describe in your paper "Unidentified Gram Positive Bacillus Associated with Meningo-Encephalitis", published in the Proceedings of the Society for Experimental Biology and Medicine, 1934, 31, 1095? We are on the lookout for cases of Listerella infection here and I would like to compare any organisms we isolate with those that you have isolated.

Yours sincerely,
E.G.D. Murray

December 19, 1935

Dear Dr. Murray:[74]

I am exceedingly sorry that the cultures arrived so badly damaged. I will immediately prepare fresh transfers of the other two strains and send them to you.

However, if you are going to be in New York for the American Bacteriological Society meetings, I would like to have the pleasure of meeting you and I could give you the two subcultures at that time. ...

My serological studies ... are serologically different from *L. monocytogenes*, but I believe the animal and human strains are of the same genus and possibly one or two other strains now classified in the *Corynebacterium* group should be placed in this group. I am at present completing this type of comparative study. Therefore, I certainly would appreciate receiving subcultures from the original strains of *Listerella monocytogenes* and also of Pirie's strain isolated from a gerbil.

I have tested for agglutinins in the blood sera of the different members of the family in which the infection was found, but could not demonstrate any antibodies. Agglutinins are readily demonstrable in the sera of rabbits, guinea pigs and monkeys after recovering from a previous infection.

Apparently, this infection is either being more readily recognized or it is actually

---

73 LAC, E.G.D. Murray Papers, vol. 10, file Listeria Cultures.
74 Ibid.

increasing, as I have been told about similar infections occurring in Boston and some in Kentucky. One of the recent members of the *Journal of Pathology and Bacteriology* describes a case of diptheroid meningitis occurring in England which I believe is this organism.

    Hoping that I may have the pleasure of meeting you next week, I am
        Yours sincerely,
           Caspar G. Burn, M.D.

                                      Jan. 29, 1936

Dear Doctor Burn:[75]

    I am sending you the following cultures of *Listerella monocytogenes:*

No. 53 XXIII

This organism was isolated on August 11th, 1924 by blood culture from a pregnant rabbit which had been sick since August 6th, 1924. The blood culture yielded a pure growth of this strain and it is the only occasion on which I got a positive blood culture from the natural disease. The morbid anatomy and histology was typical of the disease as described in the original paper (natural disease).

No. 58 XXIII

This organism was isolated on August 13th, 1924 from the mesenteric lymphatic gland of a female guinea-pig in pure culture. The morbid anatomy and histology of the disease did not differ from that seen in rabbits. The experimental infection of rabbis with this strain gave rise to the typical disease as described (natural disease).

Pirie's Tiger River Bacillus

This a culture of the organism isolated by J.H.H. Pirie and described by him in the Publications of the South African Institute for Medical Research, No. 20, 1927, page 163. I received this culture early in 1926 and as far as I can tell it is identical with the organisms I isolated from rabbits (natural disease).

    Immunologically I have not been able to distinguish between these various strains except in a slight variation in agglutinability, which I do not believe is significant as it corresponds with the agglutinogenesis exhibited by the different strains and as "absorption" of specific agglutinin is identical in all cases.

    Thank you again for your courtesy in sending me your cultures and I hope our meeting may not be long deferred.
        Yours sincerely,
           Professor E.G.D. Murray

---

75 Ibid.

Prof. E.G.D. Murray
March 10th, 1939

Dr. Mary Barber,[76]
London County Council,
Archway Hospital,
Archway Road, N. 19,
London, England.

Dear Dr. Barber,

Thank you very much for your strains of *Erysipelothrix* and the reprint of your paper.

I do not think you are justified in suggesting that *Listerella* and *Erysipelothrix* might be placed in the same *genus*, but, of course, it is a question of the value to be given to various criteria. It seems more suitable to me to consider the possibility of their belonging to the same *family*. ...

The fifth edition of *Bergey's Manual*[77] will be out soon with many improvements on the old editions and still retaining many faults for lack of information. Work such as you have published in your paper is very important. But taxonomic suggestions have to be considered very carefully before making them. The Third Congress of the International Association of Microbiologists (New York September 1939) proposes to consider the rules of bacteriological nomenclature very thoroughly. Buchanan's[78] draft for the committee to consider is a fine effort and if they can be adopted they will help taxonomic work enormously.

I hope you will continue your work and widen its scope.

    Yours sincerely,
        Professor E.G.D. Murray

Arctic Bacteriology

Institute of Parasitology[79]
MacDonald College, McGill University
Eleventh January 1938

Prof. E.G.D. Murray,
Department of Bacteriology,
McGill University,
Montreal

76 Ibid.
77 Ibid.
78 See references to *Bergey's Manual* in the Introduction.
79 LAC, E.G.D. Murray Papers, vol. 9, file Arctic Bacteriology (1938-50).

Dear Professor Murray:

The Hudson's Bay people have been interested in a sickness which has been killing Eskimos in the North West Territories. We have been collecting parasites from that area and they have communicated with us on the matter. I enclose a copy of their communication. We received along with it, a goodly supply of whale meat, preserved in formalin. This is free from any evidence of parasites; sections of the muscle however, show a heavy deposit of brown globules, possibly of a fatty nature.

The report is very suggestive to me of bacterial poisoning and I would be very grateful to have your own ideas on this subject. In case it should be of any value, we have kept the meat here.

    Yours sincerely,
        Thomas W.M. Cameron

                              Prof. E.G.D. Murray
                              January 18, 1938

Professor T.W.M. Cameron,
Institute of Parasitology,
MacDonald College,
St. Anne de Bellevue,

Dear Cameron,

I have read the remarkable account of the sickness at Crook's Inlet by D.A. Wilderspin. It is an arresting document.

In my opinion it can easily be interpreted as Botulism. It would be interesting to know whether the cases exhibited any of the following symptoms, besides those given: drooping eyelids, squints, double-vision, dizziness, blurring of speech, dilation of pupils, staggering gait, discomfort and distress, direction of spread and muscles involved in weakness and paralysis, reflexes, air-hunger, terminal increase in pulse rate.

It would be interesting if suitable material (*without disinfectant*) could be sent to me to see whether we could isolate the clostridium or any other organism.

    Yours sincerely,
        E.G.D. Murray

PROFESSIONAL ORGANIZATIONS
American Association for the Advancement of Science

Office of the Permanent Secretary
American Association for the Advancement of Science
Smithsonian Institution Building, Washington D.C.
October 5, 1933

Prof. E.G.D. Murray,
McGill University,
Montreal, Que., Canada.

Dear Professor Murray:

On behalf of the American Association for the Advancement of Science we, the undersigned, request permission to nominate you for fellowship in the Association. Hitherto scientific workers have been invited to become members, but have been elected fellows, if at all, at a later period. The council of the Association, at the Atlantic City meeting, decided that it would be desirable to elect as fellows those who are professionally engaged in scientific work and have advanced science by research. The entrance fee will be remitted and the annual dues are only $5. These include receipt of *Science* (or *The Scientific Monthly*), as well as all the privileges of the meetings and of membership.

There are now about 18,000 members and fellows of the Association. It is only by the cooperation of the larger part of American scientific men that it is possible to maintain our low membership fees and at the same time to do so much in the way of advancing science and the interests of those engaged in scientific work. All profit by the activities of the Association, and we hope that you will wish to share the privileges and responsibilities by accepting this invitation.

Faithfully yours,
(Members of the Executive Committee)

Canadian Medical Association

March 4, 1931

Dr. A.G. Nicholls,[80]
3640 University Street,
Montreal, Que.

Dear Dr. Nicholls:

---

80 LAC, E.G.D. Murray Papers, vol. 4, file Canadian Medical Association.

I shall do my best to meet your request. I suppose you wish to have a short article to balance D'Herelle's great enthusiasm for Bacteriolphage as a therapeutic agent.[81]

I think D'Herelle rather overwhelms people with a mass of detail and added to this was the difficulty of hearing what he said. Had I not known a certain amount about Bacteriophage I would have had considerable difficulty in following his argument. I dare say the majority of people will get more from reading the paper than they got from hearing it. There was not much discussion; I made a few remarks with the hope of checking a too great enthusiasm and to urge people to be a little critical. D'Herelle hardly replied to my criticism of his views and so the evening ended.

I am looking forward to meeting you.

    Yours sincerely,
        EGDM

                                    Prof. E.G.D. Murray
                                         April 29, 1935

Dear Doctor Nicholls:[82]

The paper by Doctor G.F. Laughlen on "A Rapid Test for Syphilis", which you submitted to me for criticism, in my opinion might well be amplified in certain parts and revised in others before you accept it for publication. …

There is no indication of resemblance or difference between Dr. Laughlen's Test and the standard Wassermann and Kahn tests. There is no discussion of the principles upon which he bases his test nor of the nature of the reaction. One might guess that it is a micro-method of performing a modified Kahn test. It seems to neglect the importance of zoning and relies upon rate of reaction for evaluation of differences in potency of different serums, which is a matter requiring some critical discussion. …

The interest of Doctor Laughlen's work lies in that it may provide a rapid presumptive test for Syphilis which might be useful when it is urgently necessary to obtain a result very quickly. As presented the test does not recommend itself to exclude and replace the Wassermann or Kahn tests.

    Yours sincerely,
        EGDM

---

81 Ibid. Born in Montreal, Félix d'Herelle gained international aclaim for his discovery of bacteriophages. See Introduction, note 19.
82 Ibid.

Canadian Medical Association
3640 University Street,
Montreal, April 30, 1935

Dear Professor Murray,

I have just received your comments on Dr. Laughlen's paper on "A Rapid Test for Syphilis". I wish to thank you most heartily for your kindness in going into the matter so fully. It is exactly the kind of information that I require. No one on the Editorial Board felt competent to form an opinion on such a technical subject. I am returning the paper to Dr. Laughlen with your comments. At the moment my idea is that he should amplify his paper and make it a thoroughly complete and scientific communication. Of course I shall keep your name out of the matter in my correspondence with him.

 Again with my best thanks, I am,
  Yours very sincerely,
   A.G. Nicholls
   Editor

Canadian Medical Association
184 College Street, Toronto 2,
Dec. 9, 1935

Dear Doctor, [no name – circular to EGDM][83]

May we respectfully direct your attention to some of our activities of the current year:-

The Journal continues to enjoy much favour with its readers, not only in Canada but in many other lands.

Our joint meeting with the American Medical Association in Atlantic City was a great success. One hundred and sixty Canadians took part in the programme.

Reports continue to reach us paying tribute to the excellent service being rendered by our Hospital Service Department.

Under the auspices of the Public Health Committee, popular health articles appearing in 371 newspapers have a weekly circulation of 2,074,000. Judging by the many hundreds of letters received, this plan of education is growing in interest and usefulness.

---

83 Ibid.

Room 206
Medical Arts Building,
Montreal, August 27th, 1937.

Dr. E.G.D. Murray,[84]
McGill University,
Montreal.

Dear Doctor,

The Province of Quebec Medical Association has ceased to function for at least two years, and particularly as an affiliated Branch of the Canadian Medical Association.

As a result, the members of the Canadian Medical Association resident in the Province of Quebec no longer secure proper representation on the Council or Executive Committee of the Canadian Medical Association. ...

The undersigned organization committee suggests that steps be taken to form an organization in this Province which shall become the Canadian Medical Association, Quebec Division. ...

There are about 30 members of the Canadian Medical Association resident in the Province of Quebec. If all of these will agree to the proposed arrangement, the suggested close relationship with the Canadian Medical Association may be maintained. ...

By this means the influence of the Canadian Medical Association can be extended in this Province and, what is perhaps more important, the medical profession of this Province will be enabled to take its proper place in the deliberations, decisions and actions of the National Body. ...

 Yours truly,
  Organization Committee,
  Canadian Medical Association
  Quebec Division

Canadian Public Health Association

The Canadian Public Health Association
Laboratory Section, April 16, 1934.

Professor E.G.D. Murray,
Professor of Bacteriology,
McGill University,
Montreal, Que.

---

84 LAC, E.G.D. Murray Papers, vol. 4, file Canadian Medical Association (Quebec).

Dear Doctor Murray:

I am writing you regarding the forthcoming meeting of the Canadian Public Health Association which is being held in Montreal, June 11th-13th.

The executive of the Laboratory Section are especially desirous of obtaining your co-operation in arranging the programme. We would appreciate receiving the titles of the papers which you or your group are prepared to present before this Section.

The programme is to appear in the May issue of the Journal and we would ask that you submit the titles before the first of the month.

    Sincerely yours,
        M.H. Brown, M.D.,[85]
        Chairman.

                                            Prof. E.G.D. Murray
                                                    May 1, 1934

Doctor M.H. Brown,[86]
School of Hygiene,
University of Toronto,
Toronto 5, Ont.

Dear Doctor Brown:

With regard to the forthcoming meeting of the Canadian Public Health Association in Montreal on June 11th, three members of my Department are willing to make communications to the Society, although they do not pretend that these are of great importance. The papers suggested are the following:

    Doctor F. Smith[87] – The Value of Kocto Vaccines.

    Doctor D.H. Starkey[88] – Streptococci found in the routine examinations of stools.

---

85 LAC, E.G.D. Murray Papers, vol. 4, file Canadian Public Health Association (1934-53). Brown was a member of the Toronto-based Connaught Institute, which had been established in 1914 in the Department of Hygiene of the University of Toronto. The laboratories operated as a financially self-supporting organization, linked with the University of Toronto until 1972.

86 Ibid.

87 Frederick E. Smith had studied with Joburg at Christ's College, Cambridge, during the mid-1920s. After a short stint at the Rockefeller Institute in New York, he agreed in 1931 to join the McGill Department of Bacteriology, where he soon became Joburg's right-hand man, supervising the work of the university laboratory. During the war years he served as departmental chairman in Joburg's absence, and in 1947 was appointed dean of medicine. He died suddenly in 1949. Lewis, *Royal Victoria Hospital*, 220-24. According to Bob Murray, "He and his wife Pat were kind to me as a young teenager and were a source of stimulating scientific conversation when I was at McGill." Robert Murray, "A Personal Narrrative."

88 Ibid. Hugh Starkey was appointed assistant bacteriologist at McGill in 1933. See John H. Glynn

Doctor Theodore E. Roy[89] – Organisms of the Bacteriodes Group found in Lochia.

If there is anything else I can do towards making the meeting a success please let me know.

  Yours sincerely,
  EGDM

           The Canadian Public Health Association Laboratory Section
                       October, 1934.

To members of the Section and Others Interested:[90]

… Because of the number of papers offered last year, and the increasing interest in the field of general bacteriology as related to milk, water and food, your opinion as to the wisdom of extending the meeting over two days will be appreciated.

The endeavor of the Section to serve as the Canadian society of those working in these fields of medical science and to have a largely attended midwinter meeting for the presentation of new work is being achieved in a most encouraging manner. The attendance last year exceeded seventy-five, and with the special programme that is being planned for this meeting it is expected that an even larger number of members from distant points will be present. The sessions will be of interest not only to laboratory workers, but also to epidemiologists and medical officers of health.

It will assist the committee greatly if you will kindly return the enclosed card promptly.

  Yours faithfully,
    M.H. Brown, M.D.,
    Chairman of Section

---

and Hugh Starkey, "The Cultural and Antigenic Properties of *Shigella sonnei*," *Journal of Bacteriology* 37, no. 3 (March 1939): 315-331; F. Munroe Bourne, D. Hugh Starkey, and L. J. Turner, "Brucellosis in a Veterans' Hospital, 1963," *Canadian Medical Association Journal* 91, no. 22 (28 November 1964): 1139-1145.

89 In the 1930s, T.E. Roy worked on staphylococcal research for several years in E.G.D. Murray's McGill laboratory. He also did bacteriological work at Royal Victoria Hospital under E.G.D. Murray's supervision. He eventually moved on to the Hospital for Sick Children in Toronto where he had "a very distinguished career as a clinical bacteriologist." Ted Roy and his wife (known as "Churchie" by everyone) were very close friends of the Murray family. Robert Murray, interviews by Mark Eaton, 24 July and 15 September 2005; Lewis, *Royal Victoria Hospital*, 221.

90 Ibid.

Prof. E.G.D. Murray
Nov.27, 1936

Dr. G.D.W. Cameron,[91]
Secretary, Laboratory Section,
The Canadian Public Health Association,
Connaught Laboratories,
Toronto, Ont.

Dear Dr. Cameron:

I have long intended to be present at the coming meeting of the Laboratory Section of the Canadian Public Health Association. I thought I might exhibit one or two points of technique which have long been in use in this laboratory with success and which might be useful to other workers. At the same time I shall sound members of my staff who might be able to contribute a paper to the meeting. These details I shall let you know shortly.

Dr. Orr, in his letter of November 25th, asks me to speak at the dinner but he does not suggest that what I say should be an opening for a round-table discussion.

I am not sure that it is desirable to imitate the details of the Society of American Bacteriologists, who have serious papers from early morning to night and then interrupt the dinner with still more serious matter in the form of an address to follow the dinner by round-table discussions lasting almost through the night. I think it is more desirable to encourage recreation and friendship in the evenings of the meetings.

I quite agree to papers at the meeting being limited to fifteen minutes but I think it is unwise to limit the discussion following paper to five minutes. Not uncommonly matters of great interest are raised in the discussion. It is common experience that many papers do not provoke any discussion at all and this usually is sufficient to allow of an adequate discussion of papers that prove interesting to the majority of persons.

These are merely suggestions provoked by past experience.

    Yours sincerely,
        EGDM

---

91 Ibid. George Donald West Cameron (1899-1983) graduated from Queen's University with an MD degree. He later obtained a diploma in public health from the University of Toronto and became associated with the Connaught Laboratories. In 1939 he joined the federal government as chief of the Laboratory of Hygiene of the Department of Pensions and National Health. Between 1946 and 1965 he was Deputy Minister of National Health. He was invested as an Officer of the Order of Canada in 1969.

Prof. E.G.D. Murray
Dec. 12, 1936

Dear Dr. Cameron:

... Since I have not been doing any research work worthwhile I have been at some difficulty to decide upon a topic for discussion. I think that a useful purpose would be served if I could make a contribution under the following title, "Some Aspects of Bacteriological Taxonomy". I hope this will be acceptable to you as it might provoke some discussion.

    Prof. E.G.D. Murray

December 8th, 1938

Doctor Ronald Hare,[92]
Connaught Laboratories,
University of Toronto,
Toronto, Ont.

Dear Hare,

I have just received a copy of the programme for the Laboratory Section of the Canadian Public Health Association for December 19th and 20th. It seems a pity that these were not sent out with the notice of the meeting and I had to write for a copy.

I observe that you are taking part in the Streptococcus Symposium and dealing with classification. As it may interest you I am sending you the galley proof of the new edition (Fifth) of Bergey's Manual dealing with the streps. It may give you material to criticize. By the way we have just discovered that Orla-Jensen named a family *Lactobacteriaceae* and we shall have to substitute it for the name Streptobacteriaceae we invented.

I am glad to see you have not put me down for anything special.

I wish to have a meeting of the Provisional National Committee for Canada of the International Association of Microbiologists at 12 p.m. on Monday, December 19th. This consists of twelve members and we would esteem it a favour if you could arrange for a room in which it might meet.

    Yours sincerely,
        EGDM

---

92 Ibid. Ronald Hare joined the Connaught Laboratories prior to the Second World War, working on the isolation of strains of influenza virus. After 1943 he was part of a Connaught team which developed methods for the large-scale production of the antibiotic penicillin. After the war, he returned to the United Kingdom, having been appointed to the chair of bacteriology at St. Thomas's Hospital Medical School. See Ronald Hare, *The Birth of Penicillin* (London: Allen & Unwin, 1970).

Prof. E.G.D. Murray
December 14th, 1938

Dear Ronnie [Hare],[93]

... I quite understand about the programme and realize the difficulties you must have met in making the arrangements for the meeting. I sent out 550 circulars to Canadian microbiologists and received 50 replies. The inertia is terrible.

The new Bergey (Manual) complete will be out about the 20th of February according to our fondest hopes. It will be a great relief to see it as it has involved a vast amount of work. I am sure it will be recognized that it is vastly improved though no doubt we shall get a fair measure of curses.

Freddie promises that he will bring your thesis to Toronto next Sunday.

Please remember me to Mrs. Hare.

    Yours ever,
        EGDM

International Association of Microbiological Societies
Prof. E.G.D. Murray
December 2, 1935

Doctor J.G. Fitzgerald,[94]
Director, Connaught Laboratories,
University of Toronto,
Toronto 5, Ont.

Dear Fitzgerald:

I think it is very important that the official delegates from Canada to the 1936 meeting of the International Society for Microbiology should be certain of going to

---

93 Ibid.
94 LAC, E.G.D. Murray Papers, vol. 5, file International Association of Microbiological Societies (IAMS), 1930-1936. Born in Ontario, John G. Fitzgerald (1882-1940) graduated from the University of Toronto medical faculty in 1903, followed by research in bacteriology at the Harvard Medical School and the Pasteur Institute. After becoming a member of the Department of Hygiene and Public Health in 1913, he initiated a program for the large-scale preparation of diphtheria antitoxin. The following year, with assistance from the Canadian government and private sources, he founded the Connaught Medical Research Laboratories, based at the University of Toronto and a special laboratory in the northern part of the city. Under Fitzgerald's direction Connaught emerged as a major centre for the production of various antitoxins, vaccines, and antibiotics. Paul Bator and Andrew Rhodes, *Within Reach of Everyone: A History of the University of Toronto School of Hygiene and the Connaught Laboratories*, vol. 1, 1927 to 1955 (Ottawa: Canadian Public Health Association, 1990), 13-26.

the meeting and should be of senior standing. There is no doubt that Frasier would be an excellent nominee.

The question arises whether you or I can nominate the delegates from Canada to an international meeting. I rather feel that the nomination should come from a representative society and the only Canadian society which seems to be so qualified is the Royal Society of Canada. I think the International Society for Microbiology has to be taken very seriously and I do not think the nomination of delegates would be asking too much of the Royal Society.

Thank you very much for Volumes 5 and 6 of the "Studies from the Connaught Laboratories". I am very glad to have them.

    Yours sincerely,
    EGDM

                                School of Hygiene
                                University of Toronto
                                       Toronto 5
                                          CANADA
                                December 4th, 1935

Professor E.G.D. Murray
Department of Bacteriology
McGill University
Montreal, P.Q.

Dear Murray:

I should imagine the Royal Society of Canada would be quite willing to nominate official delegates from Canada to the International Society for Microbiology if it were asked to do so.[95] My suggestion to you was not intended to imply that you and I should constitute ourselves a nominating committee, but was a specific answer to the question asked in your previous letter.

    Sincerely yours,
    J.G. FitzGerald, M.D.

---

95 Ibid. E.G.D. Murray became a member of the Royal Society of Canada in 1938.

International Society for Microbiology[96]
Second International Congress for Microbiology
London, 27 July – 1 August 1936
Section 8

March 25th, 1936

The Lister Institute, Elstree, Herts.[97]
England

Dear Professor Murray,

The sub-committee of the above section of the forthcoming International Congress for Microbiology will be very glad if you will take part in the general discussion at the session on Friday, July 31st.

The session is devoted to "The Relative Value of Antitoxic and Antibacterial Immunity in the Prophylaxis and Treatment of Human and Animal Diseases."

It is hoped that you will speak with special reference to your experience with meningitis/meningococcal. ...

Yours sincerely,
Douglas McClean

April 6, 1936

Dr. Douglas McClean,[98]
International Society for Microbiology,
The Lister Insitute,
Elstree, Herts,
England.

Dear Dr. McClean:

---

96 The International Society for Microbiology was founded in 1927, and was later renamed the International Association of Microbiological Societies (IAMS). In 1967 it became a division of the International Union of Biological Sciences, separating in 1982 to become the independent International Union of Microbiological Societies.

97 Ibid. The Lister Institute (named in honour of Lord Joseph Lister) was founded in 1891 as the British Institute of Preventative Medicine. It soon became an important international research centre in the tradition of the Pasteur Institute in Paris and the Rockefeller Institute in New York. Located in London, it also worked closely with the government's National Institute for Medical Research, Public Health Laboratories, the Royal Army Medical Corps, and the Medical Research Committee (later Council). Harriette Chick et al., *War on Disease: A History of the Lister Institute* (London: A. Deutsch, 1971).

98 Ibid.

I will be very glad to attempt to contribute to the discussion in Section 8.

You ask me specifically to contribute towards the discussion on meningococcal meningitis. Although I have done a certain amount of work on this subject in the past I have not had any opportunity to continue it during the past five years and I am afraid that my contribution would not be worth very much. On the other hand I have given a good deal of attention to the immunological treatment of staphylococcal infection, both by active immunization with antitoxins and passive immunization with concentrated antitoxin. I feel that if your list on that subject is not full I might be able to contribute to it more significantly than to meningococcus infections. ...

Yours sincerely,

EGDM

International Society for Microbiology
Second International Congress for Microbiology
London, 25 July-1 August 1936

24th May 1937

Dear Professor Murray,[99]

... As you know, Canada had not a separate representation in the old International Society, and it is clear that with the formation of the new International Association of Microbiologists this omission should be rectified.

As you took an active part in the work of the London Congress, the President of that Congress and I feel the official adherence of' Canadian Microbiologists to the new Association might well be arranged through your good offices.

If you will undertake this task and when you and your colleagues have decided what body of Canadian Microbiologists should function as your National Group, please communicate this information, together with the name and address of the Secretary of the Group and the name and address of your designated member of the Permanent International Commission *direct* to Dr. T.M. Rivers, the President-Elect of the New York Congress 1939.

Yours sincerely,

R. St. John-Brooks[100]

---

99 Ibid.
100 Ibid. Ralph Terence St. John-Brooks (1884-1963) was a bacteriologist whose major contribution to science was his curatorship of the National Collection of Type Cultures, located at the Lister Institute of Preventive Medicine, London, England. In addition to the collection and preservation of cultures, St. John-Brooks was also deeply involved in the naming of cultures and served on international nomenclature committees with the likes of E.G.D. Murray and American bacteriologists R.E. Buchanan and R.S. Breed. He was also instrumental in the founding of the Society for General Microbiology (U.K.). *Oxford Dictionary of National Biography*.

June 3, 1937

Dr. R. St. John-Brooks,[101]
International Society for Microbiology,
Lister Institute,
Chelsea Bridge Road,
London, S.W.1, England

Dear St. John-Brooks:

I have just received your letter of May 24th concerning the new International Association of Microbiologists. I shall do everything I can to help by stimulating my Canadian colleagues to support the Association.

This is not a particularly easy task as there is no Canadian microbiological society and most people belong to various American societies. On the 17th of this month I am attending the meeting of the Canadian Public Health Association and I shall endeavour to gather the bacteriologists attending the meeting and discuss the matter with them.

It is probably that the best approach would be to circularize everybody who might be interested in the Association but to do this I would require the fullest possible information on the purpose and proposed activities of the Association and its relation to existing scientific societies. The population of Canada is relatively small and widely scattered and it is difficult to form active societies interested in special subjects; therefore, the tendency is for individuals to become members of American societies. ...

From the literature included in your letter I gather that the Association is a nebula of passively interested workers centering around a permanent international commission for the organization of congresses and that this commission determines the point of condensation from time to time. ...

The easiest way out of the difficulty seems to me for the competent authority of the International Association of Microbiologists to authorize me to invite interested representative individuals in Canada to form themselves into a national group which could serve as a starting point for an organization to ensure adequate representation. ...

    Yours sincerely,
        EGDM

---

[101] Ibid.

The Hospital of the Rockefeller Institute for Medical Research,
66th Street and York Avenue, New York.
June 9, 1937

Professor E.G.D. Murray
Dept. of Bacteriology
McGill University
Montreal, Canada

Dear Professor Murray:

Thanks for your letter of June 3rd with enclosures regarding the International Association of Microbiologists.

It seems to me that there are two lines which you might follow:

First: You can ask the Canadian Public Health Association to become your national group and appoint a member to the permanent International Commission. Such action would not obligate the Canadian Public Health Association in any way.

Second: You can follow the line suggested in your letter to Dr. St. John-Brooks from which I quote: "Our situation would be greatly simplified did the Association have an enrollment of members which need not amount to more than a list of individuals who have promised their adherence. They could then appoint the national committee". Your national committee under this arrangement then could appoint a member for the International Commission.

I as President of the International Association of Microbiologists am happy to authorize you to follow one of the two plans set forth above or I shall be glad to consider any other plan you might devise.

    With best wishes, I am
        Sincerely,
            Thomas M. Rivers, M.D.[102]

---

102 Ibid. Thomas Rivers (1889-1962) was a dominant figure in the development of virology research in the United States and internationally. Most of his career was spent at the Rockefeller Institute for Medical Research. See *Tom Rivers: Reflections on a Life in Medicine and Science; an Oral History Memoir, prepared by Saul Benison* (Cambridge, MA: MIT University Press, 1967).

June 10, 1937

Dr. Thomas M. Rivers,
The Hospital of the Rockefeller Institute for Medical Research,
66th Street and York Avenue,
New York, N.Y.

Dear Dr. Rivers:
... I am very doubtful of the suitability of the Canadian Public Health Association to act as a national group. The majority of the members are health officers and sanitarians who would not have any appreciation of the problems involved. I shall see a number of the bacteriologists at the meeting of the laboratory section and will consult their opinion and let you know the result.

The second plan appeals to me as more feasible and more suitable. I interpret the name of the Association in its widest sense so that the national group will have to represent not only bacteriologists but botanists, protozoologists, helminthologists and mycologists. If I am wrong in this interpretation please let me know before I go too far in asking for the support of workers in subjects other than bacteriology.

I am very glad to offer you my congratulations on your appointment as head of the Rockefeller Hospital. This work has been one of the outstanding features of American medicine in the past and I have no doubt that it will continue to be so in the future.

Yours sincerely,
EGDM

R.D. Defries, Chairman,
Editorial Board,
Canadian Public Health Journal,
The Canadian Public Health Association,
105 Bond Street, Toronto 2, Ont.

March 5, 1938

Dear Professor Murray:
Dr. Craigie[103] very kindly showed me the circular which you sent to him concerning the International Association of Microbiologists.

... It seems to me that it would be very reasonable if the Laboratory Section in Canada could be the official body for contact with the International Association. You

---

103 Ibid. Born in Scotland, James H. Craigie moved to Canada in 1931 and played a major role in the development of microbiology at the University of Toronto until 1946, when he returned to the United Kingdom. See note 37.

know, too, that there is a directory of laboratory workers in Canada with approximately 250 names.
>With kind regards,
>>Yours sincerely,
>>>R.D. Defries, M.D.[104]

>>>Department of Medical Research,
>>>Banting Institute, University of Toronto,
>>>Toronto 5, Ont.
>>>March 14th, 1938

Prof. E.G.D. Murray,
Department of Bacteriology and Immunology,
McGill University,
Montreal, Que.

Dear Dr. Murray,

I have received the circular concerning the International Association of Microbiologists who plan to meet in their Third Congress in New York September 2nd-9th, 1939.

I would like to submit the names of those in this Department who are interested in the cancer problem:
>Dr. Bruno Mendel
>Dr. W.R. Franks
>Mr. H.J. Creech …[105]

I might also say that I have been working on the tumor problem for a number of years myself. …

The above mentioned would be supporting members of the Association, and I

---

[104] Ibid. Robert Davies Defries (1889-1975) played "a central role in the development of public health in Canada." He was long associated with the University of Toronto's Connaught Medical Research Laboratories, which he helped found in 1913-1914, and with the School of Hygiene, serving as director of both from 1940 to 1955. In the early 1920s, he played an important role in obtaining the patent for insulin for the University of Toronto. Chris Rutty, "Dr. Robert Davies Defries (1889-1975): Canada's 'Mr. Public Health'," in *Doctors, Nurses and Practitioners*, ed. Lois N. Magner (Westport, CT: Greenwood Press, 1997).

[105] Ibid. These three scientists were members of the Banting Institute for Medical Research at the University of Toronto. They had quite different backgrounds: Mendel was one of the many Jewish scientists who fled Nazi Germany during the 1930s, while Franks and Creech were Canadian medical graduates. Michael Bliss, *Banting: A Biography* (Toronto: McClelland & Stewart, 1984).

would like to nominate Dr. E.G.D. Murray for membership of the National Committee for Canada.

    Yours sincerely,

        F.G. Banting, M.D.[106]

                        Prof. E.G.D. Murray
                            March 15, 1938

Sir Frederick Banting, K.B.E., F.R.S.,[107]
Department of Medical Research,
Banting Institute,
University of Toronto,
Toronto 5, Ont.

Dear Sir Frederick,

    Thank you for your kind reception of my circular on the representation of Canada in the International Association of Microbiologists. I hope others may respond so that a Provisional National Committee may be formed. ...

    Yours sincerely,

        EGDM

                        Ottawa 17 March, 1938

Prof. E.G.D. Murray,
Department of Bacteriology and Immunology,
McGill University,
Montreal, Que.

Dear Sir:

    I have read with interest your circular referring to the Third International

---

106 Sir Frederick Banting (1891-1914) was Canada's most famous scientist during the interwar years because of his Nobel Prize-winning work in the discovery of insulin. Although he and J.J.R. MacLeod shared the 1923 award, Banting gained special recognition: he was awarded a life annuity by the federal government, appointed Canada's first professor of medical research at the University of Toronto, and knighted in 1934. During the 1930s he explored a number of research projects, notably silicosis and cancer, as well as aviation medicine and biological warfare. Bliss, *Banting*.

107 LAC, E.G.D. Murray Papers, vol. 5. Everitt Murray would later become involved with Banting's efforts to alert Canadian military officials to the threat of a German biological weapons attack. See chapter 2.

Congress for Microbiology and asking that steps be taken leading to the formation of a Canadian National Committee.

I observe that nominations from individuals are provided for by the forms you enclosed. That being the case, it would probably serve no useful purpose for the National Research Council to make a nomination in this way. We would like, however, to be represented on the Canadian Committee, and, if it can be arranged, would ask for the appointment of Dr. N.E. Gibbons[108] of our Division of Biology and Agriculture.

 Yours very truly,
  A.G.L. McNaughton,[109]
  President.

March 24, 1938

Major General A.G.L. McNaughton,
President,
National Research Council,
Ottawa, Ont.

Dear General McNaughton,

My first concern, in trying to secure representation of Canada on the International Commission of the Association of Microbiologists, as requested by the International Commission, was to secure the cooperation of all those interested in each branch and phase of the subject. It seemed to me this might best be achieved by a personal vote by individual workers and to that end I have tried to reach each of them. For this purpose, therefore, I venture to suggest that each individual on your staff qualified to be considered a microbiologist exercise his right to vote for whoever he would like to have on the *Provisional* Committee. ...

---

108 The microbiologist Norman E. Gibbons, PhD, provided instrumental assistance to R.G.E. Murray in the establishment of the Canadian Society of Microbiology in the late 1940s and early 1950s. He worked in the Division of Biological Sciences of the National Research Council of Canada for three decades. He was honorary secretary of the Royal Society of Canada from 1956 to 1959. N.T. Gridgeman, *Biological Sciences at the National Research Council of Canada: The Early Years to 1952* (Waterloo: Wilfrid Laurier University Press, 1979).

109 LAC, E.G.D. Murray Papers, vol. 5. General A.G.L. McNaughton was a McGill engineering graduate who achieved great distinction during the First World War. In the interwar years he served as Chief of the Defence Staff and president of the National Research Council, which he helped shape into a defence science organization. Between September 1939 and 1943 he was Canada's senior military officer in the United Kingdom. In 1944 McNaughton became Minister of Defence, and in 1946 he represented Canada on the United Nations Atomic Energy Commission. John Swettenham, *McNaughton*, 2 vols. (Toronto: Ryerson University Press, 1968).

The International Association is one of *individual* microbiologists and therefore the National Committee should represent Canadian microbiologists. There are good reasons why an important body like the National Research Council should be represented on the Canadian National Committee, but the same can be said for the Royal Society of Canada, the Canadian Public Health Association and other various scientific societies. Perhaps a National Committee might be permanently constituted by one member appointed by each of a number of important and representative organisations and societies, but that is for the Provisional Committee to decide.

My central concern is merely to carry out the personal request of the Central Committee of the International Association and once the Provisional Committee is formed I am relieved of responsibility.

Yours sincerely,
EGDM

International Association of Microbiologists
Third International Congress for Microbiology
New York City, September 2-9, 1939[110]
August 10, 1939

Dear Professor Murray:

By now you doubtless know that we are going to have an International Congress of Microbiology provided we do not have a world war. As a part of the Congress it seems that we must have an official banquet. As a part of the banquet it seems essential that we have speakers. So far as I am concerned it seems essential that these speakers should he amusing as well as intelligent. Inasmuch as you answer both of these requirements the Committee has chosen you to make what it considers more or less the main speech of the evening. ...

It looks as though we are going to have a wonderful Congress. At least the program is a superb one.

Looking forward with a great deal of pleasure to seeing you at the Congress,
Sincerely yours,
Thomas M. Rivers, M.D.

---

110 LAC, E.G.D. Murray Papers, vol. 5.

August 15th, 1939

Dear Doctor Rivers,[111]

I have just returned from a three weeks holiday in the North woods and therefore there has been an unavoidable delay in answering your letter of August 10th.

I am very appreciative of the signal honour your Committee confers on me in requesting that I make a speech at the banquet. I see no alternative but to acquiesce to your command although I do not consider myself in any way fitted for the task. ... I shall do my damnest though you certainly overestimate my powers.

... I look forward to a very enjoyable time, in spite of European conditions, and I hope to meet many of my British friends. Some who were expecting to be over here about this time have written to say that they have instructions not to be out of the country in August and September. This looks as if many of our distinguished colleagues we expect to welcome at the meeting will be detained. ...

Yours sincerely,
EGDM

Prof. E.G.D. Murray
September 9th, 1939

Doctor T.M. Rivers,[112]
Rockefeller Institute for Medical Research,
66th Street and York Avenue,
New York, N.Y.

Dear Tom,

I am deeply grateful to you for all your many kindnesses to me during the past week, in spite of your many arduous duties in connection with the Congress and the worries imposed on you by the interference with arrangements through the international situation. I cannot express very well the admiration you inspired in me.

I enclose a copy of my speech as you requested for whatever purpose you have in mind.

Yours ever,
EGDM

111 Ibid.
112 Ibid.

The Hospital of the Rockefeller Institute for Medical Research,
66th Street and York Avenue,
New York.
September 13, 1939

Dear Professor Murray or Murphy:

You have got me so mixed up by now that it is hard for me to tell whether you are an Irishman or a Scotchman because you have the good qualities of both of them and none of the bad.

I have received your nice letter of September 9 and a copy of your speech made at the official banquet. I think it is a delightful speech and needs no editing. Furthermore I see no reason why it should not be included in the Proceedings.

Somehow or other this Congress has made it possible for me to get closer to you. In other words I feel that I know you much better than ever before and with this increasing knowledge comes a greater appreciation of you as an individual. Despite the stress of the situation in Europe you were able to contribute much to the success of the Congress and I wish to express my appreciation for what you did in spite of your personal worries.

Cordially yours,
Thomas M. Rivers, M.D.

Attempts to Create a Canadian Society of Microbiology

Division of Bacteriology
Science Service
Department of Agriculture (Canada)
Central Experimental Farm
March 6, 1939

Prof. E. G. D. Murray,
Department of Bacteriology,
McGill University,
Montreal, Quebec.

Dear Prof. Murray,

I have delayed replying to your recent letter until I had an opportunity of consulting the members of our local Bacteriology Club on the points you brought up.

At our monthly meeting last Wednesday night the matter was discussed at some length. It was agreed (1) that the chances of successfully organizing a new, and purely Canadian, association of bacteriologists in the near future appeared to be slight; (2) that expansion of the Laboratory Section of the C.P.H.A. would still fail to provide for a considerable number of workers in industrial (including the bulk of dairy and food), soil and pure bacteriology; (3) that formation of a local branch of the S.A.B. appeared to be the most promising way of achieving our objectives i.e., to encourage closer association of workers in all branches of microbiology, and particularly to facilitate contacts between the junior members of the various staffs who rarely have an opportunity of attending the large annual meetings of the several societies. Our committee was therefore instructed to continue negotiations with bacteriologists in Eastern Ontario and Quebec, with a view to the formation of such a branch.

As to your feeling that Canadians are regarded as "foreigners" by those across the line, may I say that it is the unanimous opinion of our members, a number of whom have lived in the United States for periods up to 18 years, that this is not the case. Organization such as the Society of American Bacteriologists, American Public Health Association and others have gone out of their way to secure ample Canadian representation on their governing bodies, committees, etc., indicating that they interpret the term "American" in the "continental" rather than in the "national" sense.

To summarize the views of the Ottawa group, I would say that we favour the formation of a Local Branch of the S.A.B. for the following reasons:

1. It would mean organizing a branch of a well-established society rather than starting an entirely new one.
2. It would provide for all branches of microbiology.
3. It would extend to us the privilege of publishing abstracts of papers presented at local branch meetings in the Journal of Bacteriology, thus drawing attention to the work going on up in this area.
4. It would eventually entitle us to a representative on the Council of the S.A.B.
5. It would encourage closer co-operation between workers in bacteriology on both sides of the line, while at the same time preserving our identity as a Canadian group.

In the light of the above, it is our hope that you will agree to join with us in organizing a Local Branch to promote closer association between workers in bacteriology in this area.

    Yours sincerely,
        C.K. Johns[113]
        Associate Bacteriologist

---

113 Ibid. Gridgeman, *Biological Sciences*.

Prof. E.G.D. Murray
March 13th, 1939

Doctor C.K. Johns[114]
Division of Bacteriology
Department of Agriculture
Central Experimental Farm
Ottawa, Ontario

Dear Doctor Johns,

I still feel that the possibilities of forming a Canadian Association of Bacteriologists has not yet been explored and that it should be before a Local Branch of the S.A.B. is formed precipitately.

The discussion in Toronto during December last envisaged the inclusion of all branches of microbiology and the publication of abstracts. Naturally a Canadian association would not be entitled to representation on the Council of the S.A.B., but I see little advantage in having it. On the other hand, a Canadian association could make a sound plea for recognition, support and encouragement by the National Research Council and the Federal and Provincial Governments, whereas a branch of an American Society could not. That official support for adequate microbiological research is wanting now, might be attributable to there not being any Canadian organisation of microbiologists.

A Canadian association would secure adequate Canadian representation in international activities, such as the International Association of Microbiologists etcetera and thereby secure recognition which the restrictions of a Local Branch of the S.A.B. could not allow of.

A Canadian association would not preclude close cooperation with American workers and might give greater strength to the Canadian contribution to any such undertaking.

The Local Branch of the S.A.B. you purpose to form would seem to have little more opportunity than the Ottawa Bacteriology Club for interchange of views, since bacteriologists from Kingston and Montreal (who I understand are to be included) would seldom be present. Can you afford to neglect the large and very active group in Toronto and those elsewhere? If you do, as I see it, little is gained and the possibilities and advantages of a thorough going Canadian association are lost. I feel too that we would gain in strength by standing on our own feet without relying upon a S.A.B. crutch.

I know perfectly well that there is no tendency among individual Canadians and Americans to regard one another as "foreigners" but the official national distinction

---

114 LAC, E.G.D. Murray Papers, vol. 5.

exists nevertheless. In any case we have our own ideals and potentialities which should be developed.

I feel strongly that the possibilities of forming a Canadian association of microbiologists should be explored. It could readily be formed by the association of any existing societies (mycologists, bacteriologists, protozoologists, etcetera and the Laboratory Section of the C.P.H.A.) to hold joint meetings and to form a combined council to represent every kind of interest, attract membership and secure official recognition and funds.

I am inclined to remain opposed to the formation of a Local Branch of the S.A.B. until I am sure the possibility of a Canadian association has been thoroughly examined and found wanting; or, having been formed, that a Local Branch of the S.A.B. still has a useful purpose.

Yours sincerely,
E.G.D. Murray

Prof. E.G.D. Murray
March 13th, 1939

Professor G.B. Reed,[115]
Department of Bacteriology,
Queen's University,
Kingston, Ont.

Dear Reed,

I enclose part of a correspondence which I think is self-explanatory. I feel sure you have been approached as I have and I would like to know your views.

The International Association of Microbiologists asked me to take steps to get Canada represented on the Permanent Commission. As you know this was not an easy task and in carrying it out I realized the need for a Canadian association. This need would not be met by a Local Branch of the S.A.B.; we need more than that.

Yours ever,
E.G.D. Murray

---

115 Ibid. Guilford B. Reed (1887-1955) was one of Canada's most accomplished and respected bacteriologists. His research involved work on tuberculosis, gas gangrene, tetanus, and rinderpest. He was a professor at Queen's University from 1915 to 1954, headed the Defence Research Laboratory at the university, and worked alongside Everitt Murray in Canada's wartime biological weapons program during the Second World War. He was elected to the Royal Society of Canada in 1932, was awarded the Society's Flavelle Medal in 1947, and acted as its president in 1952-1953. See Introduction, note 233.

Prof. E.G.D. Murray
March 22nd, 1939

Doctor C.K. Johns[116]
Division of Bacteriology
Department of Agriculture
Central Experimental Farm
Ottawa, Ontario

Dear Doctor Johns,

I do not think the C.P.H.A. alone could meet what is required but it could combine with a *Canadian Microbiological Association* embracing all fields of work with great advantage. Such a united endeavour could enlist official recognition and support and perhaps financial aid. That would be a great advantage.

I believe a Canadian microbiological association could be most active and most effective if constituted primarily in divisions, each of which functioned in the way you propose for the Local Branch of the S.A.B. Then only the annual meeting would coincide with say the December meeting of the Laboratory Section of the C.P.H.A. to swell attendance, increase the range of subjects and secure united policy.

May we consider this proposal before deciding on the Local Branch?

Yours sincerely,
E.G.D. Murray

Department of Agriculture
Central Experimental Farm, Ottawa
April 11, 1939

Chairman, Editorial Board,[117]
The Canadian Public Health Association,
105 Bond Street,
Toronto, Ontario.

Dear Dr. Defries,

I have postponed answering your letter of March 22nd until the matter in question could be discussed at our April meeting.

The feeling of our members is that the proposal to form a distinctively Canadian Society of Bacteriologists at the present time is rather premature. Furthermore, the

---

116 Ibid.
117 Ibid.

proposal to form such a society by attaching to the C.P.H.A. Laboratory Section those whose work has little or no relation to public health finds little or no support here, even from those who are in the field of public health. It is further felt that very few workers in soil, dairy, food, or industrial bacteriology will associate themselves with such an organization in preference to the existing international associations. Again, the Canadian Public Health Journal scarcely appears to be an appropriate medium for publication of abstracts of papers from fields devoid of any connection with public health. (No such objection holds for the Journal of Bacteriology). Finally, such an organization does not answer the need for a small regional organization to promote personal contacts between workers in a given area. ...

 Yours sincerely,
  C.K. Johns
  for Ottawa Bacteriological Club.

         The Canadian Public Health Association
              105 Bond Street, Toronto[118]
                   April 15, 1939

Professor E.G.D. Murray
McGill University
Montreal, Que.

Dear Murray:

 I am enclosing a copy of a letter received from Dr. Johns, dated April 11th. I presume that you received this letter also but in case you did not, I thought I would forward a copy to you.

 Dr. Johns is quite definite in his desire to go forward with the plan of a local branch of the S.A.B. I discussed the matter in detail with a member of his group and this member's attitude was that fifty cents a year would secure the privilege of membership and of printing their proceedings in the Journal. Our discussion was a very cordial one and I think we have gone as far as we can.

 Thanking you for your helpful interest,

 Yours very sincerely,
  R.D. Defries, M.D.

---

118 Ibid.

The Canadian Physiological Society

The Canadian Physiological Society was organized, and its first meeting was held, in Toronto, October 19th, 1935. Three resolutions were adapted with respect to the organization:

1. "The Society shall be called the 'Canadian Physiological Society'."
2. "The object of this Society is to promote the advancement of physiology and its related branches of science, and to promote 'a friendly spirit among those Canadians who are engaged in these fields."
3. "Membership. There shall be eligible

"Those persons who have conducted researches and published papers in the fields of physiology, biochemistry, pharmacology, and the experimental aspects of biology, pathology, therapeutics and hygiene." ...

G.H. Ettinger[119]
Secretary, Provisional Committee,
Canadian Physiological Society.

Office of the Secretary
The Canadian Physiological Society
Queen's University, Kingston, Ont.
August 29th, 1939.

Professor E.G.D. Murray,[120]
Department of Bacteriology,
Mc Gill University, Montreal, P.Q.

Dear Professor Murray:

The Canadian Physiological Society is attempting to determine whether there is a need or justification for publishing a Canadian Journal of *Experimental* Medicine. A natural step is to ascertain the number of papers of a research nature published from Canadian laboratories and hospitals. Unless this output exceeds 500 pages *per annum* it would be hardly wise to attempt to publish such a journal.

Could I prevail upon you, please, to send me at your earliest convenience, a list of all papers published from your department for the calendar years 1937 and 1938, under the following description:-

---

[119] LAC, E.G.D. Murray Papers, vol. 4, file Canadian Physiological Society. George Herald Ettinger was a well-known Canadian physiologist who spent most of his career at Queen's University, where he was dean of the Faculty of Medicine from 1946 to 1962. He was also active in the National Research Council.

[120] Ibid.

Author(s) - Title of Paper - Journal - Year - Volume - Pages

Would you please star or mark in some way those communications which you regard as of an experimental nature. If you could send reprints of all papers they would be useful and much appreciated.

    Yours very truly,
        G.H. Ettinger
           Secretary

FAMILY LIFE AND ISSUES

The Province of Quebec Association for the Protection of Fish and Game: Incorporated[121]
Extracts from Charter

By virtue of powers conferred by Letters Patent under Part Third of the Quebec Companies' Act, recorded on the 23rd of January 1929, by C.J. Simard, Assistant Provincial Registrar, the Association has been incorporated with the following objects;

1. To undertake, carry out and perform all acts, measures, and things necessary or deemed to be necessary for any or all of the following purposed;
   a. The establishment, breeding, maintenance and protection of Fish and Game;
   b. The establishment, maintenance and operation of fish hatcheries and the distribution of fish or spawn for the restocking of lakes or other purposed;
2. In connection with any of the foregoing matters;
   a. To maintain and assist in all conservative, preventive and educational work in connection with any of the objects and purposed of the corporation.
   b. To promote, organize and carry on, or assist in the promotion, organization and carrying on of any exhibitions, shows, conventions or meetings ...
   e. To enter into any arrangement with any governmental, municipal, local or other authorities that may seem conductive to the corporation's object or any of them, and to obtain from any such authority any rights, grants, privileges and concessions which the corporation may think it desirable to obtain and to carry out, exercise and comply with any such arrangements, grants, rights, privileges and concessions ...

---

121 LAC, E.G.D. Murray Papers, vol. 3.

> The Province of Quebec Association for the Protection of Fish and Game, Inc.
> 1154 Beaver Hall Square, Montreal
> Aug. 11, 1933

Dr. E.G.D. Murray,[122]
Bark Lake, Barcmere, P.Q.

Dear Sir:

The directors have learned with great pleasure of your election as Vice-Chairman of our Argenteuil division and wish to congratulate you. With the support you are bringing our local organization, the association is looking forward to much improvement in fish and game conditions in the district and you may feel with confidence that the Head Office will do its utmost to cooperate and support you in all your efforts.

Sincerely yours,
E.A. Cartier
Secretary-Treasurer
Quebec Association for the Protection of Fish and Game

The Bark Lake Cottage[123]

I think it was Bark Lake, where we rented the Whitall's cottage that first summer in Canada, 1931, and soon they were planning 1932-33 'The Cottage', which was her and their job for years to come. It brought her back into contact with nature and life in relatively unspoiled, spacious territory. ...

Bark Lake is a winding body of water nearly eight miles long with an average width of about one mile and several large bays opening off the main body. The average depth is about 50 ft. though there are many deeper spots and a maximum depth of 183 ft. It receives an ample supply of water at all times from springs and from streams draining the small lakes in the surrounding hills, and is completely surrounded by well-grown mixed brush which has not been lumbered off for a considerable time.[124]

---

122 Ibid.
123 Robert Murray, "A 'Mum for All Seasons.'" Bob Murray described the Bark Lake cottage "as having two small bedrooms, with a marvelous 12 x 30 foot veranda where people primarily slept during the summer." Robert Murray, interview by Donald Avery, 10 October 2007.
124 H.W.H. Murray, "Physical and Ecological Changes and Their Effect upon Game Fish in Bark Lake," co. Argenteuil, Province of Quebec, 1946; see chapter 4.

Preparing Robert Murray for His Medical Education at Cambridge

Personal Reflections[125]

Starting a new kind of school was an adventure. Advice was sought and decision was made to put Bob in Grade 9 in Westmount High School. It was difficult and neither age nor content seemed to fit but he soldiered on to the summer of 1931. By then the private schools had been canvassed and approach was made to Lower Canada College and there the considerable differences from the Summerfields pattern of studies could be compensated for by shifting classes. Also there were masters of some academic strength and there was soccer and cricket. The next five years through junior and senior matriculation to McGill passed pleasantly. Some of the teachers were influential and several were friendly for years thereafter, including V.C. Wansbrough, Stephen Penton, Hugh MacLennan. In fact, for three years Bob and Steve Penton (later Head Master of LCC) went to the Montreal Symphony Orchestra as subscribers. ... During those years the family lived close to the mountain and in wintertime Bob took advantage for skiing for a brief hour or so after getting home from school. ...

I entered McGill as a student in September 1936 at age 17. I could theoretically have gone the year before but I was not ready although I wrote the entrance examination. I wanted to enter the science program and, having essentially no science other than a bit of physics in my schooling, I needed to start from the bottom with the usual first year courses. It was a real excitement for me even though I knew the layout of the university and some who would teach me because of family contacts. I was not pushed along the academic way by my parents although there was no doubt of university training being the route to a future. They were interested in what I wanted to do and were glad it included science. I got advice if I asked for it and I think they were assiduous in trying to avoid any semblance of interference. ...

Going to McGill and taking science subjects was an excitement for Bob because school had concentrated on everything but science. So, despite the considerable discussions of many aspects of science at home and in general reading, it was new exciting stuff. The two years 1936-1938 were a full quiver of the basic courses, a bit more than the standard number on top of the required English, Mathematics, and scientific French(!), notably two Chemistry, one Physics, basic Biology courses (Botany, two Zoology, Comparative Anatomy and Genetics). There were some notable teachers in the biologies - V.C.Wynn-Edwards, N.J. Berrill, and R.D. Gibbs. It was a good start.

---

125 UWO Archives, R.G.E. Murray Collection, Robert Murray, "The McGill Experience."

Prof. E.G.D. Murray
Feb. 10, 1937

S.W. Grose Esq.,[126]
Christ's College
Cambridge

Dear Sid:

I wish to enter Bob at Christ's for October 1938.

He has the necessary exemption from the Previous by McGill matriculation. By the time he comes up he will have spent two years at McGill and should have completed the necessary examinations to give him exemption from all parts of the First M.B. and Part 1 of the Second M.B.

From what his teachers tell me Bob seems to be doing fairly well in his studies but it is important that they are directed along lines which will be of use to him in taking up Physics, Chemistry, Zoology, English, French and Mathematics. I gather that the Physics, Chemistry and Zoology are more than are required for the First M.B. Next year he will do Botany and any other subjects that are desirable and these can be arranged according to the advice you give me.

The regulations at Cambridge have altered so much during the past six years that I must rely entirely on your advice and I hope you will not mind me asking for it. We are naturally anxious that he shall be able to go to Cambridge without any disadvantages.

Please send me the necessary entrance forms as there may be some advantage to enter him early, particularly with regard to rooms in College. ...

We are all well and busy in our various ways. Bob, as I have said, is enjoying the work and seems to be a good student. Susan is doing quite well at school and getting to be a big girl. Freda, after some three years of serious illness and a big operation, has completely recovered and seems to be her old self once more. She has the same habit of overburdening herself with household affairs and not giving time to her artistic and other talents.

My love to you and yours and my friends in College.

EGDM

---

126 LAC, E.G.D. Murray Papers, vol. 17, file R.G.E. Murray. S.W. Grose was a respected numismatist and authored, among other works, the *Catalogue of the McClean Collection of Greek Coins*, vol. 1 (Cambridge: Cambridge University Press, 1923). He is fondly remembered by Bob Murray: "There were a number of people in the College [Christ's] who were familiar to me from my childhood. Among the Fellows the closest family friendships I maintained were with S.W. Grose ... (who) was the Senior Tutor and mentor." Robert Murray, "Personal Narrative."

Christ's College
Cambridge
3 March 1937

My dear Murray,[127]

I have tried to settle down, to write to you at some length since I got your letter last week not only to say how delighted I am to think of Bob coming here, but also to give you some of our news. But it has been a busy and rather difficult time and I have been doing the normal everyday work in between a heavy succession of University and College meetings. Our particular College troubles have been the necessity of appointing a new Bursar in place of Campbell who is retiring as from the end of this term. The trouble there was that Wyatt was not at all anxious to go on to this office and has only been persuaded very much against his own desires. As to building, we have not got anywhere, a majority not wanting an architect whose plan commended itself to a very strong minority. However, no more about these topics.

You will probably know that the recent changes in regulations mean in effect that all medical students now approach the M.B. by means of the Tripos reformed so as to give the medicals a combination of subjects suited to their needs. The old 2 M.B. Part I, the Organic Chemistry, has now become part IV of the First M.B. We like our men to have exemption from part I (Inorganic Chem.) part 2 (Physics) and either part 3 (Biology) or part 4, but we sometimes take them with only two parts exempt. However, from what you say, Bob will be well up to this. But with the Honours course in front we feel bound to warn parents that it is to their own interests that their sons should be exempt from as much of 1 M.B. as possible in order to prevent disappointment later on.

Now, I think the best plan is for me to send you a reprint of the Regulations, and I am also sending some pages from the Student's Handbook. I think that perhaps the best thing is for you to study these and to put any questions which then occur to you, rather than for me to try to give advice. But there are two preliminary points which occur to me:

(1) Could you send the Certificate on which you rely for exemption from the Previous? I would submit it to the Registrar and have it accepted as in accordance with the Regulations and so save any source of future trouble on this head. As Bob is to be at McGill only two years I take it that he is not proceeding to a degree there and that exemption from the Previous will not be through Affiliation.

(2) See p. 728 s. 23, 24. Would it be possible for me to take up with the Faculty Board the question of their requirements for purposes of exemption by means of the McGill examinations? Perhaps your Department has already had this question

---

127 Ibid.

settled for other students who have come here. But it would be well for you to know what evidence they require for exemption from each several part and if there have not been previous cases they might very well ask for schedules of your examinations.

Meanwhile I send an admission form which I will ask you to return with the College admission fee of £2, and I think that the best advice I can give (in my ignorance) is contained in this second suggestion to which you perhaps already know the answer.

Ever yours v. sincerely,
S.W. Grose

We were sorry to hear of Freda's continuous ill-health and trust that the trouble is now satisfactorily cleared up. Love from us all to all of you. I heard my wife say something about writing a long letter to her this week. Perhaps she has done so.

Prof. E.G.D. Murray
October 27, 1937

V.C. Wansbrough Esq., M.A.,[128]
Lower Canada College,
4090 Royal Avenue,
N.D.G., Montreal, Que.

Dear Wansbrough,

Please write the necessary little screed on the back of the entrance form to Christ's College for Bob and return it to me here. I wish to send it on to complete the formalities.

The boy did quite well last year:

Chemistry 1	1st Class
English 1	2nd Class
English 2	3rd Class
French 15	3rd Class
Mathematics 1 (Algebra 1st Class; Trigonometry 2nd Class)	
Physics 1	1st Class
Zoology 1	1st Class

I am rather pleased with him and feel he has a chance to do reasonably well at Cambridge.

Yours sincerely,
EGDM

---

128 Ibid.

November 10, 1937

S.W. Grose Esq.,[129]
Christ's College
Cambridge, England

Dear Grose,

I am enclosing Bob's application for admission to Christ's College, together with a testimonial from the Headmaster of the school he attended here and a certificate from the Registrar of McGill University providing his entrance qualifications from McGill, which I believe exempt him from the Little-go. The certificate further gives the results of his first year B. Sc. Work in McGill where the chemistry, physics and zoology, in each of which he got first class marks, should count towards exemption for certain parts of the First M.B. The subjects he is studying this year towards exemption are botany and organic chemistry. If he passes these creditably in May he should be able to claim exemption from the whole of the First M.B. under the new regulations. Under these circumstances he could go straight on with his Tripos and general medical studies.

I think that you will agree that this certificate recommends him as the kind of student the College likes to have and I am rather proud of his achievement. He is not a boy who likes to ask favours from anybody and I think he is determined to get on through his own merit. ...

My little family are doing very well and Freda has completely recovered from the illness she suffered for two or three years. There is a great deal of hard work to do here but I hope it will result in something worthwhile in the long run. There is evidence that the University appreciates what is being done and I feel sure that developments will be more rapid in the future than they have been in the past. ...

Freda joins me in sending our love to all of you. I look forward to dining in hall and seeing many of my old friends at some time while Bob is an undergraduate. ...

    Yours ever,
    EGDM

Prof. E.G.D. Murray
June 7, 1938

Dear Grose,[130]

I enclose a certificate from the Registrar of McGill University giving the results of Bob's examinations for the past year. This together with the certificate for 1937

---

129 Ibid.
130 Ibid.

should exempt him from all of what used to be 1st M.B. or 2nd M.B. Part 1. subjects.

He did not do quite as well as he hoped to because he was unwell at the time of the examinations though there was no indication of what ailed him. However shortly after the examinations were over he went down with acute appendicitis. He has recovered by first intention from the operation and goes up to our camp on Bark Lake today to recuperate.

In July he is spending six weeks at a Marine Biological Station on the Coast of Maine, to do some more Zoology which interests him greatly. I am very pleased with his achievement and feel sure he will do well.

The rest of the family are well and delighted to be going into the woods again for the summer.

    With love from us all.
        Yours ever,
            EGDM

                              Sept. 10, 1938

Dear Grose,[131]

It is good news that Williamson thinks it probable that the Board will grant exemption for Part 4 of the First M.B. on the grounds of work Bob did at McGill. The usual notices arrived recently and Bob is sailing on September 23rd to reach Liverpool on October 1st, and will arrive in Cambridge on October 6th after spending a few days in London getting some necessary clothes. ...

The family are all very well and have enjoyed their summer, Freda and Susan at the Lake and Bob, for most of his time, at a marine biological laboratory on the coast of Maine where he enjoyed himself immensely and did some very useful work.

    With love to all of you,
        Yours ever,
            EGDM

                              Sept. 29, 1938

Dear Grose,

Bob sailed last Friday and the turmoil in Europe has been very worrying to us in consequence. Freda, who lost her only brother in 1918, is naturally very upset.[132]

---

131 Ibid.
132 Ibid. Freda's brother, Thomas, was killed in March 1918 on the Western Front.

Today's news is encouraging and it is to be hoped that sound sense may prevail in the long run though I fear it will involve unpleasant sacrifices.

    Yours ever,

      EGDM

Robert Murray: From McGill to Cambridge[133]

The two years introduced me to McGill even if I was not a sociable student. ... I knew and enjoyed talking with the friends of my parents like the Collip's, the Noble's (both later to be friendly colleagues at UWO), Fred Smith, J.S.L. Brown, the Kalz's, W.V. Cone, and many others. I often visited the bacteriology lab and talked with people at work and was received kindly for more than brief conversations; these involved a range from Willy Clarke (the Head Lab Man who taught me inter alia about glass blowing), the professorial group like Fred Smith and J.W. Stevenson, and those professionals who were there for research experience like T.E. Roy and K. MacPherson (all of whom would be friends and colleagues later in life except Fred Smith who died about 1946). There were "lab parties" at home every year and it was not a too quiet life even if a bit peculiarly centred. ...

The decision to go to the Salisbury Cove Laboratory, Mount Desert Island, Maine for a summer course in invertebrate zoology (described elsewhere) was mine and I persuaded my father of my real interest. It was independent of my father's decision that he could support my going to Cambridge, a long term goal he had for me, and talked to me about the state of affairs in Europe making it likely that it might be best to go then in 1938 rather than complete a bachelors degree at McGill. ...

In 1938 it was clear that Europe was likely to be embroiled in war and Hitler's ever more threatening take-overs and Germany's military strength was disturbing. There had been family discussions about Bob going to Cambridge as a student and it then looked as if 1938 might be a last chance. Fortunately good marks and Joburg's personal involvement with Christ's College made it possible; the College accepted him to start October 1938.

The adventure of leaving for England was preceded by a party at home that was fun but the more exciting because of high winds and rain, the bequest of the 1938 hurricane. ... I sailed out of Montreal in the SS Antonia, a Cunarder, headed for Liverpool via the Strait of Belle Isle. A large group of the family friends were there on Victoria Pier to see me off. It was a beautiful send-off in another way because the fall colours of the Laurentide hills along the north shore east of Quebec were truly spectacular. It was an exciting week because the Munich Crisis occurred while we were on the Atlantic. ...

---

133 UWO Archives, R.G.E. Murray Collection, Robert Murray, "The Cambridge Experience."

CHAPTER TWO

SCIENTIFIC WARFARE AND THE THREAT OF BIOLOGICAL WARFARE

MCGILL UNIVERSITY, JOBURG, AND THE RESPONSE TO WAR, 1939-43

>Inter-Departmental Correspondence
>Faculty of Medicine
>September 11, 1939

To the Heads of all Departments in the Faculty of Medicine:

Major D. Stuart Forbes was asked by the Principal to collect certain information ... from the Departments in the Faculty of Medicine.

What is required is a list of equipment or apparatus which might be useful for the purpose of testing or for research in the war situation. In addition ... will you also state:

a. In what ways your Department could assist in training technical workers for war service, and

b. What research work your Department could conduct to advantage?

Might I have these data at your earliest convenience please?

>Grant Fleming[1]
>Dean

>September 14th, 1939

Dear Grant Fleming[2]
Faculty of Medicine
McGill University
Montreal, Que.

Dear Mr. Dean,

It is urgent above all else to use every available possibility to the successful and rapid termination of the war. The Department of Bacteriology can and should

---

1 LAC, E.G.D. Murray Papers, vol. 29, file War Service. A.G. Fleming had been director of the Department of Public Health and Preventive Medicine since 1927. He was dean of medicine from 1936 to 1940.
2 Ibid.

contribute importantly to the military and civil emergency: because of the danger of epidemics of bacterial diseases in massed troops, because practically all war wounds are infected wounds, because the necessary depressed standards imposed on civilians exposes them to epidemic bacterial disease, and because this Department comprises a well organized and highly trained, experienced staff; also the laboratory is well equipped.

These potentialities can be utilised advantageously as follows:

1. A thorough competent personnel can be provided for army bacteriological laboratories.
2. The Department and its equipment could easily be organized as a central source for the preparation and distribution of essential supplies to army laboratories; e.g. culture media of all sorts, diagnostic agglutinating serums of all sorts, diagnostic ... suspensions of all sorts, essential reagents, etc etc. Such activities were immensely valuable in the last war, in economizing personnel, material and apparatus and in making available methods and services not available otherwise, as I can prove from intimate personal knowledge and experience during the last war.
3. The preparation of necessary vaccines for the immunization of troops or civilians on a large scale. Of this I had extensive experience during the last war.
4. The intensive training of pathologists with insufficient bacteriological experience to fit them for duties required of them in military hospitals etcetera. It must be remembered that in the last war pathologists were concerned entirely with bacteriology.
5. The maintenance of essential bacteriological service to civil hospitals could be organized satisfactorily.
6. The maintenance of essential teaching of medical students, while it is required, could be organized satisfactorily.
7. It would be possible to undertake special investigations as need arose; such as chemotherapy, causes and control of infection ... by-products of bacterial metabolism which are important solvents necessary to the production of war materials.

To carry out the whole, or the most important, of the duties outlined above would necessitate a discriminating selection of the present staff of the Department for particular duties. It would also need the absorption of certain selected members of the staff of other Departments in the University, whose usual civil duties are not so essential to the needs of the war as is bacteriology but whose special knowledge and training fits them for selected duties in this scheme.

To organize such a scheme efficiently it would be almost essential to place the Department on the footing of a military unit.

If required I am prepared to expand this report or any part of it in detail.

There follows a *curriculum vitae* of each member of the staff of the Department and a confidential brief evaluation of their qualifications for special duty.

Yours sincerely,

E.G.D. Murray

Department of Bacteriology and Immunity, McGill University and Royal Victoria Hospital, Montreal, Canada

Sept. 14th, 1939

Affiliations Outside the Department[3]

Ayerst, McKenna and Harrison (Montreal) have developed a bacteriological laboratory for the production of therapeutic rabbit immune serums and vaccines. This laboratory is guided by consultations with me and its products are only made available with my approval.

It has a competent and well trained staff. Its products are of a very high order and I believe it could be used to produce immune serums, vaccines, toxoids or special requirements with advantage.

It is licensed by both the Dominion Health Department and the American Department of Health (Washington).

October 5th, 1939

Professor David A. Keys,[4]
MacDonald Physics Laboratory
McGill University

Dear Professor Keys,

I believe there are research problems which might be undertaken immediately with advantage in the present war emergency. I take upon myself to make these suggestions, because I think it is perhaps more important for the University War Service Advisory Board to make suggestions to the Department of National Defence, to use highly skilled and learned members of the University to advantage, than it is to advise individuals on their "particular activity". There are in this and other

---

3 Ibid.
4 McGill Archives, RG 2, C-100, Principal Cyril James Correspondence, file 2724. David Norman Keys (1890-1977) was a popular physics professor at McGill University from 1922 to 1947, and from 1947 the vice-president of the atomic laboratories at Chalk River, Ontario. He was a close friend of Joburg.

Departments highly trained individuals who, singly or in groups, could undertake important investigations.

I. *Bacteriological Investigations*
 1. *Fermentation Processes*
    Work in this subject could be made a coordinated effort between the Departments of Bacteriology, Biochemistry (Professor Collip) and Cellulose Chemistry (Professor Hibbert) ...
 2. *Antiseptics*
    Work in this subject could be a coordinated effort between the Departments of Bacteriology, Pharmacology and Chemistry. ... These investigations should be along four main lines:
    a. General disinfectants for latrines, hospital wards etc. ...
    b. Antispetics for the disinfection of wounds, operation sites, surgical instruments ...
    c. Chemotherapeutic agents for specific infections of wounds such as gas-gangrene, streptococcus, staphylococcus, microorganisms of intestinal origin ...
    d. Chemotherapeutic agents for specific disease processes such as pneumonia, enteric, dysentery, etc. ...
 3. *Immunological Processes*
    This work could be undertaken in conjunction with other bacteriology and immunity laboratories.
    a. Active prophylactic immunization in enteric fevers, tetanus, staphylococcus infections, cholera etc. ...
    b. Active therapeutic immunization in wound infections, staphylococcus infections, respiratory tract infections, etc. ...
    c. Passive and therapeutic immunization in gas-gangrene and other wound infections ...
 4. *Investigation of carrier states* for a variety of diseases. This could be undertaken in conjunctions with the Department of Parasitology and other bacteriology laboratories.

II. *Other Services*
 1. The preparation and distribution of media, diagnostic immune sera, diagnostic anti-reagents ...
 2. The intensive training in bacteriology and immunity of officers and orderlies required in army bacteriology laboratories.
 3. Preparation of vaccines, toxins and serums for the prophylactic and therapeutic immunization of troops or civilians on a large scale.

I may be entirely out of order making these suggestions. It may also be that these questions have all been provided for ... In either case, I hope I may be forgiven on the grounds of anxiety to be of service in this war as I believe I was in the war 1914-18.
    Yours sincerely,
        E.G.D. Murray

                          Professor E.G.D. Murray
                          Department of Bacteriology

Professor J.C. Simpson
Associate Dean of the Faculty of Medicine[5]

                                        25 January, 1940

It is a sorry state of affairs when a department, whose accounts are in good order, cannot get a straight answer on a financial stricture put upon it without any consultation. ...

    The needs of the department until the end of this University year (May 31st, 1940), require ... \$338.00 ...

Dr. F. Cyril James,[6]
Principal-McGill University

                                        31st January, 1940

Dear Dr. James,
    I regret to inform you that, as a result of our 'balanced budget', we are experiencing considerable difficulty in providing for the general expenses of the Department of Bacteriology ...
    Yours truly,
        Professor J.C. Simpson,
        Dean of Medicine

---

5 Ibid. J.C. Simpson was dean of medicine from 1940 to 1941.
6 Ibid. F. Cyril James (1903-1973) was born in the United Kingdom and was a graduate of the London School of Economics. Following a career as an economist in the United States, James assumed his duties as principal during the fall of 1939. He held this position until 1962. During the Second World War he assumed a major role in helping to mobilize Canadian university scientists behind the war effort. Donald Avery, "Canadian University Scientists and Military Technology: The Challenge of Total War, 1939-1945," in *Canadian Universities during Two World Wars*, Paul Storr and Lisa Storr, eds., University of Toronto Press (forthcoming); see also Frost, *McGill University for the Advancement of Learning*, vol. 2, 1895-1971.

Dr. F. Cyril James[7]
Principal, McGill University

18 December 1940

Dear Dr. James,

As I informed you in our telephone conversation this morning, Professor E.G.D. Murray has asked me whether it would be possible to get a grant towards his traveling expenses in attending the Annual Meeting of the Society of American Bacteriologists, which is to be held at St. Louis, Missouri, on 27th, 28th and 29th inst.

Although I have informed several members of our staff that the University was not in a position to make grants in aid of traveling expenses ... I believe that Professor Murray's case is exceptional, and that it is in the best interest of the University that he should ... be given a grant of $63.45, which is the amount of his railway fare at the reduced Christmas rate. ...

Professor Murray has been appointed to represent Canadian bacteriology on a Committee which has been appointed to study the formation of an Inter-American Society of Microbiology ... and has been nominated for membership on the Council of the Society for the coming year. I believe it is important, therefore, that he represent the University on this occasion. ...

    J.C. Simpson, Dean

December 27, 1940
St Louis, Missouri

The Committee appointed by President Charles Thom to consider the relationship of the Society of American Bacteriologists to the Inter-American Society of Microbiology begs to recommend:

> that a special Committee of the Society of American Bacteriologists be appointed by the President of the Society to explore further the means which may be chosen to establish closer relationship between the bacteriologists of the American Hemisphere.

It is the belief of the Committee that this relationship may be established by:

a. making contact with various agencies already engaged in creating a closely conducted cultural relationship along various lines with the South American Countries.
b. creating a suitable medium in which abstracts of scientific papers as well as original research contributions in the field of microbiology emanating from

---

7 McGill Archives, B & I, RG 2, C-100, Principal Cyril James Correspondence, file 2724, Faculty of Medicine.

Latin American Countries may be published under the direct supervision of the Society of American Bacteriologists.

Signed:
>   K.P. Meyer, Chairman
>   E.B. Fred
>   E.G.D. Murray[8]

> National Research Council
> Research Institute of Endocrinology
> McGill University

Principal F. Cyril James
McGill University

> 9th December, 1941

Dear Mr. Principal,

At a recent meeting of the Executive of the Associate Committee on Medical Research of the National Research Council of Canada, I was empowered to set up a new sub-Committee under the chairmanship of Professor E.G.D. Murray. The terms of reference of this Committee are of Grade I secrecy, and its work is of the greatest importance. I am sure that you will be glad to know that Professor Murray was chosen for this important post, and that you will approve of it. It will involve a certain amount of his time, but I do not think this is a matter for serious consideration insofar as the University is concerned, particularly in view of the urgency of the work which he will be undertaking.[9]

>   Yours sincerely,
>   J.B. Collip, Chairman
>   Associate Committee on Medical Research

---

8 LAC, E.G.D. Murray Papers, vol.1, file Wartime Correspondence.

9 Ibid. During the Second War the National Research Council established a wide range of committees to investigate various aspects of military medicine. Most of this work was coordinated by J.B. Collip, who succeeded Frederick Banting as chairman of the Associate Committee on Medical Research in January 1941. Terrie Romano, "The Associate Committees on Medical Research and the National Research Council and the Second World War," in *Building Canadian Science: The Role of the National Research Council*, ed. Richard Jarrell and Yves Gingras, *Scientia Canadensis* 15, no.2 (1991): 71-87; Li, *J.B. Collip*, 148-58.

>McGill University, Principal
>and Vice-Chancellor, Cyril James
>December 10th, 1941

Dear Professor Collip,[10]

I am very glad indeed to hear of the creation of the special sub-committee of the National Research Council under the chairmanship of Professor E.G.D. Murray. If the work involves any rearrangement in Professor Murray's teaching schedule, I want to assure you that the University authorities will be only too glad to do anything that lies in their power to facilitate the work that Professor Murray is undertaking. I am equally certain that the Hospital authorities will cooperate to the limit of their ability.

With best personal wishes to yourself, I remain,

Cordially yours,

Cyril James

Memorandum by Prof. E.G.D. Murray to Dean J.R. Fraser and Dr. G.F. Stephens on the Salaries of Technicians in the Department of Bacteriology & Immunity[11]

>July 5th, 1943

The inadequacy and unevenness of salaries and privileges in this department has been a source of trouble for the past twelve years ... without reaching a conclusion. ... This does not inspire the workers with confidence but makes them feel that their services are not valued on merit. ...

Postponement of military training has been obtained for some because it would not be possible to replace them; with this they have a feeling they cannot get improvement because the alternative is to go in the army or stay where you are. They do not object to joining the forces (some have) but I have to ask them to say here because they are essential, only I cannot get them the wage they deserve.

This is not new to us, for even in 'depression time' personnel worth the training was exceedingly difficult to get for the same reasons operating now. The type of work required of some of these people needs a better class of intelligence than many other jobs and certainly better than that of many people we have been forced by circumstances to employ. ...

E.G.D. Murray

---

10 LAC, E.G.D. Murray Papers, vol. 29, file Research M1000 Project.
11 McGill Archives, B & I, RG 2, C-100, Principal Cyril James Correspondence, file 2724, Faculty of Medicine.

Dean J.R. Fraser[12]
Faculty of Medicine
McGill University

September 17, 1943

Dear Mr. Dean,
   ... Should the department suffer any further loss of technical staff it is most unlikely that replacement will be possible and the quality and quantity of work must deteriorate. The standard of work has been very well maintained in the department because of the loyalty and determination of the staff, but they become discouraged and unsettled. ...
      Yours sincerely,
         E.G.D. Murray

Department of National Defence-Army
Ottawa, Canada
12/12/1943

Dear Mr. Principal [James],
   It was kind of you to write inviting me to attend the Convocation & the reception for the Governor General. I was interested to meet him as I felt sure he would remember my father.[13]

   I was glad still to be a member of Senate on this occasion when the MD, CM degrees were conferred on my son. It was nice to be on the platform on that occasion.
      With kind regards,
      Yours sincerely,
         E.G.D. Murray

---

12 Ibid. J.R. Fraser was briefly dean of medicine until J.C. Meakins assumed the post in 1941. See chapter 1, note 26.
13 Ibid. The Earl of Athlone (Alexander Augustus Frederick William Alfred George Cambridge) was Governor General from 1940 to 1946, and was the second member of the royal family to hold that position. Between 1923 and 1930 he had been Governor General of South Africa, where presumably he met George Murray, a prominent physician in Johannesburg.

MURRAY'S INVOLVEMENT WITH GAS GANGRENE RESEARCH

National Research Council
Proceedings of the First Meeting of the Subcommittee on Gas Gangrene Research
Council Chamber, National Research Building, 7 April 1940[14]

... The following members of the Sub-Committee were present:

Lt. Col. Duncan Graham (Chairman), Head of the Department of Medicine, Banting Institute, University of Toronto, Toronto.

Dr. R. Armstrong, Mountain Sanatorium, Hamilton

Dr. G.D.W. Cameron, Food and Drugs Laboratory, Department of Pensions and National Health, Ottawa.

Dr. P.H. Greey, Department of Bacteriology, University of Toronto, Toronto.

Professor E.G.D. Murray, Department of Bacteriology, McGill University, Montreal.

Dr. G. B. Reed, Department of Bacteriology, Queen's University, Kingston. ...

4. *Report by Dr. G.B. Reed:*

DR. REED stated that he had been working along rather similar lines, although he had not been aware of the investigations of Dr. Murray. ... he had also developed a series of tests by means of which it was possible to identify within forty-eight hours all the strains of gas gangrene which have so far been described. ... He had been producing and using antitoxins as one means of rapid identification of the organisms but felt that this method would have little value ... since certain organisms were antagonistic to the production of the toxin by the gas gangrene bacillus. He felt that might have some bearing on the clinical observations that gas gangrene organisms might exist in wounds without producing symptoms. ...

PROFESSOR MURRAY agreed that the mortality rate was higher for intramuscular ... infections.

THE CHAIRMAN pointed out that a comparable situation occurred in the chemotherapeutic treatment of streptococcal infections, since the drug was found to be of little value in cases where a large collection of pus had accumulated. ...

DR. REED inquired whether the members had any information concerning the use of gas gangrene toxoid and antitoxins in the army and their production on a large scale.

DR. GRAHAM stated that the Connaught Laboratories were preparing antisera chiefly for *B. Welchii*.

PROFESSOR MURRAY said that a large supply of antisera had also been prepared in England and a polyvalent serum was available. He thought that this supply would be sufficient for the Canadian as well as the British Army.

---

14 LAC, National Research Council Papers (NRC), RG 77- 88-89/046, vol. 30, file 4-44-4-36.

After some discussion, it was agreed that a recommendation should be sent to the Connaught Laboratories that they take into consideration the method which had been developed by Dr. Reed for the production of a highly potent standard toxin for use in the preparation of antitoxins. ...

PROFESSOR MURRAY pointed out that the number of clinical cases of gas gangrene infection in Canada was relatively small and it was difficult to obtain material for examination. It was agreed that the hospital laboratories and surgeons of the larger hospitals in Toronto, Hamilton, Kingston and Montreal should be asked to co-operate by sending specimens to Professor Murray, who agreed to forward anaerobic tubes to the various hospitals for this purpose. ...

The meeting adjourned at 12:45 p.m.

New York University
College of Medicine
Department of Bacteriology
First Avenue, New York
February 27, 1942

Professor E.G.D. Murray
Department of Bacteriology
McGill University Medical School
Montreal, Canada

Dear Doctor Murray:

I am writing to ask if you would be interested in joining a small informal conference on the organisms associated with *gas* gangrene to be held during the meetings of the Federated Societies in Boston, April 1st to 4th. Our present plan is to hold the meeting either the afternoon of the first or second, (preferably the second), around a table at the Harvard Medical School. ...

Some of the questions which I think might be discussed profitably are:

1. What are the anaerobes (other than Cl. welchii) concerned in gas gangrene and just how important are they? Is anyone working or planning work on their toxins?
2. Is the lecithase produced by Cl. welchii identical with lethal toxin? How important are hemotoxins?
3. Are we using strains of Cl. welchii of maximum toxin-producing capacity? Do strains vary in the amount of toxin which they produce under a given set of conditions?
4. How much concentration is necessary to produce an effective toxoid?

We stand greatly in need of some one to lead the discussion and prevent it from straying too far afield. Doctor Meleney tells me that you are interested in the problem and I wonder if you would consent to act as chairman for the group. Please let me know what you think of the idea and any further suggestions you may have. A similar group of those working on gas gangrene has already formed and met in England to great advantage. It would please us a great deal if you would join us as chairman.

  With best regards,

    Sincerely yours,

      A.M. Pappenheimer, Jr.[15]

      (cc: Dr. Meleney)

                        March 9, 1942

Professor E.G.D. Murray[16]
Department of Bacteriology and Immunity
McGill University
Montreal, Canada

Dear Murray:

 In answer to your letter regarding the meeting in Boston in April, I see no objection to your taking part in the discussion of infections by anaerobic bacteria and associated organisms. Except for the reports coming from here which have been classified as "restricted", "confidential", or "secret", most of the work is of general knowledge, and I know that you will use judgment in leading the discussion. I also spoke to Dr. Clark of Dr. Fred's Committee, and he sees no reason why you should not take part in the meeting.

 I hope that when you are down, you will get to Washington as well, so that we will have a chance to see you. Please remember me to Mrs. Murray.

  Yours sincerely,

    Sanford V. Larkey, M.D.

    Chairman, Subcommittee on Correlation of Information

---

15 Ibid. Alwin M. Pappenheimer, Jr. (1908-1995) obtained his PhD in organic chemistry at Harvard University in 1932, and after a series of research positions became a member of the Biology Department at the New York School of Medicine in 1941. His research on pneumococcal polysaccharides brought him to the attention of Oswald Avery at the Rockefeller Institute of Medicine, while his wartime work on gas gangrene brought him into contact with Joburg and Guilford Reed. In 1957 "Pap" was appointed chair of the Board of Tutors in Biochemical Sciences, Harvard University.

16 Ibid.

March 23, 1942

Doctor Murray:

I enclose the program on the conference on the gas gangrene problem. I do hope that you will find it worth while to come. I don't know just how important the problem is in this war because of the new chemotherapy and the improvements in surgical care of wounds. However, that may be, I feel that we should pool our resources as much as possible these days and I hope that the meeting will serve to get each of us acquainted with what the others are doing.

Sincerely,

Pap (A.M. Pappenheimer, Jr.)
National Research Council[17]

March 25, 1942

Dear Pap,

... I shall be at your meeting on April 2nd. It should be very interesting and I am looking forward to it.

The question is an important one in spite of new drugs and present surgical techniques. There is still need for research. Also there is a certain danger of a false sense of security because of too much reliance on the empirical use of Sulfonamides.

Yours sincerely,

E.G.D. Murray

---

17 Ibid. The United States National Research Council had been established during the First World War as a means of mobilizing US scientists behind the war effort. During the Second World War, it worked closely with the Office of Scientific Research and Development, chaired by Vannevar Bush, scientific adviser to President Franklin D. Roosevelt.

MURRAY'S ROLE IN CANADA'S BIOLOGICAL WARFARE PROGRAM

> Canadian Military Headquarters
> 2 Cockspur Street (Trafalgar Square)
> London, S.W.1
> 9th January, 1940

The President[18]
National Research Council of Canada
Ottawa

Sir,

I have been requested by Major Sir Frederick Banting to forward the enclosed report on the necessity for research in connection with defence against bacterial warfare.

A copy of this report is being sent to the Department of National Defence, one to Major-General A.G.L. McNaughton, G.O.C., 1st Canadian Division, and one to the Director of Pathology at the War Office (UK).

If it develops that there is any kind of research work which could be usefully done in Canada in this connection, you will no doubt hear in due course.

I am, Sir,
    Your obedient servant,
        E. Brown
        Lieut. Col. G.S.
        Canadian Military Headquarters

> June 24th, 1940

Dean C.J. Mackenzie
Acting President
National Research Council
Ottawa, Ont.

My dear Mackenzie,

... During the past weeks, and particularly since I found out that I would be here, I have been thinking more and more of the necessity of experimental work on

---

18 LAC, NRC Papers, RG 77, 87/88, vol. 69, file 36-5-0-1. Dean Chambers Jack Mackenzie (1888-1984) was trained as an engineer at Dalhousie University and Harvard before serving overseas in the First World War. During the interwar years he was dean of engineering at the University of Saskatchewan, as well as being actively involved with the NRC. He became acting president of the NRC in 1939 when General McNaughton resumed his military career, and was the official president from 1944 to 1952. After the war, he supervised Canada's atomic energy program.

bacteriological warfare. I have re-read the memorandum which I sent to the Council from England and which was considered by the British [BW] Committee. I am more than ever convinced that we should be actively engaged in this research. I think the matter should be taken up with the Minister of Defence and with the Director General, Medical Services, and if the former were to request the Council, I am sure the Medical Group would do all in their power to expedite matters. ...

    Yours sincerely,

    F.G. Banting[19]

25 June, 1940

Major Sir Frederick Banting, K.B.E., F.R.S.[20]
Department of Medical Research
Banting Institute
University of Toronto,
Toronto 5, Ont.

Dear Banting,

I enclose copies of a curious correspondence which I know will interest you. Whether it has any value or not I do not know.

I am glad of our conversation yesterday as the subject has worried me and I had not determined how to broach it to the authorities. I am convinced Biological Warfare must be explored fully because it is not beyond possibilities that it will be used by our unscrupulous enemy. Under these circumstances I am absolutely certain that the only safe defensive position against any weapon is afforded by a thorough understanding which can only be gained by complete preparation for the offensive use of that weapon. This view I know coincides with your own.

I have marked this letter confidential with the meaning that it is available for any official use you see fit to make of it.

    Yours sincerely,

    E.G.D. Murray

---

19 Ibid. In 1938 Banting had convinced General McNaughton that biological weapons represented a threat to Canada's national security. He was, however, unable to convince British officials to undertake a major defensive program during his personal mission of December 1939. Nor did the Canadian Department of National Defence make BW defence a priority until 1941. Donald Avery, *The Science of War: Canadian Scientists and Allied Military Technology during the Second World War* (Toronto: University of Toronto Press, 1998), 150-175. See chapter 1, note 106.

20 Ibid.

12th July, 1940

Secretary[21]
National Research Council
Ottawa, Ontario

Re: Bacteriological Warfare

Dear Sir,

The Department of National Defence has had under consideration for some time the possibility of the development by the enemy or by enemy agents, of Bacteriological Warfare.

It is now felt that ... the National Research Council ... might be asked to carry out a preliminary investigation along the following lines:

a. The possibility of the distribution of infectious agents by bomb, shell or aeroplane;
b. The determination of the infectious agents which might be used;
c. The determination of suitable methods for the protection of personnel against infectious agents distributed by such means, including large scale production, etc.;
d. Investigation of any other matters pertaining to or arising from any aspect of Bacteriological Warfare which may be considered important. ...

Yours faithfully,
H. DesRosiers
Acting Deputy Minister (Militia)

NATIONAL RESEARCH COUNCIL

16th November, 1941

Dear Professor Murray,[22]

I wish to thank you very much for attending the special discussion group at the National Research Council on Thursday last, November 13th.

The recommendations which were made as a result of this discussion were brought to the attention of the Executive of the Associate Committee on Medical Research ... and they have authorized the setting up of a Secret Working Committee under the chairmanship of yourself, to facilitate the carrying forward of Research Project M-1000.

I shall appreciate it if you will let me know privately if you will consent to act as Chairman of this special committee.

Yours sincerely,
J.B. Collip (Chairman)

---

21 Ibid.
22 NRC, vol. 69, file 36-5-0-6.

... MEDICAL RESEARCH COUNCIL
Temporary Address: c/o London School of Hygiene
Keppel Street, London, W.C.1
16 December 1941

Professor J.B. Collip, F.R.S.
McGill University, tMontreal, Canada

Dear Collip,
I am writing to you in my capacity of joint secretary of the Bacteriological Warfare Committee of the War Cabinet. I understand that you are chairman of the committee of the National Research Council of Canada which is interested in this subject. We are anxious to exchange information with you, and I am accordingly sending you the following papers: 1) a memorandum on "Defensive Measures against possible Bacteriological Warfare" ... and Committee papers ... reporting the results of experimental work undertaken here. Further papers of this kind will be sent to you as issued and we hope that you will reciprocate by giving us any information which you have. ...

... As regards the experimental work, I may explain that it has been undertaken primarily because a practical study of possible methods seemed necessary for the evaluation of forms of attack which we must be prepared to meet. There is no suggestion that this country would ever take the initiative in using such methods.

With kind regards,
Lansborough Thomson[23]

December 19, 1941

Professor E.G.D. Murray[24]
McGill University

Dear Professor Murray:
The National Research Council in co-operation with scientific and research stations in both the United Kingdom and the United States has worked out a system to facilitate the interchange of information, reports etc. bearing on war research, and we have been asked by the authorities in both of these countries to request that all

---

[23] Ibid. Lansborough Thomson was a member of the British Medical Research Council, which assumed a major role in coordinating biological weapons research in the UK. See Brian Balmer, *Britain and Biological Warfare: Expert Advice and Science Policy* (Basingstoke: Palgrave Macmillan, 2001).
[24] NRC, vol. 69, file 36-5-0-21.

workers on war research in Canada should use the formal channels of communication. ...

 Yours sincerely,
  C.J. Mackenzie
  Acting President (C.J. Mackenzie)

                  Dec.24, 1941

Prof. G.B. Reed, Dept. of Bacteriology
Queen's University, Kingston, Ont.

Dear Reed,[25]

 Your application has been signed ... I have consulted with Collip and it is recommended that you administer grants under M.1000 yourself. It is not desirable that anyone not concerned with M.1000 sees such statements and these applications are not to be signed by the President or Principals of Universities.

 Best wishes to Mrs. Reed and your old self,
  Yours ever,
   E.G.D. Murray

                 29 December 1941

Col. A.A. Magee[26]
Senior Executive Assistant to the Minister of National Defence
Woods Building, Ottawa

Dear Colonel Magee:

 I am handing you herewith a report which I trust is along the lines which you indicated that you like to have for the Minister.

 It has been prepared by Dr. E.G.D. Murray, Chairman of the special subcommittee on Project M.1000. This Committee is quite active and will no doubt have further recommendations to make to your Department from time to time.

  Yours very truly,
   J.B. Collip
   Chairman of the Associate Committee for Medical Research

---

25 LAC, E.G.D. Murray Papers, vol. 29, file M-1000. Guilford Reed's laboratory at Queen's University was the focal point of most of the research activity on the various BW agents.
26 LAC, Papers of the Honourable R.L. Ralston (Minister of Defence 1940-44), vol. 30.

RECOMMENDATIONS TO THE DEPARTMENT OF DEFENCE OF CANADA
BY THE COMMITTEE ON PROJECT M.1000[27]

... No. 1 Obtain at once sufficient Anti-Rinderpest therapeutic serum to be able to treat adequately 5000 (five thousand cattle).

No. 2 Investigate thoroughly all laboratories (public and private), all supply houses and all importers of "biologicals". These are considered capable of producing or providing biological agents either causing disease or injurious to men, domestic animals, crops and industry. ...

NOTE TO RECOMMENDATION NO. 2

The Committee is concerned as to possible sources of infectious material or injurious material which could be distributed by enemy agents. ... The possibilities discussed are:

1. A highly skilled individual could take advantage of the opportunity afforded by a position of trust in a laboratory to prepare material which could be conveyed to enemy agents for distribution according to instructions.
2. Private laboratories, preparing and distributing biological products ... could be a dangerous source of infectious material. Such laboratories could have very opportunity to work behind a mask.
3. Persons in position of trust in firms importing or distributing biologicals products could make the opportunity to distribute to agents. ...

Such persons would use to full advantage the appearance of loyalty and transparent honesty. The higher the positions ... the greater the danger. Directors of laboratories seldom have adequate means to investigate the antecedents of all employees.

Refugees have been taken into many laboratories, and, though the fact that they are at liberty indicates they have been investigated, it would be well for the Director of the laboratory to have full information on them. ...

Certain firms still trading have been part of organizations in enemy countries or occupied territory. ...

PLAGUE

Although it is possible that Plague might not lend itself entirely to Bacterial Warfare it cannot be neglected. The Committee feels strongly that conditions which would favour the spread of Plague should not be allowed to continue. ...

RECOMMENDATION

The Committee therefore recommends that measures be taken immediately to reduce the rat population in cities to a minimum, particularly sea ports. That the Cheopis Index be watched and measures be taken to reduce it. ...

LIAISON WITH INTELLIGENCE

The Committee feels the need of a close relationship with Intelligence. In a recent

27 LAC, E.G.D. Murray Papers, vol. 29, file M-1000 (29).

visit to Washington D.C., the Chairman was impressed by the fact that qualified liaison officers of the U.S. Navy Intelligence and of the U.S. Army Intelligence are members of the U.S. Committee which corresponds to our Committee on Project M.1000 (Biological Warfare). ...

    E.G.D. Murray
    Chairman of the Committee on Project M.1000 Dec.24, 1941

                  January 2nd, 1942. At Washington D.C. (Handwritten Report)[28]
I visited Colonel James S. Simmons, War Department, Office of the Surgeon General, 1818 H. Street on his invitation. ...

The Division includes five Subdivisions as follows: 1) Epidemiology; 2) Veneral Disease Control; 3) Sanitation, Laboratories, Industrial Hygiene; 4) Sanitary Engineering; 5) Medical Intelligence, Tropical Medicine. ...

Colonel Simmons then introduced me to a number of the members of staff before leaving requesting them to inform me fully of their activities & information they had collected. In the Subsection of Epidemiology & Disease prevention I was shown a very complete & very informative set of charts of the incidence of disease in the US Army. ...

  My experience in the 1914-18 war & what little I have seen of the Canadian effort in this in no way compares with what the U.S. Army is doing. ...

                                                                          15 April 42
                            Biological Investigations
        (Memorandum for Colonel R.L. Ralston, Minister of Defence)[29]

1. The recommendations which I made to you and the Hon. Angus MacDonald (Minister of Naval Services) has led to the following proposals.
2. That a committee, consisting of Dr. J. Craigie, research member of the Connaught Laboratories, Dr. G.B. Reed, Professor of Bacteriology at Queen's, Dr. E.G. Murray, Professor of Bacteriology at McGill be appointed. Dr. Collip, Chairman of the N.R.C. Medical Committee will act as Liaison Officer on the medical side. Mr. E.L. Davies will be in charge of field work. All activities will be governed by the Chemical Warfare Directorate.
3. Dr. Reed will be authorized to visit the United Kingdom and get in touch with Dr.

---

28 Ibid., vol. 29, file M-1000 (23).
29 LAC, Records of the Department of National Defence, Chemical & Biological Warfare Activities During the Second World War, microfilmed reel 5018, file HQS 4354-33-1 (hereafter HQS).

Fildes with a view towards co-ordinating the work carried out in Canada with that under way on the other side.

4. Principal Wallace of Queen's University, without being informed of the exact nature of the work, is willing to place the top floor of the Pathological Building at our disposal in order that experimental work may be carried out there. At present this space is being used as an animal house and it will be necessary to construct an animal house adjacent to this building at an estimated cost of $17,000. The accommodation which will be made available is ideal, as considerable equipment and services are available and alternations which will have to made will be covered by the expenditure of $5,000. It is estimated that special equipment at a cost of $10,000 will be required. Furthermore, a recurring expenditure of $15,000 for salaries and $5,000 for travelling expenses would be required. This adds up to a capital expenditure of $32,000 and a recurring expenditure of $20,000.

5. For purposes of record, I may state that this project is to be carried out at the urgent request of the Ministry of Supply, U.K., and I was informed that the United Kingdom is prepared to share the cost of carrying out this work. I recommend, however, that this be considered a Canadian enterprise and the total expenditure carried out by the Canadian Government.

6. The question of dealing with animal diseases (rinderpest) must be considered separate from the above. This is a continental problem and should be undertaken in co-operation with the United States.

    J.V. Young (Major General)
    Master General of Ordnance, Canadian Army

May 20, 1942

Dear Dr. Maass,[30]

... You will remember I asked for your guidance (about March 27, 1942) on whether it was permissible to make copies of British documents on Bacterial Warfare, originating from Porton, and make them available to the W.B.C. Committee and to the

---

30 Ibid. Born in New York City, Otto Maass was educated at McGill University and Harvard, where he received a PhD in 1919. He joined the teaching staff at McGill in 1920 as assistant professor, was subsequently promoted to associate professor in 1921 and was appointed Macdonald Professor of Physical Chemistry in 1923. He chaired the Department of Chemistry at McGill from 1937 to 1955. From 1940 to 1946, he served as assistant to the president of the National Research Council, and from 1941 to 1946 as director of the Directorate of Chemical Warfare and Smoke. His various honours included being named a Fellow of the Royal Society of Canada (1922) and receiving the United States Medal of Freedom (1947). E.A. Flood, "Otto Maass, 1890-1961," *Biographical Memoirs of Members of the Royal Society* 9 (November 1963): 183-204.

Office Chief Chemical Warfare Service of the United States. Your opinion was that this was important to cooperation in the war and it should be done. ...

Yours sincerely,

E.G.D. Murray

June 11, 1942

Bacterial Warfare Project for Defence

The M-1000 Committee is informed by the Director of Chemical Warfare (Otto Maass):

1. that the Canadian Government would view the project (Rinderpest) with a great deal of interest. That it will be adopted if the Government of the United States is as interested in it as the Canadian Government is and if a suitable site can be found.
2. that it is very desirable that a definite proposal be made immediately on the project comprising personnel, equipment and expense, in order that it may be given prompt consideration with a view to prompt action.

The M-1000 Committee therefore recommend:

1. That the project be put under the aegis of Chemical Warfare.
2. That Grosse Isle in the St. Lawrence River, or other suitable site, be made available for research and production of selected projects of Biological Warfare.
3. That a joint committee composed of scientists competent to work in Biological Warfare be appointed by the Government of Canada and the members appointed by the Government of the United States.
4. That the duties of the Commission shall be to give direction to the work and recommend the appointment of the scientific director and staff.
5. That the immediate urgent project for which provision should be made is Rinderpest.

June 11th, 1942

E.G.D. Murray

Chairman of the M-1000 Committee

SECOND REPORT OF THE COMMITTEE M-1000 ON BIOLOGICAL WARFARE

12 June 1942

PREAMBLE [31]

1. This Committee was started by Sir Frederick Banting, with the permission of the Minister of National Defence in December 1940, to consider and investigate the

---

[31] HQS 4354-33-1.

possibilities and dangers of Bacterial Warfare. It was then constituted as

>Sir Frederick Banting (chairman)
>Dr. Philip Greey (Secretary)
>Dr. Donald Fraser
>Dr. J. Craigie
>Professor G.B. Reed
>Professor E.G.D. Murray
>Dr. E.A. Watson

These various members submitted written reports on the subject and met on one occasion (Dec.17, 1940) to discuss particular aspects of the problem and to devise experiments and research projects.

2. On the death of Sir Frederick Banting the project lapsed until November 13th 1941 when Professor J.B. Collip called a meeting of an advisory group of bacteriologists to discuss the advisability of initiating further action. Following the recommendations of this meeting the *Committee M.1000* was appointed on November 16th, 1941 as follows:

>Professor E.G.D. Murray (Chairman), Professor of Bacteriology and Immunity, McGill University
>Dr. Philip Greey, (Secretary) Lecturer in Bacteriology, University of Toronto and Bacteriologist, Toronto General Hospital
>Professor G.B. Reed, Professor of Bacteriology, Queen's University
>Lt. Col. A.C. Rankin, R.C.A.M.C., Medical Services, Department of Defence
>Dr. J.M. Swaine, Director of Science Service, Department of Agriculture
>Dr. E.A. Cameron, Veterinary Directory General, Department of Agriculture
>Dr. Donald Fraser, Professor of Hygiene and Preventive Medicine, University of Toronto
>Dr. E.A. Watson, Assistant Chief Animal Pathologist, Department of Agriculture
>Dr. C.A. Mitchell, Assistant Chief Animal Pathologist, Department of Agriculture
>Dr. G.D.W. Cameron, Chief of the Laboratory of Hygiene, Department of Pensions & National Health
>Dr. A.G. Lochhead, Dominion Agriculture Bateriologist, Department of Agriculture
>Dr. Frederick Smith, Associate Professor of Bacteriology, McGill University
>Dr. James Craigie, Assistant Professor of Epidemiology, University of Toronto

Liaison Members
Professor J.B. Collip, Chairman of the Associate Committee on Medical Research, National Research Council

Professor O. Maass, Department of National Defence
Colonel E.A. Flood, Department of National Defence
Representative Appointed by the W.B.C. Committee of the United States of America
THE STATUS OF THE COMMITTEE ON PROJECT M.1000
The Committee was formed on November 16th 1941 ... and funds to meet its immediate needs were made available through the War Technical & Scientific Development Committee, under Grant No. M.1000. As the very name of Bacteriological Warfare was abhorrent and the subject had to be surrounded by the utmost secrecy, the Committee became known by the number of the above grant. ...

The Committee has a strong sense of the gravity of the recommendations it has to make and fears the consequences of ignorance of the potentialities of Bacteriological Warfare. The Committee also knows that it is not sufficient to be aware of the imminent dangers and recognizes the urgent need for preparedness to meet them.

Therefore, with every respect for authority, the Committee recommends that it be instructed to report in future to an Office of high responsibility and powers. ...

The Committee enjoyed an enormous advantage through a close association of the W.B.C. Committee of the United States of America. ... The extremely complete documented reports of the W.B.C. Committee ... so generously contributed to our file, are of the utmost importance.

Certain documents from the Bacteriological Warfare Committee of Great Britain were also received. ...

PREVIOUS RECOMMENDATIONS

On December 19th, 1941 certain recommendations were made by the Committee at the request of the Department of National Defence and transmitted through Dr. J.B. Collip, December 24th, 1941. These dealt with what the Committee believed to be the most urgent need.

  a. *Rinderpest* The Committee are convinced this virus disease of cattle, sheep, goats, wild animals (deer) and occasionally swine is the most likely and most devastating possibility. The Committee recognized it would take time to provide for the preparation of Rinderpest vaccine and desired to secure some measure of protection immediately. It is therefore recommended that sufficient anti-rinderpest therapeutic serum be obtained to be able to treat 5000 cattle.
  b. *Sabotage* The Committee is concerned as to possible sources of infectious material or injurious material which could be distributed by enemy agents. The Committee therefore recommend the investigation of all laboratories and supply houses capable of producing or providing biological agents injurious to men, domestic animals, crops and industry.
  c. *Plague* The Committee feel strongly that conditions which would favour the

spread of Plague should not be allowed to continue. The Committee was disturbed by the conditions prevailing in Vancouver. ...

d. *Liaison with Intelligence* The Committee felt the need for a close relationship with the Intelligence Service, to secure information and the investigation of the dependability of information.

BASIS OF RECOMMENDATIONS

The review of published literature indicates clearly that treaty or moral obligations will give way to likelihood of military effectiveness. ...

The recognition that Biological Warfare has potentialities beyond incapacitation of military personnel is of the very greatest importance. ...

SUMMARY CONCLUSIONS

1. Bacterial Warfare (or in a wider sense Biological Warfare) is unquestionably within the range of practical use.
2. Active measures are urgently necessary to provide means of protection against the use of Biological Warfare by the enemy. This consists of the production and accumulation of immunizing vaccines and serums against infectious agents usable in Biological Warfare; precautionary elimination of circumstances and conditions which would favour the introduction and spread of infectious agents or pests; organization of a vigilance for the immediate recognition and control of any disease or pest introduced; and the recognition of methods the enemy might use and sources they avail themselves of.
3. The High Policy as to whether Biological Warfare should be initiated, or only used in reprisal, is not for this Committee to decide. But it is important that every preparation for its immediate use be completely ready in case it is required.
4. The institutions of adequate protection (defence) depend absolutely on a complete knowledge and understanding of the nature and the possible methods of using biological agents of disease and pests as offensive weapons.
5. To provide against being caught unprepared it is not enough to be aware of the danger. Personnel, laboratories and facilities should be established immediately for the investigation and the provision of the requirements for both defensive and offensive Biological Warfare. Research is necessary as very much of the desirable knowledge is not available from experience.
6. The range of Biological Warfare ... indicates that the interests of several Federal and Provincial Government Departments are involved and in addition University laboratories, Hospitals, R.C.M. Police and Secret Service are probably concerned.

  Signed

    E.G.D. Murray
    Chairman of the Committee on Project M.1000

Department of Bacteriology & Immunity
3775 University Street
Montreal, Que.
June 17th, 1942

Dr. J.B. Collip[32]
Department of Endocrinology
McGill University
Montreal, Que.

Dear Dr. Collip,

I forward to you herewith a special resolution of the M.1000 Committee on "Biological Warfare Project for Defence", passed unanimously on June 11th, 1942 at Ottawa.

The resolution arises of Section V 2, *Rinderpest* of the Second Report of the M.1000 Committee. It is deemed advisable to make the special recommendation because of the urgency of the matter, because it requires inter-national arrangements and because it can be dealt with separately.

This question has greatly interested the W.B.C. Committee of the United States of America, which passed a special resolution of similar character at its meeting on June 1st, 1942 in Washington D.C.

    Yours sincerely,
        E.G.D. Murray
        Chairman of the M.1000 Committee

Research Institute of Endocrinology
McGill University. Montreal
8 June, 1942

Dr. Otto Maass[33]
Director of Chemical Warfare
Department of National Defence
Room 502-Post Office Building, Ottawa

Dear Dr. Maass,

... For some time it has been the considered opinion of Dean C.J. Mackenzie and myself that all experimental and development work relating to projects that have been considered by the M-1000 Committee should be placed under the aegis of the

---

[32] NRC, vol. 69, file 36-5-0-6.
[33] Ibid.

Division of Chemical Warfare of the Defence Department. ...

The advisability of continuing the M-1000 Committee should shortly be considered by those concerned. It is my opinion that it should be continued.

    Yours sincerely,
        J.B. Collip
        Chairman-Associate Committee on Medical Research

Grosse Île Project[34]

Department of Bacteriology & Immunity
3775 University Street, Montreal, Que.
May 30, 1942

Dear E.B. Fred[35]
Academy of Science,
2101 Constitution Avenue
Washington DC

Dear Dean Fred,

I enclose a copy of a letter I received today from Dr. H. Barton, Deputy Minister of Agriculture, relating to Grosse Isle. It is of course a cautious statement; but a telephone conversation I had with Dr. Barton yesterday clarifies it. Dr. Barton thought it probable that any definite plan to use the island would have to be passed upon by his Department; but any plan entailing adequate precautions, such as have been outlined for the Rinderpest project, would not be opposed by his Department if they were approved by the Government as of National importance, or essential as a war measure.

It seems clear now that the Rinderpest, or other similar project would be most easily and most effectively developed under the aegis of Chemical Warfare. I have informed Dr. O. Maass, Director of Chemical Warfare, of the Rinderpest project and the suggested management of it as fully as is now possible. He in turn has been in

---

34 Grosse Île is a small island located about thirty-five miles downriver on the St. Lawrence from Quebec City. During the nineteenth century it had been used as an immigrant quarantine station, notably during the arrival of the "famine Irish." In 1942 it was placed under the jurisdiction of the Canadian Department of Agriculture, before being transferred to National Defence, which retained control until the late 1950s. Now it is a heritage site.

35 Archives of the National Academies of Sciences, Washington, DC, US Biological Warfare Program (hereafter cited as NAS-BW), E.B. Fred Files. Dean Edwin B. Fred of the University of Wisconsin had been appointed chairman of the United States biological warfare organization (code-named WBC Committee), with the support of the Office of Scientific Research and Development and the US Army's Chemical Warfare Service. See Avery, *Science of War*, 156-57.

touch with the highest Government authorities and he authorizes me to make a definite statement to the W.B.C. Committee.

I am to say that the Canadian Government would view the project with a great deal of interest. That the project will be adopted if the Government of the United States of America is as interested in it as the Canadian Government is. I am to say, further, that it is very desirable that a definite proposal be made immediately on the project comprising personnel, equipment, and expense, in order that it may be given prompt consideration with a view to prompt action.

I hope we can quickly formulate a plan which will be adopted by both Committees [W.B.C. & M.1000] and forwarded by each to their respective authority without delay. As I have informed you, I have called a meeting of the M.1000 Committee for June 9th and will bring the matter forward then.

      Yours sincerely,
      E.G.D. Murray

July 9, 1942

Dr. Otto Maass
National Research Council
Ottawa, Ont.

Dear Dr. Maass,

Under ordinary circumstances Dr. C.A. Mitchell[36] as a Civil Servant in the Department of Agriculture has to get leave from his Chief (Dr. E.A. Watson) for any absence from his ordinary duties. He also has to make a written statement of why the leave of absence is required and where he is going.

This formality might easily be undesirable for the purposes of C.1. [Committee]. It seems desirable that the Department of National Defence write to the Department of Agriculture, asking approval for Dr. Mitchell's occasional temporary absence from usual duty on the business of C-1, without the necessity of a statement on the Department of Agriculture file.

      Yours sincerely,
      E.G.D. Murray

---

36 HQS 4354-33-1. Charles "Chas" Alexander Mitchell served as chief of the Animal Pathology Division, Department of Agriculture, Canada. He was a Fellow of the Royal Society of Canada and an honorary associate of the Royal College of Veterinary Surgeons of England. For his participation during the Second World War on the Joint Canada-US Commission on War Diseases, Mitchell was awarded the US Medal of Freedom. "Dr. Charles A. Mitchell Retires," *Canadian Journal of Comparative Medicine and Veterinary Science* 21, no. 4 (April 1957): 107-108.

Lt. Col. E.A. Flood
Acting D.C.W.
Department of National Defence
Ottawa[37]

July 31, 1942

Dear Colonel Flood,
   This is to advise you that this Department is prepared to concur in the proposal that the project referred to as G.I.R. be undertaken subject to the following conditions: ...

   2. Since the project involves the introduction of (rinderpest) virus to Canada which is not permitted under the Regulations of this Department, this Department is prepared to waive the requirement of permit provided the importation of all virus is confined by admission by diplomatic pouch and consigned to Dr. C.A. Mitchell, of the Animal Diseases Research Institute, who is a member of the Joint Commission and through whom the virus will be made available to the Director of the project. ...

     Yours very truly,
        H. Barton
        Deputy Minister,
        Department of Agriculture

Report of a Visit by Dr. Fildes and Dr. Henderson to U.S.A. and Canada in November-December 1942[38]

... Visit to Canada
On our visit to Canada Dr. Henderson and I were much impressed by the strides taken by Prof. G.B. Reed of Kingston, who had spent the month of May with us at Porton.

   Dr. Otto Maass, Director of the C.W. Service, is responsible for B.W. Under him are Prof. E.G.D. Murray of McGill University as a "political" organizer and technical consultant, and Prof. G.B. Reed in active charge of B.W. explorations and technical work. On the field trial side the C.W. experimental station at Suffield is available. Suitable laboratories are in course of erection, and these will probably be administered by Major J.C. Paterson, R.C.A.M.C. ... Before I left America I heard that negotiations for permitting the use of dangerous materials at Suffield had been concluded successfully. ... Production in a (N) semi-scale plant has been underway for some time at the

---

37 HQS 4354-33-13-3 (C-5019).
38 National Archives of the United Kingdom (NAUK), Cabinet Records, CAB 106/1 (BW files).

University, Kingston, and a very large plant already existing on Grosse Isle in the St. Lawrence river has been taken over. ... (NA-UK, CAB 106/1)

P. Fildes[39]

Department of Bacteriology & Immunity
3775 University Street
Montreal, Que.
December 8th, 1942

Dr. O. Maass[40]
Dept of National Defence, Chemical Warfare
New Post Office Building
Ottawa, Ont.

Dear Dr. Maass,

Dr. G.B. Reed and I reviewed the requirements for the *"G.I.N. Project"* with the purpose of getting it started as quickly as possible and to estimate the approximate cost.

We believe the Project can be started immediately the reconstruction work is completed if the procuring of the apparatus and equipment is speeded up. ... Dr. Reed and I have listed the greater part of the equipment and supplies required and cost them as well as we can. We have also decided the number and categories of staff and their duties but we are not quite prepared to name the individuals. The members of the C.I Commission will meet in Toronto on Dec. 18th of 1942 (at the same time as the Canadian Public Health Assn. Meeting) and soon after that we hope to be able to submit names.

Judging by experience with G.I.R., the assignment of selected personnel is still difficult to secure. ... Bacteriologists are scarce and have to be hand-picked for this

---

39 Sir Paul Gordon Fildes (1882-1971), educated in medicine at Cambridge and the London Hospital Medical College, published a number of papers on the use of the new drug Salvarsan in the treatment of syphilis prior to 1914. During the First World War he explored the problem of gas gangrene and other anaerobic infections, and cerebrospinal fever. In 1920 Fildes co-founded the *British Journal of Experimental Pathology*, highly regarded particularly for its critical review of the etilogy of influenza; he also was highly regarded for his own important research on tetanus. During the 1930s he became closely involved with the Medical Research Council, warning its leadership about the threat of biological warfare. As a result, in 1940 he was asked to assume direction of Britain's BW program, and he spent the Second World War working on biological weapons problems at Porton Down. He left the MRC's bacteriological labs in 1949 for Oxford, where he joined the labs of Sir Howard W. Florey of penicillin fame. G.P. Gladstone et al., "Paul Gorden Fildes, 1882-1971," *Biographical Memoirs of the Royal Society of London* 19 (1973): 317-346.

40 HQS 4354-33-13-3 (C-5019).

work. Dr. Reed and I propose to visit some other laboratories to see what we can find.

The *"G.I.N. Project"* as we have planned it is entirely Canadian. It is designed to provide 'N' for the British Authorities immediately in suitable quantity. If practice comes up to our calculations we hope to produce 300 lbs of 'N' spores per week; this would provide for 1500 of the 30 lb bombs a week. Should the U.S. wish to join us in the "G.I.N. Project" C-I would not raise objections, but there are no indications that they are prepared to do so immediately. ...

II. Personnel for Project G.I.N.

1 Bacteriologist in charge (equivalent to Major)	$ 4,000-5000
1 Bacteriologist (equivalent to Captain)	$ 2,500-3,000
1 Assistant Bacteriologist (equivalent to Lieutenant)	$ 2,000-2,500
1 Assistant Bacteriologist (equivalent to Lieutenant)	$ 2,000-2,500
1 Technician (equivalent to Sergeant)	$ 1,500
1 Technician (equivalent to Corporal)	$ 1,000
6 Technicians (equivalent to Private)	$ 8,000
2 Animal Attendants (equivalent to Private)	
Estimated	$ 25,000

E.G.D. Murray
HQS 4354-1-23

Brig. General R.A. Kelser[41]
Surgeon General's Office,
War Department
Washington DC

January 29th, 1943

Dear General Kelser,

You asked me to send you the present state of the G.I.N. (Anthrax)[42] Project before the meeting on Feb. 6th. I hope this is sufficient detail for your purposes.

Reconstruction & Alterations

Plans have been drawn and the reconstruction of the cut-off part of the building are under way ... the sterilization holding pit into which all drainage goes is finished. ... Some heating equipment, the refrigeration equipment and the incubator room though on order has not yet been delivered. ... More men are being put on the work as the R. (rinderpest) project nears completion and Col. Dickenson thinks the N (anthrax) project construction will be finished by the 15th or 21st of February. ...

---

41 HQS 4354-33-13-3 (C-5019).
42 "N" was the code for anthrax, while "X" designated botulinum toxin. These were the two major agents being explored for offensive military use during the Second World War.

Giant Autoclaves
These have been tested and found to be in good working order. ... Bristol Recording Regulators ($900.00 approx.) have been ordered to control two ranges of temperature, 25 to 40 C and 100 to 130 C. ...

Laboratory Equipment & Supplies
Orders have been placed to the value of $2583.00. ... experiments have been made with the material recommended for trays in the (anthrax) cultures are to be grown ... (approx. 4000 trays).

The Staff
This has been estimated at four officers ... one staff sergeant, one sergeant, 1 corporal and eight privates. The military ranks are given ... for pay purposes.

Method & Procedure
The general method and procedure has been worked out in Dr. Reed's laboratory. The product of the small scale production at Kingston has been tested for virulence in small animals by Dr. Reed and on large animals by Dr. Mitchell. Sample clothing for the protection of the workers is under consideration.

Maintenance & Utilities
These will be shared with G.I.R.1

    With kind regards,
        Yours sincerely,
            E.G.D. Murray

Brig. General R.A. Kelser[43]
Surgeon General's Office
War Department
Washington D.C.

March 23rd, 1943

Dear General Kelser:

    I enclose four copies of a full report on the present status of Project "G.I.N." for purposes of the *Contract* if accepted. This is in accordance with the suggestions you made.

    Should the report not be satisfactory and a sample is required, we can supply 1 lb

---

43 HQS 4354-33 (C-5018). General Kelser was a specialist in veterinary medicine at the University of Pennsylvania prior to the war. His wartime duties at the Surgeon General's Office was to coordinate Canadian-American cooperation in the development of a vaccine against rinderpest and in support for the Grosse Île anthrax production operation. After 1945 he became dean of veterinary medicine at the University of Pennsylvania.

of "N" (100% Sporulation) immediately. If this is desired, please instruct me where and how to send it.

    Yours sincerely,

    E.G.D. Murray

DEPARTMENT OF NATIONAL DEFENCE – ARMY

Prof. E.G.D. Murray
3775 University Street
Montreal, Que.
March 30, 1943

Dr. O. Maass,[44]

I. Please get from Fildes:
 1. N from older "buns" ...
 2. Strains from cattle of accident in Scotland
 Note: We are not satisfied with the virulence of any of the (N) strains which we have for bovines. ... The strain Fildes gave Reed seems to be variant with low sporulation.

II. Fildes requires a certain amount of N. for us for "buns". Please arrange exactly how it is to be sent. ...

III. In view of the failure of "Brasso" (bomb) we are inquiring into the American M.69. We will test these as soon as the field trials at Suffield are underway. ...

IV. Peat and Attractant (Fly) baits on the way as requested.

V. Lockhead's (Queen's scientist) results on six month trials with Salmonella are excellent. Field experiments planning for trying use.

VI. "X" is in good shape. Reed producing 200 gms a week of average potency of 0.02 grams per K.

VII. We require an example of the new sampling apparatus. ...

VIII. Various points from Fildes:
 1. Altitude from which bombs are dropped.
 2. Width of front of dropped bombs and comparison with stationary.
 3. Depth of cloud to be examined.
 4. Exact preparation of acid-bleach solution (decontamination). ...
 5. Suggestions on weapons are desired.

    E.G.D. Murray

---

44 National Archives of the United Kingdom, WO 188/699 (BW files).

Notes on a meeting held on 28.4.43 to discuss Canadian co-operation in the development of Bacteriological Warfare weapons. ...[45]

General

Dr. Fildes outlined arrangements made in Canada and the U.S.A. regarding the production of N at Grosse Isle and in the United States. He emphasized that the design of aircraft bombs and concomitant N storage problems should be proceeded with concurrently in all three countries in order to reach solutions in the shortest possible space of time.

Dr. Maass agreed that this would be done in Canada. All available information and designs had been, and should continue to be, forwarded to Canada for guidance. Grosse Isle would be used mainly for production and storage problems, whilst Suffield would be confined to bomb design and proofing. ...

May 12th, 1943

Dear Dr. Fildes,

... the proposed site of the enclosed square mile ... is some thirty miles to the N.E. of the Station. About four miles to the S.W. there is a lake near which is the proposed site of the camp. A mile or so from the proposed enclosed area there is a spring of fresh water near which it is proposed to make an enclosure for the infected animals. Some thirty to forty miles to the N.E. is the boundary of the main Suffield enclosed area, so there is a safety zone of some thirty to forty miles with the prevailing S.W. wind. ... It remains to be seen whether the material (anthrax) for this summer's trials will be forthcoming. ... The ordering of all the equipment for the trials appears to be well in hand and while we were at Suffield a preliminary dropping trial of peat powder and fly bait was carried out. ...

Lord Stamp[46]
To: The Minister of Defence

June [n.d.] 1943

1. It is becoming increasingly urgent that the work at Grosse Ile, the field experiments resulting from this work to be carried out at Suffield and the extra mural research related to this work be expedited.
2. The work at Grosse Ile is carried out under the direction of the Joint U.S.-Canadian

---

45 Ibid.
46 Ibid. Lord Stamp of Shortlands was a senior member of the Biology Department, Porton, responsible for Britain's BW research and development program. In March 1943 Paul Fildes sent him to North America in order to coordinate British priorities with those of Canada and the United States.

Commission, of which Professor E.G.D. Murray is Joint Chairman with General Kelser, U.S.A. The extra mural work and the field experiments at Suffield are the responsibility of the Canadian half of the Joint U.S.-Canadian Commission, known as the Canadian C.1 Committee, of which Professor Murray is also the Chairman.

3. It is essential, in order to prevent delay, that Professor Murray devote his full time to the co-ordination of these various activities. He is at present Professor of Bacteriology at McGill University with a salary of $5,600 per year, and received at the same time a consulting fee of $2,400 for the use of his name by a pharmaceutical company; the latter entails no work. It is recommended that the C.W. Headquarters be established and a position of Colonel added, whose duties would be those incumbent on the Joint Chairman of the Joint U.S.-Canadian Commission and the Chairman of the C.1 Committee. These involve spending periods of time at Grosse Ile and at Suffield, as well as frequent consultations in the United States.
4. It is recommended that if this position is authorized, Professor Murray be appointed.
5. The rank is recommended as the incumbent will be required to deal with high-ranking officers in the United States and will be in control of establishments on which there are quite a number of Canadian and American officers of Lieut.-Colonel rank. Professor Murray distinguished himself in the last war and held the rank of Captain.

    J.V. Young (Major General)
    Master-General of the Ordnance[47]

July 23rd, 1943

Dear Dr. Maass,[48]

The C.1 Commission considered the needs for an increase in staff at the Experimental Station, Suffield as submitted by Mr. Davies, and wish to recommend strongly that the scheme outlined at this meeting be adopted. We have now reached a stage where weapons development and Field Trials are essential. Without these it will be impossible to make use of the important results of the research which has been done and the production being realized.

In addition to the purely Canadian developments which are likely to prove important, Great Britain is looking for us to carry out trials on her behalf. To do this work a selected and trained staff is essential at Suffield.

    Yours sincerely,
    E.G.D. Murray

---

47 HQS 4354-33-17-1 (C-5018).
48 Ibid.

Dr. O. Maass
Dept. of National Defence, C.W.&S.
New Army Building
Cartier Square,
Ottawa, Ont.

July 23 1943

Dear Dr. Maass,

General Kelser is very anxious to receive the financial statement of the W.D.C.S.[49] ... The contracts between the Government of Canada and the United States relating to the "R" and "N" Projects expire in September 1943. This financial statement therefore is a matter of immediate urgency.

  Yours sincerely,
    E.G.D. Murray

Dr. O. Maass[50]
Director of Chemical
Warfare and Smoke
Ottawa

30 August 1943

Dear Dr. Maass:

1. Work at the N Project at the War Disease Control Station is becoming increasingly urgent.
   a. Production is being delayed
   b. The project is under the Joint American-Canadian Commission. The American Group agree to support the project up to January 1, 1944, to the extent of three quarters of the cost and further support depends upon production. Material is urgently required for crucial Experiments which must be completed before winter. The U.K. has expressed great interest in these experiments. ...
4. This project has been delayed about eight months by the difficulty of obtaining equipment. It is undesirable to have further delays on account of manpower. ...
6. Qualified bacteriological technicians are very difficult to find and it is therefore of the greatest importance that the above men are obtained with as little delay as possible.
     Yours sincerely,
       E.G.D. Murray

---

49 Ibid. The War Disease Control Station was the official name for the anthrax and rinderpest projects carried out at Grosse Île during the Second World War.
50 Ibid.

FEDERAL SECURITY AGENCY
National Academy of Sciences
2101 Constitution Avenue (Washington DC)
September 24, 1943

Dr. E.G.D. Murray
Department of Bacteriology, McGill University
3775 University Street, Montreal, Canada

Dear Doctor Murray:

We here at the Academy and our friends of the Chemical Warfare Service, as well as Col. Rhoads, Chief of the CWS Medical Division, are very anxious to have a round-table conference with you and your associates to cover all aspects of our mutual understanding.....

Can you, Dr. Reed, Dr. Mitchell and Dr. Craigie come to Washington ... and would it be possible to have Dr. Collip also?

Sincerely yours,
George W. Merck[51]
(Director W.R.S.)

October 2, 1943

Dear Dr. Maass,[52]

That when a special trial is being carried out the C1 Committee will appoint a representative to take charge of the experiment and work in cooperation with the Biological Section.

That all field programmes dealing with pathogenic material be submitted to the C1 Committee for approval and endorsement.

Sincerely,
E.G.D. Murray

---

51 NAS-BW, War Research Service Files. In 1943 the US biological weapons program had been greatly expanded with the appointment of George Merck as chairman of the newly formed War Research Services. Merck's status as the chief operating officer of a giant pharmaceutical company greatly facilitated his executive authority in both the United States and Canada.
52 HQS 4354-33-17 (C-5018).

Biology Section,
Experimental Station,
Porton, Salisbury

Dr. O. Maass[53]
D.C.W.
N.D.H.Q, (Army)
Ottawa

October 21st 1943

Dear Doctor Maass,

    As stated in my cable,... a more or less inexplicable case of anthrax occurred 10 miles away from the explosion of the 4-lb bomb. According to our meteorological advisers, we can hardly be responsible for it, but I am not myself so certain. ... I have to assume that we are responsible from the point of view of paying compensation, although I do so 'without prejudice.' Of course, if a single 4-lb bomb is responsible for killing a cow at 10 miles, the idea of letting off fifty 4 lb. bombs at Suffield, with only 20 miles of safe distance, would seem to be associated with considerable risk, and I must say that I would hesitate before staging such a trial. Stamp tells me there is quite a possibility that you come to an arrangement with the U.S.A. for using Camp Wise (Dugway), and I hope that this can be done if you decide that the risk is too great for Suffield.

    Yours Sincerely,
        P. Fildes

November 20th, 1943

Dr. Paul Fildes
Biological Division, Experimental Station
Porton, Salisbury, Wilts

Dear Fildes,

    ... The preparations for the joint Canadian-U.S. trials at Horn Island[54] are moving fast and we hope to get started there at the beginning of January. ...

    On returning to Washington I found Maass and Murray there to negotiate the plans for the combined trials. Murray has very definite ideas as to what has to be

---

53 NAUK, War Office Records (WO), file 188/499. Although anthrax spores did not represent the threat of widespread infection, there was a serious contamination problem. And no-one really knew what level of anthrax exposure was fatal.

54 Ibid. Horn Island, Mississippi, was located in the Gulf of Mexico. Between 1943 and 1945 it was used for joint American and Canadian BW trials with both live agents and simulants. Neither plague nor anthrax was tested because of concerns of public health.

done, unfortunately not altogether based on experience and technical knowledge. He has left McGill and is working full time with Maass at Ottawa. He is not however in uniform! It is quite possible that his new position will make my liaison between Canadian and Americans not quite as easy as it has been. ...

    Lord Stamp

Colonel M.B. Chittick
Chemical Weapons Service, U.S. Army
Washington D.C.

23 December 1943

Dear Colonel Chittick,
    ... Mr. Davies is not worried about carrying out the entire procedure outlined (in Fildes Bio/2461, Sept 27th, 1943) but the Committee were not satisfied that it could be done with reasonable safety at Suffield. The contamination of a very large area would be involved and must be regarded as permanent for soil and climate existing at Suffield and are liable to be spread by the large numbers of gophers and antelope there.[55] Adequate decontamination of the area is not considered possible. It was thought the area now available at Granite Peak might be considered not large enough for an experiment of this scale, though soil, climate and animal life are more suitable than Suffield. Your views on this question would interest us very much.

    Sincerely,
        E.G.D. Murray

Review of Present Position of B.W. in Canada by the C.1 Committee
(15 February 1944)[56]

I. Achievements
   A. "X"
      The process for X has been completely worked out for a year past and is ready for translation to mass production. ... It is evident that effective lethal clouds of X can be developed from aircraft bombs. In C.1 Committee's opinion X takes first place as a B.W. weapon.
   B. *Protection against X*
      An immunising toxoid has been developed. ... Human volunteers have been

---

55 HQS 4354-33-17-1 (C-5019). Fildes circulated a number of special reports to Canadian and American BW scientists. Despite some discussions about carrying out future anthrax trials at Suffield, these never took place because of wide-scale contamination.
56 HQS 4354-33-1 (C-5018).

inoculated; they show extremely slight local reaction. ... At an estimated cost of $10,000 to $15000 it would be possible to institute a plant at Kingston to produce enough fluid toxoid per week for 65,000 man doses. ...

C. "N"

A method has been developed for the production of N spores using the equipment at the W.D.C.S. Grosse Ile, Que. This plant (Grosse Isle) is now approaching 2/3 production ... there is a present stock of over 40 litres of high grade material (better than 4 x10/10 spores). ... Field experiments at Suffield are projected to this purpose too, using W.D.C.S. product ... It is anticipated that the production will shortly be a minimum for 160 Mark I Type F bombs per week. ...

D. *Treatment of and Protection against "N"*

Chemotherapy of N infection is being investigated. ... A Prophylactic vaccine has not proved possible so far. ...

E. "LMP"

A carrier (peat moss) has been developed for the distribution of pathogenic bacteria maintaining their numbers, viability and virulence, and is ready for translation to mass production. ... The infected material is in the form of a fine dust and can be effectively disseminated as a cloud or as a ground contamination. Suitable weapons for this purpose are being developed at Suffield.

F. "LPM"

An infected bait for house flies has been developed and is ready for translation to mass production. This material combines maximum attractiveness to the flies together with prolonged survival and maintaining virulence of pathogenic bacteria; typhoid, paratyphoid, food-poisoning, dysentery. Field Trials at Suffield have proved (that) ... when dropped in a Mark 1 weapon from aircraft, liberated house flies became heavily infected by feeding on it and they transmitted the infection to food supplies at a distance. ...

G. *Low temperature food spoilage*

The spoilage of meat and fish during refrigeration using bacteria which grow at low temperatures has been developed. ... The suggested use is sabotage of meat and fish imported into enemy countries.

H. "R"

Under the U.S. Canadian Joint Commission, at the W.D.C.S., Grosse Ile, a considerable amount of proved protective vaccine against R has been prepared. Stocks are ready for complete protection of 25,000 animals. ... The possibilities of using this virus as an offensive weapon is under investigation.

I. *Projects under investigation and development*

... A Canadian team is at present at Horn Island, Miss., U.S.A. working in collaboration with a U.S. team on field trials to obtain further information for the large scale use of B.W. agents. When the weather permits more extensive field trials are to be done at the Experimental Station Suffield.

II. Appreciation of B.W. Situation in Canada

1. Several potential BW weapons have been investigated and developed to a stage where they are ready to put into production. Notably "X" and X toxoid (for protection), LP, LMP, and food spoilage. The effectiveness of each of these is reasonably well established by experiment.
2. The C.1 Committee is not advised of the policy of the Canadian Government and General Staff relative to Biological Warfare. A definite statement of policy is desired to determine whether the C.1 Committee is to continue and if so what its aim and purpose is to be.
3. Up to now the Canadian achievement has been research and development but a stage has been reached when mass production must be considered. The U.S. is preparing for mass production of "X" and "N" and "X" toxoid while maintaining research on other possible agents.
4. Much of the U.S. development has been based on or improved by the results of Canadian research, developed first by the M 1000 Committee and later by the C.1 Committee. These Canadian Committees have always had the close relations and complete exchange of information with the U.S. Committees, as well as relations with the U.K. interests.
5. Canadian action and authorization of research has largely depended upon policy adopted by either the U.K. or the U.S. But Canadian research developed the X and X toxoid projects, contributed to the N project, and is completely responsible for the LP, LMP and Food Spoilage projects together with others now under investigation.
6. The only production in which Canada has a part is "R" depending on U.S. policy and cooperation and "N" (depending on U.K. policy and U.S. cooperation) and has contributed materially to both. Of these only the R project was completely equipped with all it required. The N project resulted from Canadian initiative designing a way of using old equipment existing at Grosse Ile. Throughout the work has been hampered and delayed by low priority for obtaining personnel and equipment.
7. The field trials on BW at the Experimental Station, Suffield, depend on the research, development and guidance of the C.1 Committee.
8. The C.1 Committee consider the possible use of BW by the enemy as a very serious menace. Although it appreciates the evaluation of "N" as a weapon by the Biological Division, Porton, U.K., it considers X superior and more likely. It is also unwilling to dismiss the possibilities of LP and LPM.
9. The C.1 Committee recognises that U.S. production, once under way, will

outstrip the quantity of any product Canada could afford to provide for. At the same time, the C.1 Committee differs from Detrick and WRS in certain details of the process and product and believes it should be in a position to produce easily at least adequate amounts for Canadian field trials. It also believes it could produce amounts which would be of material value were its use adopted. The C-1 Committee does not view with tranquilly the possibility of its hard won developments of proven efficacy being set aside because they do not conform to the policy or do not appeal to the imagination of authorities of the U.K. and U.S.

10. The C.1 Committee considers production to be the purpose of its research and development. If this is not a desirable end, then it should be considered whether or not the members of the Committee and the expert staff working with them, would not be better employed in other endeavours for which their special knowledge and training equips them.

11. ... In C-1 Committee's opinion X takes first place as a BW weapon ... although it appreciates the evaluation of 'N' as a weapon by the Biological division, Porton, U.K. On the other hand, Murray made reference to the various ways Canadian scientists had contributed to the UK anthrax project ...

    E.G.D. Murray

February 2nd, 1944
Conference at Camp Detrick with Colonel Chittick, Dr. Baldwin, Commander prime, Lt. Col, Woolpert, Major Pile, Lord Stamp, Prof. Reed, Prof. Murray[57]

1. It was stated:

    The U.S. is very disturbed about the situation and the information received. The Secretary of War and the General Staff are anxious that work on X be pushed to the utmost. Instructions have been received to prepare to construct a production plant for B.W. agents and the plans have already been drawn. The Detrick plant (present capacity 18,300 gallons and could be quickly stretched to 19,760 gallons) is regarded as a pilot plant. The power to fill 1,000,000 4 lb bombs a month is aimed at.

2. A school for officers to be trained in B.W. as a division of intelligence is to start the week of Feb.7th. These officers will be used to instruct troops and for active duty in B.W. if need arises. It consists of a three week course on Offensive and Defensive B.W. and 50 officers have been detailed for the first class. ...

    1. X toxoid for active immunisation is in production. We understood about

57 Ibid.

1,000,000 doses were prepared and that 400 gallons more was being processed and would be completed in a month after which a similar amount would be started again. Negotiations are in hand to put commercial firms into production of X toxoid on a large scale. If this succeeds the Detrick plant will not continue this product.

2. *Munitions* 5000 Mark 1 Type F bombs (Porton specification) have been ordered. It will take 8 weeks to get into production. The factory is now being tooled up. This bomb can then be produced in desired quantity.

E.G.D. Murray

Major R.C. Duthie[58]
War Disease Control Station
P.O. Box 128 Upper town
Quebec, P.Q.

August 19, 1944

Dear Major Duthie,

The U.S.-Canadian Contract under which the N project has been operated terminates on August 31st, 1944. The project has been reviewed thoroughly in the light of prevailing circumstances and it has been decided to close it down by September 30, 1944.

Instructions will be sent to you regarding the safe storage of the equipment and the preservation of the plant, which is to be held in reserve in case of need. The safe storage and continued examination of the product (N spores) will be provided for in these instructions. ...

It is understood that the instructions of my letter of August 14th, relating to the sample of N to be taken to Washington, are in hand and will be carried out. I hope soon to receive notice that you are ready to proceed.

The contribution you and your Officers and men have made in carrying the N project to its conclusions is appreciated. It has been difficult and dangerous work, beset with many trying circumstances and handicapped by the use of an old plant designed for another purpose. It is recognized that the difficulties were largely overcome by painstaking perseverance and the part played by all of you in this accomplishment is acknowledged.

Yours sincerely,
Professor E.G.D. Murray
For Director, Chemical Warfare & Smoke

---

58 Ibid.

From: Lieutenant General J.C. Murchie (Chief of the General Staff)[59]
To: Colonel R.L. Ralston, Minister of Defence

23 August 1944

For security and policy reasons BW research and development has been confined, up to very recently, to a small group of experts and appropriate senior officers of the Service. It is only in the past few weeks that it has been considered advisable that definite material, from these experts, should be made available to the Service Staffs on which the latter could attempt a proper strategic appreciation. ... It is obviously not a matter which can any longer be confined to the technical officers concerned, regardless of the very valuable work which they have done, and are still doing.

September 2, 1944

Major A.J. Skey[60]
Canadian Military Headquarters, S.D. 11
2 Cockspur Street
London, W.C., England

Dear Skey,

... I have not yet received any precise account of the new U.K. set up to deal with B.W. ... I gather that the W.O. Policy committee has been formed ... It is stated that Washington is to set up a Policy Committee similar to that of the W.O.[61]

The Canadian Staff at N.D. H.Q. have lately shown a very real interest in the subject. ... This has resulted in some consultations which will develop naturally and effectively. ...[62]

The G.I.N. Project (MS 23) is being closed down, but the equipment and plant will be maintained in case of need. We have stored over 380 litres of "N mud" which is enough to charge about 1200 Type F bombs.[63] The plant would not do more than

---

59 HQS 4354-33-1 (C-5018).
60 Ibid. A.J. Skey was the representative of the Canadian C-1 Committee in London.
61 The British committee, established by the War Office, was called the British Inter-Services Sub-Committee for Biological Warfare, while the US organization was known as the United States Biological Warfare Committee. Both were created during the summer of 1944.
62 In March 1945 the Canadian Chiefs of Staff established the Inter-Departmental BW Policy Committee, to coordinate defensive responses against a possible Japanese BW attack.
63 Ibid. During the spring of 1944 the British and American governments agreed to produce over a million of the small Type F bombs (110 per cluster bomb) for retaliatory use against Nazi Germany. Significantly, this project continued after Germany's surrender in May 1945, raising the possibility that this weapon might have been used against Imperial Japan if the war had continued into 1946. Of course, the two atomic bomb attacks in August 1945 made BW weapons unnecessary.

produce for experimental purposes and as the method used is not that projected by the U.S. there is no reason to continue.

Trials with X dry soluble powder are now in progress at Suffield. We look forward to some very interesting results and I shall inform you of them when they come through. We are preparing to expand the Laboratories at Kingston to produce X powder on scale sufficient for extensive trials. ...

Yours sincerely,
Professor E.G.D. Murray
For Directorate Chemical Warfare & Smoke

8 Oct 1944

Dear Prof Murray,

This is to put you in the picture as much as possible with what has transpired in the field to date. ...

There shall be established a BW Intelligence Committee, having British, US and Canadian representation, which shall report to the Joint Intelligence Sub-Committee on all matters of BW intelligence. ... A list of targets was considered by the Committee with a view to indicating to the Combined Intelligence Objectives Sub-Committee (COIS) in order of importance the targets [German BW scientists] for investigation. ... It was also stated that the primary and most important object of the field team investigating targets should be to determine whether in fact there have been any [German] BW preparations ...[64]

I have been up to see Dr. Fildes and have asked him what he thought of your review of BW agents developed in Canada and Col. Goford's comments.[65] ... Generally, Fildes maintains that the review tends to over impress the non-expert with the dangers of BW which may be real but have not been proved. The classification of the BW into types of agents, methods and effects tends to give the impression that these methods can be used with the results intended. ... I suggested to Fildes that the classification was obviously meant as a general indication of possibilities. ...

Yours sincerely,
(Major) A.J. Skey

---

[64] Despite serious concerns that Germany would use botulinum toxin weapons against allied forces at Normandy, or in the subsequent military push into Germany, no BW incidents occurred. Allied intelligence later concluded that Germany had not developed an operational BW capability, unlike its formidable (and secret) achievements in developing nerve gas munitions and delivery systems. Avery, *Science of War*, 144-50, 165-75

[65] Colonel Wally Goford had submitted a report to the Chief of Staff in July 1944, reviewing Canada's BW policies and calling for greater involvement by the Army's hierarchy.

19 October 1944

Dear Skey,[66]

... Up to now more N has been produced by the Grosse Ile plant than in the U.S. and U.K. ... The agent produced is satisfactory when tested on sheep and on guinea-pigs but it is not effective with cattle. Canadian opinion is not convinced that N would be as effective on many as some suppose. Man is one of the more resistant animals to that disease, possibly more resistant than cattle. Quite considerable exposure at both the W.D.C.S. ... and the plant at Camp Detrick (see Monthly Progress Reports) has resulted in a very few insignificant cases of human disease. Certainly, quite extensive skin contamination does not present a danger of desirable significance in man and whether respiratory infection would have the devastating result supposed is an unanswerable question. Our respiratory tract infection experiments [C.A. Mitchell's work] with cattle have not been successful.

Yours sincerely,
E.G.D. Murray

Protecting North America: Japanese BW Balloons?

Ottawa, January 20th, 1945

Memorandum for the Prime Minister[67]

With regard to some discussions this week in War Committee regarding the discovery of Japanese balloon-bombs over Western Canada, General McNaughton has telephoned to say that he has just received a report from Regina. ...

"Begins, Rendered safe Japanese high explosive anti-personnel bomb ... Ends."

General McNaughton[68] observed that he thought that the Japanese object in releasing these balloons has more to do with a study of meteorological conditions ... [that] they hoped to get reports from our press as to the bursting of the bombs on landing in our territory. For this reason he is very concerned that no publicity whatsoever be given to these incidents.

There is also the possibility that they might have something to do with an attempt at bacteriological warfare.

Our technical experts were busily investigating these matters and there was a complete interchange of information with U.S. authorities on the technical level.

---

66 Ibid. Murray's critique of the lethality of anthrax weapons contrasted sharply with the views of Paul Fildes.
67 LAC, Records of the Department of National Defence (RG 25), vol. 5739, file Japanese Biological Warfare.
68 General McNaughton had become Minister of Defence in September 1944, following the resignation of Colonel Ralston over the issue of overseas conscription.

Minutes of a Meeting Held in the MGO Conference Room,
New Army Building, Cartier Square, at 10:30 Hours, 14 Feb 45.

Present:

Dr. O. Maass, DCW&S, Chairman[69]
Major-General C.U. Fenwick, DGHS
Brigadier W.P. Warner, DGMS
Colonel W.H. Brown, DGMS
Dr. D.C. Rose, Operational Research
Lt. Col. J.G. Collinson, DMO&P
Lt. Col. E.A. Flood, CW Labs
Lt. Col. A. Leduc, DSD (W)
Group Captain R.C. Ripley. RCAG (D.Ops.)
Prof. E.G.D. Murray, DCW&S

Secretary: Major D.J. Dewar, DCW&S

The Chairman opened the meeting by explaining that it had been called at the request of the Minister (Defence) to make recommendations regarding counter-measures against the threat of BW from the Japanese balloons; which now appear to be coming over the Continent in ever increasing numbers. He believed ... that the present organization for reporting such balloon incidents, for studying any material collected and maintaining close liaison with US authorities was working well but he pointed out that there was as yet no organization to deal with any epidemics which might arise from the use of BW. Formation of such an organization was of course difficult since no public disclosures could be made for fear of alarming the populace unduly and of giving the enemy much desired information regarding the landing places of these balloons. ...

As far as the information regarding the outbreak of animal diseases is concerned, Professor Murray believed that the information would be obtained through Dr. Mitchell from the Department of Agriculture, which already had an organization set up for investigation and control of such diseases. ...

Questioned by Lt. Col. Leduc regarding the availability of vaccines, Professor Murray advised that Rinderpest and Foot and Mouth Disease were the two animal diseases most likely to be introduced. Approximately 300,000 doses of R vaccines were available at the WDCS and production could be stepped up if desired. No vaccines for Foot and Mouth were available since Canadian and US authorities had refused the Committee permission to import the virus into the country. ...

As for human diseases as a result of BW activity, Professor Murray advised that vaccine was only available for "X" and this agent was unlikely to be used in the Japanese balloons. ...

---

69 HQS 4354-33-1 (C-5018). DC's in the quotation refers to District Commanders.

... the Chairman recommended that:
1. GOC-in-C Pacific Com and the DC's of Military Districts 10, 12 and 13 be informed of the situation regarding the possibility of BW from Japanese balloons and made responsible for counter-measures in their Districts. ...

  O. Maass (Chairman)
  Director of Chemical Warfare and Smoke
  4354-33-1

                    Ottawa 19 Feb 45

The Minister[70]

1. Following the meeting in your office on 12 Feb 45 re Japanese Balloons a conference of interested officers was held under the chairmanship of Dr. Maass on 14 Feb. ...
3. ... Dr. Maass visited Washington for a preliminary discussion with U.S. officers interested in this subject. The U.S. War Dept. is planning to send specialist officers to Western Defence Command to brief Commanders, Medical Officers and Laboratory heads regarding the BW aspects of Japanese balloons. Provision is also being made for the attached of BW specialists to BD squads.
4. These measures are very similar to those proposed for Canada. ... In view of this and the need for close cooperation between the United States and Canada, it has been decided that an international conference will be held in Washington on the morning Feb. 23 for the purpose of discussing BW aspects of Japanese balloons. Arrangements have been made for the following Canadian representatives to attend-Drs Maass, Murray, Mitchell and Rose. ... The Washington representative of the RCMP will also attend.

  J.C. Murchie (Lieut. General)
  Chief of the General Staff

                       26 Feb 45

Possible Introduction of Disease Producing Agents by Japanese Balloons[71]

1. A meeting between United States and Canadian representatives ... was held in the Pentagon Building on Friday, 23 Feb.45. ...
4. ... Dr. Murray was asked to outline the plan proposed by the Sub-Committee of the Canadian Chemical Warfare inter-Service Board for defence against possible outbreak of disease. The United States officers present endorsed this plan as being

---

70 Ibid. The minister was General McNaughton.
71 HQS 4354-33-1 (C-5018).

parallel to their own and General Styer expressed the wish that if and when the high level Canadian Committee be formed that a joint meeting with a similarly constituted United States committee be held.
5. Following the request by DCW&S that Canadian officers be permitted to attend the conference of the BW (US) group with the western Service Commands, a cordial invitation was given. ...

    J.V. Young (Major General)
    Master-General of the Ordnance

Dr. C.A. Mitchell[72]
Animal Diseases Research Institute
Hull, P.Q.

27 February, 1945

Dear Dr. Mitchell,

In view of the definite possibility of the first indication of enemy intention being an outbreak, the certainty of recognition becomes of paramount importance. This in turn may largely depend on adequate laboratory facilities and suitable staff.

It is assumed (is it right to do so?) that all laboratories for the investigation of animal diseases across Canada come under your direct supervision. ...

An interim Service Organization is being set up but it cannot cope with what might be necessary to deal with civilian involvement and the most likely real danger of animal diseases. It is therefore most desirable to review the present situation in the light of the possible threat and to indicate augmentation and improvement for investigation and control of B.W. Would you please give me a considered report on this rather urgent matter.

    Yours sincerely,
        Professor E.G.D. Murray

27 Feb.45

---

72 Ibid.

San Francisco Conference on BW
Aspects of Japanese Balloons[73]

... Dr. Murray, Chairman of the C1 Committee, has the whole background of BW aspects and it is recommended that he be sent to the San Francisco conference. It will be seen from the attached list of American representatives at this conference that a large number of US civilian experts have been asked to attend. The appreciation of what will be discussed at this conference will be of considerable value to Dr. Murray.

If favourable consideration is given ... it is suggested that travel by air be authorized.

    O. Maass
    Director of Chemical Warfare and Smoke

Ottawa, 6 Mar 45

SUMMARY TO:
WAR COMMITTEE OF THE CABINET SPECIAL ORGANIZATION FOR DEALING WITH ANY ASPECTS OF JAPANESE BALLOON ATTACK

1. *Purpose*

   To organize a special interdepartmental committee for dealing with any disease outbreaks that might result if Japanese resorted to Biological Warfare in connection with Japanese balloon attacks against North America. ...

   This sub-committee will report to the Minister of National Defence and will comprise representatives of the Departments of National Defence, Agriculture, Public Health and Justice.

2. *Considerations*

   a. On 15 Nov 1944 a Japanese balloon was discovered in Montana. Since then the following incidents have occurred up to 28 Feb 45 ...

       Balloon findings (i.e. on ground)
       Canada 7   U.S. 35

   In addition a number of balloons have been sighted in flight and which were not recovered.

   b. In Canada these findings have extended as far east as Regina, Saskatchewan, and north to Fort Simpson, N.W.T. However, with favourable winds, the range of these balloons could include Manitoba. ...

   These paper balloons with the automatic ballast dropping devices are capable of traversing the distance from the mainland of Japan to North America and it is estimated that at an altitude of 25,0000′ they would move eastward at an

---

73 Ibid.

average speed of around 30 to 60 miles per hour with the winds that prevail during the winter months. ...

Ottawa 7 Mar 45

Dr. E.G.D. Murray
Major J.L. Blaisdell[74]

1. Attached is a copy of the itinerary for your briefing session to Pacific Command and MDs 10, 12 and 13. ...
2. In briefing the DOSC concerned regarding the BW aspects of Japanese balloons I suggest that you make the necessary arrangements for the District Hygiene Officer to accompany the Bomb Disposal Officer to the site of any balloon finding to assist in the investigation, collection and forwarding of samples. ...
    Colonel
    DND &P

April 28th, 1945

Dr. E.G.D. Murray[75]
The D.C.W and S. National Defence Headquarters
Ottawa, Ontario

Dear Dr. Murray:
... (B) Should the matter become factual, there ought to be a plan of control providing for (1) scientific direction and (2) enforcement along the following lines:
1. Scientific Direction: A principal veterinary officer, assistants and clerical staff: Field Laboratory: Decontamination stations, stores and equipment: Field accommodation/transport and supply. ...
2. Enforcement: Provincial police control: Stock inspectors and range guards: Signal units: Provision for barbwire fencing and its erection: Field accommodation, transportation and supply.
        Yours very truly,
            T.W.S. Parsons
            Commissioner of (BC) Provincial Police

---

74 HQS 4354-33-1-1 (C-5018). After the war, J.L. Blaisdell was associated with the UWO medical faculty.
75 LAC, Records of the Department of National Defence (RG 25), vol. 5739, file Japanese Biological Warfare.

April 14, 1945

Dear Dr. Maass,[76]

I have sent you a copy of the Minutes of the Meeting of the C.1 Committee of 5 April 1945. The following recommendations are drawn to your attention:
1. Minute No.7
   a. That the work at Belleville Parasite Laboratory be closed down on June 30, 1945. ...
2. Minute No. 13
   The Committee approves of the grant of $20,000.00 to Dr. Reed to continue the work at Kingston for the year 1945-46. The "N" [anthrax] project also to continue separately under Dr. Reed's charge.

        Yours sincerely,
        E.G.D. Murray
        Chairman of C.1 Committee

23 May 1945

Dear Dr. Maass:[77]

Confirming the information given you at the conference in Ottawa on the afternoon of 18 May 1945, and for the record, Honourable Henry L. Stimson, Secretary of War, under date of 9 May 1945, informed the undersigned that the War Department concurs in the proposal of the Joint United States-Canadian Commission to supply avianized "R" vaccines produced under its supervision, to the British authorities for the proposed field trials in East Africa. ...

Professor Murray and Reed discussed informally with several British officials charged with BW responsibilities the possibilities of a field test and it was understood that they were interested in the same and indicated that it could probably be conducted in East Africa. ...

It was also pointed out to the Secretary of War that if the test is undertaken the responsibility of maintaining required security would devolve on United Kingdom authorities who would take the necessary safeguard measures. ...

        Sincerely yours,
        R.A. Kelser
        Brigadier General, U.S. Army

---

76 HQS 4354-33-1-1 (C-5018).
77 HQS 4354-33-16 (C-5018).

June 5, 1945

The Chief Superintendent[78]
Experimental Station
Suffield, Alta.

Ventilated Suits
1. A special suit has been devised at Camp Detrick on the principle of positive air pressure within it to protect the wearer from contamination. ...
2. This suit would protect the wearer in the chamber but there are difficulties of decontamination on leaving the chamber and undressing. Unless decontamination is very effective the wearer would be seriously exposed while taking off the clothing.
3. It is the opinion of those with whom I discussed this question at Camp Detrick last week, that your Bursting Chamber is not safe for the use of "hot" agents. The fact that it has to be entered for the collection of samplers and exposed animals introduces dangers and difficulties. These cannot yet be overcome by any decontamination process available.
4. Lord Stamp is aware of these discussions. Take it up with him when he goes to Suffield.
5. If you wish to get one of the suits you should arrange it with Detrick through your procurement.
   E.G.D. Murray

July 9, 1945

Dr. Otto Maass[79]
Ottawa

There will be a meeting of the Joint U.S.-Canadian Commission at the War Diseases Control Station on 19th and 29th July, 1945. ... Reservations have been made at the Chateau Frontenac. ...
   E.G.D. Murray

---

78 Ibid. Murray was rigorous in his insistence that laboratory workers be afforded proper protection when working with dangerous pathogens. He was, therefore, most concerned when several of the GIN personnel developed skin anthrax in 1944.
79 Ibid. This was one of the last meetings of the Joint Committee, which was disbanded in the fall of 1945.

July 25, 1945

Dear Dr. Maass,[80]

The C.I Committee have asked me to suggest that you write to Dr. R.C. Wallace, and if possible see him, to facilitate agreement on suitable arrangements between Queen's University and the Department of National Defence for the future services of Dr. Reed. ... The Committee, too, wish to acknowledge the most helpful cooperation of the University during the past three years and the most valuable services of Professor G.B. Reed.

    Yours sincerely,
        E.G.D. Murray
        Chairman C 1 Committee

July 25, 1945

Dear Dr. Maass,[81]

At the meeting of the C.1 Committee 21st July, 1945 in Quebec (City) I announced that I would resign the Chairmanship ... and I have accordingly sent you a letter of my resignation (Enclosed).

The Committee wish to recommend to you the appointment of Professor G.B. Reed as Chairman of the C.1 Committee dating 1st September 1945.

This appointment is tied up with making satisfactory arrangements with Queen's University and with the Recommendation for Canadian Organization of B.W. made by the C.1 Committee and forwarded to you. ...

    Yours sincerely,
        E.G.D. Murray

July 25, 1945

Dear Dr. Maass,

It has become important for me to return to my duties at McGill University and in the Hospitals, and I believe the Department of National Defence has agreed to release me on 1st September 1945. Recent visits to my Department at McGill lead me to believe that my duties there will require my full attention, so it is necessary for me to offer you my resignation as Chairman of the C.1 Committee and request that it be accepted as effective 1st September, 1945.

My association with this work from its inception makes me relinquish the position

---

80 Ibid.
81 Ibid.

with some regrets. However, if it is desired I am prepared to continue as a Member of the Committee.

I have been treated with the greatest consideration during the time I have been a member of the Staff of your Directorate and I only hope my services have been of some use. I believe the affairs I have had to handle will be found to be in good order.

I believe the Canadian Joint Chairmanship of the Joint U.S.-Canadian Commission is so closely related to D.C.W.&S and the Chairmanship of the C.1 Committee that both Offices should be held by the same person. ...

With kind regards and my grateful thanks for your many personal kindnesses.
    Yours sincerely,
    E.G.D. Murray

EVERITT MURRAY AND THE SITUATION AT MCGILL, 1944-46

National Collection of Type Cultures[82]
Lister Institute,
Elstree, Herts. [United Kingdom]

Prof. E.G.D. Murray, M.D.[83]
Dept. of Bacteriology
McGill University
Toronto, Canada

17th November 1944

Dear Murray,

It seems many a long day now since I saw you in Washington, D.C. and I hope all goes well with you. ...

My elder boy, who is a Major in the I.M.S., writes from India saying that after the war is over he would like to do some postgraduate work in Canada. He was educated at Harrow (Classical Scholar and Gold Medalist in Math.), Trinity College ... and St. Thomas's Hospital. He was House Physician at Thomas's in 1939 and had a good career in the Medical School there, gaining prizes in Clinical Medicine and a Gold Medal in Pathology. Shortly after getting his M.B. at Cambridge he obtained the M.R.C.P. London at the early age of twenty-six. He is now 30 years old and has been in Iraq and India since 1940. ...

I should be most grateful for any advice you could give me on his behalf. His special interests are clinical Medicine and Pathology and he has a strong inclination

---

[82] McGill Archives, B & I, RG 2, C-100, Principal Cyril James Correspondence, file 2724, Faculty of Medicine.
[83] Ibid.

towards research and teaching. I think I may say that he is extremely able and is a very hard worker.

With best wishes for Christmas and the New Year,
    Yours sincerely,
        R. St. John-Brooks
        Curator

Directorate of Chemical Warfare and Smoke,
National Research Council Building,
November 30 [1944]

Dear St. John-Brooks,

I am glad of your letter of 17th November to have news of you. First to answer your enquiry.

With such qualifications your son would be gladly accepted anywhere for postgraduate work. My wonderment is why should he want it. It is, of course, a way to get a footing in Canada but to practice he would have to take the Dominion Council licensing examinations, which would not present any difficulties to him. Various people with British or American training get teaching and research appointments in Canadian Universities and there is always a varying opportunity for people of ability and experience. However there is a dependence on vacancies and on developments everywhere and some element of chance.

With a special interest in clinical medicine and pathology it would seem necessary to go to the larger centres, which are for that reason crowded. Smaller centres may offer more opportunity in certain ways but less scope. The only two large centres are McGill University (Montreal) and Toronto University. A choice between them would naturally depend on exactly what type of work he wishes to do.

Very often it is wise to chose to work under a particular man eminent in his line. That is the usual determined choice or else an institution oriented in a special type of work. Not knowing any special choice your son may have I cannot now express an opinion. In general, Professor of Medicine at McGill is J.C. Meakins, a very widely known and distinguished physician and a great stimulator of research; and at Toronto Duncan A. Graham, very widely known and respected in Canada. Both of these men are nearing the retirement age and it is anybody's guess who will succeed them – perhaps within or shortly after the period of the war. The Professor of Pathology at Toronto is William Boyd, a prodigious writer of textbooks, an inspiring teacher and maker of museums, but not a researcher; at McGill is Lyman G. Duff,[84] a young man not yet widely known but will be for his research and teaching ability. ...

This is a rather vague review but if Major St. John Brooks cares to write questioning me more specifically, I will make suitable enquiries to inform him and put him in touch with the right people. ...

     With all good wishes,
          Yours sincerely,
               Professor E.G.D. Murray.

                                            April 10, 1945

Dr. F. Cyril James[85]
Principal
McGill University

Dear Mr. Principal,

I think the time has arrived for you to make formal application to General Hand, Master-General of Ordnance, for the termination of Professor Murray's leave of absence.

Looking forward to next year, I think it is most important to have as many heads of department who are on leave of absence back on duty as possible.

I am led to believe that the work in which Professor Murray has been engaged is now well in hand and his services might be dispensed with by the M.G.O. in the near future.

     I remain,
          Yours faithfully,
               J.C. Meakins
               Dean, Faculty of Medicine

---

84 Ibid. George Lyman Duff was born in Hamilton in 1904, graduated with honours in biology at the University of Toronto, received an MD degree with the David Dunlap Memorial Prize in 1929, and received a PhD in pathology with the Starr Gold Medal in 1932 from the same university. He was on the teaching staff of the Johns Hopkins University from 1931 to 1935; assistant professor of pathology at the University of Toronto from 1935 to 1939; and Strathcona Professor of Pathology and director of the Pathological Institute, McGill University, after 1939. His research interests focused on two main subjects, arteriosclerosis and diabetes. In 1949 he was appointed dean of the McGill Faculty of Medicine and elected as a member of the Royal Society of Canada. In 1956, he was awarded the Flavelle Medal.

85 Ibid.

General J.V. Young[86]
Master-General of Ordnance
Department of National Defence
Ottawa, Ont.

April 18, 1945

Dear General Young,

    Several Major years have now passed since Professor E.G.D. Murray, Chairman of the Department of Bacteriology of this University, was granted leave of absence for work of national importance in your department. ...

    It would be of great assistance to the University if he could return to his academic activities. It is appreciated that you might desire his advice in a consulting capacity, even although he could be released from the active direction of the work he has been supervising for the past few years. ...

    Cordially yours,
        F. Cyril James

May 2, 45

Dear Dr. James,[87]

    ... after further discussion and examination of our problems we feel that we shall be in a position to release Professor Murray by 1 Sept 45. We should, however, greatly appreciate retaining Professor Murray's services in the capacity of consultant and would request that he be allowed to remain on two committees, for one of which he acts as chairman.

    Professor Murray's services have been invaluable to the Department and we are very reluctant to release him, but we understand the importance of his civilian appointment.

    On behalf of the Army I wish to extend my appreciation of the services rendered by many distinguished members of the faculty of McGill University to the war effort, and I hope that in the days to come the Army will continue to benefit from the close co-operation of your university.

    Yours sincerely,
        H. DesRosiers
        Deputy Minister
        Department of National Defence (Army)

86 Ibid.
87 Ibid.

Dr. F. Cyril James[88]
Principal and Vice-Chancellor

June 28th, 1945

Dear Mr. Principal,

Thank you very much for your note of June 25th conveying my reappointment as Chairman of the Department of Bacteriology. ...

The members of the Department greatly appreciate your approval of the efforts they have severally made during this somewhat difficult interim period.

 Yours sincerely,
  Dr. Fred Smith

June 28, 1945

Dear Mr. Principal,[89]

I am naturally glad the Board of Governors have restored my salary to $6,000, which was what I was promised, and received for a few months, when I was invited to come to McGill in 1930. As an appreciation of my contribution to the University and Hospitals I value your personal tribute more. I suppose the formality of the Chairmanship is in order although you do not mention it.

I hope I may expect the work of my Department as a whole to receive recognition in the near future by the provision of conditions it urgently needs. The subject deserves a proper opportunity and the Department has amply demonstrated its ability to take its proper place.

I greatly look forward to my return to McGill in September of this year. It will be a real home-coming, for my appreciation has not been lessened in any way by my absence and different responsibilities.

 Yours sincerely,
  Professor E.G.D. Murray

12 July 1945

Dear Mr. Principal,[90]

I have been expecting a letter from you in view of your recent announcement in the Montreal newspapers of 30 June 1945 concerning the Chairmanship of the Department of Bacteriology. ...

I was invited to come to McGill University from Cambridge in 1930 to be Head

---

88 Ibid.
89 Ibid.
90 Ibid.

of the department at a salary of $6000.00 and to Bacteriologist in Chief of the Royal Victoria Hospital. When I arrived there was no real department, there was a difficulty over the Hospital position and shortly afterwards my salary was cut by 12 per cent. None of this was very encouraging.

During my absence on war work the University made quite proper provision for the running of the Department but on returning I quite naturally expect to have my rightful position on the Staff. I feel sure the correction was overlooked in drawing up a long list of names and I hope the correction will be made before I return.

With kind regards,
Yours sincerely,
Professor E.G.D. Murray

July 17th, 1945

Dear Dr. Murray,

... the announcement that appeared in the Montreal papers on June 30th regarding the chairmen of departments was not in error. As you may remember, departmental chairman and other administrators throughout the University ware appointed for periods of twelve months beginning on June 1st of each year, and Professor Fred Smith was appointed chairman of the Department of Bacteriology and Immunity from June 1st until such time as you should return to the University. ...

Cordially yours,
F. Cyril James
Principal

July 23, 1945

Dear Mr. Principal,[91]

... I have to accept the explanation though I do not find it satisfactory. It seems to me a pity that policies and methods are adopted which shake the confidence of members of the Staff and their sense of security.

One of the curious procedures prevailing at McGill, which I have not met elsewhere and which I consider very damaging, is that we always only learn from the Star and the Gazette of decisions which vitally affect ourselves, our Departments and the Faculties and University at large. It seems to me reasonable for the Staff to expect to be informed before such important matters are given out to the Press. The teaching staff are the most important part of the University and in fact are least informed of what is going on.

---

91 Ibid.

I shall await developments with interest,
>With kind regards,
>>Yours sincerely,
>>>Professor E.G.D. Murray

April 8th, 1946

Dear Dr. Murray,[92]

Dr. Donald McEachern has told me about the project, under the auspices of the National Research Council, to write the history of the Army Medical Committee. It has been agreed that a copy of this History should be presented to McGill University for the use of Mr. Fetherstonhaugh in the preparation of his own volume and as a permanent document in the University Archives.

I am wondering whether the Chemical Warfare Committee and the Bacteriological Warfare Work are being similarly written up under the auspices of the National Research Council. If they are, I should appreciate it if you could make arrangements for a copy of the document to be deposited with me for the McGill University archives since I think you will agree with me that this University played a significant part, under your personal direction, in both these activities.

>With renewed good wishes to you as always, I remain,
>>Cordially yours,
>>>F. Cyril James

April 10th, 1946

Dear Mr. Principal,[93]

Your inquiry concerning the Chemical Warfare history should go to Dr. Otto Maass who was Director of that division, of which Bacteriological Warfare was a part. The National Research Council was only indirectly concerned and any historical review would, I think, be done by the Army.

The whole management of releases of information has been most unsatisfactory and unduly complicated by international arrangements, which were not always adhered to by other parties. I do not know the exact situation at the moment and so cannot say what could be done, but Dr. Maass should know it.

>With kind regards,
>>Yours sincerely,
>>>EGDM

---

92 Ibid.
93 Ibid.

May 28, 1946

Dear Professor Murray,[94]

Thank you very much indeed for your thoughtfulness in sending me the copy of the Science News Letter with its most interesting account of one of your wartime activities. This is the type of thing that I am most anxious to have, and I do hope that if you have any other similar accounts, or if you know of any other staff members and their wartime activities, you will take steps to have copies sent in here. I doubt if you realize how much the University as a whole is interested in, and proud of, the accomplishments of its scientists in the last six years. ...

    Cordially yours,
        F. Cyril James

September 10, 1946

Dear Mr. Principal,

I enclose two copies of the publications of results of one of the most important projects I was concerned with during the war.

You will find the members of the 'Joint United States-Canadian Commission' listed on page 133. This worked under the Joint Chairmanship of General Kelser representing the U.S. and myself representing Canada. The Commission controlled and directed the work of the station dealt with in this publication and some not to be published. It is hoped that results of some further extensive field trials will be published later.

A large amount of the vaccine was supplied to the UNNRA after hostilities ceased, to be used in China to help save the Chinese cattle which were being devastated by Rinderpest.

    Yours sincerely,
        EGDM

September 13th, 1946

Dear Professor Murray,

Your letter of September 10th enclosing the two copies of the 'American Journal of Veterinary Research' has just come to my desk and I should like to thank you very much indeed for your thoughtfulness in sending them to me, since as you know I am most anxious to make a collection of the publications of those members of the McGill staff who were concerned with these very important projects during the war. ...

    Cordially yours,
        F. Cyril James

---

94 Ibid.

POST-WAR REFLECTIONS ON ALLIED BW RESEARCH

Defence Committee [UK] –Friday, 5th October 1945[95]

... As in the case of the atomic bomb, the Americans, working on the ideas of British and Canadian scientists, have spent large sums (40 million dollars) in the development of biological warfare and we understand that they have decided to continue B.W. work in the post-war period. ...

The new agent "U.S." [Brucellosis] promises to be many hundred times more effective than the anthrax bomb, and a very potent factor in future warfare. This new agent induces a virulent Malta fever which incapacitates the individual for periods of up to one year. ...

Office of the President[96]
The University of Wisconsin
Madison 6
February 20, 1946

Dear Joberg,

I feel highly honoured to receive such an excellent photograph of my old friend and bacteriologist. I say friend because of the sympathetic and understanding way in which you have tried to take care of visitors from the States, especially in railroad stations!

How are things going since our activities in the field of BW have quieted down? Why don't you pay us a visit in Madison. ...

Like all other universities, we are flooded with veterans and are working day and night, trying to keep up with the procession. ... I assume the same situation is true in Montreal.

Once more I want you to know how much I appreciate the photograph. It is an excellent likeness. I hope that all is going well with you and your associates in Montreal. With kindest regards,

Yours sincerely,

E.B. (Dr. E.B. Fred)

---

95 NAS-BW, files of the DEF Committee, 1944-45.
96 UWO Archives, R.G.E. Murray Collection, Post-War Correspondence; LAC, E.G.D. Murray Papers, vol. 17.

March 19, 1946

Dear Joberg,[97]

... I have just received a copy of the report of our work on the island.[98] Apparently they have decided to release the highlights. It seems out of place to see in print the various statements concerning our research project.

For my part, I enjoyed the opportunity of working with such a distinguished group and also to see something of the North country. Our trips to Quebec and down the river bring back many pleasant memories.

Once more, I want you to know how much I enjoyed the hospitality of our Canadian friends and hope in the very near future that we may have the pleasure of a visit from you in Wisconsin. With kindest regards,

    Yours sincerely,
        E.B. (Dr. E.B. Fred)

                R.A. Kelser, D.V.M., Ph.D., Dean
                The School of Veterinary Medicine
                    University of Pennsylvania
                           Philadelphia 4
                         December 11, 1946

Professor E.G.D. Murray[99]
Department of Bacteriology
McGill University
Montreal, P.Q., Canada

My dear "Joburg":

A day or two ago I received from Rolla Dyer the special medal "awarded" the members of the Joint United States-Canadian Commission. This reminded me that I had on my desk your recent letter to which I had not as yet replied. It was good to have heard from you and I especially appreciate your congratulations on my having been awarded the Gorgas Medal.

I greatly miss our Commission group and, as Dyer put it, we ought to somehow arrange to have an occasional reunion. I guess this would be a little bit difficult for Craigie as I understand that he has made arrangement to return to Scotland, if he has not already gone.

With regard to the field trial which was initiated during the last year of our joint project, I, too, am anxious to see it published. As a matter of fact, I wrote Daubney

---

97 Ibid.
98 The "island" refers to Grosse Île.
99 Ibid.

and suggested that it would be very desirable to have the results published and that I hoped some arrangement could be made for a joint publication by he and his group and Maurer and Walker. I have had no comment from him on that subject.

It may be of interest to you to know that Schoening attended a meeting of representatives from France, India, Iran, South Africa, United Kingdom, and the United States and UNRRA, which was held in London in August of this year. At that meeting the committee urged action along the widest international lines to curb the constant menace of animal diseases to the world's supply of food. One of the recommendations involves the use of vaccines, and it was pointed out that veterinary officials in East Africa and China were testing the vaccine developed by the United States and Canadian research workers for the control of rinderpest. I believe a rather full report on this meeting will appear shortly in the Journal of the American Veterinary Medical Association. ...

I was glad to know that your son is following in your footsteps, having entered the field of bacteriology. I wish for him all possible success. I am sure that you and Mrs. Murray are very glad to have your daughter back in Montreal.

With very kindest personal regards to you and your family in which Mrs. Kelser joins me, I am

    Sincerely yours,
        (R.A. Kelser)
        Dean

                                              Lincoln Avenue
                                              Rahway, N.J.
                                              January 3, 1947

Dear Jo'berg:[100]

Your Christmas card – not only the illustration and its subtitle, but the very nice message inside – gave me a great deal of pleasure. Congratulations on your grandson. My grandson was down for Christmas and has now returned to Montreal with his parents. They found a little house over at Saguenay on the back river, which has almost flooded them out, while the snows seem to be burying them. Maybe I'll have to come up and help dig them out. If I do, you shall see me.

In the meantime, all good things to you and yours, including the grandchildren, for the New Year.

    Sincerely,
        George W. Merck[101]

100 Ibid.
101 Ibid. Robert Murray notes, "George Merck, the founder of Merck and Co., was important to the war effort and the accomplishment of fermentation on industrial scale for the development of

The Foreign Service of the United States of America

Office of the Military Attache, United States Embassy
Ottawa, Ontario, Canada
9 September 1947

Dr. E.G.D. Murray[102]
Department of Bacteriology
McGill University, Montreal, P.Q.

Dear Doctor Murray:

I am happy to inform you that the War Department has approved the award of the Medal of Freedom in recognition of the exceptionally meritorious service rendered by you during the period June 1942 to February 1946 as a member of the Joint United States-Canadian Commission and later as Chairman of the Committee for Direction of Biological Warfare Research in Canada

If it is agreeable to you, I would like to make arrangements for a presentation ceremony to be conducted by the American Consul General in Montreal some time within the next few weeks. ...

Please accept my personal congratulations on winning this high award.

 Yours sincerely,
  R.E.S. Williamson, Colonel, G.S.C.
  Military Attaché

Sept.11th, 1947

Dear Colonel Williamson,[103]

I am greatly appreciative of the high honour the War Department of the United States has seen fit to confer on me. I would like to add that the association I had with distinguished scientists, military authorities and Government officials of the United States was a source of inspiration and a sincere pleasure during my office of Joint Chairman of the Joint United States-Canadian Commission and as Chairman of the Canadian Committee. The cooperation and friendship between United States and Canadian representatives was nothing short of magnificent and greatly relieved the worries of important responsibilities and was productive of valuable results ...

Penicillin for use during WWII and for help with fermentation problems to do with BW agents. EGDM always spoke of him with high regard but I don't know specifically what brought them together other than industrial expertise and advice."
102 LAC, E.G.D. Murray Papers, vol. 17, file 36.
103 Ibid.

... if I may, I would suggest any day between now and Sept. 22nd, or any day after Oct. 1st for the presentation ceremony.

 I remain, yours sincerely,

  E.G.D. Murray

Office of the Chairman
Defence Research Board
Ottawa, Canada

Dr. E.G.D. Murray
Department of Bacteriology
McGill University
3775 University Street
Montreal, P.Q., Canada

April 1st, 1949.

Dear Professor Murray:[104]

 In accordance with the regulations of the Defence Research Board, you have been retired as Chairman of the B.W. Panel, to be succeeded by Dr. Mitchell. However I hope that you will continue to serve as a member of the Panel.

 I would like to take this occasion to thank you on behalf of the Board and personally for the time and effort that you have given to the work of the Panel. Under your able guidance it has given valuable advice on the consolidation of our B.W. research programme at the end of the war and on the launching of the peacetime programme.

 I am very grateful to you for all the work that you did as Chairman of the Panel.

 With kind regards.

  Yours sincerely,

   Omand Solandt

   Chairman,

   Defence Research Board

---

104 UWO Archives, R.G.E. Murray Collection. Omond McKillop Solandt (1909-1993) planned the Canadian government's military research program in the years following the Second World War, becoming the founding chairman of the Defence Research Board (1947-1956). He was a long-time associate and Fellow of the Arctic Institute of North America, and served on its board of directors from 1960 to 1965. He was later chancellor of the University of Toronto (1965-1971) and chairman of the Science Council of Canada (1966-1972).

CHAPTER THREE

ADJUSTING TO WARTIME CHALLENGES: 1939-1945

BOB MURRAY: FROM CAMBRIDGE TO MCGILL
Robert Murray Reflects on Wartime Correspondence
(Recollections, London, Ontario, 2004)[1]

Throughout his life my father kept touch with home when he was away by writing to my mother. Those we have seen always saluted her with "My Beloved". He kept up this regular correspondence with me in England and with my mother during the war despite his busy and concentrated life at that time.

When I was at Cambridge his regular letters were very important to my peace of mind. I kept them but when it came to returning to Canada there were fierce instructions that letters or communications of many sorts had to be censored. Unfortunately, I then decided to burn them and never thought to leave them in the care of a friend for the duration.

Among the most enjoyable letters were those about the times he took off for rest and recreation. They were not many or long during the war but other than with mother at the lake they were fishing trips, usually to the Columbus Club near Nominingue, QC. These usually included mutual and, to me, senior friends that I knew well from my own inclusion in earlier trips: Charles Pascoe (Metallurgist with Dominion Bridge), Dr. Walter Scriver (actually our doctor), Bill Newman (Chief Engineer for the CPR), and Hilary Robertson (Lawyer) were particular friends among a dozen other members of the Club.[2] Accounts also included the ebullient and lively character of the "guardien" of the sixteen lakes that made up the territory. His letters encouraged me to do some fishing in the U.K. but getting permission or affording the hostelries that had riparian rights was more than a bit daunting to me. I did do some in the company of our old friend, a Don at King's (College), David Stockdale.

---

1 UWO Archives, R.G.E. Murray Collection.
2 Charles Pascoe was a member of the Columbus Fishing Club and a good fishing friend. Robert Murray remembers him as a "man of many interests and a dedicated fisherman for Bass and Trout." Hilary Robertson was a Montreal lawyer and often provided the Murray family with legal advice. Robert Murray, personal communication (hereafter cited as RGEM).

Robert Murray, 1938-1941: From Cambridge to McGill

Prof. E.G.D. Murray
November 21st, 1939

Sir William Girling Ball, F.R.C.S.,[3]
St. Bartholemew's Hospital,
London, E.C. 1,
England.

Dear Ball,

I have occasion to write to you to secure, if possible, another generation of Murray at Barts.[4] It would not seem necessary in ordinary times but I do not know existing conditions.

My son, R.G.E. Murray, is at present a student at Christ's College, Cambridge, and is nearing the time to enter upon his clinical studies. This I learn has been accelerated by the war and he is anxious to enter Barts, even as I am that he should. I have told him to write to you for advice on how to proceed only it is quite likely he will not mention who he is as he is a modest young man.

It would be a great satisfaction to me could you find time to see him or send him a note of advice. I feel sure you will like him; he has proved a good student and a dependable fellow giving us every reason to be proud of him.

Things are strange but quiet here. A serious, determined war effort is being made though I feel, like others, some impatience at not being called up at once. However, I am desired by the Department of Defence to organize research in my Department on gas gangrene problems and I hope to be able to make some contribution of use.

    Best of luck to you.
        Yours ever,
           EGDM

---

3 LAC, E.G.D. Murray Papers, vol. 17, file R.G.E. Murray. Sir William Girling Ball (1881-1945) was Dean of St. Bartholomew's Medical School from 1930 to 1945. St. Bartholomew's Hospital, "Deans and Wardens' files of St Bartholomew's Hospital Medical School," http://www.aim25.ac.uk/cgi-bin/search2?coll_id=5473&inst_id=51.

4 See the Introduction and chapter 1, "Cambridge Days and Wartime Service."

July 15th, 1941

Professor G.E. Gask, C.M.G., D.S.O., F.R.C.S.,[5]
Hatchman's, Hambleden,
Henley-on-Thames,
England.

Dear Gask,

I am writing to you about my son who is at Cambridge and is arranging for some experience in case taking and physical examination at Addenbrooke's Hospital[6] to qualify to enter the third year Medicine, McGill University. I write to you because I gather from Christie[7] that you are at Cambridge. I find it difficult to advise the boy adequately from here and I would greatly appreciate it if you could find a moment to look into his arrangements and give him the benefit of your opinion.

Robert was to go to Bart's but under the present war conditions it has been thought advisable for him to complete his clinical work here. I am not perfectly sure that that is the best plan though, of course, it would be the most economical for me. Nevertheless, I am very anxious that Bob has the best opportunity and it would be a disappointment were he not to go to Bart's as have all the members of my family.

It is to be borne in mind, of course, that he will probably work in Canada and under those circumstances it is possible that a McGill degree would be of advantage. I feel that as you know the circumstances in England and I cannot discover them at this distance that your advice would be very valuable. ...

With kind regards,
Yours sincerely,
EGDM

---

5 LAC, E.G.D. Murray Papers, vol. 17, file R.G.E. Murray. The surgeon George Ernest Gask (1875-1951) received his medical training at St. Bartholomew's and became a respected surgical teacher and consultant at the school, where he remained for his entire professional career. He also served as warden in the residence at Barts for five years. *Oxford Dictionary of National Biography* (Online). Hereafter cited as *ODNB* (Online).
6 One of the university hospitals at Cambridge University
7 Christie unidentified.

July 15th, 1941

Dr. Fred Grauer,[8]
Addenbrooke's Hospital,
Cambridge,
England.

Dear Grauer,

My son, Robert, has arranged to do some clinical work at Addenbrooke's Hospital with a view to satisfying the conditions necessary for him to enter the third year here, if possible. Please get hold of him and give him the benefit of your advice because you know very well the conditions that have to be satisfied.

I don't believe Bob is very certain of what he has to do and I feel sure you can help him greatly. I would greatly appreciate your kindness because it is difficult for me to advise him at this distance and under present circumstances.

I hope all is going well with you. With kind regards,
    Yours sincerely,
    EGDM

Hatchman's[9]
Hambleden
Henley-on-Thames

Aug 20, 1941

My dear Murray,

I was very interested to get your letter of July 15th, telling me about your son Robert.

I should be very ready to assist in any way possible, but unfortunately Christie got me wrong for I am at Oxford, not Cambridge.

So I can't see Robert and talk to him as I should like. All I can do is to put some general principles before you.

You know that the Rockefeller folk offered to take a number of our medical students and place them in the schools of U.S.A. and Canada. This was well received here and we sent some of our best men. There is no doubt that clinical instruction is disturbed here and it is more difficult for men to gain good experience than before.

On these grounds I should say that … it would be a good thing for Robert to finish his course at McGill, especially as he is likely to live his life in Canada.

---

8 LAC, E.G.D. Murray Papers, vol. 17, file R.G.E. Murray. Grauer unidentified.
9 Ibid.

I am sorry I cannot do more than offer these generalizations, but if I can do any more please let me know. ...

You have done very well – my congratulations.

    Yours very sincerely,
        George E. Gask

                Department of Bacteriology and Immunity
                            Prof. E.G.D. Murray
                                  Sept. 23rd, 1941

Professor G.E. Gask, C.M.G., D.S.O., F.R.C.S.,
Hatchman's, Hambleden,
Henley-on-Thames,
England.

Dear Gask,

Thank you very much for your letter of August 20th advising about what I should do for my boy. I am very glad that you confirm my opinion so fully and I hope that it will be possible for Robert to complete his clinical work here.

In his recent letter Robert did not feel so hopeful of being able to arrange things. He has been filling out forms in triplicate and hopes that Canada House will be able to help him.

We have one English student at McGill, another has gone to Toronto and several others to the United States. Some of them are friends of Robert's and when they pass through Montreal we see them.

Florey[10] is in Montreal this week and is dining with me tonight. It is very nice indeed to see an old friend, particularly as we worked together at Cambridge for some years. Through him I have had news of a number of my old friends...

    With kind regards,
        Yours sincerely,
            EGDM

---

10 Ibid. See Howard Walter Florey, Introduction, note 182.

December 2nd, 1941.

W. Bentley Esq.,[11]
The Bursar,
McGill University.

Dear Mr. Bentley,

My son, R.G.E. Murray, has returned from England and has registered for the remaining two terms of Second Year Medicine. [In] England he was two years at Cambridge [where] he took his B.A. (with Second Class Honours in Part II of the Tripos).

Please let me know what fees I am to pay and if I can get any allowance for him in respect of fees.

    Yours sincerely,
        EGDM

Copy to Professor Murray
December Nineteenth 1941.

Dr. J.C. Meakins,[12]
Dean of the Faculty of Medicine

Dear Dean Meakins:

I would appreciate your letting me know if Robert G.E. Murray is to be considered as a second year student for a full year or only part of a year, as this will affect the fees he has to pay.

As you know, all children of full time members of the staff are granted a bursary in their first year, to be maintained by academic ability. In the case of medical students the bursary is $250.00, and Professor Murray is entitled to this or such fraction as corresponds to the amount of fees he is required to pay.

    Yours faithfully,
        W. Bentley
        Bursar

---

11 Ibid.
12 Ibid. See chapter 1, note 26, for a description of Meakins.

6th January 1942.

Professor E.G.D. Murray,
Department of Bacteriology and Immunology

Dear Professor Murray,

 I am sending you, herewith, a certificate with regard to your son, Bob. I should be glad if you would sign this, if you regard his Cambridge bacteriology sufficient to be considered equivalent to the course in your Department.[13]

  Yours sincerely,
   J.F. McIntosh, M.D.[14]
   Secretary,
   Faculty of Medicine.

The Cambridge Experience (Recollections, London, Ontario, 2004)[15]

It was no shock to travel in 1938 to Cambridge from London by a train that seemed no different and even smells were the same in Liverpool Street Station and on board a corridor-less compartmented carriage, and the scenery and stops were familiar to one who last took that trip at age eleven. The landmarks and streets were still familiar, so it was a sort of homecoming rather than the fearsome novelty that faces most of the new matriculates. ...

 I knew that for the first year or so I would be quartered in "digs" in a lane running between St. Andrews Street and Petty Cury and not a hundred and fifty yards from the College gate. My sitting room, complete with coal-burning fireplace and gas lighting (two incandescent mantles), desk with side chairs, reasonably comfortable stuffed chair, and a side table. The bedroom was upstairs ... It was great good luck that the first person I met after introducing myself to the landlady was Arthur Harold, a second year student doing medical subjects of the Natural Sciences Tripos, as I would be doing. ... He got me clued in on the steps to take ... You could not eat in hall without wearing a gown (specific to the college) or go out after dark without cap and gown lest the Proctor and his Bulldogs caught you for a fine of 6/8d or 13/4d. One learned the simple rule that if you returned to your digs or College to be let in after 10:00 pm your lateness would be reported. ...

---

13 Ibid. Note in E.G.D. Murray's handwriting on the letter reads: "Certified Part II Trip. Path at Cambridge exempt from Course 1, 2nd year McGill."
14 Most likely "Hank" McIntosh, later Professor of Physiology. RGEM. J.F. McIntosh was also at one point honorary secretary of the Montreal Medico-Chirurgical Society.
15 UWO Archives, R.G.E. Murray Collection.

My Tutor was C.P. Snow, then still a physicist, and in overall charge of the studies of the science students while Mr. S.W. Grose, an old family friend, was the Senior Tutor and mentor to the others.[16] Dr. Snow carefully outlined the course of study that led towards medicine (decided because it had unlimited possibilities and was a primary goal) and we agreed which courses would be essential (i.e. all of them) in the first two years, Part 1 of the Natural Sciences Tripos. They were: Anatomy, Physiology (adding Pharmacology, to be taken in a "long vacation" term), Biochemistry, and Pathology. A Part 2 subject could be taken for the third year as a whole year of specialization and would have to be an agreement with both Department and College before the end of the second Summer Term. You had to take at least two major subjects during Part I and they were not fooling about the degree of concentration offered. ...

There were a number of people in the College who were familiar to me from my childhood. Among the Fellows the closest family friendships I maintained were with S.W. Grose and J.T. Saunders and social visits happened as they did with D. Stockdale of Kings College. ... Tibby Marshall (a pioneer in Physiology of sex) never failed to stop me for a chat. Fanny Saunders was a classmate of my father at the college and became a distinguished zoologist and an administrator of Cambridge University. ...

There were other interesting people in that Department and I wish, along with similar wishes for other segments of the experience, that I had made better contact with them at the time. Dr. Marjorie Stephenson was putting bacterial biochemical physiology on the map at that time – a true pioneer in a pioneer Department headed from its start by Sir Frederick Gowland Hopkins of vitamins fame and much respected in retirement. ... Joseph Needham was a member of that Department but he was seldom seen because he was immersed with his helpers in his classic work on the history of Chinese science.

Pathology was headed by Professor H.R. Dean who was a family friend dating from my father's appointment in the 20's and he still did give some of the lectures in the Part I course – an impressive figure with interests both in Pathology and the beginnings of immunology. Despite the old connection he treated me as a student and, for my part, I enjoyed him and had the greatest of respect for this senior member of the University and former Master of Trinity Hall ...[17] Dr. Spooner,[18] a bacteriologist, looked for me and greeted me on my first day in the Part I course and I was to find out that he had been one of the students in the first year that my father taught. ... When it came to the Part II the majority of the microbiology was given in lectures and labs designed by Allan Downie; a remarkably nice and thoughtful teacher and destined to be a virologist of note and Professor at Liverpool. ... Physiology was a

16 See Introduction, note 302 for information on Charles Percy Snow.
17 See Introduction, note 178.
18 E.T.C. Spooner, Department of Pathology, Cambridge.

major and interesting course with excellent teachers of which the most distinguished and later Nobel prizewinner was E.D. Adrian (later still Lord Adrian); a pioneer in electrophysiology. ...

After two years came serious examinations taking a whole week, morning and afternoon, ushered in with a written essay on one of a large number of topics listed. This was a sort of limbering-up exercise to get one into the swing of coping with real questions. There were afternoon laboratory exams with real experiments to do in physiology ... and a written test in biochemistry. Anatomy had a viva-voce exam before the end of term. ... The pharmacology course was taken in the first long vacation term and was separately examined. ... The Part II was differently organized. A whole big laboratory was assigned to us and each of us five had a lab bench for ourselves with space enough for equipment when we needed it and side benches for demonstration material. ... Every morning at 9:00 there was an informal lecture and discussion and it took place in our laboratory. We were free to visit or go to ask questions of an authority in the building or the Health Service and the Molteno Institute (mostly parasitology) nearby.[19] We attended autopsies regularly to learn "morbid anatomy" and obtain tissues to fix, embed, cut sections, and stain for all manner of purposes and to make diagnoses based on microscopy. ...

Although we were busy and well engaged in studies it was the way of Cambridge students to be seen in outside activities; the odd pint in a pub, ... some participation in College sports, and some sort of relaxation. It was inevitable I suppose, that I would find a favourite pub, The Eagle ... where Watson and Crick, as regular customers, proclaimed in 1953 that they had resolved the structure of DNA and their model turned out to be a turning point in modern biology. ...

I have to say that all this work and fun took place in a country fearing war in 1938-39 and at war from Sept. 1939 on. There was a sequence of changes in aspects of life and in the people around you. On the streets an increasing proportion of people in service uniforms and civilian folk with understood responsibilities. With war came the blackout and Air Raid Wardens wearing a black tin hat labelled in white. In London the barrage balloons appeared strung up in their hundreds. ... As a student in the medical stream I was excused military service but involvement with national defence was inevitable in vacation time and to a small extent at nights during term.

---

19 The Molteno Institute of Biology and Parasitology was founded at the University of Cambridge in 1921.

Death of George Murray, 1941

Godwan River Estate,[20]
Ngodwana
E. Transvaal

July 30th 1941

Dear Biff,

Just to let you know that Rory has received an immediate award of the D.S.O. for gallantry in action, by the Commander-in-Chief, Middle East.[21] I felt sure you would like to know of this great distinction conferred on him, and will feel equally as proud as I do. I heard from various sources some time ago that he had been recommended on several occasions, when I asked about it, he said he had only done his duty. My one regret is that your dear old Dad never knew of it, how happy and proud it would have made him. With much love to you all.

In haste,
'Nita
(Mrs. RHE Murray)

Obituary
South African Medical Journal, August 9th, 1941.
Dr. George Murray

Dr. F.H. Napier writes:-

The death of Dr. George Murray has brought to an end an almost lifelong friendship.

We were students together at Barts[22] sixty years ago, where he had a remarkable career and obtained the Bracken[bury] Surgical Scholarship.[23]

---

20 LAC, E.G.D. Murray Papers, vol. 17.
21 Roger (Rory) Murray (1895-1968) served in both world wars. After graduating from Woolwich Military College in England, he served in the First World War with gallantry, being discharged as a captain. Despite his age, Lt. Colonel Roger Murray served with the South African Army in the Abyssinian Campaign, and was awarded the Distinguished Service Order (D.S.O.) in 1941. During his wartime absence his wife Unita (Nita) operated the family farm at Godwan River Estate with great efficiency. *The Cape Argus*, 7 August 1941; *The Lowvelder*, 9 May 1968, cited in Marge Shearing, "Family Book" (unpublished manuscript, in possession of Peter Murray).
22 St. Bartholomew's Hospital, London, England.
23 The Brackenbury Surgical Scholarship was awarded at St. Bartholomew's Hospital to a top surgical student. By all accounts G.A.E. Murray was always top of his class. He took his first part of the Licentiate of the Royal College of Physicians of London (LRCP) in 1882 (at the end

He held many resident appointments at the Hospital, in all of which he showed remarkable ability.

After obtaining the Fellowship of the Royal College of Surgeons of England, the staff of the Hospital were anxious that he should join them with the object ultimately that he should be appointed to the Permanent Surgical Staff.[24]

That was the opinion of his abilities that the authorities had of Dr. Murray.

But his heart was in this country, the country of his birth, and he threw up his prospects in England to return.

At this time we had been intimate friends, and when, in 1897, I came to Johannesburg the friendship was renewed.

I found that Dr. Murray had established himself as one of the most successful medical men in the country.

In those days there were no specialists in the Transvaal; he was on the staff of the Johannesburg Hospital as a General Practitioner.

In all branches of medical work he was first-rate, but particularly in surgery.

In those early days I helped him occasionally in his operative work, and it was a lesson to me to have the privilege of watching him work. He had beautiful hands with long, delicate and sensitive fingers; he was always calm and collected, and whatever the emergency I never saw him in difficulties: he knew what to do without hesitation, and did it instantly and quickly.

He was a tall, handsome man, with a kindly and considerate manner to his patients, by whom he was much loved.

Our friendship was never broken, and it was a great regret to me when, a few years ago, he was forced to retire through bad health.

Now he has passed on after a life devoted to his work and a life of cordial relations with colleagues and patients alike.

Can there be a finer record of any man's life?

Dr. E.B. Fuller writes:-

"The old order changeth, giving place to the new." With the passing of George Murray, full of years and full of honour, one of the last of a distinguished band of general practitioner surgeons has disappeared.

They were a noble fellowship and Murray was one of the finest of the type.

He was the perfect general practitioner and a surgeon of great judgment and skill. He was beloved by his patients and, what is perhaps more rare, he was much beloved and respected by all his medical colleagues.

---

of his second year) and the second part in 1884. He was likely awarded the Brackenbury at this stage, once he had "shown what he could do ...". He took his MB in 1887. RGEM.

24 G.A.E. Murray passed the FRCS examinations with high honours in 1887.

In the medical history of Johannesburg his name will always stand for all that is best and most worthy in the medical profession.

My memory of him brings up a kindly, courteous gentleman, full of the milk of human kindness, looked up to by the public and the profession as a skilful surgeon and a wise friend and counselor.

Royal College of Surgeons of England,
Lincoln's Inn Fields,
London, W.C. 2.
4 December 1941

Professor E.G.D. Murray O.B.E., F.R.S.Canada.[25]
Department of Bacteriology
McGill University, Montreal

Dear Professor Murray

It is a part of my duty to keep a record of the Fellows of this College, and I am writing to ask you if you would be so kind as to help me with certain dates and details to make my record of the career of your late distinguished father complete and correct.

The points which we particularly need are the following:
1. The date (day, month and year) and the place of his birth?
2. His father's Christian name and profession?
3. His mother's maiden name?
4. His place in their family (e.g. 3rd child, 2nd son)?
5. Where he was at school before entering St. Bartholomew's?
6. The date of his marriage?
7. Your mother's maiden name?
8. The number of your brothers and ? sisters; it would be also of interest to know if others of them beside yourself followed the medical profession?

Any other facts that you would wish recorded we should of course be very pleased to hear from you. Of course I realise that many of the queries set out above may not be readily answerable, but shall be grateful for answers to any of them.

Believe me, Yours sincerely
W.R. LeFanu
Librarian

---

25 LAC, E.G.D. Murray Papers, vol. 17.

Department of Bacteriology and Immunity
Prof. E.G.D. Murray
February 12th, 1942

W.R. Le Fanu Esq.,
Librarian of the Royal College of Surgeons,
Lincoln Inn Fields,
London, W.C. 2, England

Dear Mr. Le Fanu,

I am very glad to answer your questions about my father, G.A.E. Murray, M.B. (1887), M.R.C.S. and L.R.C.P. (1884) for the records of the College. Having been away from South Africa since 1906 and the family records being rather scattered now, I cannot give you all the information you might wish to have. Possibly my brother, Major R.H.E. Murray (Godwan River, N.E. Transvaal, South Africa) or my aunt, Miss Annie Murray (Cypress Grove, Graaff Reinet, South Africa) could supplement it.

I shall answer the eight points in your letter as best I can.

1. Born March 18th, 1862 at Roode Bloem, Graaff Reinet, South Africa.
2. Eldest child of Walter Everitt Murray; Farmer.
3. His mother was Anne Southey (closely related to Robert Southey, Poet Laureate).
4. Eldest child (His youngest brother, F.E. Murray was also F.R.C.S.)
5. Graaff Reinet College (Where he won prizes for being first in classes, some of which I now have); St. Bartholomew's Hospital, London; Durham University.
6. Married 1889.
7. Kate Elizabeth Mary, younger daughter of Captain J.J. Dunne (The Hi-Regan). She died April 27th, 1936.
8. His children: Professor E.G.D. Murray, O.B.E., M.A. (Cantab), L.M.S.S.A., F.R.S.Canada.
   Captain T.H.E. Murray, R.F.A.
   Major R.H.E. Murray, D.S.O., R.F.A.
   A.C.E. Murray, Royal South Africa Airforce (I don't know his rank)

My father was House Surgeon and House Physician at Bart's 1884-1887 and went to Johannesburg in 1888, where my mother, who was a nurse at Bart's, went out to join him in 1889.

He was one of the founders of the Johannesburg General Hospital and was its Chief Surgeon for thirty-seven years and then Consulting Surgeon. The new Operation Theatre was named "The Murray Operating Theatre". During the Boer War he was Consulting Surgeon to the British forces and besides the General Hospital he founded and ran two auxiliary hospitals (in two large hotels) of which my mother was Matron. In 1900 my father received the Knight of Grace of St. John of Jerusalem and my mother the Lady of Grace of St. John of Jerusalem and they were bestowed by Queen Victoria.

My father was well known and much loved all over South Africa. He was noted as a magnificent shot. He had a great knowledge of horseflesh and was for many years a Steward of the Turf Club[26] and on the Committee of the Witwatersrand Agricultural Society and judge of many events. He was President of the South African Medical Association several times and was President of the Rand Pioneers for five years. His many social and sporting activities and his kindness and good works gained for him respect and love in all classes of society; I well remember how gladly he was received wherever he went. He was a very handsome man of six foot four inches and with his kindliness and dignity he made an imposing figure. ...

I hope this is what you wish to have and I would be glad to know if you receive it safely. I am sad knowing of the damage done to the Library and Museum of the College, both of which I used so much in the past and were of such great national value.

Yours sincerely,
EGDM

Robert Murray to Doris Marchand

12 Alexander St.
Cambridge
16.x.38

Miss Doris Marchand[27]
72 Dryads Green
Northhampton, Mass.

Dear old Doris,

I was damned glad to get your letter, it bucked-me-up no end. Coincidentally (good word!!) I received the azure scroll at breakfast time and devoured both it and my scrambled eggs at the same time. ...

Full term began last Monday (not semester!) and work started in earnest on Thursday. I stayed in London till last Monday morning.

Things were not dull in London as I expect I told you. Things had cleared up nicely

---

26 Ibid. The Johannesburg Turf Club organized popular horse racing events at the Turffontein racing grounds, located in suburb of Gauteng.

27 Correspondence of Robert Murray and Doris Marchand, private collection, Robert Murray.

(peace with *dis*honour). Trenches were still everywhere and city officials, corporations, etc., were beginning to think of making them permanently bomb-proof. All the parks looked like mole-hills although things looked better when they moved the anti-aircraft guns from every other corner. On the Friday afternoon of that week I went to the Zoo again and covered most of what I did not see before. It was great fun watching the seals fed with herrings – a spectacular performance that could not be equalled in any circus. The only place I could not stick was the parrot house. Perhaps my face put them off. Anyway, they made such a disgusting noise (about 400 of them) and the atmosphere was so heavy that I walked through in a hurry. ...

On the Monday I came up to Cambridge, got into my rooms (write to address above, it saves ½ day), unpacked again, got a cap and gown, saw the Senior Tutor, my tutor, arranged what I was going to do, inscribed my name on the book, went to hall, came back and read up on what I should do and went to bed. Things happen everyday which must be dealt with and my time is pretty full. I have to go around and see old friends that I haven't seen for years etc. Everyday I get an enormous mail – 75% touting letters, the rest mainly from people who want to know why the --- I haven't been to see them yet and will I come to lunch next Sunday at 1.15! Damn it all – today's Sunday and it is 12.15 and I have *got* to go to lunch at 1.15. It will be alright though – the daughter is quite nice! ...

Cambridge looks much the same as it always did except that the streets seem much smaller than they used to be. Part of it is the increase in traffic the other part is human growth. ...

Well, old thing, I have some anatomy to read so goodbye. ...

    Love,

        Robin

                              Montreal
                              14.ix.39

Miss Doris Marchand,[28]
12 Edgehill Street
Princeton, N.J.

Dear Doris:

Sorry not to have written to you before now. Things have been moving very quickly and we have all been very busy. ...

The international situation is still being a nuisance to me. I don't know yet whether I can go back or whether I shall have to stay here. There are hindrances at both ends,

28 Ibid.

as a matter of fact, which are not making things too easy. Boat sailings are vague and indefinite – passports have to be visa'ed for every possible purpose and news from the other side is so heavily censored that one scarcely knows which way up one is. In fact, if you'll excuse the expression, everything is bloody ---. And I'm horrid myself. I'm so horrid that I haven't the energy to think of interesting things to say to you.

I enjoyed the stay in Princeton very much indeed, and now fully appreciate the time you gave up to my horrid self. I enjoyed every minute of the time. My father sends his regards to you – don't forget the standing invitation. ...

From your temporarily horrid friend.
    Love,
        Robin

My Dear Doris,

19.xi.39

Warning: This letter has too many "I"'s in it and is not, on second thought, quite sound. I'm sorry I haven't time to rewrite it – Love R.

I am disappointed to hear that my letter has not yet reached you (Oct 30) but you should have it by now. It is comforting to think that you look forward to receiving them. I am, also, flattered to think that you should write to me twice in a comparatively short space of time because you think I am laggard in writing to you. ... With all due deference, I say you are utterly mistaken – though reading, crudely I admit, between the lines I would say that you were giving me an easy "way out". If ever I do wish to break a correspondence which I treasure, and take much pleasure in answering, I shall say quite plainly the why and the wherefore.

The pangs of homesickness are very unpleasant. I used to suffer severely (I was a nasty little treble) when I was sent off to a boarding school. It is said that "familiarity breeds contempt" and I am sure that when you have become more used to your new surroundings and environment you will lose that "empty feeling". I think, from a Psychological point of view, that you are feeling the absence of friends of your own age and inclinations rather than actual "homesickness". Can't you contrive to meet up with some of the younger people in Richmond (if there are any) – do a bit of scheming and become the complete social butterfly – it's a much better existence than the "old maid-ish" atmosphere in the common rooms of most schools.

I am very differently situated here. I can get as much rowdiness and gaiety as I want. And though I could, if I wanted, get female company of a doubtful character (i.e. not quite out of the top drawer) I have all too little of the company of pleasant souls I should like. The occasional binges, and the pleasant social affairs in town, between them manage to keep me out of mischief and away from the miserable feeling of loneliness and ennui.

This letter has been, so far, a miserable and soul shattering affair. On then; to brighter things and pastures new.

A week and a half ago I went down to town to attend the socialite wedding of my cousin (Nancy Heard).[29] I was in on the whole show (I was staying at the bride's house). I gave them a wedding present of two silver tankards which, incidentally, cost me £12. I fortified the bride with brandy before the show (she was wearing, this for your benefit, a gown of old lace with an Elizabethan short collar, a filmy white veil with a pearl circlet, and a bouquet of wax orchids). I arrived at the kirk in morning coat, topper etc., and acted as usher seeing that all the important people got in the right places. It was a pleasant service with some really decent music. Afterwards we adjourned to ... the reception. There I had to talk to all the people I knew (about 60 out of 225) and fortified myself the while with plenty of excellent Champagne. We saw the couple off to their honeymoon in Scotland. I went to a show that night and together with a few friends did a bit of celebrating. It was a very pleasing and pretty wedding, beautifully arranged. I only wish you could have seen the wedding presents – one of the nicest and most tasteful collections you could imagine. It did a lot of good and made a very necessary break in a long stretch of hard work.

I wish you would tell me a little more of your teaching – the sort of biology you teach these girls, what they do themselves, the interests they take (I am interested from the academic point of view of teaching) – and, of course, any amusing peccadilloes and 'boners' which you may think fit for my tender ears. How well are the labs equipped for teaching and do they allow any scope for any unusual demonstration work, or work that you may be interested in. ...

Keep your chin up and don't forget that I haven't forgotten you.

    Love,

          Robin

P.S. This, as usual, will go *"par avion"*

---

29 Ibid. Nancy E. Heard, referred to as "Nan" by the Murrays, was the daughter of W.S.N. Heard and a cousin on Freda Murray's side of the family. She died in 2006 at the age of ninety-seven. RGEM.

Christ's College,[30]
Cambridge.
27.11.39

Dear Doris

This typewriter business seems to be a craze; but, at any rate, I cannot call this one "new-fangled" 'cause it is a 1917 model. A sterling machine that has followed me through thick and thin. The real reason for the typing is not a split thumb but is that there is only a very short time left for me to write you in. I have to go to a dissect in less than an hour from now and there will be little chance for me to write again until Thursday. The end of the term approaches and so does an anatomy viva. Anatomy includes neurology and emorphology as well as the human body and goes on until you take the exams in part I of the tripos. ...

I am sorry to hear about the job difficulties. The one at Yale I am afraid, was a little out of your reach. As I have probably said before, it will come down to teaching giggling youngsters with projecting teeth a little elementary biology. They will ask for more but I doubt they will get it. Don't forget the McGill idea. ...

The term has not been without its amusements. The college has at last imported a ping-pong table (a very good one) and I play on it regularly. My game is improving slowly after the slump of the last term. I am also playing a little bit of Squash, a very fine court game. When I get a little more proficient at it I shall enjoy it very much. Most of the people here seem to have played it, and many other games, since their infancy. ...

My father is beginning to agitate concerning the possibility of my returning home during the summer. I cannot see my way clear to doing this yet, but time will show. If I do I shall try to come via New York. How about yourself? Is all the German money to be forfeited to the cause of medicine and to the detriment of travel.

I am getting tired of the term and tired of the sight of books and no one has come up to see me yet. As a matter of fact they were coming but the 'flu bug stopped them, there's a little of term left so they may come yet.

... So long for now, old girl, and send me a long and chatty letter as soon as you can. A little bit of chatter does me good.

Yours ever,
Robin

---

30 Ibid.

23.ii.40

My Dear Doris:[31]

... My letter writing is a bit irregular now I'm afraid – partly because I am frightfully busy and partly because I take my leisure moments in a very lazy fashion. ...

One amazing thing about this war is the slackening of the restrictions on jokes, gags, and displays of crude nudity (good phrase that!) which it has engendered on both stage and radio. The sort of jokes that we used to tell in "Bull sessions" are now fairly common in the more intimate stage shows. ... Much same with the radio variety shows (which, incidentally, are as lousy as ever). ...

Have you heard about the man who had no torch in the black-out and who held a lighted cigarette in each hand as an indication to other pedestrians. It was all O.K. until another fellow tried to walk between them. English Choke! Ah-ha! I know a lot of amusing stories, but they are all much too indelicate.

Well, that must end this epistle, which has, incidentally, been a good bit longer than of late – it has the flavour of an egotistical diary, written in code by a man who intended it for publication!

Love,
Robin

26.ii.40

My Dear Doris:

When I saw the cover of your letter yesterday – one of my names spelt with 3 E's, 1 V, 2 R's, and 3 T's, I knew that something must be wrong. Anyway, as I've said before sometimes, you only use *all* my names when I have not been writing regularly or in dire emergency.

I was definitely shocked when I saw the address – after a morning in the Pathology labs one is apt to imagine anything. When you stated that the cause of your illness was pneumonia I started thinking of my old friends "Red hepatisation" and "Grey hepatisation" and all the various complications. ...

You say that you were dosed with one of the Sulphanilamides[32] – did they wait a couple of days before administering it? I believe they are not really effective unless the patient has developed some active immunity during course of disease. I very well believe you when you say that the drug made you feel worse than the disease did. ...

---

31 Ibid.
32 Ibid. Sulphanilamide is defined as: "The amide of sulphanilic acid, which has wide bacteriostatic activity, has been used, esp. topically, in the treatment of infections due to hæmolytic streptococci, and is the parent compound of the sulphonamides." *Oxford English Dictionary* (Online). Hereafter cited as *OED* (Online).

Well here's to our next meeting. I'll have a good scotch for you when I am next out on the bend.

    Love,

          Robin

                              Marlston House
                              Newbury, Berks
                                       27.vi.40

Miss Doris Marchand[33]
Marine Biological Laboratory
Woods Hole, Mass.

My Dear Doris:

... Political and war affairs are well out of my scope – though over here they are discussed with acerbity and vigour. Things have happened with such alarming regularity and with such little regard for the "man in the Street" that no-one except the heaven-born can hope to understand the complexities. The sudden capitulation of Belgium, which was in effect a fateful blow to the Allies was more explicable than the sudden and shattering sue for peace by France. A peace which, I am sure, is as little understood by the Frenchmen who have been fighting. Still, this morning the news was better from our point of view and showed that we were taking a little of the offensive.

I am unable to understand at the moment where you people in the U.S. stand, or think you stand. The little news we get of your activities is not exactly heartening – in fact at moments it is downright disheartening. Because, as we see it, should we be conquered that will mean that the dominant force in the Atlantic will be German (or at least the Axis) while in the Pacific it would be Japanese. However much you increased your armaments the G's and J's would be a step ahead of you, because you would have to keep a fleet and an armed force on equal terms in both the Atlantic and the Pacific – no quick job either! Whatever the feeling in the States might be I'm sure they could, with advantage to themselves, buck up and see their real position, which is structurally far away and yet in real fact is just across the street. ...

    Much love to you.

          Robin

---

33 Correspondence of Robert Murray and Doris Marchand, private collection, Robert Murray.

26.vi.41
Christ's College, Cambridge

(Sorry it's such a short letter but duty calls. Love, Robin)

Dearest Doris:

I'm not at all sure what "hep" is – as a word I have seen it in various contexts. If it means what I hope it means your last letter was definitely hep-hep – and by that I mean damn good. It cheered me up quite a lot at a time when I was feeling very low. Since then I have read and re-read it whenever I have been feeling in poor spirits. Now, at long last, I set me down to write ... So now I must cut down all gentler pleasures and get back to the pen again after leaving it aside for some three weeks. ...

About two weeks before the big exams came up I received a very gloomy, despondent, and rather worrying letter from my father. Written as you may understand, from a very "dismal Joe" attitude, which was the first news I had heard of his being a sick man (Pyelitis with a few complications).[34] That alone worried me more than somewhat and was sufficient to put me off work. There were many other personal affairs, mainly to do with money, which also contributed to the atmosphere. The upshot of the thing is that he wishes me to go home in October. I have been moving heaven and earth ever since to try and accomplish this. Another letter arrived after the exams (unfortunately) and reassured me that he was better and is now, in fact, completely recovered. However, the plans still stand and if all goes well I shall be home before Xmas. I look forward to it although I feel very badly about leaving England at this stage in proceedings – a point you can imagine better than I can tell you. The main necessity for this change in plan is money for the Murray family at this moment is none too well supplied with boodle in any form. So it looks as if, should the stars look favourably upon me, I may be a bit nearer your abode than I am at the moment.

Then, soon after the first part of this bombshell had arrived, came the examinations. About 40 hrs of them (5 written and 3 days of practical) all within a week – covering Pathology, Bacteriology, and Serology and a few of the sidekicks but essentially along those lines. They were pretty difficult and seemed even more so in my rather depressed state of mind. Still I did my best. When they were finished there was the usual agonising wait until the results came out during which time there was some phenomenal blinds, drunks and what have you. One fine morning I picked up courage and forced my way through the crowd to the lists and found I had got a 2nd, which isn't too bad in a part II examination. The result: Just over a week ago I, and a lot of other guys, got all dolled up, gowned, capped and hooded and went through the

---

34 Ibid. Pyelitis refers to the "Inflammation of the mucous membrane of the pelvis of the kidney." *OED* (Online).

elaborate and time-honoured ceremony which gives me an honours B.A. So you see you are trying to get me qualified before my time and, incidentally, much too young. I won't be qualified for at least another *two years*.

Of course another quite important event was my 22nd Birthday – which you very kindly remembered. It was a damned wet day, and I was depressed. At exactly 12 A.M. I was standing outside a café after having some coffee and suddenly decided to go to London and immediately started walking along the road. I wanted to go see a friend at St. Mary's Hospital among many others. By great luck I was picked up by Himmelweit, one of the Virus Kings, about two miles out of Cambridge.[35] I spent the afternoon in his lab, and learnt a great deal about technique particularly of Gradocol membranes.[36] They are doing a certain amount of work similar to that you were doing last year, but a great deal else besides. So that although I went to London to get away from work, at no time was I very far from it. ...

Just at that moment a letter arrived from Canada House stating, as far as I can decipher from the official jargon of these government offices, that they are satisfied that I am returning for a just and proper purpose and that it is O.K. by them. Well, that's part of the journey anyway – but I expect there will be quite a bit of red-tape to be unravelled yet awhile.

To get back to your work – it seems to be interesting you which is the main thing and it makes me very glad to think you are sticking in the family business (at any rate in a branch of it). You should have almost unlimited opportunity to learn the practical side of Bacteriology, Immunology and, of course, many of their ramifications.

As for spring fever, you say you have (or had) a bad bout of it – well it gave me a miss and now I think I'll put it off until the fall. By the way who's this Yale guy?? A rival!

As you can understand my plans rest rather in the lap of the gods – but I shall be here for the whole summer working partly in the lab and partly in the Hospital doing routine work and wardrounds.

---

35 Fred Himmelweit (1903-1977) fled Nazi Germany in 1933 and ended up in London, where he obtained a position at the Wright-Fleming Institute at Saint Mary's Hospital, working first under Sir Almroth Wright and then under Sir Alexander Fleming. At the institute he studied viruses and influenza vaccines, and eventually he was appointed director of the Department of Virus Research. He was a fairly senior member at St. Mary's Hospital when Robert Murray met him on this occasion. Murray remembers that he probably came to the Cambridge lab to visit Alan Downie, one of Murray's teachers, who was a virologist who had to spend much of the war working on bacteriological problems. RGEM; Arthur M. Silverstein, "*The Collected Papers of Paul Ehrlich*: Why Was Volume Four Never Published?" *Bulletin of the History of Medicine* 76, no. 2 (Summer 2002: 335-339).

36 The "Gradocol membranes" were an early version of the fine filter membranes that can be used to filter out most microbes like bacteria but allow molecules and many viruses to pass through. RGEM.

It is two months since you wrote the last letter and I expect one any day now if you are sticking to your self imposed schedule. And for gods sake let me know what was at the bottom of the mysterious Xmas card of three years ago.

I must get back to work for it is only 4 P.M. and there is a lot of waiting to be done. Look after yourself.

All my love,
Robin

14.viii.41
6, Tennis Court Terrace
Cambridge

Miss Doris Marchand[37]
210 East 68th Street
New York City
N.Y.

Dear Doris:

... My plans for going home are a little more definite. I have got Canada House on my side and they are doing the arranging for me. So that after Oct. 1st I will be sitting around waiting for a short notice sailing. Of course if the war flares up into something like an invasion I shall be staying to help – but barring Hell and Highwater I'm going back as soon as I can. As far as my work is concerned it will be a good thing.

18.viii.41

Since I last wrote the theatres of war have changed about yet again. Russia seems to be having a hard time of it yet appears to be fighting well. I, personally, do not trust them an inch but I suppose in this looney war such prejudices must be cast away and everything to our advantage must be utilised. ...

I wish I could work in your place for a bit; it would be of great interest to me to find out the method of producing vaccines and antitoxins in quantity and the institute would be a good place to learn. However, I suppose I must be good and get on with my own job of work at the moment (i.e. getting qualified). They seem to look down very much, quite rightly too, on Pathologists and Bacteriologists without medical degrees. Why don't you save up a bit and go to Med. School? Good for you and would probably qualify you for a first rate job. ... You've got enough of the inbred scientific spirit to make a good job of it. ...

---

37 Correspondence of Robert Murray and Doris Marchand, private collection, Robert Murray.

I spend four or five hours a day in the Hospital where I have managed to oil in as a Medical Clerk to one of the honoraries. It involves a sort of reduplication of the house physician's duties – History and exam – of all his patients and writing up the proper sheets. Good practice and experience for me. There are some very interesting cases involving some new pathological work being done ...

Life is not "all work and no play" for there are occasionally monumental binges and pub-crawls which enable me to take a not sober, but, more reasonable view of things. Otherwise I would age before my time and become a sort of "Don" before ever taking an M.B. or an M.A. Fate, who despite everything is a very perspicacious lady, decrees that, every time two or more medical students gather together without a thought of work, good drink and all the etceteras should be rife! An expensive but pleasurable dictum. ...

Cheers, my dear, and have a scotch straight for me and at a reasonable price.

All my love,

Robin.

16.ix.41

Dearest Doris:[38]

I have started to write to you several times in the last couple of weeks and somehow I have been unable to get into the mood for letter writing. But on this occasion I have something definite to write about and it is, in my humble opinion, great news. On Saturday, in fact three days ago, I received my Exit Permit which allows me to go whenever I can get a passage. It looks probable that I should be over in Canada by Xmas time. I'm not very sure at the moment how easy it is to move between the States and Canada, but I rather fancy that it is easier for you to come to Canada than for me to come to the States. With a bit of luck we may be able to see each other in the not so very distant future. I don't know how much you have changed in the last two years, but I know only too well that I am now a very different individual from the rather odd type you used to know in me! Partly, of course, a change in years and partly through the tremendous change in experience involved in living over here in peace and in wartime. It has given me added experience and knowledge. ... [Wartime] has a very great tendency to kill ambition although I hope it hasn't quite killed my ambitions, high-sounding though they may be. In a way, I suppose, I'm returning home because I feel that I am at the moment discontented with being merely a student among people who are fighting for a very real cause – but in another sense I will be very glad to get home in order to get a more objective point of view that I am missing, or feel

---

38 Ibid.

that I am missing, over here. ... So that, all in all, I feel justified in leaving England although feeling at the same time as if I were leaving a trusted friend to fight a battle without my being in attendance. To boil down all this sentiment, I am a person of mixed feelings: I want to go home yet I'd like to stay here; I would like to work on alone and yet I'd like to have the benefit of my father's knowledge. And, my dear, I'd like to see you: so now you know.

As far as things go at the moment arrangements are made for my return. Plans for the future are a bit nebulous. I'm not altogether sure, as yet, that Medicine in its purer sense will be my future. I think that I will eventually end up in the labs – probably on some side of the medical sciences – but on what specific line of work I have little idea. I'm young and so is my learning – time will tell. But for at least another two or three years I'm going on with my present line of attack via Medicine: for in that I can do the most good in the present times and it provides the best education for any further work I may explore. You may say, and you would be right in a way, that it is a very uneconomic way of starting any career. I admit that it puts much too much of a drain on the parental purse, but he agrees and so do I that it is better to be fully qualified for a type of job than to be insufficiently qualified with only interest to keep the job. Selah! So I'm going on: If I had better eyesight I might chuck the whole project and go into the R.A.F. since it is in that service that my own type of thinking is of use, but, thank the lord, I'm not gifted with good eyesight so I have to give that pipe-dream to others. Despite this loose talk I have a real interest in the academic side of medical investigation so I'd rather stick to that.

At the moment it looks as if I may be home before Xmas but will not be likely to come by New York. So when we see each other it will probably be well after I get home and we can correspond with much less delay. Anyway, if something drastic were to occur I might not be able to return. ...

    All my love,
        Robin

                                9.xi.41
                              Montreal

Dearest Doris:[39]

Here I am, like the perennial "jack in the box", back on the continent of N. Am. again – and about time too! It's been a very long time and I must say that, now I am here, it is a very thrilling sight to see lighted cities, plenty of stuff in the shops, and a slight veneer of prosperity. On the other hand I'm astounded at the high prices that prevail in these parts – that and the amount of soldiery you see in training are about

---

39 Ibid.

the only signs of war over here. I expect you have had all my screeds and diatribes by now; written in all states of doleful sobriety and maudlin drunkenness!

So you may gather that the order of my departure from England (which I seem to have formed a habit calling "home") was as sudden and unexpected as possible – true enough, it was. I got no more than 12 hours notice to pack and report in London, to find on getting there that I had to wait two days and need not have broken my neck in trying to be punctual. As a result many of my books and papers had to be left in Cambridge in charge of friends or on gift to the Lab. So many things that have been with me for years are now lost forever. All this was late Wednesday afternoon. Having packed my goods and chattels there was left the task of saying good-bye to whole rows of people – but I sort of funked it and we went off on a binge and met most of the people that way. Money was another difficulty for I had precisely £43 in the Bank and the fare was 40. So I had to borrow £20 off my cousin when I got to London (after finding out that I didn't have to leave London till the Saturday).

While in London most of my mornings were spent attached to the end of a telephone; lunch-times and afternoons seeing people in the flesh; evenings getting drunk with some of my more reliable friends. Although it was a hurried parting it was some party while it lasted – finally I was put onto the sleeper for Liverpool where I arrived with a terrible hangover, a feeling of nausea, in one of the lousiest stations in the world, on a truly Liverpudlian day (i.e. soot and water mixed in equal proportions), at the ungodly hour of 7.15 A.M. Summoning my energies I eventually found my bags and sent them off, while I and my microscope went forth in search of coffee and a light breakfast. At midday after a lot of palaver and to do with censors, detectives, police, customs, and immigration men we are allowed to embark on a tender which takes us to our Banana Boat which looked about 2x4 on the river but turned out to be clean, comfortable, and seaworthy.

We didn't set sail properly until the Monday morning when the convoy was ready. Then on up through the Hebrides along the west coast of Scotland then up to about L 25° W x 61° N: after that we seemed to zig-zag all over the North Atlantic so that on the 8th day we were nearer Spain than N. America. At long last, a few hundred miles from Newfoundland after 48 hrs of alarms and excursions the part of the convoy bound for the St. Lawrence broke away from the convoy and we steamed at full speed. We landed at Montreal last Tuesday, after 17 days of sea.

Now after a few days getting settled I'm back at work again – in an atmosphere of unfamiliar routine, must get back to work and catch up. Will write again soon – let's hear from you.

   All my love,
     Robin

15.xi.41
Montreal

Dearest Doris,

It was grand to get your letter and to realize that it takes but a couple or three days to exchange. Quite a difference after our bi-yearly efforts! However, it occurs to me that my letter described practically nothing; what I really need is a dictaphone and a few private secretaries, and even then I'd miss the spirit of the thing.

When we were up near Iceland we were joined by some boats from Iceland together with several American destroyers (one was one of your most recent types; the others were the same as the Reuben James). From then on there was almost ceaseless activity and the destroyers (we had about 8 by now) were fussing about: every now and then, once or twice a day there would be several salvoes of depth-charges, usually a long way off from the convoy. At 5.00 A.M. one morning, just as I was about to get up and go on watch, there was a very determined set of explosions and then a number of gun noises – by the time I got on deck there was nothing to be seen – convoys are, as you no doubt know, blacked-out completely and it is as much as you can do to make out the ship next to you about 3 cables away. We were told at breakfast that morning that an American Tanker (16,000 tons) three ships behind us had been picked off. (That was all in the papers).

The previous night it had also been quite exciting – about 9.00 P.M. just as we were about to start a drinking party several of us took a turn round the deck and one of the destroyers came by quite close signalling to the Commodore "Am going into action" … – then went off at full tilt. About 5 minutes later there was a lot of flashing depth charges etc. and a Verey light[40] or two. I have good reason to suspect that it was the U.S.S. Rueben James but I'm not altogether certain. For the following 48 hrs there were a lot of alerts but nothing definite – then, at last, we were out of the danger area and the convoy broke up. I'm not sure which was coldest Lat 62 where we were still manning the guns or the St. Lawrence and gulf where we weren't. The mountains on the north shore and on the south shore at Rimouski were very bleak and had snow on them – and there was a howling wind.

It was a fine and exciting time to see Montreal again from the river. I hadn't time to be too excited for many people on the boat kept asking what various landmarks were and all that sort of thing. Then when I had had my lunch I buzzed off home, having no trouble with customs, immigration or any people of that type and arrived home some hours ahead of schedule. Now, after 10 days I'm settled-in and comfortable.

Binges here will be few and far between – somehow, there is a very different drinking atmosphere and I haven't got used to it yet. However, a week ago we held a fairly

---

40 Ibid. A Verey light is a type of flare.

large one in the house here for about 30 people – a sort of super bottle-party – which was a lot of fun.

About getting down to the States: The situation is difficult. As you no doubt realise U.S. dollars are a very essential commodity to those in the Sterling area and quite naturally Canada is a bit fussy about people who go down without essential business since it means a loss in exchange. As far as I can find out they do nearly everything except throw you in jug even when you have adequate permission. *However* U.S. citizens *with* U.S. dollars are very welcome on this side of the border – so, unless the war ends fairly quickly, our only hope of meeting each other is up in Canada. What about it?

Those friends of mine from Cambridge are over here now – but I expect that when they arrived they had to go direct to their various universities. I have not written to them yet; but I will in the near future. ...

You may be only 400 miles away but it still seems as far as when we had a broad, grey, Atlantic between us.

      All my love, Yours,
          Robin

10.xii.41
3590 University Street
Montreal

Miss Doris Marchand[41]
156 East 52nd Street
New York City

Dearest Doris:

I'm a bit late in answering this but I got a bit tied up in a bit of work – not only that, but things have been happening just a bit quickly in this world of ours. The damned Japs have done a Hitler in the end as many have feared – since they did it as a bolt from the blue most of us are a bit mixed up. The case of Radio panic hasn't helped atall. So I myself am resolved to wait a couple of weeks before I read the news atall carefully. They tell me that there was a bit of air raid warning excitement in N.Y. yesterday and all was "spoof". I suppose it is just conceivable that there could be bombing of N.Y. but I feel that it is more than unlikely for quite a while yet. Out west they are having black-out troubles and I sometimes like to think that people, like myself, who have had some experience of the system might be of some use in helping along

---

41 Ibid.

the arrangements – at the moment they are very slow and don't seem to have had any arrangements or advance plans for an authority which could possibly work quickly in an emergency like this (Applying to B.C. as well as U.S.). ...

Dad has returned from Washington ... and is now on a circular tour to Kingston, Toronto and Ottawa on the same sort of business – the old man is being kept very busy and is being very hard pressed to keep up with all his business. Soon, he tells me, he has to go to Vancouver and will at last have a chance to see the West Coast. Sometime when all this ... is over I must make a real effort to go over to that side of the continent and do some fishing. ...

I hope you can come on the 20th, my dear, for I shall be looking forward to seeing you after so long a time away from these shores. We will be able to gab about the many things that don't seem to find their way into my rather patchy letters.

All my love – and don't let war get you phased.
    Robin

Apr. 2, 1942

My Darling:[42]

... Well, gasoline rationing and the snow came in together today. ... I can't say that I've noticed any very great reduction in the number of cars on the road as yet – though mother tells me that yesterday the used car markets were clogged-up with people trying to sell. ... In the end there will have to be a great reduction in cars since there will be no sale of new tires until well after the war is over and there will be little in the way of "retreading" possibilities except for essential vehicles. ...

There are two war points that are interesting me. One was a report, broadcast in a news bulletin on the Canadian radio but not repeated in any other bulletin either here or England or in the papers, that "poison gas" had been used by Japs in Burma. Was it published down your way? (*if you didn't see or hear of it yourself don't ask others* as you may *unwittingly* start a man-sized rumour). The other was a report I have confirmed but also would not like repeated – that was in the newspapers as well (but well hidden) – that the Japanese, by some means unstated, dropped from an aeroplane infected material and caused at least 6 cases of bubonic plague. Now this is a very interesting addition to present methods of war – it is perfectly feasible – in fact it was used about 1350 in the siege of Tana (on the Black Sea) when the Tartars catapulted victims into the city and gave the disease to the Genöese, who took it back to Italy with them. Pus or sputum on clothes will remain viable for 2-3 days; and heavily contaminated clothes or bedding may have viable p. pestis for some 6 weeks or so. ...

42 Ibid.

I'll write again once the week is out – if possible before the promised binge on Friday night (or maybe I'll write under the influence again)!

Much love to you, my darling, and sweet dreams.

Robin

Sept 29, 1942

My Darling:[43]

Well, the lecture finished at last: a rather long-winded, tedious, and I think unfair, account of arteriosclerosis. So Ian Stevenson[44] and I returned here post-haste for some congou tea (very good it is too, thank you!). He is very anxious that I should have something to do with the McGill Medical Journal. This is a very attractive idea in some ways. The only great fly in the ointment is that I might be tempted to spend a little too much time on it. However, it does provide certain privileges and opportunities. Also it does provide opportunity to have a hand in profecting a little more forcibly a policy that Ian and I are much in agreement with – Viz: providing a basis in fact and policy to extend the McGill Medical Journal into a real bulletin of the researches and observations of the doctors and research men in the University and in the teaching hospitals – instead of being, as it has been in the past, merely a rather futile expression from the medical student alone. With a bit of forethought and cooperation the Journal could become very useful again. More about this later in the year. ...

Darling, I love you, and remember you always.

Yours,

Robin.

---

43 Ibid.
44 Ian P. Stevenson (1918-2007) was born in Montreal and educated at St. Andrew's University, Scotland, and McGill University. He was a classmate and close academic friend of Robert Murray in the medicine faculty at McGill University during the 1940s. Stevenson spent most of his distinguished career at the University of Virginia, Charlottesville, Virginia, where in 1957 he was appointed professor and chairman of the Department of Psychiatry. He founded the Division of Perpetual Studies at the University of Virginia in 1967 and remained its head until 2002. He was internationally renowned for research into "survival of personality after death," or reincarnation studies. He died on 8 February 2007. RGEM; "Ian Stevenson Dies at 88; Studied Claims of Past Lives," *New York Times*, 18 February 2007; University of Virginia Health System Web site, http://www.healthsystem.virginia.edu/internet/personalitystudies/.

Dec. 30, 1942

Miss Doris Marchand
445 East 65th Street
New York City

My Darling one:

I got back to this home of mine last night feeling ever so dim and gloomy – but all the same elated to have been able to troll you about for 5 whole days (think of it). Do you know that we have been together for only 35 days since Bar Harbor.[45]

The more I think of it the more I feel that this last time was the very best ever – untrammelled by colds, accidents, or revolutions we managed to be happy and close for the whole time.

I wonder how long it took the train to get out of that By-our-Lady station after I left. It was about 11p.m. when I got home frozen to the marrow feeling like a real icicle outside. ...

Another thing that makes me smile *all* over is our conversation by the ticket booth (such an original place) and the changing of the ring. Don't misunderstand "smile" – I mean it in the old sense of happiness – so happy am I that I can hardly write and it is very very hard to sit down and read. Everyone is very understanding about the whole thing. ...

Did I ever tell you *how* much I love you? Well, if I didn't it's only because it's impossible. ... I regret not having given you anything (but myself) as a souvenir of this visit. It's a damned bad show – somehow I will make up for it. ...

All my love, dearest, and don't worry yourself too much.
      Your beloved,
      Robin

19 January 1943 Montreal

My Darling:

Did you get my rush of apoplectic letters last week? Come to think of it you may have had to wait for today to get them – the last one was dropped into the post-box as David Ashdown[46] and I were going to the National Film Society show.[47] The latter was

---

45 Ibid. Robert and Doris met each other in the summer of 1938, when they attended summer courses at the Mount Desert Island Biological Laboratory, Old Bar Harbor Road, Salisbury Cove, on the coast of Maine.
46 Ibid. David Ashdown was one of Murray's fellow medical students. He hailed from Bermuda, but the two have not been in contact since the early 1940s. RGEM.
47 The National Film Society (Canada) was founded in 1935 at a time when few films could be

very good – the society is a closed membership affair and since it shows the stuff w/ 16 m.m. film it does not have to bother with censorship regulations (which, I might say, with regular film is some problem in this Catholic province).

They showed two films both of them by that bloke Flaherty.⁴⁸ The first was "The Land" which had a sound track commentary and which was all about Farms, Erosion, and the redistribution of farm workers due to mechanical aids. The second was "Moana of the South Seas" which was a chronicle in the life of a family on Savaii (Fiji) over a period of about 2 years. The photography and treatment of the individuals involved was marvelous though it was all idyllic and showed nothing of the glummer side of life (if there is any down there in peacetime). After this was over (it was a long show) we dropped in at the Berkeley⁴⁹ for a drink and somehow didn't manage to leave there till the Bar closed at 2 a.m.! So a part of your $2 was of great use to me – and sooner put to use than I expected – so thank you very much for the nice drink. It was quite amusing (I drank *very* little) for I met again, after a seven year interval, several people I was at school with. Half the fun was seeing the diversity of occupations they had got themselves into in the meantime, on the other hand I wasn't very impressed with other sides of them.

Sue is all aglitter and agog – for on Friday ... she is going to her first "Ball" (so-called). She has bought an evening dress to go with it and this she is at the moment trying on for the first time at home. Everyone is suitably impressed and, I must say, she looks very pretty in it. (For your information it is made of black corded silk with an instep length full circular skirt with a sort of 1ft rim at the bottom, puffed short sleeves – relieved with two red bows lower front R & L etc., etc.) The occasion is the Medical and Plumbers Ball (combined) and is usually quite an affair. Mother is wondering whether I shouldn't be around just for the appearance. But I wouldn't be wanting to go very much unless you could be around so ... And anyway who is going to pay the colossal sum of $5 a ticket? Anyway Sue is very excited about the whole thing and

---

viewed outside the commercial theatres. The founders hoped that the Society would "provide information and distribution services to groups of non-theatrical film users such as departments of education, adult education groups and various technical organizations." "Chapter IV: Films in Canada," Royal Commission on Development in the Arts, Letters and Sciences, 1949-1951, http://www.collectionscanada.gc.ca/2/5/h5-411-e.html.

48 Robert J. Flaherty (1884-1951), a documentary filmmaker, wrote and directed many works, and was nominated for an Academy Award (Best Writing, Motion Picture Story) for his 1948 film *Louisiana Story*.

49 The Berkeley was a restaurant and drinking establishment on Sherbrooke Street not far from the McGill campus. Murray visited there with friends a number of times. Murray recalls going there on one humorous occasion and finding it closed, with a notice reading "Closed owing to the immaculate conception." It also closed on VE day in 1945. RGEM.

would probably rather be without having me about, even in the far distance! She has got a lot of pleasure out of the mere fact of having the dress!

Well, that's just by the way – Saturday came and I did little but work in the morning and laze all the afternoon. And in the morning I went down to the M.G.H. for a quiet evening of sipping beer and serving-up drunks with Ted Rose,[50] Ian Stevenson and Clive Phillips-Wolley (an old Cambridge man, by god)[51] – in fact, a very select gathering. We talked a hell of a lot about nearly everything under the sun except sex and quoted from everything except bawdy rhymes. So it was an evening of "uplift" with a bit of beer drinking on the side! We didn't have much in the line of duty to play about with except a couple of transfusions and visits to the v. sick patients of Ted's and Clive's – but we had a negro woman who came in about 1.30 a.m. with a finger blown off by a detonator cap (she worked on munitions) which required a little attention. After an extremely pleasant session I trotted off home at 3.15 a.m. so I had two latish nights in a row and it was all of 1 p.m. when I woke up yesterday, and I struggled out of bed for a quick lunch. The best part of the afternoon and evening was spend in Editorial work – with a break for tea and a late supper (9.15 p.m.) and then Bed. ...

How are things going darling; for my part I am bearing-up bravely under the strain and I'm almost up to standard. I feel terribly lost without you and I love you more every day.

    Your loving, Robin

[Diagram of an imaginary virus with the following description]
This, believe it or not, is the anti-letter gremlin virus – gives them letters before the eyes, a severe fever, and in their delirium they tear up *imaginary* letters all day and all night long. It is incurable and no prophylactic measures are of any use.

"Doris Marchand and Robert Murray, Canadian, Engaged: Rockefeller Foundation Aid Fiancee of Student at McGill University"[52]
Mrs. Richard Werner Marchand, of this city, formerly of Princeton, N.J., has announced the engagement of her daughter, Miss Doris Marchand, to Mr. Robert George Everitt Murray, son of Professor and Mrs. E.G.D. Murray, of Montreal, P.Q.

Miss Marchand's father, the late Dr. Richard Werner Marchand, was formerly on the staff of Rockefeller Institute for Medical Research in Princeton. The bride-elect

---

50 Ted Rose and his wife, Sylvia, were close friends of the Murray family in Montreal. Ted Rose was a classmate of Robert Murray at McGill and worked as a physician at Montreal General Hospital before moving to Victoria, BC, where he continued to practice. RGEM.
51 Murray does not recall Phillips-Wolley.
52 *New York Herald Tribune*, March 11, 1943.

attended Abbot Academy, Andover, Mass., and the Bennett School, Millbrook, N.Y., and was graduated from Smith College. She is with the international health division of the Rockefeller Foundation, this city.

Mr. Murray attended McGill University, Montreal, and was graduated from Christ's College, Cambridge University. He is now completing his medical studies at McGill Medical School where his father is head of the department of bacteriology.

"Troth Announced"[53]
Miss Doris Marchand

Miss Marchand, whose engagement to Robert George Everitt Murray, son of Professor and Mrs. Everitt George Dunne Murray of Montreal, was announced this week, attended Abbot Academy ... and the Bennett School ... and was graduated from Smith College. She is a daughter of Mrs. Richard Werner Marchand of this city, formerly of Princeton, N.J., and the late Dr. Marchand, one-time member of the staff of the Rockefeller Institute for Medical Research in Princeton.

> 3590 University St.
> Montreal
> Oct. 27th (1943)

Miss Doris Marchand[54]
445 E. 65th St., New York City 21,
N.Y., U.S.A.

My very dear Doris,

I am looking forward so very much to seeing you on Sunday. It will be a real joy to have you with us and I do hope you will really feel at home and be very happy.

Your room was painted this summer; it is a nice 'gentle' shade of blue green and will 'go' with anything....

Bob is so happy to know that you will be here soon. Poor old boy, he has missed you sadly and letters aren't by any means the same thing. I too am happy to know you will be together.

Everitt is away and will not be home till Friday but I know he will be very glad to welcome you. Sue sends her very best love and can hardly wait till Sunday!

My love to you, my dear. I am so glad you are coming.

    Affectionately and in fearful haste,
        Freda Murray.

---

53 *New York Times*, March 12, 1943.
54 Correspondence of Robert Murray and Doris Marchand, private collection, Robert Murray.

Robert Murray's McGill Medical Experiences, 1943-44
(Recollections, "The McGill Experience," London, Ontario, 2004)[55]

Returning to McGill about October 20th 1941 was an almost instant return to being a busy student. Nominally I was entering third year Medicine but I had to have agreement of each of the professors regarding completion of the basic medical sciences subjects. ... Because of wartime needs the medical courses were accelerated by the expedient of abandoning any formal holidays or vacation time. Because my Class was still involved with taking Pathology and Bacteriology and Immunology I was excused lectures but demonstrated in laboratory sessions and then, as a matter of parity, took the examination with them. ...

The Cambridge experience and a touch of maturity made for better association with my fellow labourers and there were lots of happy moments despite the constant grinding coursework. The latter was soon diluted by clinical work which added the complication that some elements were in the Royal Victoria Hospital and others in the Montreal General Hospital separated by a mile and a half walk. When in full swing of clinics and ward work we added the Maternity Hospital, the Children's Hospital and The Royal Edward Hospital (for tuberculosis), and later there were a few weeks at the Verdun Protestant Hospital (Psychiatry). Somehow it worked out for our groups of four (Marksfield, Mintun, Mumford and Murray) for ward work and larger for clinics ... when at the MGH a bigger group of us adjourned for lunch and chat in a restaurant on St. Lawrence Main. In those days the MGH was on Dorchester just south of the major Red light district ...

It was the kind of medical training based on the British pattern that was brought into full form by Sir William Osler and his generation of pundits. Senior physicians gave formal lectures and among the most memorable were those by J. C. Meakins who seemed to come in with galley-proofs of a forthcoming edition of his textbook (which most of us used) and those by the more physiologically inclined J.S.L. Brown. In Pathology the teaching sessions with G. Lyman Duff were excellent both in the lecture room and in the autopsy setting. ... Some time every day was assigned to a ward in one or other of the hospitals, a number of weeks in each, and with patients assigned. That meant that we had to be prepared each day when the physician or surgeon made rounds with his retinue of resident, interns and nurses with the student group assigned to that ward. You were expected to be ready on your feet to answer questions about the patient's history and signs and symptoms of diagnostic import; equally you had to keep alert and learn as much as possible about prognosis and

---

55 UWO Archives, R.G.E. Murray Collection.

treatment. Time was also assigned or spent willingly on other kinds of tasks including attendance at the "outdoor" services.

... When we had finished, passed our exams, and at long last attained a degree we had an internship to serve. ... The internship year was contracted to about 10 months. There was no pay but you got a room with internal phone line and a set of white uniforms. ... I asked for and got a year in the Bacteriology service so that I came under the direct direction of Dr. Gertrude Kalz, and I could not have had a better tutor for learning clinical bacteriology at the highest standard of the day and still in contact with the research group of the Department in the same building (across University street from the Royal Victoria Hospital and next to the Neurological Institute). I did not have ward responsibilities but I had access to all cases of infections in the Hospital and did direct work on many of them. I had to act as relief for other interns as needed. Every bit of this experience was helpful if not crucial to my future. However, I was a bit short of some detailed medical and surgical experience when it came to my practice in the Army the next year; it was lucky that we did so much as medical students.

In the end I chose to go into the army with a touch of concern that I would end up in the set of problems that involved my father. So ... I went to the Officers Training Course ("boot camp") at Brockville along with about 40 newly hatched medico's many of them from my McGill class of '43b. It was a very good experience and we all became remarkably fit in a short time. ... I was sent to #1 Tank Training Regiment as a junior Medical Officer to gain experience with two seniors as a lowly Lieutenant. After a couple of months I was transferred to #2 CACTR with one senior colleague to enjoy a really cold Christmas on duty and then ... on my own to #3 CACTR where I stayed with a happy crew in the RAP until I was seconded to London, Ont., to teach at UWO and to be demobilized.

Freda and Everitt, Correspondence, 1943-44

Department of National Defence
Army
7/12/43

My Beloved,

This day 26 years ago you & I made a beginning of our life together & with the years our love & confidence have grown greater & closer. I know I love you to the uttermost possible. There is no greater happiness for me than to share with you the doing of things we love to do.

Our children have grown up. Bob is full fledged & on the threshold of his career. He will do well & our pride in him will grow with his achievements. I feel sure of him. He has chosen well in his girl & they will be happy as we have been.

Susan is troubled in spirit at that difficult age for a girl. She is sensitive & tries to suppress it ... I think the thing to do is to try to gain her confidence more fully & so soften her resentment of imaginary grievances. ... Don't worry my dear, we'll do all right, with patience, a little pain & increased love. She is a very dear girlie, really, but now a little disturbed. It will take a little time.

When I got in last night, Mrs. Bell had specially warmed my room before I arrived & met me with a cup of tea. Very kind. Could you find a small something to send her as an Xmas appreciation. ...

I had a letter from Ted Roy[56] today. He is going overseas soon & he opened up quite a lot. Dear old Ted. He is a good fellow & will do well when they let him. Perhaps this special job will give him a chance. He finished up saying, "and I would like to tell you that I am proud of the fact that I have (been) one of your boys." Nice of him. I too am proud of him, and the other boys too.

The pyjamas have arrived (I think); I have not opened the parcel yet. My glasses came today too & they were much improved. I am glad of them.

Tonight there is a C.W. shindig & I have been invited. A sort of Xmas time good will get together & cannot be avoided. It could be fun.

None of these things divert my loving thought from you & the years of struggle & achievement we have gone through together. There will yet be wholly happy ones for us when this turmoil is over.

    With all my love always,
        Your mate,
            Ever

3590 University St., Montreal
April 30, 1944
Professor E.G.D. Murray[57]

C/o Col. Sawyer
Canadian Military Headquarters
2 Cockspur St.,
London, England

Beloved,
I wonder if you will get any of my letters? ... I am just going ahead & writing more of them – just in case.

---

56 UWO Archives, Freda Murray Papers; Correspondence of Freda Murray, private collection, Peter Murray. See chapter 1, note 89, for information on Ted Roy.
57 Ibid.

Everything is alright & we are quite fit & well. The weather is beautiful spring like & the ice is gone out. That being the case I shall go up to the Lake at the end of the week to get that most important spring planting done. ... I think I shall go on Sunday: that early train gets in at a comfortable hour & will enable me to start work right away on Monday morning. Sue is looking forward to a week of house keeping 'on her own'...

Thank you for your cable, darling. It came on the morning of the 25th & I was so very happy to have it. ... I do hope you had an opportunity to go to C. & must have been very pleasant to visit old friends. How I wish I could have been with you! ...

    Your mate,
        Freda

                                  3590 University St., Montreal
                                              May 7/44

Professor E.G.D. Murray[58]
C/o Col. Sawyer
Canadian Military Headquarters
2 Cockspur St., London, England

Beloved,

I am all ready to go to the Lake for a week's work & enjoyment. It is 6 am & my train is at 9. The pup has a new collar & having put two & two together ... has come to the conclusion that something joyous is in the wind. He was snooping round my various activities all last evening. ...

I think I have left everything all right here for Sue & Doris to manage. ... All bills are paid – the cheque you spoke of has been paid into my account. I've put Bob's allowance into his account. ...

So beautiful to see you at the house once again – whenever that may be.

    My heart is with you as always,
        Your mate,
           Freda

---

58 Ibid.

May 16, 1944

Beloved,

Here I am home again after a glorious week's holiday. ... Weather was patchy, but there were three superb days ... the lake was very high – up to our bottom step. I found everything in good order except that R. had forgotten to dig out the drain under the house as I had directed & the house was standing in a bog! I had to spend a whole morning digging out. ...

I always miss you when I am having a good time away from you – as you know, darling. However, one day we'll be able to have a holiday by ourselves again & that will be something to look forward to. ...

Your letter arrived on Thurs. Bob phoned & read it to me. I was glad & relieved to hear Guilford [Reed][59] phoned on Monday morning & gave your message that I was not to expect you just yet. He had a good journey – and poor man, wanted his shirts which arrived here a couple of days after your departure & which I had kept as I was sure he would be coming through MTL. He arrived at midnight & left on the 7:30 for Kingston & would not disturb us by coming at that hour!

... I look forward to your return so very much: It seems ages since you left. ... We are all well & the youngsters find ... love ...

All ... to you beloved,

Your mate,

Freda

May 19, 1944

Beloved,[60]

No further letters from you. I expect you have written & letters are slow in coming. Your cable to Bob arrived yesterday. He was delighted to have it. Today I am taking a large cake up to the Lab for 4oC tea & afterwards Bob & Doris are coming in for supper & we will have

Borsch

Lobster & salad & French long loaf

Asparagus

Mince pies

-- a small celebration of his birthday!

I had news of you today from, of all people, Mrs L.M. (Le Merchant). She had a

---

59 Ibid. See chapter 1, note 115.
60 Ibid.

letter from Wax (her son) telling her that he had run into you in Bond St. – What a surprise for you both!

The old boy (B) is well but a bit tired. He could not spare time to go fishing with Hilary [Robertson] this weekend which is a pity. He feels he has had too much time off already what with mumps & ... I have given him a cheque for $20.00 from us both: better than trying to find something he probably doesn't want. ...

  Keep well, my darling & take care of yourself,
    Your mate,
     Freda

               3590 University Street, Montreal
                        June 5/44

Professor E.G.D. Murray[61]
Canadian Military Headquarters
2 Cockspur Street, London, England

Beloved,

  I was happy to have two letters from you today – the first word from you for a very long time. They came fairly quickly – 17 & 19 days respectively. ... It was nice to hear news of so many old friends & of your happiness of seeing them again. I know just how much it has meant to go & find the old friendships & affections fresh and strong after so many years absence. How I would of loved to have been with you. I have thought so much of you & have imagined you working in various Labs & Hospitals. I suppose the beard caused no little amazement (or should I say amusement!!). ...

Yes, I was at the lake at about the time you mentioned. ... Last weekend Sue and I had the offer of a lift up there ... & we joyfully took it. ... The countryside looked lovely. ... It was good to get home to the cottage again – everything there looked grand: flowers out everywhere ...

  Take good care of yourself,
    Freda

---

61 Ibid.

3590 University St.,
Montreal, Canada
June 10, 1944

Professor E.G.D. Murray[62]
C/O Department of National Defence
Ottawa, Canada

Beloved,
... As I told you in a letter some three weeks ago, I have made arrangements to have the three rooms at the top of the stairs occupied for the summer months. The housing shortage is still acute & interns & their wives are finding it very difficult to get living quarters near the hospital. ... All this entails some extra work before I can get away but it is very well worthwhile. ... I am moving Doris & Bob down to our room. ...

Doris has 2 weeks holiday starting 1st July (& two weeks later on with Bob) & will come up with us to the lake. ... At any rate, none of them will be alone in the house at any time! You have probably had Doris' letter telling you the news that we are prospective grandparents. ...

All my love, your mate,
Freda

My darling Daddy,[63]
How are you? Thank you so much for your letter. It was wonderful to hear from you again. ...

When you see all our friends (in UK) give them my very best love & tell them that believe it or not, I shall write some day!

Life here is much the same as usual – except for a bit of excitement a week or so ago: I went out on a blind date with an English flying officer who has been all over the world, both in and out of the R.A.F. Well, he was devastating (absolutely!) one of 'London's top social playboys'!!! Well we rushed around seeing & doing all sorts of exciting things – until he disappeared & I found (by method of the grapevine) that he was *Married*!!!! I could have bopped him over the head!!!

I finish school on Wednesday. ... I wish you could be here for my graduation – but I shall think of you! I have been working quite hard on my art. ... This graduating business is quite tedious. There are all sorts of lunches, picnics, teas, dinners, garden

---

62 Ibid.
63 Ibid.

parties, meetings – gosh the other night we were invited to the Old Girl's dinner & were formally 'inducted' as O.G.'s!

I miss you & hope you will be home soon. All my love,
    Sue

McGill University
Sept 8, 1944

Dear Old Dad:[64]

... The exams are all over now thank goodness. They were quite an ordeal. Although the standard is undoubtedly a bit low (!), there was an alarming difference between examiners: some hardly bothered to ask a question and some went into minutiae. Besides, no one examiner deemed to keep up a standard level of questioning. If the papers are dealt with in the same vein there must be some coarse results. I feel fairly confident about most of them but feel a bit squeamish about obstetrics, and the written in medicine.

Money may be a bit difficult for a while but we will discuss it together later.

Doris and I are going up to the lake tomorrow & I am going to stay as long as possible. Ted & Sylvia [Rose] are coming up for the weekend. Ted is ordered to work in ... the Army but has been lucky in being allowed to do it in the O.D. of the Vic. Some others have not so lucky & have to go to horrible places like Longueuil.

Ian [Stevenson], poor fellow, is being forced into the RCAMC willy-nilly. He is in better health now than he has been for years but I feel that if he is sent to some god-forsaken place in the winter it will do him a lot of harm. If the Army did ruin his health he would never be compensated.

Thank you, Dad, for your fine letter – very cheering. I certainly hope you can get a really good fishing holiday at the [illegible]. Mother is looking forward to it very much. She is planning to come down to Montreal next Tuesday (I think). I talked to her on the phone this morning. She sounds well & happy,

    All the best to you, Dad,
        Your loving son,
            Bob

---

64 Ibid.

H.W.H. "FREDA" MURRAY'S WARTIME WORK
Food Conservation Committee (I.O.D.E.)[65]

The Food Conservation Committee
Lt. Col. Scrimger V.C. Chapter I.O.D.E

3590 University Street
Montreal, Que.
February 28th, 1941.

The Disposals Board,
Ottawa, Ont.

Dear Sirs,
The above Committee has been in operation since February, 1940, and has, as its dual objectives:
1. The conservation and utilization of waste foods;
2. The conservation of waste articles from private households.

With regard to the latter:- in the course of this work it has proved possible to collect considerable quantities of waste fats, both raw and cooked. These fats have hitherto been sold to the City Renderers, who in turn sell it to the soapmakers; and until about November 1940, we received one cent (1¢) per pound, at which rate it was a reasonable item of salvage. The price was then lowered to one-fifth cent (1/5¢) per pound, which made it barely worth the time and labour involved in its collection; nevertheless, I have continued to collect it and have on hand some hundreds of pounds ready for sale.

It would be easily possible for our organization to extend this collection of fat very considerably and, under present circumstances, this seems to be desirable. I see by notices in the daily press that salvage is receiving the serious attention of the Disposals Board, and that the collection of bones for the extraction of glycerine is instanced. Therefore, with the much higher glycerine content of pure fats, it would appear that we have a valuable source readily available.

Information as to the value of extending the collection of fats, the proper means of disposal, and prices to be expected, would be gratefully received. ...

Yours faithfully,
(Mrs.) H.W.H. Murray
Convenor

---

65 Robert Murray Private Collection.

Report to the Annual Meeting of the
Lt. Col. Scrimger V.C. Chapter of the I.O.D.E (February 1944)[66]
Annual Meeting
Lt. Col. Scrimger V.C. Chapter of the I.O.D.E
Welfare Committee
February, 1944.

On the occasion of the fourth Annual Meeting of the Chapter I beg to give a review of the work accomplished by the Welfare Committee from the time it was organized in February 1940 to October 1943, when certain of its activities ceased; together with an account of the Codliver Oil Bank which succeeded it in November 1943.

As you will remember, in May 1943 I reported that since the second increase in Government Allowances it was found that the number of dependents needing our help was steadily decreasing, and it appeared that we might be able to cease our work on their behalf once those remaining on our books became firmly established on their own resources. In the case of needy mothers on assigned pay the situation was different for they received no increase, but they were few in number and their Auxiliaries were willing to take them over from us.

A very different situation faced us in February 1940. At that time there were rapidly increasing numbers of dependents of service men desperately needing help. Causes of distress were many and varied, the most usual being:

1. Large families unable to live on existing allowance.
2. Families burdened with heavy debts or sickness.
3. Delay in payment of Government cheques ...
4. The fact that recruits did not get pay or allowances for about fourteen days after joining up. As many had been on relief their families were unable to obtain credit.
5. Common-law wives with young children to support on part allowances only.
6. Widowed mothers on assigned pay, with young children to support.

Some of the Auxiliaries were already formed but found it impossible to provide enough funds for the ever-increasing requests for assistance. Some Regiments and Units had no Auxiliaries at all, and ... no existing organization ready to take them over. Civilian Social Services were not able to take soldiers' dependents in addition to the work normally required of them. From every point of view the situation was indeed desperately difficult. ... and with that in view this Welfare Committee was organized.

Briefly, its fourfold aims were as follows:
1. To provide an office to which reports of needy cases could come from H.Q.

---

66 Ibid.

M.D.4, C.O.s of Regiments; Auxiliaries; Directional Services; Civil Social Services and Welfare Organizations to whom dependents had applied for help.
2. The organization and maintenance of a Salvage Scheme whereby immediate and constant funds could be made available by the collection and sale of waste materials of many kinds; and a central Salvage depot through which such collections would pass.
3. The constant provision of considerable quantities of foodstuffs, parcels of which would be delivered to needy dependents.
4. The organization of a transport system which would make it possible for such parcels to be delivered without delay in the first instance and thereafter regularly once or twice weekly according to the needs of individual families, and for as long as might be found necessary.

The provision of adequate funds was the major problem. The Chapter was at that time able to provide a monthly grant of $5.00 but considerably more was needed. I turned to Salvage at that time as an unexplored source of revenue to all but junk-dealers, and slowly at first, and only by dint of much hard work, sufficient salvage of different types was collected and sold to bring in badly needed funds. Fats, paper, metal, tinfoil, bottles, egg-cartons, cork, string, rags, old silk stockings, stamps, coat-hangers, bottle-caps – all brought grist to the mill.

At first it was not sufficient and other means were sought to augment the considerable quantities of foodstuffs that had to be provided. Food markets were investigated, and it was found that large quantities of good and useable foods were going to waste. These were carefully collected and used and *good* meat and *fresh* vegetables were immediately available to us at no cost. For instance, in 1940, over 300 lbs. of meat was collected, and in 1941 over 900 lbs. ...

Transport was essential: It was of no avail to provide food unless it could be delivered directly to the families in the shortest possible time after receiving a request for help, for it must be understood that at that time many of these needy families were literally almost starving when reports reached us. For the transport of the year 1940-41 we are deeply indebted to friends within and without the Chapter, who so generously gave their time, their services, and their cars. During that time we had twelve cars available for rounds six days in the week and were often out within an hour or so of reports coming in. In 1941 it was seen that this form of transport was becoming increasingly difficult to maintain. I applied to the Canadian Red Cross Transport Service for help, and from April 1941 to June 1943 they provided all the necessary transport. ...

Parcels of food of approximately 20 lb. weight were provided once or twice weekly for as long as was necessary for the welfare of any particular family, and varied considerably according to their needs. It was obvious that these foods must give utmost nutritive value for weight; advice was sought from the Health & Nutrition Board and as far as possible the proper foods were provided: Molasses, canned milk, oatmeal, macaroni, wheat germ, peas, beans, canned tomatoes, bacon ends and bacon fat, beef dripping, bread, meat, potatoes and vegetables were considered essential; rice, sugar, salt, tea, cocoa and soap were added when necessary; apples, oranges, and later codliver oil were provided for sick children. In addition, used but useful clothes, shoes and bedding were collected and given to those most in need. ...

We found that those families who had endeavoured to effect economy by moving lower and lower in the rental scale were those whose morale suffered most. Bad living conditions undoubtedly contributed to this for it is difficult to struggle against filth, gloom and discomfort such as we too often encountered. There were other factors to be considered and the most important of these was the fact that allowances of that time were too low for the *average* sized "poor family" to live on (the allowance was for wife and *two* children only – certainly not the "average" amongst lower class families) and in order to try to make ends meet economies had to be effected on food and rent. ... Where a mother had to support herself and 6, 8, 10, or 12 children on money provided for an adult and 2 children a truly desperate situation ensued – malnutrition was found to exist in direct ratio to the excess number of children – the larger the family the greater degree of malnutrition – and the underfed family soon lacked the strength of will to fight against circumstances. ...

It was obvious that regular provision of carefully selected supplementary foods of high nutritive value, over long periods of time, were necessary to combat this condition. Time and expenditure were not to be considered – improvement in health and morale was the objective. Our system of regular visiting twice weekly with the necessary foods made it possible to note progress of individual families. Some little time usually elapsed before a change was noticeable but after a while the children began to look less pallid, became noisier, more friendly and cheerful. The mother would take some interest in our visits and began to tidy the house when we were expected and to keep it generally cleaner; wash her face and put on a clean apron; and by this time the children would be found at home much more frequently. Such results, although they came but slowly, were interesting and encouraging, and were on the whole very constant. It was noted that those families who had been on relief for long periods before the men joined the forces were the least responsive and took a longer time before change was observable. This appears to indicate that relief is insufficient for the needs of growing families. ...

Housing conditions in the slums were, at that time, as appalling as they are now. ... Large families crowded into one or two rooms with no sanitary conveniences other than a communal toilet in the yard several floors below. Ricketty houses with leaking roofs, walls with huge gaps in the plaster, and floors with missing boards and rat-holes in the skirtings. ... Houses where the only toilet was frozen and unusable for weeks on end and the cooking and drinking water had to be fetched from nearby houses for almost the whole of one winter. ...

Such were the conditions as we found them. In the course of over 2700 visits "below the tracks" unusual opportunities were available for comparison of living quarters, both the good and the bad, mean, filthy and ghastly. These things cannot be seen from the streets and sidewalks. One has to go into lanes and backyards, up pitch dark flights of stairs and ricketty wooden ladders in swarming backyard tenements, and down into dark cellars and basements. Hidden from the eye of resident and tourist alike, they remain a festering sore in the midst of the city. ...

That our work was useful was evident, both by the observed results and by the appreciation expressed by those with whom we worked. M.D. 4, Regiments, Auxiliaries, and Social Services all grew to respect our efforts to help those in need and they were glad to call upon our services and were confident that we would do what was required of us.

When the allowances to dependents were increased in 1942 to include four children instead of two only, we found an immediate decrease in the number of cases reported to us; and when, in 1943, two more children were included in the allowance, making six in all, the situation improved still more rapidly. The Auxiliaries were by this time all organized and in working order, and were able to help cases of emergency distress, and to provide clothes and supplementary food for those families whose very large number of children still made difficulties for them; and the Dependents Board of Trustees was able to provide assistance to dependents needing advice and financial help.

In every way the situation from February 1940 to the present time was vastly changed and improved, and we felt that the need for our services no longer existed.

When the Welfare Work, as such, ceased, I determined to continue help to the sick children of whom there are such tragic numbers in poor families, and the Codliver Oil Bank was established. ... Results have been reported as good and the health of children receiving it has definitely improved. ...

Total of Visits Made & Amounts of Food Distrbuted:
2784 Visits

42,625	lbs. meat, vegetables, cereals, canned foods, dripping, etc
5,603	loaves of bread
2,543	cans of milk
1,062	pints molasses
79	dozen oranges
113	large bottles Codliver Oil

Approximate total weight: 59,473 lbs. (including donated foods and Xmas baskets)

H.W.H. Murray
Convenor

Elan, Fleet, Robertson & Abbott
Barristers and Solicitors
Canada Life Building
275 St. James St. West
Montreal
10th March, 1944.

Mrs. E.G.D. Murray,[67]
3590 University Street,
Montreal.

Dear Mrs. Murray,

I confirm my telephone conversation with you of the 8th instant and return herewith the copy of the report which you gave as Convenor of its Welfare Committee at the recent annual meeting of the Lt. Col. Scrimger V.C. Chapter of the I.O.D.E.[68]

Congratulations on a really magnificent effort. I am delighted that the report will be published to show the extraordinary results achieved by your little group through hard work, initiative and ingenuity.

I think your description of the living conditions of the poor in this City as you saw them is the most graphic that I have come across. The publication of your report

---

67 Ibid.
68 The surgeon Lieutenant Colonel Francis Alexander Carron Scrimger (1880-1937) was educated in medicine at McGill University. He was awarded the Victoria Cross for his actions during the German poison gas attacks at Second Battle of Ypres in 1915. He joined McGill University as a lecturer in surgery in 1921. Canadian War Museum Web site, http://www.civilization.ca/cwm/media/bg_scrimger_e.html.

is bound to help in the efforts that are being made to correct these conditions that reflect so on our City.

    With very kindest regards.
        Yours very truly,
            Hilary

> The Montreal Daily Star
> "Canada's Greatest Newspaper"
> Montreal, Canada.
> May 6, 1946

Mrs. E.G.D. Murray,[69]
3590 University St.,
Montreal, Que.

Madame:

This is an interesting letter on a subject of general interest and importance. Unfortunately, it is so long that we cannot find space for it as it stands. Do you think you could (scale) it down and let us have it back?

    Yours faithfully,
        E.J. Archibald,
           Associate Editor

[Letter to the Editor]
A Sad Commentary On Our Slum Areas[70]

Sir, In your issue of May 1, I read with deep appreciation Senator Athanase David's grave and righteous indictment of the appalling conditions existing in the slums of Montreal. I can agree with all that he has said on the subject, for I know from personal experience those slums of which he speaks.

It is very necessary to bring these matters most clearly to the notice of all responsible citizens; for it is only through their righteous indignation and determination to change these things that the City Authorities can be made to achieve something more effective than a "Clean-up and Beautification Week." This, an excellent principle in itself, should do more than give an impression that we have a completely clean and tidy city to be proud of. It may be sufficient for those who are content to

---

69 Robert Murray Private Collection.
70 *Montreal Daily Star*, 17 May 1946.

view slums from the outside or to forget their very existence; but it does nothing to improve the dreadful lot of those who have to endure conditions "unfit for man."

The following extracts from a report I presented in 1944 may, perchance, add something to the growing realization of the terrible and urgent need for sweeping and drastic changes.

[See text of report above]

I have presented a harsh and unlovely picture. Believe me when I say that I have only touched upon the fringe of the matter, and that it is in the slums of towns and cities that the depths of human misery are reached. Anyone who has worked in them will recognize the dreadful miseries and discomforts of those who, through force of circumstance, must endure intolerable conditions. In many cases, "unfit for animals" as Senator Davis has pointed out.

H.W.H. Murray

MURRAY FAMILY PLANNING FOR THE POST-WAR YEARS

Robert and Doris

#1 C.A.C. T.R.
Camp Borden, Ont.
8 Sep 45

My dearest one:[71]

Another letter arrived from you today and I think with horror of the fact that I have written to you only once, and that was an absurd apology for a note. I guess I would have written to you at greater length were it not for the fact that there is virtually nothing to tell you (but) that I love you and that there is *no news* as yet. I am getting all jittery with all this waiting and I wish that something would break soon; I have almost got to the state of requiring grs½ Phenobarbital tid ac. If this teetering on the fence goes on I will have to take some to get to sleep at night. ...

We are not doing much around here except patching up the results of the odd fight. We have very few sick men. There has been no decision so far on what we are going to do with all these overseas men and I am crossing my fingers and hoping that I will be gone when they decide to board the whole lot out of the army. Nobody seems to give a hang anymore and everything is very loose and happy-go-lucky. Tonight there is going to be another small party in the mess but it will be nothing like the last one. It has been largely instigated by the returned officers who are a wild bunch of hooligans. It seems to me that they have a party every night of the week ... my room is too near the mess on these hot summer nights to make sleeping very easy when a

---

71 UWO Archives, R.G.E. Murray Collection.

couple or five fellows are whooping it up. I don't know what the hell they did overseas but from the way they talk you would think that they spent their whole time drinking and ravishing the countryside ... maybe, but I doubt it. ...

All my love to you darling and a hug to both you and Alice.[72]

Your ever-loving,
Robin

Monday 10 Sep 45

My darling:

Everything seems to happen to me. If a thing can be side-tracked, it is. This time it is not so serious. Being the day term started, I suppose all of us were beginning to wonder why the delay. Dean Hall's office[73] phoned this morning to say that they had a letter asking them to tell me to send my application for discharge "through normal channels". That surprised me a little because I had done that very thing on the 18th of August. They also said that the DGMS office had concurred in my release. So that leaves me in the anomalous position of having the people at Ottawa agreeing, and the people here agreeing, but the essential documents not having met in the middle. So I phoned the Adjutant of #15 Coy RCAMC to see what he had done. We went over

---

72 Alice Blair Murray, the first child of Robert and Doris Murray, was born in Montreal on 12 November 1944. RGEM.

73 Ibid. Born in Lindsay, Ontario, in October 1907, G. Edward Hall (1907-1972) attended the Ontario Agricultural College and the University of Toronto, where he quickly obtained a bachelor's degree in agriculture, a master's degree in biochemistry, and a doctorate in physiology along with a medical degree. By the outbreak of the Second World War, he had published forty scientific papers and had been appointed full professor in the Banting Institute of Medical Research at the University of Toronto; he was one of the youngest to hold this rank in North America. During the Second World War, Hall participated in the newly formed Committee on Aviation Medical Research, and then became a leading member of the Royal Canadian Air Force's medical services. For these achievements he was awarded the Canadian Air Force Cross and the US Legion of Honor and was made Companion of the Order of the British Empire.

Hall's appointment as dean of medicine at the University of Western Ontario in 1945 represented another stage in his illustrious career. In short order he recruited a group of outstanding medical scientists and clinicians who quickly elevated Western's medical faculty into a position of national and international prominence. After two years, Hall embarked upon another challenge: president and vice chancellor of UWO, a position he held for twenty years, during a period of momentous changes in post-secondary education in Canada. Western, for example, had under 1,000 students in 1947, and over 10,000 when Hall resigned in 1967. See UWO Archives, G.E. Hall Papers, "Dr. G. Edward Hall – Scientist, Medical Administrator and University President: A Profile of the Man and His Times," ed. Donald H. Avery (Symposium, April 11-12, 2003).

to Camp HQ and dug the thing up in the files, it having been there, approved and everything since the 20th of August. So we finally cleared it up and they promised to send it up to Ottawa. So we should hear something about it in the near future, ie next week. Another typical army bollocks-up.

Meanwhile I have been doing a certain amount of work at Bacteriology, trying to brush up on those things that I have not touched for a long time. I do not want to appear too much of an idiot. There is an awful amount that I have forgotten and it will take a little time to get back into the swing again. I am finding that the scheme of lectures that professor Asheshov sent to me is proving a great help in regulating my studies.[74] I certainly have the jitters compared to my ordinary self. ...

All my love darling.
    Your ever-loving,
      Robin.

Murray Family Correspondence, 1945

#1 CACTR
Camp Borden, Ont.
26 March 1945

Dear old Dad:
    I got your letter today. ...

I am still in the hospital but I will return to the unit tomorrow. This second bout was sort of a funny thing. No fever but considerable swelling of my cervical glands with attendant dysphagia. I had a 15,000 white count but nothing remarkable on differential count. The discomfort soon subsided but I have felt sort of played-out – cultures showed a heavy growth of "Streptococcus haemolyticus"! ... So for the last three days I have been sucking Penicillin pastilles. I will have another culture tomorrow morning before I go and see if there is any significant difference in flora. My sedimentation rate, for that's worth, has dropped from 25 mm/hr to 18 since Saturday and I think the infection has pretty well subsided.

Well, I had hoped to see Col. MacNabb,[75] but though he was supposed to come here last week he did not turn up. He is both elusive and uncommunicative.

I should be coming home this next week-end (30th) and with any luck we should be able to make it together.
    Your loving son,
      Bob

---

74 See Introduction, note 377 for information on Igor N. Asheshov.
75 UWO Archives, R.G.E. Murray Collection. Col. MacNabb was likely the O/C of the Medical Training Regiment at Camp Borden. RGEM.

#1 C.A.C. T.R.
Camp Borden, Ont.
14 July 1945

My dear Old Dad:[76]

We have been extraordinarily busy since I last wrote to you. Owing to the establishment of the pacific force we have been going through all volunteers as a sort of primary medical screening. In addition, we have transferred all our men to other training units so that we may act as a holding establishment for those coming in. As a result I have now less than nothing to do – this state of affairs will only be true for a short time – but I have to stay in because the other MO has gone on his furlough. When the others come there will be a hell of a lot to do.

The brass hats are agreed to transfer me to the lab when they have a replacement and when the present 'medical emergency' is over. That will probably not be till well on in August from the way that things look at the moment. As with all things in the army there is a good deal of chance about the whole deal. ...

That paper you sent me is very interesting and I am now returning it having digested as much as I could. Considerable work must have been put in on the problem. I wish that instead of saying "special methods" must be used to demonstrate such and such, he had come out with the method or, at least, the name of the method. Also he does not say anything of the media he used. I don't object, for I can always find out from the original papers. On the other hand it would have made the exposition clearer. I feel that some of the taxonomic affiliations he makes are a little far fetched but justifiable from the point of view of a working classification. Looking back I think that, on at least one occasion, I have grown them by accident and never identified them. ...

Doris and Alice are very well; and so am I.
    Your loving son,
        Bob.

---

76 Ibid.

Robert Murray Joins The University of Western Ontario

June 17th, 1945

Dr. H. Alan Skinner,[77]
Assistant to the Dean.

Dear Dr. Skinner,

In reply to your letter of June 6, I would like to put the following items before you for your consideration.

Since I took charge of teaching bacteriology and immunology in 1937, the arrangements of the course have undergone several changes with the object of improving the teaching. The most important change took place two years ago when it was decided to pool the courses of bacteriology and immunology. Practice has shown that such a change was not only logical but that it has also improved the possibilities of teaching and presentation of both subjects. It also gave the possibility of avoiding repetitions inevitable in the separate courses, thus saving time, which has been used advantageously in giving [a] more comprehensive course. ...

... it is my well considered opinion that the trimester system would be definitely detrimental in the teaching of bacteriology and immunology, reverting back to the arrangement which has already been found to be less satisfactory in practice. ...

I would also like to point out that from my point of view, the extension of a semester into January, breaking off the last week or two from the main part by Christmas vacation is very unsatisfactory. To have a course interrupted for 2-3 weeks and then to resume it just to give some last words on the subject for a week or two, does not bring any satisfactory results. Therefore I suggest that the fall semester is terminated before Christmas, while the first one or two weeks after Christmas vacation would be set aside for examinations. ...

    Yours sincerely,
    Dr. Igor N. Asheshov.

---

77 Robert Murray Private Collection. H.A.L. Skinner served as head of the Department of Anatomy at the University of Western Ontario from 1934 to 1964.

June 20th, 1945.

Dr. G.E. Hall,[78]
Dean,
Faculty of Medicine,
University of Western Ontario
London, Ontario.

Dear Dr. Hall:

In my opinion the development of research work in this Department, and a secure arrangement for the running of the course for medical students, cannot be achieved without further increase in personnel.

At the present time the actual teaching of about 40 students – 60 one-hour lectures and 60 two-hour periods of practical laboratory work – is done by me alone. Though in my preparatory work I am assisted by a Research Assistant and a technician, the arrangement is admittedly unsatisfactory. The Council on Medical Education of the A.M.A., for instance, recommends a ratio of teacher-student of 1:16.

The course given by this department is very compact, giving the students only the basic knowledge of the subjects and it admits no omissions. In case of my forced absence, due for instance to illness, there is nobody in the department who could replace me, and the missed part of the course could not be compensated for at any later date.

Besides this, the amount of preparatory work for the course – for demonstrations, and practical work – is such, that while it is going on, all other activity of the department, notably research, has to be completely suspended. This, of course, means not only a temporary suspension of the work but an actual set-back, usually with irreparable loss due, for instance, to alterations in culture, etc. This sets the department at a considerable disadvantage in comparison with other institutions where research is being conducted year round.

To bring at least a partial improvement and a certain degree of security to the course, I would suggest the following appointments:-

1. Lecturer – to be filled by an already experienced bacteriologist, capable of taking over a part of the course and of assisting continually in the practical laboratory work with the students.
2. Fellow – to be filled by a graduate intending to specialize in bacteriology and immunology.
3. The present position of laboratory technician should be filled by a laboratory assistant with at least 2 preferably 3 years of university training in science subjects and with a good theoretical and practical experience in bacteriology.

78 Ibid.

Though this department has another defect, which it seems impossible to correct at present – lack of space and room – the suggested increase in personnel would seem to be the only way for the long wanted improvement.

Yours sincerely,
Dr. Igor N. Asheshov.

Copy for Dr. Asheshov

No. 1 Canadian Armoured Corps Training Regiment (CA)
Camp Borden Command, 6 Aug 45

Dean G.E. Hall,[79]
Faculty of Medicine,
University of Western Ont., London, Ont.

Dear Dean Hall:

I wish to apply for the position of Junior Lecturer and Demonstrator in the Department of Bacteriology and Immunology.

The salary, I understand, is not fixed but is to be between $2500 and $3000 per annum. If it is possible I would appreciate it if it could be nearer the latter figure since I am married and have one child.

During my visit on Saturday I was most impressed with the present and future plans for the Medical Faculty. I am sure that I would enjoy the work and give my best efforts.

As you requested I enclose a summary of my training.

Yours sincerely,
"R.G.E. Murray"
(R.G.E. Murray) Capt. RCAMC.
Regimental Medical Officer,
No. 1 C.A.C. T.R.

Camp Borden Command, 6 Aug 45.

Capt. R.G.E. Murray RCAMC[80]

Qualifications for Employment as a Bacteriologist
My university education was started at McGill in 1936 where two years were spent upon basic sciences especial attention being paid to zoology.

79 Ibid.
80 Ibid.

In 1939 I went to Cambridge University where I entered the Natural Sciences Tripos taking both parts one and two. Part one consisted of courses in Anatomy, Biochemistry, Physiology, Pathology and Bacteriology over a period of two years. Part two consisted of a year's course (honours) in Pathology and Bacteriology. The degree of B.A. was taken in 1941 and M.A. was granted in Feb 45. Due to the nature of courses at Cambridge there is much opportunity to gain special knowledge and my attention was mainly focused upon Bacteriology.

In Nov 41 I returned to Canada and entered the Medical Faculty at McGill University. The degree of M.D., C.M. was taken in November 43. The medical degree was taken because of the necessity of medical training to a background for bacteriology.

Practical experience of bacteriology, apart from university courses was started at Cambridge at the Dept. of Pathology between June and October 1941. However, the most valuable experience was 9 months in the clinical Bacteriology laboratory of the Royal Victoria Hospital in Montreal, where a large volume of all types of specimens are dealt with. There was good training in the performance of Wasserman and Kahn tests, as well as in all phases of general bacteriology.

My education was planned to give a good grounding in Bacteriology. Through my father I have always been in contact with laboratories. As a result I have had the opportunity to do much work over and above that offered in university courses.

Experience in teaching has been very limited but that which I have done I have enjoyed. I have done some demonstrating as a student, both at Cambridge and at McGill, in Bacteriology, Pathology and Histology. I have given a few lectures in addition.

(signed) R.G.E. Murray
(R.G.E. Murray) Capt. RCAMC

Faculty of Medicine
Office of the Dean
The University of Western Ontario
London, Canada
August 8th, 1945

Dr. I.N. Asheshov[81]
Pike Bay P.O., Ontario.

Dear Dr. Asheshov:
Following the letter from Professor Murray re the recommendations for a young Bacteriologist, I wrote to his son, Capt. R.G.E. Murray, and arranged an interview.

---

81 Ibid.

Murray visited us on Saturday, August 4th and was quite interested in the possibility of obtaining a teaching and research appointment in your Department.

I went over his history carefully and it appeared that he had a rather outstanding training in this field and would be a definite asset to your Department. He is interested in doing a certain amount of routine Hospital work, or rather, supervising it, and also in doing demonstrating work in the laboratory with any amount of teaching which you might care to pass over to your lecturer.

He is a fine, well-built lad, appears to be capable, and possesses a rather pleasant personality. I showed him the facilities in Victoria Hospital and went through your Department. August 7th I received a letter from him applying for the position. A copy of this letter and a list of his qualifications are attached. I would have no hesitancy in recommending Captain Murray to you.

If you approve of this appointment, it would be appreciated if you would write to him directly. I would suggest a salary of approximately $2,700. If you approve, please let me know so that I might write Ottawa in an attempt to effect his release by September 1st.

I will be leaving town about August 16th so would appreciate hearing from you at your earliest convenience.

    Yours sincerely,
    G.E. Hall, Dean.

August 8th, 1945.

Capt. R.G.E. Murray,[82]
Regimental Medical Officer,
No. 1 C.A.C.T.R.,
Camp Borden, Ontario.

Dear Capt. Murray:

I have your letter of August 6th applying for the position of junior lecturer and demonstrator in the Department of Bacteriology and Immunology. I have forwarded a copy of your letter and a list of qualifications to Professor Asheshov and have asked him to write you directly and to advise me as soon as possible about the position.

If your application is acceptable to Professor Asheshov, I will assist in any way to effect your release from the service so that you might take up your position at the earliest possible time.

    Yours sincerely,
    G.E. Hall, Dean of Medicine

---

82 Ibid.

International Telegraph
August 13th, 1945
4.10 p.m.

Capt. R G E Murray[83]
Regimental Medical Officer
No. 1 C.A.C.T.R.
Camp Borden, Ontario

Letter received from Professor Asheshov indicating acceptance of your application as lecturer in the Department of Bacteriology and Immunology Would appreciate your efforts for release from service to commence academic term September 10th Advise by wire collect if we may officially request your release for this teaching position.
    G E Hall ...

G. E. Hall, Dean of Medicine.
August 1945

Major General Charles Renwick[84]
D.G.M.S.
Elgin Building,
Ottawa, Canada.

Re Capt. R.G.E. Murray

Dear General Renwick:

    For some months we have been considering various candidates for the position of lecturer in the Department of Bacteriology and Immunology under Professor Asheshov at this University. The appointment of Capt. R.G.E. Murray to this position has been approved.

    Since the academic year, other than for the final year, commences on September 10th and the course in Bacteriology commences at this time, it would be greatly appreciated if consideration could be given to the early release of Capt. R.G.E. Murray presently at Camp Borden so that he might assume his academic teaching post prior to the commencement of the academic year.

    It cannot be too strongly stressed that Dr. Murray's release at the earliest possible time would be of tremendous help to this University.

    Yours sincerely,
        G.E. Hall, Dean.

---

83 Ibid.
84 Ibid.

#1 C.A.C. T.R.
Camp Borden, Ont.
18 Aug 45

Dear Professor Asheshov:[85]

It was very kind of you to write to me. I am looking forward very much to being with you and teaching in the department.

I liked the arrangement of things when I saw them two weeks ago and I am only sorry that you were not there to show me around yourself.

Dean Hall has written to NDHQ officially requesting my release and I have tendered my request for release through headquarters at Camp Borden. With any luck the action will be reasonably prompt. It will need to be for I understand that teaching will begin when term does on Sept 10th.

My work recently has not been very arduous and they should not have any difficulty in finding someone to take over at this end.

I look forward to meeting you in the very near future.

    Yours sincerely,
        R.G.E. Murray, Capt.

#1 C.A.C. T.R.
Camp Borden, Ont.
27 Aug 45

Dear Professor Asheshov:[86]

There is as yet no news of my release from the Army. It is still under consideration in Ottawa. Since there is not likely to be very much time for me to collect my wits before teaching starts I would very much like to have some idea of the teaching schedule for Bacteriology and which lectures and labs I am likely to have to take over. This would be of help to me while I am waiting for release. If you could send me this information and also an idea of what material is projected for the appropriate labs I would be very grateful and I could spend my spare time profitably.

My wife was in London during last week. She returned a little footsore and discouraged but she was able to find a roof for our heads to last us until we can find a more suitable home. She is planning to move down there on September 5th; there is little use house hunting unless you can be on the spot when the opportunity arrives.

    Yours sincerely,
        R.G.E. Murray

---

85 Ibid.
86 Ibid.

August 30th, 1945

Capt. R.G.E. Murray,[87]
Regimental Medical Officer,
No. 1 C.A.C.T.R.
Camp Borden, Ont.

Dear Capt. Murray,

Thank you for your letter of 27.8.45. I think it is a good idea for you to have the schedule of the course while you are awaiting your release and I am sending you an outline for your information.

As to arrangement of division of work, I hesitate to make any definite statement until I can discuss it with you personally. Of course I have certain ideas and my own preferences, but before committing myself to any plans, I would like to hear your views. In correspondence it will take too much time and, besides, everything may be altered again after a personal meeting.

However, just to give you a rough idea of my plans: I thought, to begin with, of handing over to you some of the lectures – some theoretical and some factual. Our practicals I intend to carry out together, as I attach to them much more importance than to the lectures and have my own cherished ideas about them. Your most time consuming work will be to help me to improve the demonstrations and practical teaching, looking ahead in collecting material for demonstrations and illustration of the lectures and practicals.

Dean Hall went to Ottawa [a] few days ago and got a promise that you will be released as soon as possible – whatever it means. ... Let's hope for the best.

I am sorry that I was still absent when Mrs. Murray came to see our lab. I hope to have a chance to show it to her myself in the near future. If she succeeded in finding, in a strange town, even a temporary shelter for you, she must be a marvel of efficiency.

    Kind regards,
        Yours sincerely,
            I.N. Asheshov

"Capt. R.G. Murray on Western 'U' Staff"[88]

Captain R.G.E. Murray, R.C.A.M.C. was yesterday appointed full-time lecturer in the Department of Bacteriology and Immunology of the Medical School of the University of Western Ontario. His appointment was announced by President Dr. W. Sherwood Fox, following a meeting of the board of governors. Dr. Murray will work

---

87 Ibid.
88 *London Free Press*, 5 October 1945.

under Dr. Igor Asheshov, head of the department. He has already commenced his duties while awaiting his release from the army.

Dr. Murray began his university education in McGill University in 1936. After two years studying basic sciences he entered the natural science course in Cambridge University where he took his final work in pathology and bacteriology, obtaining his B.A. degree in 1941. He then returned to McGill for his final years of medicine and graduated in November, 1943, in a war accelerated course.

Dr. Murray interned in the bacteriology laboratory of the Royal Victoria Hospital in Montreal. He continued his bacteriological work in the R.C.A.M.C.

Cambridge University granted him an honours M.A. degree in February of this year.

Dr. Murray comes to Western well qualified for his post. In addition to his own special training, he has been in constant contact with his father, Prof. E.G.D. Murray, well-known professor of bacteriology at McGill.

"Lecturer in Bacteriology Appointed"[89]

Capt. R.G.E. Murray, of Montreal, has been appointed a full-time lecturer in the department of bacteriology and immunology under Dr. Igor Asheshov, head of the department. Capt. Murray spent two years at McGill in the basic sciences and then went to Cambridge University, where he obtained his B.A. in the natural science course. He returned to McGill, graduating in medicine in 1943. After an internship in the bacteriological laboratory of the Royal Victoria Hospital, Montreal, he continued his bacteriological work with the R.C.A.M. Cambridge University granted him an honour M.A. degree in February of this year.

The Move to London: Family Response and Adjustment

25 Dundonald Street
Barrie, Ont.
5 Aug 45

My dear Mum and Dad:

How are you? By now I expect that you have been doing quite a bit of tripping about the countryside and you should have had very good weather for it. How I envy you.

The day before yesterday I got a letter from Dr. G.E. Hall and he asked me to come down to London and see them. I went down as quickly as I could. I left Friday noon and got down there at 9 pm. I stayed overnight with one of my troopers at his house. Saw Hall on Saturday morning, and had a long talk with him. The upshot of it is that I will

---

[89] *Alumni Gazette*, University of Western Ontario, November 1945.

probably be going there in the near future – the nearer the better. They have a very ambitious but well conceived program for the improvement of the medical school and I feel that it should be very successful. There is a considerable building program in sight too. The labs are in the hospital just across the road from the medical school (Victoria Hospital) and are modern in construction and fairly well conceived. Unfortunately I could not meet Asheshov for he had just gone on a short holiday. Since they are in a hurry to fill the position there is no waiting till he gets back. Other than the professor there does not seem to be any other bacteriologist, except a woman who is doing the routine clinical bacteriology.[90] There does not seem to be overmuch space: this is, of course realized, and they are planning in the next few years to build another building attached to the hospital for the pathology and bacteriology labs. It seems that the post I am to fill is to be partly a teaching position to take the burden off Asheshov. This I know I will enjoy. On the research end I gather that I may be wanted to join Asheshov with his TB anti-biotic work – however concerning this I was told that Asheshov is looking for a man to do whole time work with him. The teaching is not over burdensome and consists as far as I know of about 17 weeks a year. The salary will not be much above the 2500 mark – it has not been settled yet.

The pathologists there are very nice: Fisher[91] and Col. Patterson.[92] The both of them send their best to you. Work seems to be of a high standard and very conscientious.

The university is small but seems to have a small but powerful staff of young men. The medical school is separate from the rest and is in the city next door to two hospitals. The rest of the university is out of town a little way and looks as if it were in the middle of a park – around one side of it runs a golf course. They have not very many buildings but what there are in good taste.

The city and its surroundings are very pretty. Chapman[93] (my clerk) was very kind

---

90 UWO Archives, R.G.E. Murray Collection. This is a reference to Josephine Bittner. When Robert Murray arrived at UWO she was performing the day-to-day clinical laboratory work, which provided diagnostic service for Victoria Hospital. She was trained as a physician at the University of Iowa and was originally a specialist in obstetrics and gynecology. She arrived in London in the 1930s when her husband was appointed as a professor of sociology, and instead of going into medical practice she decided to learn bacteriology under Asheshov's tutelage. She continued in the job until retirement in the mid-1950s and died around 1960 of a cancer. RGEM.

91 John Heber Fisher had been a professor of pathology since the early 1920s. RGEM.

92 Fisher was joined in the pathology department by James C. Patterson, who had been a pathologist in the biological weapons program at Suffield, Alberta, and was a colonel in the Royal Canadian Army Medical Corps. The Pattersons were close Murray family friends in London. Murray remembers the assistance Patterson provided to him, due to his experience of animal parasites, and his willingness to give lectures to Murray's medical students on the subject. Robert and Doris visited the Pattersons frequently at their Mayfair Street home. RGEM.

93 Jim Chapman was Robert Murray's clerk in the Regimental Aid Post (RAP) at #3 CACTR in Camp Borden. RGEM.

and drove me around the town and showed me a great many things. Nearly all the houses are small homes with wide lawns and gardens. This together with the wide streets and the cleanliness makes it most attractive. The shopping seems most adequate and there is a good bus service. Most of the land in the region is given to truck gardening, grapes and small fruits. All the land has a well-kept look. Outside the city there are a number of palatial homes housing the super-rich. For the rest there is a very good city park set about four miles out. I liked the whole place very much and I am sure that Doris would too.

I came back to Barrie on Saturday night. I never would have believed that there could be so many people traveling. Union Station was crowded to the back teeth and there were queues all over the place.

We are all very well. Next week Mother Grace is coming up here for a few days. Doris will be going back with her to Rochester for a few days to visit Eric and Mary.[94] I was hoping that I could go down with them but I may not be able to.

All the best to you both and have a good holiday.

    Your Son,

        Bob.

---

94 "Mother Grace" was Grace Blair Watkinson Marchand (1879-1959). Grace Watkinson studied biology at Smith College in Northampton, MA, and following a family tour in Europe decided in 1903 to do graduate work in zoology in Freiburg before returning to the United States to take an MA at Smith. She returned to graduate studies at the University of Leipzig where she met Werner Marchand, who was also a graduate student in the Zoological Institute. She studied in various institutions, including the Naples Zoological Station, and received her PhD in 1908 at Zurich (because Germany did not give degrees to women). All the while, Watkinson and Marchand kept in touch, and after a year of teaching school in New York, Watkinson returned to Europe and in 1911 she married Werner Marchand in London, UK. Mrs. R.W. Marchand taught at Bennett Junior College, St. Mary's School in Peekskill, Cazenovia Junior College, and Smith College before retiring in the mid-1940s. Dr. Richard Werner Marchand (1880-1936) was a scientist, writer on evolution, and lecturer who helped establish the Princeton branch of the Rockefeller Institute of Medical Research, where he specialized in the research of mosquitoes and horseflies. He left the Institute in 1920 and focused on lecturing and writing. He died of pneumonia in 1936 at the age of fifty-six. In addition to their daughter, Doris, Richard and Grace Marchand had two sons, John and Erich. The elder son, Dr. John Felix Marchand, was a physician, while Dr. Erich W. Marchand was an optical physicist at the Eastman Kodak Company and an amateur chess champion of the United States. Mary was Erich's wife. RGEM; *New York Times*, 25 March 1936; *New York Times*, 14 March 1959.

> #1 C. A. C. T. R.
> Camp Borden,
> Ont.
> 10 Aug 45

My dear Mum and Dad:

This has been a very hot week. Scorching hot weather with some really good thunderstorms thrown in for good measure. It has been quite bad here in camp because of the sand – which has only one compensation in that it cools off quickly at night. ...

There is no development on the London front. I got a letter today saying that all that remains is for Prof Asheshov to make his choice. So I should hear fairly soon Yea or Nay. If it works out, and there seems to be fair chance, it will offer about the only exit from this rather stultifying service.

Everyone is agog for the developments of the next few days. We heard with a lot of joy that the Japanese were considering acceding to the Potsdam terms but with reservations. I do not think that the end can be very far away – just a few more doses of the new bomb[95] should make their minds up if they are unable to be definite today. They have sustained some very mortal blows in the last three or four days.

There is little or nothing new here. We are still having a very idle time waiting for the main rush, if it ever comes. ... We do our work in a couple of hours and spend the rest of the time playing baseball or horse-shoes. At the beginning it was a very welcome rest but it has now become a bore.

All the best of love to you both,
Your Son

> #1 C. A. C. T. R.
> Camp Borden, Ont.
> 14 Aug 45

My Dear Dad:

Things are developing very nicely on the London front. I received a telegram this morning saying that they accepted my application and that they would go ahead and arrange my release from the Army.[96] This is very good news and I am looking forward to going there very much. This should come through fairly quickly since the academic year begins on September 10th. I went and spoke to Col Warren and Maj. Minshull this morning and they agreed that they would concur when the request came through. So I think that things are running fairly straight.

---

95 UWO Archives, R.G.E. Murray Collection. The "new bomb" refers to the atomic bomb.
96 Ibid.

There has been a terrific amount of news in the past week. Some of it is very momentous. The development of what my staff in the RAP calls the "Atomiser" was a great tribute to scientific tenacity. I was glad to see the names of several of your old friends as participators in the development.[97] The newspapers of course have made a terrific and no doubt rather inaccurate fuss over it and particularly since it appears to have brought the war to a very sudden conclusion. I am so glad that the Germans did not make the development first. ...[98]

My heavy work should begin here tomorrow – despite the news they are going on with their plans and I have no doubt, even if peace is declared, that they will all come here for there will be no other place for the troops to go. We may even become a discharge centre.

Before the day is out we should know whether the Japanese have given in or not. We will be glad to know one way or the other after enduring the incredible spate of empty talk over the radio for the past three days.

All my love to you all.
    Your Son,
        Bob.

#1 C. A. C. T. R.
Camp Borden, Ont.
27 Aug 45

My dear Dad:[99]

I have not yet heard anything about my release. I suppose that it is sitting upon someone's desk in Ottawa – certainly there is nothing through at this end yet.

Doris got back from London on Saturday night, footsore and a little discouraged. However, some good has come out of the trip for we will have a roof over our heads for a little while at least. It will give us an opportunity to get down there and look around in earnest – a few days is not enough. We already have some friends keeping the weather eye open and, although it will not be easy, I do not think that we will go homeless. In London apparently it is very difficult to find a place to rent – although

---

97 This refers to the McGill physicists and chemists who were involved with the Anglo-Canadian nuclear project between 1942 and 1945. See Avery, *Science of War*.
98 Throughout the war there was a concern that German scientists, given their high level of expertise in nuclear research, would develop an atomic weapon. See Mark Walker, *German National Socialism and the Quest for Nuclear Power, 1939-1949* (Cambridge and New York: Cambridge University Press, 1993); Mark Walker, *Nazi Science: Myth, Truth, and the German Atomic Bomb* (Cambridge, MA: Perseus Publishing, 1995); Paul Lawrence Rose, *Heisenberg and the Nazi Atomic Bomb Project: A Study in German Culture* (Berkeley: University of California Press, 1998).
99 UWO Archives, R.G.E. Murray Collection.

if you are willing to buy there are plenty of houses. A sort of proposition that is way above our means. The temporary abode that we have got is in the house of a retired Army man who will let us have a whole floor – virtually rent free except for the electric light bill. I thought there might be a catch in it so I enquired about through friends of mine – it is apparently on the level. It will give us the foothold that we need anyway.

I have not heard from mother yet. Both of us have had a letter from Sue – the latest one included the fact that she is believed to have a touch of lead poisoning. Is this true? It is possible of course – e.g. from some cheap lipstick. It may also account for some of the trouble that she has been having.

Time flies. I suppose that you will be packing-up in Ottawa this week and that next week you will be back in the old house again.

Doris is planning to move down to London on Sept 5th... I feel that it is important to get the house hunting done before the fall race of October 1st.

    All my love,
        Bob.

                #1 C. A. C. T. R.
                Camp Borden, Ont.
                3 Sep 45

Dear Old Dad:

It was grand to get your letter and Mother's long account (and a very good one too) of your trip to the small lakes. It must have been great fun and well worth doing and I feel most envious of the two of you being able to get away together, and fish. The both of you deserve every bit of relaxation that you can get for you have worked unstintingly and without relaxation for so long. ...

Today I received more detailed information with regard to the NRC bursaries, fellowships and studentships. I think that I would be able to qualify for a bursary at least were it not for one provision in the regulations. This states: "Successful candidates are required to devote themselves for a period of at least eight months of each year wholly to the objects of the award, and during that time are forbidden to hold any position of emolument or to engage in teaching." This seems to me a bit silly, but, if they mean it, this excludes me straight off. Another provision that amused and annoyed me was this: "The principal work ... must be a research in some branch of science, the extension of which is important to the national industries." Sounds as if they were running a patenting bureau. Aside from all this it would take a while to get an application ready even if I could qualify under these provisions for they require miles of testimonials from the people under which study was undertaken.[100]

There is no news of my release yet except that the orderly room has a letter stating that the application is being considered by the ISRB (Industrial Selection & Release Board). I also hear that Hall was up in Ottawa last week and they said it would be arranged "as soon as possible", whatever that means. I also have a very nice letter from Asheshov giving some idea of the arrangement of his lecture and practical courses, ... of course, it is all unsettled as yet since nothing definite could be arranged until we meet. Anyway I shall get a lot of experience out of the whole thing.

I got a letter from Gertrude[101] today – she sounds well and rested after her holiday. She is very glad of your return and looks forward to your direction and planning. Apparently Hugh[102] is thinking twice about returning.

Doris will be leaving (bag, baggage and infant) for London on Wednesday. We have a semi-temporary roof for our heads which will last until we have found a more permanent place. My spies tell me today that they are holding onto a much more suitable place for us that has come up since Doris was there. We will see.

I am sorry that the leg is still giving you trouble. Those things are very crippling and about the only treatment of them is rest ... so keep an eye on it ... you may be laying down a bit of calcium in your tendons.

All my love,
Your son,
Bob.

---

100 Ibid. Murray recalls that he did not end up applying for National Research Council money in 1945. In fact, there was no NRC money until 1948, and until then he relied on Asheshov's grant and "dribbles from the Dean," G.E. Hall. RGEM.
101 The routine serology at Royal Victoria Hospital was carried out by Miss Ida Ward until 1940 "when Dr. Gertrude Kalz of Prague was appointed as serologist." Much of this serology work involved performing Wasserman tests for venereal disease, which soared during the war years. In March 1946 Kalz was listed as assistant professor in the Department of Bacteriology and Immunology. Kalz is remembered by Robert Murray as a close personal friend of the family. Her husband, Fred Kalz, was the dermatologist at Royal Victoria Hospital. Lewis, *Royal Victoria Hospital*, 223; RGEM.
102 Hugh Starkey worked in the clinical laboratory at McGill and spent the war in the Royal Canadian Navy at its clinical lab in Halifax. He did return and was i/c microbiology and infectious diseases in Queen Mary Hospital in Montreal. RGEM.

#1 C. A. C. T. R.
Camp Borden, Ont.
18 Sep 45

My dear Dad:

... Things have gone slightly astray on the question of my release. I was getting pretty restive by the 10th and so was Dean Hall. He got a letter from a Major in the MPAB (Medical Procurement and Assignment Board) to say that my release was approved from that end and where was the approval from this end. Well, I had written my formal letter on the 18th August and I knew that the Col., had approved it then and there. So I got hold of the Adjutant and we went to Camp HQ. We found it after a hunt neatly filed away. ... This delay is very irritating for us all. It is now just over a month since the request went in. It is very galling to see people released to attend universities within six days of application.

We have been very busy here during the past couple of weeks. We have been discharging people to depot at a great rate and have to make out reams of forms ... in the unit: there are now four under me. I thought it would be reasonable to ask for a leave pending discharge but when I requested it they gave me a very flat "No" ... which I think was a little unreasonable.

Last week-end I was down in London. Doris is quite nicely settled in her temporary quarters – which are fairly comfortable. ... I managed to get a long weekend for I had not had the extra day that was given for V-day. I was able to spend most of Friday and Saturday in the Lab, meeting Asheshov and seeing how they work things there. I like Asheshov very much and I think that I will enjoy working there with him. He has asked me whether I want to join him on his research project which is concerned with this antibiotic which is active on TB.[103] I have not yet made my mind up on this – one advantage would be an opportunity to learn a lot about fungi and a little about extraction methods. I don't know whether it is better to go on one's own bent or to work in conjunction.

The department itself is not very well off at the time being and since Asheshov has no subsidy for his work (because he hates being tied down and making accounts!) there are certain economies that would not allow of many things that would be done elsewhere. However, facilities are fairly adequate.

---

103 UWO Archives, R.G.E. Murray Collection. Murray recalls that he was really a spectator for the work on tuberculosis but learned a lot by being useful from time to time. He found he was able to apply what he observed about fractionation of growth media and simple lab procedures in work that he later started with Lazarus J. Loeb, who did an MSc at UWO (see chapter 6, note 48). Asheshov encouraged Murray to think of things he wanted to work on and they ended up being the mucoid streptococci and the swarming of bacteria on agar surfaces that he had already written about in an early notebook from 1943-44. RGEM.

The teaching is good but, due to the preparation of the student, is far more elementary than Medical Bacteriology should be. Hall is trying to fix this by changing the premedical requisites. I shall get a certain amount of lecturing and a great deal of arranging of practical classes. This will be very good experience.

Doris and Alice are very well. Alice is taking the odd step now and then but usually feels that crawling is the safest method of progression.

All my love. I am looking forward to hearing the account of the trip.

Your son,

Bob.

P.S. Asheshov sends his regards and hopes that you may be down sometime to see his lab.

422 St James St.
London, Ont.
23 Sep 45

Dear Old Dad:

Well, the move has come at last. Somewhere about last Thursday I was getting very restive about the delay so I went to see Col. Warren[104] and asked him if he could bring himself to ring Col Zinkan at Ottawa and try to find out the score. This he did and was able to tell me that night to expect to be moved by Friday noon. I was sort of congratulating myself on the manoeuvre but I found out that Hall had been working on it at the same time for a much more serious reason. Apparently Asheshov was taken sick last weekend – he thinks it is some sort of infection – which gave him violent vomiting and later diarrhoea (but without fever) but this was complicated by a fairly serious gastric haemorrhage. The result is that he has been ordered to keep away from work for a month at least after he has been let out of bed. The result is that I am going to have to give pretty well all the lectures and the labs. This is going to be a tremendous lot of work for me – particularly since I lack a lot in experience but will no doubt be of the greatest value to me. So I am now at the depot (but living at home) getting my medical board and discharge, and will start teaching on Wednesday next. The loss of a week of teaching is serious since the schedule calls for only 17 weeks and we are going to be hard put to get it all straightened up.

I will have to try to get down to Montreal one of these weekends (while I am still allowed to wear uniform because that saves a lot of money) for I have to get my microscope and one or two other things. Also I want to talk a few matters over with you and get your advice. I may be able to get down next weekend since there is no teaching on

---

104 Ibid. Col. Warren was most likely o/c Medical Services at Camp Borden. RGEM.

Saturdays or on Mondays but I will have to see what the requirements of the depot may be. ... they may not be as quick about doing things as one could hope.

We have not got a permanent place of abode yet. The process is a difficult one and you have to get a lot of people in the know before your chances are very great. A couple of possibilities have turned up but they are all dependent upon the move of the people who are living there now – everyone steadfastly refuses to be tied down to a date. The whole rental business smacks of being a racket without any visible controls.

I am eagerly awaiting the account of the trip. I fear it may have been a little cold for you towards the end – I saw in the paper that there had been a minor blizzard in the region of the Laurentide Park. It was pretty cold weather with us too – up at Camp there was no heat and we were all back in winter dress for three days.

My love to you all.
    Your son,
        Bob.

The University of Western Ontario
London, Canada
Faculty of Medicine
Department of Bacteriology and Immunology
Ottaway Ave. and Waterloo St.
1 Oct 45

My dear Dad:[105]

I am now free of the Army and have been able to work here relatively uninterruptedly for about a week. The only trouble now is that we are still looking for a house and have not yet been successful. It has been a very hard job and does not look too promising even now. What makes it worse is that we will have to move from our present position by the 15th of this month. ...

I have an awful lot of work here getting lectures ready and getting the practicals in shape. By the end of the term I will have accumulated a lot of experience. It does not give me too much time over for the necessary business of looking for a home or for anything else. However, I am enjoying it – though I wish that I had a little more teaching experience. I do alright and have no trouble – but it takes me such a time to get everything prepared.

This worry about finding somewhere to live is not making things any easier. Despite the fact that everyone is doing all they can to help find us a place there is not a great deal of hope – although I don't think that we will be left out in the cold

---

105 Ibid.

altogether. It seems odd that a place such as London should be so crowded – a smaller city without a great deal in the way of industry.

    Your son,
        Bob

London, Ont.
2 Oct 45

Dear Mum & Dad,[106]

It was very nice to talk to you on the phone this morning. I am sorry that I have not been keeping you abreast of the developments around here but I have had little time to write. It was grand to get your long letter describing the trip to the Jeanotte. It must have been a very pleasant trip despite the poor fishing weather and I wish so much that I could have been with you.

I am sorry to be so indefinite about next weekend – I have got to get down some time. I wish that you had been in last weekend for it would have been so much more convenient for me. At the present time, with all these lectures and labs to prepare and with the worry of getting a roof over our heads, I don't really know that I can spare the time. But I will somehow for there are a number of things I would like to discuss and I am looking forward to seeing all of you.

Asheshov is still occupying a hospital bed. Everyone is a bit foggy as to the precise lesion – there is nothing abnormal demonstrable by all the tests except that he has been losing blood into the stomach and that his Haemoglobin has dropped 30%. He says that he has had it before, at which times no lesion was demonstrable, and that he thinks that it is due to a brittleness of his smaller vessels. Anyway he seems to be gaining a bit of ground now and will probably be away quite a few weeks yet. He sends you his very best regards and hopes that you will be down here one of these days.

I will probably be phoning you sometime on Friday to let you know whether I will be coming down or not. I never realized that it was the Thanksgiving weekend.

    All my love,
        Your son,
            Bob.

---

106 Ibid.

422 St James St.,
London, Ont.
6 Oct 45

My dear Mum and Dad:

... I have now finished all the introductory lectures on infection and immunity – and I'm quite glad too. Although they were really good fun they were infernally hard to prepare. Making things simple but reasonably accurate is very hard I find. Next week we start on the Enteric group[107] which will be a lot easier. I am enjoying this business of teaching and, so far, have been pretty successful. The students seem to get a little out of the lectures and they don't seem to get bored – so I reckon I have got over some of the difficulties. For some reason the "Class of '48" which is the one taking Bacteriology, had the temerity to elect me as class honorary president – why, I don't know – so I told them that if the duties involved the occasional glass of beer I was all for it. I don't think that they are teetotallers!

This house hunting business is reaching the serious stage. There is nothing much one can do about it. In most cases if a person is going to move they tell all their friends they are going to do so before they get round to the fellow that owns the place. Lately our propaganda has been directed to getting as many people acquainted with our situation as possible on the off chance that they may let us know before someone else. We also have our tentacles out to try and find someone who winters in Florida who wants a caretaker. ... Several kind people have offered to shelter us in case we should have to choose between the Police Station and Victoria Park. ...

Your son,
Bob

London, Ont.
7 Oct 45

I got a letter yesterday from Del.[108] He seems to be in great form and seems to intend to marry Josie. He has taken his California State Boards and intends to go into a Hospital Practice (giving up Pathology for some obscure reason) – He seems to have got a good job lined up and should have started at it by now. He sends you all his very best and says that a letter will soon be on its way. ...

---

107 Ibid. Enteric defined as (adj.) "Of or pertaining to the intestines." *OED* (Online).
108 UWO Archives, R.G.E. Murray Collection. "Del" was the nickname of Herbert Delwyn Mintun, Robert Murray's best friend in medical school and the best man at his wedding. Originally from the San Joachim Valley in California, he was a physician for many years in Berkeley, California, and retired to Santa Rosa, where he died in 2004. RGEM.

Everyone has been very kind and helpful – especially Dr. Patterson[109] who sends you his best regards, Dad.

Dr. Asheshov is now at home though he is still confined to his bed (for one more week). So it will be a little time yet before he will be out and around. He is very anxious to get back to work. He asks me also whether there is any chance of your coming here for a day or two. He would very much like to see you again. He is putting up a very great hunt for Colchicine, which has, apparently, almost disappeared from the market. Undoubtedly it has all been taken for the growing of Penicillin.[110] He needs it very much to continue his experiments on production of his antibiotic – he has completely run out of it. He thought that perhaps he might find some in one of the older French druggists in Montreal. It might be possible that Ayerst's might be able to let him have some.[111]

In our phone conversations I did not ask about Sue and Blake. How are they doing? I don't suppose that they have had any more luck than we have in finding a place to live. It is, from what I hear, frantically difficult to find anything in Montreal. Please give them my love. I will write when I can.

All our love,
Bob.

> 104 McClary Ave.,
> London, Ont.
> 24 Oct 45

My dear Mother and Dad:[112]

The furniture arrived this morning at breakfast time. It was all intact – except that the desk was to follow because they said that a leg had been broken when they were unloading at Hamilton and that they were having a new leg fitted. Otherwise the journey was perfect – every bit of china and glass was intact. We are beginning to have quite a furnished look about the place. It will be more or less complete when the sofa arrives tomorrow and when the desk arrives. Of course there are all the little frills such as curtains etc which will have to be added if and when we can. Doris is having a great time in arranging and re-arranging things and making plans. In spite of our advanced state we would still be classified as campers in the rooms – but we are having a lot of fun doing it.

---

109 See above, note 92.
110 Colchicine is defined as "An organic alkaloid $C_{17}H_{19}NO_5$, found in all parts of the *Colchicum autumnale*. Now used *esp.* to induce genetic changes in plants and animals." *OED* (Online).
111 E.G.D. Murray did consulting work for the pharmaceutical firm Ayerst, McKenna and Harrison Ltd. in Montreal. RGEM.
112 UWO Archives, R.G.E. Murray Collection.

Alice is also having a great time in the new surroundings and she is as full (or fuller) of beans as ever. We should report that she has now made her first honest attempt to walk – yesterday afternoon she walked, on her own hook, 54 inches by my ruler. She has made a few shorter efforts since then. This was the first real "no hands" effort and we are all very excited.

Dr. Asheshov is very pleased and very excited to have the Colchicine. $1.40 will arrive in due time.

Not too long from now we will be able to hold open house here. We are looking forward to seeing you both and we will let you know when the stage is set. It should not be long from now.

I am enclosing a cheque for 45.00 to cover the carriage charges on the furniture. I don't see why it should have been paid by you people after all the things that you have done – actually we can afford it. So no arguments.

    Your son,
        Bob

Dear Mum –

Just a note to tell you again how glad we are to have the table and everything. As Robin says we are beginning to look civilized at last.[113] We will have to live partially out of trunks for a while longer but not too much so and is only because dressers or highboys are few and far between and we are waiting until we find something that suits us and our pocketbook! But we do have a place for you & Dad to sleep when you come and *you can be* comfortable. The Hydro expects to put the hot water equipment in this week and in the meantime we are using the downstairs bathroom. I am becoming an expert on furnaces but I hope it won't last all winter.

Miss S[trelitz] expects to have an oil heater in by Christmas. A telephone is a thing of the future – the estimate is about a year from now but even that isn't so bad as it seems for Miss S is out all day and said, I could use the phone so I can order groceries and can call in case of emergency, but I said I wouldn't have any incoming calls so the phone is strictly hers when she's in and I only use it when she's out. It works out very well. There is still a lot of unpacking to be done and I have to get curtain & drape material. When these are done we should be in complete running order. I think you'll like our apartment and we are very much looking forward to a visit from you & Dad.

Robin has told you that *Alice walks*! She does it with a very experimental attitude and uses the old method of locomotion when she is in a hurry, but I guess it won't be

---

113 Ibid. "Robin" was Doris's nickname for Robert Murray.

long at all before she is all over the place with me not far behind! Alice is fascinated with my baby doll and most of all wants to know what makes her eyes open & shut. Incidentally the doll isn't very big yet the clothes Alice came home from the hospital in just fit it. It just doesn't seem possible!

Much later – well my post script has delayed the letter so I'll close and send my love to all and tell you again that we are looking forward very much to a visit from you before very long. I'm going to try to have all the unpacking and everything under control by the end of this week. Please send my *love* to Sally[114] and tell her I shall write just as soon as I get down to my semi annual effort.

    With much love,
        Doris

P.P.S. Robin & I are both very well & happy and are thrilled to have a home of our own at last. It is great fun for both of us. We feel much too lucky!

                28 Oct 45.

Dear old Dad:[115]

... Oddly enough no pneumo's except esoteric types have turned up in the routine lab. There are some rather old dried cultures that Professor Asheshov had and I spent some time trying to recover them – and was unsuccessful. The Public Health lab is in the same fix – for pneumo's seem to be very few and far between this year. So we were stuck & I telegraphed you, hoping that you might have some that are viable. It happens that we have adequate sera, for the class, of types I-VIII and that accounts for the peculiar request.

The Sera arrived yesterday and I will ship them back to you tomorrow – thanks for the promptness but we got our signals a bit mixed.

Up to now it has been a little difficult seeing more than a few days ahead – so far everything has worked perfectly but we may have to do a little fancy switching this next week. This town is too well doctored and it is hard to get even a nice specimen of pus.

The sofa has arrived and we have an almost furnished look. The desk has not yet arrived but should be along this week. I could do with it.

    All my love.
        Your son,
            Bob

---

114 Doris tended to call Sylvia Rose "Sally." See note 50 above. RGEM.
115 UWO Archives, R.G.E. Murray Collection.

104 McClary Ave., London, Ont.
3 Nov. 45

My dear old Dad:

It was grand to get your letter today – though I am a trifle annoyed to hear the "sad accident" to the cheque. Since I refuse to go on writing cheques only to have them come to pieces in your hand, I guess you win. We are glad to have both the furniture and the money.

We are looking forward to seeing you both – *whenever* you are able to come down. Things are still a bit primitive but are way past the difficult stage, and we're prepared for visits.

The cultures that Freddie sent arrived on Tuesday – by which time I had got classes switched around – and I will need them next week. The Types I & II are alright after mouse passage and are good typing strains – although culturally they are a little small & opaque. The Type III (which was in a Meat Mash) seems to be dead: both by culture and by mouse inoculation. The I & II were dried in 1943! One of Asheshov's strains (dried in 1939) grew alright but is rough & cannot be recovered – all the others we had are dead as mutton.

Asheshov has, at last, got enough Colchicine to get his expts under way – from a very comic fellow in a town in New Jersey of whom we know nothing. This following an advertisement in Science. A couple of other possibilities have turned up too.

I would like to have a strain of Leptospira biflexa[116] very much – on, or about the 21st of November I am going to be showing technique of Darkfield Examination[117] and it would be very useful. How do you keep it going?

By the way: If there is a good Darkfield condenser on the market I would like to get it for my microscopes. Also I could get the fort to pay for 2/3 of it under the Re-establishment scheme.[118] I would have to know: who has it and how much it would cost exactly. It would be well worthwhile for me to have one for myself.

---

116 Ibid. *Leptospira biflexa* is the specific name for a spirochete. Some spirochetes cause disease and some are free living. Murray describes then as "very elegant bacteria." RGEM.

117 Robert Murray describes the dark-field process thus: "Dark-field is a way of illuminating a slide for microscopic examination so that the bacteria are illuminated from all round but the light beams are at such a low angle that the cells are brightly shining against a black background because no light is coming directly through the lenses of the sub-stage condenser. It is like seeing motes of dust in a sunbeam or a spider web in a similar situation." RGEM.

118 This reference is to the economic and social policies enacted in the late war and early postwar period to facilitate the demobilization of Canada's armed forces and the transition from a wartime to peacetime economy. For information on these policies see Robert England, *Discharged: A Commentary on Civil Re-establishment of Veterans in Canada* (Toronto: Macmillan, 1943); Peter Neary and J.L. Granatstein, eds., *The Veterans Charter and Post-World War II Canada* (Montreal and Kingston: McGill-Queen's University Press, 1998).

You must be having quite a time with University politics. I am glad that bacteriologists will, at last, deal with their own part of Public Health – but I cannot imagine that you could do this without increase in staff. The teaching burden is very great even now. I am most glad to hear that you are going to have a virus division with proper equipment.

On account of a grant he has received Asheshov is on the look-out for another technician – one who has a knowledge of Bacteriology but who is fully trained on Biochemical lines: Salary Ca. $100 monthly. He prefers a female because, he says, they are easier to control! I told him that I did not think there would be any available from the McGill area before about June, if then.

Trained people must be very hard to get now and with some of the old staff going to other jobs you must be having quite a time...

You have certainly got a nice collection of equipment for bush trips – I envy you very much – and I'm so glad that you & Mum have been able to get such a good set of holidays this fall. ...

All our love.
Your son,
Bob

> 104 McClary Ave.,
> London, Ont.
> 11 Nov 45

Dear old Dad:

The bacon arrived, and in good shape, yesterday. We are very glad to have it (it brightened up the Sunday breakfast a great deal) but I still feel that it is depriving you of a pleasure that was really your own and not for others. Still, we enjoy it very much and will savour it to the full.

Excitement for the moment is reserved for Alice's birthday on Monday – she is full of beans and good promise. Yesterday, for the first time she decided that walking on the two feet was the method of choice and she spent most of the time staggering drunkenly about the room – up to now it has been a very poor alternative to a speedy crawl. So we feel quite elated that she definitely walks by her first birthday. A large parcel arrived from you people which will be opened in state tomorrow afternoon – functions to be illuminated by one candle.

The week has not, however, been devoid of excitements. Not the least of these was the illness of Miss Strelitz (Asheshov's assistant, our landlady and who lives below us) who on last Friday week started to run a hell of a temperature with considerable rigours. She refused to be seen by a physician for three days but finally on Sunday she allowed me to send for one – when he arrived, of course there was nothing to be found and at the moment she had no temperature. This was while Doris and I were at a party at the Halls'. When we returned she was having a real rigour and was very cyanosed so I phoned the doctor and we had her admitted to hospital pronto. Early blood cultures were negative but last Friday there appeared undoubted typhosus. I rather feel that she got it in the lab for we were using it less than four weeks ago. We have sent the cultures off to see whether they can be typed with phage.

The party at the Halls' was a curious one but very pleasant. When we were invited I understood that we were being invited to lunch, so did all the others who were going that I knew. So we went, ravenously hungry and well prepared for a swift drink and an attack on someone else's rations. Instead we found that it was a twelve o'clock cocktail party with seemingly hundreds of people there. The old fashioned's were very good but I could have consumed an old fashioned horse, hoof, hides and all. There were lots of people there to meet and it was a fine gab-fest. There was a Jack Lewis there (Lt-Cmdr) who brought tidings of you.

Among other things good that have arrived is a telephone. This largely through the good offices of Dean Hall who arranged a priority with the Bell. It is a great convenience for both of us. ... The number is, for your information, Fairmont 4086 J – a party line, but that is everyone's lot these days. Also the Hydro has at last installed the hot water heater so that we can wash in comfort at last. ... All in all we are ready to receive visitors – so give the word when you would like to come down.

Work in the department goes on as usual and it is still all teaching. I am a little worried at the moment about the practical class on precipitation. It is only one class and I have damn-all in the way of sera. In previous years Asheshov made his own sera in rabbits but forgot to tell me – all these rabbits have been used for other purposes so there is no chance of getting one up to reasonably high titre[119] in a hurry. He is all for my doing a Ramon flocculation test – killing two birds with one stone. I would rather do Toxin neutralization separately and have the class do a plain precipitation reaction with the other as a demonstration. How are you off for a good precipitating system suitable for the class? (This class is about Dec 7th) I would like some advice on this – also some sera if you have enough to spare. Asheshov is not by any means an immunologist and has left this up to me.

The magazines are very welcome and Doris is very appreciative. But you did not let us know whether you would like to have them returned to you or not. We are holding them in case you want them.

Winter is pretty well upon us. During the past few days there has been a definite increase in coal consumption – a thing we notice now for we have to keep the furnace

---

119 UWO Archives, R.G.E. Murray Collection. In medical usage, a titre refers to "the concentration of an antibody, as measured by the extent to which it can be diluted before ceasing to give a positive reaction with antigen." *OED* (Online).

fired. It is a great contrast to the bright warm weather we have been having. The first snows cannot be far ahead.

When I was uptown yesterday afternoon I bought a book for you about travel by canoe in the far north. It is so good that I am reading it myself and will be sending it on to you in a little while. It is an excellent commentary on bush travel especially in unmapped areas. The man writing it is an American who has evidently had a great deal of experience in all the northern territories, and there is no nonsense. It is called "Sleeping Island" after the lake that this fellow (P.G. Downes) is looking for. There is much interesting comment upon the peoples, and matters geological and biological, as well as on technique of travel out of sight of the coca-cola sign.[120]

All my love.

    Your son,

        Bob

104 McClary Ave
London, Ont.
18 Nov 45

My dear Dad:

We are looking forward to seeing you should you be coming to Toronto this next week-end. The apartment is quite comfortable now & Doris feels ready to do some entertaining. Mother Grace sent us an old rug and some curtain material that helped greatly.

The event of the week was, of course, Alice's birthday – she had a profusion of presents and a very happy day. We set them out, all wrapped up, on the floor in the morning and she had a marvellous time getting them disentangled. I think she was a little overwhelmed and could not understand why so many new things should arrive on that particular day. She walks all the time now and is getting confident despite a very noticeable stagger. She has a well developed curiosity and has to be rescued from all sorts of perilous situations. She is full of fun & is just as cheerful as ever. ...

20 Nov 45

Your long letter (17/11/45) arrived today and very interesting too. I'm glad there is a possibility of your coming to London – but you don't give me any idea of when. Is

---

120 Reference to the Massachusetts school teacher Prentice G. Downes (1909-c.1978), who documented his travels in the book *Sleeping Island: The Story of One Man's Travels in the Great Barren Lands of the Canadian North* (1943). Trent University Archives, Prentice G. Downes fonds description.

it that week of the meeting of the Lab Sectn of the C.P.H?[121] Both Asheshov & I had planned to go but I think we may both be stymied by teaching. This because of the loss of a week earlier-on when Ash was sick – which will make it necessary to use the week we had planned to take off ... the dates I can't remember.

Asheshov is now quite recovered from his gut trouble – but I very much fear that he also has a ruptured inter-vertebrae disc which is giving him some neurological symptoms. He won't go & see a neurologist and is very stubborn even though it gives him trouble. I am trying to persuade him to go & see MacKenzie[122] in Toronto.

Miss Strelitz is going through a perfectly normal typhoid fever and is now feeling a good bit better. Hers was an isolated case – there being no others for a considerable time in these parts – I am fairly sure that it was a lab infection. Her illness has not added much to my work except to add a few small details – it holds back Asheshov's work quite a bit.

The Leptospira culture arrived in good order – there were plenty in it but not so many motile forms. However, I am making up your medium and will get them "freshened-up" & put some by for keeping.

I don't think I told you about the pneumos of a few weeks back. The two (I & II) dried in 1943 recovered perfectly after mouse passage; and, when mixed with sputum killed their mice in 24 hrs (practically to the minute, three dying while I was lecturing that afternoon!). The type III, kept in meat mash after Freddie's fashion, were totally dead; even colossal amounts put into mice failed to show any effect and not even rough forms appeared air plating. By the way, I have not heard anything of that anaerobic pneumo XXXIII I sent to Freddie: it is probably languishing in some dusty corner of his desk.

I wrote to Don[123] – deciding to put on a demonstration since everyone seems to

---

121 UWO Archives, R.G.E. Murray Collection. This refers to the Laboratory Section of the Canadian Public Health Association.
122 Murray only remembers MacKenzie as a distinguished senior surgeon in Toronto at the time. He cannot recall the full name or position. He now doubts that Asheshov really needed surgery. He had dyspepsia and a gastric ulcer "which nowadays we would treat with high doses of a penicillin." RGEM.
123 In late 1945 Murray was involved in the difficult process of organizing teaching lab projects and lab demonstrations and he did not have time to develop antisera for those purposes. So he borrowed things from various places both local and distant. In this case, Murray wrote to Donald S. Fleming, a member of his father's staff, who was teaching a course for medical students and had materials to spare. Donald S. Fleming was a medically trained bacteriologist and a professor in E.G.D. Murray's department. He taught both medical and science students. RGEM. In 1950, he held the post of associate professor and was the medical director of the Child Health Association of Montreal. See Donald S. Fleming and Louis Greenberg, "The Use of Combined Antigens in the Immunization of Infants," *Canadian Medical Association Journal* 62, no. 2 (February 1950): 146-148.

be short of good precipitating sera. I wrote to Toronto some time ago for the Dip. Toxin.[124] Owing to curtailment of time as well as materials we have had to concentrate more on those things which med students will have to do (or may have to do) themselves – Next year we can plan differently. There are many very practical things I would like them to do but they will have to wait until I have them thought out & planned into the course.

You ask if I took my M.A. – Yes I did in February of this year. I have not sent any notification of my appt to. I don't know how to do that – however, I will probably write to Grose soon.

Glad to hear about Ted. The rat hasn't answered any of my letters – not that he has to but I would like to hear source of the gossip & how he is doing.

All my love to you & Mum, Sue & Blake.

Your son,
Bob

P.S. Did I tell you the desk arrived. Very welcome too. The men thought that it would never go up the stairs but Doris told them that measurements indicated it would just go up side ways – it did too with only a coat of varnish to spare!

104 McClary Ave.
London, Ont.
25 Nov 45

My dear Mum:[125]

We were delighted to get your telegram to say that you were coming next Friday – it makes us very happy and Doris is plotting feasts for both the eye and the tummy. We won't have any nonsense about hotels and you will stay here – as long as you don't mind being just slightly cramped.

Alice has been a little fretful lately – waking up several times at night. This has been the harbinger of another tooth – probably two, though only one has made its appearance so far. Her walking is progressing and is not punctuated by nearly so many bumps as before. She follows us around and responds to call like the faithful hound – her birthday toys are a joy to her and some of them already show signs of wear & tear.

Everything is going smoothly and we are enjoying ourselves greatly – as you will see. Not much time or opportunity to go out & do things but we manage to get out every now & then to see a movie. The local breweries are not up to much, from all

---

124 Diphtheria toxin.
125 UWO Archives, R.G.E. Murray Collection.

reports, and (being essentially small town AND Ontario) it is "not done" to be seen in the local [tavern]. ...

By the way, you don't say whether you are coming CPR or CNR – they both have trains coming in at the same time and the stations are some way apart. There is nothing to pick between them – the CNR is the nearer station to us. I would like to know so that I can meet you.

All our love.

Your son,

Bob

104 McClary Ave
London, Ont
7 Dec 45

My dear Mum & Dad:[126]

We are looking forward to the projected trip to Montreal for a grand family party. We are trying to arrange for someone to keep the house warm & prevent the pipes from freezing – with a little luck I think we will have someone trustworthy. We are planning on coming down on the 19th or 20th and I shall probably have to return on or before the 31st – at any rate it will provide a fine holiday for us all.

Keep an eye out for a present that you both would like as a supplement to your bush equipment. I have an idea but I can't find the thing I want round here.

This next week is the last of the teaching & when that is finished I am more or less free. Tell Don that the Precipitin Sera worked marvellously & went around, without anything to spare – made very valuable teaching material.

The pictures of Alice are finished and we have one for you. They are very good and I think we will have more prints made.

I nearly ruined my right ring finger today on a bit of broken glass & put a very deep cut over the extensor tendon – luckily did not cut it. It is annoying.

You will be off to N.Y. very soon & I bet you will enjoy the trip - both of us wish we could do it too.

All our love.

Your son,

Bob

---

126 Ibid.

EXTERNAL PROFESSIONAL CORRESPONDENCE
The Montreal Medico Chirurgical Society

Prof. E.G.D. Murray
June 7th, 1940.

Doctor E.S. Mills,[127]
 Suite 206,
Medical Arts Building,
Montreal, Que.

Dear Doctor Mills,

As you know I am anxious to do anything I can to be of use to the Montreal Medico-Chirurgical Society. It seems strange to select me as a member of the Finance Committee as it involves questions of which I am absolutely ignorant and therefore cannot contribute anything.

However, I shall do my best to deserve the confidence of the Council and take the opportunity to learn more about the working of the Society.

 Yours sincerely,
  EGDM

Montreal Medico-Chirurgical Society
Suite 206,
Medical Arts Building,
Montreal, May 21st, 1941

Professor E.G.D. Murray,
McGill University, Montreal.

Dear Professor Murray,

It gives me great pleasure to advise you that you were elected Vice-President of the Montreal Medico-Chirurgical Society at the Annual Meeting of the Society on May 16th, 1941.

 Yours sincerely,
  J.F. McIntosh, M.D.[128]
  Honorary Secretary

---

127 LAC, E.G.D. Murray Papers, vol. 7.
128 Ibid. See above, note 14.

Montreal Medico-Chirurgical Society
Suite 1020,
Medical Arts Building,
Montreal, November 13th, 1941

Dear Professor Murray:[129]

I am pleased to advise you that you have been appointed a member of the Special Committee with regard to the investigation of families closely connected with the medical profession, who are in this Country without adequate financial resources, as referred to in Dr. Grant Fleming's letter which was read at the Council Meeting on November 10th.

The other members of this Committee are Dr. D.S. Lewis and Dr. Neil Feeney, and the Committee has the power to add to its numbers, representatives from the Canadian Medical Association Quebec Division if it should see fit to do so.

The matter should be brought before the Society in general meeting.

Yours very truly,
J.F. McIntosh M.D.
Honorary Secretary.

The Montreal Medico-Chirurgical Society

Selected Lectures: Session 1941

1941

1. November 7th:

    Dr. Wilder Penfield:[130] "Impressions about things medical in Great Britain"

    Dr. Sclater Lewis:[131] "Drugs and their relation to the War"

---

129 Ibid.

130 Ibid. Wilder Penfield (1891-1976), the world-famous American-born Canadian brain surgeon, was professor of neurology and neurosurgery at McGill University. He revolutionized these fields through the development of his so-called "Montreal Procedure," which enabled patients to remain awake during surgery and describe their reactions to the stimulation of different parts of the brain, as well as through the application of this procedure to the surgical treatment of epilepsy and the mapping of the brain cortex. See "Wilder Penfield (1891-1976)," http:www.mcgill.ca/about/history/pioneers/penfield/; also see Wilder Penfield, *No Man Alone: A Neurosurgeon's Life* (Boston: Little, Brown, 1977); Jefferson Lewis, *Something Hidden: A Biography of Wilder Penfield* (Toronto: Doubleday, 1981).

131 D. Sclater Lewis (1886-1976) was a well-known and respected teacher of medicine at McGill University and the Royal Victoria Hospital. After serving in the First World War with the No. 3 Canadian General Hospital (McGill) in France, he returned to Canada and in 1921 joined the McGill/RVH staff. He steadily rose through the ranks and was named acting

2. December 19th:
    Dr. Gordon Murray:[132] "Heparin" ...

>   Department of Bacteriology and Immunology
>   3775 University St.
>   Prof. E.G.D. Murray
>   Oct. 12th, 1945.

The Secretary,[133]
Montreal Medico-Chirurgical Society,
720 Medical Arts Building,
Montreal, Que.

Dear Sir,

I wish to inform you that I have now returned to my normal duties in the University and hospitals and I wish to resume my normal Fellowship in the Society.

I wish to thank the Society for their generous consideration while I was away on war service.

> Yours faithfully,
> EGDM

---

physician-in-chief in 1943-1944. He also served at various points in his career as president of the Medico-Chirurgical Society of Montreal, first vice-president of the American College of Physicians, president and honorary treasurer of the Canadian Medical Association in 1943, and president of the Royal College of Physicians and Surgeons of Canada from 1949 to 1951. "David Sclater Lewis: Student and Maker of History," *Canadian Medical Association Journal* 116, no. 3 (5 February 1977): 314.

132 The surgeon Gordon Murray is remembered as a pioneer for his experiments using heparin as an anti-clotting agent and for his work in closed-heart surgery; he also attracted international attention in 1946 after he performed the first kidney dialysis. See Shelley McKellar, *Surgical Limits: The Life of Gordon Murray* (Toronto: University of Toronto Press, 2003).

133 LAC, E.G.D. Murray Papers, vol. 7.

Royal College of Physicians and Surgeons

The Royal College of Physicians and Surgeons of Canada[134]

Room 3018,
National Research Laboratories,
Sussex Street,
Ottawa,
Ontario.
September 1944.

Dear Doctor:

In the last ten years an active interest has developed in Canada in the matter of Certification of Specialists. The Canadian Medical Association which began to study the question as early as 1934 soon reached the conclusion that certification should be undertaken by one central body, such as the Dominion Medical Council or the Royal College of Physicians and Surgeons of Canada. This was the more desirable in that thereby the establishment and maintenance of uniform standards of qualification in the different medical specialties in the various provinces of Canada would be secured.

The Dominion Medical Council was first approached. After studying the question, it decided that it could not undertake the responsibility. Thereupon, in October 1936, a request was forwarded to the Royal College and in June 1937 the Royal College formally accepted the responsibility. In order to carry out the undertaking it was found necessary to secure from Parliament an amendment to the Act of Incorporation of the College. This was secured in 1939. ...

In 1943, a circular letter was sent to members of the profession practising the following specialties: Anaesthesia, Dermatology and Syphilology, Otolaryngology, Paediatrics, Radiology, Urology, and already a number of practitioners in each of these specialties have been certified. More recently the Council of the College has approved of certification in the following specialties: Internal Medicine, General Surgery, Obstetrics and/or Gynaecology, Orthopaedic Surgery, and Neurology and/or Psychiatry. ...

To assist the College in this undertaking of Certification of Specialists and permit the College to prepare a Register of Certified Specialists in Canada as complete as possible, Fellows of the Royal College of Physicians and Surgeons of Canada and those holding a graduate qualification in a Specialty granted by a recognized medical or surgical organization other than the Royal College of Physicians and Surgeons of Canada, are requested to return the application form simply giving the details of their graduate qualifications.

Those already certified by the College in certain specialties need not return the

---

134 Ibid., vol. 8.

enclosed application form, as their names now appear on the College Register of Specialists.

>Yours faithfully,
>>Warren S. Lyman, M.D.,
>>>Honorary Secretary.

>>>>The Banting Institute,
>>>>Toronto 2, Ontario.
>>>>March 29th, 1945.

Professor E.G.D. Murray,
Department of Bacteriology and Immunology,
McGill University, Montreal, P.Q.

Dear Professor Murray:

In connection with the Certification in Pathology and/or Bacteriology for the Royal College of Physicians and Surgeons of Canada it has been decided to add three bacteriologists to the Committee now that Bacteriology is to be included. I hope that you will be able to act on this Committee. You will receive official notification from Dr Warren S. Lyman, Secretary of the Royal College. I am enclosing a copy of the regulations for Certification in Pathology and/or Bacteriology.

>Yours sincerely,
>>William Boyd[135]
>>Chairman
>>Committee on Certification in Pathology and/or Bacteriology.

>*The Royal College of Physicians and Surgeons of Canada*[136]
>Certification in Pathology and/or Bacteriology

There shall be three classes of certificates:
- a. Clinical Pathology
- b. Pathological Anatomy
- c. Bacteriology

---

135 Ibid. William Boyd (1885-1979) was a Scottish-born Canadian pathologist. He served in the First World War with the Royal Army Medical Corps, and then moved to Canada where he held teaching positions at the University of Manitoba (1915-1937), the University of Toronto (1937-1951), and the University of British Columbia (1951-1954). He also authored numerous medical textbooks, including *Surgical Pathology* (1925), *Pathology of Internal Disease* (1931), and *Textbook of Pathology* (1932). He was made a Companion of the Order of Canada in 1968. *The Canadian Encyclopedia*.

136 Ibid.

1. Certification by Examination:
    1. General requirements –
        1. Satisfactory moral and ethical standing in the medical profession.
        2. Graduation from a medical school approved by the Council of the Royal College of Physicians and Surgeons of Canada.
        3. The applicant to devote his time primarily and principally to the practice of pathology or bacteriology.
        4. One year's general internship in an approved hospital.
    2. Special Training –
        1. A period of study of not less than three years in an approved institution or department of pathology or bacteriology.
        2. Study or practice in pathology or bacteriology for a further period of one year.

        Note 1: After the period of training, the candidate may apply for certification by examination in one, two, or three of the above classes.

        Note 2: The examination for certification in Clinical Pathology shall include routine diagnostic work in bacteriology, immunology, parasitology, haematology, biochemistry and clinical microscopy.
2. Certification Without Examination:

Notwithstanding the provisions set forth above, the College may, during an initial period of time to be determined by Council, grant a certificate without examination to a candidate whom Council shall consider to have acquired an adequate training and qualification by study and/or practice of at least five years in his specialty.

Directorate of Chemical Warfare and Smoke,
National Research Council Building,
April 11, 1945.

Professor William Boyd,[137]
The Banting Institute,
Toronto 2, Ont.

Dear Professor Boyd,

I have now received your letter of 29th March., 1945 and the enclosed schedule of requirements for "certification" as a pathologist or bacteriologist.

I wish I knew the purpose and intentions of this movement. I believe the Pathologists in Ontario have been organizing in this way for some years and in your letter

---

[137] Ibid.

you use the phrase "it has been decided to add three bacteriologist". I would therefore suppose there exists some form of constitution and regulations. There must be, too, some legal standing to allow the imposition of a qualification to enable practice, which I suppose the purpose to be. It would be a help to know the names of members of the Committee.

Not knowing any of these details I find myself uneasy and hesitant. But, on general principles, the sponsorship of the College is encouraging.

Yours sincerely,
Professor E.G.D. Murray.

Culture Collections: Bacteriological Nomenclature

Prof. E.G.D. Murray
Department of Bacteriology and Immunology
December 7th, 1940.

Dr. Robert S. Breed,[138]
New York State Agricultural Experiment Station
Geneva, N.Y.

Dear Robert,

Thank you very much for sending me the copies of letters concerning the Nomenclature Committee of the International Association of Microbiologists.

I only wish to remark that I believe it very important to elect the Judicial Commission, if for no other reason than to produce the provisional rules of bacterial nomenclature. It seems to me essential that these rules should be made available to bacteriologists in order that they may be thoroughly tested out before the next International Congress, whenever that may be.

War or no war something along these important lines could always be done and one of the primary duties of the Judicial Commission is to make available the proposed rules of nomenclature; in my opinion, at the present time this is their most important duty. Other duties might rest although it would be very useful to have

---

138 LAC, E.G.D. Murray Papers, vol. 6. Robert S. Breed (1877-1956) was an American biologist and one of three trustees of *Bergey's Manual of Determinative Bacteriology*, along with David H. Bergey and Everitt G.D. Murray. He served as the *Manual*'s principle editor from the 1920s until his death. Early in the twentieth century he taught biology at Allegheny College, and in 1913 he became head of bacteriology at the New York Agricultural Experiment Station at Geneva, New York. He was also president of the Society of Bacteriologists (1927). Harold J. Conn, "Robert Stanley Breed," *Journal of Bacteriology* 71, no. 4 (1956): 383-384.

rulings on particular difficulties of nomenclature, such as nomina, conservanda, nomina nuda, etc., which are a trouble to us in compiling the Manual. ...

    Yours sincerely,
        EGDM

                Office of the Director
                Agricultural Experiment Station
       Iowa State College of Agriculture and Mechanic Arts
                      Ames, Iowa
                    December 20, 1941

Prof. E.G.D. Murray
McGill University
Montreal, Quebec, Canada

Dear Professor Murray:

I have a note from Doctor St. John-Brooks[139] from which the following two paragraphs are quoted:

> "Murray is strongly of opinion that some means should be found of printing and publishing the Rules of Bacteriological Nomenclature which you drew up and which were approved in principle by the Nomenclature Committee of the International Association of Microbiologists of the New York (1938) Congress. He suggests a publication of about 2,000 copies which could be distributed to libraries, universities, scientific societies, etc. and later on, when bacteriologists had thus had the opportunity of familiarizing themselves with their provisions, the rules could be laid before the Nomenclature Commission, which would probably be then functioning and which could then take action." ...

    Sincerely yours,
        R.E. Buchanan[140]
        Director

---

139 LAC, E.G.D. Murray Papers, vol. 6. See chapter 1, note 100.
140 Robert E. Buchanan (1883-1973) was a contemporary of E.G.D. Murray and, along with Murray, a long-time member of the Board of Trustees for the Bergey's Manual Trust, as well as co-editor of *Bergey's Manual of Determinative Bacteriology*. Buchanan was appointed in 1910 as the first head of the Department of Bacteriology at Iowa State University, and he also served as dean of the Graduate College (1919-1948) and director of the Agricultural Experiment Station (1933-1948). "Memorial to Professor R.E. Buchanan," *International Code of Nomenclature of Bacteria* (1990 Revision), http://www.ncbi.nlm.nih.gov/books/bv.fcgi?rid=icnb.

Jan. 6th, 1942

Dean R.E. Buchanan,[141]
Iowa State College,
Ames, Iowa.

Dear Dean Buchanan,

    As you know from St. John-Brooks' letter, I saw him in Washington during one of my visits there and we discussed the present position of the Judicial Commission. I think it is obvious that very many questions which should be referred to this Commission cannot be dealt with on account of the war even though a majority of members are available. However, I am convinced that the proposed Rules of Bacteriological Nomenclature could be published and would be used by discriminating bacteriologists and their final adoption at the next meeting of the International Association would thereby be made possible. ...

    Yours sincerely,
    EGDM

---

141 LAC, E.G.D. Murray Papers, vol. 6.

CHAPTER FOUR

FAMILY CORRESPONDENCE: 1946

London
3 Jan 46

Dear Dad:[1]

    I have got the name of the company that makes the "Tygon" paint & Tubing. It is:
        United States Stoneware
        Akron, Ohio.

    Keep an eye out for those "background" and "idea" books that we were talking about – I would be much interested in any list you could give me.

    To hell with Mackenzie King.[2]

    Your son,
        Bob

London, Ont.
7 January 46

My dear Dad:

    The photographs of the charcoal drawings arrived this morning and we are most glad to have them.[3] They will make a very welcome and beautiful adornment for our walls.

    I have just read an article which I think will interest you & you may want it for your file. It is called "The Scientist & the War", written by J.H. Hildebrand. He was apparently in the London Mission of the OSRD [Office of Scientific Research and Development] – I think you would appreciate many of his comments. Anyway I know I did.[4]

    The last few days I have spent my time going through examination papers – all of which tells me that 80%+ of the population is illiterate and the remainder cannot be called educated. We have just started the orals and they don't seem to be any better. I could have wished for a better showing particularly when I did over 60% of the teaching.

    I am going ahead, on the side, with making a list of books and papers which have value in giving a broad biological view of the medical side of science – and I would

---

1 Unless otherwise specified the correspondence in this chapter is located in the UWO Archives, R.G.E. Murray Collection.
2 Robert Murray cannot recall what caused this critical remark about Mackenzie King. RGEM.
3 The charcoal drawings are of Robert Murray's sister, Susan Murray Robinson. RGEM.
4 Published in *Chemical and Engineering News* 23 (24 December 1945).

very much appreciate it if I could have your *ideas & suggestions* as soon as possible. I have got Stavraky[5] interested in it too and we are going to try to get the points over in one of the graduate studies seminars fairly soon.

Mother Grace[6] has been here since New Years day – unfortunately, she got laid low by this influenza that is going the rounds and has been in bed with a hell of a fever since Friday. She is not really on the mend yet. Poor Doris has been worried by it & has also had a lot of extra work – but she thrives just the same.

I hope Mum is well again by now and that Blake is feeling better. Let me know how things are with you.

 Your loving son,
  Bob

21 Jan 46

My dear old Dad:

It was grand to talk to you and Mum yesterday – a very pleasurable surprise for us. I'm sorry that we had not written for such a long time but we were both a little tired and not in a very letter-writing mood. Doris, especially, has had a hard time of it and is very much in need of a bit of time to herself. Mother Grace has been in bed ever since the 4th of January and for a good part of the time has required complete nursing – we would have put her into hospital but for the fact that there are no beds except for emergencies. She has had a very severe go of influenza with a great many râles in her right lung and with quite severe constitutional symptoms – also with all the concomitant depression. None of this made the task any easier for Doris. She is still in bed and is not likely to get up for a few days yet – the result is that Doris (& Alice)[7] have not been able to get out except when I've been able to get home to relieve her. Despite this she keeps very well and has managed wonderfully.

Alice is really very well although her "sniffle" is still persisting a little bit. She doesn't do anything really new except that she is becoming a little more constructive in her playing. Tonight she has the most enormous bump on her forehead as a result of a fall – Doris feels that she is scarred for life but that is a bit exaggerated.

My activities have been a little mixed. Up till the beginning of last week we were tied up with exams and reading papers. A very horrible and disheartening job. As a whole I felt the class did very poorly and since I did most of the teaching I feel rather badly about it. However, in the final analysis I feel that most of the low marks were

---

5 George W. Stavraky was a senior teacher and researcher in the Department of Physiology at the University of Western Ontario. RGEM.
6 See chapter 3, note 94.
7 Alice was the first child of Robert and Doris Murray; see chapter 3, note 72.

for students who had little background and extremely poorly organised minds. No sooner was all that over than Dr. Bittner's husband came down with acute biliary colic and consequently I have taken over the routine lab for the time being.[8] This is only right for taking either of the two competent girls for that job would discommode Ash[9] & his research far more than it would me. It is also a nuisance because I have to get things organised and ready to teach these "lab technicians" – since there is not enough routine work to go round them we really have to set up a separate little course – in addition, they have no background whatever except a smattering of techniques from Biochemistry and Pathology which are of no use to us. So for the next ten days I am going to have to concentrate more on that part.

Just lately I have been doing a good deal of reading and I have learnt a great deal. I have been particularly interested in the Burnet monograph I told you about – it is concerned with quite highly controversial matters in the production of antibodies but just the same it is valuable from that point of view alone – as a stimulus and not just as another textbook.[10] A review of that sort by a man of such caliber is of great value in that it not only presents a new view point but also points out gaps in fundamental knowledge. Much the sort of thing that I keep talking about as desirable in the "background" books we have discussed. Burnet is, I think, one of the few who constantly strive to think of things in terms of general biological functions; considers man as another vertebrate mammal and not as a sort of superior organism.

I got a letter the other day, without much in the way of detailed news, from Eugene Munroe.[11] He is, apparently, fomenting a minor revolution in the musty files of

8 Josephine Bittner performed most of the routine lab work in the department at UWO. See chapter 3, note 90, for description of Bittner's role in the department. Robert Murray recalls that by the time he arrived, she had become "adept and essentially handled the service herself." RGEM.
9 See Introduction, note 377.
10 Most likely a reference to either F.M. Burnet, M. Freeman, A.V. Jackson, and M. Lush, *The Production of Antibodies* (Melbourne: Macmillan, 1941) or Burnet, *Biological Aspects of Infectious Disease* (Cambridge: Cambridge University Press, 1940). Frank Macfarlane Burnet (1899-1985) has been described as "the greatest biologist Australia has produced." See F.J. Fenner, F.R.S., "Frank Macfarlane Burnet," in *Biographical Memoirs of Fellows of the Royal Society*, 1987. From 1944 to 1966, Burnet was the director of the Walter and Eliza Hall Institute of Medical Research in Melbourne, Australia. He was awarded the Nobel Prize in Physiology or Medicine in 1960 (with Peter Medewar). See the guide to Burnet's personal collection at the Australian Science and Technology Heritage Centre, the University of Melbourne, http://www.esrc.unimelb.edu.au/.
11 Eugene Munroe was one of Robert Murray's good friends from his first year at McGill in 1936. The alphabetical seating plans had the two seated next to each other in at least two classes, and as Murray recalls, the two "became friends doing biology together for two years." Munroe became a respected authority on the Lepidoptera and a senior researcher in entomology at Agriculture Canada. In 1966 he was elected to fellowship in the Royal Society of Canada. RGEM;

systematic entomology. He must be within sight of the end of his thesis for he says that he is "turning an exploratory mental tentacle towards job hunting" and he is listening for any gossip of openings for such as he. A good man – he should have been a bacteriologist!

This had better be all for now but I will write again soon when opportunity presents.
    Your loving son,
        Bob

P.S. Two interesting cases: (1) Pure culture of *Gaffkya tetragena*[12] from pus removed from a ureter (2) 29 year old man who has complained (not too much) of pain in one hip for four years suddenly develops acute pain and a small swelling in the region of that joint. Xray shows bone destruction in the femoral head. Pus aspirated by needle shows a pure culture of Staph. pyogenes.[13] He has negative tuberculin test and no history of previous osteo or even any moderately severe infection that might have been due to staph.

27 Jan 46

My dear old Dad:

The general circumstances that surrounded us have now abated somewhat. Mother Grace is better and has been getting out of bed for the last three days – so that Doris has not had quite the strain of previous weeks. Dr. Bittner is back doing the routine work and I can get on with some of the other duties.

At the moment I have to worry most about the teaching of this class of 7 "technicians" which is a pesky nuisance. The time we are to have them has been cut down again and I have to compress it into 6 weeks – hardly enough time to skim the surface. The worst problems are finding space to put them in and giving them the work without straining our resources or hindering Ash's research.

I am also trying to find time for a great deal of necessary reading – some of which I am accomplishing – partly in preparation for a seminar and of course to do with immunity problems.

We have been doing some "costing" in the past week or so because it has become evident that the hospital is getting far more service than they are paying for. I would

---

"RSC: The Academies of Arts, Humanities and Sciences of Canada," http://www.rsc.ca.
12 *Gaffkya tetragena* is a "species of Gram-positive coccus" that Murray occasionally identified. RGEM.
13 "Staph" is the informal term for "Staphylococcus, name of a genus of pathogenic bacteria." *OED* (Online). "Pyogenes" translates as "generating pus." Several bacteria have that character in producing disease so it is used as a specific epithet, i.e., the second name in a formal name for a species. RGEM.

be very interested if you could give me some figures for comparison – such as the charges made for culture work, TB, etc.

Ash and I had a great talk the other day about what they would call "futures" of the stock market. The plan is that I will take complete charge of the Medical course this next fall and will be responsible for a large portion of the details around the lab such as the Routine Room – as a part of this he and Hall[14] are planning to move me up a step this summer. This latter, of course, is a matter of persuading the board of governors but Ash thinks it can be done. This will mean a rise in salary but most important a rise in local *political* prestige, which will be very helpful in dealing with the hospital – not that relations are atall difficult now but it might ease up the cooperation angle.

I have some Staph titres I want to do – and although I have some toxin which was standardized 1-1/2 years ago it may have deteriorated a bit by now. Could you send me a small amount of antitoxin[15] that is standardized?

That is about all for now – I hope to hear from you soon.

  Your loving Son,
    Bob

                    5 Feb 46

Dear old Dad:

It was very nice to get your letter the other day and to hear all the news; also to get the LIFE and PUNCH that arrive so regularly (we enjoy getting them very much).

I was a little over-optimistic when I said that all our troubles seemed to be over. Mother Grace started to run a slight fever again and also developed a lot of pain in her chest. She had a small area of pleural friction in her left base – she still has a low grade fever and Dr. Kennedy[16] is going to admit her to hospital for further investigation as soon as he can get a bed for her.

Just to add artistic verisimilitude the main drain decided to have a subacute obstruction and the men had to dig for it in the front of the house. For a few days we were afraid to take a bath in case it would put out the furnace – exaggerated, but the cellar was a little damp.

Alice is in good health but is a little fractious because she is obtaining some real grinders – only one is in view but the others are not far off.

14 See Introduction, note 313, and chapter 3, note 73, for information on G. Edward Hall.
15 An antitoxin is defined as a substance which "has the property of counteracting the effect of a toxin; one of the antibodies capable of neutralizing toxins." *OED* (Online).
16 Frank Kennedy was the Murray family doctor in this period, and he attended to Mother Grace (Marchand) when she had pneumonia while staying with the family in London. RGEM.

Doris is finding things a bit much for her and getting more easily tired. But she is bearing up under the strain. I have not been feeling too perky myself for the last few days and since yesterday have developed some quite severe muscular pains in the back which are a plague to me.

All-in-all, we are a little subdued.

A very peculiar culture has turned up in the routine lab. It is from a persistent draining sinus in the costo-lumbar angle[17] which followed a perinephric[18] and pyelitic[19] abscess. In the culture there is a very light growth of micrococci[20] and E. coli but there is a heavy growth of small haemolytic colonies. These colonies are of two types and both producing a zone of B-haemolysis.[21] The smaller one is undoubtedly a small form of strep; the other is very peculiar. It grows best under microaerophilic[22] and anaerobic[23] conditions although it will grow aerobically though the colonies are hardly visible (but haemolysis is evident). ... Even in the youngest and oldest cultures the forms are most bizarre with a definite tendency to chain formation. Some of the chains look, as much as anything, like preparations of salivary gland chromosomes from Drosophila![24]

... I have not had time to study it properly yet but I hope to get something more on it by the week end. If I can maintain the thing in subculture I will send you a strain. So far I have not found any character that helps to classify it.

I find there is a paper by a fellow called Grasset from South Africa who, in 1929,

---

17 In this context, the "costo-lumbar angle" presumably refers to the shift in direction of the spine in the "small of the back." RGEM
18 "Perinephric" (adj.) means "situated or occurring around the kidney;" also formerly "of or relating to the perinephrium" ("the connective tissue surrounding the kidney; the renal capsule"). *OED* (Online).
19 "Pyelitic" is defined as "of, relating to, or of the nature of pyelitis." *OED* (Online). See chapter 3, note 34, for a definition of pyelitis.
20 "Micrococcus" (micrococci, plural) is defined as a "small germ or seed postulated as an elementary precursor to a living organism," and "Any member of the genus *Micrococcus*, originally comprising all spherical bacteria, now comprising some nine species of Gram-positive spherical bacteria that form part of the normal bacterial flora of the skin of warm-blooded animals, and occasionally act as opportunistic pathogens." *OED* (Online).
21 "Hæmolytic" refers to "destructive of the blood or of the blood-corpuscles." "Hæmolysis" refers to the "dissolution or lysis of red blood cells with the consequent liberation of their hæmoglobin." *OED* (Online).
22 "Microaerophilic" means "requiring less oxygen than that of the atmosphere; of the nature of a microaerophile." *OED* (Online).
23 "Anaerobic" means "pertaining to or characterized by anaerobiosis, anaerobic." "Anaerobiosis" is defined as "life in a medium devoid of free oxygen." *OED* (Online).
24 *Drosophila* is a genus of small fly which belongs to the family Drosophilidae; the flies are often referred to as fruit flies. *OED* (Online).

tried to immunize chick embryos using Diphtheria Toxin and failed.[25]

Some years ago I seem to remember your telling me that Cow serum was toxic to some degree to animals. Is that my imagination?

I have read the paper on the making of discs for testing penicillin sensitivity. For the usual reason there seem to be very few requests for such tests – the usual trial and error persists – and when they come up I usually do it by the cup method. I don't think I have even done one a month. The P.H. people[26] tell me that such requests are just as rare with them.

Thanks for the address of where to get the standard Staphylococcal antitoxin. I have written to them.

I have just finished reading two good books. One is Nevil Shute's "Most Secret" – which is very good. Also J. Frank Dobies "Texan in England" which is a most charming and sympathetic book with many good observation. Made me want to see Cambridge again. It is no wonder that he was so popular there.

It is good news that Blake is so much better – I hope that he remains free of amoebae[27] now.

We have received notice that there is going to be a CPH meeting in Toronto early in May and also a SAB meeting in Detroit on May 21. I suppose that you will be going to them both. I think that Ash & I will too. Apparently the university will pay expenses. What are your plans?

    Your loving son
      Bob.

10 Feb 46

My dear Mum & Dad:

You must have noted my rather depressed letter of a few days ago. Things are still not too good and I do not feel much more cheerful.

---

25 See E. Grasset, M.D. (Serologist-in-charge, Serum Department, South African Institute for Medical Research), "Diphtheria Immunisation," *Journal of the Royal Society for the Promotion of Health* 56, no. 12 (1935): 818-830..

26 This refers to the workers at the Public Health Institute, which was situated near Robert Murray's laboratory on the Victoria Hospital grounds. Murray was in frequent touch with Wes Wilson, who was director of the laboratory, because they performed all the general lab work for physicians and public health officers in the region. The Institute also performed other useful tasks including the serology for syphilis. RGEM.

27 "Amoeba" (plural, amoebae) is defined as a "microscopic animalcule (class Protozoa) consisting of a single cell of gelatinous sarcode, the outer layer of which is highly extensile and contractile, and the inner fluid and mobile, so that the shape of the animal is perpetually changing." *OED* (Online).

Mother Grace is still running a daily fever up to 100 F and although she feels pretty well there is still something going on in her chest. We are still trying to get a bed in the hospital so that she can get a more extensive investigation – but the hospitals are full to the back teeth and it may be a few days yet.

I am particularly anxious that she should get there as soon as possible, not only for her sake but for Doris also. I have been rather worried about Doris just lately – she has been getting so very tired and to make it more ominous has had a little bit of "show". ...

One thing that is not improving my feelings at the moment is this "technician" class which is composed of some mighty-fine dimwits. In addition they have no foundation in service and I doubt if they will ever learn to do every procedure with intelligence. I have only had them for a week but I doubt if I will have any occasion to change my opinion. 5-6 hours contact with them every day is souring my sunny disposition.

That comic organism I was telling you about seems to be a form of strep. The further subcultures of the original culture have changed gradually to be more like a strep and have lost that peculiar character except the colony form is maintained. (Dr. Bittner did most of the work on that original & there may have been some distortion!). However I have collected some material for my own use - & examined the primary culture yesterday. The smallest colonies are definitely a streptococcus. ...[28]

Alice is very well & cheerful. Her teeth have ceased to bother her & she is full of beans. She has matured a lot since you saw her & she is just beginning to climb upon things. Her vocabulary is not much more extensive but she has acquired some new & highly original noises.

      Your loving Son,
         Bob

P.S. I got the standard serum from Ottawa on Friday.

Montreal
12 February 1946

Dear old Son,

We got a letter from you this morning dated 10 February & one a few days ago dated 5 February & I am sorry I had not answered the first of them yet.

We are worried that Doris has had such a tiring & trying time since Xmas. The situation is difficult to handle if you cannot get a hospital bed. I suppose you have pointed out to Dr. Kennedy that getting a bed is as important to care of Doris as of

---

28 Streptococcus is a spherical Gram-positive bacteria that grow in chains or pairs. *OED* (Online).

Mrs. [Grace] Marchand. Doris is such a grand girl & so eminently practical & a doer of things that, like Mum, she will not spare herself. I hope you can get this woman to help her a bit. Mum, of course, is itching to do something to help & would be with you like a shot if it were practical, but you have no room.

It is great that Alice is well & over the bit of teething trouble. She will make ever increasing progress & at the same time need constant care. Mum is arranging to go to look after you when Doris is in hospital & proposes to travel with me as far as London on May 20th; I shall go on to Detroit and will see you there.

Your plaguey pains seem much the same as I have been having for some time. Seemingly some after-effects of "a touch of 'flu" & a deal to do with lack of proper exercise. Your worries would contribute to some extent too.

The "Technician Course" is what I anticipated. The thing to do is to put in a strong adverse report, substantiated by facts, when it is over and be justly severe if there is an examination. Make a strong bid not to have it repeated. Meanwhile do the best you can without grousing for the impression it makes & because it will very valuable experience in evaluating that kind of people & a basis to judge what preliminary scientific training is essential.

I have had various of these "Certified Technicians" (U.S., Canadian and U.K.) & their training has been no use to me whatever; I had to pick them over on the ground of personal intelligence & then train them properly for what I wanted.

The organisms you write of are interesting. One I think is most likely the "Minute Haemolytic Streptococcus" of Long-Bliss[29] (see Sherman, Bact. Reviews 1, 1937, 40) which very often shows marked haemolysis at a time when or before the colony can be hardly seen. The other, your most curious organism, seemed to me at first like a Corynbacterium possibly *pyogenes*. I would like to have a culture to see.

Serum of various species seem to be primarily "toxic" to other species of animal. Claude Bernard[30] found dog serum killed rabbits with haemoglobinuria[31] & Creite[32]

---

29 See Perrin H. Long, M.D., and Eleanor A. Bliss, Sc.D., "Studies upon Minute Hemolytic Streptococci: I. The Isolation and Cultural Characteristics of Minute Beta Hemolytic Streptococci," read before the Annual Meeting of the American Society of Bacteriologists, Philadelphia, December 29, 1933, http://www.jem.org/cgi/reprint/60/5/619.pdf.
30 Claude Bernard (1813-1878) was a distinguished French physiologist. In 1855, he became professor of experimental physiology at the Collége de France, and he made important contributions in many areas of medical science, including in the area of digestion and the functioning of the pancreas. See Times (London), 19 February 1878, 3, and 20 February 1878, 11.
31 "Haemoglobinuria" refers to "the presence of free hæmoglobin in the urine." OED (Online).
32 This is most likely a reference to Joachim Creite, who was the first to observe under a microscope the "dissolving of red blood corpuscles by the serum of foreign blood ...". See Herbert W. Rand, "Friedenthal's Experimental Proof of Blood-Relationship," American Naturalist 35, no. 420 (December 1901): 1017-1022.

found some toxicity of ox serum & thus for rabbits & still more marked toxicity of bird serums ... There is a general agreement that the worst "toxic" animalian serum for man is ox serum & the least toxic for man is horse serum. If you are specifically interested I can send more detail.

I spoke to Jim Craigie[33] about your interest in the development of antigens in the embryonated egg. He thought there were interesting things to do with it & would be very glad to discuss it with you. It would be worth while having a talk with Jim Craigie; you could write to him & sometime arrange to meet him in Toronto.

Some work done at Grosse Ile is going to be published soon, I will let you know when it is; but, meanwhile, it was observed that various embryonated egg fluids *augmented* the activity of complement in C. fixation reactions. ...

... I feel you have to pick problems which have a reasonable chance of giving publishable results fairly quickly & not a very difficult very long term problem. It is also important to pick problems which have an interest which attracts attention. However that does not mean that problems interesting per se should be thrown out of court. I am anxious for you to get some good work into the literature – I have found that what I have done here helping others during the past 15 years has not got me anywhere.

This place is beset with difficulties & I have not got them sorted out yet. Taken all round the future looks pretty black to me & I am glad you are not at McGill.

I want to read the "Texan in England"; I have known of it since it appeared. "Most Secret" is a marvelous good story very well done. People who know J.F Dobie say he is a magnificent fellow & a great power in his own bailiwick & fearless in dealing with the politicians.

I'm very glad you are going to the S.A.B.[34] and C.P.H.A.[35] meetings.

 Much love to you all,
  Love, Dad

London
18 February 1946

My dear Dad:

It is a wonder I can write at all because we are suffering an inside cold spell owing to a crisis in coal. I calculated the supply a bit too finely and the last shovelful (of dust) was consumed at 8 a.m. this morning in a vain attempt to get the place habitable. We managed to raise the temperature from a low of 47 to a high of 62 – luckily the coal

---

33 See chapter 1, note 37.
34 Society of American Bacteriologists.
35 Canadian Public Health Association.

company lived up to their promise but they only came through at 5 p.m. and we are now getting up to normal. ...

... Mother Grace is much better – she went to hospital a week ago today. X-ray showed a little remaining pleurisy but no fluid but did not shed any light on the fixed swelling on her chest wall anteriorly. They decided to put a needle into it and obtained some very bloody pus. This I found to be a pure growth of a Pneumococcus[36] (definitely micro-aerophilic on primary isolation) and which was not of any types I-XI. The[y] drained it surgically (after a saturation raid with Penicillin) and found a multilocular deep abscess one part of which had invaded. ... In the end the surgeon not only drained an abscess but did a bit of rib resection as well. It seems funny to me that a pneumo, which kills mice in short order, should produce such a chronic lesion. She does not have any signs of localisation anywhere else though she has been told by her dentist that she has two root abscesses in her teeth. I never found any pneumo's in her sputum and she has not had any consolidation of her lung.

The newspapers have been having an unwarranted field-day of guess & supposition over the occasion of Mackenzie King's announcement of leakage of secret information. Although I admit the probability of Russian interference I don't think that should be said until official announcement is made. As a matter of fact they could have kept the whole thing quiet a little longer. Tonight there is much tale acquired from the Mtl Star concerning the arrest of people from McGill, U of M, Bell Telephone etc. Also there are tales of blackmail. All very disquieting.[37]

I enclose a couple of cuttings. One important one is the impending appointment of G.E. Hall to be president of U.W.O. when Fox[38] retires in 1948. I feel very sorry for this in a way since he could have done so much more for the Med. Faculty if he were here – though, of course, he will have laid much groundwork by the time he leaves. ...

... The technicians' course seems to go quite quickly, thank goodness, but it takes

---

36 "Pneumococcus" was originally defined as "an opportunistic pathogen causing pneumonia and other infections, often in hospitalized patients," and later as "the bacterium Streptococcus pneumoniae, occurring as Gram-positive oval or spherical forms, typically in pairs, which causes the most common type of bacterial pneumonia (often lobar in distribution in adults) as well as other infections including otitis media and meningitis." OED (Online).

37 This refers to the events that unfolded in the weeks and months following the defection of the Soviet cipher clerk Igor Gouzenko in Ottawa. For sources on the events collectively known as the Gouzenko Affair, see Amy W. Knight, How the Cold War Began: The Igor Gouzenko Affair and the Hunt for Soviet Spies (New York: Carroll and Graf, 2006); Reginald Whitaker and Gary Marcuse, Cold War Canada: The Making of a National Insecurity State, 1945-1957 (Toronto: University of Toronto Press, 1994).

38 W. Sherwood Fox was the third president of the University of Western Ontario, holding the position from 1927 to 1947.

a deal of my time. I will take care to put in a very adverse report on such a course – though Dean Hall has promised that it will not happen again.

That organism I was telling you about is a Streptococcus but seems to be very variable in its microscopical appearance – the colony form would seem to be due to its very long and rather inflexible chains. ...

I much enjoyed your last letter & the information in it. I agree with you about the problems – & their choice, that is why I am holding my fire. The reading is purely interest, academic curiosity & a starting point for a longer work.

 Your son,
  Bob

        The University of Western Ontario
           London, Canada
           Faculty of Medicine
     Department of Bacteriology and Immunology
        Ottaway Ave. and Waterloo St.
             1 Mar 46

My dear old Dad:

It seems some time since I heard from you, though I don't expect it is as long as all that. I wish that we could have opportunity to meet soon. May seems a long way off, for there are tons of things I would like to talk to you about. Letters are alright for many things but they do not supplant a good chat. It was grand to talk to you on the phone last Saturday.

I am glad to say that Doris seems to be perfectly well again – looks far better, and rested, and is able to carry on all daily duties without getting too tired. There has been no repetition of the warning signs so I feel a great deal easier in my mind and so does she. Alice, too, is in the pink and is her usual cheerful self. Mother Grace is much better, almost well, but is staying in hospital.

The Technicians course has only another two weeks to cover, for which I am very glad. The seven, are without any exception, a dull lot and there are none that I would be willing to take on for our own use let alone recommend for anyone else. However, I have learnt a little from teaching them. ...

I sent you, on 25 (February), a culture of that funny streptococcus. After all the subculturing it seems to be more constant in character than it was on primary isolation. I think that the morphology of the isolated colonies is due to the very rigid chains that are formed. In fluid media the chains are far more typically cocci. I have repeated the Lancefield grouping a couple of times. ... The woman died early this week and this was confirmed at autopsy & none of the streps could be isolated from the scars or from the remains of the kidney. She died of a *miliary Tbc*; though *Tb* could *not* be demonstrated in the discharge from the virus during life.[39]

I have been working in off moments on a rather interesting "test". One of the biochemists here is working on Hyaluronidase (spreading factor)[40] in tissues & organ extracts from the chemical point of view. He was worried by contamination by organisms which I have been isolating & identifying for him. Meanwhile I remembered observing at R.V.H. that Mucoid Streps changed character & seemed to lose their capsules in proximity to a Staph. streak, and that this was most likely due to action of hyaluronidase. I have quite a good strain of Mucoid Strep (group A type 9) in the lab and I find that this is true for *Staph pyogenes* but not for *epidermidis* and other micrococci. (I have yet to try it with other S.F. producing organisms.) ... If Gertrude[41] isolates any good strain of Mucoid Strep., especially those which show little if any haemolysis, I would be very glad if you would send it to me.

There doesn't seem to be much else to tell you at the moment except that we are well, happy, and nicely settled.

   Love to you all
    Your son
     Bob.

P.S. I saw in "Nature" a while back that Paul Fildes[42] was made a Knight in the New Years Honours list. They don't seem to be giving the Canadians any recognition.

               Montreal
               3 March 1946

Dear old Son,

I gave your microaerophilic pneumo to Fred Smith[43] but I don't think he has done

---

39 Miliary tuberculosis is "a generalized form of tuberculosis in which there is haematogenous spread of mycobacteria resulting in the formation of numerous small granulomatous lesions in the lungs and other organs (esp. the liver, spleen, kidneys, and meninges)." OED (Online).
40 Hyaluronidase is an enzyme produced in tissues that require it and by some pathogenic bacteria. The enzyme "attacks and depolymerises the hyaluronic acid polymer which is a large part of the ground substance produced between cells in the connective tissues of mammals." RGEM.
41 See chapter 3, note 101.
42 See chapter 2, note 39 for information on the British bacteriologist Paul G. Fildes.
43 Fred Smith worked in E.G.D. Murray's department at McGill University. He took over the majority of the problems of the department and the teaching responsibilities from Everitt Murray when the latter's focus and energies were diverted to Canada's BW program during the Second World War. After the war, he served as McGill's dean of medicine (1947-1949). Robert Murray remembers Smith as "a very competent man ... thoroughly capable of looking after things", and "also a very good microbiologist ..." RGEM. See also Introduction, note 238,

anything with it. The other culture Gertrude examined & on our media it is a very characteristic pneumococcus. Perhaps you got your labels mixed & I shall get back the culture from Fred. I have an idea you promised me a Gaffkia tetragena you had isolated; I would much like to have it as I have very seldom found it.

Nothing is settled here yet. I have an appointment with F.C. James[44] on Wednesday & then I hope to know what "the authorities" propose to do; but I am not hopeful.

Mum is very well & so am I, though I am soft & still have a cough, and we were cheered up by the approaching time when we shall get up to the lake again. I have not yet done any work on tackle or equipment; I must get going on it soon. Mum is enjoying the training of Sue's Collie pup & is loath to part with him. He is a very nice little fellow & promises to be a good dog. He bullies Uncle Fudge[45] who takes it very well & only protests when his sleep-time is interrupted.

We had dinner with the Robinsons (Senior)[46] last evening & it was quite pleasant. Sue & Blake were here too & both look very well, having each put on necessary flesh. They have settled in to their little home & are well pleased.

Rosemary[47] is still here & continues to astonish me with her lack of general knowledge & stodginess. She does go skiing most week-ends, but they remain in compact little groups of English eschewing any Canadian contacts & so lose a lot. When they return home they will still be the perfect "little islanders".

I hope Mrs. Marchand (Mother Grace) is well again; it must be very trying for her to have been ill so long & her plans must be completely upset. Please give her our regards & good wishes. Doris' letters indicate she is feeling quite her dear self again, which gladdens us, & the little girlie seems to be in grand form. We are looking forward to the chance of seeing all of you in May.

I am annoyed that the Royal Society & the S.A.B. meetings fall on the same days. I am committed to the S.A.B. because of a symposium on Taxonomy I shall have to be chairman of. All wrapped up with Bergey's Manual problems & there are many. Have you fixed up your accommodation there? Let me know where you are going to stay as soon as you know. I hope to be able to introduce you to many of my old friends.

There may be something important in the reported Russian ultraviolet microscope

---

and chapter 1, note 87.
44 See chapter 2, note 6.
45 Uncle Fudge was the Murray family's "faithful hound," a fudge-coloured cocker spaniel who lived at the Montreal home in the 1940s and 1950s. Robert Murray remembers him as "a true family favourite (who) would sell his soul for asparagus," a can of which he would receive every Christmas. RGEM. See also Introduction, note 262.
46 This is a reference to the parents of Blake W. Robinson, who married Susan Murray on 9 June 1945.
47 Rosemary Pamplin was a British evacuee who stayed for a period at the Murray family home in Montreal. RGEM.

but the newspaper reports lack specific information. There is an American microscope in process of development which by report has definite possibilities ...; it is said that such things as phage particles are easily observed with it in the living state. I don't know the details & have not seen any report as yet.

Hall will have a couple of years yet as Dean (with additional interests) & with much influence, during which he can achieve much for the Faculty. He will probably be judicious in selecting his successor, so it is probably not necessary to worry. I hope the scheme for your advancement is in good order. I hope it will give you the best possible chance for research, which is the most important thing for you to do. Don't worry about finding large & seemingly important projects, work at whatever presents itself & squeeze what you can out of it & publish anything informative. Often the most trivial seeming matter brings out unexpectedly important information leading to great advances. You can't judge until you follow them up. The important thing is to get & keep the habit of investigating & not to let administrative & other occupations deflect you. Profit by the big mistake I made on coming here, of letting myself be kept from research on however simple questions. I am determined to get back into research, even if it is only on very simple ordinary things. Make everything possible into research by the approach you take to it & you will find contentment & progress; administration (etc.) are sterile fields full of worries & dissentions & jealousies & whatever you may do it is all destroyed by a single stroke at any moment leaving nothing atall. It may appear to have an iridescence at the time but in the end it is merely a froth of bubbles easily dispelled, but once you are trapped in it it is hard to escape again to freedom & solid work.

I picked up an interesting book of exploration the other day. "When Fire was King" by H.J. Moberly, dealing largely with his work & travels for the H.B.Co. & giving many details of geography & hunting starting in 1853 & written in 1929 when he was 94, helped by W.B. Cameron. It is a chronicle of his journeys & doings & he was quite a doer of things.

Yesterday & today we have had quite a thaw. It is to be expected as the Kandahar[48] was set for today & that is as influential on the weather as a flower show & fete & gala. The running water undercutting the ice in the streets has quite a spring-like appearance. I have heard that an early break-up is foretold for this year. Have you made any enquiries about your canoe? It is in perfect condition except for the perishing of the paint.

---

48 The Kandahar trophy was awarded to the winner of the Akademische slalom ski race, held annually in Switzerland. The McGill Redbirds Ski Team won the event in 1933, and participated often thereafter. RGEM; McGill Athletics Web site http://www.athletics.mcgill.ca/varsity_sports_player_profile.ch2?athlete_id=1789.

Max Sauer[49] has just called up & is coming to see us this afternoon. He only recently returned from overseas. Sue & Blake are coming to tea too.

    Love to you all,
        Dad

Montreal
11 March 1946

Dear old Son,

    Yesterday I at last got down to straightening out things & tying a few flies. Some evident loss of skill & I must put in some work to get it back again. I enclose some "Old Gold"; a fly (streamer) invented last year by Brian (Bill) Gulline[50] & tried out with great success by him last spring in the Mastigouche River. Charlie Pascoe[51] was there too & Bill's fly beat all comers there. Hilary[52] gave me one last fall at the Columbus (Fishing Club) & I found it did very well. …

    … Mike Carmichael[53] gave me a reel he had made in his plant of Magnesium. Quite good but the design could easily have been improved though the metal is nice. I don't know if they are going to make more. It's too small for one thing & has no room for backing. It will be alright for Mum's rod.

    Mum has been using the frying-pan you gave us & says it is splendid. We greased it & baked it well first & she has used it since for various things including the tricky Omelet & everything is a great success. I thought you would like to know.

    Your cultures arrived this morning & I have handed them over to Gertrude. I have no facilities for real work yet. I am still "reorganizing" & trying to get budgets re into line.

    I had an interview with the Principal (F.C. James) & Dean Meakins[54] the other day & it was fairly successful. There is some increase in the budget (about 50%); not quite enough for my requirements but I can do something with it and there is a chance of getting some space though still rather uncertain. In any case I want some men; two lecturers who will be mainly concerned with the clinical work under Gertrude & two Demonstrators who will also be able to work for a Ph.D. if they wish to. The finding

---

49 Max Sauer was "a very talented photographer," who visited Bark Lake and took "some memorable photos." RGEM.
50 The Gulline brothers ran a fishing shop in Montreal selling tackle and specializing in beautifully tied flies for fly-fishing. Everitt and Robert Murray purchased many items from the store and made many "conversational visits as well." RGEM.
51 See chapter 3, note 2.
52 Ibid.
53 One of the Bark Lake fishing companions.
54 See chapter 1, note 26.

of men is going to be difficult & important. I would be glad to hear of anyone you & Ash may know of.

Give my regards to Ash & ... Hall when you see them. Did you hear the C.B.C. "Sunday Night Show" last night? It was a very fine bit of satire, full of fun & caricature. The theme was a supposed Secret Treaty with the Indians which required that Canada be given back to the Indians if we made a mess of it. The various Provinces were canvassed for their opinions very amusingly; P.E.I. didn't care about the rest but could not tolerate that "the Cradle of the best Canadians" should be handed over; when Ontario came in it was introduced by "Nearer my God to thee" & so on. I liked the discussion over the Canadian Flag too: It was suggested that it be a red ensign with a large beaver & nine small beavers, but Ontario objected to an ensign with the big dipper because it was too suggestive of some drinking bout I forget the detail of; Quebec wanted a "Godbout couchant & a Duplessis rampart on a field of split peas"; etc. The whole thing was clever & amusing, with a good grain of truth.

You will soon have finished with your Technician's Course & I hope you can persuade them (the University) to abandon it. The entire thing is on a wrong footing. Anyway when it is over you can devote your time to your own work & I hope you can get a start on a good problem. The Hyaluronidase problem looks to me good. Have you the Biochemical Reviews (I think it is) in your library; there are a number of references to Hyaluronidase in relation to bacteria in the last volume issued (1945). Let me know if you have not got it. Gertrude & Co are on the watch for mucoid forms of anything that crops up & they will be sent to you fresh, stat! ...

Gertrude is finding great advantage in the blotting paper disc method of testing organisms for Penicillin sensitiveness; it was described on the 2nd last number of the J. Path & Bact (Brodie 50 No 2 p157 & Moreloy 57 No 3 p. 379). It simplifies the test considerably and you can make it roughly quantitative by (writing on the disc with pencil before sterilizing them) impregnating with various concentrations before drying them. A great advantage is that they are always ready for use. ...

... Let me know what flies you are short of & any special ties you want. I need practice in tying & would like to tie them for you.

My best love to Doris & Alice & best of luck to you all.

Love, Dad

3590 University Ave.
March 11/1946

Dear Doris & Bob too, of course:

... Your last letter arrived on Saturday & I fully intended to write ... but decided to go out to supper ... to Desjardins Seafoods on Dorchester. There we had a very excellent supper – yes! Oysters, followed by eels & including wine & good cheese & excellent coffee. Afterwards we prowled our way home through a snow storm ... which rendered roads impassable for an hour or so.

On the way we browsed a while in a bookstore on Dorchester ... that seemed to remain open all night ... full of good books ... & we see the queerest people doing just what we do - browsing ...

Sunday was a lovely bright day. Dad tied his flies all day very happily & I dragged Rosemary off down to see some real Montreal & really walked her off her feet – down to ... square ... & from there east through all the delightful old squares & queer lanes & old streets past dockside warehouses & odd old pubs with 'John's House', ' Ed's Place' sailors ..., monuments to by gone adventurers whose sweat & endeavour have been forgotten & of wealthy benefactors whose money is ...?

Bill is back in town, having given up the job at Chalk River in despair of ever getting a place there for the two of them ... but he immediately got an equally good – if not better job here with a group of consulting engineers. He is happily thrilled by it. The sad fact that they cannot get a house of their own is somewhat mitigated by the fact that they are at least together in the same town! ...

I am simply delighted to hear that your mother is recovering so well, & that Erich[55] was able to come up to fetch her home ... I think that was a very good idea & much better for her traveling alone. ...

I hope you still have your nice 'sitter' & that you & Bob get out to the weekly show & to see your friends. It does make such a difference if you can. Dad and I haven't seen a show a show for some time – they have been a poor selection – but we did go to one of the MTL. Branch of the Nat. Film Soc. with Mrs Turner (who is a member). There were two shows –one very good, the other appallingly bad. The good one was 'Carnival in Flanders' & in spite of bad lighting & poor sound reproduction we enjoyed it immensely. ...

    A hug to each one of the three of you,
        Mum

London, Ont.
13 Mar 46

Dear old Dad:

It is a long time since I wrote to you and I seem to have a pile of your letters in my pocket unanswered. It was very nice to get your letter this morning with the flies. "Old Gold" looks a very good fly and I look forward to giving it a try. It is good to hear that Bill Gulline is going back into the fishing business and I'll see if I can promote

---

55 Erich Marchand, brother of Doris; see chapter 3, note 94.

some customers for him – there are a couple who might be interested & Hall is one of them. ...

... The mix-up of cultures was most unfortunate. I wrote to Gertrude at some length about the matter – I don't know that the labeling got mixed up, I am sure it was a mixed culture for one of the stock tubes made on the same day grew that pneumo and the Strep in about equal qualities. My fault for not doing it all myself. I am pretty sure the 2nd batch was alright. I also sent a slope of the *Gaffkya Tetragena* that you asked for ...

... Once I devote some time to it the hyaluronidase problem should show something promising – at any rate, a useful method of determining that an organism does produce it – if it works out to be quantitative so much the better. So far, the organisms which have shown the effect have all been those known to be hyaluronidase producers: Pneumococci, some haemolytic streptococci, staphylococcus pyogenes; but not any others. I need more strains of streptococci of various groups, also some strains of Clostridia, particularly *Cl. Septicum* (a known producer) & *Cl. Welchii* (a non producer).[56] Some organisms produce a most marked effect, particularly the Pneumococci, but with all of them there is zoning shown also with hyaluronidase from tissue extract. ... Information about mucoid strains of streptococci is a bit meager in the literature though a little work has been done on them of late years. I have not seen the reference in Annual Rev. Bioch.; we have it in the library.

I am glad to hear that they are doing something to increase your budget. I hope that you will get some of the things you want. I am very sorry to hear that the virus division is not to be – it is very foolish of the university not to start now.

Mother Grace left to Rochester on Sunday – Erich came up Saturday to take her back and it was nice to meet him at last. He is extremely nice and seems very capable.

 I must to bed.
   Your son,
    Bob

The University of Western Ontario
London, Canada
25 Mar 46

My dear old Dad:

 It was grand to talk to both of you last week – a great pleasure to us. After we rang off the Operator rang up & told us you had asked to reverse the charges and "was it

---

[56] "Clostridia" is a general term for members of the genus Clostridium. The generic name may have a short form, Cl., when repeated frequently; Cl. septiicum and Cl. welchii are species of the genus Clostridium. RGEM.

alright". It would have caused too much agony for her if we had argued about it so we gave in!

... The Streptococcus equi is very interesting and I shall certainly write a note about it. The monograph arrived but I have not had time to delve into it yet. You mentioned that there had been some possible isolations from humans – have you got the references to them? I am eagerly waiting a note from Gertrude concerning her observations – for I have not got the sugars for some of the described fermentation characters.

I am running some tests on the mucoid streps to see whether there is a quantitative phase – so far, it has been uncertain. The chemist who is working on these substances is going to be away for some weeks and I probably will not be able to get the problem properly settled before June. The things that remain to do upon it are fairly simple once we get organised.

I have another organism I am busy trying to identify which I grew in pure culture from an abscess on the back of a rabbit (subcutaneous) which was apparently chronic & which contained a pus that looked just like dough. Of course the rabbit belonged to the pathologists, they brought the pus to me and meanwhile killed the rabbit & incinerated it. ... It is aerobic and easy to grow. So far I have not been able to fit it into any definite group.

I have not written for some time because I have been saturating myself with various aspects of developmental Immunology – which has, I must say, proved to be a fascinating field but very hard to gather any concrete information. There is no sure way to gather the information because most of the observations have been made in the course of other work. I finished writing the paper this weekend – but the seminar is postponed until next Monday because of competition from a university lecture this afternoon given by a Psycho-analyst! I didn't go to hear him. When I have given it I will send it on to you for your amusement. It was written with a view to providing as many good discursive points as possible for those that come to the seminar – people all working in different fields. But as a paper I am not satisfied – too many unwarrantable conclusions, too much flimsy evidence, and I had to borrow too many ideas from others as half-baked as myself (Joe Needham for instance).

To provide some relief from the above indigestible bit I have read Cannon's "The Way of an Investigator" which is a very charming book I am sure you would enjoy.[57] He must have been a fine man.

... I saw in 'Nature' that the post in Botany at Auckland was filled by a Cambridge

---

[57] Walter B. Cannon, The Way of an Investigator (New York: W.W. Norton, 1945). Walter B. Cannon (1871-1945) was an influential American physiologist and an expert on the physiology of human emotion. Theodore M. Brown and Elizabeth Fee, "Walter Bradford Cannon: Pioneer Physiologist of Human Emotions," American Journal of Public Health 92, no. 10 (October 2002): 1594-1595.

man named Chapman – so I guess poor Gibbie has been disappointed again.[58] There was a nice write-up there too about Wynne-Edwards taking the chair at Aberdeen.[59]

I will write again soon.

    Your loving son,

        Bob.

                27 March 1946

Dear old Son,

... Gertrude will write to you about the *Strep. equi* as soon as she completes the sugar reactions. They were not satisfactory in peptone water[60] & she has them growing well in serum-water now. I don't know the references to the human cases which were unsatisfactory, it is mentioned in Bergey 5th Ed, but a note to Dr. J.M. Sherman,[61] Cornell University, Ithaca. N.Y. USA would get them I am sure. He is a good friend of mine & if you say I suggested you write to him he would tell you.

I sent you a strain of a "very mucoid strep" Gertrude gave me for you. She is watching out for all such things for you.

Your rabbit organism seems interesting & it is a pity you did not get the rabbit. Many years ago I isolated a G +ve organism[62] from a subcutaneous dissecting abscess in a rabbit; the pus was a dead chalky white & very curious tough stringy cheesy material without smell & the condition was chronic. The organism turned out to be an Actinomyces of unknown species which produced a definite soluble pigment (as far as I remember); grew in small colonies very slowly. I will look up my old notes & tell you more of it. I

---

58 "Gibbie" was R. Darnley Gibbs, professor of botany at McGill and a friend of E.G.D. Murray, who taught Robert Murray a first course in botany. He wrote the ground-breaking work Chemotaxonomy of Flowering Plants, vols. 1-4 (Montreal: McGill-Queen's University Press, 1974). RGEM.

59 Vero C. Wynne-Edwards (1906-1997) was a British zoologist and a leading behavioural ecologist. From 1946 to 1974 he held the post of Regius Professor of Zoology at Aberdeen University. From 1944 to 1946 he was an associate professor in the zoology department at McGill University. He was elected to the Royal Society of Edinburgh (1950) and the Royal Society of London (1970).

60 In biochemistry, "peptone" is defined as "A mixture of proteins made soluble by partial digestion or hydrolysis, now widely used in microbiological culture media." In microbiology, "peptone water" refers to "a solution or broth of peptone used as a culture medium." OED (Online).

61 J.M. Sherman (1890-1956) worked as a microbiologist at Cornell University and in 1937 served as president of the Society of American Bacteriologists (later the American Society for Microbiology).

62 "G +ve" is shorthand for "Gram-positive." Nearly all bacteria can be designated as either Gram-positive or Gram-negative, as a primary step in determining characters for identification. RGEM.

sent the strain to the National Type Culture Collection at the Lister Institute.[63] It was not obviously Actino like at first, but developed branching filaments later. I will enjoy seeing your paper when you have given it. I would like to be there ...

I have not read Cannon's book. I was very tempted to buy it as I knew Cannon & liked him very much indeed. A wondrous fine fellow.

I'm sorry for Gibbs' sake that he was not selected. It's tough for him here.

Ted Roy[64] is leaving & is taking the job at the Children's Hospital Toronto. My recommendation for his promotion was without result & now I expect they will say they would do it, though I have not been able to get an answer to any enquiries. I don't see how I can find a replacement for him nor candidates for the other jobs I have. The salaries here are not near good enough.

I am tying a selection of useful flies not usually obtainable here & will send you a set when complete. I shall also send a set to Hall.

Our love to you all,
Dad

Department of Bacteriology and Immunology
3775 University St.
Dr. T.E. Roy
March 28th, 1946

Dr. R.G.E. Murray,
Department of Bacteriology and Immunology,
University of Western Ontario, Ottaway Avenue and Waterloo St.,
London, Ont.

Dear Bob,

A few days ago your Dad asked me to send you a series of strains of Clostridia. It was not clear to me whether you wanted only perfringens for hyaluronidase tests or a general selection for class purposes. I am sending some 30 varied strains and a list accompanies this letter.

... If you are busy preparing potassium hyaluronate maybe you can tell me whether one buys umbilical cords by the pound or by the yard.

Yours truly,
TER (Ted Roy)

---

63 This collection was founded in 1920 and today is the longest operating bacterial culture supply service in the world. Information on the collection can be found at the website of the Health Protection Agency (UK) at http://www.hpa.org.uk/.

64 See chapter 1, note 89.

London, Ont.
6 April 46

My dear old Dad:

Neither of us seem to have written for some little time – as a matter of fact I have just put through an unsuccessful phone call, thinking you would be in, and got Rosemary just as I was about to hang up – vague as usual. We both of us thought we would like to have a chat.

You were asking a while back whether the plans for my promotion were going through without any hitches. Ash told me the other day that it had been approved and would become effective on July 1st.[65] This will be no less of a pleasure to you than to me. Doris is very bucked.

The seminar I was supposed to give has suffered from a chapter of accidents. As I told you it was first cancelled due to a lecture by a psychoanalyst; last Monday they cancelled it because the University was showing one of the documentary films of the war – a sort of glorified newsreel. As I result I as supposed to be giving it this next Monday. Having been "keyed-up" twice already and having found some additional interesting information some of the original sparkle will be gone & there will be a tendency to be lost in a *welter* of detail. Anyway I enclose a rough typescript of the things I could say – if you would like a bibliography I will send it later when I have time to make it up. Many things in it are not stated in the way that I would like and many arguments are weak, ill-supported and even unwarrantable. The idea, of course, it to make people think but I fear that most of the stuff is both above my head & their heads!

The experiments with hyaluronidase & mucoid streps is going rather slowly at the moment. I have been able to show that effects are demonstrable to a dilution as great as 1/26 of one of McCleans Viscosity Reducing Units but I don't think we will be able to make this test of any value in determining the amount of enzyme. I have not yet set up against Clostridia because we suffer from having only one anaerobic jar. (It would not be feasible for us to have ones made like those you have; but can you recommend a jar, made commercially, that would be worth investing in??) The mucoid strep Gertrude isolated arrived in good shape but it is a devil to work with. …

I have not heard from Gertrude about the *Strep. equi* yet. She is doing so much that would be hard for me to do here that I feel it should be published jointly – what do you think?

I have also been extracting phospholipids from BCG to try their effect on TB & growth. Dubos[66] reported getting more homogenous growth by growing them in

---

65 The promotion was to the rank of assistant professor, a title Murray obtained after less than one year at UWO.

66 This is a reference to René J. Dubos, the world-renowned microbiologist, "humanist-

fluid media with phospholipids (soya bean) but this did not work well in our hands – together with detergents (Tweens, which are *Mannitans & Sorbitans*).[67] I have some other ideas which may be worth trying.

We have had a busy week and have not done much out of the ordinary. Doris remains very well though she finds that there is very little room for a large meal. Alice is her very good self and is doing new things all the time. Her latest is to point at things of interest to her & make exclamations of interest & joy. New words are now ... appearing.

   All our love to you.
     Your son,
       Bob

                 7 April 1946

Dear old Son,

  I was glad to talk to you & Doris this evening & to know you are all well. We were very sorry to miss your call yesterday. Saturday is a bad evening because Mum & I usually go out to eat oysters at Desjardins & may be late. In the summer too, as we hope to spend the weekends at the lake.

  The news that your promotion is confirmed gives us enormous pleasure & we are so proud of you. All the way through you have done well in fitting yourself to take responsibility & even as a little chap you showed you had it in you; you were able to take charge of things in emergencies as you did when I was in Baltimore & Mum was taken ill & taken to hospital, also on other occasions. It is fine that the good job you did is recognized materially by the University, but most of all our gladness is in

---

philosopher," and for over fifty years professor at the Rockefeller Institute for Medical Research in New York City. According to James Hirsch and Carol Moberg, Dubos's "best known and most remarkable achievement" of the microbiology period of his career (1927-1944) was the discovery of gramicidin and tryocidin, "the first antibiotics systematically cultivated from bacteria and produced commercially." See James G. Hirsch and Carol L. Moberg, "A Biographical Memoir of René Jules Dubos, 1901-1982" (Washington, DC: National Academy of Sciences, 1989).

67 "Tween" is a "proprietary name for any of a class of polyoxyethylene derivatives of fatty acid esters of sorbitan, several of which are extensively used as emulsifiers, solubilizers, and surfactants." "Mannitan" is defined as a "cyclic anhydride of mannitol, $C_6H_{12}O_5$, occurring as crystals or a light syrupy liquid." Sorbitans are "any of a number of cyclic ethers which are monoanhydrides of sorbitol", "a colourless crystalline solid," and a frequent attribute "in names of fatty-acid esters of these compounds, which are used as emulsifiers and surfactants." *OED* (Online). As Robert Murray explains, the Tweens "are chemical combinations of some sugar-like molecules (such as mannitol and sorbitol) with a fatty-acid like oleic acid. They are surfactants like soaps are and have useful properties."

your ability to do things well & properly. I know you can & will not only do the job required but will do research that matters. Only, I want to impress on you to publish it & not do as I have leaving a lot of significant work buried in your notes.

I have just packed up some flies for you. If you let me know those you would like to have I'll tie them for you. These are not as well tied as I used to tie them but the knack will come back with practice. Later flies can be expected to be better but these will catch fish: I hope you will have a chance to use them. I am told there is fishing to be had not too far from you. It would be wonderful if you could join us during your holiday.

The lakes & the woods are very much in our minds with the wonderful spring weather we are having. We are expecting to hear any time that the ice is out & you may be sure we will be up at Bark Lake the first week-end after.

I'll send you some Pneumo Broth tomorrow.

I am hunting everywhere for staff. The appropriation for the Department has been increased quite a bit & the work required of us is greatly increased so that staff is vital. There is a scheme to make alterations in the building to give us more space; largely by altering the Pathology class-room so that we will share it with Pathology, when our class-room will be cut up to give us a smaller class room & 4 rooms. The autopsy demonstration theatre is to be cut across on the B floor & I have planned out the space to make a very nice unit especially designed for the clinical bacteriology. I hope too to get a bit more necessary space if some other things can be arranged.

No notice was taken of my recommendation to promote Ted Roy, in spite of my repeated pushing it & Ted has now accepted the offer made him by the Sick Children's Hospital, Toronto. I cannot see how I can replace him. Experienced men are rare in Canada. There are 2 or 3 places after Don Fleming[68] too & I am trying to manoeuvre things to keep him. I am giving a few lectures but my time is chiefly taken up by organization & trying to get the Department on its feet again. The new work required of us & the difficulties in getting supplies immediately needed & staff & space for the future keeps me hopping & troubled. ...

Mum has written a very good report on the fishing in Bark Lake for the association. She has given an historical review & a critical analysis of the present situation with the factors involved. It is very good. She is to be elected President of the Association next week at the Annual Meeting. They can be quite sure she will keep things moving & bring understanding into whatever is necessary. Delia Lichte is Secretary & she & Mum get on very well together.

I am looking forward to seeing you at the Detroit meeting of the S.A.B. I shall have some jobs to do there but we will find time to be together quite a lot I hope. You will

---

68 See chapter 3, note 123 for information on E.G.D. Murray's colleague Donald S. Fleming.

enjoy meeting a lot of the people too. Unfortunately the S.A.B. meeting clashes with the Royal Soc of Canada (some days) which is meeting in Toronto & I would like to be there. I am committed to some business at Detroit so I cannot even consider a choice between them. ... Unfortunately some friends I would like to see will not be going to Detroit.

It is very cheering that Doris is well; give her my fondest love. I am looking forward to a couple of days in London on the way back; I will look up the dates & let you know so that you can you book a room for me at the hotel. I want to see the three (or four) of you.

  All my love to all of you,
    Dad

                    10 April 46

Dear Bob,[69]

I feel very guilty for not having written earlier, but I was feeling very miserable and everything seemed to happen at once. ...

About your Staph: I am sorry to say that the fellow died before I had finished what we wanted to do, I have not been able to revive him. ...

I just heard big news in the dining room, Barbara Barker is going to marry Ted's brother Stewart. I was told it came from the horses mouth so I think it should be true. Ted has not mentioned anything, by the way he feels very guilty for not having written to you but he claims there is so much to write he would never finish. ...

I am sorry that you are not coming to Montreal for Easter – it would have been nice to see you all. How is Doris getting on and when is the happy event expected? I am sure Alice is soon going to be a young lady. Have you finished with teaching and made time now to do other things? We have had a lot of interesting cases. ...

Give my love to Doris and Alice and all the best from the lab crowd to all of you. Your father and I are enjoying your letters a lot. You have quite a gift for it. Have you ever tried your hand in writing articles (non scientific ones, I mean).

  Much love.
    Yours, Gertrude

---

69 LAC, R.G.E. Murray Papers, vol. 28 (KALZ, Dr. Gertrude, 1946-1970).

>   Faculty of Medicine
>   Department of Bacteriology and Immunology
>   Ottaway Ave. and Waterloo St.
>   14 April 46

My dear Mum and Dad:

I may be a little long-winded so I have decided to do this on the infernal machine. Maybe you should get one of these, Mum, if you are going to have further trouble with your thumb!

Things seem to happen thick and fast – all these engagements and marriages seem to indicate that there is something in what they say about spring. ... Spring is more than definitely here. I suppose that it is with you too since you are planning to go up to the lake so early this year. We have a definite blush of green on the trees and the plants are well on the way up. Fishing times are not far away but I don't see how I can get any unless I can persuade some other enthusiast who has a car to take me with them. I have been working on this problem but I have not had any definite offers yet. ...

We were just looking at the map last night and London is right in the middle of just about the most populous part of Canada. There is little real promising territory for at least 200 miles. There [are] tons [on] the Great Lakes but I hear that they are not very much good until you get up to the Bruce peninsula. The only real solution to travel around this part is to get a car and I don't see any sign of that for a few years yet. A car would certainly be a solution to Doris getting out and about a little more than she can for the present.

The paper that I gave to the Seminar group last Monday went over very well. A lot of people came to hear it and I got a number of nice things said to me by some of the senior men present. It was much the same as the rough typescript that I sent to you but I feel that it was very much improved by being presented. There is something about being on your feet and having the subject matter at your finger tips that adds to an argument. Things crystallized a bit better and the summation and integration of ideas from other fields was much better presented. One of the things that most people commented upon was that I did not READ it – which I am afraid is what most of these people do. God only knows why they do it – a most vile habit. The discussion was very good but did not add very much. ...

Another event of the week was the Third Year class party, of which I am hon. President, held on Thursday night. It was a pity Doris was not there but she would have found it very tiring after a usual days hard work. It was remarkably well behaved even though the drinks were much in evidence, and everyone had a good time including me. A lot of talk, dancing and general banter made it a pleasant though long evening. ...

Life is looking very good and the income tax is a far smaller burden that I had calculated! We keep our eyes open for some permanent abode of suitable size (we *could* fill even the biggest one if we had to) but I don't see any hope of one within our means for some time to come. Since there is a land boom everyone is selling on the rising market, and that is not the market for us. We were looking as we walked, at some of the palatial homes around here where you could make a golf drive on the front lawn, speculating on the number of people inhabiting them. Some of them look terribly empty.

All our love to you both and we do enjoy your letters.

    Your loving son,
        Bob

Montreal
14 April 1946

Dear old Son,

I like your "Seminar" & it provokes thoughts of various possibilities for work & I am sure you had a lot of fun working it up. Gordon[70] & I used youngish rabbits to make our mennigo serum when we did our typing (the first typing of anything); we did so because the serum was more specific though of lower titre. This seems to be forgotten (1915). It would be interesting to see if there is an orderly selectiveness of antigens according to the age of the animal immunized (after globulin formation becomes possible). I would have liked to have been present when you gave the paper.

Mum was elected President of the Bark Lake Protective Association at the annual meeting. She took office with a very graceful little speech & a nice dignity. At the meeting she read a very good report on the fishing in the lake; a very thorough analysis of the factors influencing the conditions [see below]. Very well received. I have no doubt she will exert a good influence with the usual thoroughness.

I enclose the news of the Osler Society Dinner, also a fanciful diagram to represent the chronological acquisition of knowledge on the thyroid; this was the background of a very good review by Dr. J.H. Means.[71]

We had a very poorly attended meeting & dinner at the Columbus Club last night. Adelard[72] made his usual extravagant demands & ended, as usual, with a grateful acceptance of the terms proposed by the club. I don't suppose he will do any better

---

70 See Introduction, note 127 and chapter 1, note 7 for information on Mervyn Henry Gordon.
71 James Howard Means (1885-1967) was a professor of clinical medicine at Harvard University Medical School and chief of medical services at Massachusetts General Hospital. *New York Times*, 6 September 1967.
72 Adelard Gregoire was the guardian for the territory and facilities of the Columbus Club. The others mentioned in the paragraph were members in that fishing club and good friends. RGEM.

this year than usual.

Mum & I hope to go to the lake next Thursday afternoon (18 April 46) & return on the following Tuesday morning (23 April 46) if the ice is out by then as it promises to be possible. It will be nice to see the old place after so long away from it. The trout season opens this week but I suspect the water will be too cold, I will have a try at Twin for fun.

I posted some flies last week to you & Hall. I have not heard if they arrived & I hope they were not lost in the mail. I am sending you two more this time.

1. My *"Pelican"* which is a very useful fly in the very early spring when the trout are taking dragonfly nymphs & when all other flies fail. ...
2. *"Jungle Hornet"*... I have found it an attraction for big trout in the Jeanotte River when they won't come up for anything else at the time. I have never done any good with it in any lake.

The photos Doris sent of Alice are nice though out of focus. She has grown a lot & and looks splendid; Doris in the background looks very well. We are glad of them. It will not be long now till we see you all again.

Gertrude tells me she wrote to you recently about your "C Strep".[73] In some ways it is not quite convincingly *Str. equi* though it is definitely closely related to it. A "human" species & probably worth a description. I hope the media got to you safely. ... You could mention that Gertrude checked certain reactions, but there is no reason for joint authorship. It is interesting that you may have found another strain – I hope so.

Did Dubos send Ash a copy of his observations (typescript – not published) on the medium to give rapid diffuse growth of TB in fluid medium? Some who have seen his cultures are very impressed by them. A selective method of growing TB certainly from small inocula is badly needed. We find culture still to be vastly inferior to ls-pig inoculation. I'll look up our figures & send them to you if I can remember.

I thought I was getting some sense into the affairs of the department & seeing a little daylight at last. But Meakins made some remarks to me last night that seem to indicate trouble – I think of his making. I hope to know what it means sometime tomorrow.

J.W. Stevenson[74] is coming to me as Assistant Professor, replacing Ted Roy; he has been working with G.B. Reed[75] at Queens all the war. A very good fellow & I hope I can make him happy here. I have drawn 1/8" scale plans of the suggested alterations in the building but still wait to hear whether they will be carried out – they are absolutely necessary.

---

73 See chapter 3, note 101 for information on Gertrude Kalz.
74 J.W. Stevenson was a bacteriologist and an associate professor in the department at McGill, and he served as dean of medicine at the school in the 1950s. RGEM.
75 See Introduction, note 233 and chapter 1, note 115.

The old pup is groaning & fidgeting round & Mum just said to him "Hi! Stop fussing about! You're just like an old grandmother." At which I laughed heartily.
Love to you all,
Dad

PHYSICAL AND ECOLOGICAL CHANGES AND THEIR EFFECT UPON GAME FISH IN BARK LAKE, CO. ARGENTEUIL, PROVINCE OF QUEBEC
by
H.W.H. Murray
Montreal

Based upon past history and present conditions of Bark Lake, this Report includes observations made by Professor Murray and myself during the past fourteen years – observations made primarily from the fisherman's point of view, but in addition, arising from our deep interest in Zoology and Ecology. Notes on past history were gathered from fisherman residents of long standing, whose interest and intimate knowledge of the fishing are well known; those on present conditions are carefully observed facts.

Bark Lake is a winding body of water nearly eight miles long with an average width of about one mile and several large bays opening off the main body. The average depth is about 50 ft. though there are many deeper spots and a maximum depth of 183 ft. It receives an ample supply of water at all times from springs and from streams draining the small lakes in the surrounding hills, and is completely surrounded by well-grown mixed brush which has not been lumbered off for a considerable time. ...

SUMMARY

The serious deterioration in the Bass and Pike fishing which we have just considered in detail may be summed up as follows:

1. The raising of the water level made conditions a little too difficult for the successful breeding and survival of Bass.
2. The closing of the outlet by the permanent dam excluded the Suckers in the spring migrations, thus affecting the food supply of the larger game fish.
3. The flooding of swamps, reedbeds and muddy shores by the increase in water level limited the production of green frogs and their tadpoles – further affecting the food supply.
4. The introduction of motor boats, and the greatly increased motor traffic on the lake in the past ten years, resulted in a marked decrease in the number of nymphs. This has further reduced the food supply of adult Bass which take fly

or nymph very greedily: but a more important [fact] is this destruction of the food supply of Minnows and young fry of all kinds.
5. The removal of "floaters", stranded logs, and brush from the shores limited the breeding space available for minnows and destroyed the shelter which Bass fry *must* have if they are to survive the early stages of their existence.
6. The increased depth of water over submerged reefs and shallows caused a decrease in aquatic vegetation, thus limited the production of nymphs and water insects as well as reducing available shelter and feeding grounds for minnows and small fry.
7. The decimation of the minnow population by the fungus epidemic of 1934-35 brought the food supply of the game fish to a very low level.
8. An unsatisfied hunger for minnows has caused the Bass to make great depredations among both Bass and Pike fry.
9. Underfeeding has caused a very marked slowing down of the growth rate in Bass – so much so that large Bass taken from the lake are apparently not replaced by the succeeding generation. It must be understood that this is due to the fact that Bass are no longer growing to full size (up to 6 lbs. in this lake) and that they are growing very much more slowly: and, in addition, there are many fewer Bass than in former years.

These changes in condition have brought about the following results:
A. A marked decrease in breeding activity.
B. An increase in cannibalism due to shortage of food supply.
C. A marked slowing down of growth rate resulting in a gradual disappearance of big Bass and Pike.
D. A decrease in resistance to infection and disease as a result of underfeeding and wrong conditions: possibly contribution to the Trematode or "Black spot" infection of the Bass and the Fungus infection of the minnows.

17 Apr 46

Dear old Dad:

I got your letter today and enjoyed it very much.

I am glad you enjoyed the 'paper' – it was rather fun to work-up and my information of the subject is far from complete. This is largely due to the fact that it has never been properly collected from that point of view and consequently it requires exhaustive search of the literature without much help from reviews or from the various index books. It might be worth publishing as a review sometime for the help of others when I have more information and have more organization of certain parts of it. I meant you to keep the typescript for I have my own copy – I will send it back to you, if you like, when I have appended a bibliography. I agree that it allows of some fascinating speculation although meager information cannot substantiate any far reaching conclusions.

Mother's report on the state of the lake is very well done and worthy of a better informed audience than the Protective Association. A fine piece of work and well put together.

The batches of flies have arrived and they are fine – I wrote about my general prospects of fishing in my last letter.

The other possible strain of Str. Equi was disappointing. When I saw it (that Sunday) it looked very like it on anaerobic plates but it is not AB or C Lancefield and is one of those odd bugs that produce haemolysis aerobically & B haemolysis anaerobically. Not the right organism but still a peculiar one. I have formed several like it & so has Ash. I fancy that some of them may be group B – but I don't know.

I am glad you have got [J.W.] Stevenson. I hear he is a very nice fellow & should be good acquisition.

 All my love.
  Your Son,
  Bob

I enclose some references which may be of interest; also a sample of Kum-Kleen labels which I think are worthwhile for a lab.
  R

Montreal
18 April 1946

Dear old Son,

Mum & I are off to the lake this evening & will not return till Tuesday morning. There is an Easter Holiday for that time. We have just finished all our teaching & the exams start Tuesday. There is no lack of work to do & I am very worried trying to find candidates for 3 lectureships at a salary $2,400. There are numerous other worries too & a lot of work to be done on new courses for D.P.H.,[76] Dentists, Graduate nurses. There are others (coroners) too they are trying to get us to undertake which I am trying to avoid.

… I do hope you can get some fishing. There must be some enthusiasts who would be kind to you & give you a lift in their car. There must be some possibilities; many

---

76 The Diploma of Public Health (DPH) was awarded for completion of a McGill program within the Medical School's Department of Public Health. RGEM.

New York State streams are much fished & run through populous districts. I am glad you like the flies & will tie some more for you.

I'm delighted the Seminar went so well. I wrote to you about it recently & was tempted to keep your notes. Yes there is a lot on the "living voice" when the presentation is good.

The 3rd year class party seems to have been good fun. A pity Doris could not take part in it, but wiser not to.

With luck you will find a suitable place one day. Perseverance will be rewarded & no doubt friends will help if you remind them.

I have booked our reservations to leave Montreal the evening of May 20th. Mother will get off at London on 21st May about midday (CNR) to stay with you & I will go straight on to Detroit. Please reserve a room for me in the hotel for the night of May 24th & I will stay there over the weekend. We can fix up minutiae when we meet in Detroit. ...

Best love to you all. I'll write at greater length when we get back from the lake.

Dad

Montreal
23 April 1946

Dear old Son,

We have just returned from the lake. It is cold enough there to want thick things & jerseys etc in the evening, but here it is summer heat & oppressive. We had a grand time. The train was rather late arriving & fairly crowded - when we got to the lake we found Mrs. Lichte & Mrs. Carmichael staying at Rosie's[77] because they could not get through a 200 yard ice jam at Thomson's Point. ...

We had lunch around 12:20 pm Monday, put tackle & tea things together & paddled over to the swamps having a chat [with] Charlie & Fred Moore on the way. I carried the canoe in with 2 rests (up hill & on top of largish lunch) & met Ricky Owens on the trail; he has grown into a fine & handsome, upstanding young fellow. We quickly got my tackle together & paddled down the S. shore of the lake fishing every inch without any luck until we came to the bay carefully & got 9 small trout 1/3 to ½ lb; 4 on Silver Doctor & 5 on Pelican. ...

This trip illustrated another thing about very early spring fishing. The fish are in markedly limited area feeding, in this case about 50 yards of shore line, but I have

---

77 This is a reference to Rosario Miller, who lived close to the landing at Bark Lake. He would meet lakers at the train, ferry them across the lake when needed, and deliver supplies such as wood and vegetables. As Robert Murray recalls, Rosie was "an indispensable helper and friend" to the lakers. RGEM.

often seen it very much less. The sport is not predictable, but I fish carefully along the shore until I find a fish then fish that region thoroughly & patiently. ...

We were tired when we got back to camp & cursed because we forgot to bring some Rye in our stores. There was a little brandy in a bottle on the shelf so we shared that & felt better. I cleaned the fish & put canoe, tools etc away while Mum got supper so that we would only to have to shut doors when we left this morning.

I wish you could have some of the fish. It is too far to send & the weather is warm. In fact it looks & feels like a storm; a wash would do Montreal a lot of good. Sue & Blake will have some & they are coming this evening to fetch them. ...

24 April 1946

The camp is in very good shape. We will have to paint the cottage this year & it will be a tough job & one I don't care much for.

I have just finished going over my income tax return & it will dry us up completely. It is very disappointing as I had a plan to take Mum on a trip to the West but it will take everything we have saved to meet the $1500.00 still to be paid & I will have to put everything possible aside for the insurance premiums due later & for next year's income tax. I thought we would get by comfortably.

Your letter of 17th arrived yesterday afternoon. Thanks for the references & the note on the Kum-Kleen labels. ... I would like to have the copy of your paper.[78] If you polish it up it might make a good review for the *Am. Journ. of Med. Sciences*. I want one for Jan 1st 1947 & you would have time to work it up by then. I cannot remember the author but there was some work done, I think at Boston, on the immunity response of infants which has a bearing on the question; I'll try to find it for you.

Mother's report is very good, but, as you say, wasted on the Association. I don't know where it might find a wider interest.

Pneumococcus constantly gives greening aerobically & very marked haemolysis anaerobically. This is not usually recognized; nor that Pneumo grows better anaerobically.

I have a spare copy of "Respiratory Enzymes" by various authors from the University of Wisconsin. I will send it to you if you would like it, let me know. It is a review monograph of the subject dated 1939, but I don't think much has been added since then.

Let me have your lab phone number, it might be useful.

I had a nice letter from Dean Hall to whom I sent some flies.

---

78 Robert Murray does not recall exactly which paper this refers to: "Anyway, it did not survive or get written up." RGEM.

Our best love to you all, it is good to know, from Doris' letter as well, that you are all fit & happy.

Dad

London, Ont
28 April 46

My dear Mum:

It was a most pleasant surprise to get a telephone call from you on Friday night – we do so enjoy talking to you although the time is so short and we never seem to be able to say all the things that we want to say. You both sounded very cheerful despite your trouble with the lumbago.

Doris keeps reminding me that tomorrow is your birthday and that now I am too late to get a letter to you in time. Anyway even if it is late I wish you a "hippy happy birthday".

I very much enjoyed reading your report to the B.L.P.A. [Bark Lake Preservation Association] on the 'state of the Lake' and I consider it a very well constructed report – a very worthwhile effort. Also congratulations Mrs. President upon your position – maybe you can make them stick to business.

We have not done anything very constructive of late – just gone out to a couple of movies and had some people in for the evening. Doris keeps very well and is slowly learning not to do too much. Sometimes she gets out of hand and does a flurry of work, only to regret it for the next couple of days. Doris, I am sure, has written you a long letter about Alice and her doings … she is as full of fun & devilment as she ever was.

At the beginning of the week I had a long letter from Nan. They are settling down in Scotland and have bought a house at Rhu, somewhere near Helensburgh. It sounds like quite a big place that has been modernized, having a few more mod. cons. than the average Scottish abode. They are going to move in July 1st. Their present address, in case you don't know it, is Glenfenlan, Shaudon, Dumbartonshire. They sound happy though Jimmie, as a steel man, does not look forward to being "nationalized" and converted into a form of civil servant.[79]

I am now released from bondage to the furnace. During this week an oil burner has been installed – no more coal & no more ashes to be manhandled. The whole thing is fully automatic and operates from a thermostat. A great blessing. …

Your trip to the lake sounded like fun and I do so wish we could have been with

---

79 See chapter 3, note 29 for information on Nancy "Nan" Heard. She married James "Jimmie" Stuart Abercrombie Pearson, who became an executive in a large steel firm in Glasgow. They lived north of Glasgow near Helensburgh. RGEM.

you – especially on the trip to Twin for the nice catch of trouts. Please write & tell us all about the property & the garden and all the little things going on there.

All my love,
Bob

28 April 1946

Dear old Son,

It was fine to speak to you both & know that you are all well & that everything is going nicely. We are looking forward to seeing you soon.

I enclose the flies you want: Teal & Red & Silver Doctor. The S.D. is the correct tie, which is not usual in the shop flies; I find it quite effective at times.

… Mum's lumbago is better though she still has twinges. She had a sudden bad agony which crippled her completely for some 30 hours. She probably brought it on by moving a lot of heavy things in the basement & carrying some upstairs; at any rate very quickly afterwards she was stricken. She is feeling the strain now in other muscles due to unaccustomed misuse while saving the affected groups. It stopped us going to the lake for the weekend but we will go next week if Mum is alright.

… I hope very much you will get the promised fishing. I am told there is some to be had in the region centred on Orangeville which is high ground. I'm told they have dammed the streams there to make it possible. Perhaps that is where it is proposed you shall go. I have no doubt you can do well with a fly; but the long rod may be awkward if they are very small streams & much wooded. If they are reasonably open you'll manage alright. Any way it will be fun to try.

I have no particular lab news. The alterations scheme is still in the stages of estimates & not near approval yet. I wish they would get a move on. The only change is the suggestion to build the Clinical Lab Unit I designed on the roof rather than floor over the P.M. Theatre.[80] As far as I am concerned it is an improvement. I still have no news of candidates for the jobs I have available. This is a serious obstacle; the other difficulties are manageable if the space is provided.

… Sue & Blake were here yesterday. Both are very well indeed. Sue looks marvelous; if anything she is handsomer than ever. She is very cheerful & happy; greatly enjoying running her home, which she does extremely well. They have done various paintings & arranging & have improved the place enormously. …

Best love to you all,
Dad

---

80 The Post-Mortem Theatre was a room at the McGill Medical School, with raised seating around two sides, for demonstrations to medical students. RGEM.

104 McClary Ave.,
London, Ont.
6 May 46

My dear Mum and Dad:

Since I am now recovering from the effects of a large mess of trout for dinner I feel that I should write and tell you about the trip. Yes, I have been on a fishing trip and I am now feeling very healthy and full of beans.

It all started on Wednesday afternoon when I ran into Dr. Ivan Smith[81] in a corridor of the hospital. He asked me if I would like to go fishing. When I answered appropriately he said that they were leaving at five o'clock – the time at the moment was 3:30. This was a bit short on notice but it was an opportunity not to be missed since I have not had any spring fishing for a good three years. So I spoke to Ash and arranged about a few cultures, then sped for home. Doris was out at tea but I phoned and she was a little surprised ... So I spent a very hurried 40 minutes finding things, getting dressed and cutting down the bulk of my baggage. ... We were to foregather at Ivan Smith's office at five, but, as is usual with such things people arrived in driblets and ... we did not really get out of London until after 7. We made good time up to Listowel where we had supper (very late). Then up through Mount Forest and Durham where we turned off to Flesherton. This last was our goal – an old hotel on the main corner of the town called Munshaw House. A good hundred years old and rather ramshackle and gone to seed, but a family affair with plenty of good points. The owner is a fisherman (fly only) and so has attracted a fishing clientele. The hotel restaurant was not in commission so we had to eat in a greasy spoon across the way.

All the country in the region is pretty high up (around 1500 feet) but is not the same sort of hilliness as the Laurentians, much more rolling. The rivers are fairly fast and they have cut steep little valleys into the land. Some of these, particularly the Beaver valley, are very handsome. The land is, or rather was, agricultural. Most of it is pretty poor stuff and Grey County is more or less renowned for the number of rocks that push out of the fields with the frosts. Many of the hill sides look like Quebec province farms; some of the farmers have been very persistent and have built great stone walls around their fields, and by the looks of things are still building.

On the first day we did quite a bit of traveling around. Two of us went to Eugenia Pond, which is a Hydro pond on a branch of the Beaver River. It contains a lot of big trout but we did not rise or see one (I know they are there because I saw three beauties that were taken out of there on Sunday). I think it is a place that would require a little

---

81 Ivan Smith was professor of radiology in the McGill Medical School and chief of radiology at Victoria Hospital. He was a pioneer in the application of high level radiation sources and he supervised the beginnings of the Cancer Clinic in that hospital. RGEM.

bit of study before one could fish it adequately. A great part of it is a hurrahs nest of dead trees and stumps and simply full of hidy holes for trout. ...

The next day (Friday) after a breakfast of trout (and eggs and ham and fried potatoes etc), country appetites had appeared, we set out on a much more exciting expedition to the Rocky Saugeen River. Going over to Markdale and then turning SW for about four miles. This is a beautiful river, fast and with the clearest water bubbling over a fine rocky bottom. There are plenty of weeds and watercress, logs and deep spots for the trout to feed and hide. It is thirty to forty feet wide down there and there is plenty of room over a great part of it to cast a fly with comfort, though a fairly short line is in order. I had borrowed some waders so I was very happy and comfortable fishing the stream. As with most of this part of the country the fish run pretty small and you catch about six of seven inches for every one of legal size. Being in fast water they fight about twice as hard as their lake brethren and they are in very good condition. Not being in good condition myself I was very glad to have convenient cedar trees lying into the water on which to sit every now and then the better to contemplate the beautiful water. It really was fine, extremely clear and pretty. ...

... Altogether it was a very successful fishing trip and we brought back a nice mess of trout for our respective families. The country is very pretty as far down as Palmerston, the rest is typical SW Ontario and very flat. On the way, near Mount Forest you cross the Saugeen river and it looks very fine but I understand that most of it is private water (not that it would stop you from fishing a lot of it). The land changes from the rocky sandy upland soil to good stuff about Yeovil and Holstein.

So I got back to the lab on Monday morning feeling a lot better and with a beautifully blistered nose (which has been the butt of a lot of boisterous comment which I attribute to envy). ...

So I had better stop now even though there are lots of things I should answer in your marvelous letters –

 Your loving son,
 Bob

104 McClary Ave.,
London, Ont.
10 May 46

My dear old Dad:

... I am sorry that things concerning the expansion of the Lab are going so slowly – I suppose that it is only to be expected – but you can be thankful that this time there is some possibility of it being carried out as opposed to some of the previous reverses. The manpower situation is serious and I don't see what you can do about it. The only further possibility that I can suggest (you may not like it) is to advertise in Science. I have tried to contact some of the people that I thought might be possibilities but they are all fascinated by absurd things like Surgery. Somehow we have to have a way (in addition to more money) to attract people into the basic biological sciences – all the other subjects seem to be in the same boat. One reason, I think, is that too many places give a very short premedical training and rush them through medical training before they have time or opportunity to get any guiding interest – so that nearly all of them are only aware of clinical and dramatic (to them) subjects. The people that have to be encouraged are those that take a science degree for the love of it and go into medicine to get the opportunities.

I would be very glad to have the book on Respiratory Enzymes. I have read it and found it useful & would like to have a copy if you do not want it.

As far as I know Ash and I are planning to go to Detroit on the afternoon of the 20th and will be staying at the Book-Cadillac. You will be coming on the following day. The only thing that bothers us is whether the "banquet" will be a 'tails' affair or not – we would be very glad if you would let us know what you think. It does not make much difference to me for I have none, but Ash is well provided and would like to know.

In case you want it here is the Lab phone number:
Metcalf 3808 (i.e. Victoria Hosp), Local 58.
I will write again soon.
    Your loving son,
        Bob

                Montreal
                16 May 1946

Dear old Son,

This is to wish you well on your birthday & many years of joy & success. Through the years you have always brought us happiness & our memories are full of interest & enjoyment with but few anxieties. Now with your own little family you add to our happiness and as we grow older our contentment increases.

We look forward keenly to the few days we will be with you soon. But as soon as it can be managed we must arrange a fishing trip; often we say on the trail or lake or river "the old boy would love this." It happens so often that one of us says it when the other has just then thought it too, and we know it is true. I think of you a lot as I do little jobs in the workshop at the lake as there is so much in it to remind me of you & Mum treasures & uses things you made for her; I cannot imagine her parting with the dustpan you made from a syrup can.

May you & Dorry have the happiness in your children we have had in ours. We are sure you will.

(FREDA)

All my good wishes, too, darling, for a happy day, & for many years of happiness & success. Dad has said all that can express our pride & delight in you & Doris & your little family: you are, all of you, such a great source of joy to us; & this, with your achievements, is something we have looked forward ever since May 19th, 1919, when you were a "very small potato" – if I may so allude to a scientist & solid citizen!

We look forward most eagerly to seeing you all quite soon, & only wish we could have all been together on your birthday. We shall think of you, & will drink to your very good health at 6:45pm on Sunday.

This cheque is for you to buy something for yourself you particularly want.

 All our love,
   Mum & Dad

                London, Ont.
                 5 June 46

My dear old Dad:

Somehow I did not have time to write to you before you went off on your fishing trip. I wish that I had, for I never thanked you properly for my birthday present which was so welcome, but purchase not yet made. ... Also I did not thank you for all the nice things that you said in your letter. ... I might say that I feel that I received a great deal more than I could catalogue from you and Mum, hence the credit is really yours. However that may be, I am very gratified by the way everything is going though a little surprised and – I don't know what – about having such a high-sounding title at my stage of development. I sometimes think that I would rather earn it by hard work than have to justify it after its establishment.

Doris is very well – though has slight signs of a cold – and has refused to show any signs of impending labour though they may appear at any time. She is finding it a great help to have Mum here and we are both enjoying having her more than we can say. Alice is a handful but a very cheerful one.

No more fishing yet – but I have some offers for the future and I think it will work out. Jack Pawlitska (the florist) is planning to take me to their private stream and also to the good fishing parts of the north branch of the Thames. Every day, now, as I cross the river I look for my three carp friends – one of them is very big as far as I can judge.

The *Streptococcus equi* cannot be reported as such for the blighter ferments lactose though it seems to conform in all other characters except the pathogenicity for mice.

I have been experimenting with various media to see if I can enhance its virulence but it won't kill them – though it makes them sick for a few hours the organisms do not seem to appear in the blood. I am also trying to identify a Strep, isolated from CSF[82] by the Public Health people, from a case of purulent Meningitis. The culture seemed pure that they gave me, but after a lot of screwy results I find it is a mixture. One produces greening both aerobically and anaerobically, the other produces greening aerobically but good B-haemolysis under anaerobic conditions. They may be interesting and I will tell you later what I find, also the screwy results I got at first which were very peculiar, even for a mixed culture.

I have run into a funny snag in the TB work. Though I can get good reproducible results in 10 cc amounts of Medium, trying to do it in bulk with exactly the same medium in various shapes of bottles is harder to do, in fact it is not easy to get the organisms to grow at all. I am now proceeding to try and untangle it; it would seem to me probable that it is something to do with both the surface exposed to air, the rate of oxidation of the detergent and the relative rate of multiplication of the organism. The inoculum seems to have to be enormously bigger, in proportion. Soon I will be able to extract more phospholipids and repeat my previous observations on a larger and better scale.

Ash is getting some very interesting results on the antiphage agent – for which I am very glad for it is giving some hope of a better solution than the other problems that were worrying him. Schatz[83] (of Waksmans lab) is publishing, apparently, the results that they reported at the meeting. ...

Your loving son,
Bob

Montreal
9 June 1946

Dear Family,
It was grand talking to all of you this morning. It was too late to call up last night as I only arrived back about 2 am. I was very tired – a bath, a drink & bed, after

---

82 CSF is an abbreviation for cerebrospinal fluid, "which can be drawn by needle by lumbar puncture for diagnostic purposes." RGEM.

83 Albert Schatz was a graduate student in Selman Waksman's laboratory (soon to be the Waksman Institute) at Rutgers University, New Brunswick, NJ. He was intimately involved in the discovery and the early characterization of streptomycin, for which Waksman and others at Merck & Co. received the Nobel Prize in 1952. Schatz never forgave them (particularly Waksman) for not giving him the credit he thought he deserved. RGEM. For details see William Kingston, "Streptomycin, Schatz v. Waksman, and the Balance of Credit for Discovery," *Journal of the History of Medicine and Allied Sciences* 59, no. 3 (July 2004): 441-462.

unpacking the fish into the fridge. They traveled very well in Sphagnum moss & a little ice. Today I parceled them out for different people some of which I have delivered & others at the lab tomorrow.

Sue & Blake & Fudge walked over at noon & are all very well. They are coming in the car this evening to fetch the fish for themselves & Dr. & Mrs. Robinson. I only wish I could give you some; but I hope Bob will get some more fishing, as he mentioned the possibility this morning.

We, Tom Matthews & I, left here 31 May /46 in Ernest Trott's Ford V8 at 7:30am in rain & quite cold & it continued so to l'Annonciation where we ran into snow. The snow became heavier & whitened the ground & stuck on the evergreens. We arrived at l'Ascension at 12:30, stopped there a little to make some enquiries & then drove on to the Sawmill on Lake Shibley where the club has a boathouse & a garage. There we met up with George Wood who completed the party. We loaded a mass of boxes & gear onto a rig with two very fine well cared for black horses, had a sandwich lunch then rowed ourselves the length of L. Shibley to the trail. A two hour walk to Matheson Lake which we again crossed in a boat. Then about ¾ hour walk to Green Lake. The trail was wet but not too bad, there were no flies as it was too cold for them & it was comfortable going though I was glad of warm clothes & boots. There was nothing at the camp, no tea etc, though they had some tinned food which however they seemed unwilling to open up. The team & gear arrived about 5:30 or 6 pm & we unpacked the grub stake & stowed it "according to custom" & then had a good supper. ...

Green Lake is a very beautiful lake, surrounded by low hills with mixed woods but very few Hemlocks, but lots of Spruce & Balsam, no Tamarack, lots of Cedar & Birch (mainly yellow). I saw tracks of 3 separate bears, one moose, numerous deer, one wolf, one fox & smaller tracks I thought were skunk & mink & other smaller ones I do not know. There was one porcupine & one Partridge. Birds were plentiful ... A pair of Herons (blue) have a nest on a broken white birch & are busy hatching. A seagull nest (3 eggs) on a bare bit of rock. (The others said there were very few birds – they could not see or hear & were up too late for the best of it) Two rabbits (snowshoe) were very tame around the camp. I tried hard to see the bear but though I could find absolutely fresh tracks most days I seemed to be just too late.

... The camp was very comfortable & its site well chosen. The view from the camp was magnificent, especially at sunset. In the evening the silhouette of the hills, with the reflections in the lake, was gorgeous.

... The coming out on Saturday was another matter. A party to occupy both camps was due in by plane that day & so both camps had to go out. There was only one rig available ... We took the road & it was bad going being very wet. We got to the loading place on Shibley but there was no boat; if it had been brought back it was at the other trail. So we started to walk a further 2 or 3 miles. ... I found the slow pace they walked & the frequent short rests very tiring & would have been much happier at our usual faster pace.

When I got back here (2 am) I found some biscuits, some cake & with these a little rye & ginger ale. I felt better. Then I read your letters (2 from Mum & one from Bob) had a hot bath & went to bed. I slept soundly & felt fine this morning. ... I fried a trout & made tea & that was my breakfast ...

All my love,
Dad

104 McClary Ave.,
London, Ont.
21 June 1946

My dear Dad:

I realized with horror, a day or so ago, that I have not written to you since the great event when I have been under the impression all the time that we had exchanged several letters. This was I think due to the phone calls ... which created such an impression.

Mum will have told you most of the news about us and the family. Doris came home yesterday and she was very pleased to do it. She is in the very best of health and does not seem to be fatigued very easily – which is a nuisance because it is hard to get her to rest properly. However, she is well and looks marvelous. His Lordship (Peter) is being very good too, though I have not got back into the habit of being able to wake at two and six in the morning – practice, undoubtedly, will do it. ... Alice greeted Peter very sweetly with the most appropriate noises and does not yet show any signs of acute jealousy.

I suppose that I should do something about sending word to Christ's [College] about all this. But it is rather foreign to my nature to do such things and, anyway, I don't know who I would send the information to and in what form. I think I included some information on previous events to Grose[84] in a letter but I don't think that any of it ever got into the Magazine. In Mrs. Grose' letter to Mum I saw that Grose is retiring this year.

There have been no opportunities for fishing as yet but I live in hopes that someone may take me out soon. The bass fishing opens this weekend, and Dr. Ramsay murmured something to me a while back about going out around July 1st.

Evelyn's[85] lectures must have been very fine. The précis is very good if a little bald. I wrote to him a note of thanks. I think that the radioactive isotopes will bring a lot of

---

84 See chapter 1, note 126.
85 K.E. Evelyn was a "brilliant Canadian researcher" who worked in what was then the Donner Research Institute attached to McGill's Faculty of Medicine. RGEM.

important knowledge to the surface. Though they may be available from the NRC I don't see that such as I could use them for they require rather expert handling and the use of techniques about which I know nothing. Some of the reports on therapeutic use are very promising and may come to something when they learn more about control.

You should try and get hold of the Chemical and Engineering News Vol. 24, #10, May 25th 1946. In it there are some very excellent general articles, especially one by Linus Pauling on molecular architecture and biological reactions. Also in there is one by G.W. Merck called "Peacetime implications of Biological Warfare".

I am so glad to hear that they have started on the structural alterations. It will be a great blessing to you and I hope that they get on with the job and don't have it hanging on over the beginning of term. It is also very welcome news to hear that you have filled some of the vacancies in the staff and I am hoping to hear who they will be. I don't think you would be able to find any first class men at the moment and it seems to me that everyone is looking. I am sure that (J.W.) Stevenson will be a great asset to you.

Talking about Stevenson reminds me that I had a letter from Ian (Stevenson)[86] the other day who sent you his best regards. He has left Arizona and has gone to New Orleans where he is a Fellow in Medicine at the Ochsner Clinic. I gather from his talk that it is a diagnostic mill which spews forth a slightly battered patient and a diagnosis at the end. But he seems to think that he will learn something there. He says that his health is much improved.

I also had a letter from Ronnie Hare[87] who is going to be here for a day next week. I couldn't read his letter very well and don't know why. He said that he was in the process of dismantling his lab and that he was coming to St. Thomas to see and be seen. What the hell there is at St. Thomas I don't know except a few railway lines. He also knew about Peter and I can only guess that you told him.

A week ago today I took a trip with Jim Patterson[88] to Guelph to have a talk with MacNabb. We stopped off for lunch on the way at Kitchener at Walper House which has quite the nicest dining room I have seen in Canada and which serves very good food. That part of the country is all square heads and it is rare to see a good old Anglo-Saxon name – they are all Kalbfleishes and Schwartzentrubers. However, it is pretty rolling country and worth having a look at.

Guelph is a pretty town and all the houses in the centre are built of stone and have

---

86 See chapter 3, note 44.
87 Ronald Hare was a microbiologist who worked on viruses. He was trained in St. Mary's Hospital, London, with distinguished teachers such as Alexander Fleming. In the 1930s he came to Canada for a research appointment in the Connaught Laboratories, University of Toronto, where he worked on influenza virus. He was the author of an important account of the environment of the discovery of penicillin by Fleming in the late 20s, *The Birth of Penicillin and the Disarming of Microbes* (London: Allen and Unwin, 1970). RGEM.
88 See chapter 3, note 92.

a very old world look. The OAC (Ontario Agricultural College) is outside about a mile and has very fine grounds with really beautiful lawns and well kept gardens. I did not have time to look around very much for we were there only an hour and a half. The Veterinary College is very small and as MacNabb realizes badly in need of a transfusion. I was trying to see whether I could arrange with them for embryo sera from the pig. I think that I may be able to get it but with a bit of delay. He asked after you.

We have had some wild and wooly weather the last few days – though there have been no tornadoes in London there have been some remarkable wet and gusty thunder storms. Alice was very fascinated by them and not a bit frightened.

I was very glad to see the cutting concerning new appointments at McGill. It is good they have given Freddie (Smith) a raise – I must write him a note. If they were going to do that why the hell couldn't they have raised Ted (Roy) to Associate rank? (I don't know what sort of job he is going to at the Sick Kids or the opportunities it presents). Also they have finally made an appointment in Zoology – just as Wynne Edwards is leaving too!

Two things I want to ask you. (1) Are you using Penicillinase in the clinical lab? How are you using it & what brand is it? How effective is it? (2) What do you think about preservative for the serum or do you think it is better not to use any? The latter is alright if you do it yourself & know what you are up to – but I was thinking of these veterinaries? Do any of these preservatives have adverse effect on animals or the formation of antibody?

In the last week or two I have been a bit distracted & not been doing very effective work in the lab. Not much to tell you except I have found out for the nth time that green streps can be hard to identify; that paper disks seem to absorb Hyaluronidase; and that I've got a lot to learn.

Will write again soon
Your loving son,
Bob

104 McClary Ave.,
London, Ont.
26 June 46

My dear old Dad:

We were awfully sorry to hear that you have been feeling poorly, and such an annoying form of sickness too. Mum was very worried that you might not be feeling up to going out to get your meals and might, conceivably, need to keep quiet for a few days – we felt it best to persuade her to go home to you – which she did at noon today. Now don't go getting worried that we are being left in the lurch. Nothing of the kind.

Doris is very well and as long as she does not get an urge to do any unnecessary work we will get along very well. We both of us have missed her – and we are very appreciative of the great work that she has been doing for us during the past five weeks. ...

Yesterday and today I have spent a lot of time looking at Entamoeba histolytica.[89] There is a good case of recurrent dysentery on the ward in a flyer who has been stationed for a long time in Nicaragua. There were a fair number of very typical vegetative forms, they remain active a long time in this weather, in the fresh passed stool... Today there were immeasurably more amoeboid forms and they were very active. Today I have "subcultured" this trying the added refinement of a drop of blood. I found also, after a dimly remembered recommendation, that the amoebae are more easily found and identified if the suspension is tinged with Neutral Red as a vital stain.

I was very sorry to hear in this regard that poor old Blake is again troubled with his dysentery. After such a thorough treatment at Xmas time it is a bit disheartening for him to have trouble again so soon.

It has been very hot here for the past few days and we have been much relieved by the possession of a fan, which we have had for about a week. Our place is a good few degrees hotter than is comfortable on many sunny days, even though they are not excessively hot, and the fan makes all the difference between comfort and misery.

Mum will give you all the news of Alice and Peter. Needless to say they are both flourishing.

Doris sends you her fondest love and hopes that you will soon be better. She says also that she will be sure to write to you soon.

      Your loving son,
         Bob

Montreal
27 June 1946

Dear old Son,

Mum walked in on me last night at midnight just when I was addressing an envelope to her. Took me by surprise. I said over the phone not to change plans that I was alright. I hope very much it is not going to wear you out & make it too exhausting for Doris.

It is good to know you are all so well. The heat will worry the little fellow. You might try the old India trick of hanging a wet sheet in front of the window, dipping in

---

[89] Entamœba is a "member of a genus of amœbas so called, including several species parasitic on vertebrates." *Entamœba histolytica* is "the cause of amœbic dysentery and liver abscess." *OED* (Online).

a bath of water, to cool the air by evaporation. It's hot here too & for the first time in my memory I am feeling the heat.

I am still a little giddy but steadily improving. McNally cannot find anything wrong with ears, sinuses etc & thinks it must be circulatory. Probably my low blood-pressure is letting me down a bit & he wants me to see a physician. ... I have been taking salt tablets since Tuesday & have benefited.

Selfishly I am glad to have Mum home. We are going to the lake for the weekend tomorrow. Sue & Blake are coming too & Sue will stay up a couple of weeks; she needs it as she is pale & feeling the heat in town. I hope to get up there for a while soon. Mum is looking well but I am sure she needs a spell at the lake to set her up properly & she is longing for it. The old pup gave Mum a joyous welcome. He was surprised & delighted to see her.

Blake's stool frequency has abated a bit today with "Vioform". The Entamoeba is not so numerous as in December last but there are fairly numerous trophozoites & there was a lot of blood mucus. The stool is not typical & the onset was much too sudden, so I suspect some complicating thing. Gertrude has found a few Lact. mg colonies but it is too soon to know what they are. I hope we can find the real trouble & get it under control.

Roddy Heard[90] is here & he & I dined pleasantly together last night. We arranged for Mum to stay with them on her way home but that's shot to pieces now. He gave me a card introducing Blake to one of Imp. Oil higher executives & I have given it to Sue. No harm in Blake seeing what there is to offer & trying to better himself. His present job does not seem to offer much of a future. I hope something good comes of it. ...

... About your serum problem. I have asked Don to look up what he can find & he is interested in knowing something about it himself. Some preservatives are found to give altered specificity. ... One would expect many disinfectants, especially heavy metals, which kill bacteria to act in some way on proteins. It would seem that such types should be avoided for your purpose & choose rather those acting on bacterial enzyme systems such as sulphonamides. ... I'll think more about this & will try to find some evidence.

One by name Crabtree who was at Lower Canada Coll. with you was in here a while ago & wishes to be remembered to you.

Some building operations have started in the department – mostly destructive. I suppose constructive action will start sometime soon.

  Love to you all,
    Dad

---

90 See chapter 3, note 29 for information on the Heard family. Roddy was the brother of Nancy Heard, cousin of Robert Murray on his mother's side.

Mum gave me the book you sent me. It looks very interesting & I shall read it on the train tomorrow.

104 McClary Ave.,
London Ont.
30 June 1946

My dear old Dad:

It was good to get your letter & hear that you are feeling better. I hope that by now you are completely well after a week-end at the lake.

Although it would have been a great help to have Mum here we are getting along very well & have a scheme for division of labour that works very well. Doris seems to have regained her strength very remarkably and I think that getting up early has a lot to do with it. Also, for the present situation, it is a good thing that she was on her feet & able to do things at two weeks. Doris is really well and is not being tagged out by the work even though it is strenuous. Peter thrives and does not show any alarming tendency to vomit; he eats & sleeps, doing both of them well. All this despite the heat, which seems to worry us more than it does the kid.

It has been really hot here for the last week. Our apartment is pretty poorly adapted to extremes of both heat & cold. This would not be so bad if it was confined to the daytime but for a considerable part of the night the roof radiates and, if anything, it gets hotter. Suitable clothes are just about impossible to find – certainly there are no odd trousers to be bought. However, I managed to find a reasonably cheap suit made of a rayon tropical fabric – even though it makes me look like an ice-cream vendor it will be very useful for everyday wear. No one at the medical school has the guts to start turning up for work in shorts so I fear we have to be dressy and suffer the heat.

I hope that Blake's trouble is coming under control. I have found a reference that may be useful & interesting. "The treatment of Amoebiosis with special reference to chronic amoebic dysentery" W. H. Hargreaves The Quarterly Journal of Medicine (New Series) 15; 1-24 January 1946. I am in no position to evaluate it but it provides what seems to be a good survey of the literature & of recent experience.

Thank you for the information on preservatives – I think I will use Merthiolate if & when I have to. It seems odd that these effects on antigens are not well known. ... (Even sulphonamides are suspect because some of them will attach to protein to a considerable degree).

Yesterday I got a copy of the 2nd Edition of Landsteiner's "Specificities of

Serological Reactions"[91] into which he seems to have incorporated a lot of the recent work.

I think you will enjoy "Goosefeathers" – there is some fine writing in it and the descriptions seem to be very authentic.

The pipe arrived safely, in a relatively enormous package, and I am pleased to have it even though you should not send me so much of your own stuff. I think those pipes smoke well or better than the most expensive briar – much as I like them.

There were two Crabtree's at L.C.C. – the one you refer to is probably the one that went into Chemistry. Last I heard he was working on sulfite mashes for the paper industry. I never thought much of him but he may be improved.

This isn't much of a letter but I will write soon – Love from all of us & we hope to hear from you that you are all better.

  Your loving son,
    Bob

4 July 1946

Dear old Son,

... I am better than I was a week ago but I am still troubled by giddiness & lassitude. McNally cannot find anything amiss in ears, sinuses etc & Brow[92] cannot find anything wrong with my heart & circulation. He did an electrocardiogram etc & went over me thoroughly. Brow says some 10 or so people have come to him lately with giddiness & nothing to be found obviously wrong & he cannot account for it. It's a plaguey nuisance & disturbing, also I tire easily.

Your letter of 30 June arrived this morning. It is grand that Doris & the kids are all so well & that Peter is putting on weight so well. The hot weather is trying for you & I wish you had more comfortable quarters. However, you could easily be worse off. Your rayon-tropical suit is as good as you could want for comfort which is important.

Blake's trouble seems to be coming under control & he is better though not right yet. Gertrude found a Lact neg. motile G -ve rod ...[93] It does not fit anything described but it may contribute to his trouble. We are going to give him vaccines ... & see what

---

91 Karl Landsteiner, *Specificities of Serological Reactions* (Springfield, Ill., and Baltimore, MD: C.C. Thomas, 1936). Landsteiner (1868-1943) received the Nobel Prize for Physiology or Medicine in 1930 for the development of the ABO system of blood typing. *Encyclopedia Britannica*, http:www.britannica.com.
92 Ray Brow was a senior cardiologist at Royal Victoria Hospital. EGDM went to him for a real cardiac consultation. RGEM.
93 This is a rod-shaped bacterium that reacts negatively to the Gram stain, is motile (swims in fluid media), and does not ferment the sugar lactose, which are the first few diagnostic clues in the identification of such a bacterium. RGEM.

happens. Thanks for the reference – I'll look it up & see if I can learn something new from it. I feel sure there is a complicating factor in his case.

... I'm glad the pipe arrived safely. It is one I smoked a lot one time. It is made from the outer wood of the Cree tree while your other one was made of the red heart wood & is harder. I shall write to my relations in Graaff Reinet & try to get some more of them. I bought these in 1909 when I was there last & they cost 1/6 each, as far as I remember.

... I am worried that you are not arranging a holiday; even a short one is necessary for you. Let me know if you can get a line on a place you could go for 10 days or so & get some reasonable fishing. Enquire from your friends. It's very important as you have a lot of work coming up next fall.

I'm getting my staff lined up. I still have the teaching fellowship to fill & have place for two Clin. Lab Technicians & 2 general lab technicians (Lab boys of some capability). The teaching fellowship is worth $1200 a year & must be qualified to work for a PH.D.; it's a good chance for a young fellow. ...

Best love to all of you,
Dad

104 McClary Avenue,
London, Ont.
7 July 1946

My dear Old Dad:

This should get to you by the time you get back from the lake – I hope that you are going to give up bothering about city affairs for a while and take a decent holiday. I gathered from Mum's note to Doris that you might be doing that in a week or so – it strikes me that it is even more important for you to take a bit of time off at the present than it is for me. I will admit that there is something in your arguments but I cannot leave the city for a while yet, even if we could afford it, and I have a number of things going at the lab now that cannot be left for a month. Anyway, I have to hold the fort for Ash until about the 15th of August. Even if I took time off I don't think I could stand being within sight of the lab and not go there every day. I am happy and not overworked and, under the circumstances, I could not ask for more.

You make matters much worse (in a nice way) by sending that marvelous sleeping bag. It is a really fine affair and I am very glad to have it. But it gives me the itchiest of feet and I wish that I could go off with you for a trip into the bush....

I am glad that Blake's trouble is quieting down. It may be that the peculiar Lactose negative organism is causing some ancillary effect. The vaccine is probably well worth a trial. Multiple factors in gut infections seem to be terribly hard to disentangle.

You seem to have got some way towards filling the vacancies in the department. I fancy that it is just as hard, if not harder, to find technicians as it is to find qualified people. You have an advantage over most places in that there is the Course III class to pick from, even if they are mostly girls. The only sort of technician grown domestically here is of the Association of Lab Technician variety and of no earthly use. You have not told me about any of the people you have got except for [J.W.] Stevenson. ...

... I have started growing eggs at the lab again and hope to be able to get some rabbits immunized before the summer is out. In that way there will be work ready to be done by the time we are finished with the fall class. I have also started some rabbits for class work on precipitins. I think that by a year from now I will have things more or less ready for decent teaching without having to do too much hectic work in preparation for classes. The main thing seems to be to get ready well in advance and to have some of the snags anticipated. I will bet I miss quite a few anyway.

Pearce[94] and I have been spending the last week on an intensive bit of work on my test for Hyaluronidase presence. It looks as if it might be definitely useful in that it will detect to at least 0.02 of a Turbidity Reducing Unit which is definitely in the range where other methods are uncertain without having to do a complex concentration procedure. Also it seems to provide a useful quick test for production by an organism. You can't have everything but it would be useful if it was definitely quantitative. I am not sure that it would not be a good thing to include Hdase production as a character in Bergey's Manual one of these days. May be worth noting concerning *Staph. pyogenes*, Pneumococci and some of the Clostridia. It is interesting that the Testicular and the Pneumococcal enzyme do not act on Group C capsular substance while they both do on Group A. (I have not tried other combinations) ...

I don't know whether it is worth burdening the literature with a description of that Group C organism that resembled *Strep equi* in morphology and a few points. It seems to be a definite lactose fermenter. I don't know about Trehalose, I have had that sugar on order for about two months and there is no sign of it yet. It is interesting that it was in combination with a Group H Strep, but I am not particularly convinced that it is worth writing up unless it is definitely an equi. What do you think?

I have not heard from Ash but he is down at Cold Springs Harbour. Mrs. Asheshov told me that he is finding it a good meeting.

Enclosed is the letter I received a while ago from Ian Stevenson. I thought you might be interested in seeing it. Don't return it to me, I have replied and I have written down the address.

---

94 R.H. (Dick) Pearce was a post-doctoral student in Earl Watson's Clinical Chemistry Laboratory in Victoria Hospital, interested in the biochemistry of the polymers between connective tissue cells. One of these polymers is hyaluronic acid, and Robert Murray collaborated with Pearce on this substance in the late 1940s. RGEM.

We are doing alright as a family group even if the going is a bit hectic at times. We split up the work as well as we can, though I am physiologically unadapted to some of it. ...

We love to hear from you about all the doings at both the lake and at home. Soon we hope that you will both be up at the lake for a good spell – do you both the world of good.

 Your loving son,
  Bob

              Montreal
              11 July 1946

Dear old Son,

 We came back from the lake on Tues, 9th, & your letter was here as you expected. We were very glad to see it & know all is going well.

I have finished drawing the detailed ¼" scale plans for the extension over the roof, I completed the plans for the other alterations involving the old classroom etc. a while ago. If it is all carried out completely it will enormously improve the department. When it is finished it will be a real pleasure to show it to you; but presently I shall send you a copy of the plans to see & you will have some idea of the changes.

Briefly: the extension over the roof will comprise a complete unit for the Chemical Division (lab for Gertrude, Serology Lab, large general lab for 6 technicians, lab for 2 lecturers, incubator room, cold room, cloak room, 2 animal rooms, operating room) media room, washing up room, store rooms, plate pouring & sterile distribution room, library. The alterations in the old classroom gives 4 labs like [Cliff] Kelly's[95] a course 3 classroom & frees the present media room & washing up room as labs for staff & research students. The path classroom is being altered for the use of both departments with time table arrangements. I'll give you fuller details & of staff & class etc when I send you the plans.

I have also got the class schedules & who is to do what teaching outlined. Freddie [Smith], Don [Fleming], Cliff & [J.W.] Stevenson are busy drawing up the class practical sheets & I shall go over them when they are done; I hope next week. I can now consider planning a holiday – probably starting in a week's time.

My dizzy troubles wax & wane a bit, better some days than others, but on the whole better than they were. A rest may help besides being pleasant. I quite see it is difficult for you to leave until Ash comes back & while you have experiments cooking. Still I

---

95 Cliff Kelly was a general bacteriologist interested in non-medical topics who was brought to McGill from R.S. Breed's dairy bacteriology lab. RGEM.

think you must have a bit of time away before term starts again. You will have a busy time next year & better be fit & well to tackle it. I'll take this up again at a better date.

... Blake says he is back to normal. This evening we bled him to see if he has any antibodies against his queer organism & also gave him 1/20cc *Korto* vaccine intradermally as a test of reaction. He did not like it much but it is well to find out about it. I propose to give him a course of vaccines to see what happens. ...

You are right to get class material ready well in advance. One of my annual concerns is to get essential preparations on the way during the summer. These things cannot be done quickly just before they are wanted. There is nothing so unsatisfactory & humiliating as meeting a class with no material for them to use.

I have an idea that there is some published work touching on your Hyaluronic Acid & Hyaluronidase work. Someone was able to digest off capsules with extracts of other organisms. I'll try to get a line on it for you. You & Pearce seem to be making good progress with it & there would seem to be a useful application as a fairly delicate test. ... I think you might write to Jim Sherman & ask his opinion on whether your group C strep is worth a note. He is reached at the Dept of Bacteriology Cornell University, Ithaca N.Y. USA. I believe you met him at Detroit but I'm not sure. You can say I suggested your writing – he is a good friend.

I did not feel inclined to do much at the lake last weekend. I messed about a bit in the workshop & chopped a little wood. Mum & I had a paddle in the canoe each day with the pup & I had a few casts with a plug off the wharf. There seems to be very few Bass, rather small & wormy. Nothing like the old days. I hope to do a bit more this coming weekend – perhaps go after a few trout. ...

  Love to you all.
    Your loving, Dad

104 McClary Ave.,
London, Ont.
11 July 1946

My dear old Dad:

Tonight we had a small meal of Bass – I had almost forgotten how good they were to eat, and I am again convinced that they are among the best of eating fish. As a fishing trip it was not of very great interest but it took me to a new spot and was important from that point of view.

Dr. Ramsay and Dr. McLachlin[96] arranged it as a trip to a spot on Lake Huron

---

96 G.H. Ramsay was a London physician. A.D. (Bam) McLachlin was chief of surgery in Victoria Hospital and the professor of surgery in the Faculty of Medicine. RGEM.

just for an evening's fishing. We left here, after a lot of waiting for people to turn up, at about 4:30 and we reached the lake at 6:00. The place is about 50 miles from here and the place we went to is supposed to be called Cedar Point, it is about 4 miles south of Kettle Point. The nearest town is Forest and it is about 5 miles from there to the lake. There were five of us and Dr. McLachlin's 7 year old son out to catch his first fish.

The shore-line is only slightly indented in those parts and you have an impression that you ought to be at the sea. There are no cliffs but there is a sandy bank about fifty feet high which is pretty steep. The water is very shallow for three hundred yards or so from the shore and the bottom is very rocky. It is so shallow that you can wade for a long way out into the lake and many people fish in this way.

... As you might gather the trip was not very effective as far as good fishing technique was concerned – conditions were rather against it. But it was a nice trip for an evening and I now know what sort of tackle is needed for that area and the general lay of the land. The others did not do very well though everyone got at least one fish – I had easily the greatest number of strikes but was not very effective in hooking them securely. The main thing was that Doris and I had a bass each and Alice had a Perch just her size (she enjoyed it very much). I don't know what other fish there are to be caught in that area but I did see a pike taken by another party and I was told that pickerel are taken also. The bass bellies were empty so I don't know whether the main part of their food is minnow or not – from the look of the land I would say that it was.

Our fishing was cut short by half an hour because of a ... storm blowing up. They certainly do come up very suddenly, and the wind licked up waves in no time.

Other activities have not been many. Doris has been very busy and I have been home as much as I can to give her a hand. She gets the occasional bad night and it takes a little time for her to catch up sleep afterwards. Alice and Peter are both very healthy – Peter, though well, has been doing rather more than his fair share of yelling lately – he certainly does not have the quiet disposition that Alice had as a baby.

We got a very cheerful letter from Mum this morning. I am glad that her neuritis is clearing up a bit and I hope that it may be due to the parenteral Thiamine.[97] She may not have been absorbing any of the oral doses that she had for so long. We also had a letter from Sue who sounds cheerful and seems to be getting a lot of good out of her holiday.

---

97 "Parenteral" is defined as "Designating or involving the introduction of a drug or other substance into the body by a route other than the digestive tract (esp. by injection, infusion, or implantation); designating a substance so introduced." "Thiamine" is defined as "a water-soluble, heat-labile, sulphur-containing compound that is present in many foods (esp. whole cereal grains, pork, and liver) but absent from fats and is necessary for carbohydrate metabolism, its dietary deficiency resulting in disturbances of the nervous system." *OED* (Online).

Mitzi and Bob Beale[98] were married in Toronto last Sunday and dropped in to see us yesterday and today during their honeymoon tour of western Ontario.

After a series of thunderstorms things are a bit cooler and sleeping is far more comfortable. The roof of this house is entirely without insulation so that it absorbs heat all the day and radiates it all night. An appreciable and annoying effect.

All my love to you both.

Your Son,

Bob

London
15 July 1946

My dear Dad:

Your long letter came this morning and I found that I had not posted the letter I wrote to you about last Friday. While I am waiting for Alice to finish her breakfast and for Peter to wake up for food I will make this addition to the news.

On Saturday evening I went fishing again with Ivan Smith. We took food and rods and went over Dorchester way. Dorchester is a very pretty little village about ten miles (perhaps a little less) from here and on the Thames. On the way we drove through some of the country a little south of here and looked at the odd pond there is – all of them jealously preserved and most containing Bass and Pike. Ivan has a pull with a Miller near Dorchester and we had a very pleasant swim and fished for a while without result though I know there are big bass in there. After a picnic supper we went to a "hole" in the Thames about a half mile this side of Dorchester. ... I fished below into the deep water first and had a good strike but he got off after a half minute and before I had a look at him. Then I cast into an eddy back of a rock at the foot of the riffle and got a hell of a bang and hooked into a good fish. He gave me a hell of a fight going both up and down river also picking up quite a bit of weed when he tried to rub the hook off the bottom. But I got him into the bank and landed him. He later proved to be a 2lb 9oz Bass when weighed on Peter's scales about two hours later. ...

I have been meaning to ask you to keep an eye on a spot of hyperplastic skin on one of mother's calves. She said something to me about it and I wanted to show it to

---

98 Mitzi Leopold was one of a group of friends Robert Murray made in Montreal during his years as a medical student. She lived in an old house on Union Street (one block east of University Street) and they enjoyed having a fairly large influx of young and interesting people even late into the night for "a lot of innocent fun." One of the group was Bob Beale, and another was Margaret Scott, who Murray "gave away" in marriage to Bill Fear. "So," writes Murray, "there were various consequences!" RGEM.

Ivan Smith but he was away at the time. I think that perhaps it would be a good idea if it were cut out. She says it has been growing slowly.

Ivan Smith had quite a time out west and caught some trout and Salmon (cohoe). He has a long and amusing story about a monster salmon that he lost just about over the epicenter of the 'quake that they had while he was there and the peculiar association of the quake with the struggles of this fish on his line.

I must rush off now but will write again soon.

    Your son,
        Bob

Montreal
17 July 1946

Dear old Son,

Your two letters, of the 11th & 15th, describing your two fishing trips arrived this morning. We are glad to know about them & to share the fun of them by your description. Yes, Bass is a good fish to eat & good sport to catch.

Your first trip to Cedar Point seems to have been like a traffic jam & it must have been difficult casting. Still it was fun to get out & at it. … Your 2nd trip, Dorchester way, is an interesting contrast. You can easily make your plug more or less weedless with a piece of thin springy wire which is easily compressed by the Bass' bite but feeds off the weeds quite successfully. You had good fishing for only an hour's opportunity. I hope you will get a lot more.

I'll arrange to have the hyperplastic spot excised. I don't like it either & I will feel better when it is removed; though it is quite free at present you never know what these things will do. We cannot take risks with Mum anyway. Mum's neuritis is definitely better though we have not been regular with the doses & must improve.

Sue's holiday did her good & she looks fine again. That tooth removal gave her a very bad time shortly before her holiday. Blake is much better. He does not like our vaccine treatment – he objects to being "bayoneted". The intradermal "Kocto" gave quite a reaction though there is no agglutinin in his serum. The doses of Kocto give him no trouble. The last stool, after dosing, was full of proteins & difficult to judge, but we may find something in others. Blake is enquiring about another job with a big paper company; he is going to Ottawa to see about it on Saturday – the Gatineau Paper Co 9 miles outside Ottawa. It may well offer better prospects – I hope so.

Mum & I were at the lake last weekend. We kept fairly quiet. Sat was a gorgeous day & we spent it at Twin Lakes. We got 2 keepers & put back several small ones, but we did not fish systematically. … We are going up on Friday & will spend a week at the camp & then I propose we go to the Columbus for a while; Mum & I propose to explore some of the distant lakes & to camp a bit on the island on Iberville. We'll let you know about it & we wish you were coming with us.

We're having quite a time trying to fit our teaching time table together. Working in 9 courses, sharing lecture room & classroom with Pathology & the complication of 3 different faculties don't make it easy. We will fit it together if Arts & Science cooperates a bit.

Our love to you all. It's fine to know all are well & that Doris is not finding it too hard.
  Dad

Faculty of Medicine
Department of Bacteriology and Immunology
Ottaway Ave. & Waterloo St.
21 July 1946

My dear Mum and Dad:

I am so glad to hear that you are taking your holiday at last. It will be grand for you to get some time at the lake as well as a fine trip up at the Columbus. I so wish that I could go with you for it is a trip that I have wanted to take for a long time. ...

I have a bad case of itchy feet myself. Dr. McLachlin (the professor of Surgery) is a canoe tripper from way back and he was asking me whether I might be able to go with him to the Algonquin Park for a week when he can get away. He can't set a date and, for that matter, nor can I. It would have to be after the 27th August and that is a bad time because of the proximity of the new term. Oh well, it is a faint hope to hang on to while the summer passes. I have not been on any more trips but they will probably crop up again and at very short notice as usual. Most of these people are very busy and plan these things at the last moment when they are fed-up with the sound of the telephone.

... Work in the lab goes on. I have run into a snag in the egg supply but I think I have that cleared now. I have found it rather difficult to learn to bleed the embryos efficiently and find it a chancy business so far. If others can do it so can I. Another snag is the extracts of whole embryos. I can get a beautiful extract but it is such a good colloid that you cannot clarify it with our centrifuges nor by filtration with all grades of sintered glass filters. (I don't want to use other filters because of their absorptive powers). With potent sera, however, one should get good results using complement fixation techniques.

The kids are thriving. Alice is improving her vocabulary and her devilment day by day. Peter has been a bit fractious lately but he seems to be better in the last few days. With everything Doris has been a little busy and, though she is a little tired, she is very well. She has not many energies to spare at times and she says that she will write to you when she can but transmits her love.

Enjoy yourselves and catch some big trout.

    Your loving son,
        Bob

Bark Lake
23 July 1946

Dear old Son,

... We are delighted at the possibility of a canoe trip to Algonquin. We are convinced that you must do everything you can to take it & arrange work so that it can be left. You will have lots of time to do it & besides needing a holiday badly the season is short. I am convinced that it is all important to doing good work to have a good holiday. To help you to have one without too much difficulty we send you this cheque with our love. Please tell Dr. McLachlin that you will certainly go.

By report Algonquin is wonderful territory away from the hotel & the tourists.

... We have mainly done jobs around the place since we have been here. Mum's garden is wonderful & we are enjoying the produce. It has never been so fine as it is this year. A rabbit was doing great damage to the peas, he ate 2 rows & was starting on a third. While we were sitting on the kitchen step having breakfast this morning I saw him coming across the clearing towards the garden. I took the air-gun & opened the bedroom screen & when he appeared, making for the peas again, I shot him through the back of the head & dropped him in one.

Fudge was terribly excited & leapt on it. I hung it in the apple tree & he is camped underneath it. ...

When we came in this time we forgot the keys & had to cut our way in with a hacksaw, cutting through the staple & ruining two good ones it will be hard to replace.

On Friday we were among a huge crowd waiting for the train at a temperature of 95°F in the shade. Presently Norry Owens[99] came to the edge of the crowd & called us out so we drove up with him. He was coming up alone so went to the station to see if there were anyone he could give a lift. Few people would think of it & much less do it....

    Love to you all,
        Dad

---

99 O.N. (Norry) Owens was a good family friend together with his wife, Eva, who was one of the Whittall family with a cottage at Bark Lake. In the First World War he sustained a severe head injury in an explosion and lost a patch of his skull, which Robert Murray remembers as being fascinating to a twelve-year-old boy. Murray remembers him as "a very handy man and a great canoeist and competent in many ways, ingenious and thoughtful and kind." RGEM.

St. Bartholomew's Hospital Journal (August, 1946)

Would Yer Believe It!

They have asked for a gossip column.

A little time ago one of the most able and truly intelligent members of the Hospital suggested, in a brilliant letter to the Journal, that the proper study of mankind was man. He asked for Hospital Gossip in a Hospital gossip column.

Immediately a collection of polymorphous phrenopods (with brains in feet, etc) rallied round the Editor in calling him a scandal-monger. This just showed lack of erudition. Scandal is gossip made tedious by morals – and who is Editor to quarrel with Oscar Wilde or who would suspect the original letter-writer of an urge to wax moral?

More recently the true worth of this suggestion penetrated the ivory editorial domes. And the lot fell upon me, as upon Jonah, in several senses.

How shall we begin? With the story of the preclinical who is to get an S.S. Jaguar for passing 2nd M.B.? Or by telling of the proportion of Vicarage spirits which is removed to Hill End by be-spectacled Chief Assistants? Or shall we borrow a trick from Mr. Agate and have a gossip-column bearing no discernible relationship to tittle-tattle?

Shall we find sufficient material to make this a regular item? I doubt it. We shall see. But the collection of the material should be fun.

 Evelyn Tent

              Faculty of Medicine
          Department of Bacteriology and Immunology
              Ottaway Ave. & Waterloo St.
                    5 Sep 46.

My dear old Dad:

The other day I managed to finish the account of the trip to the Algonquin and I fear it was a little inadequate, both from the literary and general detail point of view, but I sent it anyway and hope that you enjoy it. A fine trip and we enjoyed all of it and all the weather.

Since getting back I have been infernally busy – at first because there was a heavy routine load (Dr. Bittner was on holiday) as well as checking materials for teaching. Pearce and I were trying to finish our bit of work but had some technical difficulty with our enzyme preparation as well as with one of the other tests. Despite this we should be able to fill in the gaps fairly easily. Now everyone is back and I should have more time for my own things but teaching problems keep cropping up: Lack of supplies which have been on order for months and trials of fitting large classes into small space of both lecture rooms and labs. I have finally got things arranged so that I can teach them all without having to duplicate lectures or demonstrations. You must have this problem about 10 times magnified. ...

 Your loving son,
  Bob

104 McClary Ave
London, Ont.
9 Sept 46

My dear Mum:

... You are going to be a busy grandmother! The news of Sue was given us by dad & we have had a couple of letters since. However encouraging the new job is in Ottawa they sound a little discouraged by the housing – and I don't blame them. Boarding houses of that character would lower anyone's morale. But I expect they will find something fairly decent soon and have a roof for the baby as well as themselves. Doris is very keen to see Sue & to talk with her on the matter of supplies – they have been writing back & forth of a possibility of Sue coming to stay with us for a while – we would love to see her.

We are so pleased to hear how well you & Dad are after your summer of holidaying – such a change for Dad after all these years – though the dizziness must have been a great bother to him it seems to have passed away so that he could get real benefit from sun, trips and fishing – I hope he can find time for more. ...

After a summer at the Lake I'm sure you are very well – and Fudge must have lost some of his waistline! You seem to have done a lot there – and a lot of wood cutting too. I would love to see the place again.

We are very well – taken all around. Doris has had no relief from cares of family but seems to keep well despite this. When Peter's colic was bad she got very little rest & was correspondingly tired but now that she can get a reasonable nights sleep she is very well. The week's trip in the Algonquin did me a lot of good and I am pretty well refreshed for the strenuous few months ahead. I wish that we could have got away for a while. ... Next summer we must get away for a real holiday.

Alice has changed a lot since you saw her last – she seems to have grown and she is far more articulate. All we regret is that she is discovering how to be naughty & is up to some devilment most of the day. She is great fun, though, and is surprisingly good compared to a lot of kids we see. Peter of course is still small but is showing some signs of intelligence – Doris will probably give you a lot of detail ...

Teaching starts tomorrow and will go on till term ends at a 4 days a week clip. I will be good and busy most of the time.

London has been almost entirely free of Polio (only one case to my knowledge) though Sarnia has had quite a number. There has been the usual psychological flapdoodle among the populace & much fuss which the M.O.H has dealt with quite successfully.

All my love to you, Mum, and write again as soon as you can.

    Your loving son,

        Bob

                              Montreal
                       10 September 1946

Dear old Son,

    I have not written to you for a long time; not since before our trip to the Columbus. We greatly enjoyed your account of the Algonquin trip. It was a fine trip & you enjoyed it so we are sure it did you a deal of good & it was necessary. Doris must be having a strenuous time with the youngsters & it is good to know she is keeping fit & enjoying it. We would love to see you all.

    ... I expect you are getting very busy now with teaching started & the hospital work to supervise & some work of your own. Lack of supplies is a serious matter; we have a number of orders dating back to the beginning of April & it's a curse. Some orders I have cancelled & replaced with a firm in Chicago (whose address I shall send you) which will produce better results. ... The local branches of Central Scientific, Fisher etc don't do their job & I can't put up with their ways.

    The alterations in the building are coming along slowly. The new classroom seating 130 students is finished & in use for Course I (Med). It is very good except for the ventilation, which is vile. We have 132 students in Course I but can use 4 of them as demonstrators as they have had more advanced teaching or a lot of experience & that eases the seating.

    ... This classroom & the lecture room we share with pathology so our timetable has to be integrated with that of pathology. We have also to integrate the timetables of the Faculties of Medicine, Arts & Science, Dentistry & Graduate Nurses, which has taken quite a bit of doing. We have nine courses in Bacteriology this year, we could not undertake the 10th & will introduce that next year. Thus you will see that when October comes around we will be really busy. There are few hours in the week when the lecture room & the classroom are not in use, & most of it is our work.

    ... I had hopes of a Division of Virology. That was scuppered but I am trying to revive it. The quite serious epidemic of Polio should push some sense into people here.

    I am sure you will enjoy seeing the place when you can. Perhaps when the C.P.H.A. Lab Division meets here in December.

    ... Gertrude is having quite a time. She has 6 new people to teach now and one more coming in October. It will take a while for her to lick them into shape – it's a bit hard on her now, but later all will be well & she will be able to get onto her own work.

I have plans for the development of the Clinical Division into more than a diagnostic service.

I have quite a problem to solve in the C.M.H. & Alexandra.[100] I have not tackled them yet but must do so soon. The trouble of greatest moment will be finding staff for them & for the C.M.H. space.

The Department has become quite large & with varied activity.

... I have reread your trip for the nth time. Splendid & I hope you will do many more. Perhaps when we can get a car (a dream we conjure up sometimes) Mum & I can join you on one of them. We would love to go through that country. Perhaps we could arrange a trip going from here by CNR, but that would take more thorough planning to make the return by rail. Things can be done if thorough enquiries are made.

Mum is the chronicler of our trip & she is writing a tale for you which will be sent in due course.

We did not expect to go to the lake last weekend. I had a lecture on Sat. morning. Norry ... rang up & offered us a ride up Sat afternoon & return Sun evening so we jumped to it. Got there about 5pm, paddled over with Norry in his canoe & that evening did odd things about the place like splitting logs. Rosie is replacing some of the foundation posts & it is necessary as some of them are mere shells; I have told him to get on with it on the "stitch in time" principle.

Sunday was a wonderful day. We took lunch & paddled to the end of Green Bay (20 min) portaged through to the end of the lake opposite Jarvis' island (20 min) & we had lunch on the rock after fishing about a bit with a plug. After lunch we went up the inlet & had a look at the trail to Green Lake & then back the same way to be in good time for Norry.

Do you get the Christ's College Magazine? Your appointment as Lecturer is on p. 126 of the last number. You should send them your promotion to Assistant Professor.

This is a scrappy letter. I have not written for so long & am so busy just now that all is disjointed.

It brings our love to you all.
    Dad

London, Ont.
15 September 1946

My dear old Dad:

---

100 This is a reference to the Children's Memorial Hospital. The Alexandra was a "fever hospital" for infectious diseases, which according to Robert Murray closed in the 1950s "largely due to being able to control some of the diseases and their complications with antibiotics ..." RGEM.

It was grand to get your letter and to hear all the news. I am very glad to have the old Brock cartoon – I remember that the original hangs in the reading room at Xts.[101] It is nice to see it in more or less of its old state for they have now taken down all the creeper in the extensive face-lifting just at the beginning of the war. The only thing that is missing is the sun-dial on the wall of the Hall – but that may have been put up since 1918. No, I don't get the magazine and I have not seen the announcement. I would like to see it. I have not sent them any word though I did write to Grose a good while back. Who is going to be Senior Tutor now that Grose is retiring?

I was glad to hear about the changes in the lab. It must be a very difficult task getting the business reorganized on such an extensive basis and with a lot of new men too. There are so many new classes that I don't see how you have managed to fit them in at all – and it can't leave any of you with very much time for your own work. I certainly hope that I can see it not too far from now. I have not heard about the CPHA meeting but if it is on in Montreal I will certainly try to be there. I hope that it is after December 20th for that is when our term ends.

We are pretty busy. We have 58 medical students and a couple of others – since our class room holds only 42 this made a bit of a problem. We would have had to split the class but I found that by juggling with the timetable of the final years I could free another class room (though not very satisfactory) on our floor. So we are housing the overflow there. It makes demonstrations rather difficult and I have had to change some of them and it also spreads our demonstrating forces out a little thin. But it provides a better solution than any other. What we are going to do next year is an even greater problem for the class as it stands is over double the capacity of our class room and there is no lecture room (except the auditorium) that will hold that many. So far they have been taught as two sections and I don't look forward to that one little bit. As you say something will have to go and I suppose it will mean that the more interesting (to us) demonstrations and asides will have to be cut.

There are not very many changes in the course from the practical side though I have made some changes in the organization of lectures. An attempt, that may not be successful, to put more lectures on applications which are not well treated in the books. A large amount of spoon-feeding in the lectures seems to be inevitable but I would certainly like to cut it down a bit. However, I am told that these veteran classes are very good and that some of the top men have beaten all previous records. Not very surprising because they are in earnest and the competition must be very keen. Dean Hall was telling me that there are three who have almost impossibly high average marks (87-92).

---

101 The reading room in the old library at Christ's College, Cambridge.

... During the summer I collected a lot of odd references that might be of interest to you but I have not put them together yet. I wrote a rough outline of a paper to cover the hyaluronidase problem and I now know the gaps we have to fill – there are not many – if only we can find a bit of time to get cracking we could have it finished in a few weeks. One damn thing after another holds us up. I told you that I had to abandon a preliminary canter on the embryo protein business because the egg supply ran out in the middle of the first series. A damn nuisance.

In a week or two I should be well into the swing of teaching and be a little less harried and hurried.

All my love,
Bob

Montreal
15 September 1946

Dear old Son,

We have Dr. John Ipsen, his wife & daughter of 11 yrs, staying with us; they arrived on Thursday & are going back to the States on Tuesday. He is chief of the Biological Standards Division in the Staats Serum Institute in Copenhagen. You met him in Detroit. I could not find a hotel reservation in Montreal so invited him here; they are nice people.

We have decided that I shall go with Bill Newman to the Columbus on 25th of Sept for the last of the trout for this year. All my gear is at the lake so I shall have to go up next weekend to fetch it. ...

... I have just finished my turn of the first 5 lecture in Course I (Meds) & Freddy takes over for a bit this week. When I come back various of our other courses will start. However there are enough of us to share the burden. Things look pretty well on the whole.

... We hope very much you & Ash can arrange to come to the CPHA Lab Section meeting which is to be in Montreal this December. Perhaps you could stay over a few days then. I expect we shall have quite a time arranging a reasonably good meeting & visitors will find it hard to get accommodation. The accommodation problem has not been looked into yet & we better get moving.

Mum & I saw Lawrence Olivier's production of Henry V a day or two ago. It is marvelously good & we would much like to see it again. It is one of the best films I have seen & easily the very best attempt at producing Shakespeare on the screen. I don't think it could be improved & it effects a marvelous transition from the stage of the Old Globe Theatre in Elizabethan times to that of King Hal & back again. As if you were transported back in imagination while watching the play. Very neatly done. Every part is splendidly played with due restraint & most convincingly. The scenes &

colours are very beautiful. I hope it goes to London.

... We called you by phone this evening & like to now & then because it is next best to seeing you all. With Sue gone from Montreal too, Mum feels a little at a loss. Sue & Mum used to see one another most days & used to shop together & phone frequently. We get great pleasure from letters & the occasional brief phone talk is a delight. Mum is going to join in with Mrs. Maass[102] & some others to try help with students housing by providing a means for them to get information about rooms etc. It will give her an interest to follow up.

It gets late & I am sleepy.
  Love to you all,
    Dad

London, Ont.
30 September 1946

My dear old Dad:

 I am reminded that this is the last day of the trout season in the Province of Quebec and that you should be returning to Montreal and the hurly-burly of work again. Somehow I have not been as attentive to my correspondence as I should be and I am still way behind in all the things that I intended to tell you about and it seems hard to catch up with them. We are all pretty busy; Doris has her hands good and full but things have eased up a bit for her and we have managed to get a little time out together though still at very infrequent intervals. I am up to the eyebrows in work of all sorts. The class provides most of it but I am squeezing a little of the investigative side in at the same time.

 It was a great pity that Sue could not come to see us. We were so disappointed but she had best stick to orders. I agree that the midwives seem to be absurdly careful – but she may have good cause to be a little careful at this time. I must say that the difference between journeys to Montreal and to London do not seem to warrant the discrimination.

 Peter has been growing at a very great rate and is now quite a recognizable young man. He has not advanced very far with his muscular coordination but spends his waking time pushing up on his hands and taking a good long look round at everything. ... Alice of course has been at the door to take her to new adventures.

 ... I am now getting into the swing of classes and it is all going on quite easily. This class is largely veterans and they are a keen bunch. Some are quite interesting: among

---

102 Carol Edna Robertson, a former student of the renowned Canadian chemist Otto Maass, earned a PhD in physical chemistry and married him in 1926. *The Canadian Encyclopedia* (Online). For biographical information on Otto Maass, see chapter 2, note 30.

them is a biochemist, a statistician, and a wing-Cmndr of the air-force who wrote the standard book on flying for the RAF. Having the two separate class rooms is a bit of a headache but it works out alright (although it is a bit hard on the feet, being 225 feet from one to the other).

There is some interesting reading in the volume of the Advances in Protein Chemistry (Academic Press, 1945) Vol II. Particularly one by Paul Cannon on protein metabolism and the production of antibodies – would be good reading for the Course III.[103] Vol VI of the Advances in Enzymology has one or two articles which form good reviews of phases of bacterial enzyme systems.[104]

If I had a couple of clear weeks I could clear up the enzyme work with Pearce. As it is we do it when we can fit it in. We are really clearing tag ends and trying to get some good photographs of the effect. The latter has stumped us so far and we have had a series of failures. We have something to present now – but I suppose we should wait until it looks good on paper.

Don Fleming is I hear in charge (or Secy) of the December meeting – how is it arranged? I suppose that I could present this test at that time if it is worth it – but I don't think that anyone is working on H'dase in Canada. I will have to find out whether Pearce has any other ideas. ...

Your loving son,
Bob

Montreal
6 October 1946

Dear old Son,

Sue has been here since Tues 1st & is staying on at least another week. She looks splendid, seems very well though tires a bit. However she recovers quickly with rest. I wish they could find a reasonable place to live. Blake came for the weekend yesterday & returns tonight. He seems fit.

Yes the sundial was put up, I think, about 1920. Brock's drawing is quite accurate & very clever – it is perfect of Hobson & McLean.[105]

103 Paul Cannon, "Antibody Production and Resistance to Infection," in *Advances in Protein Chemistry*, vol. 2, ed. M.L. Anson and John T. Edsall (New York: Academic Press, 1944).
104 F.F. Nord, ed., *Advances in Enzymology and Related Subjects of Biochemistry*, vol. 6 (New York: Interscience Publishers, 1946).
105 This is a sundial on the east wall of the first court of Christ's College on the second-floor level. E.W. Hobson and Norman McLean were Fellows and therefore had the right to walk on the sacred grass of that court. There was a cartoon in *Punch* by H.M. Brock, with a World War One cadet saying, in reference to Hobson and McLean: "They think they own the place;" which Murray remembers "they did in effect." Norman McLean succeeded A.E. Shipley as Master of

The alterations in the old classroom region are not quite complete (shelving, cupboards, sinks & locks on doors are still wanting) but the rooms are in use. It is a great improvement & looks well. Course 3 has 18 students, the medical courses 1 & 2 have 134, the dentist course (5) will have 40 (I hear) & we expect about 60 for elementary sciences course 4, the D.P.H. course has 5 students, the graduate nurses has 90 & there are others. We all have a lot to do.

... I am sure you find the split class a nuisance. Overcrowding is bound to cause deterioration of some of the teaching. We have had to cut out our demonstrations for lack of space, as you have. I think the veteran classes are good & for that reason I would like to be able to give them better teaching. I found them good to teach after the first war too.

I'm glad you are finding some time for your own work. I hope you will not allow yourself to be submerged in the teaching & hospital routine. It is extremely important for you to publish good research. I wish you could make the opportunity for a talk with Jim Craigie. It would be a good thing for you to present your test at the CPHA Lab Section. It does not confine you to publish in the CPHA journal, but it brings you in contact with other bacteriologists in Canada appropriately. I asked Don (Fleming) to write to you.

Thanks for the references. I must find time to do some back number reading. There is a great deal I have not kept up with. ...

Bill Newman and I had a pleasant trip ... We got 6/7 fish to take out, which was much more than others got. There was not the usual the fishing crowd, only Tom Williams, Hilary & Tom Robertson, Bill Newman etc. Very quiet & pleasant but not profitable for the club. The club is not in good shape & something drastic must be done to put it right – it is not being seen right & Adelard is not working for its good. I suspect he is plotting a scheme to acquire territory for his own purposes.

... I have to go to New York at the end of this week. Tom Rivers[106] & Rene Dubos are planning a book & they want me to contribute to it. I'll let you know more of it when I come back. ...

  Love to you all,
   Dad

London, Ont.
6 Oct 46

My dear old Dad:
 I am looking forward to hearing from you how the fishing was at the Columbus this

---

the College in 1927 and held the post until 1936. RGEM.
106 See chapter 1, note 102 for information on the American virologist Thomas M. Rivers.

year. They should have been running-in fairly well with the short spell of cold weather we had in the middle of September. Mum said in her letter that the Owens boys got more than a little sport up at Barbot – it is a pity you could not have given that a try.

... Do you happen to know of any books or collected papers on diseases of lab animals? I find I am very ignorant on that score & want to know something about it. I think it is very important. I have lost one of my precipitin rabbits – he had a gradually increasing flaccid paralysis starting with the hind limbs. Nothing to be found on P.M. [Post-Mortem] and I did not want to do any animal inoculations in case I started something. Many of the animals we get seem to be of uncertain origin & from very amateur breeders, I have since heard that two other people lost some rabbits in similar manner. There have been no more cases in the last 3 weeks. The new animal house is not ready yet & unlikely to be so until the New Year. It would be a nice place if they had only made it bigger or given the proper proportion of space between Physiology cats & dogs and our guinea pigs & rabbits. Too many functions to be served!

Teaching is going along very well. The larger class & the use of two class rooms certainly increases the work out of proportion to numbers. However, we are able to do pretty well and have given some pretty good classes. They are a keen lot but with a fairly usual proportion of slow thinkers!

I am afraid the seminars the staff employed so much last year have been changed in character. There are now 24 graduate students and Dean Hall wants them all to get a chance to present later. That could have been arranged without sacrifice but M.L. Barr[107] has taken over from Stavraky (who is very heavily burdened with teaching) and he never quite agreed with the rest of us on the function of those hours. We will have to leave it be for this year and hope to regroup our forces next year.

I don't remember if I told you that two new professors have been appointed: one to replace Macallum[108] in biochemistry and the other to take over physiology from Miller. Both sound to be good men & both from England. The biochemist (name is

---

107 Murray Barr (1908-1995) taught in the areas of histology and neuroanatomy at the University of Western Ontario. Barr worked with Robert Murray to secure the funding from the Bickle Foundation for Western's first electron microscope. He published the popular textbook *The Human Nervous System: An Anatomical Viewpoint*. In 1968, he became an Officer of the Order of Canada. R.G.E. Murray, interview, 17 June and 24 June 2005; Canadian Medical Hall of Fame, http://www.virtualmuseum.ca/Exhibitions/Medicentre/en/barr_print.htm; Paul Potter and Hubert Soltan, "Murray Llewellyn Barr, O.C., 20 June 1908 – 4 May 1995," *Biographical Memoirs of Fellows of the Royal Society* 43 (November 1997): 32-46.
108 Bruce A. Macallum was appointed in 1924 as the first professor of biochemistry (with department status) at the University of Western Ontario. He served as dean of the Faculty of Medicine from 1927 to 1934. Meanwhile, he continued as head of the Department of Biochemistry until 1947, when he became a research professor until retiring in 1954. *A Short History of Biochemistry at the Western Medical School*, http://www.biochem.uwo.ca/history.html.

Rossiter, I think)[109] is at present lecturer in biochemistry at Oxford. The other whose name I have forgotten, is physiologist at the post-graduate school at Hammersmith. Both good posts & Hall must have offered them good opportunities to come here. Coming from those places those fellows should offer some opportunity to change teaching methods from the usual stereotyped courses of N. America!

It seems to me that two things are needed to put this medical school higher on the map. One is more room for graduate research students and the other is the right to grant Ph.D. to attract some of them.

We are all of us pretty busy and have not had very much time or opportunity to do much. We have been lucky in getting tickets to see the productions by the London Little Theatre – an excellent local amateur repertory. No doubt there will be a few university functions as well.

I hope all goes well with you. This year you should be in good physical shape after being able to take adequate holidays.

All my love to you both.

Your son,

Bob

I enjoyed the commemoration number of the Barts Journal. I agree that the description of the ceremonies were unnecessarily brief. Do you want it back? I expect you do.

Faculty of Medicine
Department of Bacteriology and Immunology
Ottaway Ave. & Waterloo St.
8 October 1946

My dear old Dad:

I was very glad to get your letter this morning. It makes me think of a lot of things

---

109 The biochemist Roger Rossiter (1913-1976) was born at Glenelg near Adelaide, Australia. He attended Thornburgh and Wesley Colleges and the University of Western Australia, and upon receiving a Rhodes scholarship in 1935, he moved to Merton College, Oxford. The University of Western Ontario recruited Rossiter from Oxford to head its biochemistry department, a post he held from 1947 to 1965. His research focused on neurochemistry and the degeneration of nerves, and he established a neurochemistry study group very early on in his time at Western. He played major roles in establishing the National Cancer Institute and the Medical Research Council, and was the founding Chairman of the International Society of Neurochemistry. Robert Murray and Rossiter became close friends and shared a common interest in camping and fishing. RGEM; W.C. McMurray, "Roger J. Rossiter," *Journal of Neurochemistry* 27, no. 4 (1976): 827–828; "Rossiter, Roger James (1913-1976)," *Australian Dictionary of Biography*, online ed.

that I want to tell you. It sounds as though you had a pleasant trip to the Columbus – in such company it would be so, and I wish that I could have been there for I enjoy all those people. It is curious that the members are taking so little interest in the club – it cannot be purely that people are too busy, a state almost physically impossible where fishing is concerned. ...

I am very sorry to hear that Norman Williamson[110] is so sick and with such serious implications. His doughty spirit will be much missed.

When I heard about the CPHA meeting I was going to write to Don [Fleming], then in quick succession I got a note from Don and a mention from you. I have written to Don saying that I would like to present a paper on the test. I have been doing work on aspects of it all the time when I have been able to fit it into the days work – unfortunately not the best way to have it working at peak performance but good enough in the circumstances. About the only free time for such things that I get is between 4:30 and 6 in the evening and it is sometimes hard to feel like working. However I have worked out a fair approximate of a good medium and have determined most of the optimum conditions...

I am arranging to see Craigie some Monday in Toronto before he goes and I am looking forward to it. I would certainly like to hear his views on the subject. One trouble for me is that some of the problems presented are not suitable for solution here for space reasons. I don't think that any reasonable scale immunologic experiment could be done here until after next year after looking at the animal house and its facilities. Also there will need to be some animal man who can be trusted to feed the proper diets and give one reasonable assurance that the animals will live as long as they should. I am not very happy about the animal situation.

There are plenty of other things to work on. I am toying with the idea of doing some work with the Corynebacteria.[111] As a group they seem to be in a hell of a mess – and those that we continually find in disease processes deserve much more attention than they have had. I am beginning to collect them as they come up.

The labs and lectures keep me busy and I have to do a lot of reading all the time. A terrific amount of the classical work to say nothing to lots of contemporary stuff seems to have escaped me.

The classes you have to teach are horribly large and I don't see how the university can have the brass to undertake to give opportunities to such large numbers of students. The added keenness of the students does not make the matter any easier for it only impresses you with how little you can give them. It must be quite encouraging

---

110 Norman Williamson was a lecturer in the Department of Pathology at McGill University and taught Robert Murray during his Part II course. RGEM.
111 Examples of Corynebacteria include the diphtheria bacillus and other animal and plant pathogens and parasites. *OED* (Online).

to see that the expansion of the department is being carried out in some measure at last. I am really looking forward to seeing the place in December.

Your loving son,
Bob

London, Ont
17 October 1946

My dear old Dad:

It seems to me that I have not written for some time – although nothing very special has been going on there doesn't seem to have been much time left out of the day for relaxing in writing. Just recently the lecture & lab preparations have kept me busy. Things seem to be going along alright – but I feel that I could work a lot of improvement upon lectures and lab procedures. These things take a lot of thinking about. Rather foolishly, perhaps, I got myself roped in to give a talk to the "noon-day Study Club" – a group of doctors of the town who meet once a week throughout the winter (they have an interesting bye-law that insists all officers & speakers of the club should be under 40 yrs of age.[112] Dr. Seaborn says that this should really be interpreted as "apparent age" – he is an old, but very bright, codger). I had a free day last Wednesday & offered to do it then. It was a very pleasant session (after it was all over!) and they seemed to enjoy my talk. It is hard to get some angle of approach that might be new to them and condense it so that it will balance in 25 mins. I think I made a successful stab at telling them something about bacterial enzymes and how knowledge of them affects concepts of disease. It certainly helps to fix ones reading to have to pass it between the teeth.

We had the first of the winter series of graduate seminars this week. It was about the work that Stavraky has been doing (with Drake,[113] one of the MSc students) on the excitability of neurons (partially) isolated from the influence of more central neurons. He has shown some very interesting things especially in the line of an almost non-specific motor response of epileptic quality to such things as Adrenalin and Acetylcholine, metrazol and other convulsants as well as to some other non-specific things as cations.[114] It is extremely interesting that the effect on decerebrate cats is almost exactly the sequence of events in generalized tetanus though, of course, it is of

---

112 Robert Murray remembers giving a talk on measles and its complications to this assembled group of practising physicians at the YMCA building in downtown London, Ontario. His long-time friend, Douglas Bocking, persuaded him to give the talk. Bocking was a practising physician and secretary for the group, and a future dean of medicine at UWO. RGEM.

113 Charles G. Drake (1920-1998) later became a world-famous neurosurgeon at the London University Hospital, UWO; he was known for his work on aneurysms.

114 Cations are the positively charged ions of metallic elements and dissociate from salts of those elements. RGEM.

a temporary nature with the dose of these drugs that were used. We were wondering whether it would be worthwhile to try Tetanus toxin and see whether this method of study would give any information as to site of action of the toxin. In view of the effect of some non-specific substances it would be better to use one of the partially purified preparations – would this be obtainable or would we have to make it ourselves for a trial? One pest in this would be the variable incubation time (if there is one under these conditions) and the necessity of round the clock observations on a complex animal preparation.

By the way did I tell you an odd thing about Tetanus toxin that I found in the literature. In all the books it is said that Tetanus toxin is ineffective in cold blooded animals. Grassett (in South Africa) tried it in reptiles & repeated that part; then he showed that if you incubated the animal a typical picture of tetanus action is obtained. ...

We have not been out very much. We were lucky in being able to get season tickets to the London Little Theatre (Local Amateur Repertory) and last night we went out to see the first play of the season. I think it was one you saw in London – called "When the Sun Shines". An amusing "drawing-room" comedy and very well done. As a matter of fact I think the cast would have got high praise in a big city for its job, and there was little one could criticize, or want to. They are supposed to be very good & they have listed some good plays for this season's performances.

Well, I have some work to do before I get to bed & must end now.

    Your loving son,

        Bob

P.S. Have not heard from Sue for a while. Is she alright?

P.S.II I nearly forgot to thank you for the windproof lighter that arrived yesterday. It works very well and I am very proud of it.

Montreal
22 October 1946

Dear old Son,

I have two letters from you to answer. I should have written before but last week I was not very well. I had to go to New York on Oct 11th to a meeting of those who have been invited to contribute to a new book on Medical Bacteriology to be edited by Rene Dubos & financed by the National Foundation for Infantile Paralysis. It is to be a sister volume to one on virus diseases of man, edited by Tom Rivers. I have been asked to be responsible for two chapters (1) Historical & (2) the cultivation & identification of pathogenic bacteria. Neither an easy assignment.

... I'm glad you are giving a paper at the CPHA Lab Section. It sounds interesting. Mum is delighted at the chance of seeing you & only wishes Doris & the babies were coming too. You are doing well to get some research done while you have so much else to do. It is very important to get some good work published; it need not be a big problem ... pertinent work can use a simple approach without losing any importance.

Your animal room problem is a difficulty hard to overcome; also animals are hard to get & very expensive now. Also, the problem itself is a large order.

There is great need for some good work on Corynebacteria. I think I have some interesting specimens collected & if they are still alive I would send them to you. The present classification of them is not satisfactory.

... We are making up the programme for the CPHA Lab Section. There are a lot of interesting papers offered & I think we cannot avoid a double session on at least part of one day.

The hyperplastic spot on Mum's leg proves to be an area of hyperkeratosis with extensions of the papillae & "round cell" accumulations. There is no sign of malignancy. The dermatologists may have a name for it.

... The day seems to slip by with little to show for it although it is taken up with things it is essential to get done in order to keep the department going. At the same time there seems to be so much still to be done.

I'm sure it was a good thing for you to give a talk to the "Noon-day Study Club". I would have liked to have heard you; I'm sure it was interesting. I think you would probably have to make your own tetanus toxin. Some work done in the war on botulinus toxin might be used for purification of tetanus toxin. However, you might get a lead with a potent crude toxin. When you & Reed are here together it might be possible to talk it over.

It's good to know you & Doris are having some recreation by going to the London Little Theatre & that the players are so good. There is a similar group in Ottawa & they are very good too. There is often remarkable talent in such amateur groups & many little towns in Ontario seem to have them. Mrs. G.B. Reed is a very enthusiastic promoter of those around Kingston.

Miss Cox gets married tomorrow afternoon. We had quite a party at the lab last week to give her a present & it went off very well. She was delighted with it & the present, a very nice silver plate teapot, sugar-basin & cream-jug. Mum & I will go to the wedding & I suppose many from the lab will go too.

I must work up tomorrows 9am lecture a bit.

  All love to all of you,
    Dad

London, Ont., 24 Oct 46

My dear old Dad:

It seems an awfully long time since I have heard from you – but it is probably mutual for I have not had very much time for writing myself. I have been doing a great deal of teaching lately and also there seem to have been rather a lot of diagnostic problems in the last couple of weeks. This has been a nuisance for it has interfered badly with some work I have been doing – however, I hope to get back into the work again this week-end.

There must be something funny about this town – today we confirmed the 4th Friedlander pneumonia for this year. Seems a lot for a comparatively small hospital but, perhaps, having a big TB sanatorium nearby probably increases the possibility of cases being sent for investigation. In this regard it is annoying that Lederle[115] is no longer making Friedlander typing sera. ... I suppose this is a late effect of the antibiotics & sulfonamides! For class work in future years, if this practice spreads, we may have to make our own – which is not necessarily simple.

Teaching goes on much as last year. There are some things that need change & modification but I don't see how we can effect them until the classes get back to normal size & shape. Though this seems to be quite a decent class I am not quite satisfied that they are getting good enough foundation and I have not been able to put my finger on the trouble.

I am very glad that "Henry V" is coming here next week and we are very much looking forward to seeing it. From what I hear & read it is really a magnificent production & done in a manner quite novel compared to usual movie practice. I hope it makes Hollywood & some of the other studios feel that it is worth putting out the odd film that is not a pot-boiler.

We have had a McGill man named Brown[116] working in radiology for some time. I discovered only the other day that he is the son of Prof Brown of engineering at McGill. Discovered it by accident when talking with him about the Laurentians & the region of Morin Heights. He is a very nice fellow.

All of us are well – despite lots of hard work. Peter is progressing very nicely and he is quite a young person. Alice is her usual self and in perpetual motion from dawn to dark. Her words are getting better but it is not quite strung together into sentences.

Must away, but I will write again on the week-end.

    Your loving son,
    Bob

---

115 Lederle Laboratories was a major pharmaceutical company located at this time at Pearl River, NY.
116 Lewis identifies N.N. Brown as a radiology resident at Royal Victoria Hospital starting in 1946. Lewis, *Royal Victoria Hospital*.

London, Ont.
27 Oct 46

My dear Mum & Dad:

This is the most extraordinary nice fall weather I have seen. Today might just as well be the very nicest kind of summer day except for the falling leaves. Although London has plenty of trees the colours here were not as good as I had hoped – very pretty but most of them the yellows and russets with the occasional scarlet. I expect that the colours up north were good for the leaves have stayed on the trees rather longer than usual. There has not been the usual rain and high wind to strip the trees.

The other night they had Hamilton Bailey[117] – surgeon and writer – give a special lecture in the medical school. Instead of the usual form of lecture he had Kodachrome slides of a series of cases seen in the London Hospital outpatients and more or less carried out a facsimilie of an English outpatient clinic which was, I think, most instructive to all. I was invited to go out to Dr. Ramsay's house afterwards with some others of the staff and we had a great talk. He is a keen fisherman and it is a pity he came when the season for trout is over. He says that the English fishing is in a poor state. My goodness, they are having a hard time over there. He said also that TB is getting much too common – largely due to the closing of most sanatoria – that tuberculous adenitis is a common thing in clinics; and that the vast majority of cases are due to "hominis" (80%+ in Kent) – which is a sad commentary.

28. 46.

Dean Hall certainly can get things when he goes after them. He has obtained a grant to set up a unit for fundamental research on cell physiology – he got it from the cancer research fund and there is no mention of cancer or anything resembling it in the project! I think this is a bit of a record. His idea is to get various people together with different lines of training and get them to attack cellular problems from a fresh angle. Very praiseworthy, I think, because up to now the various attacks that have been made have been along totally stereotyped research paths. Parts of the plan are a bit ambitious, I feel, but there should be no difficulty in finding something new & may be good in the course of such work. I have an idea, that I might play around with sometime, that there may be a lot of information obtained by working out the characteristics of various molecules foreign bodies on cells – I have seen many times in Ash's work that many substances have a stimulating action in a certain narrow zone

---

117 Hamilton Bailey was one of the most influential British medical personalities of the first half of the twentieth century, and the author of three very influential medical textbooks, including *Demonstrations of Physical Signs in Clinical Surgery* (1927), *Emergency Surgery* (1930), and *A Short Practice of Surgery* (1932).

of concentration – I feel this may be a general rule. ...

I am not sure that anyone has really investigated many aspects of this – though there has been something related to it such as *Harper's* macromolecular theory in Atherosclerosis. I noted with some interest in a recent review (JAMA) on streptomycin that in one set of invivo-experiments a similar thing was noted – using a (mouse) typhoid protection test they got them in the controls. Another argument against homeopathic doses. Analogous too are the effects of toxins on the RE system. There may be something in the suspicion of some that there is a relationship between toxicity and antigenicity. (All this aside from the rather exhaustive work on the carcinogenetic hydrocarbons).

... Organizing the C.P.H.A. meeting must be quite a chore – since there has been no meeting for quite a time I suppose there are many people who have things to offer. It is a pity that you may have to run a two ring circus.

We are very well and getting out & about as much as we can. I am glad to say that Doris has been able to have a little more fun than she has had time for most of this year, and she is beginning to find a number of congenial friends. We both went out on Sunday night when a brave friend put on a dinner for 10! And very successful it was too. On Friday night we are going to Henry V and next week to the Little Theatre. We feel quite gay though we are not likely to kick over the traces until they legalize drinking in Ontario!

Alice & Peter are disgustingly healthy. Peter is growing in stature as well as cubic & we feel that he may be a throwback to the very tall Murrays. ...

... I have enjoyed the Barts Journal – the number will be sent back to you. The observations on emphasis to medical teaching were very pertinent. The Countrymen came yesterday – I have never read it before and I am surprised at the astonishingly large collection of material it contains. Rather good!

    Your loving son,
    Bob

Proposed Presentation of A Copy of His Portrait in Oils by Captain Oswald H. J. Birley, M.C., R.O.I.

To Professor Henry Roy Dean[118]
M.A., D.M. Exon., M.A., M.D. Cantab., Hon. LL.D.
Western Reserve University, D.Sc. Liverpool

Doctor Dean, previously Professor of Pathology in the Universities of Sheffield and Manchester, became Professor of Pathology in the University of Cambridge in 1922 and from that time one of his chief concerns has been the welfare of medical students, especially those who have passed through his department. In 1924 Pathology became a subject for Part II of the Natural Sciences Tripos and those students who have taken that subject owe much to the personal interest which Professor Dean has taken in them. Many who were his pupils at Sheffield, Manchester and Cambridge have become his friends and colleagues, and now it well becomes his pupils and friends to show their appreciation of the lively interest he has taken in them.

The members of Trinity Hall, of which he is Master, have recently had his portrait painted by Captain Oswald Birley, and some of his friends, who are not members of Trinity Hall, think that it would give him much pleasure if a copy of the portrait could be made and presented to him. The Fellows of Trinity Hall have given permission for a copy of the portrait to be made and Captain Birley has recommended an artist to copy it. The pleasure of receiving such a present would be greatly enhanced if the Professor knew that past Part II students had contributed to the reproduction of his portrait, and it is certain that they would wish to do so; accordingly it is felt that all past Part II students should have the opportunity of contributing.

The cost of the gift is estimated at about L300, allowing for contingencies, and if the subscriptions exceed the amount required, it is suggested that the surplus shall be used by the Library of the Department of Pathology.

Subscriptions (not generally exceeding one guinea) may be sent to Dr Raymond Williamson, Department of Pathology, Tennis Court Road, Cambridge. Cheques should be crossed "a/c H. R. Dean, Presentation Fund". The names of subscribers will be printed and circulated with details of the arrangements for the presentation.

October 1946.

---

118 See Introduction, note 178 and chapter 1, note 25.

Montreal
3 November 1946

Dear old Son,

We got your letter of 27 Oct. Yes this is a wonderfully mild fall & we wish we could have made use of it. Various things, such as lectures, the meeting in New York, sudden emergencies & things to do with Sue's housing difficulties, conspired to prevent us taking weekends at the lake. The place has to be put to bed for the winter so we must get up there soon. Mother is going up on Wednesday when I have to be in Ottawa & I hope to go up on Friday night & return on Sunday as I have to lecture on Monday. I suppose it will be our last trip until spring. Delia Lichte[119] says the colours were wonderful.

... Hall is a great fellow & the cell physiology project has great possibilities. I have long maintained that physiologists have neglected intracellular functions & so have the pathologists. I believe too much stress has been placed on nuclear histology & the more important cytoplasmic functions have been neglected. To my mind the future of pathology lies in the study of functional pathology (derangement of physiological function) & that largely intracellular.

The last number of "American Scientist" (Vol 34, No 4, Oct 1946) will interest you; it has a paper by Michaelis on "Oxidation & Respiration" & one on "Complement Functions".

I have no doubt that there is a definite zoning of a variety of substances in relation to biological activity. I found it so in relation to media for virulence of meningo, Fred Smith showed it for streplolysin, Strean did for growth of lactobacillus from the mouth. Immunity reactions are zoned in the body as well as *in vitro* as Horsfall[120] showed with pneumo. When making meningo serum I had many samples which in high concentration or in any concentration gave a higher mortality than normal serum or no serum; this I believe still was due to bacteriolysis in the absence of sufficient or any antitoxin. I have seen what I think to have been similar action in patients.

The idea of transferring certain (or all) Corynebacteria to propionibacterium seems to me no more than shifting the emphasis on criteria; as [in] this case giving

---

119 Murray remembers Delia Lichte as "a lively, interesting and helpful summer resident at Bark Lake" who along with her husband "did good work for the Bark Lake Protective Association." RGEM.

120 Frank L. Horsfall (1906-1971) was a microbiologist at the Rockefeller Hospital from 1934 to 1960. Subsequently he served as director of the Sloan-Kettering Cancer Research Center. He devoted much research time to the study of viral diseases. Although he was born in Oregon and received his BA at the University of Washington, he attended medical school at McGill University, graduating in 1932. He also spent a resident year at the Royal Victoria Hospital in Montreal. "Finding Aid to the Frank Lappin Horsfall Papers, 1940-1971," United States National Library of Medicine, National Institutes of Health, http://www.nlm.nih.gov/hmd/manuscripts/ead/horsfall.html.

prime weight to a difficult & impractical criterion, the detection of production of propionic acid among other fatty acids ... New criteria are badly needed in many groups. However it is necessary to have criteria which are not too recondite; they must be within the scope of reasonable laboratory conditions.

The programme for the CPHA meeting is "oversubscribed", we have to refuse several papers & that in spite of introducing a special session. There are a number of interesting papers & it should be a good meeting.

... It's fine to hear the babies are so well. We would love to see them & will have to arrange a trip as soon as we can.

I suppose your class comes to an end before too long. I hope you have not got to take on the so called "technicians" this year. You must be careful not to be harnessed too much, to an overload of teaching & hospital work; it is vitally important for you to have time to devote to research. The only thing that brings progress & recognition is research. I made a great mistake & have paid heavily for it & don't want you to repeat it.

... I would think Hall might be interested in the remarks on teaching in this Bart's Journal as there are one or two important points raised.

... Entin, who you will remember as a student, was in to see me the other day. He asked me to remember him to you. He is working with Baxter[121] & under a NRC research grant is doing an interesting bit of work on skin grafting.

Ian Stevenson's brother is a student in our present medical class. He is a nice fellow, much taller I think than Ian who he says is doing well. ...

 Love to you all,
  Dad

             London, Ont.
             3 Nov 46

My dear old Dad:

Time certainly flies by and it is very little over a month before I will be seeing you. ... This odd pre-winter lull still continues and we have some quite warm though wet weather. ... It is rather putting off the evil day so I suppose that we will get a rude and cold shock one of these days.

---

121 Martin Entin was doing research under Dr. "Happy" Baxter as part of his work towards a fellowship in the Royal College of Surgeons. He was involved in tissue culture research, and Murray consulted him about his techniques and his availability because Dean Hall was interested in possibly finding someone to initiate a tissue culture laboratory. Eventually an authority in the area, Raymond Parker at the University of Toronto, was consulted. Martin Entin became a professor of surgery at McGill and chief of surgery at the Royal Victoria Hospital and "did good things for that institution." He became a good friend of the Murray family and Robert often met him on his visits to Montreal. RGEM.

... Doris is managing to do a little socializing – for which I am very glad though it does not ease her work atall. She has been to a couple of the university women's bunfights and is taking a hand in something to do with the student's wives. A nucleus of very pleasant friends is forming and she will have some opportunity for a cheerful time. We have had some people in odd evenings which has been pleasant. Jim and Louise Patterson were in last night. They are great fun. They have moved into their house though it is not yet completed. There are no stairs and they have to use a ladder affair, but they have the furnace and a bed and a few chairs so they are living better than some poor souls. ...

... We have become tied up in the help problem again. One of technicians has left us (Louise Butler) and right in the middle of the busiest time for preparations. We will be able to manage alright but it will mean a heavy burden for the others with all the media etc for the class as well as the extended work of the lab. Luckily she was about the least essential of the people we have. It means partly that Elizabeth Hall[122] will have to do more of the preparation for Ash and Miss Strelitz – and leave her less time for the very excellent work that she is doing. Ash is extremely pleased with her and says that she is about the best he has had. She certainly has a brain of her own and can use it, being capable of planning and doing her own experiments.

Dean Hall has had an inquiry from one *George Ling* who requests an opportunity to do an MSc in Bacteriology. You may have run across him. He is I believe doing Bacteriology in the Montreal Military Hospital and is still in the army. Elizabeth Hall knew him in the army and she tells me she is a little dubious about his tale. He claimed to have a BA from Cambridge and to have been in Medical classes at McGill – the reasons for his leaving them are a little obscure but rumour had it that he tried to practice before his time. I gather also that he is a peculiar racial mixture, being half Chinese and half Negro. From the type of letter that was written it would seem likely

---

122 Elizabeth Hall was a native of Windsor, Ontario. In the Second World War she worked for a short time in Guilford Reed's laboratory at Queen's University until she joined the Army and was posted as a laboratory technician. Robert Murray first met Hall at Camp Borden, where she was one of his bacteriological technicians. After the war, Murray next met Hall at Wolsey Barracks in London, Ontario, where both were being demobilized in late 1945, two months after Murray arrived to teach at UWO. When Asheshov needed another competent bacteriological technician for phage work, Murray suggested contacting Liz Hall and she accepted the offer. While doing the phage work, she registered as an MSc student, became engaged to Asheshov, and left with him in mid-1948 for the New York Botanical Gardens, where they worked together on a March of Dimes research project. She completed the MSc under Murray's supervision in 1950. The March of Dimes project ended in 1953 and the Asheshovs, who by then were married, moved to London, England, where Asheshov continued his work at the Lister Institute, while Elizabeth began and completed a PhD on staphylococcal bacteriophages at the Central Public Health Laboratories. Eventually she headed a section on staphylococci and the typing of strains in the CPHL. RGEM.

that he has more or less circularized various universities and may have applied to you. If you have, or could get some information on him, would you let me know as soon as possible. Even if he has good training or ability he may still be undesirable.

Work goes on as usual but is fairly strenuous. I certainly wish I had a little more time to put a polish on the Mucoid Strep question. Though I am a little encouraged by it I still feel that there may be some pitfall in it – it still suffers from being a slightly complex test as are all the others for the enzyme. ...

We had a very nice long letter from Sue the other day. She sounds very cheerful despite the adversities of life in Ottawa. Their "new" car sounds like a vintage bone-shaker guaranteed to keep you amused for short journeys only!

All our love to you both
Your son,
Bob

> Chateau Laurier
> Ottawa, Ont.
> 5 November 1946

Dear Bob & Doris,

I am here for a meeting & had to disarrange things a bit to do it & I'm not sure it is worth while. I go home tomorrow 4:10pm.

Last night Blake called Sue by phone, she was in Montreal & he here, to tell her he had lost his job. She was frightfully upset & wept bitterly. It was a real shock to her & there was not much Mum & I could do. However she decided on action & calmed down then; she came here with me this morning & Blake met us. He, poor fellow, was very depressed.

... They both dined with me here tonight & feel much better. They are planning out what is to be done now. It is interesting what a good head our Sue has & fine pluck too. Blake is going to look round here to see what there might be doing & they are getting ready to move to Montreal. They had a hard time in a foul lodging, rat-ridden, dirty & uncongenial people to deal with. No doubt they have learned a lot but they do not deserve such a buffet.

I met Sir Astley Cooper, Governor of the Hudson Bay Co, in G.D.W. Cameron's office this afternoon. An interesting fellow in several ways ...[123]

Love to you all,
Dad

---

[123] See chapter 1, note 91.

London, Ont.
7 Nov 46

My dear old Dad:

I got your letter today and I am very shocked at the news of Blake's losing his job. It will be quite a blow to them, poor people, but if a firm permits that sort of thing it is not really a good place to work. It is hard experience but I am sure he will have gained something from having learnt about that aspect of work and it may put him in a position to get an even better job. I certainly hope so. ...

You seem to be having a lot of traveling about to do these days. It must take quite a cut out of your time. I certainly hope that things will get suitably settled in the lab so that you will be able to have some time for your own work and interest so that you have to spend less time looking after the interests of others. ...

Ash had a very interesting letter from Dr N.A. Bulgakov who is at the Laboratoire du Bacteriophage in Paris. They are having a hell of a time with supplies. He has apparently done a great deal of classical work on phage and with Sertic[124] has headed that lab for some time. He is apparently looking for a job outside of France and would be interested in Canada. He is in his forties Ash says, and is a very good research worker. You might like his address for the record, it is: 75, rue Olivier de Serres, Paris 15.

I have had quite a busy time the last week or so but I have managed to fit in some of my work as well by dint of doing long hours. I have found out some interesting but useless things! As well as the usual lectures and labs I was invited to give a talk to the interns on Wednesday evening. It was rather fun and good from the point of view of academic evangelism. I gave a fairly short talk on applications and we had a discussion, and a good one, that lasted about two hours. They are a pretty keen bunch. I was glad to notice that it has borne a little fruit already.

I will have to get this paper ready for the meeting soon. I don't know what the usual habit of the CPHA meeting may be. Is it a ten or fifteen minute paper? I think that I can make it fairly interesting. I am still a bit stymied by difficulty in getting a photograph of the effect – we did some more trials today.

It is very hard to fit in the work that interests you when there is a lot of preparation and teaching to do. It seems to me that you can only do it by extending hours and that can be carried a lot to far for comfort or well-being.

I am afraid I will have to give up the chance of seeing Craigie before he goes. I can't really stand the expense of a special trip to Toronto at the moment and also I have not really got the time at the moment. I wrote to him and sent him a copy of

---

124 Nikolai Bulgakov (1898-1966) and Dr. V. Sertic were old friends of Asheshov. Ash had worked for years on bacteriophage with Sertic in Dubrovnik, Croatia, and they published papers together on phage biology supported by the Rockefeller Foundation. RGEM.

the seminar that I wrote with some amendments, but I have not heard from him. It is a pity for I know that he could give me a lot of good sound advice. Anyway I think I am going to keep most of that in the back of my mind at the moment until things are clear enough to begin any sort of immunological experiment. I will have to go to Toronto later on in the month, but too late for Craigie, but my expense will be paid – I have been asked to go up with Murray Barr and (Jim) Hamilton[125] to look at the tissue culture rig they have there, see Parker, and see what can be done about setting up a smaller set-up here.[126] Should be interesting.

... Do you remember any Toxin that has a selective effect on vascular endothelium?[127] Does Staph Toxin have such an effect as well as all the others *in minor dosage* (or over long periods of time).

I am still thinking very seriously of attacking the Corynebacteria. ...

It is getting late and I had better get to bed.

Love to you all.

Your son,

Bob

Royal Victoria Hospital
Montreal
17 November 1946

Dear Old Son,

... Last weekend Mum & I went to the lake (after I returned from Ottawa) & put things to bed for the winter. I cut down a birch (medium size) & cut it up into firewood, both for stove & fire place. The weather was a bit damp but we got done most of what we wanted to; though not all the gardening we would have liked. We had a couple of short paddles in the canoe, but most of our time was taken up by work round the camp. The days were short, it only became light about 7am & was dark about 5pm. It was not cold but the lake was too cold for bathing. I don't suppose we will have much chance to get up to the lake again until spring.

It is good to know that both of you & the babies are well; also that Doris is getting

---

125 Jim Hamilton was a research fellow under Alan Burton, professor of biophysics at UWO. RGEM.

126 Raymond Parker was at the Connaught Medical Research Laboratories at the University of Toronto. Barr, Hamilton, and Murray made the trip to consult with Parker about the requirements for establishing a tissue culture laboratory. RGEM.

127 "Endothelium" is "the layer of cells lining a blood-vessel or serous cavity" and lies between the circulating blood and the rest of the blood vessel. It is involved in different aspects of vascular biology, including playing a role in the control of blood pressure, blood clotting, and the generation of new blood vessels. *OED* (Online).

out with friends & having some fun. It is good for both of you to enjoy your friends. We are greatly looking forward to seeing you all together as soon as we can. Meanwhile it will be great to see you on Dec 15th for as long as you can arrange to stay. Mum is laying in beer for you, though none has a name "to be on every infants tongue".

John MacIntosh (Secretary of the Faculty) has written to you about George Ling. He was a medical student 1940-43 & was turned out by the Dean because he was irregular in attending classes being much occupied with his own social affairs & amusements. He seems to have been very unstable & very easily imposed upon by undesirable acquaintances. He did well enough in his work that he was as good or better than many who get their degree. Nothing brilliant, I saw no indication in his record of a Cambridge B.A. He is well spoken of by those he served under in the army, but I do not know what that is worth – I have no reliance in the army laboratories. Whether or not Ling has learned sensible behaviour in the army & acquired some judgment of men, I do not know. He only got a C in the medical bacteriology examinations at McGill; no indication of striking ability on the subject.

I feel the time is well past when the laboratories need take cast offs & misfits. Bacteriology has suffered greatly in Canada by the taking of inferior people into the laboratory & the results of such policy in the army are clearly indicated by the lack of any work worth while. I would not take Ling myself; though I think he might very reasonably be readmitted to medical school at McGill – worse have been given the degree.

I enclose a cutting about Wansbrough[128] which will interest you. Since being got rid of by the Governors of L.C.C. he has done a fine job & done very well for himself. He must be quite prosperous & I have met him from time to time in various places to find him always the same good fellow.

Blake is still hunting a job. I am not knowledgeable on business methods & opportunities so I cannot give him any worthwhile advice. However I urge him to get advice from friends of his father & some of our friends to find out what sort of line to go for, in order to be in line for a reasonable chance of advancement.

The work on the extension of the lab is starting, with some amount of noise but no very obvious progress yet. However, starting work has some promise. The working of the lab is in pretty good order & the teaching is in a satisfactory state. There is a little difficulty brewing with technicians in our lab too! Unfortunately it involves a couple of useful men, one thinks he is worth more money (may be, but it needs proving) &

---

128 V.C. Wansbrough was the head of Lower Canada College in Robert's final year there. He subsequently embarked on a successful business career, serving at one stage as vice-president of the Canadian Metal Mining Association. The Hon. Edgar J. Benson, Minister of Finance from 1968 to 1972, consulted him during the process of tax reform he initiated in the late 1960s. RGEM.

the other is not pulling his weight. If they go it will be difficult to replace them, especially at this time of year. ...

I have not caught up with a number of things yet & there is a lot to do for Bergey's Manual & for the 2 chapters I have to write for the Medical Bacteriology book which I must get started.

Yes "Darkness Falls from the Air" is a good novel with very evident knowledge behind it. "The Little Back Room" is marvelous good & I can place people I know in place among the characters & parallel the circumstances.

There are a number of people I have heard of & some have written to me, who would like to have a job in Canada. There are jobs in Canada for bacteriologists & some of them good jobs & we have not candidates for them. How wise it is to import people who are dissatisfied because they have to face difficulties, presents a problem I don't know the answer to.

The CPHA Lab Section ought to be a good meeting as there are good papers & quite attractive circumstances. I hope there will be a good attendance. We are allowing 15 minutes for the presentation of papers. The trick in doing such a presentation is to give the essentials (leaving out historical review) emphasizing your own work & its implications. The idea is to promote a discussion & then you can put in some of what you did not have time for.

I'm sorry you cannot manage to see Jim Craigie before he goes. ... Parker is a good fellow & has done good work. What he is after is primary intracellular physiology & is most important. His work is worth watching. His sort of setup is very expensive & requires space.

We met Ted & Sylvia [Rose] in the train last weekend.[129] They got out at Morin Heights & spent the weekend walking. I gather they enjoyed it. Ted looked a bit worn out & in need of a rest. Mum had tea with Sylvia last Friday.

Yes Staph toxin has a direct action on capillary endothelium, as John Glynn[130] showed in his bit of work on it. ... I am a long way behind in my reading; the war set me back 5 years & I shall have difficulty in catching up, if ever I do so.

What sort of effect do you want to get? Haemorrhage? How do the Pasteurellas[131]

---

129 See chapter 3, note 50.
130 Identity not known. There was a John P. Glynn from the Laboratory of Chemical Pharmacology, National Cancer Institute, National Institutes of Health, Bethesda, MD, who contributed articles to the *Canadian Journal of Microbiology*. See, for example, vol. 7, no. 6 (December 1961).
131 "Pasteurella" is defined as "A genus of bacteria (family Pasterellaceae) comprising small, Gram-negative, usually aerobic, rod-shaped or coccoid forms which are parasites of mammals and birds and cause systemic and localized infections including fowl cholera and haemorrhagic septicaemia; (also pasteurella) a bacterium of this genus." *OED* (Online).

produce the haemorrhagic septicaemia effect?[132] I suspect there is a toxin involved. It is probably not easy to get out & quite likely is of the "endotoxin" type.

... We went to see "Meet the Navy" film yesterday afternoon. We found it a most satisfying show which deserved better photography. It was very good fun with a lot of real talent. There was a lot to provoke a good laugh too. Afterwards we had some first rate oysters at Desjardins.

... The Public Health Bacteriology (D.P.H) class has started. Six students (including Jock McKenzie who was Medical Superintendent of the M.G.H.) who seem to be quite keen. It is nearly as long a programme as Course 3 though not as detailed in many things but covers quite different ground because it is confined to public health interests. It competes with Toronto which is well established & with the new well financed "Institut d'Hygiène" of the U of Montreal (Govt. sponsored) & I don't know how it will fare in future years. I'm sure the teaching of bacteriology will stand comparison (but that is not everything) with the time allowed: some of what we wanted to do was pared down in time available.

It's getting late & I must give the pup a little run.

 Love to you all,
  Dad

       104 McClary Avenue
       London, Ont.
       20 November 1946

My dear old Dad:

Your grand, long letter arrived the other day and we both appreciated it very much. I also got the parcel containing the tump lines[133] and they are very good too – though you do not say how much they cost! I will grease them & treasure them. ...

I have become committed to a couple of activities in the New Year which should be quite fun. One is to give a radio talk on the immunization of children over the local station sometime in February. The other is to conduct a few evening sessions on how to cast the fly and plug. Dean Hall is very keen to foster hobbies among medical students on the excellent theory that hobbies are essential to the well being of any hard

---

132 "Septicæmia" is defined as "septic poisoning." *OED* (Online).
133 Tumplines, originally used by Aboriginal peoples, are used to make carrying heavy packs and canoes easier and less tiring. They are made with strap or rope with a wider (2-inch or so) piece of leather or cloth that is placed high on the forehead with the lines tied to the pack or the gunnel and centre thwart of the canoe. As a result, the weight bears straight down the spine instead of at an angle. The Murrays always used a tump to carry canoes and as Robert Murray remembers they were "a blessing on long portages." RGEM.

working person – not excepting medical students. I think these are interesting and they will not interfere with any work I may be doing to any appreciable extent.

Thank you for the information on Staph Toxin – I have looked up John Glynn's paper and possibly he may not have looked at sections of arteries in organs other than the kidneys or in larger arteries. Jim Patterson and I are going to try the effect of Staph Toxin in Cockerels – the basis for this work I can tell you better when I see you. Anyway, Jim has observed some funny lesions in the vessels of a large proportion of cockerels involved in one of his experiments and pretty well all these birds had acute & subacute Staphylococcal infections a while back. I have a strain of this bug and it seems to conform in all respects to the usual *Staph. pyogenes* though I have not gone into fine points on the toxin (I know it has α haemolysin). It also produces hyaluronidase & is coagulase +. These infections are not too uncommon in birds & I have seen it in the literature a few times referred to as "Staphylococcosis" – portal of entry seems to be the *feet* (which is very interesting) and not uncommonly the localization is in the large joints of the leg. Luckily for Jim he did not lose many birds in his epidemic possibly because of energetic treatment with penicillin. ...

I think we have got some decent enough photographs of the general effect of the enzyme on mucoid colonies – after a lot of trial & error in lighting positions – I have not got prints yet but the negatives look good. It would be easier if the whole change were matt – but unfortunately that effect is only given by the highest concentration of enzyme & I think is expression of almost total destruction of the capsule. The partial effect, expressed by flattening of the colonies but still with shiny surface, is much wider but luckily sharply demarcated.

Thanks very much for the observations on *George Ling*. I got a letter from McIntosh (I have thanked him) and it gave the same impression of intellectual capacity being possible but application to work being very variable. He has been refused (kindly) for MSc work here. I, too, feel that he might be worthy of readmission to medical studies. I agree entirely with the cast offs & misfits being refused laboratory opportunities – they are a waste of time.

I am glad the extension to the lab is about to materialize – it is good that there was not further delay. I look forward to seeing it. We are arranging to come down a little earlier – both Ash & I. We plan to arrive on the early morning train on the 13th – but we will have to leave on the night of the 17th. Ash has written to the Windsor Hotel for reservation – I hope the hotel situation is not as bad as it was.

Wansbrough seems to be doing very well & I am glad of it for he is a very good fellow indeed. A great loss in Education. As a matter of fact that announcement was in the local paper with a picture of him as well.

We got a nasty [note] from the Income Tax people the other day – a request for some $82 on account of Doris' earnings in 1944 – a most unreasonable state of affairs and though we have asked for detailed explanation I have no doubt we will have to pay it. It will not make our financial situation too happy, I'm afraid.

Alice is talking, to all intents and purposes, and repeats everything we say. She is gaining an enormous vocabulary of descriptive nouns but hardly has a verb to her name. ... Doris is well and thriving. She goes up to the library one evening a week with Louise Patterson & draws – they refuse Jim & I to accompany them on suspicion that we are interested in the model! God forbid!

I have much more to write but I have terrible difficulty in getting everything into the 24 hrs at the moment with teaching & trying to get the problem into good enough shape. It is late so good night – your son Bob.

P.S. ... Haven't heard from Mum in ages – she must be busy – Doris would appreciate hearing what she can do for Sue.

Montreal
Royal Victoria Hospital, Montreal
24 November 1946

Dear Old Son,

This has been a working week & there seems to be little news interest. There is a lot to do in niggling ways to keep the department running & various nascent activities from dying out prematurely. These little things take up a lot of time & have little to show for it, but without them a lot would go aground. I greatly look forward to your coming, & to Ash coming too, as there will be lots to talk over & I think you will be surprised to see the department regaining its feet after its wartime prostration. There is a total staff of 37 & one or two more yet to be found.

Your very welcome interesting letter of 20 Nov '46 arrived to our always intense pleasure. The rapid progress of the kiddies must be fascinating to watch. We wish so much we could see you all reasonably often. It is good that Doris is getting fun out of social activities & the drawing classes.

... Mum is very well & busy chiefly with family doings & preparations. She does a little on some McGill women's activities but is not very interested. We must arrange for her to visit you soon as it would delight her greatly.

... The investigation with staph toxin in fowls you & Jim Patterson plan to do sounds interesting. Jim might immunize his birds with toxoid to prevent being troubled when he wants them for other purposes.

Joan de Vries did Course 3 some years ago, then took her MD, married & now has returned to Montreal & is a lecturer in the department. I am getting her onto what seems an interesting possibility in cooperation with J.S.L. Browne & Co.,[134] tying in metabolic disturbances with immunity. I'll tell you about it when we meet.

The "roof extension" does not show much progress yet but there is quite a deal of hammering. We hope to see more positive progress soon.

The hotel situation here is bad – very bad. The CPHA has made an arrangement with the Windsor Hotel & if Ash's attempt has not been successful he had better wire the hotel. ... At worst we will do something, though the house will be pretty full. Anyway I hope we will see him on Dec 13.

Mum is getting up a party on the 14th Dec (Sat) & I hope Ash will accept an invitation to it. It will be my staff & some of your special friends.

I am sorry to hear of the income tax bomb shell. That department is a curse because it is so damned inefficient that it is over 2 years behindhand in its work. It caught me most unpleasantly, unfairly & unjustly for a considerable sum for two years in succession. For years (some 10 or so) they had allowed me certain exemptions, then without any notice they suddenly disallowed them & only notified me two years later. Not only that but they imposed interest for 2 years on each count. I took the matter step by step right up to the Minister in Ottawa; in effect injustice was admitted but it would have taken a Privy Council Order for my special benefit to rectify it. So in the end I had to pay, interest & all. It is an inefficient department & I wish my curses could take effect.

... I have not got my workroom into order yet. I may do some of it this evening. Once it is straight I can make use of stray moments. Ten minutes can mean a good fly & is worth while.

I am damned annoyed about the article on BW in Life.[135] It is unwarrantably misleading, makes unjustifiable claims, is quite wrong in places & in one very important matter tells a deliberate lie. Our Government directly forbade the release of very carefully prepared Canadian statements so Canada gets no credit for the leading & important part Canada played. The US is playing politics with the subject & besides the policy being wrong they have made a tactical blunder.

I have not got my own lab straight yet. It has been a sort of storeroom & is in a mess.

      Love to you all,
          Dad

---

134 J.S.L. Browne was one of R.G.E. Murray's instructors at McGill Medical School.
135 See *Life Magazine*, 18 November 1946.

London, Ont.
26 Nov 46.

My dear old Dad:

We got your letter today and one from Mum a few days ago and we are very happy to have them. These may be "working weeks" but they are none-the-less interesting. It has been pretty busy going for me, too. Since the practicals are serological at the moment there is a fair amount of checking & cross-checking to be done – to say nothing of the usual lectures. As well as that I am trying to get as far on as possible with the biological assay. In the latter I have two very annoying stumbling blocks – (1) A variation in the medium that throws out 2 sets of experiments. (2) The trials of statistical & mathematical analysis of curves. This latter is causing me more headache than anything else at the moment, in fact I have a very bad case of statistical indigestion, and I somehow doubt if I will have it licked before the meeting – though not essential it would be nice. Though I did get an introduction to this form of mathematics in 1938 I never got enough practice to fix it – so I thought the haywire figures I was getting on analysis for a linear curve were due to my ineptitude. Well two people who have some experience had a go & got just as extraordinary figures so I feel that they are making the same mistake – for you can't get good fitting linear curves by chance time after time. The photographs are disappointing too – the two better ones are not in sharp focus! Hells teeth. The greatest consolation is that I am learning a lot all the time even if I do not make very great progress.

I am certainly looking forward to seeing you – there are going to be so many things to talk about and see in a few short days that we will probably have to leave a lot undone. It will be very nice to have a party in the house again – like the old staff parties. Doris is terribly disappointed at not being able to come down. I'm afraid the exchequer couldn't stand it – the only reason I can do it is that the university will pay my way.

... It is terribly late and I think I am past seeing $E(XY)=aE(X) + b(EX^2)$ instead of sheep-

    Your son,
        Bob

P.S. Peter has just developed two teeth – but only just!

Montreal
Royal Victoria Hospital, Montreal
1 December 1946

Dear old Son,

We got your letter of 26 Nov '46 & are happy to have it with the news that all goes well. I'm sure you are having a busy time. Immunity classes always involve a great deal of work & the intricacies of the balanced interactions are so important in the applications to medicine & so seldom appreciated by clinicians.

The importance of the medium & the difficulty of making any two batches anywhere near alike has been a theme of mine for many years. For delicate work medium made by technicians cannot be relied on.

I confess that statistical methods are beyond me. I don't feel inclined to put absolute reliance on them, in any case, because they must leave out of account many important factors in the complicated systems we have to work with, because we cannot measure some of them accurately enough for mathematical treatment. ... The spurious accuracy of statistical methods are too much admired & people overlook the fact that you get no more out of them than you put in; the only advantage is that at times you can transpose information to a more useable form. It is useful & important to master the method to use it where it is valuable & to be able to fully appreciate the value of work done by others who insist on using it.

I am glad you are getting time to do some of your own work because it is so important for you to publish & to maintain the habit of research.

... My time is still frittered away by necessary adjustments, pushes & stimulations of people & things in the Department. It is hard to get other people to take responsibility reliably. I am eternally grateful to Gertrude; without her the burden would be dreadful.

We have plunged suddenly into winter this last week. Various small falls have today been capped by some 5 or 6 inches of snow & more is still falling. The forecast is 10 below tonight, the past week having ranged about 22 above.

This morning the pup & I went out to clean snow & shut ourselves out without the door-key. Much ringing of the bell eventually brought Mum out of bed; it looks like a good scheme as we got some breakfast (brunch) by 11am which we were in need of by then. The imputations laid on us by the rest of the family are not particularly complementary.

Mum & I went out last night to eat an oyster with our usual success. Dr. & Mrs. Cleghorn came into Desjardins looking for a seat & we invited them to join us. We had a very pleasant evening of bright conversation & friendliness. Cleghorn has joined the staff of the Allan Memorial Institute. We hope to see more of them in the future. He seems to have known you somewhere, I think at Borden, & enquired after you.

The new construction over the roof is not showing much obvious progress yet. ... I don't suppose we shall get into the new quarters before next September or October.

The arrangements for the staff party on Dec 14 are well under way & it should go well. There are a lot of nice people at the lab & with wives & husbands the lab people alone add up to over 40. There will be a few other friends too. Quite a job for Mum but she seems happy about it; I wish we had someone like old Nellie,[136] who enjoyed "occasions" so much, to help.

I have to think of a theme for a talk to the Undergraduate Medical Society. I have no ideas at the moment. It presents difficulties because it involves all years. I suppose I shall wriggle out of the difficulty somehow but I would like to interest them.

Love to all of you,
Dad

104 McClary Ave.,
London, Ont.
1 Dec 46

My dear old Dad:

The winter seems to be starting in earnest at last. For the past while we have needed to wear the winter coat and now we have the first real snow flurry of the season. Alice went out to play in it for a short time this morning and enjoyed it hugely. Doris gave her the big corn broom and she was as happy as a sandlark doing a total ineffective job of sweeping the front porch. As a result she has been running around all the afternoon shouting "NO" which is her version of the beastly white stuff.

... It is certainly good to hear that Blake has got another job. It may be a good one though I do not know what the firm does or how it operates. I had a phone conversation with Roddy [Heard] the other night and he had a letter from Blake telling him of it ... he thought it would be a good thing to keep the weather eye out all the time. I am glad Sue is well but it must be a plague to her to have a constant leg pain. It is a good thing that it is not anything worse.

On Thursday I made the projected trip to Toronto, together with Murray Barr and Jim Hamilton. We drove down (or is it up) and it is quite interesting – much more so than I had expected. Some of the country is quite pretty. We arrived about noon and after lunch went round to see Raymond Parker. He is quite a lad and with a pleasant

---

[136] Nellie was the cook-general and maidservant at the Murray home in Montreal after Robert left for Cambridge. RGEM.

sense of humour. He has a very nice lab which is, I suppose, small by the usual tissue culture standards but seemed fairly ample to me. It is a nice though expensive set-up but I think that it would be possible to set up a smaller one or two room outfit here in London without breaking the bank provided that these people map out fairly accurately what they want to do.

The chief trouble is technical help and that was partly what we were looking into – I somehow think that they have an idea that it is worthwhile to train a technician. It was and still is my opinion that it would be much better if they were to train someone with much broader background (biological plus special understanding of histology and bacteriological techniques) who would be in a better position to train others to the work here. Since it takes longer to train the ignorant this would be a more economical plan but where can you find such people. Parker, of course, stresses the fact that he is willing only to train the competent for it is a drain on his resources to train people for other labs – even if it may be flattering to have pupils strewn around it is no good if they are second rate. I found it very interesting to see their lab – and sort of wish that I could learn more about the business – someday. We spent a long time in the lab and it was more or less too late to do much when we left there.

We went and saw some friend of Barr's in anatomy – I popped over to the Sick Kids and found Ted [Roy] still there. My goodness that is a terrible lab and even if there was a prospect of a lab built to your own prescription four years hence I would have thought twice about it. It seems small, more or less impossible to clean or reconstruct, without storage space, grossly overcrowded and with a considerable load of work – in fact all a difficult matter. Ted looked a little exasperated but maybe it was because he has been bickering with architects. Future prospects seem good and they have very ambition plans – but it is future.

It was a bad week other than this. I am still a little fogged about the details of some of the work especially about the errors of method ... and the mathematical businesses are a headache but I can't find any other method of showing the significance of an assay. Anyway it will be no more than an index for I would have to do a great many for a good analysis – and I have not the time at the moment.

I can quite imagine how annoyed you were at the BW article. It all shows the foolishness of the government in not making a proper release before. Without proper publication they cannot expect proper recognition. Of course the political angle is a criminal shame. ...

    Your son,
        Bob

Montreal
9 December 1946

Dear Old Son,

You will be here with us in four days time & we are greatly looking forward to it, with the only regret that Doris & the babies will not be with you.

I received a wondrous "mock" medal, made & sent by R.E. Dyer,[137] Director of the National Institute of Health, Washington D.C., which will amuse you when you come. I'm very pleased with it & it makes a good memento.

... You seem to have had an interesting trip to Toronto. Parker is a very good fellow. He has a neat setup there which was originally designed by Jim Craigie for his work on Smallpox & Vaccinia & other viruses. Parker has changed it a little here & there. Quite expensive. I think your ideas on the suitable person to train are right. The only other thing is this, that it is a long training of a person with proper background & it is not worth taking a woman for it. They get married & all the training is lost. Parker is right about the drain it is training people for other places & that second raters are not worth while.

I am sorry about Ted [Roy] having a tough time. Most places over here (Canada) are rather like that. This place was awful when I came & Ash had a tough job too very tough, when I first saw it. Many others I know are not better. Unfortunately it will take up the time of good men struggling to get them straightened out.

Louis Greenberg of the laboratory of hygiene in Ottawa & a girl from University of Minnesota are coming here next month to work for Ph.D. The girl is going to work at a problem in immunochemistry with Catherine MacPherson & Greenberg will work at mixed antigens in immunity. Others are getting busy on various things & I hope to have the whole lab working at good problems before too long.

Mum is very fit but had a bit of tummy trouble. There seems to have been a lot of it about without obvious cause.

I wrote to Graaff Reinet for some pipes & I hope they will come along one day. I'm sure they will send them if they still make them there. They were peculiar to that little town. I rather hoped they might arrive for Christmas. ...

It will be great seeing you.
Dad

---

137 Rolla E. Dyer was a respected authority on infectious diseases, particularly endemic typhus. He served as chief of the Division of Infectious Diseases at the National Institutes of Health, Washington, DC, between 1936 and 1942, and as director of the Institutes from 1942 to 1950. Jeannette Barry, "Notable Contributions to Medical Research by Public Health Service Scientists," *National Institutes of Health Almanac,* Public Health Service Publication No. 752 (Washington, D.C.: U.S. Department of Health, Education and Welfare, Public Health Service, 1960).

Royal Victoria Hospital, Montreal
19 December 1946

Dear old Son,

It was grand seeing you & I am proud of the way you presented your paper & handled the discussion. A number of people spoke most pleasingly of it. We selfishly wish we could have kept you a bit longer.

... I'm glad you like the changes, present & to come, being made in the lab. When all is completed & everyone is settled in their work you will find it even more interesting. I expect Doris will like hearing of it all.

... Sue's leg is less painful today than it has been the past two days after the injections MacNaughton did. Whatever he injected greatly increased her pain & now it is about what it was before, but it may have an ultimate benefit it is to be hoped. She will have no more of such handling and cannot be blamed for it.

I don't believe I showed you the "Medal of the Joint US-Canadian Commission". I cannot remember doing so.

It's late & I must to bed.
 Our love to you all,
  Dad

London, Ont.
26 December 1946

My dear Mum and Dad:

We have had a marvelous Xmas – thanks in no small measure to you. We were particularly thrilled at having a little party of our own with Alice just old enough to take some part in it. There was nothing to mar our holiday and even the weather played up: sunshine in the morning and 2 inches of snow in the afternoon.

Such a lot of things you sent us – my goodness, I arrived back from my trip with a load and then came the largest express parcel I have seen since we moved from Barrie, to say nothing of many things that came separately. Most of all the very munificent Cheques: so much but so welcome and you may be sure they will be well used. As is always true, despite the fact that we should have enough money, there never seems to be enough to spend on ourselves and especially clothes. I am especially proud of my "bush shirt" and have been wearing it ever since I opened the parcel yesterday.

It has been a happy time for all of us. Alice was very thrilled with the tree, the lights but was a little mystified at first by the rash of parcels that appeared yesterday morning. She soon caught onto the spirit of things and opened parcels right left and centre – making it very difficult for us to keep track of things. She loves all her toys very dearly and can hardly bear to be parted with them to sleep. Despite her enormous selection she casts a covetous eye on some of Peter's things. They are all lovely and we really thank-you for them.

The food was almost too much for me (if you can believe it) and I could barely stagger about during the evening. I think I will have to limit myself to one meal a day for a while – though I admit to feeling a little seedy for the last day or two.

Judging from the scratching of pens Doris must have told you all the details – there is little I can add to her account. Suffice it to say we have had a very nice & happy Xmas thanks to your efforts.

    Your loving son,
        Bob

McGill University, Montreal
26 December 1946

Dear Bob & Doris,

First, thank you for the book, "Upstream and Down". It promises good reading & the illustrations are splendid. I wish you had written in it. Mum gave me "A River Never Sleeps" by Roderick L. Haig-Brown, which also looks good.

We had a happy day & the talk to you was most cheering & made us all feel the better. We missed you & drank your health in Groot Constantia at the appropriate time. We fed enormously on a fine dinner Mum & Sue prepared & shared it with Blake's father & mother. In the evening Ted & Sylvia, Stewart & Barbara Rose dropped in for a short one. All very well & in good form. Otherwise the day was spent quietly digesting.

Mum is still delighted with the little talk with Alice – it gave her tremendous pleasure & I wish you could have watched her face while it went on; it would have done you good to see the sheer joy of her expression. She is so proud of being called "Agee", it's something all her own.

The weather has become very cold today, it is now zero & the forecast is 10° below tonight with a lot of snow. We have quite a coating of snow now & it snarled up the traffic quite a lot for a day or two because the streets were so slippery. ... It is quite full blown winter today. However the shortest day is over & I had better give a little time to tying flies & getting tackle into order soon. I have so many other commitments that must be tackled that time will pass quickly.

... Mum has just come in from one of her "brisk walks" with the dog & says she is "roasting". I came in a while ago & was frozen. I don't know how she does it.

... The pup was most amusing over his present from the tree. He was full of expectancy & Mum made a great game hunting for his presents & he & his old sniffer were busy in on it all. When he got it he was delighted & did great unwrapping, the tail going like a flail. By the end of the day he was as lethargic as the rest of us.

Sue's neuritis seems a little better but not enough for her to be comfortable. Bill Cone[138] was to see her but got snowed-in... Another time has to be arranged. Blake is a bit sorry for himself; we cannot find anything other than a few Entamocbac. ... He is a bit of a nervous fussy type. However we will search some more specimens to make sure.

Our love to you all & astonishing luck in the New Year.

Dad

London
29 Dec 46

Dear Fudge:

... the family thank you for your kind remembrance of Xmas. Peter uses his rattle all day (when he has to defend it from a certain female predator). I have worn my tie; it fits my neck and my clothes well.

I heard you slept well after dinner – so did I!

Bob

X Peter (His Mark)

London
29 Dec 1946

My dear old Dad:

... The trip to Montreal was very enjoyable and it was grand to see all of you again. I must thank you and Mum for all the things you did and tell you that it was of the greatest of pleasure to me. You both seem to be very well and happy. Doris was so disappointed not to be able to see you herself but I was able to tell her a lot of things and it made up some of the leeway. I am afraid that with one thing and another I was not able to spend as much time at home as I would have liked but we did manage to have a fairly good visit. It is kind of you to say all the nice things about the paper, I enjoyed giving it – there is little left in that phase to be done and I must get it written up and published. I still do not know where to send it – Ash is rather against the Cdn. J. of Research – and since most of the workers in the field are in the U.S. and England in might be wiser to choose some more generally used publication. There is quite a lag on all Journals at the moment.

Many other things in the trip were fun and good for me. I particularly enjoyed seeing and hearing about the plans for the lab – they are fine, and I think that when

---

138 Bill Cone was a gifted surgeon at the Montreal Neurological Institute and worked closely with its director, Wilder Penfield (see chapter 3, note 130).

they are finished you will have a very fine show. But please try and arrange it so that you have a little time to yourself. Your newly arranged staff is good too – I particularly enjoyed talking to Steve. It was grand to see Gertrude again and to have a long talk with her – she is a really grand person and as good a bacteriologist as there is. So many people to see and talk to in such a short time that I really needed a good deal more time – at any rate I learned a lot.

I have had a few days off but I am now back to work again and have to look after the routine work for a week to give Dr Bittner a bit of time off. ... Starting Jan 6th we have examinations and it will be a fairly strenuous week with almost 4 hours of orals every day. So I won't be really free until somewhere about the middle of January. There is a lot of reading to be done on various new projects and a lot of preparation for them if I decide in favour of them. It promises to be an interesting year.

When I got back I saw Hall about this certification business. He had not heard of it and the last thing he had in his file was over a year and a half old and did not even mention bacteriologists. Fisher had some stuff on it but he had not had anything directly but had heard of it at a meeting and had got the forms himself. So Hall picked up the phone and called this fellow Plunkett in Ottawa and they sent the forms post-haste. From the way the forms are put together I would say that they are more interested in the medical graduate who has a slight interest in bacteriology than in a bacteriologist who also has a medical qualification. Anyway it is a bit doubtful if I can qualify for it except as an argument on their rules. ... It depends what they really mean by five years in the study and/or practice of the specialty. From the way I went at it could be argued that I was studying and learning the practice of the specialty from the time of the Part II onwards. I don't know how they interpret their own rules but I could be completely disqualified if they insist on a year's general internship which I have never taken. They may be a bit flustered by what is meant by the Cambridge degree and they may take it to you as a referee. I believe that Hall has written a couple of letters too, to those supposedly in charge of the thing.

The monograph and Mervyn Gordon's paper arrived safely and I am very glad to have them. Especially the former for it has a lot of material in it that needs digesting for me to attack some of the things I have in mind. I hope that if you see any references to things that are along my lines you would make a note of them for things are easily missed and it can lead to a lot of unnecessary work.

I am glad the ball-pen is working out well. I must say that they all seem to have the characteristic of being ever on the brink of running out of juice. ... At the moment I have enough pens to last me a lifetime though I expect that I will break down in favour of some new gadget one of these days.

Yes, I saw both Graham Smith's[139] letter & drawing and the grand Mock Medal with the smutty ribbon. Enjoyed them both.

We are just getting over the post-Xmas slump and are beginning to sit up and take notice. We have an awful lot of letters to write still. ... I wish you could see them both they have changed such a lot and are changing so fast. Peter has his own individual form of crawl, quite elephantine, and Alice is constantly improving her small talk.

I will write again soon – when the writing horizon is a little clearer. All my love to you and to Mum.

Your son,
Bob

P.S. I forgot all about my new rod. It was given me by one of the local doctors for whom I did a small job during the summer. Very nice of him.

Royal Victoria Hospital, Montreal
29 December 1946

Dear old Son,

Doris' long & splendid letter came last evening by special delivery & gave us all great pleasure. She is a great girl to trouble to give us such an interesting & full account of your doings & happenings. It brought us nearer & we loved it all. She makes too much of our little contribution; only, all of it carried our full love, & that you all enjoyed it & love us is all that matters.

On Friday Mum had an inch long fine sliver of wood, from the broom handle, run into her left thenar eminence. She got out ¼ in. of it & I could not reach what was left when I came home. John Armour[140] came all the way from home at 10pm & cut it out neatly. Such a good fellow. I put the splinter into [illegible] meat mash, fortunately no result yet, but it is well to be sure. It is not giving her any trouble.

The book Mum gave me, "A River Never Sleeps", is very fine reading on matter & in manner. When I finish it I shall lend it to you as I am sure you will enjoy it. The author is R.L. Haig-Brown & worth taking note of. I have seen a copy of his "Return to the River" but would like to get his "The Western Angler" which appeared in a limited edition & probably is scarce.

We had a fierce blizzard yesterday. Subzero wind & driving snow which was very uncomfortable & interfered with traffic quite a bit. Today is still cold & snowing but

---

139 G.S. Graham-Smith (1875-1950) was a reader in preventative medicine at Cambridge University (1923-1940). He instructed Robert Murray at Christ's College (subjects included toxins and venomous animals and insects). Robert G.E. Murray, "A Structured Life," *Annual Review of Microbiology* 42 (1988): 1-34.
140 John Armour was a senior surgeon at Royal Victoria Hospital. RGEM.

without wind. Real winter has set in now & I suppose we will be snowbound 'til spring. I am glad to live near the lab. Freddie got both ears frozen yesterday waiting for a streetcar ... Others had quite a job getting to the lab, for it was not only cold with snarled up traffic but many people could not or would not take out their cars & so added to the streetcar & bus passengers.

I suppose you have seen the paper on B.C.G. vaccination[141] of nurses in Saskatchewan in the last number of the CPHA Journal. As far as I can know it is the first controlled & convincing experiment. The results are very striking indeed & very important.

I am starting some of the writing I have to do & I shall have to work hard at it for the next few months to get it in hand. I also have to think out an address to be given at the dinner of the Ontario Veterinary Association at the end of this month & an address to the McG Undergrad Med Soc. I don't like those things but cannot always refuse them. I have not hit on subjects for the two addresses yet.

We are having a short break in the teaching, but there is quite a load to take up again when term starts again. The examinations will be a load this year with the numbers of students & 9 courses. The mere setting of proper papers involves quite a worry & then the marking will be exhausting. However, it has to be.

R. St. John-Brooks[142] is coming here for one night on his way back to Washington from Ottawa. We will have much to discuss concerning the International Association of Microbiologists, rules etc of Taxonomy, & various other international activities that are in the wind. He has retired from the National Type Culture Collection in England & is concerned with the scheme of developing an International collection. 1st & 2nd Jan will be chiefly taken up with his business. I am much involved in these things & at the disadvantage that I cannot afford to go to the various meetings I should go to.

   With all our love,
    Dad

---

141 Bacillus Calmette-Guérin is a tuberculosis vaccine that was developed by the Pasteur Institute in France in the 1920s.
142 See chapter 1, note 100.

CHAPTER FIVE

FAMILY CORRESPONDENCE: 1947

London, Ont.
5 January 1947

My dear old Dad:[1]

Thank-you both for your greetings of the New Year and the nice things you said in your letters. We had a very quiet time and did not celebrate at all except to have a swift drink before we staggered into bed more or less tired out. There seemed to be a fair amount of semi-wild celebration for Ontario.

… I have not had time to do any work and I am not likely to be able to get back to my own stuff for a little over a week. Next week we have examinations and I will be tied up for most of the time with orals. Even after that there will be plenty of little things to take up time but not so much as to interfere. The cultures arrived in good shape but I have not yet checked them to see that they have retained their characters. I have written to Gertrude[2] asking her to let me know how they grew on your media – it is important to know because several people have asked me for cultures and it is not much use sending them unless they have the medium to maintain them or a suitable substitute. Directly I can get at it I will work out these matters.

Thank you for the references. Some of them I had not seen although I keep a fairly careful eye on the literature. I had seen the article by Parkes on anti-tadpole sera and it is very interesting.[3] Amazing how they can live in such solutions of serum. Makes one think that some of the early observations may be a little inexact but it would need checking. I have found another reference for you on chronic meningococcemia; Biol. Abstr. 17982 (1946): Berk, Lionel. (U. Cape Town). Chronic meningococcemia with fever resembling malaria. Clin. Proceed. 3 (10): 469-472 (1944). A case observed for seven weeks with joint involvement.

I am glad you enjoyed the Ms. on Immunological Reactivity. It was fun to write and it brings up some fundamental problems but I now feel that they are a bit big to tackle with the present facilities. It was given at a graduate student seminar in the medical faculty on 8th April 1946. With more work and digestion of further information that I have dug up it could be published as a review in something like the Proc. Of the Cambridge Philosophical Society – but there have been rather a lot of reviews on

1 Unless otherwise specified the correspondence in this chapter is located in the UWO Archives, R.G.E. Murray Collection.
2 See the Introduction and chapter 3, note 101 for information on Gertrude Kalz.
3 See A.S. Parkes, "Anti-Tadpole Sera," *Nature* 157 (1946): 164-165.

contiguous matter in the past year. I agree that it is worthwhile to have such ideas – it does something to one's enthusiasm and is a useful stimulus to reading. I sent a copy to Craigie,[4] with a letter about other matters, before he left but I expect he was much to busy to answer it or to read it.

Common antigens are very important and will I think be of particular importance to human physiology as more substances are found in bacteria that are common in humans – Hyaluronic acid is only one that I know for certain though there are probably other polysaccharides. They should be an important basis for study of mechanisms. Certainly there are quite a large number of common antigens between different species of bacteria, both capsular and those more deeply seated.

All my love to you both,
Your son,
Bob

Royal Victoria Hospital,
Montreal
5 January 1947

Dear old Son,

The Xmas holiday is just at an end & classes start again tomorrow. This year is upsetting as everything seems to go by bits & pieces & nothing gets properly finished. The alterations in the lab are not completed but every now & again some kind of workman (painter, carpenter, electrician etc) appears & does a little bit without finishing it off. Judging by noise some work is being done somewhere towards the extension over the roof but there is nothing much to see.

The only positive thing is the work of the department which has to carry on. There will be a whale of a lot of examinations before the year ends.

There is some stirring of interest again in the possibility of starting a Virus Division. Hints of money being provided & the Dean has initiated a meeting on Thursday to do with it. There is no indication of other than an exploratory discussion so far as I know.

[Ralph] St. John-Brooks[5] while here for a couple of days raised questions of various International activities I have to play a small part in, though I cannot see how I can get to meetings; I cannot afford to go to Copenhagen etc at my own expense.

I have not worked out a theme for the address to the Ontario Veterinary Association, which preys on my mind a bit; also I have to work up something for the

---

4 See chapter 1, note 37.
5 See chapter 1, note 100.

Undergrad Med Soc. Now I have just had a request from the Univ de Montreal to address some society of theirs which I think I must refuse.

The actual running of the department is being a little difficult because small things go wrong & have to be put right, things which it should not be necessary for me to worry with at all.

The general disturbed state of things give an atmosphere of uncertainty which makes it hard to get in the swing of progress. There is rather too much new in hand at the same time.

There is too, Bergey's Manual & two chapters to be produced for the new book on Medical bacteriology.[6]

The game is a little too fast & furious but I'll get through somehow with a little luck.

Mum & I went to Desjardins last evening for some oysters & lobster. They were very good & we thought of you & wished you were there. Sue did not feel like going out so she & Blake stayed home.

We have heaps of snow & the city is not doing so well with it this year. The streets are in a mess, snowshoes would be useful on some sidewalks. The temperature varies a bit too.

Don't get one of these Eversharp ball-point things until they have improved them quite a bit more. They dry up at critical moments, though when they do work they are nice. The uncertainty is an aggravation ...

This is a gloomy letter. I'll write again when things seem more cheerful or, at least, when they don't seem to matter so much.

  Love to you all,
    Dad

Montreal
8 January 1947

Dear old Son,

We are glad of your letter (5 Jan '47) with the full news of the kiddies. They must be a host of fun. We are looking forward very much to a chance to see you all as soon as possible. Alice's conversations must be delightful & Peter's antics amusing.

... I made a mistake when I told you your cultures had changed on our medium. Gertrude says they grew just like very good Type III Pneumo & she got a fine result with them. ...

---

[6] This refers to the later completed chapters, E.G.D. Murray, "A Synopsis of the History of Medical Bacteriology," and E.G.D. Murray and Gertrude G. Kalz, "The Cultivation and Identification of Pathogenic Bacteria," in *Bacterial and Mycotic Infections of Man*, ed. René Dubos (Philadelphia: J.B. Lippincott, 1948).

Thanks for the reference on Chronic Meningococcaemia[7]; that sort of thing is not uncommon & I have seen quite a number with interesting variations. Joint infections are not rare, only they are most often overlooked because the patient never complains of them. Usually the patient does not notice them until his attention is drawn to them. The pus in the joints is bright "apple green" & at early stages crammed with meningos which quickly disappear. Resolution is complete without residual trouble. A very curious thing is how often the chronic meningococcaemias are not suspected & often it is a long time before a blood-culture is asked for. ... I agree that you could make your "Immunological Reactivity" into a good & stimulating review. I think it would be worth doing ... I would think Bacteriological Reviews might like it ...

I know people are a bit diffident about publishing in the Can. J. of Research because they say its circulation is small. However if Canadians publish important papers in it then they will compel the increase in its circulation. I am all for Canadian institutions & not depending on the USA – let's "build a better mouse-trap". At present American Journals are snowed up with proffered papers & British Journals are too & lack paper as well.

Yes I think there is a failure to the study of common antigens & one effect will be to clean up some existing muddles – including the Salmonellas.

... Ronnie Hare[8] writes contentedly (from Dept. Bact. St. Thomas's) & seems happy to have left the Connaught. He is glad of my practical class books he asked for. That is the usual story & many have used them as a foundation & reference.

We still have lots of snow & sometimes quite a snap of cold in the zero region. Taking Fudge for a stroll round the block is quite a freshener at times; the trouble is that those "quick-frozen" messages he has to read take quite a bit of thawing out.

The Star is editorially very pleased with the appointment of your new Chancellor, as might be expected, though I believe it is good.

    Our love to you all,
        Dad

12 January 1947
London

My dear old Dad:

I have had two letters from you this week but have had little inclination to write letters for one reason and another. We have spent pretty well the whole week on examining the class and it has been quite a strenuous time. With 57 of them it is quite a chore

---

7 This is defined as "The presence of meningococci in the blood; meningococcal septicaemia." *OED* (Online).
8 See chapter 4, note 87.

to put them all through an oral examination but we did it. There was also the written, which I invigilated giving a total of 23 hours spent on examinations alone during the week. Rather a disappointing show in many ways – some of the students were very good, one of them outstanding, but an excessively great number of weaklings. We can pretty well equal your record for we had one who was convinced that typhoid may be isolated from a throat swab – there were several other things which were of equally annoying quality.

I have not yet got any real work done but I am clearing the decks in preparation. I have had some correspondence with D.T. Fraser[9] concerning the Tetanus toxin and he is going to send me as much as I want.

There is not very much to report on the home front. Doris was not feeling very well at the beginning of the week, and we have all had a mild sort of cold but all is well now with Peter the only one showing the relics of the sniffles. Nothing much new. Alice is giving herself a little tea party, we have just found out that she has been eating, with gusto, a mixture of butter, salt, pepper and sugar that she made up all by herself with materials filched off the table!

As relief to examinations I have read a couple of very good books this week. Directly after I had written my last letter I heard from one of the students that he had a copy of Negley Farsons "Going Fishing" which I promptly borrowed. It is really magnificent and I think you should really get it. ... The other book is about 'down north' and is called "The Land of Feast and Famine" by Helge Ingstad. ... It is an account of four years spent almost entirely in the regions east of the Great Slave and is excellent description of the country, the inhabitants and the problems of travel and living. I am most excited by it.

... One thing I wish you would do for me and that is send me a parcel of some of your selected fishing books – including that fisherman's handbook by Knight and an old Hardy catalogue, and any other suitable book. If I have to give this sort of hobby lobby on fishing methods this winter I want to have a few good reference books and it would help a lot if you could lend me a few.

I see that they had the gall to make Camillien Houde[10] a citizen of Canada! There

---

9 The bacteriologist Donald Thomas Fraser (1888-1954) joined the University of Toronto's Antitoxin Laboratory (later renamed Connaught Laboratories) following service in the First World War, and eventually held the post of assistant director. In 1932 he was appointed full professor in the Department of Hygiene and Preventative Medicine in the Faculty of Medicine. He also assisted in the establishment of the university's School of Hygiene. His research focused on vaccines and antitoxins, and among many contributions his work helped to improve insulin production. He also participated in collaborative research that "perfected diphtheria toxoid," and worked on tetanus, scarlet fever, and whooping cough problems. University of Toronto Archives and Records Management Services, Finding Aid, Fraser Family Records (B1995-0044).

10 Camillien Houde (1888-1958) was a long-time and often controversial mayor of Montreal (1928-

was an editorial in the Free Press drawing attention to it in no uncertain terms. I wonder that even Mackenzie King or St. Laurent allowed that.

You seem to be a bit snowed under with writing and speaking commitments and I don't blame you for being a bit depressed by them. No doubt the department is taking a little more than reasonable time for administration but I expect that in a short time you will be able to lead some of the senior men into taking over some of the burden. The development of the department is most laudable and not to be decried atall.

I am glad to say that Hall is very keen to get Martin Entin[11] to come here to set up a tissue culture outfit from the start – rather a gloomy start from the lack of previous spadework but it should be a most promising job for the future. We are expecting him to be coming up here to talk it over with Hall at the end of the month. I hope that he has not become too deeply enmeshed with the surgeons and that he will join us here.

On the matter of "building a better mousetrap" I agree with you entirely. But I will have to think the matter over.

From what I hear, the new chancellor is a very good man. He is rather a friend of Ash's[12] and he says that it is an excellent step.

All our love.

  Your Son,
   Bob

"An Incongruity"[13]

It does seem a little ironical, to us at least, that Camillien Houde, mayor of Montreal, first citizen of the French-Canadian metropolis, the man who told his fellow citizens not to obey the law calling for national conscription – and went to prison for it – should now be the first person in Quebec to be honored with our new Canadian citizenship. Mayor Houde received the first citizenship certificate from Chief Justice Tyndale during a colorful ceremony in the ornate marble-lined grand salon of the city hall.

"Happy is the man who can call himself a Canadian citizen," said the chief justice. This is the proper attitude to take towards our new citizenship. It may have been necessary, in order to comply with civic punctilio, that Camillien Houde should receive

---

1932; 1934-1936; 1938-1940; 1944-1954). His mayoralty was suspended on 5 August 1940 and he was imprisoned (until August 1944) for his open opposition to conscription and for calling on French Canadians to resist the National Registration Act. *The Canadian Encyclopedia* (Online edition: http://www.thecanadianencyclopedia.com).

11 See chapter 4, note 121.
12 See Introduction, note 377 for information on Igor N. Asheshov.
13 *London Free Press*, 10 January 1947.

the first parchment, but we still believe it to be an incongruous twist of fate and circumstances that it should be he of all men who now stands as the prototype of Canadian citizenship in French Canada.

Montreal
12 January 1947

Dear old Son,

Yesterday I had an accident which is most inconvenient & painful. Going to the lab after breakfast I fell & twisted my left ankle rather badly, it is swollen & painful still but not as bad as it was all yesterday. The city in clearing away the snow dug the ice on the sidewalk into two levels (some 5" difference) & the fall of snow during the night leveled this off so that it could not be seen. I unfortunately put my foot on this edge & stepped off & in falling with my leg under me hurt my ankle. This happened only a little way from the house so I limped painfully home & as the pain did not abate I stayed there. I bandaged it, but had a miserable day as there was no position of rest & a lot of pain. Today it is swollen & sore, painful to walk on; but not troublesome at rest. It will quickly improve but be a nuisance for a while.

I have finished reading "A River Never Sleeps", which Mum gave me, & greatly enjoyed it. Mum wants to read it & as soon as she finishes it I shall lend it to you. I know it will appeal to you and it is very attractively written.

I still seem not to get anything worthwhile done. The days are filled with various fiddling things which have to be done or seen to in order to keep things moving or in preparation for necessary things. Someone has to keep an eye on things & unless I do they seem to be neglected. I suppose it will have to go on a while longer, until I can get people to take some reasonable responsibility.

Mum's knee (semilunar cartilage) is still troubling her. She gave it a very insignificant twist one slippery day when we were out; not enough to trouble her at all at the time but which had results later. ... Otherwise she is very well. The year must have turned because she has started dreaming about the lake, camping & fishing etc. Sue seems to be having less trouble with her neuritis. She saw Bill Cone[14] the other day. He said the only thing that could be done was to inject round the exit of the nerve to get temporary relief – 2-4 days. She did not think it worthwhile – probably right. He says it is a pressure effect. She seems to be very well otherwise.

Blake[15] seems to be enjoying his job, which affords him quite a bit of variety through the inspection of all sorts of premises insured against fire. Ralph Sketch[16],

---

14 See chapter 4, note 138.
15 See chapter 4, note 46.
16 The Murray and Sketch families were related on Freda Murray's side. Ralph Sketch's

who was here a while ago, said that Underwriters training led to good jobs & men trained by these people are always in demand. I hope he is right. The way the world wags at present, it seems likely there must be a "depression" before very long, forced by the present demands labour unions are making.

Some more talk last week on the possibilities of a Virus Division. Where it will get I don't know but Meakins[17] & Vivian[18] are taking part in it as far as the scheming goes. I suppose that has to be & I shall have to struggle to try to get the effort directed into proper channels.

I had a letter from H.R. Dean.[19] They are taking active steps to form a school of Veterinary Medicine at Cambridge. Their first concern is to find a first class man for the Chair of Animal Pathology, which is the key position. He wrote to me for suggestions. If they can't find a Veterinarian of required character & qualifications they will hunt a human pathologist. They quite realize the difficulties & are determined to do it right. I think it is a sound & necessary move, one which I urged years ago.

This will get to you just before your Wedding day. It brings you both all our love & good wishes for many years of happiness & success. We shall be thinking of you on 15th.

Our love to you all,
Dad

London, Ont.
15 January 1947

My dear Mum:

It was lovely to get your long letter today; we had very much looked forward to an account of the Christmas festivities and the things that went with it. You must have had quite a time and a lot of work but it sounds as if it was the greatest fun. We so wish we could have been there – despite that we had ourselves a fine time with our little celebration. I had almost forgotten how gargantuan the staff party was in December. The figure of 3-1/2 gallons for the punch that Dad and I brewed that night seems awfully big but I suppose that it is about accurate.

... We have been having a happy time being at home and looking after the kids and

grandmother, known by the family as "Aunt Nell," was the sister of Robert Murray's great-grandmother Julia Ann Griffiths. Julia Ann Griffiths married William Esau Heard, and their daughter, Florence Emily Rose Heard, married Thomas Hardwick Woods. Therefore, Robert Murray and Ralph Sketch were second cousins (once removed). Ralph Sketch worked as a general manager at the head office of Phoenix Assurance in Toronto.

17 See chapter 1, note 26.
18 R.F. Vivian was head of the Department of Public Health at McGill. After 1945, he worked closely with Joburg in providing proper bacteriology training to students in his program.
19 See Introduction, note 178.

getting them back to their pre-Xmas behaviour. They are quite a pair, and, since both of them have had colds for the past week, they have been playing together a lot. Alice is very good with Peter and only occasionally lets her feelings get the better of her. ...

... I am sorry to hear that your knee is back to bothering you again. It is the devil but you should be careful on the icy sidewalks – maybe you should get some 'creepers'!

... We are leading the quiet life our only excursion together lately was to a movie the other night. Doris goes to the library and draws once a week – which is a very good plan and she enjoys it very much. She is very pleased with the pencils, and the paper. The latter is good but makes a noise like a thousand rats in a woodpile when you start drawing! It has a funny form of surface. We are going to try and fix things so that we both can go up on Mondays when I can browse around in the library while she draws.

All our love,
Bob

Royal Victoria Hospital, Montreal
16 January 1947

Dear old Son,

... You are no doubt weary with the examining. I have not encouraged viva-voce[20] examinations here, they were not in current practice & I did not suggest them. I did much of it for matters of 100-140 candidates at Cambridge for years. It has some merits & some difficulties, as a method, but so has any form of examining. We shall have a weary time this year with all we have to do; 140 medicals, 40 dentists, 92 graduate nurses, 54 elem. arts & sci, 18 Course 3, 20 degree nurses, 6 D.P.H., with all except the nurses having written & practical exams.

... Don Fraser is a very good fellow & will help you all he can. I had a nice letter from him today on matters of International & Canadian microbiology.

... Yes, we were disgusted at Camillien Houde receiving the No. 1 Citizen Certificate; but we got inured to some things in this Province; even to miscarriage of justice. I don't suppose W.L. Mackenzie King cares so long as he gets votes.

... I go to Ottawa tomorrow morning & return Sat evening. Nothing important but a lot of the old gangs will be gathered & it should be amusing.[21]

Our love to all of you,
Dad

---

20 Latin for "by live voice," referring in this context to the oral defence of a scholarly examination or dissertation.
21 The Defence Research Board had organized a special BW Committee to study postwar threats.

104 McClary Ave
London, Ont.
Jan 16th 1947

Dear dad –

... This is mainly to send you the enclosed picture for your wallet in case anyone demands proof that you have one thriving grandson & smiling granddaughter. Gertrude would probably like to see what Alice has turned into.

They are contentedly playing together now. Alice reading Xmas cards to Peter.

We *have* Mum's marvelous & gargantuan effort about Xmas and appreciated every bit of it. I guess you had about the best Xmas ever too!

... This brings lots of love to all and I hope your ankle is functioning again without too much trouble. That was an unfortunate accident. Accidents usually are.

Alice is handing me all her toys. I suspect it is a bid for attention.

With very much love,
Doris

P.S. Also sent the rest of your Xmas pies. Hope it arrives safely. We think you & Mum will like them. D.

Royal Victoria Hospital, Montreal
19 January 1947

Dear old Son,

... I still don't feel sure of what I shall talk about to the Veterinarians at the end of the month. It is most annoying not to have a good theme to work up.

The possibility of forming a Canadian Association of Microbiologists has cropped up again, with interest in it by those who would not support the idea several years ago. I shall have to take a hand in it again but I shall do so chiefly by encouraging others to move more actively this time, keeping in the background as much as possible.[22]

I shall have to continue with the maintenance of contact with International organizations until a definite National Canadian Organization can take over.

Anyway I have some involved problems to deal with though there is not much encouragement to some of them. I started some & accepted others because they seem to me to need doing; perhaps I can help them along until others join in. I did not expect them to pile up as they have & I shall have to put time & effort into them now.

... Things in Ottawa seem to be in a confused state & the "beam" guiding everything just now seems to be the Cabinet's instruction that everything that has been

---

22 This project would be picked up by Robert Murray in earnest in 1949, resulting in the establishment of the Canadian Society of Microbiology in 1952.

planned has to be cut 25%.[23] I'm glad I am not concerned with any of it. Economy at any price is not a good note though I am all for effecting economy by a thorough examination of the worth of every proposal or project. I do not envy Solandt[24] his job as Director General of Research & Development. The Directorate I was in has been disbanded & I do not know more than hints of what may replace it in the future. However none of it is any concern of mine any more, though there are hints of me being invited to be on some advisory committee.

It seems that the "Cost of Living" is jumping up. ... I suppose labour will come out with demands to ease things for them though the rest of us will have to take the medicine. People retired on pensions determined on old rates will be terribly hard hit & those of us approaching retirement cannot provide for better. It's a dingy outlook.

I hear of such a state in Britain too. Hartley came to retiring age last year & he immediately had to look for a job because his pension was now inadequate. It seems a sad end to a lifetime of valuable work.[25]

Meakins said as much the other day in Faculty that he would not be there next year, but there is no hint as to who might succeed him. It seems to me stupid to be so secretive (the University not J.C.M.) & it would be much wiser for his successor to be taking part in things now in order to be worked in by September. I suppose Hall's[26] successor has not been announced yet, though he is probably getting some insight into his own new job. It will all come out in the wash, but it does not make things any easier.

... There is nothing new on the Virus Division possibility.

The temperature has fallen rapidly this afternoon & evening & the forecast is rain tomorrow. It promises a slippery time. I think I shall have to invest in "creepers" as the city does no sanding, or very little.

Well it's getting late.
    Our love to you all,
        Dad

---

23 E.G.D. Murray was a member of the National Research Council at this point and he is most likely referring to its budget. RGEM.
24 See chapter 2, note 104.
25 Percival Hartley was an esteemed English biochemist. In 1922, he was appointed director of the Department of Biological Standards at the National Institute for Medical Research, a position he held for the next twenty-four years, until his retirement in 1946. He is perhaps most known for the development (at the Wellcome Physiological Research Laboratories) of the so-called "Hartley's Broth," which greatly facilitated "the regular production of potent diphtheria toxin, and therewith of antitoxic sera of high value." Henry H. Dale, "Percival Hartley, 1881-1957," *Biographical Memoirs of Fellows of the Royal Society* 3 (November 1957): 88.
26 See chapter 3, note 73.

London, Ont.
19 January 1947

My dear old Dad:

Your letter arrived on the morning of the 15th and was very welcome. Thank you very much for your good wishes. We are very happy and we enjoy being here in London very much; we feel that we have been lucky in finding our feet so soon and having such a substantial place to start our lives together. ...

20 January

... Today another letter arrived from you – the one you took to Ottawa in your pocket. The CPR Express charges on the parcel were very little and came to less than a dollar. The books were not damaged at all and are in good shape. It would be nice to have others but not yet – I have more than enough to read and think about for some time.

... I have finished my part of the marking and I can get started on things. Last week and over the weekend I have been getting on with things and preparing the way for an assault on various of the angles that have interested me with the Streps and with Hyaluronidase. I am in process of producing pneumococcal Hyaluronidase in fair amount, which both Pearce and I need for our separate work. ...

... A long letter came from Gertrude today, which was very nice to get. She says she has not been feeling well lately which is a pity. I am glad that the streps are growing well and that she has been able to get some result from it though she does not seem to have tried any wide variety of organisms yet.

... Another letter that I received today was from Arthur Harrold (you may not remember him but Mum will – RNVR Surg-Lieut) who has married and settled down to practice in Australia and seems to like it. Doris persists in referring to him as a "war-groom".

All my love,
Your son,
Bob

25 January 1947
Montreal

Dear old Son,

The news of you all & the photographs of the babies please us greatly. Your friends at the lab were glad to see the print Doris sent me & were very complimentary & we feel with good reason.

I'm glad the books arrived safely & were what you wanted. Some of them are very good & should help you to do what you want to adequately. There are of course other fishings than with the fly & some ways not known over here which are matters of great skill & much knowledge. They have to be truly experienced to be appreciated.

... "The River Never Sleeps" is a fine book & the author must be a real fellow to fish with & talk it over with. He gives you lots to think about.

I am glad to hear you will now get some time to give to your own work. You have some interesting stuff on hand worth the doing. Any problem has many varied points & opens several "avenues" & cannot be absolutely completed in reasonable time. However all problems allow of natural pauses & convenient divisions & these present the occasion to publish results. Keep in mind that publication is important. I know I should have realized it more in my younger days when I was doing real work.

Gertrude is working hard as usual, but I have a suspicion is not quite as well as she should be. She has an attack of neuritis somewhat similar in character to what Sue had but not due to the same supposed cause. I hope it passes off soon.

I shall have to think out my dinner speech to the Veterinarians these next few days. I think I shall try to develop some ideas on the functions of societies. I'll have to try to carry it off somehow.

We are having most uncomfortable weather; jumping in a day from freezing even subzero to over 40 degree F. The result is that the sidewalk is a rink & the City seems to be doing a poor job of sanding. It makes getting about a difficult and unpleasant process, especially after a light snowfall. I am glad not to have far to go to get to the lab.

The old pup is not getting much of walks under these conditions with Mum & I both a bit lame and shy of the ice. In consequence he is becoming Aldermanic and Mum threatens the poor fellow's meals. It will be a good thing for him when the Summer comes.

David Keys[27] has just been made Director of the N.R.C. Laboratories at Chalk River, that is in charge of the Canadian Atomic Research. I expect he will do a grand job there. He will be very sorely missed here where he has done the heavy work of the "willing horse" without adequate recognition. He will be missed greatly by many of us as a grand friend. I don't suppose the teaching of elementary Physics will be maintained to such a good standard and it will be a strain on Norman Shaw[28] whose coronary circulation is none too good. I don't know what the younger men are worth in that department. Keys always asks after you whenever I see him & is interested in your progress.

It is a noisy night as the Frat next door has some sort of party going. I suppose it will run into the early hours. Are Frats encouraged at Western?

Our D.P.H. (Dip. Public Health) bacteriology course is going quite well. The students are very enthusiastic about it & compare it most favourably with what they get from other departments. I wish I felt the rest of it was really worth while, but I am not convinced that some are qualified to handle it. At present, at least, it would be

---

27 See chapter 2, note 4 for information on David Norman Keys.
28 Norman Shaw was a physicist who succeeded David Keys as professor and head of the Physics Department at McGill.

far more worth while getting some Medicals interested in becoming bacteriologists. There are some quite good jobs going without any candidates for them. I don't know how to induce them to consider bacteriology as a career, nor do I know why it does not appeal to them. There seem to be some interested in Pathology & I can't see why.

All our love,
Dad

London, Ont.
28 January 1947

My dear old Dad:

... After deciding that Xmas was a bit much from the point of view of entertaining and being entertained we have been spending a very quiet time "resting-up" and going nowhere in particular. Doris has been attending her drawing evenings very regularly and is having a fine time. She has shown a great deal of improvement and is getting back some of her old skills at portraiture in pencil. It is good for her and she enjoys the change very much. ... So last week we lapsed from grace and went to a party and also to the play. The latter was a very good effort and London turned out in nearly all of its finery to go and see John Gielgud and Co., do "The Importance of Being Earnest" which is a very good play and magnificently done. I don't know if they will be going to Montreal in the course of their tour but if they do it would be well worth your while to go and see it. It being the North American premiere there was quite an elite crowd and many of them went as far as dressing up to it – which is a little unusual.

Then on Saturday night we were invited to a party given by one of the older and married medical students and it turned out to be one of the nicest that we have attended here. Very pleasant and we met some people who are nice to know. I am glad especially that Doris met Norm Brown and his wife, who are very nice. (He is the son of Prof. Brown of Engineering at McGill and is in radiology here). They have a daughter a month older than Alice who is much in need of a play mate as is Alice.

It must be a pest to have to think out dinner speeches – the character they must take can be changed so much by circumstances. However, I think that it would be a very good thing to consider some of the functions of societies. I have to get going myself soon and construct a 13 minute talk for the local radio on the immunization of children. It is part of a long series of talks by members of the university staff – as a matter of fact they put me down to do it without asking my permission, despite that I rather look forward to doing it as a new experience.

The research work has not been going too fast lately for there have been a multitude of interruptions and also I have spent a good deal of time in preparation and drawing up schemes for various of the projects. It seems to take a very long time to get things together nowadays and some of the delay has been due to the non-arrival of chemicals and biologicals.

There are many "avenues" for work that are opening up for me at the moment and I am very keen to get at them. At the moment I am running various trials to find a suitable standard method for following the production of the enzyme by pneumos in order to get a good potent enzyme preparation. Some of the materials I need have not yet arrived but I expect them soon.

I am glad to see that David Keys has got some recognition and that he has been selected for such a responsible job. He is a grand fellow and I am sure that he will be very much missed at McGill. I agree that they did not seem to appreciate the tremendous load of work that he was doing – especially in teaching and I don't suppose they will realize it fully until he has left.

The D.P.H. course is probably very valuable even though it means a lot of work. You probably have some chance to interest people in Bacteriology during that course and they have the advantage of being already medically trained. It is probably worthwhile to carry on some behind the scenes evangelism on them in the course of conversations to get them aware of the possibilities – or to offer the good ones definite jobs. It is certainly a problem to get men to go into it, and I feel just as mystified that so many choose pathology rather than bacteriology.

I must get back to work now, but I will write again rather sooner this time.

 Your son,
  Bob

London, Ont.
28 Jan 1947

My dear old Dad:

A marvelous long letter from you which we all enjoyed very much. It was grand for me, with all the information in it. I have already put the tips you gave on fly tying to great use and find that they work out well – the solution to the tie problem was simple & I don't know why I did not think of it myself.

There is a reference I must give you before I forget it. The paper is by D. Keilin and Y.L. Wang in the Biochemical Journal *41*: 491 (#4, 1947). It concerns the stability of hemoglobin & certain enzymes during prolonged storage in vitro. It contains reference to you since some of the blood used was collected & sealed by you in 1922! Maybe Keilin has sent you a reprint but he may have omitted it.

... I have just been speaking to Walter Johnson[29] on the phone and he is going to

---

29 Robert Murray recalls that Walter Johnson was originally a zoologist who went into physiology in graduate school, and possibly he was involved in research into aviation physiology with the RCAF. RGEM.

hunt up some references (and reprints too, if possible) in regard to the effects of pollution of various kinds. From what he says I imagine you will find them useful. As a matter of fact we met Walter many years ago. Do you remember the two McGill students who were doing a sort of survey up at Bark (and all over the Laurentians) while we were in the Whitall cottage? Well he was one of them! They made a tracing of the aerial photo map at the cottage.

Things have been very busy. I have about finished the trials of examinations and all that remains is the meeting to discuss the results & make final recommendations. As a whole they were pretty good with a more than usual number of excellent students. However there were some disappointing failures and I fail to see how a few of these have got as far as us. We also have budget & other annoying meetings coming up.

The past week I have been having a lot of fun looking at & through (!) the phase microscopy equipment put out by the Bausch & Lomb people. It is a great deal handier than the American Optical staff & a good deal less expensive – however, you don't get quite the range and it is all arranged as contrast on a bright background. It is amazing how much more clearly you see cells & intracellular structure, with good definition & resolution. However you have to be prepared to spend a little more time in setting up, centering etc, and have first class illumination. For my work I'm not sure that it adds a great deal but it has confirmed one thing for me and that is that the colonies move in their own drop of water – i.e. carry their own skating rink. It shows up well and when there is contact between the meniscus of one colony and another you can see the characteristic sharp fusion and the movement of organisms. ...

... Could you please send me Greenwood's "Epidemics & Crowd Diseases"[30] and any other book you think worthwhile on the subject of epidemiology that we are not likely to have here. If you have by any chance, Burnet's "Biology of Infectious Disease" I would like to borrow it as well.

Very glad to hear of your projected trip out west – I am sure you & Mum will have a lovely time. You had better take your fishing rods! Though I don't know what might be in season. Doris & I are both very envious.

I saw reference in a magazine article to a place that might be worth visiting if you are ever in New York City. It's called the Angler's Roost and is in the Chrysler Building – apparently a meeting place for fly tiers.

All our love
>Your son,
>>Bob

---

30 Major Greenwood, *Epidemics and Crowd Diseases: An Introduction to the Study of Epidemiology* (London: Williams and Norgate, 1935).

McGill University, Montreal
2 February 1947

Dear Old Son,

The meeting of the Ontario Veterinary Association went off very well & my address at the dinner seemed to be liked by the majority & afterwards many were complimentary. However it is difficult to get a general realization of the situation. The O.V.A. is quite an important body with a large membership & is one of those many to be included.

I didn't do anything else than attend the meeting. There were a couple of interesting papers: W.A. Hagan[31] gave an account of a curious cattle disease which has broken out in New York State, caused by a virus & accompanied by very definite symptoms & lesions; F.R. Beaudette[32] gave a good general picture of Newcastle disease of fowls in the USA.

... Montreal seems to be doing better with plays of late. It's a pity they don't build a good theatre. Last week Mum & I went to see the British film "The Way Ahead" & enjoyed it enormously. It is a good play & excellently well done. If it comes your way, go to see it.

I now have to think up an address for the McGill Undergraduate Medical Society. As it is to all years it must be of general interest & I would like to give it some purpose, without having to put too much work into it because I have so much to do at present & find it hard to spend time on looking up details. What I have to develop is a theme.

I have just recently heard of a portrait to be presented to H.R. Dean by his old pupils & colleagues. I did not receive any notice of it; did you? Fred Smith[33] got one. It seems that Williamson is the secretary of the subscription committee. Perhaps that is why it is not quite what it should be.[34]

I hope Mum & I can soon make an opportunity to go to London to see you all. I am very anxious to see Peter. He will probably inherit the family propensities & you will have great fun teaching him to fish. I expect you will have with him all the fun & happiness you have given me. Our old fishing waters give me much pleasure these days by the memories they awaken. Time & again particular spots recall moments especially associated with you. I hope we can fix up some more trips together.

---

31 W.A. Hagan was from the Laboratory of Comparative Pathology and Bacteriology, New York State Veterinary College, at Cornell University in Ithaca, New York.
32 Fred Robert Beaudette (1897-1957) was an influential American avian veterinary microbiologist. William R. Hinshaw, "Fred Robert Beaudette: Distinguished Avian Microbiologist and Pathologist 1897-1957," Avian Diseases 1, no. 1 (May 1957): 2-17.
33 See chapter 1, note 87 and chapter 4, note 43 for information on Smith.
34 Norman Williamson was a member of the Department of Pathology at Cambridge. "So," Robert Murray recalls, "my father was a bit miffed that Fred got a notice about the portrait and he did not." RGEM.

I don't remember whether I told you that General Pannet[35] called me, the other day, inviting me to be a Director of the P.Q. Fish & Game Organization.[36] Yes, I did. I suppose there will be a meeting one day & I'll let you know about it when I know more of it.

   Love to you all,
    Dad

London, Ont.
4 February 1947

My dear old Dad:

  I am glad that your address at the dinner of the O.V.A. went so well – as I think it deserved to. I very much enjoyed reading it and would like to keep it. Many of things you say in it have been most improperly stressed or have been missed entirely in the present scheme of publicity. I enclose the report as it appeared in the local paper. It looks as if they have put it in much as it came over the C.P. wires & without comment, as yet.

  ... Is the P.Q. fish & game control a private or government project? It sounds like a good thing & I am very glad to hear that you have been asked to be a director. You should be able to contribute a lot of experience & good advice to such an organization. Maybe they conduct surveys & you could go on a trip into the north! I believe Pannet is a very good man.

  Where the time is going I don't know, but I don't seem to be getting very much done. Many things I had intended doing seem to be still piled on my desk. I will have to start very soon on constructing exactly 13 minutes worth of talk on immunization which will, I know, require several rewrites & more then a few hours of work. It is due for Feb 24th.

  A most curious organism turned up as a contaminant in one of Ash's big culture bottles the other day. It seems to be a gram –ve (Gram negative)[37] sporing rod which is motile & exhibits a most peculiar form of swarming that I have not seen before – nor, I think, have I read of it. The colonies are extremely phomorphic though the organisms seem to be of fairly constant morphology. ... It may belong (I think) with some member of the bacillus brevis group – or may fit into a picture with one of the "Miscellaneous group". These "trails" are peculiar and may be broad & branching but if they are any length they seem to be most often single. With further standing the trails have a beaded appearance because of minute colonies growing up. We have

---

35 Robert Murray has no recollection of Pannet.
36 The former Province of Quebec Association for the Protection of Fish and Game.
37 See chapter 4, note 62.

almost decided to call it the "Ha-Ha" bacillus because these habits have caused us some merriment!

Hall got a letter from Boyd[38] in which he said I would have to take examination to get this specialist certificate, which is what you thought would happen but it was worth a try. The damned things cost money & I'm not sure that I feel like spending $25 for an examination & $25 for a certificate – but it might be wise to get it written off soon. Do they give any information about their examinations, places of examination & examiners – or are you as much in the dark as we are?

... Fishing has occupied my thoughts in many spare moments & I too remember many pleasant occasions & places in great detail. We have much to remember & I am glad you remember them as pleasurably as I do. We learnt many things together & you laid me a good foundation. However, I do sometimes wish we had learned some of the pleasures of the bush much earlier. I often think of what we can do in the future – Doris & I discuss the possibility of joining you at the lake this year & we can do some exploration together. It would give me more pleasure than any present to go fishing with you.

Your son,
Bob

(Enclosed)

Dear Dad – both letters to Mum are sealed but in your letters to Bob today I see that you are contemplating a trip our way after all and I just want to say how much I hope you can soon. Peter has an extremely favourite picture in the house guaranteed to learning a happy smile when all others fail. That is the one of the drawing of Mum. He looks at it again & again, so it looks as if he's going to approve of his "agee"!

... With much love Doris.

London, Ont.
4 Feb 1947

My dear Mum:

I am afraid this must be a hasty letter for I have spent much of the evening writing to Dad & I have still work to do. We have enjoyed your letters very much & thirst for more. Thank you very much for sending the socks: as you say they are valuable articles these days but I understand how they escaped me this time!

---

38 See chapter 3, note 135.

... I am glad Sue is better & is in good fettle.[39] Please give her my love – she had better make hay while the sun shines for she will get precious little sleep in the morning from March on!

All my love,
Bob

McGill University, Montreal
9 February 1947

Dear old Son,

I'm glad you like my address to the Ontario Veterinary Association. I sent you the copy to keep if you wished to. Thanks for the cutting from the Free Press, which is reasonable.

... I too find the time slipping by with very little to show for it. The running of the department has its complications & difficulties & the day seems to be absorbed by innumerable adjustments & organizings. Some of them are a weariness & seem out of proportion to the time they take.

... I don't yet quite know what the organization is that Gen. Pannet invited me to belong to. I am waiting for a meeting to be called. But if Pannet is in charge of it, it cannot be far wrong.

Hilary Robertson[40] is getting some of the members of the Columbus together on Wednesday to discuss its affairs, which seem not to be in good order. Something drastic will have to be done to bring order to a bad state of affairs. Adelard[41] has not played square for quite a time in looking after the club's interest & effects. I have long suspected that he has allowed, perhaps encouraged poaching. This season he was arrested & fined for shooting deer out of season, $100. I hope it chastens him satisfactorily. It seems he had a lot of pals hunting with him on out territory not of season. Then too the finances seem to be out of order & not properly kept & reported.

... The curious culture you describe sounds like bacillus alvei (or a near relation) which has "motile colonies" which spread in a way that you describe so as to migrate about on the surface of the medium. It is the cause of European foul brood of bees. This genus is in the 5th Ed. of Bergey, [which] though an improvement on previous attempts, is unsatisfactory, & you will find this genus dealt with very well in the 6th Ed. when we can succeed in getting it printed. The publishers have let us down badly & even old Robert Breed is angry.[42]

---

39 Sue gave birth to her first son, Douglas, in March 1947.
40 See chapter 3, note 2.
41 See chapter 4, note 72.
42 See chapter 3, note 138 for information on Robert S. Breed. Williams & Wilkins was the

I do not know how important it is to have the registration with the R. Coll. *JP* & *S* as a bacteriologist. I understand I was on the committee but it has never met & it seems [William] Boyd does just as he likes. I suppose you would have to take the exams if you have to register but I think the thing should be enquired into & brought into the open & justified. I don't know yet what its purpose really is. I think they have local examiners in various centres; I suppose you would have to go to Toronto for it. It would be a good thing if Hall were to ask for a full explanation from the college; I don't think he would get anything understandable from Boyd.

The suggestion that you might all come to Bark Lake for the summer has the most enthusiastic reception by everyone here. Mum & Sue promptly broke out into complete & satisfactory plans covering everything, as if it was all settled. Mum would love to have Doris & the babies for the whole summer & you for as long as you could manage. You & I could certainly do some wandering. We hope it can be managed.

Yes, it's a pity we did not start camping expeditions earlier. It is stupid that we did not think of it. You will be able to start your son on the right tracks & get great fun out of it.

Mum has a slightly husky voice today, otherwise everyone is very well. We are in the midst of a real blizzard – masses of snow & high wind & fairly cold – & it is difficult to get about, especially as the snow has been badly handled this year here ...

Dad

London, Ont.
9 Feb 1947

My dear old Dad:

We are in process of having a simply beautiful blizzard. It has been snowing hard all day and, since it is fairly cold, the snow is dry and the crystals are perfectly formed and sparkle in the light. We ventured out on a veritable sled trip of 3 blocks this afternoon to have tea with Norman Brown and his wife, with both of them (the kids, not the Browns!) in the same sled; Peter on top wrapped in his sheepskin. It is a funny sight to see them like that – both of them enjoying it hugely except when the wind drives into their faces (then we put a loosely woven scarf as a mask for them). We had to buck the drifts and it was quite hard work though plain sailing – until one particularly tough drift upturned the sleigh & decanted them both into a foot of powder snow. Much picking-up and rearranging was needed.

... I have been thinking hard at intervals of the peculiar locomotion of that bacillus I was telling you about in my last letter. Yesterday I happened to run across an article by

Baltimore publisher of the *Bergey's Manual*.

Stanier[43] on the Cytophaga – they seem to have a most peculiar locomotion and only on solid surfaces. I must do some careful observations of this peculiar bug of ours & see how it does it. I had forgotten about the possibilities of the myxobacteria. Although most of these organisms are cellulose eaters there are others of less defined taste and no reason why there are not more (R.Y. Stanier, J. Bact 40: 619-634 (1940) – but I think there have been other more complete descriptions of their motility since then.) Have you ever worked with them? They are a gram –ve rod – various shapes among the species but many of them not dissimilar to this one. I must read about them tomorrow.

... I have some minor hunger pangs to attend to & after that work to do before bedtime – I will write again during the week, though it promises to be a busy one.
 Your Son,
  Bob

McGill University, Montreal
16 February 1947

Dear old Son,

We have been hoping to hear your answer to Mother's $128 question. At the beginning of the month I posted a jar of conserve to you which Mum made specially. We hope it has not been lost on the way. You have not mentioned receiving it.

· We enjoyed your account of the sled trip with the kiddies & Alice's delight in playing in good snow. Mrs. Ernest Brown told Mum she had heard from Norman Brown of enjoying friendships with you & Doris.

I did not think of cytophaga in reading your description of your peculiar bacillus. As I remember you said it forms end spores. I have not worked with the myxobacteria[44] or a number of other peculiar fellows described in Bergey. When the new edition comes out I shall send you a copy for yourself & you will find many interesting groups dealt with better than before. There is a lot of good stuff in the new edition. We are furious with the publishers who have let us down badly & unjustifiably, giving preference to other things quite wrongly.

There is an account in "The Star" of the successful treatment with BAL [Brit. Anti-Lewcyte] of a child who swallowed a phenomenal dose of mercuric chloride. Very strange. It opens the question whether B.A.L. might not be very valuable in the lab to neutralize many disinfectants in materials to be cultured; to determine sterility,

---

43 See Introduction, note 510. See correspondence between Stanier and Robert Murray in Appendix E.
44 The Myxobacterium is defined as "Any member of the order Myxococcales (also called Myxobacterales), comprising rod-shaped, motile, slime-secreting bacteria that are capable of forming fruiting bodies containing resting cells, and which are mainly soil-inhabiting saprophytes and predators of other microorganisms." *OED* (Online).

bacteriostasis etc; it might be applied to differential killing with advantage. There is room for careful work along those lines, with many applications.[45]

Adequate methods for the certain isolation of TB is very urgently required.

... We had a couple of most interesting meetings on tuberculosis with Dr. R.G. Ferguson of Saskatchewan. His evidence of the efficacy of B.C.G. is most convincing.[46] His work is the first properly controlled experiment I know of & I feel that its application has to be properly extended now. The achievement in the Western Provinces is marvelous when all is considered & the reduction of incidence of infection in nurses & others in hospitals, especially sanitoria, by B.C.G. in Saskatchewan is more than that. In Quebec the whole problem must be tackled & vaccination alone could not succeed. There are new problems which the development of a susceptible population imposes; that is something which Saskatchewan & Alberta will have to watch & relaxation of thorough precautions would have dire effects. Ferguson presents his work very modestly & very well & he is worth getting to speak to a society. They muffed it a bit here in not getting him to speak at a general meeting. Anyway, this is something you should look into.

We have almost finished marking the 140 Medicals exam & have the several other courses to deal with at the end of the year. I am glad we will be rid of the medicals because it will be very heavy work in April & May.

I have just received from M.H. Gordon[47] three volumes of Southey's Minor Poems. It appears to be a first edition & has the book plates of Sir Claudius Hunter (Bart) & penciled in some most interesting notes on the Hunter family.

Love to you all,
Dad

London, Ont.
16 February 1947

My dear old Dad:

I was not too surprised to see that you have returned to black ink and a pen. You have had a lot of trouble with the ball-pen and it seems to be the experience of everyone

---

45 British anti-Lewisite (BAL) was developed during the First World War as an antidote for Lewisite, a poison gas (a chlorinated arsenical). RGEM.
46 See chapter 4, note 141 for information on the Bacillus Calmette-Guérin (BCG) vaccine. R.G. Ferguson (1883-1964) was known for his work in tuberculosis control and for using the BCG vaccination to protect student nurses at the Fort Qu'Appelle Sanatorium in Saskatchewan. His expertise was recognized when in 1948 he was appointed to the World Health Organization Subcommittee on TB and the BCG. See "People Profiles: Dr. R.G. Ferguson (1883-1964)," http://www.lung.ca/tb/tbhistory/people/ferguson.html.
47 See Introduction, note 127 and chapter 1, note 7.

else. I hope they make one that has a sure action for it would be very useful in a lab since they will write on most surfaces and will not wash away – would be excellent for labels etc. I don't know about the "Parker 51" but I have heard of many satisfied users.

We had a very pleasant visit of Martin Entin during this past week and it was a pleasure to show him round and introduce him to the various people here. We hope that he may come and set up a lab for special cytological study. He is rightly cautious about it for the plans are for the future and there is nothing to build from. He created a fine impression on all the people he talked with and they all hope he will join us – however, he has his interests and no doubt has some ideas as to his field of work. With his visit and other urgent matters this has been a poor week for work. Hall is conducting a general survey of all departments of the University in respect of staff & material needs and the next important matter of increased space, so we have been trying to prepare a detailed study of requirements and possibilities. Of course what we would like and what we will get are in two different realms and we can only hope. There is still the possibility that in the next few years there will be built a separate building for ... bacteriology, pathology and pathological chemistry. As far as I can gather the project is about 3rd on the list but is complicated by the necessary cooperation with the hospital building plans – since the latter is city owned there is much cautious politics involved.

An uncomfortable spectre for this year is how we are going to be able to teach nearly 100 students in the next class with our present facilities – the solution is not yet apparent. It will involve some "White Elephant" spending for it is unlikely that we will have such a class again.

The case of tetanus that has been worrying for us for the past week seems to have turned the corner and is getting better. There was a 10 day incubation period following a gash on the left index finger while sawing wood on a farm. He received no treatment until the first suspicion of tetanus was noted. Once started, however, they gave him adequate doses of antitoxin even if they did nothing about the primary lesion. I must say I was not in agreement with their reasoning on leaving the primary lesion alone (secondarily infected and draining well, notwithstanding). There seems to be an impression among surgeons ... that adequate antitoxin and saturation with penicillin is enough – I don't think this is a satisfactory statement for there is no guarantee that either penicillin can diffuse into a damaged area nor, that being the case, that toxin cannot diffuse out.

... I have not got much further with "the Bug". Thank you for the information on B. alvei; before now I was not aware that any organisms had "amoeboid" colonies. A very interesting state of affairs. This particular organism seems to fit best into the B. circulans group. ... I am going to try & see if I can observe it on agar block cultures.

... The report on BCG & its use in nurses in Saskatchewan is very encouraging and it looks as if the vaccine may be of very definite value. The people at the Byron Sanatorium here are very enthusiastic and are in fact proposing to give the vaccine to all tuberculin negative medical students. I don't know whether it has got to the stage where one may unreservedly recommend its use for children. Would you think, for example, that it is worth inoculating Peter & Alice? One of our technicians, who is tuberculin negative, came to me the other day and said she wanted to be given BCG vaccine (she is tied up with one of the med students). What would you say? The questions are beginning to roll in and I am not sure just how one should take it – since so few of the groups have been adequately controlled or followed. I believe the U.S. Public Health Service are beginning to use it but only in areas where the incidence of TB in the population is above a certain figure.

I have not done anything about the Royal College of P & S and I will not until I hear from them directly. I transmitted your suggestion to Hall but he is too busy & concerned with other things for the time being. I would have thought you would be in as good a position as anyone to ask for details & explanations of the rather peculiar state of affairs.

... Of course we would love to see you here for a visit – it would be a joy to us and you could see the children. We quite understand that as things are you may not be able to make it for some time. The possibilities of the lake in the summer, I must say, are in my thoughts a great deal. It would be lovely but we will have to leave any decisions until a bit later.

     Your son,
          Bob

                          McGill University Montreal
                                  22 February 1947

Dear old Son,

... Martin Entin kindly came to see me with your message when he returned. He seems to have enjoyed his visit to London, but he rightly wants assurance of a future in the scheme.

Hall's survey is sure to bring out needs & if he can find money for them it will be well. Money is the difficulty & the costs of everything are stupendous today. What I have seen of the building costs here astonishes me, they are completely out of proportion to the performance. It is unfortunate that your building prospects are tied up with the hospital – that is if your hospital resembles those I know here, & more so if it involves City politics.

I got some increased budget last year for increased teaching & larger classes & I have applied it so that it will benefit the department when the classes return to normal size. The expendable materials of course get used up. I have avoided "temporary" structural changes.

We have had several cases of tetanus, I forget the exact number now (that I have looked into the work) but my impression is seven, & isolated Cl. Tetani in all of them. In two it required great persistence & insistence, both in procedure & specimens. A great deal depends on a proper specimen, especially in a small wound with little embedded foreign body. It is then important to have complete excision of the wound both for treatment of the case & culture.

I think it of the greatest importance to remove the entire site of infections when possible & when not possible to open it up as freely as possible & remove all foreign material, clot & necrotic tissue. Then saturation with antitoxins is also carried out ...

Unless there was no possibility of proper surgery I would not like to rely on Penicillin & Autotoxin alone. Penicillin penetrating adequately might suppress growth sufficiently & Fred Smith is getting some very interesting results on suppression of staph toxin production by penicillin which may have a wider importance. I don't know how near to publication this work of his may be. I don't think penicillin has any action on the formed toxin. ...

... As soon as I get the page proof of Bergey on Bacillus I'll lend it to you. The new treatment may help you to run down your Ha-ha bacillus to find whether it is a known species or not. Still, how he does his tricks would be interesting to know, whoever he is.

One of these days, when I can get together some of the information, I'll tackle Boyd on the registration question.

The B.C.G. question is a very serious one. There is no doubt that nurses in sanatoria should be protected by it, if so why not nurses in Gen. hospitals & medical students. If it is true, as it seems to be, that BCG vaccine affords a real measure of protection to the most heavily exposed then it should be used to protect those any way endangered (Med students & nurses) & should be available as an option to the general public as a recommended procedure. Whether it should be the policy to inoculate all infants in a community seems to me to depend on the TB rate in the community – I think it might well be advised in Quebec Province. I think the children in a family in which there is a case should definitely be inoculated. There should probably be some selection of nurses & students, inoculating only the Tuberculin negative, or doubtful reactors ...

There is need for serious consideration of the method to be used. Subcutaneous inoculation gives rise to a definite incidence number of abscesses, which are a nuisance.

Intradermal inoculation gives quick & a larger number of tuberculin positive reactors (if that is an index of protection – which is by no means certain – at least it indicates response) but it gives rise to abscess & ulceration in a certain percentage ... Oral vaccine gives the lowest % of tuberculin positive reactors & the least trouble with any sign of infection – in fact no known ulceration or abscesses – & least trouble in administration. It does give a definite number of positive reactors. It might be argued that good results could be obtained by oral vaccine in the children not open to dangerous exposure & scarification or intradermal in children of tuberculous families. ... You might consider oral vaccine in your children & then after proper delay test them with tuberculin & consider what to do if they fail to become positive reactors.

If your technician wishes to be inoculated with B.C.G. there is no reason to prevent her from having it. The only question is by what route. I think you might write to Prof. Armand Frappier,[48] l'Institut de Microbiologie, University of Montreal, for reprints of papers by his group on B.C.G. inoculation.

Later
Lately we have been out a bit. We all went to "The Yeomen of the Guard" put on by Harry Norris & a very creditable performance. The tenor who did Fairfax was extremely good with a splendid strong voice – quite up to good professional standards & made the other amateur voices seem weak. It was very well dressed & well staged & made a pleasant evening.

... Then again Mum & I went to see another English film at the Palace "The Notorious Gentleman" with Rex Harrison & other good players, based on a sort of modernized Rakes Progress idea. A clever play very well played indeed but sordid, too much good work lavished on a waster & real enough in theme & presentation to seem true & thus like an unpleasant experience.

Last night we had supper with Ted & Sylvia Rose.[49] Oysters mainly & very good. A very pleasant evening indeed. Various topics & much friendliness as usual. Ted is working too hard. They are toying with practicing in S. Africa, Durban.

This evening we supped with Bill & Mrs. Newman[50] & some of their friends. A quiet pleasant evening. Bill sends you his regards. They are soon building a cottage

---

48 The physician and microbiologist Armand Frappier (1904-1991) was a prominent figure in Quebec and Canadian microbiology. His particular interest was the study of tuberculosis. He reorganized the Department of Microbiology at the University of Montreal in the 1930s, and in 1938 he founded l'Institut d'hygiène et de microbiologie de Montréal, which was renamed l'Institut Armand-Frappier in 1975. In 1945 he founded the world's first French-language school of hygiene and served as its dean for the next two decades. In 1969 he was made on Officer of the Order of Canada. "Qui etait...? Armand Frappier," http://www. prixduquebec.gouv. gc.ca/eponyme/s-frappier_armand.htm; *Canadian Encyclopedia.*
49 See chapter 3, note 50.
50 See chapter 3, introductory reflections.

on Dorval Island for summer use & moving into an apartment for winter use. He is rising in the C.P.R. to more importance & working hard, but keeps well.

Love to all of you,
Dad

London, Ont.
26 February 1947

My dear old Dad:

... Much of my time last week and last weekend had to be given to the serious matter of getting a radio talk into shape. It was rather a job to get the thing into shape – the hardest job being to reduce the subject of the immunization of children down to very simple language. I spent a good many more hours than I would care to estimate on getting material in shape for only thirteen minutes of talk. That, of course, is an old story for you for you have had to do it yourself before – as a new experience, all I can say is that I will know how to do it better next time. As a matter of fact it went off very well and I have had a good many compliments on the suitability of the Murray voice for Radio! It was rather an extensive field for I was told that they wanted a simple explanation of what immunization was all about, what can be done for prevention of disease in kids, a plug for the university and the desirability of research, and anything else that would fit in! Just the same it was an interesting experience and I am glad of it. After my stint I had a good look round their plant and I must say it was a good deal more expensive than I had expected for a small place like this – though they are right in the middle of adding a good bit to it.

Just the other day I had a letter from Spooner[51] – sending you and Freddie (Smith) greetings – and a day or so later the portrait of you arrived and seems to be in good shape. Apparently Spooner is going to London soon and will be taking up a professorship there – he did not say which one. With the portrait there was a note from Hudson,[52] which I enclose.

It looks as if I am going to be landed with two more speaking jobs, but not over the radio this time. They run semi-annual "lectureships" for the general practitioners of the district and one is coming up in a month or two which will be largely on burns. Hall has asked me to take up the matter of secondary infection. Then in May, just before the SAB meeting, they have cooked up a course for practitioners in Windsor on Gyn & Obs in which I have been asked to contribute two lectures on the bacteriological aspects. I think it can well be justified as useful evangelism as well as the whole

---

51 See chapter 3, note 18.
52 Hudson was the head lab man in the Pathology Department at Cambridge. RGEM.

being a good plan for keeping practitioners up to the mark. Besides I will be able to use some of the material I dig up in teaching undergraduates.

One of the new professors arrived a week or so ago. He is Rossiter[53] from Oxford – who is a pleasant fellow and knows many people you do in England and elsewhere. He is a biochemist and I would say a very good one. He has an excellent manner and on the one occasion that I heard him talk I am sure he will be a very good teacher. He certainly fits in with the spirit of the place. The other, Edholm,[54] a physiologist should be here very soon.

The medical students have got together a show which they are putting on at the Grand Theatre to capacity audiences – there is a lot of talent among them and by all accounts they have a really fine little revue. We are going tonight and we expect to have a good time.

Work does not seem to have made a lot of progress in the last few weeks and I hope to get time a little more free of time consuming interruptions. I have not gone much further with the identification of the peculiar Bacillus: except that it does not conform in all respects with any of the described species. It has many of the characters of B. alvei but by no means all especially in some important respects of habits in spore forming. By the way, why is B. alvei put in the Subtilis group – I don't see any good reason ...

2 March 47

The medical student show was marvelous and we wish that we could have seen it again. Quite the best student show that I have ever seen. It was fairly ambitious, remarkably clean for meds, and they had a tremendous amount of talent. The girls were reinforced by a contingent from the nurses and they made a most attractive chorus line and well trained. The singing, solo, quartet and chorus, was of the very highest caliber and probably the very best thing in all the show. It was a revue in show-boat style and strung together with a light story. It is the first show that they have done and they all say that they had so much fun at it that they would like to do it again. Although I don't think they did very much studying for three weeks or so, putting this show on probably did them all a lot of good. Anyway they were much

---

53 Chapter 4, note 109.
54 The physiologist Otto G. Edholm was educated in England, where he obtained his degree at King's College, London. He moved to Canada in the early postwar years to assume the post of professor of physiology at the University of Western Ontario. In 1947, he became head of the Division of Human Physiology at the Medical Research Council's National Institute for Medical Research, London, England. Waldemar Karwowski, *International Encyclopedia of Ergonomics and Human Factors* (London: CRC Press, 2006).

appreciated and they played to the public for three nights in a proper theatre and gave a most polished performance.

I am having a lot of fun and good instruction by reading the fishing books that you lent to me. I am particularly impressed by all the good observations there are in Haig-Brown's book and I am reading it over again with as much pleasure as I had the first time.[55] There is much good stuff in all the other books and I am absorbing as much as I can. You are altogether too generous with your own things ...

... The departmental surveys conducted by Hall, the Vice-President, and the comptroller are to be an annual affair which is a good thing. I fear that our appropriation for apparatus etc is to be a bit smaller this next year though we will be able to sue for more if we need it. Hall argues that the salaries of the staff have to be raised not only to bring them more in line with other universities but also to help people with the increased cost of living and they will have to do it by decreasing some of the other costs. However, he says that it does not stop you from getting important things for you can put up a much better argument when you really need a thing than when you think it may be useful.

Doris is having a busy time for Peter has no use for pens or other restraining things and is only really happy when he has a whole expanse of floor to play on. Alice for her part has cast aside all restraining bonds for she is now expert at climbing out of or into everything. So when she is awake someone has to be alert for devilment.

... It is good that you are getting out a bit. It must have been fun to go on an oyster feed with Ted and Sylvia (Rose) – enthusiasts like yourselves. We have not been able to get free as much as we would like to. The main trouble is the problem of "sitters". They are hard to find: particularly steady ones who have a reasonable idea of what is fair to charge. The medical students have a 'sitter service' as they call it but they are expensive and charge a flat rate of 25 cents an hour. It makes you think twice about going to movies. However, we have some friends who will hold the fort on occasions and we have got along alright so far.

... I have been hampered for the past week because Dr. Bittner[56] and Jean MacKinnon have both been laid low by the form of flu that is going the rounds. Jean has been quite sick and is likely to be abed for a while longer. Since Barbara has her right arm in a cast with a fractured scaphoid it has meant a little difficultly carrying on the

---

55 Roderick Haig-Brown (1908-1976) was a renowned English author, committed conservationist, and fisherman. He moved to British Columbia in the early 1930s, and published works on numerous subjects, including sports fishing. His most influential works include *A River Never Sleeps* (1946), *Measure of the Year* (1950), and *A Primer of Fly Fishing* (1964). *Canadian Encyclopedia*.

56 Chapter 3, note 90.

routine media making and routine work together with research.[57]

All the best to all of you and I will try and keep my letter writing down to a more regular schedule for the future.

    Your son,
        Bob

                      McGill University, Montreal
                                  2 March 1947

Dear old Son,

Time slips by & I seem to have little to show for it. So many things which are essential to maintenance of work take up a lot of time without leaving any mark. The department keeps going by being held together almost by force & its extension tires me. The most surprising difficulties keep developing. Much could be done with a more satisfactory wages scale to secure more stability & satisfaction of staff.

... I have to go to Ottawa on Tuesday for a couple of days (Defence Research Board meetings). I'm not looking forward to it because I do not know what it may develop into, since it involves the question of the continuation of the old work & my views won't be popular. I shall have some arguments on my hands.

There is more & more reason why I should go to the Copenhagen Congress (I.A.M.S.) but I see no way of being able to do so. I cannot pay such a trip myself & there is no way to provide for it. I'll have to appoint a deputy, someone who is going for certain.

I have made some further changes in the reconstruction of the department. I am giving up my lab & office & rearranging the space to make a media-making & technician service & washing up labs grouped in relation to the autoclaves. This will centralize those services to the great advantage of the work as a whole. I shall probably move myself to the corner lab where media are made at present. In some ways it will be inconvenient for me, but it can be adjusted & I will become accustomed to it.

We are all well. Sue seems to be fit & cheerful & much occupied with multitudinous sewings & knittings. Mum's old sewing-machine has been adjusted & seems to be humming all day. The event is not far off now. She is a little troubled by slight swelling of her feet. Blake seems well (occasional slight trouble but we don't find any cause) & he is keen on his job.

... Mum & I are fit but get a bit tired more easily than we used to. I find I haven't the energy nor the "fight" I had. I find myself feeling that it is not worth exhausting

---

57 Jean MacKinnon and Barbara Blay were technicians at this time in Robert Murray's laboratory. RGEM.

myself over things & take more time to try to ease them into place if I can. I'll be glad when this reorganization which is going on is completed so that I can arrange my own spot a little more comfortably & try to do some of the things that interest me.

I have just finished reading "The Great Lone Land" by Capt. F.W. Buller F.R.G.S., written in 1872. Much of it is prosy but it gives a very truthful account of things in the West at that time. His report to the Lieutenant-Governor is wonderfully good & far seeing. He gives, in his Report (reproduced as an appendix) & in the book a very interesting & instructive account of smallpox among the Indians & its ravages. He must have been an observant man with a capacity for gathering information.

The "Fisherman's Bedside Book" arrived. I have not read it yet, but I notice some old friends in it, & quite naturally, regret others are not there. It is worth having & I shall take it on my trip to Ottawa tomorrow. I also bought Negley Farson's "Going Fishing" & like it immensely.

All love to each of you,
Dad

6 March 1947
The Lord Elgin
Ottawa

Dear old Son,

Your phone call this morning was a surprise. I am sorry I did not mention in my letter that I was going to this hotel as it cost you extra to trace me.

There seems to be a tendency to appoint young men to important positions. I am not in the confidence of those who make the decisions so I do not know what directs this policy. You have McLaughlin in Surgery and my first impression was that it was the selection of an exceptional young man (I don't know his experience). At McGill they are in the process of making young David MacKenzie professor of Surgery.[58] He is well educated ... & has had special post graduate training, but I do not know that he has had real experience. A lot of money is to be spent for a building to house Experimental Surgery (Research) & his salary & other expenses are to be quite a lot.

Meakins retires from the Deanship & also from the Chair of Medicine. It will be interesting to see the selections for those two jobs & I shall not be surprised at anything.

I am confident that Hall would not make the offer to you unless he was sure of his ground & had considered all sides of the possibilities. The suggestion that you

---

58 David MacKenzie headed research in surgery at McGill University in the early post-World War Two period. See McGill University Health Centre Web site (http://www.muhc.ca).

work two years under his immediate direction and supervision would be a necessary period of training & trial. It would give you rapid instruction & methods & policy & procedure. Intensive experience upon which you could build. It would give you a chance to decide whether you like the job or not. Under such circumstances I see no reason whatever for you to fear accepting the job. Hall is a fine fellow & would be a good instructor & has a real responsibility in planning the offer & selecting you. Since this move puts such a responsibility on Hall you can be certain he has taken every precaution to be sure not to make any mistake. The conclusion of this observation is that you need not hesitate to accept the offer.

Of course a lot depends on how your colleagues will take it & cooperate with you. You don't want to be left with a constant struggle against the current. You don't want to have a fight on your hands with every move. Hall can tell you what you can expect & I suspect the vast majority will be prepared to play fair & wait to see how you do & with diplomacy & efficiency, sympathy & encouragement you can & will win everyone's support. I am confident you will make it a success of it.

An important consideration is what does it mean for your future. This you must discuss with Hall & Ash. It is probably the most important aspect to be considered. You are young & all I know consider you have a great future in research & teaching in Bacteriology on the slight showing you have had a chance to make. The two years probation will allow you to test the possibilities. If you can make & carry through an arrangement for some available *time* for your own use & laboratory facilities to do a reasonable amount of your own work, you can keep on some important research. You could do some teaching but research should definitely dominate. This would give you the living touch, the sympathy, the appreciation & very importantly a relief mechanism essential to your happiness and wellbeing. It is, in my opinion vitally important never to become an administrative hack. It is essential for you to be one of the workers in the research team of the University, making your own essential contribution & part of your own scientific world. I repeat, therefore, that you must obtain & securely fix ... a definite agreement of time, opportunity & facility for your own research. You will thus more quickly gain the admiration & cooperation of your colleagues & you will be happy yourself.

The offer is an honour & a tribute. I don't see how you can easily refuse it a trial. I am sure using it properly & well you can contribute to the development of the University, now well under weigh, & can make for yourself a real position in Canada & in bacteriology.

I am immensely proud of you, my boy, for what you have done & the way you have carried yourself to gain this tribute. I admire Hall & his achievement in such that his selection of you is a distinction to be proud of. Your mother will be no less proud & happy than I am. I am sure Doris is happy and proud & I am sure you recognize she has contributed to your success in a thousand ways. Give a hug for me and my grateful thanks.

> Love to you all,
> Dad

McGill University, Montreal
9 March 1947

Dear old Son,

Your letter was here when I returned from Ottawa but you did not include the note from Hudson. I had not heard that Spooner was going to London – there are a number of vacant jobs in the U.K.

I'm sure you made a good radio-talk but it is discouraging to have to compress things too much because there is always a danger that it becomes too abstruse to have any effect. There is always a danger at the other extreme (too much time) to so cover up the point that it is overlaid. After all you cannot expect to be able to make the general public understand the reasons for everything. I'm sure you did it very well.

The proposed lecture tours for you are good & useful. There is need for a proper realization of the importance of bacteriology & that "wonder drugs" only intensify the need for bacteriological control. Hall seems to have done some good picking & things should hum around your University.

You must have exceptional talent among the students. I have never seen anything here that could pack in the general public. They are not always intelligible to the University audience.

... I cannot bring to mind more than you mention yourself about glycogen storage by bacteria. I shall try to remember to see if I can easily find something. I don't think there is much known about carbohydrate synthesis by bacteria.

I'm sorry you are having technician trouble too. I don't know what is to be done for the future, it is largely a matter of wage discrepancy. Completely untrained & hardly educated youngsters ask for a wage approaching what we can offer to medically qualified people. The situation is ridiculous whatever way you look at it. It is not possible to attract medical graduates to do bacteriology at the salaries offered.

My trip to Ottawa did not help me much. The scheme put up by the Director-General of Defense Research [Dr. O.M. Solandt] covering such things as I am interested in provides for almost all requirements & details can be worked out. I cannot escape cooperating in the way I am asked to, though I do not wish to give the time involved. I shall have to try to get the system working efficiently as quickly as possible & then ask to be allowed to retire. It may be a compliment that I am wanted back as Chairman but I thought I had wriggled out of it all.

I gather the "Province of Quebec Association for the Protection of Fish & Game" also have a scheme to make me Chairman of their "Committee on Fish". From a talk I had with Percy Nobbs[59] it does not look like an easy job & I am not very happy about it because I already have a great deal too much to do, especially during the next few months.

It was fine talking to you three this evening & we wish so much we could see you. (Parts of the Bk & white film of our trip last summer is quite good. I'll send it sometime)

With all love,
Dad

9 March 1947
London, Ont.

My dear old Dad,

It has been rather an exhausting week – not so much physical as mental exhaustion. The decision itself is not so bad but the reorganization of one's ideas seems always to be a long and rather sleepless process. This was made all the more difficult because it was a development that came "out of the blue" and felt, & still feels, very strange. I must admit that, for almost a week before this bombshell and prompted by some comments in one of your letters, I had been thinking of the place of administrative knowledge & ability in a career such as you & I had planned – but thinking in such high terms had not occurred to me and made the strangeness of the offer much greater. I must admit that one of my first thoughts was of you – for the unstinting help you have given me, the advice & encouragement, and, much more recently, the things we had planned-out together. These facts made the phone call very important to me – because I feel that you should know about such things and, although decisions are now mine, I value your teaching, your experience and your advice. Your letter was awaited anxiously and read with great care & joy.

I will recapitulate the broad outline of Hall's plan. He proposes to submit to the Committee of the Faculty that I should be appointed as Assistant to the Dean on or about April 1st. That he will remain as Dean until July 1st 1948 leaving most of the presidential duties in the capable hands of Landon, the vice-president, for that year so that he will be able to spend at least half of each day on the Faculty of Medicine and training me in the duties involved. In fairness to all, it was laid-down that appointment as dean would not be final until after Jan 1st 1948 giving time for Hall, myself

---

59 Percy Erskine Nobbs (1875-1964) was a professor of architecture at McGill. He designed many buildings on the McGill campus, eight of which are still in use. He also won a silver medal in fencing at the 1908 Olympic Games in London, England. "Architect Percy Erskine Nobbs," *McGill News*, 7 December 2007; RGEM.

and the faculty to be sure of the fitness of things. So that, if all goes according to plan, I will have 15 months of training & experience. Now, of course, all this is contingent upon the approval of the Nomination Committee of the Faculty. How likely this is I can only guess – but if Hall goes after something he usually gets it.

Undoubtedly it is a serious step and as you say it would be difficult to refuse to give it a trial. I must admit that I am, and have been for some time, very interested in Education and particularly Medical Education and this is no mean opportunity to learn and practice. There would seem to be some "imponderables" behind the offer but in my talks with him, Hall has been very open and apparently certain of his ground. He says that I have many of the qualities that he thinks desirable. As you say, and he says it too, he is rather intimately concerned with the matter of his successor and would not make an offer lightly. We discussed this matter of suitability at some length and I was unable to shake him from his stand. For my own feeling on this matter I don't know what to say: my youth and inexperience are against me but I think I am capable of learning matters quickly. I get on well with the majority of colleagues and I probably owe a portion of thanks to them for this offer, in that most of them seem to like me. I have taken an interest in what they do, attend their meetings and when I could help them I have helped them. Hall did say that there were others who thought as he did on this matter. Certainly there would be some who would say ill-advised things but I think the vast majority would play fair and would give their whole support if I were worth anything. It would require constant alertness, sympathy, understanding & a lot of diplomacy but it would be necessary and I might be able to do it – but there is no way of knowing this without trying. Reading between the lines of your letter I would say that you think that a chief difficulty would be my age – despite the tendency to appoint the young to high places. With this I entirely agree, it has worried me not a little, and I feel it can only be overcome by developing prematurely mature qualities of judgement and understanding to overcome the varying degrees of antipathy and envy which would probably develop.

The important matter of my future cannot be dismissed lightly – in Hall's view, as well as yours and mine, this particular administrative position requires someone still active in research & development of the medical cause (a little teaching) apart from all the other considerations in the job. As you rightly say this would give the living touch as well as a relief mechanism which I would greatly need. As Hall sees it, this is possible. He says that the reason he has been unable to do research himself since he has been here is because of the horrible administrative muddle he had to disentangle & reorganise, and, but for the presidency, he would have been able to devote approximately half time to it from the beginning of this year on. As I see it such a program would only require determination and a little help – it would require either an excellent technician to carry things on through inevitable interruptions or an assistant for collaborative research. But it can be done, and as Hall says when you are interested enough you will get it done through hell and high water. I don't think this can be done by agreement or contract but is entirely dependent upon your own willing. In the case of the Assistant job it will then be a matter of agreement & encroachment upon only by important contingencies.

The offer is certainly an honour and is a tribute to such small bits of work as I have so far done. Even if I refused I should probably ever regret it and it is still a tribute if the plan fails. A refusal based on any reason other than complete devotion to research would be an admission of mental ineptitude and implying lack of ability to lead, and disinterest in the cause of Education, therefore unworthy of any position in a university.

On all these grounds I have told Hall that I would accept the offer if it is approved and would do my level best to justify his faith in me. The only thing I can do to quiet any misgivings I may have over my age & inexperience is to show myself that they can be eradicated and overcome – that I will do. In this way I may hope to show that your confidence & love is not misplaced despite the fact that I may attain those ends by a somewhat different route than we had planned.

Doris was just as astonished and dazed by the news as I – though she said that she would expect anything & claims she is "hitched to a shooting-star". This plan would take with it some responsibility & social duties for her shoulders which she is willing to bear though quite naturally apprehensive. Of course, to make things easier for her there would be money to provide help & furnishings, and the university would aid materially in getting the tangible necessities such as a house. Through all advantages & disadvantages, and in all events, we will have something to show for ourselves, to remember & to look forward to.

There certainly seems to be a tendency for the appointment of young men to important positions – the reasons are obscure except in the cases of undoubted ability, even so it seems hard to justify in many cases. In the case of McLachlin I do not find it strange (anyway he is about Hall's age) for he is a first rate teacher, a good administrator, and an excellent surgeon who has gained great respect from his elders. His training is good, both in research and surgery & I think he is an excellent choice for such a post in a medical school. I think I met David Mackenzie once a long time ago but I don't know anything about his abilities – anyway, the teaching of surgery is so bad & the teacher-student relation so haphazard that he could hardly do anything except improve things in that part of McGill (ask Ted about it!). The successor to Meakins will be a matter to be awaited with interest. ...

Jim Patterson[60] told me that you looked well and were in good form in Ottawa. I hope that the meetings were satisfactory but I also hope that you won't ey too tied

---

60 See chapter 3, note 92.

up with it. Your phone call & and opportunity to talk to you & to Mum was a joy for us. I expect Jim told you about some of our activities – the "Sportin' Lectures" are postponed a while, which is a good thing, but I expect that we will have something arranged in a few weeks. They will be fun, I think, and perhaps they will be appreciated by some.

I found in the latest number (#5, 1946) of Acta path et microbiol. Scand. a paper on a member of the *Bacillus* genus with motile colonies which would seem to be similar if not identical with ours. They propose it should be called *Bacillus vagans* (ex vagare). They describe the morphological appearance well and have excellent pictures of the colonies and their tracks. They state, as have others, that the colonies rotate as they move – (is that the origin of "circulans" in the case of *B. circulans*?).[61]

It is getting late & I have to go to the lab before I can go to bed so I must close now –
 Your loving son,
  Bob

McGill University, Montreal
12 March 1947

Dear Old Son,

Everything you say in your splendid letter of 9 Mar is right & your attitude to the whole situation is correct & admirable. We are immensely proud of you, that you have done so much so well & that you can write such a fine letter with command over the situation.

You are correct in reading a thought of your age into my remarks. It is true that you might seem young for the job but you are mature for your years & ready to learn & be advised (but be careful to weigh advice carefully) & make thoughtful decisions. I put against any qualms on this score that Hall & others will consider all sides of the question & the 15 months training he will give you. The only one possible difficult situation would be if colleagues would not play fair on that account; but I don't fear it as it would not involve many & others would outweigh them & you could win them round with consideration & care.

I am convinced you have nothing to fear & a great deal to gain. Plans have to be changed because they cannot be made with full knowledge of what may happen. Opportunity must be taken advantage of because it never knocks twice.

Hall's plan is sound. He has confidence in you & you will justify it. He is certain to

---

61 "Bacillus" is defined as "A genus of Schizomycetæ, microscopic vegetable organisms of the lowest grade among what were once called Infusoria. Separated from Bacterium, with which it agrees in its rod-like form, and characterized by its larger size and mode of reproduction." *OED* (Online). With regard to *Bacillus vagans* and *Bacillus circulans*, Murray writes: "They are species names in which case both the generic and specific name is italicised to indicate that it is a formally described species and a description and type species specimen exists." RGEM.

have consulted those that matter. It is an important job in which you can do a deal of good. But I still urge the necessity of research & a little teaching to keep the common touch & relieve tedium. All that you can arrange.

Doris has good right to be proud of you. She will have some responsibilities too & will have to "be nice" to many she would rather not. Simple entertaining & a well placed little touch of friendliness can help matters a great deal. I am sure she will do it gracefully & well. She is a grand girl & all will be well & a lot of fun.

J.A. Ryle, Professor of Health & Social Medicine at Oxford was here yesterday.[62] He was Regius Professor of Physic at Cambridge after Curly Brown & resigned to go to this chair at Oxford. He seems to be doing a good job there & has some very interesting results.

Quastel[63] is here for a few days & was in the lab for a while this morning. He has not changed much.

We are all well.

Love,

Dad

McGill University, Montreal
13 March 1947

Dear old Son,

I think the enclosed will surprise you as it has everyone else. A few days ago Freddie told me he had been asked by the Principal if he would accept the appointment of Dean of Medicine if it were offered to him. He had just come from seeing the Principal & his loyalty made it necessary for him to tell me, otherwise I would not have known it until this morning's *Gazette* published it.

In view of the disorganization it must cause in my department it would seem necessary & even common courtesy for the Principal to have told me.

I think it is a good appointment & it seems to be popular, as far as I have heard. He is in for a tough time with the various schemes brewing here, especially those relating to the hospitals. I don't think he can do it as a half time job as is proposed it shall be. It

---

62 John A. Ryle (1889-1950) became the first chair of the new Institute of Social Medicine at Oxford University during the Second World War. His lectures in the United States, Canada, and Britain between 1946 and 1948 have been credited with defining social medicine as a new academic discipline. These lectures were published in Changing Disciplines: Lectures on the History, Method and Motives of Social Pathology (London: Oxford University Press, 1948). His son, Martin Ryle, won the Nobel Prize in Physics in 1974. ODNB (Online); Nobel Foundation Web site, http://www.nobelprize.org.

63 J.H. Quastel had worked in the Department of Biochemistry at McGill University. See http://www.mcgill.ca/biochemistry/department/history/.

is very certain my department won't get half his time. The situation does not promise to ease my difficulties, though I am sure Freddie [Smith] won't add to them himself. It will make some things a bit touchy for him too, but he can allay suspicions with tact. There are "traditional" jealousies here as you know.[64]

There is to be a meeting of the Columbus Club on Sat 22 Mar & the order of the day is REFORM. It needs shaking up & putting in order, but it will take quite a bit of doing. It may be a tough meeting & requires serious discussions made in a sober state. I have never known a sober meeting, so there has never been any real business done.

Love to you all,
Dad

McGill University, Montreal
17 March 1947

Dear old Son,

I enclose a copy of the address I shall give to the McGill Undergraduate Medical Society tomorrow night. I don't know how they will like it, nor am I at all sure that it is worth the trouble it has cost me.

I am irritated because it has been very badly typed & so full of errors, some of which I may not have picked up in running through it since. Why it is necessary to alter quotations to make them wrong & alter punctuation to alter my meaning beats me. Barbara does not know better yet. But Irene is supposed to do my work & shifts what she can onto the other; at times it annoys me greatly.

Sue started pains at 4AM this morning & by 6:30 they had set up a rhythm & improving. She had a light breakfast & was full of fun & laughter, chaffing Blake who was all of a dither. She went to the Mat about 8 & the baby boy was born about 2PM, as far as I know.

Stewart Henry[65] rang up this evening to say she was a model patient & behaved splendidly; she & the boy are fine & he is "a big fellow" (not weighed yet – (the rule is x days after birth there!)). Nice of him to call up. Mum has seen Sue this evening & says she looks fine, good colour & had a good supper. "Spinal" wearing off. It is a relief that it is over & in good order.

I am exercised with too much on hand. Examinations for 5 courses all different in character, written & practical. Yards of proof (Bergey) to read & needing some alteration with considerable care. Several engagements of various kinds this week. And ever threateningly the two chapters of the book on medical bacteriology to

---

64 Fred Smith was dean of medicine until he died in 1949 of a sudden heart attack. See chapter 1, note 87 and chapter 4, note 43 for information on Smith.
65 Stewart Henry was a much respected Montreal obstetrician and gynecologist. RGEM.

write. Building alterations. Alterations to plan in an unsuitable space at the C.M.H. to move this lab to this summer & to draw up lists of equipment needed for it & to try get personnel for them to staff it. How to cope with the effects of Freddie becoming Dean & provide for the work he will not be able to do. Varied annoyances & problems which crop up in the department – especially technician trouble & how to get the ever piling up glassware washed & back into use, since Mary & Helen left. The place is becoming very difficult to run. Then there are several outside things – committees etc. Perhaps I'm aging & loosing the grip & stiff back I had.

Uncle Fudge knows there is something on, he feels the air is changed & he does not know what it is. Funny old fellow & so affectionate.

I must look over & arrange my 9am lecture for tomorrow.

  Love to you all,
    Dad
I am sending you some duplicate reprints.

London, Ont.
17 March 47

My dear old Dad:

Such exciting news. It came as quite a shock to me to realize that I am at last a real proper Uncle! I can quite see you & Blake toasting the young gentleman – a most suitable christening! Doris rang me up at the lab this afternoon in a state of high excitement. Unfortunately we have not celebrated his arrival properly for Doris went out to her drawing & I had to devote myself to finishing the reading of these seven essays for a prize in Medical History. But we will drink to "himself" tomorrow night.

Among other surprises, that of the appointment of Freddie Smith is not the least. I imagine that it might work out well but I must say it surprised me greatly. I enclose the cutting as it was in the local paper. ... As you say, it will not make things any easier for you – the burden being great enough as it is. I suppose the appointment begins immediately just in time to leave you shorthanded for the last push and examinations.

Bert Collip[66] was here last week but I only had a few words with him. He was doing a most hurried conducted tour of the researches going on here and was very pressed for time. NRC hurries I presume.

I had heard that Ryle was over here. I met him a few times at Addenbrookes when he took ward rounds – though he was away most of the time then (just before his appointment to Oxford). Quastel I never met though I know some of his work – he

---

66 See Introduction, note 317, and chapter 1, note 56.

has done a lot. I heard a while back, on the grapevine, that he was shopping around for a job over here – maybe!

Speaking of grapevines, I must put my ear to it again. We have a place on our budget for a fellow in bacteriology at the princely stipend of $1200. This should be filled as soon as possible (if possible) for it has been vacant for a while (I didn't know this) – apparently when I came as lecturer part of my money came from this & part from the rest of the budget.

Now that I have a full appointment, Hall has somehow got this one by the board of governors! Ash has an idea he would like to have someone with good biochemical training though he remains open to my choice of someone who would like to work on clinical material of which there is a lot going to waste. Another possibility should be someone able to do some mycology – but they are special types & would likely be above the price. If plans went through there would be an opening in the future for a good man.

Your letter concerning my future possibilities seems to be assuming the plan will work out & is a foregone conclusion! This is *not* so – for I am sure there are others to be considered & the result rests on a committee. So far it is only a rather flattering possibility.

We are still having technician trouble. Jean is not well yet & Barbara is still hampered by a cast on her right arm. It holds lots of things back & slows my work to a slowish walk so that I have had to switch tactics. Despite this I am having an interesting time, in off moments, with the immigrating bug – of which I sent you a subculture last week. I certainly wish that the taxonomists of the genus bacillus had laid down a scheme of study & stuck to it. ... Anyway I have now most of the morphological criteria including sporulation & germination fairly straight – I spent a tiring couple of days watching block agar mounts. It hasn't given any information on how the beggar moves. I am now trying to see if I can cut sections of these migrating colonies in the plane of the colony to see if there is a characteristic arrangement. (Do you know any good ways of cutting colony sections?) (on agar surfaces?). I have not yet thought of a suitable technique of following this business – though I have several impractical ideas! Some fellow, years back, studied moving pictures of this form of "migration" and decided that over 90% of the rotating colonies did so clockwise – he should have gone south of the equator to see if they reversed!

It is very late & I must go to bed – we are so glad & proud of Sue's & Blake's son and I am sure you are a very proud grandfather. Both you & Mum sounded very happy & I'm sorry we couldn't stay on the line longer specially to talk to Blake but we have overworked long-distance this month & *had* to limit the call. I do hope he was not hurt that we did not talk to him.

    Your loving son,
        Bob

... I gave Hall the copy of your speech to read – he returned it with a note of praise & asking if it was being published – if not, he asked if it could be used for the Undergraduate Journal? He was obviously impressed by it.

London, Ont.
23 March 1947

My dear old Dad:

I very much enjoyed reading the typescript of the address to the undergraduates. I think it must have gone very well and I heartily approve of its sentiments. I know these things are a bother to get together and are often not so pleasant to give but they are really worthwhile in the end.

We got a very cheerful letter from Blake this week. It is hard to believe somehow that the boy was born less than a week ago. Sue seems to have had a reasonably easy time and we are very glad that they are both so well. It is a curse that poor Blake has such a cold and that he is a virtual outcast at the Mat [Maternity Ward]. They would miss a lot of visits that way but I hope that he is better now and has been able to see the son and heir.

... I can just imagine Fudge scenting that there is some new excitement in the air and that it has something to do with Sue. He is a wise old bird and seems to have a real feeling for the family. He is a real good hound.

Thank you very much for the reprints – I am glad to have them though I have made little arrangement so far to store them in a reasonable order and usually put those that are of use with my own filed notes on the various subjects. Even the notes seem to be in unsatisfactory order – I don't seem to have the strength of mind to alter things about too much though I can see that many things would be better collected on a physiological rather than a taxonomic basis (for me at any rate). I also received a reprint on Active Immunization by a fellow called Edsall (Harvard, I think) which is useful.[67] I suppose you asked him to send it for I did not.

I think that I forgot to tell you that I have had a little correspondence with Pirie[68] about Hyaluronidase inhibitors. He told me that you had seen him and asked him to write. Thank you very much for I have got a couple of good ideas from him that may be of great use. He sent me some small samples of his acetylated and sulphonated

---

67 The Harvard Medical School graduate Geoffrey Edsall was assistant director of the Division of Biologic Laboratories of the Massachusetts Department of Public Health (1940-1942) and its director from 1942 to 1949. He was also for many years professor and chairman of the Department of Microbiology at Boston University School of Medicine. In 1951 he was appointed director of the Division of Immunology at Walter Reed Army Institute of Research.

68 The biochemist N.W. Pirie (1907-1997) collaborated with the British bacteriologist Ashley Miles in the early 1930s on research into the bacterium causing Malta fever. W.S. Pierpoint, "Norman Wingate Pirie, 1 July 1907-29 March 1997," *Biographical Memoirs of Fellows of the Royal Society* 45 (November 1999): 398-415; *ODNB* (Online).

products of Hyaluronic acid for me to try. ... So far I have nothing except ideas on the matter but hope to have some real information soon. I am glad to say that Jean (technician) is back again and I will be able to get on with a little more speed from now on.

Rossiter, the new professor of biochemistry, is a very good fellow and very helpful. He is hoping to get to work on leucocyte enzymes and is casting around for methods. I seem to remember you telling me that you had found an extract of the salivary gland of some insect (larva?) which produced an almost pure monocyte response. I think that such a manoeuvre would be of great use to him if you could remember the details of species and extraction.

Time seems to be going very quickly and already I have to start worrying about supplies for the class in September. Since there is not yet any decision on how the class is going to be handled this is a little difficult. I am afraid that we are going to have to take a few short-cuts and teaching will not be what it should be.

I am sorry that things are piling up on you so much. It is unfortunate that there is so much going on at once for it would not have been nearly as bad if the reorganization and changes could have been made a bit more gradually over the past years. I can understand the advisability of getting the media making, washing up and autoclave rooms as close together as possible – even though it will be a little inconvenient for you with your office at the other end of the hall. I hope that you are planning provision for a lab for yourself in this scheme. Are classes next year going to be as extensive as they are this? With Freddie more or less incapacitated it will be a burden and you will have to get some of the others to the state where they can take more of the load.

The troubles they are having in England are very disquieting and I feel uncomfortable whenever I read or hear of it. I had a letter from John Gibson[69] who is living with his wife in what he calls a "luxury caravan". Oddly enough he says they are better off in that than most of their friends who have real houses. He says that "in this country of mad plumbers" the pipes are either frozen, burst or both at once. Circumstance and nature have both been unkind to the UK and it seems very undeserved.

We are all very well and the kids are flourishing better than ever. They are very cheerful and happy and a joy to us. Alice sits up at the table with us now and her appetite and ability to shovel food has improved in proportion.

    All my love.
        Your son,
           Bob

---

69 John Gibson was one of Robert Murray's close friends at Cambridge and was also in medical studies. He was at Queens' College and distinguished himself as a rugby "blue" playing on the university team. They often met for a beer at The Eagle and shared a set of rooms during the summer vacation of 1940. Gibson later became an obstetrician and gynecologist and after a few years went to Southern Rhodesia to practice in Bulawayo. RGEM.

Royal Victoria Hospital, Montreal
23 March 1947

Dear old Son,

Sue & her boy are very well & she has more than enough milk for him. She is very happy & is longing to get home. She had a bad headache for 4 days, no doubt due to the spinal-cord anaesthetic she was given. I'm glad it's over as I'm scared of spinals; because the hospital's technique for treating the outside of glass vials is unsatisfactory & every one we have examined is contaminated.

... Blake has quite recovered from his ordeal. I have not seen the young fellow yet. He is said to be large & to have character. Started at 8-1/4lbs.

I saw in yesterday's paper that Joe Barcroft[70] died in a bus on his way home. He was 74 & was in the middle of some of the best work he ever did. I saw him when I was over for "D-day" & we made tea in his lab in a flask. He was full of enthusiasm & energy. When I was an undergraduate Joe was a Demonstrator in Physiology & once a week used to put on the most wonderful practical demonstrations of classical experiments in physiology: secretin, salivary secretion, blood coagulation etc. He would bring in the animal to the small lecture theatre (now long since gone) & start from the beginning & do the whole thing (dissection & everything else) before your eyes. Never did a single one fail. His accompanying talk was marvelous & inimitable.

When Bert Collip came back from London he rang me up & said he had seen you. I gathered he was rushing through business there but he was immensely impressed with the progress of your medical school. He was concerned with the N.R.C. research grants.

I think $1,200 is quite good for a fellow; it is after all pay to get a degree, presumably a Ph.D. The difficulty is to get a medical man. It might be possible to get a woman, but I am tired of training women. They are a dead loss as they soon become unavailable for jobs. Even if their family affairs allow them to work they are limited to the place where the husband has his job. A man's family follows him. I don't know where you are to find a candidate. You will have to circularize & advertise. There are a lot of jobs going for bacteriologists & among them I have to get someone to part replace Fred Smith.

I gave a sub of your bacillus sp. to Cliff Kelly[71] to play with but he will not do anything with it; I doubt if he will even use it for the class with advantage. I must get a

---

70 The physiologist Sir Joseph Barcroft (1872-1947) joined the Cambridge Physiological Laboratory in 1897 and remained there, with small interruptions, for his entire career. His early work on blood gases was the beginning of a lifelong interest in hemoglobin, and particularly its combination with oxygen. During the First World War, Barcroft worked on the medical and physiological effects of poisonous gas at Cambridge and at the British government's Porton Down experimental station. After the war he returned to Cambridge as a Reader and much of his research focused on high-altitude physiology. He returned to the CW operation at Porton Down briefly during the Second World War. He was knighted in 1935. *ODNB* (Online).

71 See chapter 4, note 95.

copy of the Galley (Bergey 6th Ed.) on bacillus for you; you will find the subject in very much better order than in the 5th ed, which was unsatisfactory.

... The speech to the Ont.Vet. Association is to be published in the J. of Comp. Med or whatever it is called.[72] Charlie Mitchell[73] has the M.S. & I shall send you a reprint if obtainable. It is nice to know Hall thinks well of it, I suppose, if he wanted to, it could be reprinted in another journal by arrangement. That was done with one of my papers years ago.

I look forward to your comments on the talk given to the McGill Undergrad Med Soc. It could have done with more thorough preparation & I am not entirely pleased with it as it is.

Love to you all,
Dad

Chester Keefer[74] gave a talk to Med Clin. on Friday on streptomycin. Good & conservative not claiming undue possibilities but will not restrict clinical excesses.

Columbus Club meeting last night. Got some good reorganization. Bill Newman President, A Gutherie V-pres, Hilary R. Sec & J. Heeney Treas. – definite plans for improvement & increase members to 40. Adelard to be kept in order properly.

London, Ont.
29 March 1947

My dear old Dad:

The tent came the other day and it is really magnificent. I have only been able to lay it out on the floor and have no real idea how it will pitch but it is just the same as the one that McLachlin has and they look good and serviceable. Directly the frost is out of the ground I will erect it in the back garden and have a real look at it. You really should not have spent the money on it – even though I really love to have it – I thought I had told you to go slow on it, especially after your blow from the income tax people. I am getting together some of the impedimenta necessary to bush travel ... for the projected trip this spring. It looks as if we will make a two canoe, two tent, trip into the Algonquin ... At present Hall, Rossiter, Angus and myself are the members of the trip – it should be fun though I have some misgivings about traveling in such a conclave, almost a fleet.

---

72 E.G.D. Murray, "Value and Function of Scientific Societies," *Canadian Journal of Comparative Medicine and Veterinary Science* 11, no. 2 (February 1947): 47–52.

73 See chapter 2, note 36.

74 Chester Scott Keefer, MD, DSc (1897-1972) was chairman of the Division of Medicine and director of the Evans Memorial Department of Clinical Research and Preventive Medicine at Boston University (1940-1955). Boston University School of Medicine, *The Academies of Advisors: Six Distinguished Figures from Our History*, http://www.bumc.bu.edu/www/busm/aa/distinguished_figures.htm#CSK.

It has been rather a bloody week and hardly any time to get on with research. Most of the time I have spent on advance cost accounting for the course next year to show Hall that we need some cooperation and help as well as to try and din in the idea that even a class of 60 medical students is a little heavy for the facilities at the present time. I have also had to give some extra time to revamping the projected timetable for next year which [H. Alan] Skinner[75] submitted in an almost impossible form. I have it now so that the three heaviest courses of that year (Path, Bact and Physiology) only have to duplicate their labs and can give their lectures all together in the auditorium.

... Edholm, the new professor of physiology, arrived this week and he seems to be a very good type, amusing, scholarly and full of energy. He should be an excellent stimulus around the place. I think Hall has made some very good choices in filling these posts.

I hope that you are able to find some reference to a good way of cutting sections of colonies. It would be easier if I wanted to cut cross sections but I want to get a section in plan. However I ran some blocks through at the beginning of the week and have been able to get some sections. I have them on slides but have not yet stained and mounted them – they don't look too promising.

... Fraser Gurd[76] is coming to give some of the lectures in this three day lectureship. I have to give a half-hour talk on the role of bacteriology in surgery which will give me an opportunity to say something of the place of the lab in practice but means that I must give up some much needed time in polishing the points into respectable language!

I am very sorry to hear of Joe Barcroft's death. He was a very great man and will be much missed. Edholm told me of it, and he saw it in the Halifax paper the day he arrived – it was not in our local rag at all.

I am glad the Bacillus sp. arrived alright and I hope that you can make some use of it. I have had some photographs made and if they are good I will send them to you for it is a good example of a peculiar form of swarming. ... The bullet-like colonies if you find a good one move fast enough to appreciate the movement while you are looking at them under a dissecting microscope, which is going some for a whole colony. I think I have almost analyzed the sequence of events going to form the "circling" colonies (as opposed to those moving in a line) but will tell you more when I am more sure.

It would be good to have a decent setup for streptomycin and penicillin assays. But it is not worth my while at the moment to do either for the requests are so few and

---

75 See chapter 3, note 77.
76 The surgeon Fraser Gurd was the head of the surgery department at McGill University and surgeon-in-chief at the Montreal General Hospital. Douglas Waugh, "Profile: Fraser N. Gurd," *Canadian Medical Association Journal* 135, no. 3 (1 August 1986): 1409-1410.

far between. I have done penicillin assays when asked using a standard staph. and as for streptomycin it has been used only three times to my knowledge in this hospital. They are not so much backward as lazy and careless.

... We are so glad that Sue and their boy are getting on so well and that he is so bonny. You and Blake will have to resign yourselves to being second fiddles for a while.

Peter is still getting on well and is putting on weight fast. We have no certainty of what his trouble is but it is some relation of infantile sprue and they grow out of it sooner or later. We hope it will be sooner. Alice is getting a great talker, improves day to day, and is fining down to be quite a handsome little girl. Doris is in great health and spirits, despite the unspringlike weather, and says that now the family is on a reasonable schedule she is a new woman!

 Your son,
   Bob

          London, Ont.
          4 April 1947

My dear old Dad:

... One of the busy things of the week has been the "Alumni lectureship" which was on surgical matters this time and the guest Fraser Gurd. He was in good form and it was nice to see him & hear him again. My talk on "The Role of Bacteriology in Surgery" went over very well I am glad to say and I received many compliments & few criticisms. It was frankly evangelism for better cooperation and understanding but I doubt if one effort would make much noticeable difference. Dean Hall approved of it and only criticized the fact that I walk when I talk! There are worse habits. I would send you a copy if I had one but it was delivered from the sketchiest of notes and I would have difficulty in reproducing it as it was given. At any rate it is a *blow* for the cause – worthwhile, if even a fleabite – and I will be able to reinforce some of it when I go down to Windsor.

Concerning the Windsor business it is a little difficult for it comes directly after the S.A.B. meeting in Philadelphia – (to which I hope you are going) – the 17th & 18th May.

By the way Fraser Gurd mentioned that my 1938 episode of "appendicitis" wasn't all appendix – but when he was talking about it we were in a crowd & I could hardly hear a word. For the record, what was going on then?

The Reid strain of [B.] subtilis and the two bottles of media arrived safely but I have not had the opportunity to look up the method & try it yet. It should be good.

... I am sending along a copy of the radio talk I gave on immunization of children for you – it is an extra copy for you to keep if you wish. Like the typescript of your talk to the undergraduates it is full of absurd typographical errors especially considering it was taken from my original & typed up. We have just acquired a half-baked girl to act as secretary technician & this was her primary effort. I doubt if it is excusable.

Ash has a terrible habit of requiring things more than a little different from the standard – even in typewriters. He wants so many gadgets on the typewriter he has ordered that I doubt if we will get it this year & it will cost a lot extra. Same with an ordinary rubber stamp! ... It is impossible to get annoyed with him, and even his gadgets make sense but they complicate matters (esp. $) and make for terrific delays.

I have just isolated an extremely mucoid strep from a bronchiectasis[77] more so than any of my others. Remains to see if it maintains and if it has other suitable stabilities. One interesting thing was that it was together with H influenza (& very little else) and that the strep colonies caused a slight satellitism of the H influenza. They may then have coenzyme I or a close relative.

I am very glad to hear of the awards to Bert Collip & Otto Maass.[78] I must have missed seeing it in the local paper ...

    All our love.
        Your son,
            Bob

                                    4 April 47

Dear Dad:

Saw Hall the other day concerning the proposition he had made. As you know it was only a general proposition for further submission to committee, but with possibilities for the future.

He tells me that a choice has been made & an outside person asked to accept the post. He told me who it was and I must say it would be a most acceptable choice to us – with only a very minor deficiency. Age & experience are necessary & in this case it is well filled with scientific accomplishment to boot. I am very glad of this – he also told me that I would likely be asked to take on part-time, as assistant to the new man, with the likelihood of there being two assistants to the Dean – one Public Relations & one academic (me!).

Anyway that is the way things are shaping – I am not sorry.
    Bob

---

77 Defined as the "Dilatation of the bronchial tubes." *OED* (Online).
78 See chapter 2, note 30.

Royal Victoria Hospital, Montreal
6 April 1947

Dear old Son,

I asked Watson to send you a copy of "Professional Fly Tying and Tackle Making" by G.L. Herter. It is a very useful book I have just found. I can add some useful tips & tricks in places, & it lacks delicacy in places of some importance. It is influenced somewhat by "professional" mass production. Still it is a useful book.

I am delighted to hear you are planning a spring trip to Algonquin. The proposed company sounds fine & there is lots of room. ... Your tent should be fly proof. But it is a good thing to take along a spray & some insect killer. I have a small nebulizer made by "The Canadian Nasal Spray Co" (cost $1.00 or thereabouts). You can spray inside the tent & it is even useful to spray your clothes when the flies are bad. Some people take a "flit Gun" & have good reason to be glad of it (even in the boat!). This year I am going to treat the tent with D.D.T. – the tent seems to attract the houseflies & stomoxys & they sit on it while waiting to plague you.

... Time-table trouble here is terrific, especially in the Fac. of Arts & Science. In that Faculty they seem to think it unnecessary to allow any time for practicals. The Med. Fac. is more easily arranged & works quite well but it does not give a damn for anyone else – so it's easy.

... The penicillin susceptibility using the paper discs with 0.5, 1 & 5 unit is easy & the titration by the Reed & Brewer method is easy too. We do a large number & it is used by the Clinical men, though not by all. Streptomycin may prove more difficult even though it is more stable.

Sue is coming down some of the day, even most of it, & so is the boy. She is very well except she is greatly troubled by her coccyx (which she hurt several years ago when she fell on the ice skating) & has troubled her much ever since delivery. She cannot sit ordinarily with it. I don't know who should see it; I have an idea that Moseley is probably the best man to see it & most likely to be helpful. The baby flourishes marvelously.

I hear on the radio that there are floods at London, Ont. & that people living on the ground floor are advised to move upstairs. I hope you are high-up enough that your house is not in danger of being flooded out.

I may be going up to Toronto to look for a virus worker before long. If so I shall take Mum & try to pop over for a couple of days to see you. There is quite a pressure of things here & I don't know how to get away, so I don't promise anything yet. I shall have to go to Kingston, probably 27th but only for a couple of days.

It is good to hear that Peter is getting a hold on his trouble & is putting on weight. I look forward to seeing him. I am sure Alice is a handsome little darling & full of chatter. It must be easier for Doris now the family is getting old enough to be easily managed. It's good to hear she is well & cheerful.

    Love to you all,
       Dad

London, Ont.
8 April 1947
15 April

My dear old Dad:

It was grand to talk to you on the telephone the other night and to get your letter today. It now seems an awfully long time since we were able to have a long conversation together – there are many things that we need to talk about and I hope that you may be able to drop in here, as you suggested you might if you go to Toronto. Doris and I would be overjoyed if you and Mum could do it. A pity that you cannot combine it with your trip to Kingston.

The Easter weekend was quite a dramatic one – and one of discomfort for a small proportion of the population. The river rose very suddenly and was already very high by Saturday noon. By late that afternoon all the low lying land was flooded especially that across the river from the hospital and over by the Adelaide bridge (where you and Mum explored when you were last here) and in many other spots. The river rose very rapidly and from what I thought were very insignificant rains. We had about two days of steady but comparatively light rain but it fell on ground that was already soaked from the melting snows of the week before – so that everything that fell ran off the land. When I went to the lab in the morning the water was only about three feet above normal and when I returned at noon it must have risen another two to three feet. By five o'clock that afternoon the water was within inches of the top of the wall of the little house by the bridge with all the chain and iron frogs. We walked over to the spill at Richmond street and there was a fine rapid going over it and all the land was flooded on the other side. That evening we went to dinner with the Browns. A more formal affair than usual for Norman's parents were there for the week-end. We had a very pleasant chatty evening with them. They were very impressed with the flood and have probably told you all about it if you have seen them. (Sorry, we went there on the Sunday night, after the shouting was over).

... The flood caused a terrific outcry for proper conservation and every man-jack seemed to have his own ideas of what should be done first despite a pretty good scheme which has been worked out in some detail being already on the cards. They broadcast the meeting of the city council on the next Monday when all these things were discussed and there was much motioning and counter-motioning. In the end I expect that the conservancy scheme will go through, involving well planned control of the headwaters and the drainage area. But the flood goes on down the river and the people in places like Chatham swear that most of their floods originate from the areas and streams below London and that they will not share the burden of the cost to deal with the headwaters properly. The usual wrangle.

The flood was not as bad as it might have been for it did not quite reach the top of the West London breakwater (though it lapped over a bit at one point with the wind) and so some thousand homes were reprieved.

All this is almost history now and we have had, in the past few days, some really rather nice weather that has made life worth living. Trees are showing signs of budding, Croci are almost over, the birds are coming in, and I can imagine the trout beginning to stir. It won't be long to fishing time.

... Sorry about Sue's trouble with her Coccyx. I remember only too well the several years of discomfort I had after I damaged mine in the Gym at school. A very painful business and rather hard to treat and symptoms take a long time to disappear ... I would say that it would be a good thing to ask Moseley to have a look at it – he seemed to me to be a fairly good man though his manner is not all that one could wish.

It is a pest that Mother feels that it is necessary to do so much work for Sue. I have no doubt that she enjoys doing it but she is undoubtedly doing too much and that alone will probably be enough to ruin her enjoyment of being an active grandmother.

... Last Saturday night they had a party up in the convocation hall at the north end. Quite a pleasant party was had by all and nearly all the faculty were there. Rather fun to climb into dress clothes again and both Doris and I enjoyed ourselves. We had a little get together over drinks both before and after with the Pattersons and a couple of other friends. The University is officially "dry" and being Saturday night no dancing or frivolity could be tolerated on university property after midnight. Anyway, as someone remarked it is better that a party should be a little too short than too long. It was nice to meet some of the people from the other parts of the university that we so rarely see.

Other than various alarums and excursions it has been a busy time. Preparing for things that don't come off, getting over the day to day work, and trying to get the research to pay a dividend every now and then. This latter has been a bit difficult betimes, various rather involved troubles have rather impeded making great strides though I am now getting things into relatively good order. I am incidentally finding out a lot of useful points that will help me with my intertwined study of the formation of capsules in M streps. I have got a lot of side information and have got some useful methods. My interim researches (or rather, the numerous little ones that help to fill in the gaps of time) are rather in suspension at the moment though I go on sniping at the moving colony as he comes by. I would have had something on the Tetanus toxin by now except for a plague of cat distemper which has held up the works.

... I am glad to say that Ash is getting some very encouraging results and is getting to a good stage in the anti-phage work - I am sure that this will pay good dividends. Miss Strelitz[79] has been doing some very sterling chemical work and has got a few Mgms of almost pure substance of activity $1/10^7$. Analytical work is the next headache for them.

Today is (or was) Doris' birthday. She was most happy to get all your letters (and enclosures) and she swears that she is going to write to you tomorrow. Unfortunately for her she has not been feeling too well today and unfortunately I was not able to be home for most of the day and some of the evening. A purely temporary upset but annoying. Otherwise she and the kids are very well.

... Spent two and a half hours at a meeting today to discus the set-up for the proposed researches into fundamental cell physiology. It will be praiseworthy if it works out properly, and I don't see why it should go wrong except that it is so very hard to find personnel. A great pity that Martin Entin could not see his way to become interested in it for he would have made a very good member of the proposed nucleus.

I suppose you have received a questionnaire from Gibbons[80] concerning the type culture collection meeting in London. Ash has a couple of interesting ideas that I will send to you in a separate letter.

I would be interested to hear what you have to say about the matter of collections.

It is very late and I should have gone to bed long ago. Sorry this is such a poor letter. A bad substitute for shorter and more frequent letters in which I can go into details properly.

All my love.

Your son,

Bob

Thank you for the "Wilderness". I like the fly very much ... What it imitates (if anything) I don't know. ...

---

79 Miss Strelitz was Asheshov's assistant.
80 The microbiologist Norman E. Gibbons provided instrumental assistance to R.G.E. Murray in the establishment of the Canadian Society of Microbiology in the late 1940s and early 1950s. He worked in the Division of Biological Sciences of the National Research Council of Canada for three decades. He was honorary secretary of the Royal Society of Canada from 1956 to 1959. N.T. Gridgeman, *Biological Sciences at the National Research Council of Canada: The Early Years to 1952* (Waterloo: Wilfrid Laurier University Press, 1979).

Royal Victoria Hospital, Montreal
13 April 1947

Dear old Son,

I got your letter & the enclosures. I am not sorry for the altered scheme. Since you are satisfied with the choice made & he is someone of experience & distinction, it is likely to be better for you to gain experience of administration working under guidance. You will learn the intricacies of administration & it will be valuable to you in the future. I shall be interested in the development when it is made public.

The other enclosure was your radio talk. I like it very much indeed; it is very well conceived & attractively put together & carries conviction. It must have been informative to listeners & at the same time easy for them to follow the argument. Sue & Blake read it with great interest too.

Fraser Gurd told me he had seen you & he told me of how well liked you are by your colleagues & that you are doing a fine job. He was pleased to tell me this too. I'm glad your talk went over well. I gather from what you say you are going to Windsor to talk there.

... About the appendicitis. You had an acute appendix alright. The addition Gurd refers to was his diagnosis of "Terminal Ileitis", which he is keen on. That is an inflamed condition of the last inch or two of the ileum where it runs into the caecum[81] with enlargement of the regional lymphatic glands. He removed a lymphatic gland & I cultured it in various ways & it was sterile, as usual. I don't know what it means but there is some literature on the subject, which, so far as I know, does not explain anything.

I'm glad you like the look of "Lord Iris"; I think it looks good. I have asked Rex Carthew[82] to get the exact description from Preston Jennings, who belongs to the same New York Anglers club as he does. It will be interesting to get it right & not guess at several possible ties. I send you four "Wilderness", which I know you like to use. It is a curiously good fly on occasions. Last fall, when Bill Newman & I were at the Columbus, there was one day when Wilderness was the fly without shadow of doubt.

... Sue & Blake's boy is flourishing exceedingly. He has filled out amazingly; holds his head up, has frogs-mouth on his knees already & is interested in things round about & is very good. He still gets an occasional small staph spot (folliculitis) but it is much less. I gather there is lots of it in the Mat [Maternity Ward] (among other things).

---

81 Defined as "The blind-gut; the first part of the large intestine, so called because it is prolonged behind the opening of the ilium into a cul-de-sac. It is present in man, most mammals and birds, and in many reptiles." *OED* (Online).
82 Rex Carthew was a well-off New York businessman and family friend who enjoyed fishing for trout in the more remote Laurentian lakes. RGEM.

The only thing that worries me is the amount of work Mum has to do & she won't take any time off to rest.

It's good to know you are all well & flourishing. I am looking forward to seeing you as soon as possible but I can't make any plans just yet. There is still a pile of things I have to do here & it is hard to get them done. I am struggling with budget at present & it is no easy matter & takes a lot of time.

  Love to you all,
    Dad
Let me know what flies you want.

                   London, Ont.
                   19 April 1947

My dear old Mum:

We got your telegram this afternoon. I was very sorry to hear of Gup's death though considering his age and the troubles that have beset him for the past few years I was not too surprised.[83] He lived a long life & had the pleasure of hearing of, though not seeing, a good crop of great-grandchildren. I would telegraph to Aunt Betty myself if I had the address but though I used to know it I find I can't recall it – thank you for making the cable come from all the family, there is little else we could do from this distance. Doris and I both hope that you are not too distressed by the event and we send you our fondest love.

I hope to be able to write you a longer letter very soon and send you all sorts of news.

  Your loving son,
    Bob

                McGill University, Montreal
                   22 April 1947

Dear old Son,

Mum was glad of your very nice letter today. Old Gup was having not too good a time & Mum seems to feel it is for the best. She is not upset.

The enclosed will interest you from tonight's Star. It is not quite what I would like to see but is what is to be expected. It is not praise enough for Collip and there is no criticism of McGill for losing the most distinguished man on its staff. I'm sure he will

---

[83] "Gup" was the family name for Hardwick Woods, Robert Murray's grandfather on his mother's side of the family. RGEM.

be a stimulating man at Western & I hope he will find ways & means to improve your lot, among others, to allow you opportunity & time for significant research.

... I got a brass lantern off a corvette for Mum's birthday. I'm sure she will be pleased with it to decorate the fireplace at the lake. It looks handsome & will be ornamented though will shed very little light.

Best love to you all,
Dad

London, Ont.
22 April 1947

My dear old Dad:

... The exciting news of the day is, of course, the appointment of the new dean. This will have been in the papers today in Montreal and probably may have made the headlines for it will be a more considerable loss to McGill. I think that Bert Collip is a most marvelous choice from all angles except possibly his notorious ineptitude as a public speaker – however, that is now a minor matter. I can imagine that many of you at McGill will miss him badly. I imagine that it reflects no great credit on Cyril James[84] and the McGill administration that they could not offer sufficient advantages to make it worth his while to stay. Another thing is that he will be bringing with him as well Noble[85] and Heard – in essence a whole department. Everyone seems to be very enthusiastic about the development, even the clinical types. I don't know where they are going to be able to put themselves but I gather that the general plan is to displace the quarters at present occupied by Burton[86] who, having expanded into reasonable space, will have to contract himself again. As Burton puts it, the only advantage of this is that he will be able to clean house. Anyway it is a great thing for

---

84 See chapter 2, note 6.
85 Robert L. Noble (1910-1990) came to Western and became internationally renowned for his cancer research. His work contributed to the discovery of the anti-cancer drug Vincristine. He was inducted into the Canadian Medical Hall of Fame in 1997. The Robert L. Noble Prize is awarded by the National Cancer Institute of Canada annually to researchers who have made significant contributions to cancer research. R.D.H. Heard remained at the biochemistry department at McGill where he achieved the first synthesis of radio-labelled steroid hormones. RGEM; Canadian Medical Hall of Fame Web site, http:www.cdnmedhall.org/; National Cancer Institute of Canada Web site, http://www.ncic.cancer.ca; McGill University Department of Biochemistry Web site, http://www.mcgill.ca/biochemistry/department/history.
86 Alan C. Burton (1904-1979) has been described as an outstanding pioneer of modern biophysics. He arrived at the University of Western Ontario in 1945 and organized the new Department of Biophysics. He served as professor and chair of the department from 1948 to 1970. "Alan C. Burton, 1904-1979," Department of Biophysics, http://www.uwo.ca/biophysics/Burton/Burton.htm.

this University and will do a great deal to strengthen its position in the eyes of the world at large and makes it an even more attractive place to be in at the moment. No announcement has been made at the moment of the plans for the assistants to the Dean.

I am glad to hear your comments on the matter of a type culture collection. It is certainly true that we should have an adequate National Collection which is well endowed and staffed. However, these collections are more often than not merely museums and the staff have not the time nor often the opportunity to take advantage of the collection from the point of view of research and taxonomy - this side is still left, in the main, to the individual worker or group. There seems to be a place still for the collection, in constant use and surveillance, by the research group. In many places, where research has stayed within groups for many years and where disciples are trained, I don't see any reason to doubt the possibility of survival of a collection provided that the NRC or some such body took sufficient interest to encourage the collection with money and aid. Such collections maintained by the active worker would remove some of the disadvantages of national collections as a source of cultures of special organisms and groups of organisms, and would encourage adequate research on the biology and taxonomy of a group as a whole instead of having work spread thin over a whole phylum. I am making up a brief of Ash's idea which I will send to you for your comment, which we would like very much as well as your help in case you approve of it.

Last Saturday night we went to a very pleasant party. First of all to a buffet dinner at the John Lewis'[87] – some 18 people and all very cheery and full of fun. After that we went to a playlet (in four acts) produced, written and directed by the students wives club which was really excellent and very well done on the theme of present living and working difficulties of married students. It might have been terrible but was in actual fact very good indeed. The star of the evening was a very talented seven year old daughter of one of the students. After that they had a very pleasant little dance in the gym at the med. sch. We had a very good time.

Another event of last week was the arrival of a son to the Rossiters. Doris has been and is all excited and has been talking nothing but very small babies ever since! Talking of such things, I came home a little earlier than usual and found Doris entertaining seven small children (including ours) from three other university families. They were unusually well behaved but there were toys scattered everywhere. All seemed to have good time - Peter played aloof and by himself most of the time sitting (like the

---

87 John Lewis was a practicing physician and cardiologist in London, Ontario. A lot of his early work was with veterans at Westminster Hospital. His wife, Betty, and Doris Murray were good friends. RGEM.

Big God Nquong in his bath)[88] in a box busy putting things in and putting them out again. ... The flies and the Pyrethrum extract have not yet arrived but I look forward to having them.

    Your loving son,
      Bob.

P.S. 23 Apr: Ash is rather infecting me with the idea of working with phage. It seems to me aside from the interest of the field that I would be wasting my time if I did not learn all that I could of the business from him, since he has so much experience and ability in the field. This could not be done without actually working with them and doing some investigation. I think that as soon as I have finished what I have on hand that I will go into the thing in detail and learn & do as much as I can.

The University of Western Ontario
London, Canada
24 April 1947

Dear Dad:

Here is a rough outline of the ideas we have talked over concerning special collections. It seems to me a bit of a large order but I can see that it could be done "without pain" if started from the right basis. I certainly think they would be useful & worthwhile. I would like your comments.

... I will probably be going to Philadelphia in May – stopping first in N.Y. to see Mother Grace & John (Marchand)[89] – returning via Windsor for this Obstetrical Junket. Approx: 10 – 18 May, inclusive. The canoe trip is planned, now, to start from Monday 26 May for 1 week approximately ...

    Your son,
      Bob

McGill University, Montreal
25 April 1947

Dear old Son,

Collip's appointment has caused quite a stir here. It is quite a shock to the academic people, who feeling it is a tremendous loss to McGill recognize it an indication that all is far from well. That a man of such international distinction should find it worth while to leave McGill after 20 years there makes one wonder how wrong things

---

88 A reference from Rudyard Kipling's *Just So Stories*.
89 Chapter 3, note 94.

may be. It is, of course, a tribute to Western that it can offer the attraction to him. The enclosed editorial from the Gazette is taken as a justly severe criticism.

I am going to Kingston tomorrow & shall return on Monday (28th). I hope the meeting will go well & accomplish some useful work. I shall get back in time for a meeting of the Alexandra Hospital Medical Board to consider the hospital becoming largely concerned with children's tuberculosis. A very important matter here.

I shall get back the day before Mum's birthday. I have got a fine brass lantern for her which I am sure she will like as an ornamental piece at the lake. It comes off a corvette & is of the characteristic workmanship required for the sea. I don't know what it is designed for on the ship. Anyway it looks fine.

Today agreement was reached at a meeting in the Principal's office (between the Canadian Legion and the University) for the gift of $15,000 to start a virus laboratory. Some details have to be fixed up within the University for suitable space & I shall now have to hunt staff for it & try to get the equipment needed. It will give me a lot of additional work & responsibility. One thing of some importance is that it stymies some moves that Vivian & Meakins were busy with which seemed to be designed to get virus work into Vivian's hands, which is ridiculous. Old Cushing[90] has been of the greatest help & steered things my way.

... Yes there are things to be done with phage as with other things. Craigie believes there are fundamental things concerning viruses to be learned from phage. Anyway you must do what interests you.

... Breed is very insistent on my going to Philadelphia (S.A.B.) so I shall have to try to fit that in too & there is a job to do in New York on the way. I am going to the Royal Society in Quebec too. So there is a lot to arrange, without leaving out examinations, building alterations, lab organization & that damned book. I must get a bit of fishing in sometime too.

  All my love,
   Dad

London, Ont.
1 May 1947

My dear old Dad:

... The money for the Virus lab is a good thing but I fear that it will be quite a chore for you to set it up and get it going. Not the least of the troubles will be to find

---

90 H.B. Cushing, a member of the McGill Faculty of Medicine, did some early work on the treatment of diphtheria. See H.B. Cushing and E.V. Murphy, "Treatment of Diphtheria at the Alexandra Hospital, Montreal," *Canadian Medical Association Journal* 6, no. 9 (September 1916): 817-822.

personnel – a hard thing for any lab. Still it is a good thing that you have the matter in hand for it will mean a lot to Canada to have more virus labs.

... I am sure that Mum will very much like the lantern. It sounds most imposing – it is probably a stern light.

I have a room booked for myself and Ash at Philadelphia. Now Ash is not going and if you are to be there I would far rather share it with you than with any other odd type that may turn up. It is in the Warwick Hotel and booked from the afternoon of the 12th May. I hope to go to NY first to see Mother Grace and John – others if I have time – and I hope to leave on the Saturday night arriving there on Sunday morning and staying until leaving for Philadelphia sometime on the Monday afternoon. We might meet there if you are going to be in NY as you mentioned in your letter.

The clippings from the *Star* and the *Gazette* were interesting. They quoted from them in the local rag but did not have them in full. I am also very glad to have the picture of part of the Staff of the lab. ... Everyone here is more than a little annoyed at the policy of the new animal house. They have just put into effect a schedule of charges some of which are pretty steep – e.g. such that a colony of 100 rats maintained for a year would cost $1825 at a rate of 5 cents per day per rat!

For the animals that concern us the figures given are:

	initial cost	*maintenance per diem*
Rabbits	$2.00	$0.10
G-pigs	1.50	0.05
Mice	0.25 (0.45)	0.02

We are all sure that it need not and does not cost as much as this to maintain animals in an animal house – otherwise labs would never be able to maintain an animal house of their own. It is probably right that in a shared animal house one should pay for what is used but it seems unfair to make it a profitable organization. Grants for even small bits of research would have to be relatively enormous to pay for the animals and their keep at these rates.

... Jim Patterson told me that he saw you and that you were well. I am glad to hear it for you have a great load of stuff on your shoulder. You will have to make a little time for fishing. I want to know in this regard whether you are planning to take a proper holiday this year – you must. I hope to be able to take off most if not all of July so that we may see you. Please let me know about this.

In getting supplies for the bush I am finding it impossible to get dehydrated vegetables locally and have not been able to enlist the interest of a local merchant in troubling to find them for me. Can you get them in Montreal? Especially appreciated would be potatoes. Another most important article is nesting billycans – not to be found. Do you know of any? Any other suggestions.

I have looked over my flies and think I have enough to get by on but I am short on a few. Of the streamers about the only one that I would like is Mickey Finn, unless you are going to be tying more of Preston Jennings streamers like Lord Iris (which I am prepared to bet will be a good fly).

... I am afraid that I have missed the noon post and I hope that this gets to you before the weekend. We are all well and my snivel seems to be on the mend today. Give my love to Blake and Sue and the boy.

    Your son,
        Bob

                McGill University, Montreal
                3 May 1947

Dear old Son,

Your letter of 1 May '47 arrived this morning as you wished it to. I'm sorry you have a cold & I hope you will shake it off quickly. It slows you down.

... I was not going to the S.A.B. & so did not book a room. Breed is worried that I might not go & I also find I am on a couple of committees so I have considered changing my mind. Now that you propose I share your room I am quite tempted & will go if it is certain Ash is not going. I would go straight through to Philadelphia arriving there on 12th & meet you there. I would go straight to the Warwick Hotel & deposit my things; so if Ash is definitely not going you might let the hotel know I am going instead & let me know so that I can make my railway reservations.

Thanks for the references to Corynebacterium pyogenes[91] infection; I'll look it up. I have seen two perfect cases & treated them successfully.

I have never yet seen a "communal animal house" that worked satisfactorily. The care of the animals does not suit everyone, in experimental infections their observation & notification of illness & death is not done properly, immunizations are often not reliable because of mix-up & the attendants have no loyalty or responsibility to any department, and finally a curious greed develops & they become much too expensive. I am sure it is essential for any department of bacteriology to have its own animal rooms & the care of them training its own attendants. I have never yet seen healthy animals in a biochemistry or physiology animal room or the appreciation of the situation. The prices per diem for the animals is much too much. I'll send you my figures.

... Mum is going to try get the dried vegetables you want. There are some to be had here though not all we would like to have. I have not seen any nesting billy cans for years.

---

91 See chapter 4, note 111.

I shall take a proper holiday & will arrange it to coincide with yours if you come to the lake. You need not worry about it. I feel I could do with some time away from things & people. ...

    Love to you all,
        Dad

                London, Ont.
                4 May 47

My dear Mum:

It was very nice to have your long letter the other day and to hear how happy your birthday was. All the things you got sound very exciting. My imagination is very much exercised by the rum-pots! They sound fascinating but since I have not seen them I can only imagine them to be a swag-bellied (excuse the expression but you know what I mean) fine copper pot obviously meant for the measuring of drink. ... they rather sound like the definition of a Demagog (or sumpn) in that book of Boners by Hunt.[92] The enormous lantern also sounds a worthy object and should look well at the lake – it must be enormous.

Your trouble with the hard boiled eggs must be rather like the trouble that Doris claims to have over the toasting of toast – especially on Sunday mornings. She has to have two or three trial runs before she can get the toast the way she likes it. We are a sort of carbon factory! (I shall have to get this letter out to the post before she can censor it) Anyhow, she claims that the toaster runs a lot hotter on Sunday mornings than at any other time and that this is the reason that she has so much trouble on that day of all days – I can get no admission out of her that it may be sluggish reflexes or unawakened liver bile or what have you. Much indignation is expressed at all my helpful diagnoses.

We think about the summer holiday rather frequently. As things go I think that we shall be able to get away for most if not all the month of July (Plans may change due to other circumstances, but I hope not) – June would be too early for the kids and I have to be back for August to prepare for the teaching session as well as clear up other work. We would like to hear how all these thought fit in with your plans.

It is good to hear how well Sue and her baby are – they sound well from your letters and Dad is full of talk of the baby. ... However, it sounds as though you have been doing far too much work – and I hope that we may take you literally when you say that you are now "off-duty". Doris is very thuggish on this subject.

... No signs of houses yet – especially for rent. The Rossiters by some stroke of luck were able to get a five bedroom house and rent it. They have not moved in yet and are

---

92 This may refer to *Wit and Humor* by Leigh Hunt (1784-1859).

still staying with the Halls, baby and all. The Edholms had to buy a place and hope to move in soon. We are looking and hoping. There was a ray of sunshine in the classified ads yesterday when there were all of five places to rent (unsuitable for us, but there) which has not happened since the year dot. We hope that it is an indication of space to come.

... Must now write a short note to Dad and then on to other work for there is a lot to do before I can clear off.

    Your son,
        Bob

                              London, Ont.
                              4 May 47

My dear old Dad:

Here it is May already and the fishing season almost upon us. ... in actual fact it is open here for trout now but I have not been out to open it, several of the others are away this week-end on sundry excuse but really to trap an early trout.

... I expect that you saw in the paper the other day that Almroth Wright died recently.[93] I also saw in a recent number of Nature the list of new F.R.S.'s included in the list was James Craigie. A very good choice. There were several other names a couple of which I knew but I have forgotten them now. You might be able to look back to it. Someone should send it to the S.A.B for they may miss it.

I hope to hear very soon what your plans are for the SAB meeting and perhaps your time in NY before hand. As it is I plan to leave here on the night of the 10th and stay in NY till sometime on Monday afternoon (12th) and then on to Philadelphia. Leave Philadelphia on Friday evening for Windsor where I have to give a lecture on Saturday and two on Sunday. (Makes me feel like an itinerant evangelist!) ... If you will be at the meetings I will be very glad for we have not met for many moons.

The page proofs of the *Bacillaceae* for the new edition of Bergey arrived safely. I have not had time to study it carefully but it really looks as if the group has been put in pretty good order and also as if it is a classification that has some practical value, which the old one certainly did not have. It is straight forward and not too confusing. I think that my organism may be B. circulans but it differs in one respect (I think) in

---

93 Sir Almroth Edward Wright (1861-1947) has been described as "Britain's first academic immunologist." He was a physiologist, immunologist, and bacteriologist, and was a pioneer in the study of vaccines and immunity to infectious diseases. He was known for developments in vaccination through the use of autogenous vaccines, and particularly the development and introduction of anti-typhoid inoculation. Frank Diggins, "The Life and Times of Almroth Wright," Biomedical Scientist (March 2002): 274-276; ODNB (Online).

that it does not ferment sucrose. The organism that the two Scandinavians reported and wanted to give the name of B. vagans would seem to fit in with circulans almost exactly.

I am glad that you are not against the ideas in the notes I sent you. I sent them to G.B. Reed[94] in case he would be going to that conference and might be interested anyway. Is the conference in London on type culture collections an international thing or is it commonwealth only? It makes quite a difference. ...

We had a very cheerful letter from Mum telling us all about her birthday, the rumpots and the Corvette lantern. She seems to have enjoyed herself tremendously.

I expect you have a tremendous pressure of work but I hope that you can get away to the meeting so that we may have a talk. I see that you are down in the SAB to do something with the taxonomy discussion. The Corynebacteria interest me and I hope someone may have something to offer as a rational way of classifying and specifying them.

    Your son,
        Bob

P.S. Doris has one of her drawings in an exhibition at the art gallery! She has been reading the paper from cover to cover looking for press notices! It is among an exhibition by Western Ontario artists but is a part of the sketch club section – no price tag attached.

London, Ont.
7 May 1947

My dear old Dad:

This morning there arrived a nice little box of flies. They are very nice and just what I want. ... I sent you a telegram today to tell you that I had written to Hotel Warwick saying we were arriving on afternoon of May 12th & putting Ash's reservation in your name. I very much look forward to seeing you and talking over many things.

At the moment, after a quiet spell dictated by many small reverses, I am working long hours and have to, after a long day, spend a long night getting the more extra-research work done such as preparation for these lectures in Windsor & I'm learning a lot!

Must be a short note.
    Your son,
        Bob

---

94 See chapter 1, note 115.

The University of Western Ontario
London, Canada
20 May 1947

My dear old Dad:

Thank you very much for your nice letter for my birthday. As to the cheque, I don't know what to say! It means a lot to me and is very much appreciated; on the other hand your presents in aid of my birthday started with a tent, progressed via a pack-sack reel and line, and a nylon head-net to this cheque. I feel that you have given me far more than a birthday warrants aside from the damages to your purse. They are all things I want and can use, all very much appreciated. Also many thanks for your joint telegram.

There was not much exciting after you left Philadelphia. It was grand seeing you there and being able to have a talk with you – though there were many things we left out – I wish you could have stayed a day longer. I left on Friday afternoon in company with Sandford Hooker[95] for N.Y. where we met John Marchand & had an amusing evening together. I then caught my train (midnight) for Windsor. My lecture was not til evening so I went over to Detroit for a visit with Perry & Louise TeWalt[96] and saw their daughter aged 3 months. By the time the evening session of lectures was over I was good & tired – got some sleep despite the fact that there was an advertising & sales managers convention – very fluid & exceedingly noisy – going on in part, as far as I could make out, in the room next to mine. I more or less struggled through my two Sunday lectures – but all of them went over well though I myself was not very satisfied with them. Got a drive back from Windsor Sunday evening with Frank Brieu – really tired by then.

Though the papers at the S.A.B. were not too good I did get a lot out of the meeting as a whole – definitely valuable. I saw N.R. Smith[97] & had a good talk with him. Also got some ideas from Stanier who played with allied problems.

I will have quite a rush this week – getting things ready for the Algonquin trip among other things. We are looking forward to it and expect to be leaving next Sunday – an advance of a day or so.

Norm Brown & Jean are coming in tonight – for they leave for Montreal at the end of the week. We are sorry to be loosing the company of such pleasant friends.

Doris is well but has been fighting the fates in the week I was away – came in the shape of colds, unusual persistent activity on the part of the kids & rather wearing.

---

95 Sanford B. Hooker was a medical scientist at Evans Memorial Hospital, Boston, Massachusetts, specializing in antibodies and immunological reactivity. RGEM.
96 Louise TeWalt was a lifelong friend of Doris. They met while attending Smith College. RGEM.
97 Nathan R. Smith was an American bacteriologist who worked with Everitt Murray on the Board of Trustees of *Bergey's Manual of Determinative Bacteriology*.

Things are under control though. She is talking of the lake trip & will contact Mum on details.

    Your loving son,
        Bob

                           McGill University, Montreal
                                      21 May 1947

Dear old Son,

    On returning from Philadelphia I had a meeting to finally convince Meakins & MacFarlane[98] that certain alterations are essential for them to put it up to the Principal. It is alright so far. Meanwhile the building extension is taking shape & the partitions are now almost completed. There are some silly things which I cannot stop. I think the improvement will be a great advantage but it could have been so much greater for very little more.

    That afternoon (Fri 16th) Mike Carmichael[99] offered me a lift to Bark Lake & as Mum was there already I put off most other things & went. We had a pleasant drive up in clearing weather. Mum was fine & so was Fudge & it was lovely to see the old place & a good log fire. Saturday was a lovely day & Mum raked her garden – & did a few odd jobs which showed how soft I am. Later in the afternoon we paddled over & went into Trout Lake. ... We stayed a bit over an hour (during part of which Mum & the pup went over 2 mountains & 2 valleys looking for Mirror Lake, without luck, but found where quite a stream gushes from a hole in the ground) we took 4 trout on the Pelican, one on the G.R. Hares Ear, & lost 2. We then boiled some tea, packed up & went home. ...

    ... On Monday we sent you a wire but you were probably in Windsor. It was a terrible day of Faculty meetings (3 diff faculties) & various interruptions have continued since so that I cannot get some things done. Everyone seems to have some sort of bleat & several people are or have been ill. We are trying to move everything out of my lab & office to allow of the alterations to that part & putting all technician work into the classroom. I feel as if some sort of a curse has been put upon me which purposes to delay everything.

    Mum & Fudge came home today very cheerful & looking well (except that the pup had a most gloomy countenance) & pleased that she had planted everything. She is full of hopes of a fine garden for the summer & no doubt it will be.

    ... This is to wish you a grand fishing trip.

        Love to you all,
             Dad

---

98 MacFarlane was perhaps a faculty member attached to the Dean's Office, or because the meeting involved building extensions, he may have been with the physical plant of McGill University. RGEM.

99 See chapter 4, note 53.

London, Ont.
25 May 1947

My dear Mum:

Many thanks for the veg. & the birthday greetings. ... Anyway we should be off early this afternoon (wet!) and will probably not set off on the waters of Opeongo (Algonquin Park) until tomorrow morning. However we propose to follow a similar route to last time – being one of the few well suited to a leisurely 7 day trip – with a side trip or two if time & energies permit. ...

    Your son,
        Bob

Royal Victoria Hospital, Montreal
9 June 1947

Dear old Son,

It was good to hear from Doris that you had returned looking well & having had a good trip. We thought of you a good deal & were especially disturbed because we were having such bad weather & the general weather reports were unsatisfactory. I hope you were lucky.

We returned last night after a short weekend at the lake. Mum's garden promises splendidly for this summer. ... The black-fly are pretty bad but the mosquito & sand-fly have not come yet. The black-fly drove us from the garden & any sheltered place & the portages into the wind on the lakes. I hope you escaped plagues of flies & that the dope helped. We only got bitten in places free of dope; such as hands, frequently wetted, unrecognized holes in shirts etc which they never fail to find.

... I wrote to Ash about such possibilities of candidates for your jobs; not very much to offer. Competition is high & the supply light. I am looking for two people with suitable qualifications for Assistant Professor at $4000 to $4500 according to qualifications. One teaching & research as a replacement for time lost to the dept by Fred Smith becoming Dean & Don Fleming[100] becoming Secretary of the Faculty; the other trained in virus work in relation to the Virus Division I am to start. Both will be very difficult to find & attract here & this is late to be starting that sort of search. I thought I was set for the time-being when I filled my appointments last year but my calculations are entirely upset. The alterations to the building are progressing & do not make this sort of rearrangement any easier.

I am going to Alberta for a three day meeting, leaving here by air on 19 June & returning the end of the week. That does not ease the situation & I am worried to find

---

100 See chapter 3, note 123 for information on E.G.D. Murray's colleague Donald S. Fleming.

time to do the chapters for the book on medical bacteriology. Changes in the Clin. Lab assistants & various technician troubles & necessary changes add a bit of worries. Also I have an itch that I would like a bit of fishing & a rest, to which the weekend at the lake contributes.

Sue & Blake are fit & Blake is pleased with a small raise in pay, very small, only $300 a year making $2700. However it is a slight mark of appreciation. Their baby is a very fine fellow & grown greatly in size, very quick & good ...

The photos Doris sent of Alice & Peter are splendid to have. Peter is very like some we have of you about that age.

    Love to you all,
        Dad

                London, Ont.
                17 June 1947

My dear Mum and Dad:

Thank you for your nice letters. We are gradually beginning to see more clearly how we are going to stage manage this "move" to the lake.

First of all, where are you all? From previous letters I gather that Dad will be away, from the 19th but no mention anywhere of the date that Mum is going to the lake for the summer. Important to know this for communication purposes. Sue and Blake I take it will be in Montreal at the time we intend to go through and will be at the house.

Although you say it will be "alright to come" it will be a big crush in the house with 6 adults and 3 kids and will be a lot of work and little holiday for someone. There will have to be a fair division of labour. Also I am keen that Doris have a reasonable chance at a holiday for a change.

... We will look after the blanket situation and also bring sheets. We will have to express a crib and a mattress ahead of us to the lake – and hope that it arrives in time. I will have my sleeping bag and prob. tent – so we can camp if nec. We have also been thinking about the control of the kids up at the lake. I will have to get some life vest – or something with similar properties for Alice and Peter (esp. if he walks).

... We had a fine lunch birthday party for Peter ... the place was a madhouse and he had a fine time. So did everyone else without regard to birthdays or any such excuse. In future I can quite see that all presents will have to be given in duplicate. Alice has semi-appropriated most of his toys but Peter, anyway, seems to enjoy his dirty old ball and a 25 cent set of boxes better than anything else.

Must rush to post this – we have people coming in a few minutes from now.

    Your son,
        Bob

London, Ont.
25 June 1947

My dear Mum and Dad:

Our plans are progressing very nicely and we are all looking forward to seeing you next week and to having a holiday together. Doris has been in a frenzy of housecleaning, re-arranging etc and I seem to have carried a small ton of reckonings down to the basement. ... In case you are wondering we will be coming on the CNR train and we have good accommodation so the trip will not be too much of a trial – I had funny visions of being in the awkward position of having to pot Alice and not being sure whether to take her to the little boys room or the little girls!

... I will be bringing most of my fishing gear and look forward to using it a bit. I hope that you, Dad, will be able to bring up your basic fly-tying equipment so that I may have a few lessons in methods and then I will have a basis to buy some of my own.

Doris asks me to thank you, Mum, for the note on the cheese situation – it should be fine though it sounds as if it is going to be an expensive feed. It is certainly a damn nuisance that we have to be continually fussing with Peter's diet but it seems to be necessary and I think you will agree when you see him that he does not look as if there were anything wrong with him. He is as active as a trout these days but he has such a good means of progression that he has so far refused to do any walking much to Doris' disgust and Miss Strelitz' disappointment (she wanted to see him walking before we go away for she reasons that he will be walking by the time we come back)...

All our love to you both.
 Your son,
  Bob

Royal Victoria Hospital, Montreal
31 July 1947

Dear Old Son,

I hope you had an uneventful journey home & that Doris & the babies were not overtired.

It was great having you at the lake & delightful to visit some of the old places together in something like the old stile. So some happy hours recalled past happy ones.

... I'm going back to the lake for the weekend only tomorrow. Mum returned yesterday after suffering a very hot day here. I'm sure she was glad to get home to the cottage.

 Love to you all,
  Dad
  (Oupa to Peter.)

London, Ont.
31 July 1947

My dear Mum & Dad:

... Our holiday was grand & we all feel very fit. I must say that I feel fitter than I have been for many years. It was grand to be with you & it was grand of you to do so much for us.

This is just a short note which will, I hope, get to you while you are both together at the Lake for the weekend.

    Love from us all,
        Bob

Royal Victoria Hospital, Montreal
4 August 1947

Dear old Son,

We got your letter at the lake this past weekend & were glad. It is fine you are feeling fit & well. The time passed so quickly & there are a lot of things we did not do. However it was fine to have you with us again.

... When you get a new abode I hope it has a garage. I am sure you would find the canoe worth having & could enjoy some good weekend with it on the river. Also one day you can get a car to widen your scope. The projector lamp has not arrived yet so I have not seen the film. I hope it will be here in a day or two.

... I am trying to hurry our building operations. Nothing was done while I was away on the media rooms & the new part is some weeks behind time. The staff situation is still in abeyance as I have no candidates for the job. I fear this will be another difficult year & I am not looking forward to it. ...

    Your loving,
        Dad

London, Ont.
6 August 1947

My dear old Dad:

Your letters arrived safely. I am afraid that I have not felt much like writing with this hot weather. It has been rather sweltering here in the house for Doris and the kids and for us all in the evenings but it has been made bearable by having the garden and Labatt's brewery fairly handy! The kids have had a great time with a tub of water on the grass, and today Doris and the two of them went out to Springbank Park where they paddled, played, and had a ride on the miniature railway.

The lab is rather hectic and I seem to have tons of things to attend to and, with the heat, find it a little trying. The lab gets terribly hot – damn these architects – and the temperature has been in the 90's in the media room and pretty near that in the rooms across the hall. Despite this our work on the motile colonies is going very well and we have some very interesting data. Since the weekend I have been able to show that the colonies can be directed in their movement but there are so many possible variables in the conditions imposed that I don't expect to disentangle it for some time.

A great many people are on holiday but there seems to be no lack of work! Collip has not turned up yet and Noble is said to be in this part of the country but has not been seen by us as yet.

... As a small present to you both I have got three books to add to your collection – I hope you have not got them but I could not see them in your shelves when I had a look. They are: Winter Studies and Summer Rambles by Anna Jameson, 'T Ain't Running No More by Sherwood Fox,[101] a nicely illustrated small history of the Aux Sable region of Ontario, and a small booklet on Audubon's water colours of birds. I will be sending along your fishing books and will put these in with them and hope that you will enjoy them.

Your staff problems are a curse and I wish I could help you. We have had difficulty, thus far failure, in filling the Fellowship one possibility having just fallen through. I wish we could get this filled soon for the university will get a bit annoyed at having the money earmarked for so long without visible result.

The kids are very well and Peter is very active, standing up much more but is not yet walking.

  Your son,
    Bob

P.S. You had some idea on the motile colonies but I did not make note of it and so have forgotten. We have found purely incidentally that Brom Thymol Blue inhibits migration although growth seems to be reasonably exuberant.

Royal Victoria Hospital, Montreal
14 August 1947

Dear old Son,

... It has been a real hot week. The worst of the year, 90° with 95% rel. humidity, making work difficult. The lab is terrible. I'm glad Mum, Sue & Douglas are at the lake. I go to Baltimore on Sunday & will be away until Friday. It is a stupid time to call a big conference. There is not all that hurry about things & it will be sweltering there & added to it great intensity of earnestness – an awful prospect.

The books you have bought for us sound exciting & we have not read them. I see

---

101 See chapter 4, note 38.

there are some new books on Canadian country & on Canadian history which I must look into them while they can be got.

... I am full of hope that I may have found my man for the Virus Division. It seems to be almost certain but I have been disappointed before & so prefer just to hope for the present. The other job is still wide open & without candidates.

Your answer to the University is that it costs more to live these days & the pay offered does not attract. That there are good jobs going begging & ready to take even the inexperienced with just bare degree qualifications. Young untried new graduates of D.P.H. Courses are offered $5,500.00 as Health Unit Officers all over the Dominion. Lab jobs offer much less & call for more training & more experience & get no applicants. The answer is clear.

... On Sunday last Mum & I went into Pike Lake. We got two bass 1-1/2 & 1-1/4 lbs & several small ones not kept (the smallest I have seen in that lake). The 2 larger fish came to Mum's jungle hornet & the smaller ones mostly to the plug. We did not see any fish like you & I got; nor did we see a pike. ...

 Love to you all,
 Dad

London, Ont.
14 August 1947

My dear old Dad:

After all this heat I don't know that I am capable of remembering if there is anything interesting that has happened in the last week. ... We have been existing but not much more. The apartment gets fiercely hot, and because of the lack of insulation the roof radiates heat most of the night.

... Mother Grace has been with us for a week and a half now but, unfortunately must be returning to Rochester this week. It has meant a lot for Doris for it has given her some degree of freedom and relief in this weather. As a result we were able to go to Lake Huron a week ago Monday and to Detroit last weekend.

Sunday, a week ago, was a very hot bright day and we went with Jim Hamilton and his wife to a place near Port Franks – on Lake Huron between Ipperwash and Grand Bend. It is where the Aux Sables river comes out – the one in (the) Sherwood Fox book (that I have yet to send you).[102] It really is Aux Sables for the country for a ¼ mile inland is all dunes and there are the most wonderful sand beaches, hard and without a pebble in them, stretching for miles. The water is crystal clear and has a definite

---

102 The book by Sherwood Fox was *T'aint Runnin' No More: The Story of Grand Bend, the Pinery and the Old River Bed* (1946).

bluish tinge – the lake in the sunlight has an extraordinary clear light blue – almost aquamarine. The water was warm and I spent nearly all afternoon in it. We added to our suntans and felt much refreshed.

This last weekend was refreshing in a different way – certainly not for temperature. We went to Detroit to see Perry and Louise TeWalt and also for Doris to get some shoes and a hat. On all counts a successful expedition although it was damned hot in that city both day and night. Like a dutiful husband I helped shop and even bought myself a shirt (of which there are plenty down there) wearing my legs down to a short stump. Doris enjoyed herself hugely on her first visit to the US in a long time.

... The lab has been a terrible place to work. I thought it was more comfortable today (maybe it was the contrast from Detroit) but it was 32°C on my bench at 9am. I certainly hope that it gets better as promised for it has put me rather back in my preparation work for the classes as well as hampering other things. I had hoped to get a little writing done in the evenings but it has been almost impossible. ...

> Your loving son,
> Bob

27 August 1947

Dear old Son,

Your letter which arrived today did not surprise me. We have had a hot & humid August & I heard it was as bad or worse in Ontario. ... This house is habitable but your apartment must be fearful.

Your trip to Lake Huron must have been good but I cannot imagine Detroit being pleasant. I can't find the places you mention on the map (Port Franks, Ipperwash, Grand Bend) & the Aux Sable River in Michigan (about ½ way between Thunder Bay & Saginaw Bay) seems a long way to go there & back on a Sunday from London, Ont.

... I start lectures on 5th Sept. I am not looking forward to this year for a number of reasons. I have not got a replacement for the [Fred] Smith-[Don] Fleming loss. The alterations to the lab are not completed though they have started putting in the fittings; we will be cluttered up with workmen for some time & then will have to move the Clinical Division. There is quite a mess of technician trouble in the junior grades & the washing up & no satisfactory solution in sight. The new media rooms are almost completed & I am moving the work in as a way of pushing the workmen out. The changes in the animal rooms are nearly finished & there is much to reorganize & much arrangement & time to settle in. The new Virus Division must get started.

... I have not had a trout-fishing trip this year & I do not feel sure of getting one this Fall, though I shall try to. Mum was hoping we could do a small camping trip but the chances seem very slight. She is going to stay at the lake with Sue & Douglas through September. The summer slips by so quickly …

I am rather weary & must to bed.

    Love to Doris & the kiddies.

        Your loving,

           Dad

London, Ont.
31 August 1947

My dear old Dad:

… Research into "swarming" & "migration" goes on apace and we have assembled large amounts of data which do not seem to mean anything very much! However ingenious our experiments seem to us, they turn out gloriously inconclusive as far as the bugs are concerned. All we can really say is that swarming & migratory colonies are each stopped by the same things; that the migratory types will swarm if conditions are right; we have not been able to make swarmers into migrators. … It is interesting that despite the fact that you can make these migratory species give the appearance of swarming – under the microscope the swarming is different from that of proteus in that there is still active, but unorganized, movement of groups of cells.

I wrote to Richards at the American Optical Co, to see if we could borrow a phase microscope – but he replies that they have no spare equipment for loan & only suggests that we go down there to work with his own setup. Not very practical at the moment. However, I saw Collip a week ago and suggested that one he obtained & he agreed – so we may see one in the New Year – I hope!

Collip is now actively in charge and seems to be happy. He has bought a house up in the north end and is satisfied with it. I have no doubt he will do a good job even if he may be a hard man to find at times. We spoke briefly about my proposed part time administrative job & agreed to leave it over to January when the big class is done with, Dr. Skinner carrying on until then. It will be better that way for both of us. We spoke about a research assistant for me, to which he agreed, but for the life of me I don't know where we would put him or her even if available! Space troubles are terrible in all departments here – and could not be more acute in ours.

I had an extraordinary phone conversation yesterday am with the head man (Priest!) of Ottawa University asking me if I would like to be Professor of their Department of Bacteriology! A flattering but impossible offer – even if I wanted to leave here – I feel that I must get properly set & do work here before entertaining a change by which time I hope there may be far better ports! The thought of a Catholic institution run by the priesthood leaves one rather cold.

As you say, summer slips by too quickly – we too will be immersed in teaching in just over a week. Necessary changes are only just being completed after months delay, by the works department – and I have not spent all the time on preparation that I should. I shall be damn glad to see Dr. Ash, Dr. Bittner & Miss Strelitz on Tuesday for it will mean that a heavy load can be skirted giving me time to get at the last minute essentials.

We all of us send you & Mum our fondest love.
   Your son,
    Bob

              Royal Victoria Hospital, Montreal
                    4 September 1947

Dear Old Son,

Your letter of 31 August '47 and the books arrived. The three extra you included look most interesting & I look forward to reading them; especially "It Ain't Runnin' No More" by W.S. Fox. It's nice of you to think of it.

… Lectures start (by me) tomorrow at 9AM. The lab is still in some confusion & there is too much to stir up, push along or provide for to make it fun just now. Things are coming along but it takes time & patience.

The Ottawa University job has been shopped a lot & I don't feel that it is a good opportunity. It would be a big gamble though they might treat you well. I don't know any details. I wish I could recall the suggestions about migration & swarming. I have an idea it was an application of absorbed flagella & absorbed somatic sera.

… Have you seen papers by C.W. Hall on Hyaluronidase problems in the Biochem Journal? 38 No 5 1944 p362 & p368. It's work done at the Lister Institute.

It's cooler here too, in fact quite pleasant. Up at the lake there is a tinge of fall in the air & even Mum admits it. The hot spell during August was very trying but I regret the rapid passing of the summer & do not look forward to winter.

It's getting late. I am glad to hear the kiddies are in such good form & flourishing.

  Love to you all,
    Dad

                      London, Ont.
                      6 Sept 1947

My dear old Dad:

… I have sent a letter of refusal on the Ottawa business. I, too, think that it would be a risky business. I told Collip about it – that was his view. George Stavraky[103] had an

---

103 See chapter 4, note 5.

offer from them too and went down to have a look at the place. He says that there are at least three factions in the medical school: priestly, French and English-speaking.

Collip also promised a phase microscope for us (i.e. the school generally, but we would be the greatest users) but it is becoming quite a lot of work to find out what to order – just ordering a phase microscope would not be of any use – so we are having an ordeal by correspondence.

Did you ever get your copy of the new edition of Topley and Wilson?[104] The library here has had orders in for a hell of a long time, in fact ever since the first notice was sent out and have not had a smell of it. The same thing seems to have happened with a number of books - apparently Canadian libraries are having a very raw deal …

How are you getting on with the chapters of the book? I don't suppose you have much time to deal with them. I had meant to write to Gertrude about a couple of points but my notes are at the lab and I have forgotten what they were about.

I am pretty sure that I have that reference to C.W. Hale's[105] work for I have all his reprints on that business. He has done very good work but spread pretty thin and from a rather narrow point of view.

Since all this talk in the papers about the Sterling Bloc and such details as they give, I have been wondering about the value of that insurance that you are holding for me in England. Certainly if the money were needed it does not look as if the money could be got out of the country. Do the people it is taken out with have a representative in Canada? Perhaps they should be asked about the situation.

We start teaching next week and there still seem to be many last minute administrative points coming up, even at this late date all because of some muddle up in the office. We are going to be full up to the gills, having only two student spaces left over after doubling the class – and it looks as though we are going to have a fight again over the laboratory technician courses. Blast.

Concerning the migrating colonies. I have been having a philosophical argument with myself over whether one could compare motility as such, in a fluid, with motility on agar. In the first case the organism is living in the depth of medium; in the latter he is living very near or in the surface film where there is greater or different concentration of various substances of surface activity. … Why does increasing the agar concentration have such a profound effect? Is it that the gel takes up more water and nutrients and makes them completely unavailable to the organisms?

… We have not heard a word from Mum or Sue for a long time and would like to hear how the lake is and all the details of Douggie etc, please give them our love.

---

104 This is a reference to the popular book *Topley & Wilson's Microbiology and Microbial Infections* (1947 edition). First published in 1929, it is currently in its tenth edition and has grown from one to eight volumes.

105 C.W. Hale of the Medical Research Council, Antibiotics Research Station, Clevedon, England.

... Alice is going to a very nice little nursery school. She loves it and feels very 'portant and it seems to be doing a lot of good to her behaviour. We felt that it was justified to get her playing with nice kids for at least part of the day, and away from the tendency to play "in the gutter". Doris looks almost as if she had lost a chick or were at least sending her to university!

Your loving son,
Bob

P.S. Since yesterday there are a couple of bulletins: (1) Peter definitely started walking today – no argument and plenty of witnesses! (2) The canoe arrived yesterday – Angus and I unpacked it this afternoon and it is in perfect shape – we portaged it to the North Branch and paddled around a bit. ... It will be a lot of fun for us and well worthwhile having here.

Montreal
11 September 1947

Dear old Son,

I got your letter of 6th & we were glad of the news, especially the Late Bulletins that Peter definitely walks & that the canoe arrived safely. Mum came home here on Wed ... She seems to have had a nasty fall this morning, grazing off some epidermis, bruising & straining other places – not very bad but annoying & a trouble.

... The chapter is coming along & I shall send you a copy for criticism presently. I wish I felt sure nothing important will be missed. Send along your suggestions in any case.

I don't know what value anything is in the Sterling area. It will surely recover & stabilize in time. In any case the University Life Assurance Association is thoroughly sound. Roddy Heard[106] got his inheritance out of England alright but Mum has not learned anything whatever of hers, what it amounts to or whether she can get it out. The S. African money is still out there.

I enclose a copy of a snapshot taken at Suffield Alta this past spring which you might like – it amuses Mum.

... I had a great surprise this morning which I am not publishing abroad & just keeping in the family for the present. I received a letter from the Military Attaché of the U.S. Embassy (Ottawa) telling me that "the War Department has approved the award of the Medal of Freedom in recognition of the exceptionally meritorious services rendered by you" etc. He also wished to know in order to make arrangements for the presentation ceremony etc & his personal congratulations. It is bound to appear

---

106 See chapter 4, note 90.

in the back of the newspaper in due course but it is gratifying & may help to restore my prestige here a little bit. It is very nice to have.

The "Lectors" are going alright so far & it seems to be a reasonably good class. We have only Medicals at present so we are not in full swing. However we are almost completely prepared for all nine courses we have to give. The media making region which was reconstructed is occupied though not quite completed yet. The new part is coming along but will not be completed yet awhile. It promises to be good except for a couple of major errors upon which the architect was insistent & had his way in spite of my vigorous protests. I shall have to get our most serious one changed & am still battling over it – that is the stupid fixed windows.

It is arranged that Pascoe & I are going with Mike Carmichael to Lake Marcotte (N of Shawinigan) for 4 days on 23rd Sept. Otherwise I was going to go to the Columbus for a few days. I expect it will be a nice trip & will be fun.

Gertrude is back from her holiday in Mexico. She seems to be very fit & had a good time. There's a lot for her to do now she is back. I have not got a replacement for Fred [Smith] & Don [Fleming] yet!

  Love to you all,
    Dad

London, Ont.
25 Sept 1947

My dear old Dad:

I am afraid that it is a very long time since I have written to you. We have been very busy and it has been fairly hard going for the last two weeks. Classes are going full swing and four hours a day of labs and the preparation work, and some research if there is opportunity makes for a long day. Things are going fairly well for the class and they seem to be a good bunch despite their numbers.

The seminars have started again and they will probably be good fun. I had to get the staff seminars organized this year which meant a little extra running around. It should be fun for there are some good topics and people seem to be more willing to come to them than they were two years ago. They are also having some other do's for graduate students that I am trying to attend to a certain extent. One is a short course of practical statistics which may be useful – when I have needed them I have needed them badly and always find that it is quite a job to read it up and apply it without much basic knowledge. My only training was a little we learnt in Huskins' course in Genetics years ago.[107] Roger is running quite a discussion group in Biochemistry

---

[107] Leonard Huskins was professor of genetics at McGill and taught Robert Murray in the late

that I find fun – they have a scheme of topics and meet once a week. It is quite stimulating.

The first seminar was about the present state of cancer research and was essentially a report of the recent congress at St Louis. It was very interesting and the developments since I last got myself up to date on the subject in 1940 are definite though not profound. Among the most interesting facets are the so called co-carcinogens involving two or more substances which are not active separately but which in concert may produce the effect. Situations like that would seem to give good chance of solution. The virus angle does not seem to have produced very much as yet. ... I don't know the evidence though the suggestion that virus is inactivated, accounting for inability to isolate it, due to the formation of antibody is interesting.

Thank you very much for sending along the cultures and the Dahlia. An experiment with the latter is being set up tomorrow. I wish that three of my migratory species would grow better in simple nutrient media. It is unfair of the soil organisms to require such complex and expensive diets.

... I have returned the manuscript today to Gertrude with such comments as I could make. I must say that with all other things go on I have had little time to give full attention to it – though I did manage to read it through. In general I think it is a good but rather matter of fact account. It will probably look a good deal better in print. It is always hard to read in manuscript.

... I saw in the paper the other day a list of US decorations and was sorry not to see yours listed yet. I hope it will be in soon. Glad to see that Charlie Mitchell got something.

... I have not heard anything yet about the Xmas meeting of the CPHA lab section meeting. If I can get a decent record of these bugs I may give a paper.

Give our love to Mum, to Sue and Blake, and of course Doug.

Your loving son,
Bob

P.S. Suddenly remembered your fishing trip. I certainly hope that it was good September fishing and that you had good fun. Look forward to hearing about it.

1930s. As Murray recalls, he "gave an excellent basic course in genetics with a great practical laboratory component that gave ... excellent training in *Drosophila melanogaster* genetics." RGEM.

Montreal
28 September 1947

Dear Old Son,

    Last 14-20 Sept I hurriedly rewrote the M.S. from Gertrude's notes & she sent it on to you for criticism which I hear you did very promptly. Many thanks. I shall get busy on it again tomorrow.

    … Mum stayed on another week as I was going with Mike Carmichael to Lake Marcotte, NW of Shawinigan. … This is his fishing club & he is an original member … & had invited me to go with him for a number of years.

    … The territory is about as big as the Columbus, with a number of good lakes & good fishing. The club house is log, squared inside & for joints, & very well built & very comfortable. The fishing is mostly done from canoes & they have a flock of them, mostly heavy.

    … It is good to have such good news in Doris' letter to Mum. I wish you could find the place you want to live in.

    Love to you all,
        Dad

London, Ont.
30 Sept 47

My dear Dad:

    I was so glad to get your letter today and to hear of your trip to north of Grandmere. It sounds like a most marvelous place …

    … I have been having an infernally busy time and have not been able to go out much. The teaching, especially keeping the labs running fairly smoothly, takes up most of my day and other activities have to be squeezed in outside. So far, the teaching is going alright though I feel that the strains on our facilities has reduced rather than increased our teaching efficiency this year and has prevented many things we had hoped to do or demonstrate. Big classes are a curse and we may, according to current rumour, have to deal with more of them – where & how, god only knows.

    Great and rather grand plans are in the air – the present lead of discussion seems to be long range discussion and affects our immediate facilities quite directly. The general plan seems to be to build afresh together with a university hospital on the ample ground at the north end – to have the university all together. It takes time & if that is the decision then they are not going to put a lot of money down the drain in new buildings in our present site. Still very formative ideas but revealing – and, according to the "Presiding genius", quite possible and probable!

    The new influences are being felt in the training of the graduate students and discussion groups & seminars are going 3 times a week. Many of these we attend &

contribute to, which is fun and stimulating even if an added burden at the end of the day. Good for us too, since they are widely attended by the staff.

The observations on swarming are coming along – and I think that I have got a lot of interesting stuff. The latest bombshell is that both Proteus and Kurthia do not spread across a plate by expansion, as I at first thought, but have means of spread entirely analogous to those exhibited by the spr. of bacillus having the so called migratory colonies. ....[108]

The investigation of dyes is not getting us far – it is at the moment hard to think how to control variables properly & I am beginning to think we are making the wrong kinds of observations. At any rate Dahlia is pretty toxic and also is not very effective at stopping swarming. The investigations of these, & pH, surface tension, effects of salts etc will have to wait till I have incubated the problem in my mind. Anyway I am gathering some interesting stuff for a paper on the "Anatomy of swarming". We have tried taking some cinephotomicrographs with very moderate results – many negative – but I now am more hopeful & await the return of our latest trials with some impatience.

About the manuscript – if I could have kept it a bit longer I would have had time to think about the various parts both individually and generally but it was Tuesday am when I got it & Gertrude said she wanted it back by the end of the week. I am afraid I did not do it justice and that I should have been more helpful. But circumstances were difficult.

All our love to you & write soon –
    Your loving son,
        Bob

London, Ont.
2 October 1947

My dear old Dad:

I hope that you are now completely recovered from your siege with the streptococci. You were lucky you did not have more trouble from it.

Just now I sent you a NLT [likely a 'night letter', or overnight telegram] for information – this Australian fellow sent me a very nice polite letter but for some, no doubt

---

[108] Robert Murray provides the following explanation of this passage: "Some organisms inoculated at a spot on the surface of a solid nutrient medium will spread over the surface as they grow. Some just expand on that surface; others actively move (swarm) either in groups or individually over that surface. The movement is usually due to activity of flagella which act as propellers. There are other bacteria that move over the surface by gliding and are without flagella by a direct cell body mechanism. The swarming I was working with was of flagellated bacteria." RGEM.

absent minded, reason forgot to sign it or otherwise identify himself. Despite this he sounds very nice!

... I am very pleased today for we got our last length of film back from the processors and it is very good. Cinephotomicrographs seemed hard at a distance but they are a lot simpler than one would expect. We have quite usable results and only a few details have to be rectified – I intend to take representative shots of one species & only a few to illustrate main differences among species & groups. Ash was most impressed and is insistent that it should be shown at the lab section meeting. About which, incidentally, I have heard nothing.

In the last number of the Yale Journal of Biology and Medicine there was something that might interest you. I have forgotten the authors name but it concerned the use of glycine in lysing bacteria.

A.J. Rhodes[109] from Toronto was here yesterday and gave me a very excellent talk on poliomyelitis and he spent some time in the lab with us afterwards. A very interesting fellow and we had a most stimulating talk. He seemed very interested and impressed with Ash's anti-phage agents, and most willing to help with the trial of these on various animal viruses.

We are still busy as blazes and there seems to be little time left in the day for many of the things I would like to do.

> Your son,
> Bob

> Montreal
> 8 October 1947

Dear old Son,

I have not written my usual letters & don't quite know where we were last. I wrote last on my return from Lac. Marcotte & I am about three letters behind you.

... Sir Henry Dale[110] is here. I saw him at the Medico-Chi. Socy[111] dinner last night

---

109 Andrew J. Rhodes (1911-1995) was a research associate at the Connaught Medical Research Laboratories, and a professor of virus infections at the School of Hygiene, University of Toronto. Beginning in 1947, Rhodes directed the polio research program at Connaught. Christopher J. Rutty, "Dr. Robert Davies Defries (1889-1975): Canada's 'Mr. Public Health'," in Lois N. Magner (ed.) *Doctors, Nurses and Practitioners* (Westport, CT: Greenwood Press, 1997).

110 Sir Henry Hallett Dale (1875-1968) was a British physiologist and pharmacologist. He spent the first ten years of his professional career as the pharmacologist and director of the Wellcome Physiological Research Laboratories in London, England. Dale was named director of the department of biochemistry and pharmacology at the Central Institute for Medical Research, which was renamed the National Institute for Medical Research in 1920, and in 1928 he was made the Institute's first director, a position he held until his retirement in 1942. He was secretary of the Royal Society of London from 1925 to 1935 and its president from 1940 to 1945. ODNB (Online).

111 The Montreal Medico-Chirurgical Society.

& he said he is going to the Physiological Soc. Meeting at Western. He would like you to introduce yourself to him as he is a good & old friend of mine & a grand fellow marvelous to meet.

The course in practical statistics will be useful; I have often wished I had some knowledge of it. The seminars & discussions must be fun as they seem to be lively.

It's good to hear that your cine-microphotographs are coming out well. You must show me how it's done. Perhaps my *Bewi* light metre might be useful to you; I have found it quite good for microphotographs, it suits the diameter of the eyepiece. I'll send it along if you would like to try it.

Many thanks for your comments on the M.S. of the chapter Gertrude & I are responsible for. It must have been a nuisance to you but you picked out useful & important points to improve & add to. We have worked at it some more & I hope improved it & sent it in. I feel it could be a great deal better, but it is a difficult assignment.

I have not heard any more about the decoration & I did not see any list of the others nor have I heard anything. I know that G.B. Reed, Chas Mitchell, J. Craigie & I were recommended. I suppose I shall hear something more some day as the letter came from the Military Attaché of the U.S. Embassy in Ottawa.

The late set for the C.P.H.A. Lab Section is most inconvenient for us. Nearly all of us on the teaching side have classes in one or other of our numerous courses. It will be hard for any of us to get away. Gertrude will be able to. It is annoying.

I hope you saw something of North & that he had a good time in London. He is most interesting on his impression of different countries & is a quick observer.

... Last weekend when Mum & I went on a voyage ... We set out from the cottage about 10AM, paddled to the end of Green Bay, portaged (a bit over a mile) through to the end of the Lake Bay. We paddled through to the old lumber road landing right up the inlet & from there portaged through to Green Lake (Lac de la Verdure) 3-1/2 miles... All told we portaged a bit over 9 miles & must have paddled about 15 miles all in about 9 hours with rests & lunch & investigations. We were a bit tired & hungry when we got back & slept easily after a good supper. It was a very satisfactory day & Fudge enjoyed himself greatly too.

Thanks for the note on the solution of bacteria in amino acids. I am glad too to have your impression of A.J. Rhodes; it agrees with other reports. It's good to hear that Ash is getting interesting & encouraging results – give him my love.

I seem not to have any time to do things I would like to do. My time seems to be fully taken up by a multitude of infinitely varied troubles. Nothing interesting but at the moment each is important to the department or students or the requirement of some member of staff. I am now short of staff & technicians & both are hard to find

– supplies are short too. The new building is standing still though nearly completed – makes me furious. ... I'll be glad when the year is over.

It's getting late. Love to you all,
  Dad

London, Ont.
10 Oct 47

My dear Old Dad:

... I am glad that the notes on the MS were of some use to you. I was afraid that in my hurry I had not done a thorough enough job or failed in giving the sort of criticism that you wanted. I am glad that you have now got it out of the way and that it is now Dubos'[112] headache. However, I suppose the chapter on history will be equally difficult if you are going to attack it in any way different from the conventional angle.

Dr. North came and went so very quickly that we hardly had time to realize how much we enjoyed seeing him. He was great fun, informative and seemed to have a very genuine interest in all that we are doing. He arrived one evening and left the next noon – since his main mission was to see Crombie at the San[113] we only had him in the lab for an hour which was nowhere long enough. We had him round here in the evening and gave him beer and onion sandwiches which he seemed to appreciate very much. We enjoyed seeing him and hearing his many observations on countries – also found the work that he and Keogh have been doing on the haemagglutinating properties of H. pertussis and influenza strains to be most fascinating. I think he enjoyed himself too.

I look forward to meeting Sir Henry Dale when he comes here, though I know he will be a hard man to see for he will not be here long and I think that the powers that be will have most of his time taped. It will be rather nice to have a meeting here for a change but I don't expect to be able to go to very much of it because we will be teaching as usual. Dale will be staying with Collip while he is here.

I had a quotation the other day from the American Optical people on suitable phase microscope equipment. What they term a fairly complete outfit, without microscope stand or eyepiece and binocular optics, comes at a price of approximately $1100! Though it would be possible to have the essentials for my purpose for about half that minus the stand etc, Collip says he wants one for the school and is willing to get the lot.

I have some new and interesting data on swarming and that is that calcium salts

---

112 See chapter 4, note 66.
113 Crombie was the superintendent of the Byron Sanatorium and worked with TB patients. The Sanatorium closed in the next decade as the disease became treatable with streptomycin and soon other antibiotics. Robert Murray has no recollection of Dr. North. RGEM.

are inhibitory, without seeming to interfere very much with growth and this effect can be neutralized and reversed with citrate or oxalate. This would seem to indicate the importance of surface films for Calcium is notorious for its effects on surface active films, particularly proteins.

… Sorry that you may not be able to go to the CPHA meeting. As a matter of fact I have still not heard when the meeting is being held nor have I received any request to give a paper. I hope that the dates are not as inconvenient to us as they are to you. I would be most disappointed if you were not able to go.

Got a nice letter from Gertrude today. She sounds well.

… The account of the trip around the end of the lake sounds like a strenuous affair but a great deal of fun. I have never seen Earl Lake and it sounds like a nice one …

Much love to you all.

Your son,

Bob

McGill University, Montreal
15 October 1947

Dear old Son,

We got your letter when we got back from the weekend at the lake yesterday. Dubos is quick off the mark & wrote about the M.S. the day after he got it. He seems to like it. He found some obscure bits & we have improved them. He also suggests there should be a formulary of media & of staining methods. I think we can do it for him reasonably on the basis of our list in the Course 3 lab sheets. That is what we are suggesting to him.

The historical chapter can't well be on conventional lines. There is not space for other than a tracing of trends. Anyway it seems to me a difficult proposition.

I knew you would enjoy North & I wish you had seen more of him. I am sure he enjoyed meeting you.

I hope you meet Dale. I am sure Collip would see that you meet him. He is most approachable & you could introduce yourself & be sure he would be pleased & friendly.

I don't know anything much about phase-microscopy. It seems to have possibilities worth trying out. I'll send you the Bewi light meter. I expect you will have to calibrate it for your purpose.

It's interesting that calcium salts inhibit swarming. Perhaps you can balance it with Na, K & Mg as I did with the growth character of meningo. But you are likely to take it deeper.

The C.P.H.A. Lab Section meeting is on Dec 15th & 16th in The Royal York, Toronto.

If they have not sent you a notice you should write to Wishart,[114] at the Connaught Labs, who is Secretary. My difficulty is that I have a lecture on 16th at 9AM.

... Mum says she is writing you the account of our trips last weekend. We had a strenuous time. We covered our 64 miles by canoe & portage on the three days, & over about 31 miles we carried the canoe & pack. Something like 5 miles we pushed through bush without a trail to follow, going by bearings; very tough as in places the footing was very bad & in other places the gradient (up or down) steep. The pup was tired too & has worn off the leather on his legs & feet.

As usual Mom showed her most remarkable bush craft & wonderful knowledge & recognition of country. She really is wonderful. The fallen leaves hid the ground so no trail could be seen & we had to depend on the feel of the ground to the feet & to spotting the old blazes. This was, of course, more difficult in the dark. Not too many people could guide the way & come out slap on the point aimed as she does. She goes slap through the bush with extraordinary certainty of where she is & how she is going. It always surprises me how she recognizes every rise or mountain & says exactly where it is & its relation to lakes & other contours.

... The canoe took a hell of a beating, especially through the bush, & hardly shows a scratch for it. We, too, came out with few bruises & scratches & not too tried, so our condition is good & hard for our ages.

  Love, Dad

Montreal
22 October 1947

Dear old Son,

... We have read the books you sent us & have enjoyed them very much. They are each most interesting & worth having of their kind ... Mrs. Jameson gives a vivid & interesting picture of Upper Canada of 100 years ago. I am very glad indeed to have it ...

Fox's book is charming on its own merit, as one would expect.

I came across an old note to tell you what I think about the registration as a specialist. I think the only virtue of the procedure lies in the protection it probably would give if "Socialized Medicine" or whatever it may be called is introduced. The threat is spreading everywhere. Rock Carling gave a nicely spoken & clever apology for the British scheme becoming law shortly; but I found what he said unconvincing & I felt it was not sincere as he seemed to be making the best of a bad job. Should State

---

114 F.O. Wishart worked in the Connaught Laboratories and the School of Hygiene at the University of Toronto.

Medicine be forced here, as it may well be in your time, recognition as a specialist may prove very important. There seems no other reason to recommend it at present.

... Last weekend Mum & I spent quietly. My chief occupation was making the shutters & doors fit & the bolts run better against when we have to shut up the cottage. It was fine weather but would have been muggy in the bush. We had a few paddles about the lake. We put the boats into the workshop & greased all the tools. Nothing exciting but a number of useful or important small things.

We cannot go up this coming weekend as I have lectures etc & some essential things to do. We have offered the cottage to Ted & Sylvia [Rose] & I believe they are going up.

The new construction of the lab is coming along very very slowly & with some silly things I tried hard to avoid but which the architect & engineer contrived to introduce nevertheless. Some of them annoy me but we will have to put up with them. The ventilation worries me & I am sure it will have to be changed. I would like it changed now, before we move in, but there seems to be little hope of getting it done.

... The Clinical Bact. Lab is ever busier & the demands greater as well as more numerous. Either the amount will have to be cut down or we will have to have more staff. It will mean an argument. I have not met the new Superintendent but I am sure we will miss George Stephens & his understanding consideration.

It is strange you have not heard from Wishart about the C.P.H.A. Lab Section meeting. You should write to him to make sure you are on the list of members. The Torontonians are rather offhand. The notice I got says the meeting is on Dec 15th & 16th. Not much good to us here, as I think I told you. Very disappointing.

Williams & Wilkins[115] have now promised to have the 6th Ed of Bergey's Manual out by Jan 1st. I'll send you a notice & you can have an order placed for your lab. I'll send you a personal copy when I get mine, which ought to be early. I'm sure you will find it helpful & most of it good.

 Love to you all,
  Dad

London, Ont.
26 Oct 47

My dear old Mum:

It was grand to get your long and detailed account of the trips that you and Dad took in the region of the lake. How I wish that we could have taken a similar trip. It sounds a terrific effort and ... I marvel that both of you manage to keep in such good

---

[115] The Baltimore-based publisher of the *Bergey's Manual of Systematic Bacteriology*.

condition and able to undertake such strenuous voyages. I must say that I would have bogged down very early, especially in the trip round Balsam Lake and region. The country must have been very beautiful just the same and well worth seeing this year.

You did not say very much about the colours this year. Round us they are not very good. Somehow the weather has not been most conducive to the formation of the colours – perhaps it is the remarkably warm weather and the comparative lack of gentle frosts to cause the gradual shutting off of the leaves.

... There has not been very much going on in London although we have been going out a little as Doris has no doubt told you. The Little Theatre has started up again and we went the other night to see Thurber's "Male Animal", a satire on college life which was amusing and entertaining though not as well received as I would have expected. As for the movies they have been foul and we have not gone near them. We do not go to the football games, which seems to be the main interest here at this time of year as well as a main topic of conversation, although we might well do it for the Western team is very good and seems to play quite a spectacular type of game. They have been beating all comers without any trouble atall and play an open game with plenty of passing, which endears them to the crowd as well as seeming to get them points.

This week we attended the Fall Convocation which was a colourful and pleasantly short affair, with an excellent speech which few people were able to hear. I was luckily in a position which allowed me to hear very well. They gave honorary degrees to Sir Henry Dale and to Weed[116] who is head of the NRC in the US. Despite all the evident preparation I feel that there are a lot of incongruities in the ceremonial of presentation of degrees which could easily be smoothed out and made more acceptable to look at as well as saving time.

30 Oct.

Doris and I will be very sorry not to see you before Xmas. We certainly wish that you both could come down to see us, as well as see the kids. The meeting in December would have provided a good opportunity and would have been fun for us all but the schedule of teaching always does interfere. We are able to get away from that problem by a stratagem.

... The news that Dad is being presented with the Medal of Freedom on *Saturday is very exciting to us. We very much wish that we could be there. Have a good family

---

116 During the Second World War Lewis H. Weed was director of the School of Medicine at Johns Hopkins University and Chairman of the National Research Council (U.S.) Division of Medical Sciences. Rexmond C. Cochrane, *The National Academy of Sciences: The First Hundred Years, 1863-1963* (Washington, DC: The Academy, 1978), 401.

celebration in *Ottawa and we will think of you all in that setting. (*Misunderstanding on my part. When I re-read Dad's letter I saw the date and realized it was today)

We had the meeting of the Canadian Physiological Society at the University last week and I saw a lot of old friends there and heard the odd bit of news. I also met Sir Henry Dale, who is a most charming and alert old gentleman and had an opportunity for quite a chat since he came to visit us at the lab. J.S.L. Browne[117] gave me quite an interesting bit of gossip when he told me that Ian Stevenson was married but he had no details.

... I have not heard any news from England or from others, such as Roddy [Heard], for a very long time and would very much like to hear any news that you have.

We both send you our fondest love.

   Your son,
    Bob

           Royal Victoria Hospital, Montreal
               27 October 1947

Dear Old Son,

... This morning the U.S. Consul's office rang up to arrange that I go there on Thursday, 10th Oct, for the presentation of the Medal of Freedom. It will be a brief & private affair as they asked who I would like to be there as there is room for "4 or 5 people in the office". Mum & Sue & Blake will be there & I wish very much you & Doris could be. I'll let you know what happens. I hear that G.B. Reed's Medal will be presented to him at a University function & recognized by the Queen's Univ. authorities as a high honour. It would have been pleasant if mine had been presented when General Marshall was here for an Honourary Degree from McGill recently. Well! No odds.

The lab construction still trickles on very slowly. There is not much to be done but it takes a very long time doing. ... Meanwhile it is difficult to get things into working order in the department. I don't believe I will get order into the place until next year. I have not found a man for the Assistant Professorship vacancy & we see only barely the essential minimum of Freddy & Don.

... On Sunday I did a tidy up in my workshop. Things are in order but some things will have to be given a bit of attention to take care of them through the winter. I found an extra pair of original Anderson Scissors, one of three I bought in 1909. I shall include them with the fly-tying equipment I shall be getting together for you very shortly.

... Have you heard yet from Wishart about the C.P.H.A. Lab Section meeting? I am

---

117 See chapter 4, note 134.

sorry I cannot be there but Gertrude will & Kay MacPherson[118] will give a paper & so will Louis Greenberg. No one else can go, most unfortunately. It is hard to think of a generally suitable time for the meeting – the best compromise I can think of is the week between Xmas & New Year.

Young Douglas is coming on apace. He now has at least 4 teeth, they say 6. He sits up firmly & even has pulled himself onto his feet several times. He is quick to recognize & makes conversational noises with wide grins. Sue is too thin & not up to par. She has seen Brabander who cannot find anything definitely wrong ...

Mum is only troubled with the "Neuro-vascular" trouble affecting her hands, especially the left. It is a curious affliction, somewhat painful, at times causing swelling of the hand & at others local sweating. Very annoying to her & not associated with posture. I don't know who to get to try help, not knowing who would be intelligently interested in it. I am still having some discomfort from painful tendons which plague me walking. Otherwise Mum & I are very well.

Enough of complaints and complaining.

... What of the Hyaluronidase work? I have been expecting to see it in the Cdn J. of Research or somewhere else. Don't let it slip too far behind, else you will not write it up.

I am anxious to hear whether you met Henry Dale.

It's getting late & I must to bed.

    Love to you all,
        Dad

                London, Ont.
                30 Oct 47

My dear old Dad:

... Doris and I wish that we could be with you on Saturday when you get the Medal of Freedom. We are very proud of you and very glad that you are getting recognition, conspicuously absent on the Canadian side, for the long service that you gave during the war. On re-reading the letter I notice that the date is today, and not Saturday, so more congratulations. To implement this I have just phoned you and very glad to be able to say this "person to person".

The meeting of the Canadian Physiological was last week and it went very well and was heavily attended. Saw several Montreal people and chatted with them and heard a certain amount about what is going on there. It was fun to hear at first hand. It was

---

118 Kay MacPherson came to Montreal following post-doctoral training with Hattie Elizabeth Alexander (1901-1968) at the Columbia-Presbyterian Medical Center. E.G.D. Murray gave her bench space in his personal lab. She later moved to London and worked in a clinical department at University Hospital. RGEM.

a very great pleasure to be able to meet Sir Henry Dale and to be able to have a chat with him. Ash knows him well for he was head of the National Institute when Ash was there. When he arrived, because of sundry mix ups we more or less received him and he was able later to come to the lab for a half hour and talk over the work that we were doing. He is a marvelous fellow and as keen intellectually as a young man. He was most impressed with Ash's work and its significance. The other distinguished guest at the meeting was Weed, who is apparently the head of the NRC in the US. I must admit that I had never heard of him before and did not know anything about him. Hall introduced me and I found him a very nice man to talk to. They had a convocation at the time of the meeting and gave both Dale and Weed Honorary D.Sc's. The latter made a very excellent speech, which was not heard by most of the audience because of a faulty audio system in a hall with foul acoustics, advocating that governmental systems of aiding research should be revised and giving a very properly pointed perspective to the place research must take in the university.

I don't think that I could say that I am "outrageously busy"! Although a good portion of the week is taken up with teaching and much of the rest of the week is clouded by preoccupation with teaching matters, that is by no means all that I manage to fit in. We are still trying to do a little research and although this is interrupted in a rather frustrating fashion much of the time we do get a little time for light scientific entertainment. The result is a longish day but quite worthwhile.

A new sort of private society has arisen here, and since it arose spontaneously from about three foci I think it may persist and be a good thing. It is like many a society at Cambridge or Oxford, where a paper is read upon more or less any subject with somewhat philosophical flavour and the subject matter opened for discussion suitably oiled with a modicum of beer. At present it has no name but is supported by some twenty staff members of the university from all sorts of departments. The first meeting was a week ago Monday and for some reason I was asked to start the ball rolling and give the opening paper. That was another thing that prevented my writing two weekends ago! I don't know what the title of the talk would be but it was a sort of synthesis of the development of parasitism, in as simple terms as possible and giving as much scope for argument afterwards as was possible. If discussion is a gauge of success it was a success. I spoke for about 35 minutes and the discussion was still going hotly an hour and a half after I finished and the chairman broke it up. They seemed to enjoy themselves, and those outside the biological sciences not the least. In fact some of the most ingenious and most accurate arguments were brought forward by a classicist and a geologist. However, it meant a lot of work and forethought on my part and some intensive reading of Theobald Smith, Zinsser and Co. I am afraid that there was not a great deal of original thinking on my part but it was good for me. I rather look forward to future meetings for it gives one a chance to meet people from other departments, whom one would rarely meet under ordinary circumstances. It is intended to have monthly meetings.[119]

The teaching is about half over and it is all going a lot more smoothly than I had expected. The preparation side has been the main difficulty but with more help than usual we have managed quite well though we have had to make some improvisations.

Jeanne Faughnan works very hard and is reasonably satisfactory as technical help. Her academic knowledge is astoundingly patchy. Any routine things are picked up quickly and reasonable accuracy attained but when intelligent cerebration is required she is rather at a loss. It is probably unfair to judge, but her ambitions and desires would seem to outstrip her performance. We shall see.

... I got a letter from Wishart yesterday in answer to my letter of inquiry concerning the omission. Rather a peculiar explanation. However, I have written back and given a title for a paper. I want to talk about the microscopical structure of bacterial swarms since this seems to have been rarely described even for Proteus, and never, as far as I know for the other groups. The Torontonian sense of red tape may veto the paper on the ground that it has little if any reference to public health. If they do, we will be forced to form a microbiological society of our own!

I agree with your dictum "Don't let it slip too far behind, else you will not write it up". That Hyaluronidase work should be published and I think that we will have to do it in its present state. It has a useful application even though it makes a poor form of assay. It was only a perfectionist sense that prompted holding it over. I have just recently talked it over with Pearce, who is just as tied up with extension of his work as I am and we decided that it would be best to complete writing it up in its present state.

Saw Hugh Starkey[120] last week for a half hour or so – quite the executive. He did not have time to look over the lab or to come home and have a drink!

    Your son,
        Bob

Royal Victoria Hospital, Montreal
31 October 1947

Dear Bob & Doris,

Mum & I are just off for the weekend at the lake. This is to enclose the cutting which gives my citation which will interest you.

It was grand of you to call last night & we were glad of a word with you & I was

---

[119] The Chiron Club was organized by Roger Rossiter, Robert L. Noble, Robert Murray, and other like-minded faculty members. It held informal meetings until the 1960s. See UWO Archives, R.G.E. Murray Collection.

[120] See chapter 3, note 102.

delighted with your reason. I was sorry you were not at the ceremony & your phone call comforted me on it.

     Love, Dad

Royal Victoria Hospital, Montreal
3 November 1947

Dear old Bob,

Mum & I went to the lake for the weekend & just before going I wrote a hurried note to you & Doris enclosing a cutting from the Gazette. ... The paper gives the exact citation so I have not had the official one copied. It should have said I was Canadian Chairman of the Joint U.S.-Canadian Commission, to be exact.

Today I received a cable from the Registry, University of Cambridge, asking me to represent the University of Cambridge at the installation of the Chancellor of the University of Toronto on 21st November 1947. So I shall go to Toronto on 20th & then do some business there 21st & 22nd & propose to go on to London on the 22nd staying until Monday 24th afternoon. I must be back here the morning of 25th to Lecture to Course 2. I am proposing that Mum comes with me & I shall let you know exact times when I have worked them out & know what is required of me in Toronto.

When we come we will bring the films taken this summer & the medal. Can you raise a 16mm projector or shall I bring ours & will it work on your current. Mine requires 120V A.C. 300 W. & I suspect it will not work with your current.

... I was sure you would enjoy meeting Sir Henry Dale & find a great deal to admire in him. I know Lou Weed quite well too. He was Professor of Anatomy at Johns Hopkins & is a very fine fellow. I would like to run into him again.

Your new Society will prove fun. It is a compliment to you to have been asked to give the opening paper & to have succeeded in provoking such a good discussion. It is always good to know those of other faculties & departments. The intimate mixing of those of all different learnings at Cambridge & at Oxford is one of their great features. There is not enough of it here & the chief medium is the Faculty Club.[121]

Jeanne Faughnan may have been handicapped by her deafness. I recommended her as a technician & not as a research worker. However she had the courage to leave home & Montreal which is much more than most of them (men or women) have the guts to do.

... For years I tried to get a Canadian Microbiological Association formed but Toronto & a few actively stupid ones at Ottawa opposed it. Some of these now wish we

---

[121] Joburg was an active member of McGill's small and unusual club known as the "Greenhouse Follies," which met every Monday evening to drink beer, eat meatballs, and engage in lively and diverse conversation. See J.B. Collip, "Professor E.G.D. Murray, An Appreciation," Canadian Medical Association Journal 92 (9 January 1965): 97.

had it. I'll support a genuine effort to revive the idea but do not feel inclined just now to lead the attempt again.[122]

You should certainly publish the Hyaluronidase work in its present state. No work is ever complete but it can be most useful nonetheless.

Yes Hugh Starkey has become the Executive & brings high lights into the discussion in the form of men of influence in distant parts mentioned with casual familiarity. Very amusing. However he has done a very good job & deserves full credit for it. He had the guts to tackle the job & others have not the same tenacity or vision.

... Old H.B. Cushing was buried today. He died quickly on Friday from a stroke. I am extremely sorry. He was by no means anywhere near senile, though 74, & was to help me with getting the virus division working. I shall miss him greatly.

 Love to you all,
  Dad

London, Ont.
5 Nov 1947

My dear Dad:

Your letter came today and we are elated to think that both of you will be visiting us so soon. It is a very happy happening. I am also very pleased to think that the University of Cambridge thought to appoint you its representative – definitely an honour.

... We got the mislaid letter – as I believe Doris told you – it must have been picked up & posted for it arrived very promptly. The full citation was a great pleasure to us – I showed it to Ash and he was very impressed and sends his congratulations.

I was very sorry to read in the paper the other day & to read in your letter of the death of H.B. Cushing. I remember him very well as a wise and friendly physician of great general knowledge. Remarkably active right up to his death.

We are busy – both of us. Doris has been doing some "socializing" (she won't like the expression!) and has started her Monday evening drawing sessions again. All very good for her. I am up to the ears with various bits of work – in fact too many. The teaching is a bit of a strain for we are chronically short of supplies of various sorts which necessitates last minute stratagems – most of which have worked quite well I'm glad to say. The class is a good one to teach and it seems a great pity to have to makeshift for them. In addition to the teaching I suffer constant misgivings about the nature of swarming and I am, at present, trying to organize the large body of observational data that I have so that I can present it at the staff seminar

---

122 Joburg later became the third president of the Canadian Society of Microbiology.

next week.

... We are still in the same state as far as fundamental knowledge is concerned and I can't see that I have added an iota – some of the observations are interesting but they are not capable of interpretation. What I need are some ingenious methods – I don't seem to be able to provide them ...

 Your son,
  Bob

104 McClary Ave
London, Ont.
Nov 5th, 1947

Dear Dad,

Congratulations again on the award. We are thrilled. I'd love to have spoken to you on the phone, but the time was limited and I knew Robin could tell you how glad we are.

I'm very glad to hear of the proposed visit and would love if Mum could come on to London on the 20th unless you specifically want her in Toronto. Then we could have a longer visit with her. Most of all be sure she comes and for as long as she can. I'm already counting the days!

Robin is working very hard and we are all very well including Peter. Alice will be especially glad to see you. She talks about you & Mum & Uncle Fudge & tells me how Gampa didn't get on the train. (Why I don't know!)

This is as usual just a note but brings lots of love from all of us & a kiss for Dougie,
P.S. Someone kindly mailed your note & clipping!
 Doris

Royal Victoria Hospital, Montreal
6 November 1947

Dear old Son,

Mum & I & the pup are going to the lake for the last possible weekend & we shall close the camp for the winter this time. The spring seems a long way off just now & most desirable. We will make full use of Armistice Day, which has been declared a holiday.

I told you I have been asked to represent Cambridge at the installation of the new Chancellor at Toronto on 21st Nov. There are things I wish to do so I shall go earlier. ... I would like to see Collip & Hall & Ash if it can be arranged, so tell them I am coming.

You will hardly believe the latest adventure. I have been appointed by the University to be one of its three representatives on the Montreal City Council. It seems silly to me as I am not a politician. Some think otherwise.

Douglas is slightly fractious for him because he was vaccinated (Sm. Pox) on Monday last. He feels not quite too well; but it won't last long & he is naturally a very cheerful fellow & very fit. He now crawls & pulls himself up onto his feet holding onto his pen.

... I have no doubt you have some interesting observations on swarming & migrations of colonies. I don't think you need be depressed because exact & reliable observations of fact is most important & explanation by hypotheses of small matter. It is seldom, if ever, that a complete explanation can be given of anything. I think there is too much tendency at present to write lengthy explanations of very small observations & to belittle observational experimental science. I think if you describe what you find you cannot go far wrong. I expect you have done pretty well with your photography; cannot do superlatively without expensive equipment & money to spend.

I wish I could get back into the lab. I don't suppose I ever shall. It seems hopeless as I become more & more involved in administration & have less chance of delegating some of it.

The weather threatens a change but I know Mum & I can find things to do whatever it is. The only thing we do not care about is the large number of indiscriminate hunters who wander in the bush. Everyday last weekend there were one or more lost & firing signals for help. Mum fears they will mistake the canoe for a moose!

  Love to you all,
    Dad

I just received a nice letter from F.C. James about the Medal of Freedom & I am glad of it.

London, Ont.
9 Nov 1947

My dear Mum:

After much intermittent hunting I have at last found the clipping from the "Star" of your letter on slum conditions. I apologize for its long disappearance from view and enclose it.

I am at present immersed in trying to sort out my large bundle of notes on the research we have been doing on motile colonies and it is a hell of a job. A terrific amount of data but which is hard to correlate and make into any form of intelligible whole. I have to give a seminar on Wednesday and still have the worst part of the job to do.

Alice had an acute sickness this week but, thank goodness, it did not last long and she seems to be perfectly well again. She was stricken with very persistent vomiting for about 5 hours starting (as usual!) at 3am Thursday and with some fever for the rest of the day – nothing else. It seems to be some bug floating around the populace for the same thing has happened in many families.

We are very much looking forward to seeing you in two weeks time – it seems a small age since the summer and there are many things to show you & talk about.

Winter is here at last and we are beginning to be glad of warm clothes and plenty of central heating.

  All my love,
    Bob

            The University of Western Ontario
                Faculty of Medicine
                14th November, 1947.

Dr. E.G.D. Murray,
3590 University Street,
Montreal 2,
Quebec.

Dear Dad;

It struck me that you may not know that the Chicago train passes through here on the way to Montreal, and that you can leave here in the evening and be in Montreal by breakfast-time the next day. Sleeping cars go right through, and it may be much more convenient to you and give you more time on your short time with us.

I have seen Hall and Collip. Hall regrets that he will not be here that week-end, since he has to pay some visit out to the west. Collip, however, will be in the city and looks forward to seeing you, also Jim Patterson.

  Your son,
    Bob

            Royal Victoria Hospital, Montreal
                15 November 1947

Dear old Son,

... I'm sorry I shall not see Hall but it will be nice to see Collip & Paterson. I sent a note to Collip telling him I would be in London.

... On Friday Mum & I went to the Arts Science Commerce Ball. It was a crowded affair but nothing attractive about it. The band was merely a very loud noise (Holman), the supper was miserable & there was no sociability of any kind.

Lab affairs are becoming increasingly difficult. William is leaving to make more money as a carpenter in the building labour market. He wanted a raise of $500 a year which is beyond possibility & even if conceded would be the start of a general demand. The building operation moves very slowly ... Supplies are still short & we work from hand to mouth for test-tubes & many other things. So my days get frittered away with trivialities & interruptions which others should look after but don't. This is turning out to be a bad year!

The weather report has just told of snow in Southern Ontario, which I suppose means you, & it has been anticipated by it snowing here too. It seems a long time to wait for spring again, for open water & the canoe & fishing.

Love to you all & we are looking forward to seeing you next Saturday.

    Dad

London, Ont.
1 December 1947

My dear Mum and Dad:

... Mum's very nice letter arrived this morning and we are both so very glad that you enjoyed yourselves when you were with us. As you say it was a short and crowded weekend and there is a limit to what can be done. We loved having you and we hope that when you come again it will be for long enough to really enjoy a leisurely stay. All credit for organization and what not goes to Doris who made the extensive & smooth working plans.

Time is very tied up with the tag-end of teaching (Immunity: which requires strict attention) with plans for Xmas, for the Toronto meeting – and so forth. We also have some entertaining and going out to do, and that will take care of the rest.

I don't know if Doris has written but the day after you left Doris went wild with a pair of scissors & sheared Peter. She claims it has ruined him – I disagree, I think it makes him look a very nice grown-up little boy. I think she really misses the curls!

This is only a lunch time note – will write properly when I can.

    Your loving son,
      Bob

McGill University, Montreal
3 December 1947

Dear old Son,

... Your letter came this morning & we were very glad to see it. I am glad that Mum wrote to thank you for giving us a good time. We enjoyed the visit immensely & only regretted that it was so short. I had to be back on the Tuesday to lecture & I have one or two lectures a day now till Xmas holidays start, & several days seem not to have a free hour from morning till midnight. My time at the lab is a series of interruptions & at the end of the day I seem not to have done anything. Well! Enough of bellyaching.

Winter has set upon us; there has been quite a bit of snow & moderate cold, though thawing yesterday & today. It is hard to regulate the house & in any case, Sue wants it much warmer than Mum & I like it.

I must find time to do my Xmas shopping before the heavy rush. The time goes quickly & I must watch it...

... The new part of the lab is nearly finished. I am going to try to move into some of the labs next week. The moving is going to be a curse in the middle of a heavy load of work so that all technicians are heavily occupied. Freddie's group are not going to be easy to handle either. This Deaning business is not the most satisfactory situation & I find some of the circumstances difficult to stomach. There is little I can do about it...

 Love to you all,
  Dad

            London, Ont.
            7 Dec 1947

My dear old Dad:

I am afraid that my letters for the moment are a little scrappy for we have had a hectic time since you were here and spare time and energies have been limited. I have had to catch up on many arrears of correspondence for Xmas, as well as teach, get ready for my paper etc, none of which matters seem to be completely in hand. As a matter of fact I have had to skip Xmas shopping for the moment though Doris is doing a sterling job of looking after things I miss out.

Serological practicals of the past two weeks and for next week keep me hopping mostly with anxiety. They have worked well so far and the supplies I got together have been just enough, the immunized animals produced reasonably potent sera, and gremlins did not inactivate complement etc!

As far as my paper is concerned everything is under control except additions & subtractions to the film. Somehow preparations have not grown well or done things at awkward times etc etc so that I have nothing more to show than I demonstrated to you and I have not been able to replace the parts that are bad.

... I am looking forward to seeing old friends at Toronto and have written Ralph S[t. John-Brooks] & Ted Roy[123] to warn them of my impending descent.

Last week we went to see the University Revue – talent drawn largely from the

---

123 See chapter 1, note 89.

medical school esp. the veterans and it was an excellent show and very good entertainment. I think they put on the best & most artistic & polished university show I have seen with some excellent comedy & no reliance whatever on crudity for effect. The singing was remarkably good & the male chorus excellent – largely due to a medical student who was brought up by the Salvation Army & who is an excellent musician!

The Sunday lamb is roasting and my saliva runs apace – must go & see if it is really edible.

 Greetings to all.
  Your son,
   Bob

P.S. How are relations with Camillien Houde?!

Royal Victoria Hospital, Montreal
21 December 1947

Dear Old Son,

... I'm sorry I could not go to Toronto for the meeting but I hear your paper was very well received & was very good. They seem to have crowded too much in & not allowed time for discussion, which is a pity.

We have moved into the new labs even though everything is not completed. Everyone is mighty pleased & satisfied, so I feel I planned it fairly well. Moving was not too hard a job but there are some adjustments to be made.

The girls put on a party on Friday in the library (which is still a bare room) & it went off very happily. Lots of fun. Mum & I hope to have a staff party early in the New Year.

I wish you could walk in to see the new wing & the other alterations, even though everything is not yet in order. There are many improvements though some things are still too cramped. However it is working well.

On Wednesday last (17th) Mum & I celebrated the 30th anniversary of our wedding. We went out & had a quiet little dinner together. We are both pretty fit; though I still have the remnant of the cold I left London with. I can't shake it off completely & it plagues me at night.

I went to my first meeting of the City Council on Friday. It was not remarkable as there was no contentious business; but the appointment of officers (Executive Committee, etc) gave opportunity for congratulations with Gallic eloquence. I expect other meetings (Budget etc) will be more active & partisan; though there was some petty politics evident in attempts to gain office. It is just as well to be pretty well conversant with French as most of the talking is in that language.

This brings you love & every wish for your happiness & success. I am sure you will

have a grand day & we will be drinking your health about 2pm.
  Your loving Dad.

Canada
Department of Agriculture
Animal Diseases Research Institute
Hull, Que.
December 22, 1947

Professor E.G.D. Murray,
3590 University Street,
Montreal, Que.

Dear Joburg:-

 This opportunity is taken to drop a line to wish you, Mrs. Murray and the family the Compliments of the Season.

 Your presence at the recent meeting of the Laboratory Section was missed and also your leadership in the discussion. However, the father was well represented in the son, who gave an extraordinarily interesting talk, supplemented by a home-made film on 'bacterial swarming'. Have you seen this film? If not, you should. I was very impressed with what he was able to do with home-made apparatus. Photography is certainly an aid most useful to the teacher. We are going to London next year and I tried to have your son placed on the Executive. Insofar as efficiency is concerned, the other members of the Council agreed but they felt that Asheshov might not appreciate being by-passed. With kindest regards to all,

  Yours sincerely,
   Chas. A. Mitchell,
   Dominion Animal Pathologist.

104 McClary Ave
London, Ont.
Dec 26th 1947

Dear Mum & Dad – We were tremendously cheered by our talk with you all on the telephone yesterday and as if the mountains of lovely gifts wasn't enough yesterday Dad's check for me arrived in this morning's mail! You people are far too good to us and keep us in constant debt for again this year we wish we could have done more on our part.

 ... I think Christmas started last Sunday for us. We got everything we could done while mother was here & children had their bad bronchial colds then. Then just when we were ready to put our feet up & do some entertaining mother left for Rochester. I was hoping she could stay longer but they needed her there. On Mon. we were invited out and on Sun. Wed & Christmas we had gallons of friends in and were at last able to really entertain a little after a very busy fall.

Alice & Peter helped decorate the tree which went up Sun. & both treat it with great awe, admiration and respect! Peter is very thrilled with the lights & points & squeaks at them. He says "drink of water" now & many other things including asking to go to the bathroom! Alice was just able to get back to school for her Christmas party on Friday though it tired her & I also took her to see Santa just before Christmas. She was so impressed she forgot to ask for her shovel, but he brought it anyway, which was nice!

Christmas eve both A & P hung their little red socks up by the Xmas tree and Alice & Daddy & I sang Christmas Carols before she went to bed. Her favourite is jingle bells which she knows very well. She is tremendously sweet & happy & such a companion. If I ever indicate I'm tired she brings me a drink of water so I'll feel better.

Well – that was Boxing Day! It is now the 28th and the tea we attended was a "soaring" success to put it mildly. They had all the noisy toys she could dream up & all going at once! However – eventually all the small toys were removed for a party by the nanny & we had ours in uninterrupted peace.

… On Xmas day we got up around 7:30 and all piled on our bed to open stockings, which was fully okay & also Christmas carols on the radio. Then we had breakfast of sorts in the kitchen & then opened the door to the tree. Alice spotted her shovel immediately others began the organized confusion of opening, recording and keeping papers more or less under control. Needless to say it took most of the morning. We gave them a table & chair which has proved a tremendous hit indeed. Peter especially loves to sit at the table and Alice uses it for her painting (put on water & bring out the colors book).

The things you sent were lovely and I don't know where to start thanking you! You must have been impressed with our garment shortage here! For we all feel well outfitted now and many thanks.

… Our tree is especially lovely this year & everyone has noticed it. Robin seems to have an eye for them. We have it decorated with care & not too much and it features the beautiful chain Alice made at school. Also Barbara at the lab loaned us some sweet little houses for it! I got a few balls & we have your decorations from last year, lights & tinsel. It is very very pretty. The children love it.

… I must get us dressed for the blizzard now or would write ad infinitem. So I'll close by saying that we all feel very happy & full of good spirits and send you & Dad & Sue & Blake & Dougie *and* Uncle Fudge lots & lots of love & kisses and hopes for the best New Year ever. And I'll write to Sue & Blake next & meantime send them our

love & thanks for far too much for Xmas.
    With [illegible] love from myself, Robin, Alice and Peter.
        Doris

London, Ont.
28 Dec 1947

My dearest Mum:

Doris is in process of writing you a full account of Christmas – she will tell you many details I have missed. We had a lovely time thanks in no small measure to your grand efforts. At our end Doris' planning was so efficient that there was little or nothing to be done at the last minute and we were able to take things as they came and enjoy every minute.

The children had a marvelous time – not the least were the preparations before the event – raising the tree, trimming it and turning on the lights. But when Xmas morning came they were full of fun opening presents on all sides starting with a little table and chairs that we took over and we, as recording angels, had quite a time getting a note of all the presents.

The things you sent us were lovely and we do and will enjoy them enormously. I have a distinct feeling that, when you were here, you took a careful inventory of what we needed. Shirts and handkerchiefs were certainly among the necessities and the fine ones you sent me are already in use. Fudge's present of Pyramid handkerchiefs were grand – reminds me of the old advertisements in the underground with drawings by Fongarre – they are still very nice.

… Doris seems to have done me out of a lot of the news. She did a grand job of planning and is still keeping things under great control being immersed at the moment in baling mountains of paper for the Salvation Army to collect. The kids, bless them, have never been better. Alice is more than her sweet self and carries on the most complex conversations – Peter for his part, though short on vocabulary, strives to keep up and does not do so badly.

We had (and have) mountains of food and little specialties – even though we seem to be doing more than the usual amount of entertaining. Tomorrow night most of the lab is coming in for the evening and other things seem to be cropping up in the agenda.

It was so nice to talk to you on the phone on Xmas day and to hear what a happy time you are having.

During my visit to Toronto I managed to have a nice visit with both Ted & Churchie [Roy], and Ralph & Mary Sketch. I spent a very pleasant evening at Ted & Churchie's little house – such a nice place – and they treated me to a royal dinner. And on the Sunday I went out to Oakville to help christen Ralph's son Michael. We went afterwards to their house and spent a very pleasant day. It is a very nice place they have with nice, though affluent, neighbours.

...I have had a grand holiday and I'm glad to say that Doris is not looking tired – we have managed to get plenty of sleep on the average. I go back to work again tomorrow to a much less arduous schedule than before, and very much refreshed.

    All our love from your loving son,
        Bob

London, Ont.
28 Dec 1947

My dear old Dad:

Christmas is over and we have all pretty well recovered from the festivities, which were great fun and went very well. For a few days before the event we were, due to Doris' industry, fairly ready and had to do little at the last minute. Mother Grace had been with us for a week which had given Doris time and opportunity to get things done. It was very nice to see her and we were all disappointed that she could not stay with us for the festivities.

Since things were more or less under control we were able to ask various friends in and do a little entertaining. I found I was pretty tired after a tough term of teaching and much other work and I was very glad to take the whole week off. ...

We love you both very much & thank you for everything.

    Your son,
        Bob

1 The first medical Murray: George Alfred Everitt Murray in Johannesberg (1892)
Peter Murray Collection

2 The Murray residence in Johannesburg in the 1890's
Peter Murray Collection

3 Kate Murray and her four sons during the Anglo-Boer War
Peter Murray Collection

4 Captain Everitt Murray on "Old Soldier" during his 1916 visit to South Africa
Peter Murray Collection

5 Bacteriologist at work: Everitt in the Cambridge Pathology Laboratory during the late 1920's
LAC, Photographic Division, E.G.D. Murray Collection, E008406489

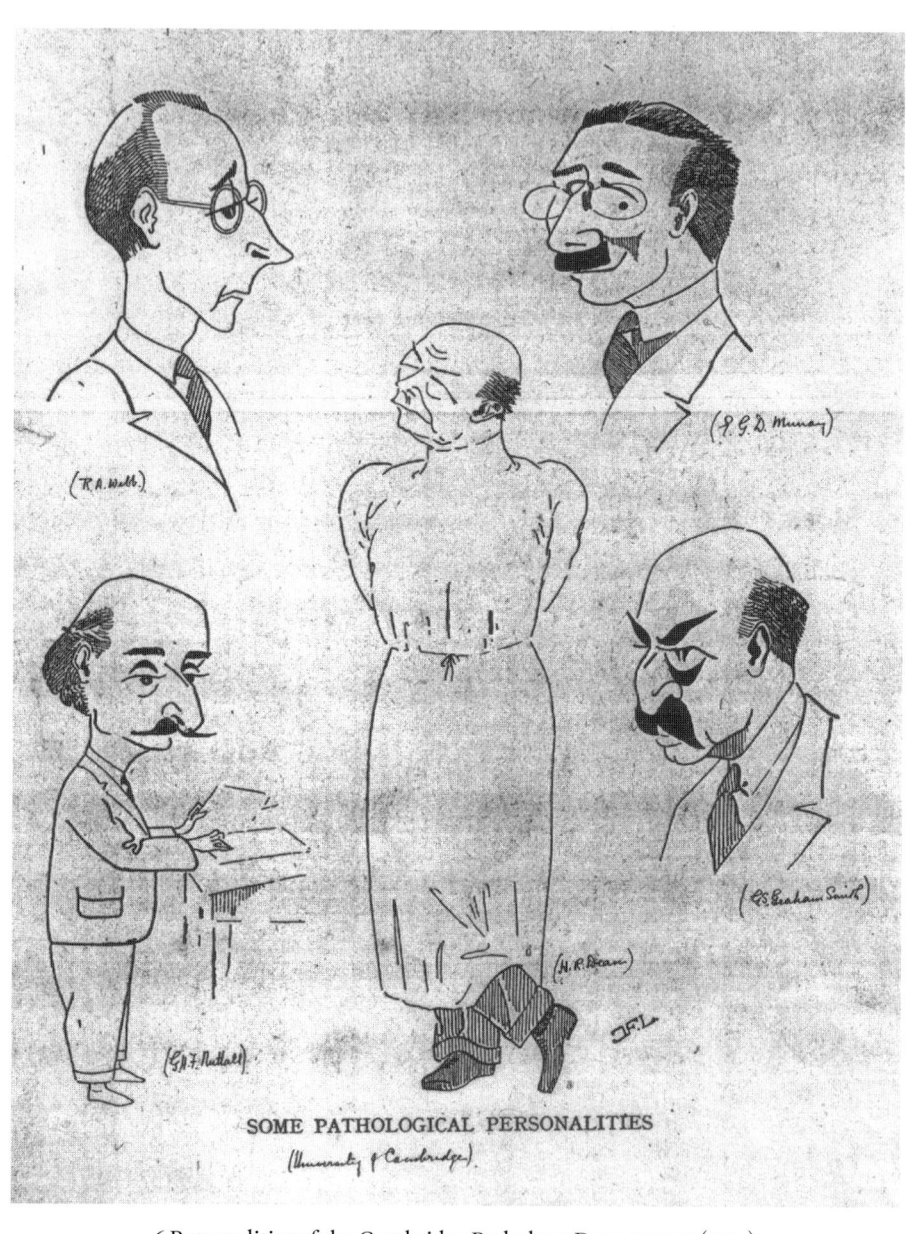

6 Personalities of the Cambridge Pathology Department (1930)
Robert Murray Collection

7 Joburg in his McGill laboratory (1939)
Peter Murray Collection

8 Joburg and Fred Smith: Scientific cooperation from Cambridge to McGill
Peter Murray Collection

9 Bob and Joburg share their love of fishing at the Columbus Club, Quebec
LAC, Photographic Division, E.G.D. Murray Collection, E008406498

10 "Soup's on." Freda and Joburg in their Bark Lake cottage
Peter Murray Collection

11 Robert and Susan enjoy the sun at Bark Lake (1942)
Peter Murray Collection

12 Scientists meet on the eve of war: The Third International Congress for Microbiology, New York City, September 1939
LAC, Photographic Division, E.G.D. Murray Collection, E008406497

CHAPTER SIX

FAMILY CORRESPONDENCE: 1948-1949

London, Ont.
3 Jan 1948

My dear Mum:[1]

... The kids are beginning to recover from the strenuous excitement of the holidays – so are we. They are enormously pleased with all their toys and play with them very happily. Peter looks like a small blue Eskimo in his "Snow soup" and it keeps him beautifully warm – when he is properly done-up about all you can see of him is his nose and a pair of beady eyes.

Us adults have not done too badly either and have had a lot of fun playing with a spinning top, several mechanical cars, and a toy ukulele (which drives me wild). The presents, more seriously meant for us, are grand and we are enjoying them to the full.

Such a nice long letter from you describing your Xmas doings. We love to hear all the details and your letter gives us a very vivid description of the good time you all had. We are very glad that you like the cigarette case and that it is something that you need.

All our love to you.
Your loving son,
Bob

104 McClary Avenue,
London, Ont.
4 Jan 1948

My dear old Dad:

... We were very glad to get your letters and to hear how things were. Mum sent us a most full and enjoyable account of all your doings on Xmas day and thereabouts. I found the book on the Mounties in Toronto and thought that you might enjoy it, although I knew that you had several other books that touched on the subject. The first edition of Mackenzie's Voyages that Mum gave to you sounds very fine indeed and I am sure it is a good addition to your bookshelf. You must be getting a very creditable collection by now.

---

1 Unless otherwise specified the correspondence in this chapter is located in the UWO Archives, R.G.E. Murray Collection.

When I saw Gertrude[2] they were due to move into the new lab the next day and were looking forward to it. I am very glad to hear that they approve of everything so well. It is a great tribute to you that the job is finally finished (or nearly so) and that it fits the needs of the lab so well. I certainly wish that I could just walk in some day. From seeing the plans and the rooms in an unfinished state I would say they must make a grand series of labs. The "house-warming" party must have been fun.

I had a very good time in Toronto and wish that you could have been there. It was good to see Gertrude, Ted [Roy],[3] Allin[4] and a number of others. But I was sorry also that Guilford Reed could not be there. It was quite a good meeting though not quite as interesting in many respects as the one in Montreal last year. However, there were some good papers and the meetings went off smoothly. It is good to hear that people enjoyed my paper – I am glad in that it represented some rather patient and painstaking work. The film added a lot to it and gave people an idea of what these things really look like. Charlie Mitchell[5] was very kind in letting the discussion go on a bit and in giving a little latitude in time. I am glad to say that Ash's[6] paper was a great success and was undoubtedly the best of the meeting as far as original work is concerned, and it caused some considerable discussion. He was very pleased.

Ash has just returned from Chicago where he attended the section on Antibiotics at the AAAS meeting.[7] From his account it must have been a really good session. He also managed to get to a meeting on the phage. He was very bucked because Waksman asked him to join a discussion group and said a lot of nice things about his work. There seems to be some very fascinating work going on in that field but I must say that half of it is Greek to me.

We spent a very quiet time over the New Year. By accident we stayed up to midnight hour and drank a new year's wish to all of you. We will be quite content to spend a quietish month or two recuperating. It won't be hard to do in view of all the talk about prices that have been in the papers.

Next week we start the chore of examining and I imagine that it will be close to three weeks before we are shot of it. During the past week I have been fairly busy. One very interesting organism has turned up in pure culture from a lung abscess (I think). It is a streptococcus which is indifferent aerobically but produces a trace of greening anaerobically. (Sheep blood) ... It does not seem to require either X or V factors for it

---

2 See chapter 3, note 101.
3 See chapter 1, note 89.
4 A.E. Allin was head of the Ontario Public Health Regional Laboratory in northwestern Ontario, most likely at the Thunder Bay–Fort William Laboratory. RGEM.
5 See chapter 2, note 36.
6 See Introduction, note 377.
7 The American Association for the Advancement of Science.

shows just the same characteristics on a yeast extract-agar without blood or serum. Seems to be quite odd ...

I am having a lot of fun tying flies but having a little trouble getting wings to set properly. I am having more fun with your present than I have ever had from a present before.

   Your loving son,
    Bob

P.S. There is a projector available so you can send down the fishing & camping films anytime you are ready.

                Montreal
               9 January 1948

Dear old Son

... I sent you a copy of my M.S. on the history of medical bacteriology which I sent in to Dubos[8] for the book. I am glad I have finished both the chapters I undertook to write. They were both about the most difficult required & least likely to gain any credit.

Dubos seemed satisfied with the technique one & I hope he will be with the historical one, though I feel it leaves much to be desired. Gertrude likes it & I am sure she means it. I am looking forward to your impression.

The new lab is working quite well & will be much better when it is all complete; there are still unfinished little jobs which are annoying. I think you will like it when you see it. Some things the architect foisted on us fester my soul but we will get used to them in time. Taken all round it is good.

I am still plagued with some technician trouble originating in too low wages. I have been lucky to solve the major trouble of media maker extremely satisfactorily.

Staff troubles begin to loom. I want a Lecturer as assistant to Gertrude starting in July when Denton leaves.[9] Salary $2400, can be green but ready to learn & keen. If you have someone send him along. I also have the Assistant Professorship to fill at $4000 to $4500, depending on experience, on the teaching side. I also have to find a virus man & cannot get a lead on that either. These I fear are going to be real troubles.

... Mum & I are thinking of writing up Pollution & Contamination of Rivers & Lakes. If you know or come across any useful references let us have them.

8 See chapter 4, note 66.
9 Denton was a research assistant in the clinical labs working under Gertrude Kalz. Joburg was commenting on the common problem of finding skilled people to replace post-doctoral (MD) assistants who frequently moved on to new positions after gaining experience in the bacteriological labs. RGEM.

I am sending you a left-hand crooked knife which has just come from H.B.Co. in Winnipeg.

    Love to you all,
        Dad

London, Ont.
14 Jan 1948

My dear old Dad:

It was lovely to get your letter, and the various parcels were very welcome: both the knives, & the books, & the chapter from the book.

I was very glad to get the MS. of the chapter on History of Bacteriology and I enjoy it very much. I must say that I have only read it over once and have not had time to read it critically. I may have missed it but I did not see any mention of the primary recognition of viruses and their characterization – with the exception of bacteriophage. In the case of the phage I am not sure [about] the credits that you give & feel that the more basic work was done by others than those mentioned as using phage for determining race & type differences. I much enjoyed your general attack and the tone of the presentation. ...

21 Jan 48

Well, I started to write this over a week ago and I have not really had much time, opportunity or energy to finish it. We have been examining pretty solidly since term began – first with a colossal series of orals and then after the written a mountain of examination papers. Now I have finished (just finished the last book) I can breath a semi-complete sigh of relief & start real work again.

Perhaps I missed also a mention of the role played by studies of bacterial physiology in relation to medical bacteriology – especially perhaps the epoch making (to me) realization of the competitive inhibition of enzyme systems through mimicry of substrates by Woods, and later elaboration by Fildes,[10] Woolley & others. This has such wide general application not only in bacteriology that it would seem to be a milestone in investigations.

... I'm very glad to hear that you & Mum are thinking of writing up Pollution & Contamination of rivers – I will try & find some source of references here & let you know.

Sorry you are having such staff troubles and I'm afraid I can't help you with them. As a matter of fact I often wish we could work together for a while – and still think

---

10. See chapter 2, note 39.

about it every now & then in relation to my present position. However, I will keep an eye out for someone to replace Denton. I might have someone for a year from July but not immediate.

We also have peculiar economies & such like evidently a part of higher strategy. Often very hard to decipher. When it happened last year they put salaries up. What next?

In between other things we have had a little relaxation and have been out a couple of times. Last Friday we went with friends to the Dublin Gate Theatre production of a slightly mad & definitely Irish play – well done – an experience, but not much more. ...

... Our wedding anniversary – the glorious 4th! – was spent quietly at home drinking, quite appropriately, rather poor brandy. You may not remember but all we had to drink on our honeymoon (hard times) was rather poor brandy! Very nice of you to remember & thoughtless of me not to congratulate you & Mum on your 30th in December. Times pass quickly.

The kids are very well & progressing by leaps & bounds. Peter is getting very gay & quite chatty. Doris competes with the chores & vicissitudes of the day with great cheerfulness and is very full of good spirits. She has just got back from some hen meeting & is positively hoarse from chattering!

Many people have asked me to give you their best regards: among them Collip, Hall.[11]

    All our love to you both.

        Your son,

            Bob

P.S. We are having a spot of trouble with an illness in G-pigs – apparently not bacterial? ... I don't know anything about lab animal diseases (except some) & I should know more. Any ideas & references???

                                        McGill University, Montreal
                                                        18 January 1948

Dear Old Son,

I have been examining the crooked knives & other knives actually made & used by two tribes of Indians & which are in the McCord Museum.[12] There is a lot of variation

---

11 See Introduction, note 317; chapter 1, note 56; and chapter 2, note 9 for information on J.B. Collip, and chapter 3, note 73 for information on G.E. Hall.

12 Located on the campus of McGill University, the McCord Museum initially housed the Canadian historical and cultural collection of its namesake, David Ross McCord. The museum was established in 1921 and operated by the university into the 1960s until its privatization.

in the curve of the blade & the workmanship. There is one constant & most curious feature that the handle is crooked too. [Image 1: whipped with raw-hide, shrunk on wet, slightly hollowed, hole through handle through which the raw hide goes & is held with a stopper knot; Image 2: one had the blade turned at a right angle; Image 3: one had the bend right close to the handle, could thereby get work on the curve most effectively ...]

... If you know any good references to pollution & contamination of rivers & lakes I would like to have them. The public health side of the question is easily found but I want information on the other effects. There are hints of information in paragraphs, here & there, in articles on oncology & many general statements, but I wish to get a line on actual observations.

Dubos acknowledged the receipt of my M.S. on the history of medical bacteriology but made no remarks about it so I don't know whether he thinks it suitable or not. There is an enormous amount of material & many ways of dealing with it. It would take a book to make a real job of it.

... We thought of you both on your Wedding Day (15th) & drank a toast to you.

This year continues to be difficult. Two more Lab Helpers have left to earn more money in offices filing papers. They are replaceable by the same class of girl, of course, but they have been in the lab 3 or 4 years & know this job & it is the devil to have completely green hands to train when you are most busy & need things to be running smoothly. At the same time the "Administration" wants departments to effect economies, that is cut down on the budget requirement for next year without reducing the amount of work & efficiency. The situation is ridiculous.

 Love to you all,
  Dad

24 January 1948
Montreal

Dear old Son,

... I am glad you like the MS on the History of Medical Bacteriology on the whole. Your comment on the comment on the omission of the contribution of bacterial physiology is just. I overlooked it & it is probably too late now though I will write to Dubos about it.

I left out the viruses because there is to be a similar chapter in the volume devoted entirely to virus diseases. I only mentioned bacteriophage in relation to bacteria to bring in the recognition of types by that means. Phage will be treated as a virus & I had no concern with it. There are many ways the subject could have been treated & if developed one of them might well have proved better than the way I chose. In any case much had to be left out unless it was made into a book of quite a size.

We have had our exam for some 120 students (practical) & their written will be soon, but we will have a terrible time about May. Thank goodness we do not have orals.

I hope you will have time now for your own work. You should be given better facilities (a better share of what there is). There are various jobs & it may be that you should look at some of them while the demand is great & some bargaining for terms is possible. I know Ash's work is good & important but you should have a fair chance to work too & it is essential that you get some papers published. I keep hoping to hear your hyaluronidase paper is in the press & I hope to hear soon that you have written up swarming. Hyaluronidase is quite prominent just now & your paper would be opportune & noticed – I'm sorry it has been delayed so long. I am sure you could get a grant towards your work from the N.R.C.

I have often wished we could work together; though I have had to give so much time to other things that I often feel I am out of it now. I still can help others in their difficulties but I have no chance to work myself just now. I still hope I can arrange things better next year. Anyway the present circumstances seem hardly worth while.

I would like to see your position greatly improved.

... Yesterday afternoon Mum & I went to see the Dublin Gate Players in "Where Stars Walk". We enjoyed it very much & admired the splendid rendering of a very difficult subject. Every part was excellently done. We would like to see it again. It reminded me of much I had seen in Ireland. Recently we went to see a revival of an old film "Wings of the Morning" & enjoyed it again. It is a Derby winning horse film & appeals to us; the real scenes of the race, the tipsters & touts & some real people riding like Steve Donahough & not actors, good horseflesh & real scenery, not sets, appeal to us strongly.

... It was good to read in Doris' letter that Peter is now on a full ordinary diet & doing well. That is the best news we have had for a long time. It will make a great difference his being able to join in with everyone else. Douglas knocked his right cheek against a corner the other day & gave himself a real black-eye. The "shiner" makes him look quite a thing. He is busy pushing through the back teeth.

You don't say what signs or symptoms your G-pigs show so I cannot suggest anything.

... We have been having some bitter sub-zero weather with a biting cold. We still have much less snow than usual. I shall be glad to see the end of the Winter.

I am arranging to go with Mum to Vancouver in June to the Royal Society meeting. She will enjoy it greatly never having been West ...

    Love to you all.
        Dad

The Lord Elgin, Ottawa Canada
29 January 1948

Dear old Son,

My two copies of the 6th Ed. of Bergey's Manual arrived this morning and I promptly posted one on to you for yourself. I am sure you will find it useful in many ways & that there are many improvements as well as new parts. The Index of Sources & Habitats should be valuable. The book is larger but not clumsy.

I have just arrived (10pm) here & will have a couple of busy days, to return on Sunday morning. I could do without these jaunts as there is still so much to get straightened out at the lab.

(A.J.) Rhodes[13] was in Montreal yesterday & gave quite a good survey of neurotropic viruses to the Neurological Section. Nothing new but I hope it sunk in that a diagnostic service, such as they get with bacteria, is not yet possible with virus diseases.

The typing of Staphs with bacteriophage is coming out very interestingly. The vast majority of the babies in the most affected nurseries are all the same type as the permanent nurses most concerned with the babies. Some of the staff have it too. This has annoyed the Head Pupper exceedingly & he behaved thoroughly badly about it. Many obstructions were put in the way of the investigation & cases were concealed. It is difficult to understand their stupidity.[14]

There was a fatal case of tetanus last week too. A clean operation on a knee-joint which was "immobilized" in plaster. No specimen was sent to the lab until the day before death, when there was a panic because of well developed symptoms & characteristic tetanus could be seen in the pus sent over. It grew quite easily from the wound with various other things. The source has not been traced; we could not get it from the faeces of the patient (P.M.). It's hard to get co-operation from the clinicians & I want to examine the plaster used.

I'm reading "Owl Pen" by K.M. Wells. Mother liked it very well but I am not quite so keen on it. It is well written but it lacks real substance & seems to me a rather watery kind of "We Took to the Woods". What I dislike in it most is a kind of sympathy seeking exaggeration of incompetence. Nevertheless it is readable.

This has been a bad week. The City Council from 3 to 6:30PM Monday, 3-6:30PM Tuesday & I did not attend the 3PM onwards on Wednesday. It was an exhibition of petty municipal politics & Camillien Houde[15] on a vote catching stunt (very clever)

---

13 See chapter 5, note 109.
14 Robert Murray comments on this passage: "For 'Head Pupper' read The Obstetrician-in-Chief! I think that would have been Dr. N.W. Philpott who was Director of the Women's Pavilion of The Royal Victoria Hospital, 1945-1956. I remember taking lectures and rounds from him when I was a student. My father could be quite sarcastic when irritated."
15 See chapter 5, note 10.

working up a feud. They spent in all 5-1/2 hours "debating" whether or not to raise the salary of one of the most important City employees. I don't want much more of that sort of thing.

We are all pretty well & busy.

Love to you all,

Dad

"Mayor Houde Suddenly Assumes Firebrand Role: Serves Notice That No Group Is Going To 'Put Anything Over on City"[16]

"It's just like old times"

With that wistful yet jubilant quote an elevator operator at City Hall today summed up the reaction to Mayor Camillien Houde's outburst last night which placed him squarely in the role of leader of the vocal opposition at City Council and Metropolitan Commission meetings – a role he hasn't filled during three years of quiet and almost conciliation towards the centralized government of the City of Montreal.

At last night's meeting of the City Council it was – as the elevator operator expressed it – "like old times" as an irate but alert Mayor Houde stepped down from his rostrum and led the opposition from the floor.

Today, in commenting on this abrupt termination to three years of acquiescence to leadership by the Executive Committee and the Metropolitan Commission, the Mayor intimated that his hat was in the ring when he gave formal notice that he would attend all Committee and Commission meetings...

His reason, as he put it, was to attend to see that neither body "put anything over on the City of Montreal." ...

The Mayor's action caused a flurry of speculation to run through City Hall, and today the general opinion appeared to be that Mr. Houde was reacting against what he considered to be the fading power of the mayor and councillors with most of the important public offices under the control of a central administration instead of City Hall.

Royal Victoria Hospital, Montreal
1 February 1948

Dear old Son,

... I am too ignorant about the Phase Microscope to make any observations on it. It is interesting that it has helped you to show that the colony transports its own puddle

---

16 Although occurring several days after the incident E.G.D. Murray is referring to, this article indicates that this particular week witnessed a shift in Houde's approach to Montreal politics. See *Montreal Gazette*, 28 January 1948.

to swim in, but it is still strange that it can push it about the surface of the medium. I wonder if you could not make your medium more suitable by absorbing the colour out of it with charcoal. Many years ago when I wanted completely colourless gelatine I boiled it up with active animal charcoal & got glass clear colourless medium.

I'll try to find something to correspond to your G-pig disease. What are they fed on? Any chance of something growing in their food & producing a toxin?

I shall look for the books for you tomorrow & send them.

I have heard of "The Angler's Roost" but have not been to it, I shall look at it next time I am in New York. I have just packed up for you some Teal flank feathers, some English Partridge breast feathers & back feathers, and some Pearsall's floss silk for bodies, tags etc. ... It is about time I started readying tackle & renewing necessary flies.

I expect Jim Patterson[17] will tell you we met in Ottawa yesterday. He had not seen you recently so could not tell me much news of you. I suppose you see less of him now he has moved out to Collip's Institute.

You do not mention writing some papers on your work. It is most important for you to do it & it is more profitable to write several short papers than one comprehensive one. Short papers are more readily accepted & you can get more into several short ones than one long one. A short paper is read & the point appreciated whereas the several important things in a long paper get overlooked. My papers were too long & had too much in them & material which should have been made into a whole paper is expressed in a short paragraph, the significance of which is not appreciated. You must publish to get recognition & opportunity so get on with it.

  Love to you all,
    Dad

McGill University, Montreal
4 February 1948

Dear old Son,

I have sent you the books you asked for & some others, except Burnet's "Biology of Infectious Disease" which I was not able to get at any time. I did not know what you were after so did not know what to pick out.

... I have not been able to look up some of the possibilities of the G-pig disease. It may be neuritis due to Vitamin C deficiency. G-pigs are also very much affected by phosphorus deficiencies & that can be fatal ...

  Love to you all,
    Dad

---

17 See chapter 3, note 92.

London, Ont.
5 February 1948

My dear old Dad:

Two letters from you and two packages to acknowledge. All of them very pleasant.

I was very glad to get and to look at Bergey's Manual. It seems to be a very much improved edition and though it is larger it does not seem so. It has been a long hard struggle to get it out but it is well worth the trouble. I have not looked into detail yet but when I do I will let you know of errors or suggestions for future reference. I liked the Index of Sources and Habitats very much and think it is a very worthy addition.

... The typing of staphs seems to be turning up dividends even though it is an investigation made in the face of absurd obstructionism. I wonder if the staphs producing the troubles in babies have any physiological peculiarity. It would be most interesting to know of their "spectrum" of toxin production.

The city council seems to be excessively long winded – it must be quite a trial to you to have to sit through so much of it. Houde's little tricks even got into the local paper and, it would seem, in just the light that would pleasure Houde the best!

I got my last 100ft of movie film back yesterday and in some ways it is quite good. The sections taken through phase microscope are of good exposure but the effect was slightly spoiled by the focusing difficulties, dispersion, and small local variations in thickness, transmission etc. However, it is quite illustrative of one of my points.

I must rush off now to the lab – been doing a little baby sitting! Will write again as soon as I can.

    Your son,
        Bob
P.S. We are all well.

Montreal
15 February 1948

Dear old Son,

... I have not done anything this week but fuss round to help the lab running & sometimes I get fed up with it. I would like to do a bit of bacteriology myself.

... I have one very difficult job to tackle now; the application for the appropriation for next year is to be in soon. It is quite a labour analyzing the accounts & estimating for next year & it is not made easier by having to split it into University & Hospital. I gather that there will be an economy drive with purpose to cut down as much as possible. I don't see how it is to be done with increased work and the higher wage scale outside & the increase in cost of everything.

Ted Rose[18] says you are a Rat & have not answered his letter.

We are marking the Medicals exam. I have not seen anything particularly good yet.

>Love to you all,
>>Dad

>Royal Victoria Hospital, Montreal
>23 February 1948

Dear old Son,

Today's paper tells of Fraser Gurd's[19] sudden death in a train from Chicago, where he was at a meeting. I am sorry & he will be much missed here as a doer & stimulator of good work.

There has been a McGill Winter Carnival & I don't know whether it has been a success. They did some good snow figures; the sphinx of the boys next door is still in fine condition & very well done, better than the photograph shows.

Ted & Sylvia [Rose] joined Mum & me last night at the "Smorgasbord" of the Ritz last night. We fed largely amid a number of most queer people …

Mum & I have written a screed on "Pollution", a sort of introduction but nothing startling, to serve the purpose of a report. I suppose I should have found time to work up something better. I do not feel inclined for "homework" after the disappointments of the day. We are still beset with technician troubles & it is hard to get it straightened out unless better wages can be paid. The burning point now is the washing-up once more. Media-making is in hand though under strain. Sterile glassware etc is in a muddle because of unskilled workers only newly employed. Replacement for two lecturers looms for July & probably an Assistant Professorship not yet filled. This has been & promises to continue a hell of a year.

I spent the entire morning in the City Council, & it was lucky not to be all day & more. I suspect the next meeting may be a tough one. I do not see that I can do any good there & I do not find it interesting either.

I hope you are getting to the time when you have some leisure & can write up your papers. You should write the Hyaluronidase before it gets too stale. You will want to try out the phase microscope possibilities; I would like to know what it is capable of.

>Love to you all,
>>Dad

---

18 See chapter 3, note 50.
19 See chapter 5, note 76.

104 McClary Ave.
London
26 February 1948

My dear old Dad:

It must be an age since I last wrote to you and I have several things to thank you for as well as at least two letters. Times have been busy and you will be glad to hear that at long last I have buckled down to writing up the Hyaluronidase work. This latter is a stinking job and I certainly wish I had done it ages ago – however, I finished the first draft today. I can't say I am proud of the way the whole thing was worked out but it may be a useful thing for special purposes. This has kept me busy all spare minutes during the day and we have also had a certain amount of going out and entertaining to do to fill in the other chinks.

Thank you very much for sending the epidemiological books. I have been giving a few general lectures on epidemiology for Hobbs[20] and have quite enjoyed reading up the subject in greater detail than I ever had before. I will be sending them back to you in a couple of week's time.

I would very much like to see your screed on "Pollution". I must say I haven't the faintest idea of what it is for & to whom it reports. Are you going to send it to a magazine?

I saw Walter Johnson[21] in the library the other day and he said that he had written around to reversal workers in the field for reprints & lists of relevant works. He says they are damn slow in answering but hopes for the best. He said he much appreciated a nice letter from you.

From the pictures you sent, the fraternities seem to be putting a lot of effort into their snow sculpture – I suppose it has all melted by now. McGill must be taking the winter carnival seriously to cancel two days classes! Sounds like a big advertising stunt.

… All of us seem fairly well and recovered from our bout of colds. Doris, however, is having tooth trouble and apparently the dentist is not very hopeful about the future of some of her teeth. She is a little upset about it.

We were both very upset about Fraser Gurd's sudden death. He will be very much missed. Angus (Graham)[22] saw him in Chicago the day before and said he seemed remarkably hearty and full of his usual stories.

---

20 Ed Hobbs was both a psychiatrist and an epidemiologist and was professor of preventive medicine as well as holding a position in the Department of Psychiatry at Western. He played an important role in Bob Murray's professional life, as he was assistant dean of medicine for a number of years. Murray remembers him as "an excellent problem solver and helpful colleague with a range of experience." RGEM.
21 See chapter 5, note 29.
22 Angus Graham worked at the Connaught Laboratories at the University of Toronto.

What with writing etc I am not doing anything dramatic in the lab. I must get some more movies and I hope this phase equipment comes along soon – the delay is holding me up. I hope to show the movie (improved) at the SAB in May – the only reason being that you can't publish a movie film in a journal! I hope N.R. Smith[23] may be there.

There were two very interesting & significant papers in the Lancet in January (I forget the numbers) one by Burnet on mechanisms of virus infection[24] and the other by a fellow whose name I forget on the easy reversible nature of penicillin fast (artificially produced) staphs by growing them together with certain organisms (streps) or a filtrate of a culture of the latter. Very interesting indeed.

I must write a note to Ted [Rose] – I am not as much of a rat as he is. At least I write a little more often.

>Your loving son,
>Bob

>Royal Victoria Hospital, Montreal
>29 February 1948

Dear Old Son,

We are glad of your letter of Feb 26th after quite an interval.

I am glad to hear you are writing up the Hyaluronidase work; I think it worth doing as there is quite an interest in the subject. I have wished several times I had written up some of my work at the time it was done. Several quite interesting things in my note books of 1917 on are cropping up in papers just coming out. It is hard to write up old notes. Because I made the mistake of not writing up quite good work I keep urging you to write yours. No work is ever complete but it can make a useful contribution.

Mum sent you a copy of our screed on pollution which we hope will prove interesting to those it is meant for. There is nothing new in it but it purposes to show attention to the need for control of pollution & contamination. It will be published in the Report of the P.Q. Association for the Protection of Fish & Game.

Have you seen Vol. 1 of the Annual Review of Microbiology (Published by Annual Reviews Inc. Stanford Cal. U.S.A.)? You should recommend it to your library as it has quite useful stuff in it & helps to keep up to date.

We have had two cases of tetanus lately. One is still in hospital & was the ordinary story of neglected infection of contused wounds through insufficient knowledge. The other case died & illustrates carelessness & ignorance where it should not exist. A clean operation on a knee joint was put up in plaster using plaster bandages bought

---

23 See chapter 5, note 97.
24 Burnet, "The Initiation of Cellular Infection by Influenza and Related Viruses," *Lancet* 1 (1948): 7-11.

(as they do now) in tins & used as supplied. Some 20 days later the man died of tetanus which was not diagnosed but was thought to be hysteria. The wound was known to be infected, as seen through the little window in the plaster, but no specimen was sent to the lab until the evening before the man died. ... Unfortunately the original plaster was removed & destroyed before we knew of the case & none of the original plaster bandages were available.

... I'm glad you are giving a paper & your film at the S.A.B. I thought you had your phase equipment already. The B. & L. representative was here the other day & I understood you had the equipment.[25]

I'll hunt up the papers in the Lancet.

I am fixing up our trip to the West. It works out that we will be away a month. I want Mum to have a real good time. I don't know what your plans are this summer & I would like to know if you have any idea of it as I want to include visiting you on our way back. As I have it planned we would arrive in London mid-day 3rd July & stay until 7th July. I hope it fits your plans but if it doesn't I'll arrange it differently. I'll send you the full details later but in general we are flying to Calgary, then by CPR to Vancouver sleeping the night at Sicamons in order to go through the mountains by daylight. After the Royal Society (Vancouver & Victoria) we will go by boat to Prince Rupert & from there by train to Jasper to stay one night & then Edmonton, Winnipeg & then to you at London.

The lab is still difficult to keep going. The washing-up is the chief trouble. I forget how many changes of women we have had since Mary & Helen left & none of them any good. I am trying to find Veterans who would do a job of it but cannot yet find the way to get in touch with them. The facilities have been greatly improved & will be better still presently.

Tomorrow I have to spend at the City Council or else be fined $20. I don't see what good I can do there with the time I can spare & the time it takes I cannot spare from what I have to do in the lab. In any case there seems to me little reason to make great efforts to work up an influence in the Council as the only effective body is the Executive Committee. Once that is elected by the Council all that can be done is pass resolutions asking the Executive Commerce to do this or that or look into some question. Then it is possible to refuse to pass resolutions brought forward by the Executive Committee which have to come to the Council. This little is accompanied by a hurricane of talk much of which is futile when you can hear it. Tomorrow is occupied with the Budget Estimates & I suppose it may go on for days on end.

  Love to all of you,
   Dad

---

25 Bausch and Lomb, Inc. developed popular versions of phase microscopes.

Royal Victoria Hospital, Montreal
7 March 1948

Dear old Son,

This has been a very trying week of urgent requirements of the department & continuous meeting of the City Council to debate the Budget. The coming week will be similar. The Budget has to be passed by midnight on 15th March & I am sure it will be debated until the last minute of that hour. The council meets everyday from 3 to 6pm & frequently again from 8 to 11pm.

. I told you of the case of tetanus & this suspicion I put on the un-sterilized plaster bandage because we have isolated Clostridia & various other organisms from new unopened bandages & could not find Cl tetani in the patient's faeces. Now we have isolated Cl tetani from another new bandage as well as various other gas-gangrene Clostridia. These have killed guinea pigs typically & controls with appropriate serum survived. I think the situation is perfectly clear & it is essential to sterilize the bandages we tested without in any way impairing its effectiveness as a plaster.

... Sue & Blake have found an apartment on Mountain St. north of St. Catherine. Mum has seen it & says it is quite nice. The rent is astonishingly low $65 a month I think. They will be moving about 1st April & are naturally pleased about it. Mum will miss the young fellow very much for a while & I am sure Sue will find she has more to do than she has here & will have to organize her day for it. Fortunately Douglas is amazingly good. Anyway it is the right & proper thing.

I have two troublesome & difficult tasks to face very soon: writing the Annual Report & making the budget estimates for next year. I wish I could have some time to work in the lab myself instead of doing all administration.

I hope you are getting on with your own work. You must do research now & it is perfectly reasonable for you to require time & opportunity for it. Otherwise they must give you adequate compensation in lieu of it. Lab people, & it seems bacteriologists more than others, have neglected insisting on their rights. It is time they took a firm stand. Ash should give you one room to yourself.[26]

I am anxious to hear of your summer plans. From your remark about giving a paper to the S.A.B. I suppose you will get expenses to go to California. It will be an interesting trip for you & I am very glad of it.

      Love to you all,
          Dad

---

[26] In the late 1940s, E.G.D. Murray was very concerned that Bob concentrate on research and publishing his results.

London, Ont.
9 Mar 1948

My dear old Dad:

... Your description of the case of tetanus and the subsequent investigation of the plaster bandages is most interesting and important. It just shows how important such details can be. I hope that you will get this into the literature for it should get some recognition.

The city council business must be a most terrible waste of time for you and most disheartening to have to spend time that can be ill afforded from your proper duties and interests. The troubles you are having in the lab are a symptom of the times, and have to be taken most seriously. When the governmental institutions have put up their salaries, it is more than time for the universities to wake up & do something constructive. With us we have the same trouble. It may be that the university will put up salaries – they are giving a salary bonus at the moment squeezed from all our departmental budgets from things we can hardly afford to lose – but I doubt if it will be much and probably to the detriment of something else.

It has been a weary couple of weeks. We have had to go out a little and I have but various things to do & write. I hope to have at least one paper ready soon and have a little progress on another. The latest duty is to make an after dinner speech at the banquet of the Science Club of the university (students) which will mean active & careful preparation at rather short notice (for the 17th). I also have one more lecture, the last, next week.

For the past week or so I have been making a more detailed investigation of the effect of levulose on the growth of mucoid streps. It is rather a fascination and maybe, in the long run, important business. Dextrose is definitely stimulating to growth & capsule formation. Levulose would seem to be the same in the earlier phases of growth but it seems to accelerate the auto-destruction of the capsules. It also does this to a lesser extent even in the presence of dextrose. ... This may be important from several points of view but it is rather amusing to speculate on the interesting association of dextrose & levulose in the factors and probably in such glands as the testis. May be of some importance in Hyaluronic acid metabolism.

I saw Walter Johnson the other day and he asked me to send you the enclosed letter since it may interest you. Looks as if you are doing a pioneer piece of work. I have read the manuscript through once but have not had time to go over it again. It seems to me a good introductory & guiding note – no doubt you will be developing some parts of it further in another report. It would seem to be worthwhile to make up a bibliography as you go.

Of course the big shindig of the past week has been the official installation of Hall as President of the university. ... it was a good show and a pleasantly dignified affair.

Hall gave a fine presidential address at the convocation. Parry,[27] Vice Chancellor of the University of London brought greetings, a scroll from the public orator and a fine silver mace to grace all further proceedings. In all a most satisfying show. I went up to hear David Thomson's speech yesterday – it was good stuff and delivered in his usual fine style.

Several people ask after you. Jim Blaisdell,[28] particularly, sends his best and says that he appreciates very much what you did for him & the war effort, during the war. He is a most remarkable fellow & I would judge a very able one. Tony Brown[29] also sends greetings.

We are not very sure just yet what we shall be up to in the summer and don't seem to be able to make up our collective minds. It is pretty certain that we can't afford to rent a cottage in this region – they are remarkably expensive. We were wondering whether we might be able to go to the lake and perhaps be up there when you come back from the west. What do you think?

I wish the SAB meetings were in California. Unfortunately they are in Minneapolis which is definitely Midwestern! All I got from the SAB was an acceptance of the paper & I will not know until after April 1st where it will be in the program. There is a dinner here I would like to attend on May 9th but I fear I will have to miss it for it is a fair sized journey out there.

  Love to you both.
    Your son,
      Bob

London, Ont.
22 Mar. 1948

My dear old Dad:

I have been more or less incapacitated, since we spoke together on the phone, by a particularly virulent cold or 'flu or something. It was particularly annoying because I had various jobs and researches started and had to abandon them to be begun again with more labor. Not only that, I had contracted to give a speech to the University Science Club banquet on Wednesday night and although I felt like it I did not want to pull out. As a matter of fact I had not fully prepared it and had to go ahead, running about 50% efficiency, and finish it off. It wasn't a good speech but fair enough and

---

27 Sir David Hughes Parry (1893-1975) was vice chancellor of the University of London in 1947-1948.
28 James Blaisdell was the pathologist/bacteriologist in charge of the laboratory at the Byron Sanatorium and he was primarily interested in tuberculosis. RGEM.
29 Tony Brown was head of the Department of Zoology at the University of Western Ontario.

well received – perhaps, delivered well enough to offset some of its other deficiencies. It was a sort of commentary on present day modes of research and its effects on thinking, with a main (focus) on the necessity of universities planning their departmental researches along lines apart from the industrial pattern which is becoming so influential. ...

... Today I ran across an interesting piece of the history of Bacteriology in an article (leading) in Nature (Feb 21st 1948, #4086, *161*: 266-267) called "Sir Charles Sherrington and Diphtheria Antitoxin". ...

I certainly wish I could come down and visit you. I don't see any prospect of paid travel for the University though you never know your luck. There are several things I would like to talk over but cannot yet. It looks as though there may be a bit of a departmental upheaval in the near future but I don't know all the detail and I am not at liberty to talk about such as I do know. Your remarks about keeping an eye open for possibilities are not lost but I would like you to amplify your ideas a little so that I may think about them.* ...

    Your loving son,
        Bob

*TOP SECRET!

P.S. ... Collip sends you his best regards. Says he has inaugurated his billiard table.

Montreal
4 April 1948

Dear old Son,

I have not written for a couple of weeks because of a pressure of work, meetings etc & when I got home I was tired & lazy.

I sent you a copy of the "new" Topley & Wilson (2 Vols) which I have been waiting for a couple of years. I have not looked at it carefully yet as there has not been time but it is sure to be useful, though probably I won't agree with some of the classification (but I don't like some bits in Bergey either). Let me know that it arrives safely. I suppose you got the original copy of your birth-certificate I sent on Mar 9th by registered post and the book "Fishing & Flying" sent later.

I have just finished correcting the Galley of my chapter on the history of bacteriology. It seems alright, though more work on it would improve it. Such a thing cannot be complete & never would be satisfactory. The note about Sherrington[30] using Dipth.

---

30 While professor-superintendent of the Brown Animal Sanatory Institute in London, England, Sir Charles Scott Sherrington (1857-1952) was one of the first scientists to produce diphtheria antitoxin. He shared the Nobel Prize in Physiology or Medicine with E.D. Adrian in 1932. *ODNB* (Online).

Antitoxin in England in 1893-4 is interesting; It does not alter my note for the chapter because Behring & Wernicke[31] used it first on a human case Dec 1891.

I have my Budget application in with statements etc. I don't know when I shall hear the decision but I suppose there will be quite as much delay as usual. I am still waiting for an answer to a letter I wrote the Dean [Meakins][32] in December, in spite of three reminders. These things are quite a lot of work because I have to partition all expenses between the University & Hospital & have no say in the final decision. I now have to write the Annual Report which I do not believe is ever read.

The Columbus Club Annual Meeting was held on Saturday. Some progress is being made in getting some order into its affairs. Tom Williams was there, in spite of having been flown out last fall with a heart attack. Most of the old crew (what's left of them) were there and some new members. We had a staff party last night & wish you & Doris had been there. All told there were about 60 people. It went very well, although there was not anything special provided.

Sorry you were not well when you had the speech to make & I hope you are perfectly well again. Doris, in her letter which came yesterday, said you had taken a couple of days off over Easter & had tied some flies. Tying wings low depends on holding them down tight with the thumb & finger, after arranging them on each side of the body but with the pressed together flies on top of the hook but tilted towards the direction from which the tying silk is coming so that when the silk is pulled down tight the wings are held down on the top of the hook.

... There are some jobs going which you are very well qualified for if any of them appealed to you. I think Western is a good place with good potentialities but I don't feel you are getting quite the share of it. I may be quite wrong & I don't know what seems to be in the wind. I have kept quiet & not spoken to anyone but I was disappointed in the distribution of available space & facilities when I was in your department last. I have not heard that it has been improved. However, I will wait until we can talk things over.

Doris' news surprised me a little.[33] You will have to put an effort into finding more suitable quarters but it may be hard unless your salary is improved.

    ... My regards to Collip, Hall & others and love to you all.

    Dad

---

31 Emil Behring (1854-1917) and Erich Wernicke (1859-1928) are credited with developing the first effective diphtheria serum in 1890.
32 See chapter 1, note 26.
33 Joburg is referring to the news that Doris was pregnant with her and Bob's youngest son, Thomas Everitt.

London, Ont.
7 April 1948-
11 April

My dear old Dad:

It is again a long time since my last letter – which is inexcusable since you have sent me many things as well as two letters – but things have been too busy and muddled so I have been conserving my energies. As a matter of fact I feel rather worse than usual tonight after a longer session than usual at his microscope – Doris aches too so we are a happy pair! ...

Doris' news may have surprised you but it was no surprise to us! We have had such a plan on the agenda for a long time and did not feel disposed to be stopped by situations, housing or otherwise. It may be a fraction difficult for a while but it will all work out. I expect some improvement in salary as well – I think Collip will see that it comes through. ...

    Your loving son,
    Bob

Royal Victoria Hospital, Montreal
11 April 1948

Dear Old Son,

I hope the various books have arrived safely; I have rather lost count of them now.

... I have not yet tied any flies & the season has opened, though the ice is not out of the lakes. However it won't be long now & I am in no way prepared.

There is such a lot to do in the lab & in the evening I don't seem to have the energy to do much. I must get started & once I do I know I will continue. It is just laziness.

We are preparing for a host of examinations & the authorities have left all too little time before they want the results returned. Those departments with great numbers of students have what appears to be an impossible task. As usual, it will be done but at an unnecessary sacrifice by the staff of the University. I think we can manage as our biggest lot is seventy odd and we have things arranged. I don't see how any kind of arranging can accomplish what some of the other departments in Arts are asked to do.

I have, by great good luck, got two good people for junior jobs, lecturers, to replace Denton & Nunes.[34] One is Pierre Masson's[35] daughter who is very good indeed & the other is a young man Leduc who is with Bertrand at Notre Dame Hospital. Leduc was

---

34 Doris Nunes, like Denton (note 9 above), worked as a research assistant in the clinical laboratories under Gertrude Kalz.
35 Pierre Masson was a professor of bacteriology at the Université de Laval. RGEM.

in the R.C.A.F. during the war & has been with Bertrand a couple of years who wants him to have training with us. I am sure it will work out very well & is good policy too. I don't know how I am to fill the Assistant Professorship. I wish Denton desired to stay as he is turning out well; but he has an inkling to return to medicine though he is vastly tempted by lab work & can't make up his mind.

We are having quite heavy rain today, &, though it is cool, it will take out the ice if it is general. Anyway it will give the city a much needed wash. The maple in front of the house is in full bloom & buds are breaking or swelling on other trees & the bark begins to look alive. Grackles & robins followed the crows so spring is on the way.

    Love to you all,
        Dad

                                London, Ont.
                                22 April 1948

My dear Mum and Dad:

Thank you for a lovely weekend. It was a grand thing to see all of you again and to have such a good talk with you. The only trouble was that time passed so quickly – the weekend was finished almost before it began. Alice had a lovely time and I do hope it was not a trouble for you to have her. I must say she behaved remarkably well. ...

I haven't got much work done, except writing of papers, since I got back – and various troubles such as the course for nurses, the sickness of our wash-up woman, etc. However, things seem to be straight now. I had a letter from Gibbard[36] yesterday and will probably go to Ottawa and look over the situation – but I can't leave until the end of next week. I have written to that effect. I have seen Collip and had a semi-satisfactory talk without any promises either way but with consent for action – but I will be seeing him some evening soon and will be able to get more of his advice. There is no need to be worried about his attitude. He says he will be in Montreal again soon for a visit. ...

Many thanks to you both, my dears, for the love and help you gave me. I wish we could all make visits a little more frequently. I will be writing again soon.

    Your loving son,
        Bob

---

36 James Gibbard was chief, Laboratory of Hygiene, Department of National Health and Welfare, Ottawa.

Montreal
28 April 1948

Dear old Son,

It was grand seeing you & Alice & we wish Doris & Peter could have joined the party. I wrote to you of this.

We were glad to learn that the return journey was good. I am sorry you were not feeling more relieved & rested on your return & I hope things will shape better & not worry you so much. I think the worry should shift to Collip & Hall by it being necessary for them to take active steps to make it worth your while to stay. Don't promise anything quickly but insist on looking round at other things with the option of taking what suits you.

Writing papers is an important part of work & you must be prepared to give time to it. It is very important for you to write now.

I hope Collip will let me know if he comes to Montreal. There are other things I would like to see him about & I would leave it to him to open this subject himself.

I hope you get some trouting in a few days time. Mum & I went to Bark Lake last Friday & had a splendid weekend. The weather was magnificent. We arrived after dark & soon went to bed. I was more weary than I thought, with months of continuous strain & no time off for quite a while, & slept eleven hours, which I have not done for many years. ... In the afternoon we went to Trout Lake & caught 5 very nice fish – missed 6 other rises rather clumsily. We slept well except for a porcupine who would eat a carton under the house with astonishing crunching noise. After a while we drove him off – a chilling performance.

... The old pup was exhausted with the Hart Lake trip & was quite ill on Monday. Too much for the old boy to start with. Possibly he was ill without it but won't be left out of things.

 Love to you all,
  Dad

Royal Victoria Hospital, Montreal
2 May 1948

Dear old Son,

... I hope to hear soon that you have been trouting. I am very anxious to hear how the rod works. It will have a different timing to your "Viscount Grey", which will take a few casts to work out & it may not lift such a long line. I expect you will have to pause longer on the back-cast. Good luck & happy fishing, old boy.

Yesterday evening Mum & I enjoyed seeing "The Bishop's Wife". Very amusing & very well done, with every character well cast. It is the first good film that has been here for months, & the first we have been to for some months. There were some curious disconnected breaks in the story which we interpreted as interference by the Quebec R.C. Censors & a couple of Angel characters mentioned in the cast did not seem to appear.

 Love to you all,
  Dad

London, Ont.
4 May 1948

My dear old Dad:

 I got back from Ottawa at noon today and found your letter and one from Mum awaiting me. Very nice to hear all your news.

 ... Collip should be down your way about now and I hope you get a chat with him. I was not able to give him your message before he (& I) left but he had said before that he wanted to see you when he was in Montreal. We were up to see them a week ago and saw the new annex to the house. It is very fine and the billiard table looks quite small in the big room. I would say there would be room for at least 3 tables. Alice had a marvelous time with a player piano and quite tired herself pumping it.

 I went to Ottawa on Sun. night and returned last night. I don't like two nights in a row on the train but it was about the least painful way of making the Journey. I spent the day at the Lab. Of Hygiene and saw the place ... and talked with most of the people. The plans for the place as outlined by Gibbard are quite good in many ways. The offer he makes is quite indefinite in some ways and might be summed up as being part of his policy to let people work out the direction they think should be taken (within limits) and plan the lab around it. However, it is a national Hygiene Lab; has a lot of routine; has to provide and make services for provinces, government et al. and in a sense rather restricted. Despite all the protestations to the contrary I would doubt if much very fundamental research, if any, will occur there unless staff can be made and becomes available and unless the pressures of coping with various forms of testing, assaying and typing become less instead of more. There are other factors such as Civil Service, Politics, etc which take some of the gilt off the gingerbread. The salary possibilities are not bad viewed from this instant – but, probably slightly inflexible, and probably not as good as I could expect in 10 yrs + in Academic life. Altogether I am not very attracted to it and would, I think, only consider it if the academic outlook were very bleak – which I don't.

 I have had a letter from Guilford Reed[37] – what he might have to offer is not more and not less attractive than the Lab. of Hygiene one though Guilford himself is the

---

37 See chapter 1, note 115.

attracting star! A little marking of time is indicated. ...
All my love to you.
Your son,
Bob

London, Ont.
17 May 1948
23 May 1948

My dear old Dad:

... The trip to Minneapolis was quite a pleasant one and I really quite enjoyed the meetings. Just before leaving I had a very busy time getting many things straightened up and particularly in editing my film and getting in the recent and good shots I had taken. The prospect of the rail strike was not pleasant however it seemed likely not to last or perhaps not to occur atall – and it worked out alright. ...

My paper, or rather film, went very well and seemed to interest and fascinate people a lot. In fact I gave it twice – once at the general session and later at an evening session on Bacteriological Motility. Quite a success. People were interested enough to ask questions and many expressed a wish to be able to get the film. Morton is interested in having it in the Society collection.

Of papers in general there was quite a lot of interesting stuff but I have not been over my notes and culled out the really good items. Nothing world shattering I think. The electron microscopists had some fine pictures to show but have little to base some of their interpretations. Stuart Mudd[38] and Hillier[39] produced some really fine pictures of *E. coli* infected with phage which pose some philosophical problems. Robinow[40] from the Strangeways lab was there and I had a lot of very informative chats

---

38 Stuart Mudd (1893-1975) held many positions during his long career at the University of Pennsylvania, where he was professor in the Pathology Department of the university's hospital, chair of the Bacteriology Department, and chief of the Microbiologic Research Program at the Pennsylvania Veterans Administration Hospital. His research work on the freezing of blood plasma and fighting hospital infections was internationally known and respected. He was the first American president of the International Association of Microbiological Societies. University of Pennsylvania, University Archives and Records Center, Mudd Family Papers (Guide).
39 At the University of Toronto in 1938, the Canadian research physicist James Hillier (1915-2007) co-invented (with Albert Prebus) the first high-resolution electron microscope in the western hemisphere. He was named an Officer of the Order of Canada in 1997. Hillier Foundation Web site, http://comdir.bfree.on. ca/hillier/hilbio.htm.
40 The German-born microbiologist Carl Franz Robinow (1909-2006) was a noted researcher of bacterial and fungal cytology. Bob Murray recruited Robinow (on leave from the Strangeways Laboratory, Cambridge, UK) for the Department of Bacteriology and Immunology at the University of Western Ontario in 1949, and the two quickly became close associates and friends.

with him. He is a real biologist among bacteriologists and makes very good observations. He really knows how to use a microscope which is one good reason for his good work. Very stimulating fellow. ...

Having just phoned you, you have the general idea of the developments. My birthday was ushered in by the first definite news that Ash had got his visa and things could proceed. There were immediate conferences all round for it had to come up officially at the faculty meeting which was Thurs. Before then Collip got the [illegible] arrangement straight and made me the general offer. Apparently I was to be made an Associate Professor anyway but now carrying the position of acting head of the department, with the proviso that they would keep their eyes open for a really top notch person to bring in as head should the opportunity present. I don't think they are too hopeful. Also promised that I would be briefed immediately as to anyone they had in mind. I don't think, honestly, that they could have acted differently.

I have written to Gibbard concerning his offer, which was a very generous one, and refused in as graceful a manner as possible. I do not think that the job was just the sort I would wish for my future interest although it would be a fine line for someone. Nor was I impressed with what Guilford (Reed) had to offer.

R.

Royal Victoria Hospital, Montreal
19 May 1948

Dear old Son,

Just a short birthday note to bring my love & good wishes. There was not a letter for you because I arrived from the Columbus last night after a few days quite pleasant fishing. The party was [Hilary] & T Robertson, F Hanken, W Scriver, G Routhwaite & ... one of Hilary's partners & myself. [Charlie] Pascoe was there (with a friend) but

He arrived after the departure of I.N. Asheshov as Murray was assuming full control of the Department. Murray remembers the period as one with many responsibilities but few personnel to carry them through. The two shared common interests in microscopy and microbial cytology, and Robinow (the senior scholar) served as one of Murray's most crucial mentors and collaborators. As Murray later wrote: "To have Carl Robinow come to us that same year as a close colleague with real cytological expertise and research experience was a truly formative event, a great pleasure, and provided a degree of support in day-to-day operations and in coming to grips with cytological bacteriology that was essential to real progress and my future." Robinow's many honours included the Harrison Prize in 1957 (awarded jointly with Robert Murray) and election as Fellow of the Royal Society of Canada in 1960. He also served as president of the Canadian Society of Microbiologists in 1962. See Robert G.E. Murray on Carl F. Robinow in *The Best Teacher I Ever Had - Personal Reports from Highly Productive Scholars*, ed. Alex C. Michalos (London, ON: Althouse Press, 2003), 178-181; Canadian Society of Microbiologists – Obituaries, http://www.csm-scm.org/english/Obituaries_det_ robinow.asp.

he & I had one day together & was the only real fishing day I had & the best catch.[41] The others were more or less sociable. ...

The fish ran rather small for the Columbus, but good eating. They were not yet up in the usual spring places, but there were enough with a bit of work. I thought of you many times, reminded by places we have fished together, & wished you were there. I hope you get some fishing soon.

I hope the University makes up its mind what it will do about your department & does so quickly. The usual is to delay to the last minute making a most unfair embarrassment.

    Best of luck old Boy & many happy years.
        Love, Dad

                Department of Bacteriology and Immunology,
                            346 South Street,
                                London, Ontario
                                  26th May, 1948

Dr. Gertrude Kalz[42]
Department of Bacteriology,
3775 University Street,
Montreal,
Quebec.

Dear Gertrude:

I have been through all the cultures you have sent to me and have tested them for hyaluronidase production, using my plate method. Of the 94 strains all except one were definite hyaluronidase producers, although these produced a little less than the others as estimated by the width of the zone. The one strain was apparently negative and was rather interesting – it produces a diffusible inhibitory substance with no evidence of a hyaluronidase effect. If it does produce hyaluronidase it would be in smaller amounts than any of the other strains. I enclose a list of the strains tested so that you can check them off with your results.

We have been isolating coagulase positive staphylococci from the routine nose and throat cultures of nurses, doctors and war maids on the obstetrical floor. Out of a total of 63 specimens we obtained 17 strains of *Staphylococcus pyogenes* from different individuals. All of these produce hyaluronidase although two of them produce

---

41 See "Robert Murray Reflects on Wartime Correspondence" in chapter 3 for information on some of the Murray family fishing companions.
42 LAC, R.G.E. Murray Papers, vol. 28.

inhibitory effect as well. I attach a list of these strains and we will be sending them on to you very soon.

I hope that everything is going well with you and yours. As you may imagine from what Dad has probably told you, things are more than reasonably hectic around here at the moment. However, come hell or high water we will be taking a holiday in July and will process to Montreal, bag, baggage and kids.

Yours ever,
R.G.E. Murray, M.D.

London, Ontario
27 May 1948

My dear Dad:

It was very nice to talk to you on the phone and to be able to discuss the possibilities of getting some help this year. I have brought the matter up with Collip and he has a few objections which I should let you know about.

With regard to Doris Nunes.[43] She would be a fine person to have in the lab since she is pretty well trained. However, there are a couple of worries in her regard. First of all she has been given an NRC Fellowship. There are some regulations about the fellowships and one of them is that they should not do more than four hours a week of "routine" work and they are supposed to spend whole time on research. It might be possible to transfer her here but the justification would be slight and would have to be taken up with the NRC. I fancy they would want very definite assurance that she was going to get techniques she could not get elsewhere – and there would be very distinct suspicion that the reason was that father wanted to help out the son! Collip was rather definitely of the opinion that if she were to come here she would have to give up the fellowship – which is absurd, and would not be in her interests. Another difficulty which should be raised is a social one: being a creole she might have difficulty in a place like this since it has almost a small town mentality, and I don't think there are any West Indian students in the university. In a big city it does not matter but here I think that it is probably an important consideration. Collip knows the girl and has good regard for her. ...

As for the two men that made first class in course III, I am myself receptive to the idea. Collip is rather against them, not because of their academic qualifications but because they are Hebrews. He says that this place is practically free of them and he would like to keep it so if it is possible. I understand his view but, I don't agree with

---

43 Doris Nunes (see note 34 above) was a potential candidate for post-graduate work at Western. RGEM.

it completely. He wants me to look around a bit more before committing myself to them. However, it is quite likely that others may not be available and I would not like to close up the possibility. There is no doubt that they could do an MSc, and the regulations would allow them not more than 6 hours a week of demonstrating etc. How soon will you know whether one or both of them have been accepted for medicine?

The lab is a flurry of packing and cleaning up. Probably a good thing for there is an awful amount of junk cluttering up cupboards and shelves. There is a lot of work to be done in straightening things out but I think I will be able to let up on it soon.

Your son,
Bob

London, Ontario
31 May 1948

My dear Mum and Dad:

... The upheavals have upheaved! Elizabeth Hall[44] left for N.Y. on Saturday and Ash tonight. Miss Strelitz[45] has got her visa but is in the process of selling her house under us and probably will not leave for a couple or three weeks. The lab seems a little less crowded but in all the flurry of cleaning up we have not noticed it very much. There is an absolutely incredible amount of junk stored away all mixed up with good apparatus. I seem to have spent the day up to the ears in bits and pieces, useless bits of paper, and the dirtiest dust. The trouble is that everything has to be looked at for possible worth. I am very lucky to have Fred Heagy[46] around who is a good capable fellow. ...

As for help, I have only promise of getting one assistant if I can find one. The grade would be either demonstrator or instructor, for the post of lecturer is a permanent appointment here – this would be only binding for one year. The salary would be up to $3000 according to experience. However, if the man were really worth his salt we would be able to make a good case to keep him. As a matter of fact I could find space for a graduate student as well – but I am not sure about the money situation and I have not got the OK from the hierarchy. The trouble would be to find the money to defray research at this time of the year. The NRC is oversubscribed and don't want to dig into their kitty except perhaps for a project of emergency and extreme importance – and our apparatus and supplies money is absurdly small. The grant we are asking is a sort of replacement of Ash's grant which lapses and the work has to be somewhat (remotely) along the same general line. I will be seeing Collip this week

---

44 See chapter 4, note 122.
45 See chapter 5, note 79.
46 Fred C. Heagy obtained his PhD in 1950 (supervised by R.G.E. Murray) and subsequently joined the Department of Biochemistry at the University of Western Ontario.

about it, and try to get some authority for most of the high muck-a-mucks seem to be traveling the world this summer.

Ash was very pleased with the letter that you sent him. ...

Times seem to be pretty busy for me at the moment and will probably continue so. I should be busier for we are slightly worried about the housing situation and should be more active about it. It is still a selling market and rented houses are uncommon.

The kids seem to be very well and extremely talkative, not to say slightly noisy betimes. We have escaped the measles so far although there have been a number of cases about the town and on the street. We cannot escape them forever but it would be nice to put it off for a little while. Doris is well but a little tired – we have not been out for a long time, which is bad, because of the sitter problem. ...

Final exams are on for all except the final year and they will continue for the next two weeks. I have a little responsibility for one of these, but will be more concerned with the meetings that follow.

    Your son,
        Bob

Department of Bacteriology and Immunology,
346 South Street,
London, Ontario
16th June, 1948

Dr. Gertrude Kalz[47]
Department of Bacteriology,
3775 University Street,
Montreal, P.Q.

Dear Gertrude:

Today we are dispatching the staphylococcus cultures that we promised you and the details concerning these cultures were sent with my letter on the 26th May. It is true that we obtained seventeen strains from sixty-three specimens but two of these strains have been mislaid so that, in all fairness to statistical figures, we have to say that fifteen strains were isolated from fifty-five nurses. One of the strains in the list I sent you was noted as "1047". This should be "1747", otherwise everything is pretty straight.

You remember that I told you in my last letter that three out of all the strains of staph. examined produced an inhibitory effect on the mucoid streptococci. I tried two of these on twenty different strains of streptococci, Groups A and C. Interestingly

---

47 LAC, R.G.E. Murray Papers, vol. 28.

enough, there was no inhibition of Group C strains and inhibition of only some of Group A strains. Group A strains inhibited were mucoid but some mucoid strains were not inhibited and the pattern of inhibition for the two strains of staph. were different. This was interesting to me because it showed the non-identity of two pairs of mucoid streptococci that I had considered to be identical. These observations may be quite interesting.

If you should want then we can isolate staphylococcus from the nurses each month. I suspect, however, that your study must be nearing completion and that more strains would be burdening.

We will be coming to Montreal somewhere close to 6th July, and I look forward to seeing you at that time.

Yours ever,
R.G.E. Murray, M.D.

London, Ontario
20 June 1948

My dear old Dad:

I seem to have had a very busy time and I don't seem to have a great deal to show for it except routine business out of the way and perhaps most encouraging some research underway too. What happens to the hours of the day I don't know but they disappear very rapidly. Now I look forward more than anything to a bit of holiday and I feel that I need it.

This is a lousy time of year to get money matters straightened up. Most of the hierarchy of the university seem to be away and one can only get things done by luck and waiting. However, the NRC grant came through and we are already ordering the things that we want for it. Collip was back from the west and we hardly saw him before he was off again. ... while he was back every thing was tied up with meetings on marks for the various years. You may well say that they cannot expect me to carry the whole works and should give me proper assistance – but where are you to find people to assist? I know damn well that part of the reason that there are no people coming up is that the treatment we get is pretty scurvy but part of the fault must be ours. We don't seem to be getting people properly interested in the subject.

I am waiting for Collip to come back from whatever meeting he is at so that I can get the final word on Loeb,[48] and then put it up to him. I would rather have a graduate

---

48 Lazarus J. Loeb (d. 2005) came to Robert Murray as a MSc student following his graduation with high honours from the McGill honours course in Bacteriology and Immunology. See R.G.E. Murray and L.J. Loeb, "Antibiotics Produced by Micrococci and Streptococci That Show Selective Inhibition within the Genus *Streptococcus*," *Canadian Journal of Research* E.28 (1950): 177-185; and

student than none at all. To make the bait more attractive for medical graduates we would have to practically dynamite the country.

Miss Strelitz has sold her house and immediately buzzed off to Toronto for a week. I don't know who the people are who have bought the place nor even what their name may be. However, they cannot throw us out without getting into trouble with the law and we can hold on here until we find a better place. I have plenty of spies out looking for an opening and have many of the people up at the university doing what they can. There are places to be bought at exorbitant prices but places for rent are a scarcity. We may have better luck in the next week but the last couple of weeks efforts have been very disappointing and discouraging. Even if we have not found a place we are coming on a holiday and will be on our way in the first week of July.

I have been waiting to hear from you, which would be the best day for us to arrive in Montreal for we have to be able to make a booking.

Everyone is still very well and we keep our fingers crossed that the kids do not get some dreadful disease just before we leave.

I will write again very soon.

All our love,

Bob

McGill University, Montreal
24 June 1948

Dear old Son,

... After my go of flu (or whatever it was) Mum & I spent a week at the lake. I had no energy & tired so easily I did not do anything. I caught 5 bass off the wharf & we kept 2 to eat (1lb & 1lb 6oz); they were good eating but had not spawned yet on June 18th. The season should not open until 1st July.

I finished my ash paddle & it seems good. It is strong, a good spring to it & has a handsome grain. It is a tough wood to work & it would have been nice to have had an English boxwood spokeshave. ... I have saved a similar piece of ash for you if you want it, to make a paddle while you are at the lake. ...

We are greatly looking forward to you coming to the lake & it should be splendid conditions for Doris & the kiddies as this year promises to be very free of flies. There were very few of any kind last week.

L.J. Loeb, A. Moyer, and R.G.E. Murray, "An Antibiotic Produced by *Micrococcus epidermidis*," *Canadian Journal of Research* E.28 (1950): 212-216. Following his masters he entered medical school at Western and subsequently practiced in Ottawa, before moving to Fort Worth, Texas, where he was an influential and respected allergist. RGEM; *New York Times*, 4 October 2005; *Canadian Medical Association Journal* 174, no. 2 (17 January 2006); *Globe and Mail*, 7 November 2005.

The garden looks well, a bit late this year, but it will develop rapidly now it has started. It is most important as so little can be bought.

The new canoe is fine; very stable & easy to paddle & well balanced. I'm sure you will like it too.

Be prepared for a trip to the Columbus. I don't expect the fishing to be good then but we can try. Mum will stay to be with Doris & the kids.

Mum sends her love & will write to Doris tomorrow.

    Love to you all,
        Dad

                The University of Western Ontario
                        Faculty of Medicine
                              29 June 1948

My dear old Dad:

... Mother Grace[49] arrived yesterday. She is going to stay while we are away and for a while after. It is very nice for us and will give Doris a freer hand for getting ready to leave and will also keep our things safe and sound.

There are developments in the housing situation but I wish I could say that we have a house. I have one on a string that may turn out. ...

Last weekend I went off on a trip to Rondeau with Charlie Thompson[50] in hopes of catching Large Mouth Bass. No luck with the fishing but had a good time. I learned a new method of fishing and will tell you about it when we meet – very interesting. Got quite a sunburn on Saturday which was hot as hell. ...

    Your son,
        Bob

                The University of Western Ontario
                        Faculty of Medicine
                             9 August 1948

My dear old Dad:

I've had a marvelous holiday and we are all, as you well know, looking fit and well prepared for the time ahead. We want to thank you and Mum for all you have done for us and your many kindnesses. I am so pleased to see the family looking so brown and well. Almost more than anything I appreciated being able to be with you again,

---

49 See chapter 3, note 94.
50 Charles Thompson was an ophthalmologist. He had a family cottage at Rondeau Park where he took Bob Murray fishing for largemouth black bass, a very sporting fish. RGEM.

to be able to fish together, and make things and do all the odd jobs and trips that we both enjoy so much – the more so that we can do them together. I feel very fit and in better condition than I have been for a very long time. I hope that you too are feeling more equivalent to the tough going that is ahead. ...

 Your son,
  Bob

McGill University, Montreal
18 August 1948

Dear Old Son,

 We have the impression that your housing problem is not yet settled & we are most anxious to know about it. Doris' letter suggested it was still unsettled.

 The colour movie of the kiddies doing various things is quite a success. I'll send them to you before long to see. I am waiting to finish off the black & white spool to send them all together.

 Have you received a copy of "The Lord King's College of God's-house. 1448-1948" by A.L. Peck? My copy arrived yesterday, with "A greeting to all members of Christ College" signed by J.C. Smuts & the Master.

 Last weekend Mum & I went into Hart Lake. We took only the trolling rod & regretted leaving behind a fly rod because the fish started rising in the evening. I wanted Mum to learn trolling & she got two, a small one about 1/3 lb & a very nice fish of 2 lbs exactly. She missed 4 or 5 before she got onto the necessary steady delay before striking. ...

 I suppose the kiddies are well settled down at home again & back to routine. The country & sunshine will have done them great good. You & Doris looked much better for it too.

 The voyagers from here to Britain have returned so I suppose Collip & Hall have too. Perhaps you will learn something of what they found there.

 I am sending you a reprint of the last paper by Fleming & Greenberg.

  Much love to you all,
   Dad

1559 Pine Ave. West
Montreal, Quebec
September 4, 1948

Dearest Family,

 London must have been frightfully hot; what a blessing you have a nice and shady garden. In which the oversized bath must have been a marvellous boon to the kids. ... You appear to have had quite a measure of success with vegetables etc., more than can be said for my poor efforts at the Lake. I never saw such a magnificent crop of weeds as that which greeted me on our return from the trip West.

... Town is very stuffy and disagreeable just now ... and I do not enjoy it one bit; but since Dad has to be here now without chance for weekends ... it is better for me to be here to look after him now, and go up again later. Unfortunately the hunting season will interfere with my chances for walking.

... Dad is going to Geneva for a meeting next Saturday and returns the following Wednesday ... perhaps he can manage to take a few days off then and stay over for the last of the season's trout fishing. We'll see how things go. He has a very full lecture schedule ahead and we shall not have many more chances of weekends the ways things are arranged. ...

   With Love,
    Mum

                                  9 Sept. 48

My dear old Dad:

... We were very glad to get Mum's letter and to hear that she would like to come down and help look after us when the event events. We very much agree with her plan of waiting until there is definite knowledge that things are starting.

... You will have heard from Doris that we are definitely moving – in fact we will be making the move tomorrow morning. This is none too soon for the teaching starts next week. Mother Grace has come up to help and look after the kids while Doris is trying to get the things straight. It is most fortunate that Alice's little school is just across the street for it will make a great difference during the winter. Doris is very pleased with the arrangement and so am I – though I could wish that it could be a longer term lease. However, it will get us over a most difficult time.

I was very glad to get the mimeographed booklet of the procedures of the clinical lab. I have read it through and think that it will make a most useful addition to the other notes you put out. It should help Gertrude a good bit in the breaking in of new help at the lab.

I had a very pleasant visit with a Cambridge contemporary of mine who is going out to Saskatoon as Assistant Professor of Bacteriology. Guy Richards[51] by name. He did not do a part II but he did a two year postgraduate course at Cambridge after he got out of the army. Apparently he heard of the post through G.S. Wilson.[52] He is a

---

51 Guy Richards was a fellow Christ's College student when Robert Murray attended Cambridge. He later moved to western Canada after his postwar medical training. RGEM.
52 See Introduction, note 196.

very nice fellow.

Next Monday I expect a Miss van Iterson to visit us at the lab. She is an Electron microscope expert from Delft who is most interested in motility problems. She is coming to talk over the swarming problem. I will be very glad to see her for she is both nice and competent.[53]

Loeb arrived yesterday. He is a good fellow I think. He will take a little time to settle down but he seems very adaptable and I am sure will fit in well with the spirit of the place. ...

All my love,
Bob

293 Oxford Street
13 Sept. 1948

My dear old Dad:

It was very nice to get the phone call from you and Mum on Sunday morning and I look forward to getting a letter from you to hear how you are and, I hope, that you are fully recovered. I fear that between us we seem to have a lot of trouble from the streptococci. ... I am attacking the thing from both ends with penicillin and feel a good deal better already. The main plague is the pains in the joints, as they were in the previous attack. ...

Your recurrent attacks of erysipelas are worrying and I wish there was something that we could do with it. I think you would have been wiser to have ceased lecturing for the week at least to give a chance of recovery.

... The moving went very smoothly and we are now almost comfortably settled. Most of the small stuff we brought up in a friend's car the day before to provide the essentials – bedding, cooking utensils, etc. Doris had everything planned for location so that the movers did all the hard work, and by lunch time on Friday the place was more or less liveable. It is a nuisance being laid up for there is still a certain amount of stuff to be unpacked from trunks – but it can wait. Doris did far too much but she is taking it easy now and realises that top shelves are not for her. Mother Grace being here has meant a lot to her for it has allowed her freedom to plan and look after things.

Alice and Peter are quite excited by their new abode. Now the surroundings are no longer strange and they are returning to their usual schedule.

There is not much more use my writing more until I feel better. Which will be I

---

53 Dr. Woutera van Iterson, Laboratory of Electron Microscopy, University of Amsterdam, The Netherlands.

hope in a couple of days.
>Your loving son,
>>Bob

>McGill University, Montreal
>14 September 1948

Dear old Son,

I am very distressed to hear you are hard up with a strep throat again. Your letter arrived this afternoon. The strep problem is far from settled & there are many things about it which cannot be interpreted. Their persistence in the mucosa of the throat is more understandable than their persistence enabling the recurrence of erysipelas. These recurrences indicate an inability to produce immunity to strep & this, to my mind, is the most cogent argument against strep being more than a concomitant opportunist in scarlet fever. There is an important question in this.

I too have annoying joint pains – knees chiefly. The erysipelas has died down again leaving me a bit washed out & quickly fatigued & easily put out of breath. The change of site is difficult to explain. I have had two facial attacks ... in the last 12 months.

Perhaps a combination of sulphonamides & penicillin would be more effective than either alone. Combined actions affecting different functions are often most effective & if an immunity could be used the most effective method would be available. However the right antigen has to be determined. ...

The lab here bristles with troublesome problems, some of which I do not know how to solve & cannot count on helpful cooperation from the financial dictators. There is no reason to trouble you with them however much they are a curse to me. You surely have your troubles too until the position is stabilized & you know where you are & can, if you are given the authority, arrange things as you wish.

I'm glad Loeb seems to promise well. I hope he turns out as well as seems possible from his record. He should be able to do quite a bit of back work for you but will need instruction & supervision.

The booklet of clinical lab methods should be of use to you as we find it a help here. Any suggestions for its improvement will be very welcome indeed.

I heard some time ago that Guy Richards was going to Saskatoon & I hope he will be happy there. I hoped he might drop in on his way through Montreal.

I hope your illness has not interfered with the visit of Miss van Iterson. I hope you are writing some of your papers. Make Loeb work & set aside some time for yourself as so much depends on putting out some good papers. It is the most important thing for you to do, much beyond any amount of good teaching & good organization. I have proved this contention beyond any question.

Mum returned to the lake on Sunday soon after phoning you. I manage after a fashion & don't care about it but it is so much better for her to be at the lake.

I am going on a short fishing trip to the Columbus with Bill Newman[54] on the 24th. I look forward to it.

On 17th I have to make a speech at the Veterinary Conference Luncheon. I have not thought of a theme yet & must do so immediately; I do not feel inclined for the job & my effort to escape it failed. I don't mind speaking when I have a subject but I am at a loss this time.

Love to you all,
Dad

London, Ontario
19 September 1948

My dear old Dad:

It is very good to hear that you are better. Such an illness takes it out of you a bit, as I well know, and I am very glad to hear that you are going to go off for a few days to the Columbus. I certainly wish that I could be going with you to help wind up the trout season for 1948. With any luck you will be getting some good fishing and I will think of you in many old favourite haunts, as well as some new ones I hope, while you are away. …

My fever and sore throat subsided with treatment – using this procaine penicillin stuff, which seems to be pretty good. They are a terrible nuisance these strep infections and I hope that I don't get many more of them. I seem to have got rid of the bugs again – but also so I thought last time. Maybe it was the fact that I swept out the old apartment a couple of days before and they were lying dormant (as in Maxton's classical experiment). On the other hand we have been isolating a lot of streptococci from throat swabs at the lab and there may be a high carrier rate at the moment.

Lectures and labs have started – a couple of days late in our case due to my being sick – and the busy period begins. We seem to have enough help to make things go smoothly though it is a pest having to split up the practical class again because the university will persist in taking more students than the classroom will hold. It does not seem to me to be either a desirable or a necessary policy. I am very pleased with Loeb and I think that he will do good work – he seems to be a good and enthusiastic demonstrator.

We are extremely pleased with the house which is working out well in convenience and in space – I think Mum will agree when she sees it. It is such a difference for us – Doris is particularly overjoyed. As you say we will have to have our plans laid extra

---

54 See "Robert Murray Reflects on Wartime Correspondence" in chapter 3.

early for next year in order not to get caught short by the housing.

Miss van Iterson came through here a couple of days late and I was able to have a long talk with her about motility problems. Elder[55] made up a demonstration for her of the swarming business – so despite difficulties we had a profitable time. She had a lot of stuff on electron microscope studies on flagella but was looking for some way of getting better results.

We look forward to seeing Mum and wish that we may see you not so long from now. Maybe you will be able to come down for the Lab section meeting in December.

All my love to you,
Bob

McGill University, Montreal
19 September 1948

Dear Old Son

... Mum returned to the lake this morning & I am going to the Columbus with Bill Newman on Friday 24th.

I tied a few flies this evening. Charlie Pascoe came back with a tale of 5 & 7 lb trout from Lake Nipigon & that the real fly for that region is the Cocatouche which I had never heard of. ... The photograph Charlie has of the Nipigon fish is astonishing.

Mary Peto (Secretary stenographer) is thinking of leaving, she told me yesterday she was offered a job at $200 a month which is $80 a month more than I can offer. She said she was undecided as she is happy in the lab! Alice Gilett left this month to take a job in a commercial lab at quite a bit more money than I can offer.

Meat (ox heart) has gone up in the past three weeks from 15 cents a lb to 55 cents a lb, this puts my estimate completely out. Everything is mounting & delivery is uncertain & late. Running the lab is a perpetual worry which is multiplied by having so many people some of which are Premadonnas & difficult to keep in line.

I'll send you a copy of the newly revised course 3 book. I cannot find time to do anything I want to do in the lab as there are so many difficulties & problems brought up every day that there is not any peace for me wherever I go.

I had to speak at the luncheon of the College of Veterinary Medicine of the Province of Quebec on Friday last; it seemed to go alright. I discussed the "Future of Veterinary Medicine". Charlie Mitchell was there from Ottawa & several others I know & Mum was there. Last night Mum & I went to their banquet with far too many voluminous speeches – it took about 4 hours. Even a number of the French were bored. I have to

55 R.H. Elder graduated from the Western medical school. He subsequently studied microbiology at McGill and became microbiologist and chief of the laboratory in the Ottawa Civic Hospital. RGEM.

talk to the Dentist Convention later on. It is impossible to avoid all of these things.

I suppose you are affected by the electricity shortage in Ontario & will be limited in some way. I suppose it is the low water. But it is astonishing how there is an ever increasing shortage of everything, even the most ordinary commodities.

 Love to you all,
   Dad

            McGill University, Montreal
            23 September 1948

Dear Old Son,

Bill Newman & I go to the Columbus tomorrow. He will pick me up about 9:30AM. I wish greatly you were coming with us ... I'll let you know all about it when I get back on Oct 1st.

I hope you don't get any more strep infections. Streps do not establish any immunity as I have seen it, in spite of the theory many hold about Scarlet Fever on what I think is flimsy evidence. No satisfactory antigen has been developed, partly I think because we do not grow them the right way on a proper medium.

It is absurd to take more students than the space & equipment can manage. We are back to 120 entries this year, I am glad to see. Something has to give way & it is, of course, quality. I am very glad Loeb promises so well; I think he will prove reliable & he has brains.

I am exasperated because I have lectures to give on Dec 9, 13, 14, 15 & 16, on some of the days two, so I cannot go to the C.P.H.A. Lab Section. I missed it last year for the same reason but I am much more annoyed this year because it is in London. I was so looking forward to seeing you & hoped to go a bit early & stay a bit late. The gaps in my lectures come at the wrong time & as the time of the meeting was only known to me too late to try juggle things a bit nothing can be done.

I am glad the house is such a success. I am sure it must make a huge difference to you all, but you must start active work on a new one for next year. If your higher-ups would decide what they will offer you in the future & when you could plan things. You cannot do so otherwise as it might not prove worth staying there.

I am pleased to know you had a good session with Miss van Iterson & that Elder came over with a demonstration. ...

Well my pack & tackle are done up ready to go & I have only to put on the old clothes & cook my breakfast & I am ready to go. I shall forget the troubles until Oct 1st.

 All my love,
   Dad

London, Ontario
25 Sept. 1948

My dear Dad:

I am busy getting arrangements tidied away for the CPHA meeting – admittedly looking after only the purely local things. I have not relinquished, yet, the responsibility of a speaker for the dinner. I think that it should be a better effort than last year. I am not so much in favour of having footling speeches – rather not have one at all. Have you any ideas on who might be asked??

I will write again this weekend. Hope you have a good time at the Columbus.

Your son,
Bob

McGill University, Montreal
3 October 1948

Dear old Son,

I expect you have quite a bit to do for the C.P.H.A. Lab Section, even though you are limiting it to the local arrangements. I don't know who to suggest for a speaker at the dinner, much depends on the sort of talk you wish to have. At our time they need to have a scientific address; in that case you might get someone like (A.J.) Rhodes to talk on modern trends in virus research or some other general review; for a more general subject you might ask G.D.W. Cameron (Deputy Minister of Health, Ottawa)[56] to give a talk on, say, International Health Organization (United Nations Organization) in relation to the laboratory. Cameron has been the Canadian Representative at the International meetings & could give an interesting & important review. There must be other possible reviews & appropriate speakers & if I can think of one I will let you know. I am very disappointed that my lecture schedule will prevent me attending the meeting.

I had quite a good trip to the Columbus with Bill Newman. The weather was extraordinary – fine & warm, almost like summer & except for the colour & lack of insects & birds it might have been summer. I fished all day with my shirt off & did not even take my rain shirt with me. ... There had been a fire at Bandy which burned the boathouse & a large part of the point round the portage. The fire control people had been in & supposed it was out. ...

The McClaren Co have built a camp on Kelly's Bay & have started lumbering through to Trudel. They have made a fine mess of the trail to Trudel. The Singer Co are building a lumber camp on Zouave just beyond the Pallasades to lumber that

---

56 See chapter 1, note 91.

region. The lumbering is going on for the next two years. The lumbering is to be expected but it is sad to hear "timber" & the crash of the big trees.[57]

There was not such a crowd as there used to be for the last week. Several of the new members were there & they are a very good lot. The prospects of the club look better than ever before. Keen new members, most of them fly-fishermen, who are all for the development of the club & very much impressed by the territory. ...

Mum & I saw Olivier's "Hamlet" last evening. It is magnificently done & we greatly enjoyed it. The only fault was in the over use of panoframing in the photography. I have never seen Hamlet played so convincingly & the only part which I was disappointed in was Horatio.

This letter is very disconnected. I thought of you & repeatedly wished you were with me. So many places reminded me of our fishing together. I wished very much we were together with the canoe.

  Love to you all,
    Dad

            Department of Bacteriology and Immunology
                   October 19th, 1948

Professor E.G.D. Murray
Dept. of Bacteriology
3775 University St.
Montreal, Que.

Dear Dad:

You will remember my telling you that Mr. R.H. Elder has been working with me for the past two summers in the swarming problem. He is now in his final year of medicine and will be graduating in the spring. He has to make some decision as to his further training. He feels that he should do a year's internship to consolidate his medical studies and when he spoke to me a couple of weeks ago he was seriously considering the possibility of doing a D.P.H. [Diploma of Public Health] following his internship ... and the training that that course gives would allow him a wider range of decision later on. When he was talking to me today he still had some doubts whether that particular plan is the best and was asking me about an internship involving Pathology and Bacteriology. He is anxious to get his further training away from

---

[57] Trudel and Zouave were among the fourteen or so lakes in the Columbus Club territory. Kelley's Bay and the Pallisades, pair of cliffs, were landmarks on Lac Zouave. Bob Murray remembers photos on the wall of the main room at the clubhouse showing part of the club territory which had been clear-cut in the late nineteenth century. RGEM.

London which is, I think, a wise decision. He is also worthy of careful consideration because he is likely to become a professional Bacteriologist.

I would like very much to hear what you think would be good advice in his case for as you know I am interested in him and he might be someone you might be able to take on in your lab. at some later date. In addition I think that advice on such matters as these should be from someone more experienced than I am.

While I am writing to you I might as well quote a rather horrible paragraph from an advertising booklet on Canadian wines.

"It is interesting to note that a recent independent and nation-wide survey showed wine to be the alcoholic beverage least blamed by Canadians for intemperance."

    Love,
        Bob

            Montreal
            31 October 1948

Dear Old Son,

...I hope to be able to send you the Virus & Rickettsial Diseases of Man[58] & the Bacterial & Fungus Diseases of Man, soon. I have not my own copies yet but the library copy of the former was here some time ago. I am writing to Lippincott about it & hope to get some action. I got a copy of the Am. Pub. Health Association book edited by Tommy Francis on Diagnostic procedures in Virus diseases; it is very good & I see in the J. of Bacteriology an advertisement offering to supply a great variety of virus antigens. This sort of work must be introduced into the lab work.

I enclose a copy of a paper I gave last week to the Montreal Dental Club. This is a modest name for a very active society to which men come from quite a way off; something over 400 registered for this meeting, so it is not just a little local club. They seemed to like the talk.

I had a very busy week & this next week is building up too, including a City Council meeting on Tuesday with a long agenda which I hope will not carry through to Wednesday. The lab bristles with difficulties & as fast as I solve one others crop up from the most unexpected sources. Money is, of course, the reason for many of them. Anyway it is a nuisance because it is preventing us from getting caught up with back work & finding time to do some lab work for my own amusement.

Elder could come here as an intern in Path & Bact & on your recommendation. I can assure him of acceptance for next year. If he decides on that let me know & I will

---

58 Thomas M. Rivers, ed., *Virus & Rickettsial Diseases of Man* (Philadelphia: J.B. Lippincott, 1949).

send him an application form & you can write a letter supporting him & he could get others too, just to keep it regular.

For the Specialist registration it is necessary at present to have done a general internship; I hope this regulation will be contended & altered but that cannot be relied on. I don't know how important such registration may yet become. It would evidently count should "State Medicine" ever come into effect here & the cult is spreading for the College of Physicians & Surgeons of the Prov. of Quebec is about to introduce it too.

I am sure that a year spent getting a D.P.H. is time well spent for several reasons. It is useful knowledge for a bacteriologist, it opens the way for P.H. appointments & there is a growing demand for Medical Qualifications in P.H. labs which is as it should be, there is a demand everywhere for Public Health officers. Nevertheless, he would do well to get additional fundamental training in bacteriology; he could get this here both in the Clin labs & in the teaching labs too & if he is going to lay foundations of internship & D.P.H. the question of an available appointment does not arise immediately. Everything depends on exactly what he decides to do & I would be prepared to give him the best opportunity I can but it would require plans ahead to make appointments available etc. It is good to know of a MD wishing to be a bacteriologist.

I must do some work on my lectures & I must take the dog for a little walk. Also I feel the need to be a bit lazy.

I will try to write a more satisfactory letter next week.

  Love to you all,
    Dad

London, Ontario
7 November 1948

My dear old Dad:

It was grand to talk to you last night. It served to remind me very forcibly that I have not written to you for a very long time and that we have been corresponding largely by telephone, which is very satisfactory but also very expensive. It is good that you are feeling fit and I hope that you manage to avoid illnesses this winter. Maybe you will be able to take one last fling at the lake before the ice sets in – I hope you do, and wish that I could be with you.

Doris is very well and is coping very well with added burdens of looking after the kids. We are both very grateful to you and to Mum for all the things you have done for us. Mum made all the difference to the event for Doris was able to relax and not worry about any little details, and that means a lot. I hope that her absence did not worry you too much – we were so glad to have her we wish we could have inveigled you both for a stay. Maybe we will be able to arrange something later on.

I have so many things going on at the moment that I have been a little preoccupied and not doing all the work I could wish. Labs and lectures take a lot of thought – and though they are going well I am not quite satisfied. With two pretty active minds working with me I have to hope to keep up with them – and advise them as I can. This side of the work is going well and we are beginning to pile up results. I wish I could give a little more time to swarming problem but I will have to be patient. Those particular experiments have to be watched over all the time and cannot be done on an occasional basis. I am not too happy about it because it is not flowing well when I try to write it up – the work seems to be alright, just my ineptitude. The thing it wants most is a little leisure for trials – for there are many aspects not even touched because of lack of a method. In addition to the CPHA meeting there is a meeting here of the Toronto Biochemical Society at which I am setting up a demonstration of the biological assay of enzymes – esp. hyaluronidase. This latter will be quite an interesting meeting even apart from the fact a University of Toronto society is visiting us!

I went to Toronto last Monday for the program committee meeting. There were only 19 titles submitted. In characteristic fashion the Toronto group (Connaught included) had not yet submitted their own papers. There were none from your lab and none from (Armand) Frappier,[59] as well as a couple of other places. Some of the Toronto people had an acute attack of cold feet (notable Defries[60]) but a little plain talking settled that – they were on the point of calling the whole thing off. I hope that you will be able to submit at least one paper from the lab if not more. I could submit a couple more but, despite the fact that the work is interesting, I do not think the work is complete enough to warrant a paper. I want to keep something in reserve for the SAB meeting in May.

I am still in the market for a really good technician as a research assistant (and to help in getting the teaching practicals ready). The girl I have had, who is excellent, has been getting more and more serious allergies over the years and last month had her first attack of bronchial asthma, practically a status asthmaticus since it lasted four days. She is going to have to leave for the environment is too dangerous for her. She is sensitive to things in the lab, which makes it worse.

… Thanks very much for your comments about the possibilities for Elder. It is indeed nice to have one MD who is thinking of Bacteriology. I have talked it over with him and he has decided that he should do a clinical internship – he hopes to get

---

59 See chapter 5, note 48.
60 See chapter 1, note 104.

in at the Montreal General – after that he will be free for real training. He seems to be very attracted by the possibilities of working in your lab and he would very much like to meet you and talk it over. There is however no immediate hurry.

... I thought the address to the dentists was very good and very stimulating. It is a very interesting disease but a brute to work with. ...

All my love,
Your son, Bob

London, Ontario
21 November 1948

My dear old Dad:

... Doris has had her hands full lately and has had little spare time or energies. Thomas seems to be settling down quite well, is gaining satisfactorily and generally looks a most healthy specimen. Everyone who sees him makes some remark about the appearance of red hair! As yet I have not been able to decide that they are right and I stoutly maintain that it is merely a shade of brown. Alice and Peter are very well. Peter has developed a great deal since the summer and is a most chatty individual now, and seems to have grown inches.

Despite this business we have managed to get out a couple of times. We went together to a performance of HMS Pinafore by a travelling company – it was quite pleasant and fairly well done. It is a company from Los Angeles who do G&S (Gilbert and Sullivan) all the time. They are a small company but keep fairly well to the traditions. Other than this we have not been able to get out together although we have both been to functions individually. We had a very pleasant concert here the other day – the Toronto Symphony – among other things they did Beethoven's seventh which was very nice. Hamlet will be coming soon and I hope that we will be able to go. The J. Arthur Rank group have built a new movie theatre here which is, apparently, very modern and comfortable – even allow smoking in a section of it – with room for the longest knees. Doris has been there but I have not.

I have to go to Toronto again tomorrow for another meeting about the CPHA meeting. I hope things will have crystallised properly by then. I gather that they have enough papers now but I am sorry that you are not able to be there and give yours. They tell me that you are sending a couple of people down which is a good thing. Planning for this is going to give me a few headaches in the next four weeks.

Teaching keeps me busy although I have some willing helpers. You still have to keep a check on every little detail to see that things work properly. As a matter of fact it has gone quite well although I would like to be able to do some more things differently. They will have to wait for another year. For the last week or so I have been lecturing on the viruses and have had to do a great deal of reading for I changed the attack somewhat over previous years. There seems to have been a tremendous lot of work in the past few years and it is quite a job of selection. I have not yet seen van Rooyen[61] & (A.J.) Rhodes new book[62] but I have read quite a bit of Rivers new book. I am very impressed by the quality of the latter and think it is a pretty good book. Too much for the average medical student as a text but a very excellent reference book. I have also been looking at the Bacterial and Mycotic diseases volume – I have not yet had time to read enough of it to criticise the overall editing bit I think that it has distinct advantages as a text and is worthwhile. However, I do think that in many places it has given too much space to the discussion of technological detail (eg Streps). I am also inclined to think that in many places there is too much general discussion of individual organism without proper emphasis on the overall characteristics of the group. Perhaps inevitable with a text written by a board of experts. One grand thing is that the texts have seen subsidised and that the cost is more than reasonable for textbooks these days.

... I am a little worried by a fungus that I have isolated from a skin lesion of a graduate student. He had a low grade inflammatory lesion, about two inches square, on the lateral aspect of his leg which showed small focal areas of abscess formation – draining every few days. It was subsiding when he brought the problem to me. It had been diagnosed by a skin merchant (by inspection only!) as a streptococcal infection and treated without success. The pus was very beautiful polymorphonuclear exudate in which no micro-organisms were seen, no mycelium, but quite a number of oval or spherical bodies staining like nuclei but more homogenous and surrounded by a narrow "halo". Not definitely identifiable as anything. ... We have sent a subculture away to Connaught. We are treating it with the utmost respect. The student is perfectly well and the local lesion has now cleared up – I only saw it in its closing phases – it was in active state for about six weeks. He has not been in the California area for two years – and previously served with the US forces over most of the Pacific.

The cost of living is appalling – particularly how much it costs to eat. This last month seems to have been an expensive brute for us and it is an infernal nuisance and worry to have to be forever thinking about price and bank balances.

61 C.E. van Rooyen was trained in the UK and had varied experience as a medical officer in the RAMC and as head of a laboratory in Egypt during the First World War. After the war he joined the Connaught Medical Research Laboratories at the University of Toronto, where he worked on polio, influenza, and bacteriophage projects. In the early 1950s he was appointed professor of microbiology at Dalhousie University, where he continued his distinguished career as an important figure in the development of virology. RGEM.
62 A.J. Rhodes and C.E. van Rooyen, *Textbook of Virology* (Baltimore: Williams and Wilkins, 1949) was a pioneering work in the field of virology.

Jim Blaisdell asks me to send you his best regards. I saw Hall the other day and he sends the same and asks after you.

    Your son,

        Bob

McGill University, Montreal
28 November 1948

Dear old Son,

I have not written for a very long time. Quite unreasonably – but you are frequently in my thoughts for very many reasons. ...

I have just finished writing the paper on Tetanus from plaster of paris & I am sorry there is no time to wait for you to criticize it. Gibbard was very anxious to get busy on the subject & I wrote him an account of it. He promptly examined the products of various makers & I gather he fully confirmed our findings & has sent a paper to H.P. McDermott for publication suggesting that mine & theirs go into the same number. I, therefore, had to get busy & I think it is sufficient though there is more to be done, which must wait. I have nowhere & no time to do real & proper work. I'll send you a copy.

I sent you a copy of each "Virus & Rickettsial Diseases of Man" & "Bacterial & Mycotic Diseases of Man". I have not been able to do more than look them over but they seem to be good in general. Horsfall[63] was here recently & was very complimentary on my historical review. I do not much like Swift's[64] chapter on Streptococcus & I think the chapter on Immunity by Treffers[65] is useless in this kind of book. The Virus book seems to me very good.

... I'm sorry about the C.P.H.A. Lab Section. However, the date of the meeting is always at a time that most of my people cannot get away. Gertrude, Anne Masson & Joan de Vries are going but I would be glad to send almost everyone if the time was suitably chosen. Several of us could then contribute papers but none are keen to just send along a M.S. to be read. This is an old fight & the answer from Toronto is that this time suits best the people from the West, but they had better add up the numbers & weight. Look out for A.E. Allin & talk with him about the character of the contributions. Look out, too, for R.M. Shaw from Edmonton who is one of my exceedingly

---

63 See chapter 4, note 120.

64 Homer F. Swift (1881-1953) was an associate member of the Rockefeller Hospital. He worked there for twenty-seven years, retiring in 1946. He was known for his work on the treatment of syphilis, rheumatic fever, and streptococcus. *New York Times*, 26 May 1953, 4 June 1953; *Britannica Encyclopedia Online*.

65 Henry P. Treffers was at this time at the Section of Immunochemistry at the Department of Bacteriology and Immunology, Yale University School of Medicine, New Haven, Connecticut. He later served as chair of the department (1950-1961).

good friends.⁶⁶ I am sure it will be an interesting time for you even though very busy; I hope Collip will give you some means of entertaining the party.

I shall have to go to Ottawa 2 or 3 times during December & I hope that Collip may be there for one of these visits.

We are going to have a semi-annual meeting for the Columbus Club & they want me to show a film. I shall have to find time to do a bit of editing for the purpose. The club is in good shape; much better than it has ever been because the officers are doing their jobs.

Hilary has done many kindnesses to Mum & me & we have worried how we could show our gratitude. I found a splendid copy of the Le Gallienne edition (1897) of the "Compleat Angler" to which is added Charles Cotton's How to Angle for Trout or Grayling and "The Fisherman's Almanac" by Hi-Regan. The illustrations are by Edmund H. New & are wonderful. There are appended Walton's & Cotton's poems etc, Bibliographies & pages of marvelously interesting notes. We are very pleased with it & hope he will like it. It is worth the price.

One of the weekends since I last wrote Mum & I went to the lake & put the cottage to bed for the winter. We hoped Rosy had done the painting but he said the fog was too bad & everything was wet. The truth is he had put it off & missed the fine weather so now it must wait till spring. Well we have the paint. We left everything tucked up, oiled where desirable etc.

Please send me the dimensions of your paddle as I want to make one like it. I have ordered some ash for it.

How are your inhibiting Staphs going? It is a surprising & interesting phenomenon & I hope you have time to work on it. The cellophane technique is good. They have not told me of any mucoid streps yet – but I shall remind Gertrude. If there is a reasonably simple test we could treat all staphs we isolate. Perhaps it requires a particular test strain. I wish I had a lab to work in, I must arrange things better very soon.

I hope the writing is almost or quite done, especially the Hyaluronidase.

There are no girls available just now. There may be when course 3 is over though I think there will be a shortage this year as most of the class are men. Perhaps one of those from the Clin Lab may want something but it depends on the opportunity & pay offered.

When does Collip propose to decide about your department so that proper development & organization can be planned?

I don't think it is essential for Elder to do a general clinical internship unless he wants to. (Lyman) Duff⁶⁷ had a man accepted for the Certification Exam recently

---

66 Robert McLeod Shaw, BA, MD, CM, DPH, FRCP(C), P.H.S. Edmonton.

67 See chapter 2, note 84. He was dean of medicine at McGill from 1949 to 1956, and a good friend of both Everitt and Bob Murray.

who had not interned atall. It might be well to write to Boyd (Path Toronto),[68] who seems to have the entire thing in his hands, to make sure. If Elder was changing his mind he had better let me know at once as we have to pick next year's interns now for Path & Bact. Anyway I am very anxious to make a place for him to give him as good a chance as possible. ...

We are having some fun out of the Arctic Institute.[69] Meetings are held monthly & are interesting & are attended by good people out of the ordinary. They have changed quarters recently & Mum has enjoyed herself helping unpack & catalogue their museum etc. She has examined some very interesting specimens & has helped to interpret the meaning of some of them, as you would expect.

It is good the family are well & flourishing & that Doris is not finding the sledding too hard. It is a blessing that you got the house but that question must be watched still. It will be a relief to you when the University decides finally on your department. They have very good reason to be grateful to you.

There is a young Latvian doctor of medicine who came to see me on Friday who had a bad time during the war. He was chiefly interested in T.B. & had good chemical training in the subject in Germany etc before the war. At present he is employed as a mechanic in a factory from 4pm to 12pm & it will take him not less than 5 years to get a license to practice in Canada & with difficulty. Gertrude is taking him on to test him out working in the mornings to see what he is worth. If he proves worth the training it may be a good thing to bring him on as a lab man for a job somewhere.[70] It seems impossible to find any of our own young men who wish to go into the lab.

I must go to bed now.

    Love to you all,

        Dad

I hope you are being exceedingly careful about the cultures of possible coccidioides imitis[71] & that you are not growing it in Petri dishes. The distribution of the diseases seems to be much wider than used to be supposed. I believe cases were found in U.S. recruits which were subacute or chronic & seemed to have come from a variety of states. It is not confined to the San Joaquin valley by any means.

---

68 See chapter 3, note 135 for information on D. William Boyd.
69 The Arctic Institute of North America.
70 Bob Murray does not recall this particular episode or individual. However, he points out that accommodating such foreign workers was made easier because "Gertrude was particularly good at making people comfortable in the service, as they say, and it helped that she was competent in several languages." RGEM.
71 A pathogenic fungus found in certain soils that can cause the respiratory disease coccidioidomycosis.

London, Ontario
December 6th 1948

Dear Dad:

Immediately after your phone call I got hold of Elder and put the possibility up to him. He thought that although it would present him with an excellent opportunity he should take a clinical internship because if he once started on the laboratory side he would be unlikely to feel inclined to undertake a year of clinical work should that be necessary. He has, I believe, written to you to explain the circumstances to you. I should add that he would very much like to work in your lab. starting the summer of 1950. ...

I received the copy of your paper concerning Contamination of Plaster of Paris. I have no very definite comments to make about the paper and I think that it is a very important piece of work. I am glad that Gibbard has taken an interest in it and that he has confirmed your findings. I certainly wish that you were giving the paper yourself next week.

The Laboratory Section meeting is giving me a certain number of headaches but I think that they will probably go fairly well. I wish that he could provide a better background for the meetings than the Hotel London offers but I am afraid that we can't do much about that since the university is still in session. Concerning the time for future meetings, I will take up the possibility of holding them at Easter although I fear there would be some resistance from certain quarters. Although there are comparatively few people that come from the west it would be a pity not to have them present. Please write to me or have Gertrude well primed on your view concerning a setup concerning the whole field of microbiology.

... We are in the process of instigating charges for bacteriological work done on private patients in the hospital. I have made out and submitted a scheme but would be very grateful if you would send me an outline of the charges made for the various categories of specimen at the R.V.H. ...

    Your son,
        Bob

McGill University, Montreal
8 December 1948

Dear old Son,

Charlie Pascoe's daughter wishes to spend the summer doing marine zoology & I thought Mount Desert would be a good place. I promised to write to you about it. She is in her first year in McGill & aims at medicine. She is a charming girl & very clever, she got seven first classes in the Matriculation Exams & an entrance scholarship. She is very interested in her courses, especially the biological sciences & works hard. I think it would be best if you wrote direct to Charlie (C.7. Pascoe 4826 Victoria Avenue, Montreal) advising him how & to whom she should apply. I gather there are two marine biological labs at Bar Harbour.

I don't suppose anything can be done about the Lab Section meeting time. It is impossible to fit everyone's convenience.

I know the mycologists & the phytopathologists have formed some sort of societies. I suggested long ago that either a microbiological association be formed or else that separate societies be formed & that they each elect representatives to a Canadian Microbiology Council. My ideas & efforts got very little support & I got very few answers to numerous letters. I have urged the matter of Canadian representation in international affairs & the proper representation of microbiology at Ottawa in Committees & etc as the chief purpose of a Microbiological Association or Council. I have quite a file on the question.[72]

I shall ask Gertrude to send you a copy of the charges we are making for outside work. The schedule could stand revision as it is old. The R.V.H. made its own scale of charges without consulting us & it is absurd, even stupid in places; it was compiled by their lay office staff (accountant etc) & I will have none of it.

I am about to go to Ottawa till Friday when I have to be back for a City Council Meeting; among other things.

Sue has been laid out with a severe attack of Flu & Mum has been looking after them. She is better today but knocked out by the fever etc.

A young bacteriologist, Dr. Jack Wilt[73] from Winnipeg has been here a few days & will be at the meeting in London. Look out for him & introduce him to a few people & have a chat with him.

   Love to you all,
     Dad

---

[72] Everitt was in the lead in promoting the study of mycology in the McGill Faculty of Medicine, notably through his encouragement of the work of Dr. Blank in 1952-1953. See Appendix E.

[73] "Jack Wilt" was J.C. Wilt, head of the Department of Medical Microbiology in the Medical School of the University of Manitoba. He was a virologist but had a wide appreciation of all aspects of medical microbiology. Early in his career he arranged to spend a bit of time in Joburg's laboratory for experience and became a good friend to both Murrays. He worked on poliomyelitis and the virus during the major post-Second World War epidemic that afflicted Winnipeg. He also did recognized work on the virus diseases that afflicted the Inuit in the Arctic and Subarctic. He died in the late 1990s. RGEM.

3590 University St.
Montreal, Quebec
16 Dec, 1948

Dearest Bob,

It was grand to talk to Gertrude on her return from London & to hear that the P.H.A. meetings were such a success & particularly that your paper & film were so good. ...

We are having the senior staff of the Lab in to drink some sherry with us on Wed- not with wives this time ... and we want to have a little sherry party for our friends of the Arctic Institute. ...

Lovingly,

Mum

McGill University, Montreal
19 December 1948

Dear old Son,

The account given me is of a successful meeting & that you acquitted yourself extremely well; of which I am sure & proud. I wish I could have been there. The Montreal papers had names & places a bit mixed up but I gather you were elected to the Council, of which I am glad. Gertrude & Co enjoyed their visit very much.

The enclosed photograph was snapped to give to the College of Physicians & Surgeons of the Province of Quebec & I thought you might like one.

I saw Collip briefly in Ottawa & again at Macdonald College at a meeting of a N.R.C. sub-committee. We had a short chat but I did not gather very much. The situation & procedure seems to be exactly what goes on here & in others of our universities & is rather less honest than horse-trading. I think you will have to ask for a straight-forward plain statement at the end of the year. The fact is that the job deserves the position & age does not enter into the question. Other people, such as Guilford Reed, would be glad to get you; he told me so a few days ago. The Western jobs are not too attractive at the moment but they will probably be forced to improve them.

There are some funny things going on here with a bit of scheming I do not like. The difficulty of getting the truth is hampering. I think I shall play them against one another to explore the plot & then if necessary play the ace of trumps which I hold.

We had a good meeting of the Columbus Club, chiefly social & a report of action by the executive committee. I showed my films taken at the club & they seemed to please. There are a fine lot of members now with the right ideas & the club is in better shape than it has ever been in. The projected improvements will make a difference if Adelard could be relied on to do his job. Two Provincial Game Wardens went in during the fall & imposed heavy fines & confiscations on a number of poachers which will do some good. The club has to prosecute seven others, one of which is a local policeman, *which* will cost a bit & may go nowhere, but it will show we mean business. It is all the more necessary now with two independent lumber camps on the limits & likely to be there for two years.

There are various Xmas parties planned, 2 in the lab, & it is astonishing how many choose the same day. Of course the suitable time is limited but I often notice with ordinary meetings when time is not limited they will pile up on a particular day even when they are not all in Montreal.

The C.M.H. Interns put on quite a skit last night & took off the staff quite cleverly & did not pull any punches. It is the only one of its kind left at McGill now.

I hope you will write to Charlie Pascoe advising him about Enid going to Mount Desert. They wish to get the application in early to be sure of getting a place in the course & find it difficult to get information.

 Love to you all,
  Dad

        The University of Western Ontario
        Faculty of Medicine
        20 Dec 48

My dear Mum and Dad:

This is a little late for a Xmas letter and I fear it may not get to you in time – but it brings you both our fondest love and best wishes for a very happy and pleasant Xmas, to say nothing of good fortune in the New Year. I wish we could be together. ...

 Your loving son,
  Bob

        McGill University, Montreal
        26 December 1948

Dear old Son,

It was grand having a word with you & Doris yesterday but Mum rang off before I could say a last word. Mum is delighted with the print, as she will tell you & it will be framed at once.

... Your presents have given us great happiness & will continue to do so.

Fudge had a grand time & was tremendously interested in parcels, he unwrapped his own unaided & very carefully, with evident delight. He knew what it was all about & was like a child with expectancy from the moment the tree was installed.

Douglas came in on Christmas Eve for a moment & when he saw the tree he stopped dead in his tracks with delight in his face. They had a little tree of their own in the morning & then some more with us which the little lad enjoyed thoroughly. Sue, Blake & Dr. & Mrs. Robinson & ourselves & Douglas made up our party & the day passed happily. We drank to you all at 2pm E.S.T. & wished you were here. ...

It's grand the kiddies are so well. I expect they had an exciting & happy day yesterday. I wish we could have seen them. Mum had much pleasure finding & sending little things to them with anticipation of their pleasure. The tree is great fun for them & to hell with psychiatrists.

I kept an account of a psychiatrist's claim that Vincent's gingivitis & other gum inflammations start by being an "inflamed emotion". Honest to God, I think there is more sense to "Abraham's Box" than in all the psychos.

The London "Vitamin E's" are making a splash in the news & will cause a lot of false hopes & make a lot of money.

All our love to all of you,
Dad

The University of Western Ontario
Faculty of Medicine
27 December 1948

My dear old Dad:

We have had a very happy Xmas and in no small measure this is due to you and Mum. You gave us such a galaxy of presents which are all much appreciated. Your good wishes in a letter, enclosing a very magnificent cheque, touched me deeply: I can only say that my love and pride in you is at least as great. Your phone call did much for us too, and we were able to think of you and drink to your health a few minutes later with even greater feeling. ...

The C.P.H.A. Meetings were an awful lot of work for me; though I enjoyed doing it. I was already busy with other matters. I thought the meetings went quite smoothly and it is very gratifying to hear that others thought so. ... There were a number of old friends there, and some absent. Many had seen you the week before in Ottawa and were able to tell me of you. There were not very many outstanding papers but some good quality stuff among the chat. My own efforts went over well but did not provide any useful discussion. The new edition of the film is much improved and the paper is, I think, a well documented summary of an unexplained phenomenon. I will send a copy of the latter in a couple of weeks. ...

Collip has not said anything to me and I have not pestered him. I suppose I should enquire about the matter soon but I don't see what they expect to be able to do about it. No one has gone out of their way to come and see me in the lab and see how things are going (except perhaps Ed Hobbs and Rossiter).[74]

Well I will write again soon. Many many thanks for helping Xmas to be merry and very good luck in the New Year.

  Your loving son,
   Bob

The University of Western Ontario
Faculty of Medicine
28 Dec 1948

Dearest Mum:

You are too good to us. It was a lovely galaxy of presents you sent. The kids had a marvelous time and had a lovely day (even if their total calorie intake was made up of candy!) The pudding was excellent and the cake, though not yet cut, looks a beauty.

You have not sent me a cheque as well as the presents for you know Dad *always* outdoes himself – but it was lovely of you and I bought tobacco, and a new pipe and (I think) a portion of a bottle of good Scotch with it. ...

Doris is looking very well. Mother Grace being here and being able to help with the kids and join in the fun has meant a lot for her. Almost as good as a holiday and I think Mother Grace is enjoying herself too. Despite Mother Grace being willing to act as sitter we have been home most of the time and enjoying it a lot. ...

A & P are blooming – never in better health and in the happiest of spirits. Peter has changed a lot even since you were here and is a most talkative and ingenious young man with a passion for trains.

All my love to you, my dear, and many thanks for all the good things. Very best wishes for a good and fortunate New Year from us all.

  Your loving son,
   Bob

London, Ontario
28 Dec 1948

Dear Fudge:

The shaving bowl came in the nick of time – I use it daily from now on. Thomas appreciates his "long legged pants" (Doris' notation) though he has not yet given them a field trial. Peter loves his "tinkertoy" very much and invents the most extra-peculiar games to play with them – he has christened* his lovely overalls, and was most proud

---

74 See chapter 4, note 109 for information on the biochemist Roger Rossiter.

in them (strutting around as if they were his first long pants!). The little Mexican (I think) dolly is a much used toy with Alice and is most carefully looked after.

I trust that you have not been severely rationed this winter – that the carving has been suitably handled in my absence, due warning given of all meals, etc. No doubt Dougie has been doing his best for you.

      Yours affectionately,
          Bob

P.S. I hope the asparagus hit the right spot!
* Don't misunderstand me!

          The University of Western Ontario, Faculty of Medicine
                                        January 7th 1949

My dear Dad:

A long time ago we talked about the possibility of reopening the work that you did with the effect of the salivary extract of *Gastrophilus* larvae on the migration of leucocytes. As you know, Rossiter is very interested in the enzymes of leucocytes,[75] particularly polymorphs, and this extract might be of particular interest to him. Through a veterinarian here I find that I can probably get a good supply of larvae so that we could try out very easily to see how selective it is as a chemotactic agent.

Some time ago you wrote me some information on the method of getting the salivary gland and the general properties of the extract, but I seem to have lost these notes. If you are willing I would like to reopen the investigation and in particular to find out the specificity of the effect. I enclose a copy of the paper I gave at the CPHA meeting – you may find it interesting. We are examining but it is nearly over, by the end of the week I will be free.

We are all well.
      Love,
          Bob

                                          McGill University, Montreal
                                                      1 February 1949

Dear Old Son, .

We are delighted at the invitation you have received to be Visiting Professor at Seattle Washington. It is a very nice compliment which cannot be refused & I am sure you will carry it through with flying colours. Hall & Collip should be pleased & impressed. We are anxious to know more about it. I am sure Evans[76] will see to

---

75 A leucocyte is a "colourless corpuscle, e.g. one of the white blood-corpuscles, or one of those found in lymph, connective tissue, etc." OED (Online).
76 Charles Evans became the first chair of the Department of Microbiology at the University

it that you have a good time & you know him already. It will be worth your while to take along some fishing tackle & be prepared for strong fish. Remember that Haig-Brown[77] described "Lord Iris" as his favourite fly for Steelheads. I hope to hear of wonderful fishing.

On Sunday morning I was suddenly stricken & that is why there has been a long delay in writing to you. The main symptom was drowsiness & no inclination to do anything, with 2 to 3 loose & copious stools, slight abdominal discomfort & very little rise in temperature. (There have been 3 other cases similar at the lab) The lassitude is extreme & I do not yet feel inclined to do anything & am easily fatigued. I went to bed & am only up today; I intended to go to the lab but did not do so. I am supposed to go to Ottawa tomorrow.

Sue has been in the Mat. a week today. Stuart Henry curetted her on Wednesday & nothing more has been done until today when some X-rays were taken & perhaps some other tests such as basal metabolism.

... Blake & Doug are staying here & it is a full time job for Mum. Doug is a very good boy & amazingly well behaved but extremely active. Everything is done at the double so he is round the corner & out of sight in no time. He understands a great deal & has quite a vocabulary for such a little chap. Mum is tired by the end of the day. He is a most lovable little chap. ...

I suppose you will have to make some arrangement about a house before you go to Seattle. It looks as if you had better come to some clear understanding with the University. There is a howl all up & down about the scarcity of bacteriologists & still the salaries offered are absurd. It's up to all of us to insist our proper recognition & treatment. The situation is our own fault for accepting & putting up with such treatment & now is the opportunity to improve it. It is absurd to pay more to pathologists & still more ridiculous to radiologists.

We have two Spalding badminton racquets. Do you want them? They are not likely ever to be used by us again. Let me know & I will send them along at once. Glad to hear you are playing.

 Love to you all,
   Dad

---

of Washington, Seattle, when it was established in 1946. He held the position for twenty-four years. Robert Murray accepted an invitation to be a visiting lecturer in Evans's department in the summer of 1949. "Standardizing Clinical Microbiology," http://www.washington.edu/research/pathbreakers/1960h.html.

[77] See chapter 5, note 55.

The University of Western Ontario
Faculty of Medicine
February 3rd, 1949

My dear Dad:

... Have you seen the very excellent pictures of flagellated organisms by (Kingma) Boltjes (see below) in the April, 1948 number of the J. Path. & Bact (page 275)?[78] Robinow sent me the other day, for my inspection, some electron photomicrographs that he and Miss van Iterson took of *Proteus mirabilis* showing very definitely that the flagella pass through the cell wall and end in bulbs well within the cytoplasm. Robinow also told me that he thought there was good evidence now of there being a plasma membrane, showing birefringence,[79] which can be demonstrated by several different methods. The cytological differences between bacterial and other cells seem to be growing less and less.

Another article that should interest you a lot, and the pictures, is by Hampp, Scott and Wyckoff in the December 1948 number of the J. of Bacteriology.[80] Particularly interesting, I think, is the definite impression in figures 1 and 2, which are pictures of a small oral Treponema, that there is a very fine axial filament. An observation which oddly enough they do not seem to mention in the text of their article.

We are very busy with requests for grants and writing reports. I am writing separately on other matters.

Your son,
Bob

293 Oxford Street,
London, Ont.
3 Feb 49

My dear Mum and Dad:

I am very sorry to delay writing to you for so long. We have had lovely long letters from both of you and done nothing about writing back. This has been far from intentional and has in fact been largely due to extreme business. This week I have not been

---

78 T.Y. Kingma Boltjes, "Function and Arrangement of Bacterial Flagella," *Journal of Pathology and Bacteriology* 60, no. 2 (April 1948): 275-287.

79 Double refraction.

80 Edward G. Hampp, David B. Scott, and Ralph W. G. Wyckoff, "Morphologic Characteristics of Certain Cultured Strains of Oral Spirochetes and *Treponema pallidum* as Revealed by the Electron Microscope," *Journal of Bacteriology* 56, no. 6 (December 1948): 755-769.

so well and have suffered something similar or identical to Dad's seizure, and at the same sort of time. I felt fine on Monday and was full of beans, played badminton in the evening. Woke up at quarter to six with extreme urgency, and a lot of nausea. The diarrhoea did not last more than about three hours but I felt absolutely lousy. ... with mistaken heroism went to the lab for an hour or two and managed to see Collip. ... A number of other people have had it – all with much the same tale and most of them without fever. Rossiter was sick this week too with IT. It has come at a most annoying time too and I am frantically busy.

The offer from Seattle has bucked me up no end. We are pleased with the prospect for I hear that the State of Washington is lovely, Seattle is a nice town, and the University is an up coming concern (with plenty of money!). Well, I had this very nice letter from Evans inviting me to come for one or two months as visiting lecturer. The lecturing consisting of a couple of lectures a week, labs apparently well taken care of. The most important duty of the visitor is to help in the exchange of ideas with graduate students and staff. He said that there would be plenty of time to enjoy the "recreational opportunities offered by the area". So I shall go armed to the teeth for the steelheads and rainbows, which I look forward to very much.

I saw Collip and he is very pleased and puts nothing in my way and I can arrange things alright at this end so that there is no gap in essential work. They are generous about money and offer $645 a month – and since my salary here will continue should allow enough leeway for us all to go about there and enjoy ourselves. Doris is very pleased. I hope that it is for the two month period for we would like to be able to make the best of the opportunity and from the financial point of view, going out with the whole family, would be in a better position. I wish we could meet out there and have a little steelhead fishing together. I have, of course, a lot of things to find out such as income tax matters, necessary permits etc. It is really a very fine offer and a nice feather in the cap. I liked Evans very much and I spent a very pleasant evening with him and old Sherwood when I was in Minneapolis. I suppose it is one of the good things about going to meetings. Robinow was out there for six months and has just come back (to New Haven) he says it is a marvelous place and he left there with regret.

We will certainly have to solve the housing problem in some manner and I would hate to have to solve it by buying at the present moment even if we did have the necessary collateral to lay on the line. This house, although it suits us very well at the moment would not be suitable in a few years time for larger kids because of the layout of the rooms. I really don't know enough about houses to trust myself on such decisions. Since we might leave for Seattle as early as the second week of June we are going to have to get cracking on the problem.

We are at the moment faced with another problem. The Sinclairs (whose house we rent) have bought and moved into a house in Ann Arbor and are moving out their furniture *this week*. Rather short notice, but there you are. I don't blame them really but it is going to make the house rather bare for a while. Luckily we can buy a few of the essentials right away. Makes things a little awkward.

Sue's long stay in hospital is a bit of a worry and it must be a financial hazard to them, poor things. I hope that there is nothing seriously wrong and await a report quite anxiously. I am sure it must mean a lot of work for Mum but Doug sounds a most good child and full of fun.

We have done little but get through the days work since Xmas and about the only relief has been going to play badminton on Monday evenings. Probably good for us.

... What with trying to send in requests for Grants-in-aid, write annual reports and write papers to say little of the other common problems I am up to the ears. By the fifteenth, when most of it has to not only to be finished but at its destination. After that I will be able to give you a more coherent account of the work we have done. The Bacteriophage metabolism problem is going well, even if on a rather unexpected line. Loeb is doing well, but I have some snags to clear up on the Staph. antibiotic purification. ...

    Your son,
        Bob

3590 University St.
Montreal
Feb.7, 1949

Dearest Family:

... I can't begin to tell you how delighted I am to hear the grand news about the probable visit to Seattle as a "Visiting Lecturer." Bob; as well as being a definite feather in your cap it has many grand possibilities and will be most interesting in every way. It will, for instance, be rather intriguing to see just how a University is run when there is lots of money to run it on! From the point of view of recreation you will indeed be in clover—what a chance to get some really superb fishing in a marvellous country. I am so glad they have made it possible to consider taking the family. ... Both Dad and I are very happy indeed about this very complimentary and pleasant invitation – in every way it is going to be a splendid and enjoyable change for you. ...

    Lovingly yours,
        Mum

The University of Western Ontario
Faculty of Medicine
13 Feb 49

My dear Dad:

This has been a most busy week and a half. This is the time that we have to put in applications for grants to the NRC, send in annual reports, worry about staff, and the budget for next year. Lots of paper work and worry, sometimes to no good purpose. I have put in for three grants and hope that they go through alright. The largest is one to continue the work that Fred Heagy is doing on metabolism of phage-infected cells. I think the work is going quite well and hope that the NRC look kindly upon it. Fred and I had to work quite hard on the annual report during the last week – will send you a copy to look over and see how it is going. I am very keen that this bit of work get some continuity for the future – for it is no short project. Fred will probably be leaving in a year and a half and I need someone to work with him and learn the techniques so that he may take it over. I think that I have found the man – although financing him is a bit of a worry. I am also putting in for special equipment to do a cytological study – especially phage infected cells. Some characters in France have observed that special lesions can be shown up in phage infected cells which are characteristic of the phage and *not* the host cell. This needs to be confirmed and extended if it is true. So I am hoping to get some good optical equipment and the necessary photomicrographic stuff to get on with it. We don't have the necessary stuff in the school. Also I am asking for support of the work on micrococcal antibiotics.

Had the annual meeting of the budget committee last week. Won most of the general battles and got more or less what I wanted. They are still in the position of wanting you to get all the money you can out of grants in aid which I think is for many things just a trifle dishonest. However, space or no space we are going to have to get more staff if things go on as they are. The clinical routine is as heavy or heavier than ever and we should expand our research work a bit. With everything we need one more fully trained hand. I have a sort of blank order for finding one but chances of success would seem to be slim.

I have not heard again from Evans so I have nothing new to tell you about the western visit. We are looking forward to the possibility with pleasure. I will have to get things well organised and running smoothly for the summer.

The hierarchy will not … say how they intend to handle the affairs of the department after July 1. However, everything they say indicates that they have not found anyone and don't think that they are looking very hard. They seem to be putting a certain amount of trust in me which is encouraging. On the other hand I wish they would say something encouraging about the way things are being handled (or the reverse). All very negative.

Love, Bob

P.S. Van Rooyan is going to send you the membranes.

McGill University, Montreal
22 February 1949

Dear old Son,

There is an interesting paper by Cuttino & McCabe on "Granulomatous Nocardiosis" in the American J of Path, Jan 1949, <u>25</u>, (No 1).[81] A somewhat similar case was found by Wigglesworth in a patient which had been in C.M.H. for 12 days & died at home a couple of weeks later. Child 5-1/2 mos which had 6 weeks fever & vomiting 99°-100° 5 wks & 100-103° during last week. Sulphas & penicillin without result ... The autopsy was done at Valleyfield & very little tissue was collected. ... Unfortunately no cultures to be able to examine the organism. ...

I saw Boltjes' paper & wrote to him encouraging his disagreement with Pijper. I think Pijper talks a lot of nonsense.[82] I also saw the paper on spirochaetes, at a glance, but did not read it. I doubt very much their growing Borrelia. ...

Sue came home on 17 February 1949. She is cheerful & making good progress. There was some sepsis, a small leak in the wound from which I took a specimen in which was a faecal strep. The situation was hidden up by 3 lots of penicillin in spite of avowals that there was no sepsis. Slack technique is being bolstered by antibiotics & sulphonamides. ... She had 22 days in hospital & the bill came to just over $400 (hospital only) of which 2/5 was met by "Blue Cross". None of this agrees with the way I was brought up & I don't

---

81 John T. Cuttino and Anne M. McCabe, "Pure Granulomatous Nocardiosis: A New Fungus Disease Distinguished by Intracellular Parasitism: A Description of a New Disease in Man Due to a Hitherto Umdescribed Organism, Nocardia intracellularis, n. sp., Including a Study of the Biologic and Pathogenic Properties of This Species," *American Journal of Pathology* 25, no. 1 (January 1949): 1-47.

82 Adrianus Pijper (1886-1964) observed live bacteria with dark-ground illumination at high magnification using sunlight instead of a lamp. He argued that the tails that streamed off the moving cells were made of materials pulled from the surface of the cells, and were not active organelles or propellers. He contended that bacteria swim by undulating the cell body and that flagella were products of the movement trailing along with other material. The main contender on the side of flagella was W. van Iterson, who was with Kingma Boltjes at the lab at the University of Amsterdam. Kingma Boltjes believed that the flagella were active propellers in their various arrangements on the bacteria coalescing together in activity. E.G.D. Murray supported the latter and believed that flagella had a function in propelling the cells. As Murray remembers, "Proofs were yet to come," and we now know that flagella are rotating structures and cooperate together as propellers. RGEM. For more on Pijper, see J.P. van der Walt, "Adrianus Pijper M.D. D.Sc 1886–1964," *Mycopathologia* 24, no. 4 (December 1964): 377-379. For more on the bacterial flagella debate see James Strick, "Swimming against the Tide: Adrianus Pijper and the Debate over Bacterial Flagella, 1946-1956," *Isis* 87, no. 2 (June 1996): 274-305.

wonder the general public are anxious to have "State Medicine".

I too am worried about the policies of University Administrators. I don't think it is quite honest. The big & wealthy USA Foundations are now refusing to finance research projects. They are using their money to finance training of men instead. A curious & interesting shift of policy.

I look forward to hearing more about the offer from Seattle. I am surprised at Hall & Collip. In my opinion if you can do the job you deserve the position, which is the principle on which I work my department & recommend appointments. I think you will probably have to force the running in the long run.

Mum is having quite a busy time & no rest. Doug is quite a fellow & amazingly active. He is very good indeed but is never still a moment. ...

There is a paper on Hyaluronidase in rheumatic fever in the last number of the Am. J of Med Sciences (Feb 1948). There is an interesting tetanus toxin paper in the last Brit J of Exp Path. using the anterior chamber of the eye & paralysis of the iris.

Enid Pascoe tells me that they are not taking students at Mount Desert[83] any more. ... Woods Hole[84] only takes 16 & they must be seniors & intending to go on with the subject. In any case she is keener on Botany than Zoo. The Zoo & Botany depts. here are not very helpful nor encouraging. ...

Love to you all,
Dad

293 Oxford Street
London, Ont.
27 Feb 49

My dear Dad:

After having a pretty healthy winter we have suddenly had a plague of colds and suchlike. All of us have had some sort of a sniffle but Doris had quite a siege. She had a bad cold two weeks ago, which seemed to clear up in the usual fashion and was not accompanied by fever. Just as the symptoms cleared she started to spike a daily fever up to 101. Nothing particular to be found, though Frank Kennedy[85] came to see her. ... We had to keep her quiet (a damn hard job). Frank recommended use of penicillin, on the basis of his experience with similar cases of the moment, and it certainly seemed to do the job judging by the drop of fever and relief of malaise. Five days after her temperature came down she suddenly threw a spike of fever with a chill (97 to 102 in two hours); since then she has had no fever and has been feeling fine. ... Except for

---

83 Mount Desert Island Biological Laboratory, Salisbury Cove, Maine.
84 Marine Biological Laboratory, Woods Hole, Massachusetts.
85 See chapter 4, note 16.

the apparent response to penicillin it had all the earmarks of one of these so called primary atypical pneumonias, of which there is a lot about. Just in the last couple of weeks we have noticed a great increase in the number of isolations of pneumococci, streps etc. Looks as if the bad respiratory disease season is going to come a little early this year.

Thanks for all the data on *Gastrophilus*.[86] It is just what I wanted. Rossiter is away at the moment (on a junket out to Saskatoon) but I will take it up with him next week. Since his interest is in the enzymes of the leucocytes the apparent damaging affect may not be of importance. An interesting effect is the exudation into the serous cavities not injected, and that in this situation monocytes predominate.

I am sorry that Enid Pascoe is having trouble finding a summer course in invertebrate zoology. I was under the impression that Woods Hole had a larger and more elementary course. ...

The badminton racquet arrived safely as did the press. It will be fine. It has been a lot of fun playing and I find I feel better for a small modicum of exercise. Glad of it the other day for we had a staff versus students volleyball game, and but for a bit of training would have been mighty stiff afterwards. The players in the badminton group are a very mixed lot and are everything from the most uncoordinated to good players. Good chance to learn. Next year I may be able to play some squash again for they are building a great big gym affair on the campus which is to include squash courts and a swimming pool. (That is the sort of things these millionaires give their money to -!)[87]

The "Damsel" fly is quite handsome but a little opaque on transmitted light. It may be worth a try and I will tie a few. I have not done much tying lately and have spent most of my time on writing. Must get to work and find out what patterns I should get ready for the west. ...

You ask about Vol XIII of the Cold Spring Harbour Symposia. I am not sure about this particular volume. The use of tracers is particularly important now that they are available and with them there is going to be a lot of very important work done in the next few years – in fact some is being done now. I have not seen the volume yet but three of the articles look very interesting and a few of the others might be informative. I think it would be important to have some work of this character. In general the Symposia are very good and a most useful reference for years after – I wish I had several of the old ones. I notice that volumes XI and XII are still available for both of them contain some very useful material. Have you got one or both of them? ...

---

86 Genus of botflies, several of which infest horses and rarely humans. Merriam-Webster Online; http://www.merriam-webster.com.
87 Western was fortunate that a number of prominent London patrons – notably the Little, Smallman, and Ivey families – donated extensive funds for University infrastructure projects.

When are you going to come along and see us? I had rather hoped that you would be coming this way in the near future. I don't think we expect any visitors until perhaps Easter when Robinow may be coming up from Yale (I hope he does, I want to learn a lot from him).

The kids are very well and progressing nicely. Thomas is a sturdy, cheerful and good little boy and you would like him. His red hair is not too obvious yet and I have hopes that it will calm down.

Thanks for the reprints, glad to see them. Thought we might have a use for ... method the other day when a H. Influenza (fatal case of meningitis) refused to type directly or on the first subculture. It did later from Fildes.[88] I had kept the manuscript of the address on Fuso-spirochaetal disease. I will be sending you some manuscripts for comment in a day or two.

I have a lot of work to do and catch up with so I had better post this.

All my love to you.

Your son, Bob

London, Ont.
4 April 1949

My dear old Dad:

... It was very encouraging for me to hear from you that Collip thinks that I am doing an honest enough job. I was about getting to the point that I felt an odd word would do me a lot of good – having run things in silence for quite a while. As a matter of fact I saw him last Friday. He said that there was little doubt that I would be given the whole show, which is nice to hear even though it is not in black and white. I was also glad to hear from him that my three grants from NRC came through unscathed (some people were apparently cut down quite a bit), so that means that some of the work that we have planned will be adequately supported. I am also glad to hear that the man I have lined up to work on the phage problem with Fred Heagy, and eventually take over, got his fellowship alright at a slightly higher figure than might have been. This is encouraging both for the problem and for the training of workers.

I have not heard definitely how long I will be in Seattle, or for how long. I expect to hear sometime this month. However, I was sent the university bulletin, both the annual and the summer quarter one, and I find that my name is down in black and white – it seems to be official from that end. It is a fantastic place – the campus is about 1x½ mile and seems to be largely devoted to the schoolteacher population

---

88 See chapter 2, note 39 for information on the British bacteriologist Paul G. Fildes.

though there is a large graduate school there. They are building a new medical school and dental school, which may be informative from the point of view of plans etc. The normal student population seems to be about twenty thousand!

In a recent number of Nature I saw that old Tibby Marshall is dead.[89] Died in February after an operation in Cambridge. There was a short but excellent obituary. The old Guard is passing, I fear. Tibby, despite his bachelorhood and funny ways was a giant in his field. Did I ever show you my collection of his notes of excuse for not attending student meetings in College? Semi-legible but priceless.

A couple of graduate students from here (Physiology) are going up to Southampton Island for a couple of months this summer in a research project. One of them asked me if there was anything that he might collect for us while he was up there. I suggested he might get a snow knife and any other implements he thought might be of interest. Any suggestions.

Carl Robinow (from the Strangeways, field of bacterial cytology) is coming to visit me here for a few days this week. I am looking forward to it a lot, and expect to learn a lot from him. Have been doing some of that work lately – it is very fascinating. You must be having quite a run with TB specimens. We are in a similar position, though not so extreme. The main trouble is that we are having an enormous number of specimens of all kinds. For the past three months we have averaged over 500 specimens a month, as opposed to about three hundred last year. Makes a lot of work for a small staff and does not allow of the careful work I would like to institute. The present financial setup and relation of university-hospital is absurd despite the changes we were able to make. There is an agreement that dates back to the twenties that gums up the whole work. ...

I have not yet tried [Guilford] Reed's method for viable counts. We have been using a loop method that gives almost comparable results (a little less accurate, but not much) but which is infinitely simpler once you have a bit of practice. It is a method Ash developed years ago for counting phage, and is equally applicable to bacteria.

I have not had any time for fly-tying for the last three weeks. Thanks very much for the material. Dr. Bittner[90] has been sick for the past four days and I have had to work like a slave keeping up on both sides.

Mum sounds as if she is having a great time at the Arctic Inst. I am sure she enjoys

---

89 Known affectionately by students as "Tibby," Francis H.A. Marshall was a Cambridge-educated scientist whose long-term research interest was in the physiology of reproduction. He served as lecturer in agricultural physiology and then reader in agricultural physiology from 1908 until his retirement in 1948. His personal and professional life centred on Cambridge University, and particularly Christ's College and the School of Agriculture. A.S. Parkes, "Francis Hugh Adam Marshall, 1878-1949," *Obituary Notices of Fellows of the Royal Society* 7, no. 19 (November 1950): 238-251.

90 See chapter 4, note 8.

it. When are you going to come and visit us???? Must close now, already late and Doris will be needing help with the kids.

    All my love.

        Your son, Bob

McGill University, Montreal
6 April 1949

Dear old Son,

We got your cheerful & encouraging letter yesterday evening when we returned from seeing Hamlet again.

It is good to know that things are far enough along for Collip to go so far as to say there is little doubt you will be given the whole show. I hope it will be fairly soon & will bring a proper salary with it. I am glad, too, that your grants have come through. It is all a tribute to the work you have done. It is satisfactory too that the Univ. of Washington has printed your approaching visit there. I have no doubt you will have a very interesting time & I suppose it will chiefly have to do with graduate students. They have [loads] of money and are branching out greatly, but it would be interesting to see how wisely.

The need you have is for an adequate staff, in number & in training, for the work to be done. Be careful to have help that allows you time to do your own work – I did not take care of that for myself and regret it now. You will eventually have to improve the set up for the hospital work if the contract is continued. I would think the old contract could be redrawn & brought to meet present needs & conditions.

I am sorry to hear of Tibby Marshall's death. We were very good friends. I know what his writing was like & your specimens are interesting in several ways. Yes he was a great man.

I don't have any other suggestions than you have made to be collected from Southampton Island. Modern transport facilities have so altered things that original crafts have died. Possibly a real Ategi would be worth having – but the Eskimo is now using woven materials bought from the trader. Similarly for Mukluks & other beautifully made and watertight footwear.

It will be very interesting for you to have a visit from Robinow. There are some interesting & important things to be done in that line. I enclose a reprint I have just received from Boltjes. If you have it return this, but if not keep it as I have the original journal. …

I want to see you & will arrange as soon as I can. Just now it is impossible. I am right behind with my work & finding it difficult to catch up.

    Love to you all,

        Dad

McGill University, Montreal
10 April 1949

Dear old Son,

Two very interesting maps arrived from you & the Eskimo pictures which I had intended you to keep. The French River map is particularly interesting with Alexander Mackenzie's description of the difficulties it presented. Do you want them returned?

Today is Blake's 32nd birthday & he did not look a day older when I saw him yesterday. Sue seems quite well again but I do not know whether the operations did her any good. It is hard to see quite why they should. The 22 days in the R.V.H. & their fee to Stuart Henry cost them $660. The hospital certainly charges all the traffic will bear & subdivides things to charge good funds for each ... a charge for pathology & for cytology. The abdominal operation revealed a true "phantom tumour" which I thought no longer happened; the lump which was felt was not there. I have not said anything to Blake & Sue. The hospitals are certainly forcing the issue of state medicine & I expect every party to have it in their platform next election. ...

Doug is quite a lad. The other day, when Sue was phoning to Mum, Doug was very quiet so Sue went to see what he was doing & found him in the kitchen draining the beer bottles. As she entered he looked up with a smile & said "Nice!"

I bought a copy of Theodore Cordon's letters & articles & they are quite interesting reading. He was one of the earliest of the real fly-fishermen of the United States. He designed the Quill Gordon which is a proved pattern. ...

I spent all Friday at the city council & it meets again tomorrow afternoon but I don't think the agenda can be completed then because there are some topics on which a number will want to talk their full time merely repeating what others have said already. Then there is an unusually large list of private motions.

I recently re-read "Sleeping Island." It is a grand book.

 Love to you all,
  Dad

McGill University, Montreal
17 April 1949

Dear old Son,

It was good to talk to you both on Friday. We are very sorry our mail for Doris is late but only because Mum's illness put us out. Our thoughts were with you.

... Mum is markedly better though not herself yet. She tires easily & so far I have succeeded in making her breakfast in bed so that she does not get up till late.

 Love to you all,
  Dad

> The University of Western Ontario
> Faculty of Medicine
> 18 April 1949

My dear old Dad:

... It is a long time since last summer and you have not yet seen Tommy. I suppose there is no chance that you will be going to the SAB meeting, with all the work that you mentioned when we talked?

Up to the middle of last week I had a rather hectic time. This was aggravated by Dr. Bittner being sick and leaving me to do the main part of the work for the hospital routine. This is not an inconsiderable load at the moment for we have been running a little better than 500 specimens a month. While she was away sick Robinow came to visit for a couple of days – it would have been longer only he had a little trouble with his visa (peculiar US immigration red tape) and had to leave early to be sure of getting back to his wife and family! This was a great nuisance for I wanted to be able to spend some unhurried time with him at a bench and play with some of the things that interest us. As it was we managed to do a little work in the short time available and I learned quite a lot. He gave a very excellent seminar to the graduate students – much enjoyed by all on the structure of bacterial cells. This work is very interesting and the techniques are relatively simple. A little time could profitably be given to such studies in the Course III – it is fascinating; teaches people to use their eyes, and the microscope, to the best advantage. Also it is an aid to making people think in terms of the cell and its development. The apparatus required is practically nil. Anyway we had a pleasant if hectic visit. He was also able to tell us much of what was going on in other labs he has visited, and especially the University of Washington.

Thank you for sending Boltjes' reprint. As a matter of fact I got an envelope full of his reprints the other day and that one was included. I will return the other.

The maps that I sent were intended for you to keep. I knew that you would be especially interested in the French River one for it is an important landmark in Canadian travel. I am glad that it will be appreciated. I will be sending an American publication called Fishing Tackle Digest which is labelled as a complete guide to fresh water fishing tackle. It is by no means complete in many places and information in others is plain misleading. Still it has some interesting information in it. ...

The housing situation does not look too good. The one place that might have been a possibility has fallen through because the people decided to sell rather than rent. That is the general happening. I don't blame them because they are hardly likely to lose by the deal even if they have to buy a house again later. The trouble with buying at the moment is that in general a rather large down-payment is required. On the other hand the trouble with renting at the moment is that places coming vacant now are no longer controlled so that the rental can be and is fantastic. I don't know what to do. As a matter of fact I have no idea what money I could raise at the moment even if I thought I did want to buy. Damn little I fear.

In this regard the insurance policy, which you are keeping up for me (and should not be, for I should be able to maintain it), may have some borrowing value in an emergency. The whole damn housing structure is cockeyed. We are still running advertisements and hope that we may get some possibility that way. We have had some answers but they have been either impossible for rents, or so damn small that we couldn't fit in.

Sue's hospitalisation in all its phases seems to be utterly fantastic. It must have been a great blow to them to have to pay such a high fee. Not at all fair. I can't think why [Stuart] Henry kept her in so long. There does not seem to be much indication. Most of the surgeons here are getting their patients up and out of the hospital in an incredibly short time, even after very extensive operations. What is most important is that the patients seem to be better for this early exercise. I always thought that Stuart Henry was a bit of an old woman.

Laz. Loeb has been doing good work and I am pleased with him. He has been a little depressed lately, with good cause, for his brother (an MD in Toronto) has been found to have an inoperable Ca of the bowel with liver metastases. We will all be going down to Cincinnatti. Fred Heagy will give a paper on the interesting inhibitors of mucoid streptococci. It will probably be a worse melee than last year – but I hope to see some of the people who interest me. Fred has been having a bit of trouble with the T-2r phage – it refuses to behave itself. I wish I could explain it for it is holding things up – seems to be an ecological problem not coincident with the known blocking mechanisms.

Easter provided a pleasant break. I did a little fly-tying and some reading. I have been putting off for quite a while. Doris had very pleasant birthday and in the evening we went to see "Oklahoma" – a very nice musical with plenty of colourful songs and dance. The kids had a good time but Alice developed a cough and some fever on Saturday, which seems to have quieted a bit now.

I must get this to post for I do not have opportunity to add to it for a while.

All my love, Bob

The University of Western Ontario
Faculty of Medicine
1 May 1949

My dear old Dad:

Yesterday there arrived a large parcel containing your large reel, wrapped in a most expensive fashion. You should not do such things! I am very glad to borrow the reel for the summer provided you can spare it and I am not sure that you can. You left a line on it which you should not have done for you may need a spare line and I have two, myself, which is plenty. ...

The N.R.C., despite the fact that they allowed my grants in toto, are being a little pesky about doling out the money. They would not send the advance I asked for on one grant, which makes things awkward and time consuming in making arrangements. It is not really their fault. It is because the damn government were more interested in politicising and a new election, than in passing the estimates.

The housing situation is not solved yet. I saw a house last week that is large enough but it is not in a good district and the people want a very fancy rent. ...

Mum says you are working too hard. I hope that you will manage a trip to the Columbus this spring. The fish should be in good form this spring and everything is ahead of time.

  All my love to all of you,
    Bob

                Montreal
                2 May 1949

Dear Old Son,

I am glad to see by Doris' letter that the Mumps have not developed among the children & I hope it is not to be a 30 day incubation period.

I did not write last week because I was working against time. I let myself in for writing an article on fishing for the "National Home Monthly" (whatever that is); the editor called me long distance from Toronto & said Gregory Clark had recommended him to get me to write & gave him some guff about me as a fisherman. It was hard to refuse but I found it hard to write the article as too many & varied things crowded my head & my style is too heavy. However in the end I concocted the enclosed but would not be surprised if it is pronounced unsuitable. ...

I glanced at the "Fishing Tackle Digest" & I do not like it. Obviously it is merely a trade journal for manufacturers. I don't like the articles, especially "How to tie flies". I have never fished in the West but I am not impressed by the article on "Western Flies" & I would rather rely on my own 43 years of fly fishing. ...

Mum & I heard the ice was out a couple of weeks ago but I could not get away. This weekend Sue & Blake went up & were having a lazy time when they heard yesterday afternoon the bush was on fire at Raab's (Merill's old place) & they spent the rest of their time fire fighting. It seems Raab thought he would burn his grass & it got away from him to catch the thick carpet of dry leaves & in a matter of minutes (not more than 30') was up over the hill with trees burning & old stumps well alight. Blake & Sue drove back here to get home well after midnight & they are still worn out today. They worked like blazes to stop it coming our way; it reached the top of the cliff behind us. Thank goodness it rained a little today – I wish it had been more. Raab phoned this evening to apologize for the anxiety he had caused us & evidently he had a good fright & feels rotten. He wanted to assure us that the fire is out & there is no more danger (he left there late this afternoon) & I hope he is right. I've seen them smoulder in the ground for weeks. ...

The old pup has been none too well & we got Dr. Baker (the vet) to see him, his treatment seems to be doing him some good. Baker is a very decent chap & he & Mum got onto the old topic of the bush & fishing. The upshot was, after a short consultation on the dog at each visit & a long yarn on the bush, that Baker brought in his friend Harry Wheeler (Gray Rocks Inn & flying tourists into the North) who happened to be in town. He seems to be a fine fellow to whom "tourists is tourists" but Mum is the real thing & we are invited to make free of his territory. I don't know more than that but Mum will no doubt give you the whole yarn.

Thanks for the maps; I like them very much & they illustrate things that interest us quite a lot. French River is a devil & no wonder people had difficulties with it.

I hope to arrange to go see you & I would like to hear about Robinow & his methods. I can see I shall have to do a lot of spade work myself to improve some of Course 3, & I shall have some inertia to overcome in places. I should go to the S.A.B. but I have not arranged it yet. Where are you going to stay?

I wish you had good news to give of the housing situation. I think Collip & Co. should let you know definitely what the situation is regarding the Chair so that you can decide what to do. Mum & I have a little salted away & we are ready to lend you about $3000 to help with a down payment & you could eventually pay it back by degrees without any interest, which Banks etc would exact from you. You could consider the wisdom of telling him you have to decide whether to stay in London or not. I shall be glad when you have some papers published – it is very important indeed & a number of shorter ones which add up to one good one in the end is better in many ways than monographs. I know because I made the mistake.

I'm sorry about Loeb's brother. It seems to be a bad outlook. What are Laz Loeb's chances of getting into the Med School?

I have been looking into some Staph trouble the neurosurgeons have been having. We have not completed the work yet but at present the indication is that it is probably the patient who is the source. Last time I tackled a similar (though worse) state the trouble was the Instrument Nurse was the carrier. These things take quite a lot of doing.

I am trying to force the Alex Hosp. [Alexandra Hospital] to reopen its lab as they are becoming a bit of a burden to us, especially with the T.B. specimens.

I got a packet of documents from England the other day which are illuminating. I am still on the register there. This is the beginning of forming a Medical Trades Union, in exactly those words, & included was a copy of the Trades Union Laws in general. I am not taking any part in it, but it is a sad situation when the rights of medical men have to be protected in that way & documents circulated to prove it.

We may well have State Medicine to contend with here before very long but it is hard to see what form it will take. The more successful the Red Cross Blood Service is the stronger argument the Government will have. The C.M.A. seems not to be doing much that matters about it. I think a time will come when you will have to consider Specialist Certification, especially if State Med. looms large. One thing is certain that the hospital situation & the general cost of illness cannot go on the way they are & the public cannot be satisfied with the platitudes one hears at the Annual Meeting of this & that hospital. Group insurance must be making fortunes for the Insurance Companies.

Hell. Think of it, the trout season in Quebec has been open since April 1st & I have not wet a line yet.

  Love to you all,
    Dad

London, Ontario
8 May 1949

My dear old dad:

The weather is simply lovely. Alice, Peter and I are sitting in the sun on the back step and it is as warm as toast. ... We feel much more cheerful about the housing for we think that we have found a place to rent. It is not a whole house but is a lower duplex, with three bedrooms. Perhaps not ideal but very convenient and only about two blocks from here. It is right opposite the school on Waterloo where Alice would be going to kindergarten next year and is of course only two blocks from Mrs. Bullers where Peter would be going. Another convenience is that it is heated by a gas furnace which requires no attention from one year to another. The timing is most convenient too – for the people want to decorate the place and we will be away for the summer.

Your long letter is a great joy to us and we both enjoyed your article on the intangibles of fishing very much indeed. I have never heard of the National Home Monthly but I think it is very nice of them to ask you to write. You will probably find yourself writing more – and I think it would be a good thing. By the way where did you get your closing quotation? I know it must be hard to write that sort of thing – it is hard not to be heavy – but I think you have managed a pleasing light touch. If they don't want it I am sure that you could find a publisher without too much trouble.

I have been so damn busy that I have not had time to do much about the fishing plans for the summer which I should do before it is too late. I have had a great many things on my mind. Next week I shall be leaving for Cincinnati. I will be at the Gibson Hotel where I have a single room – maybe we could rig things up for you there. If you want to save some money and go down come here by Sunday. Fred Heagy is driving Loeb and I down by car and we will be leaving early on Monday morning and coming back the Friday afternoon. It will be a longish drive but should save us a lot of money. We would love to see you – Mum might like to come along and stay here with Doris or even along to Cincinnati. Doris says she, Mum, could come and sit around and LOAF! What do you think?

Fred Heagy and I have done a couple of very interesting experiments with phage. He has been following the effects or dinitrophenol on T_2r and T_2rplus infected cells and I have been doing cytological studies in parallel with his experiments. So we have full data on the infection from start to finish. Despite the few experiments there are very definite lesions [in] progression during the course of the infection which are easy to see though very hard to interpret. The internal structure with respect to nuclear structures (deoxyribonucleic acid) is completely changed in less than ten minutes – this corresponds with chemical data – after that there is synthesis in a regular pattern.

Unfortunately my photomicrographic equipment has not arrived and I have to record results in drawings which are too easy to make fanciful. Anyway it is most stimulating and we look forward to some interesting work. Rossiter is very interested too and we may be able to do work later following the nucleic acid side quantitatively.

The trouble is that there are so many interesting things to do. Our high speed centrifuge arrived yesterday. A Sorval SS-1 which attains 20,000 g (about 13,000 rpm) which should be adequate for what we want. It is a pretty bit of machinery but we will have to wait a while before we can use it for I am afraid that they sent the wrong voltage adjusting transformer (our bloody 25 cycle juice) despite the fact that I wrote 25 cycle all over the order. Incidentally we may be changing back to 60 cycle in about a year which will bring a whole host of new headaches until the change is complete.

The fire at the lake must have been very worrying. Raab must be an idiot to let things like that happen but he has probably learned a lesson. I suppose that he has lost a lot of his own trees.

We are very sorry to hear that the old pup is sick – but then he is getting quite an old dog now. I am glad to hear that Baker is managing to make him better. Baker must be a nice old stick and I look forward to hearing more from mum about her talks with him, and Wheeler.

Laz Loeb will be able to do his medicine here as a special student starting next year. He has acquitted himself very well and he is well liked by staff and students. He will stay on with us, help a bit with the teaching and will also get a small but helpful salary from the university at the same time. Collip is very much persuaded that there is a shortage of bacteriologists and that those wishing to stay with the subject are to be encouraged. His brother is not improving and I fear has not long to live.

Collip has been away for a couple of weeks and I have not been able to see him since he came back. I expect to see him at the beginning of the week and to see whether things can be clarified a little. They will have to make their decisions soon.

The hospital routine is still a headache and we are getting still about 500 specimens a month. This is a lot for us and it makes so much work that it is hard to be able to improve the quality of the work. Pressure puts the quality down.

I don't think anything of the Fishing Tackle Digest either but it is useful to have for it gives an idea of the tackle made in the country.

No mumps yet and I think we can cross it off the list unless they get properly exposed again. The kids are very well and the last week of sunshine has tanned them up a lot. Young Thomas has one tooth through and looks very proud of himself. He loves sitting up but has to be helped. He should be sitting by himself very soon. He is a most cheerful baby and loves company and chat – grins all over his face on almost all occasions.

All our love goes to you both,
Your son,
Bob

P.S. hope you can fit in with the Cincinnati plan.

Seattle
25 June 49

My dear Mum and Dad:

It was very nice to get your letter the other day – it got here in good time.

I had a good journey out here – especially on the Streamline train from Chicago here. They make excellent time and very comfortable. ...

All the houses here are very neat with lovely gardens, green as Ireland, profusion of flowers. Many old familiar plants. ... Many unfamiliar birds – all colours. Mum, you would love it. All over the place there are magnificent trees – big trees, much larger than any of ours are common and they grow so fast. I counted the rings in one B.C. fir 27 inches diameter and there were about 80. The city is very spread out and you have to travel around quite a bit. We have a very convenient bus service. ...

The university is huge and this department fairly big. Well equipped but poorly arranged rooms. Very nice, generous and kindly people and very stimulating. I am sure I will get a lot out of this visit. And I'm sure Doris, Mother Grace and the kids will enjoy the town and country.

I am getting settled in, hearing the interests of everyone and finding my way about. The lecturing is not hard work though demanding in that it requires very simple explanations.

Will write again next week after Doris arrives.

All my love.
   Your son,
    Bob.

                     Seattle
                    July 5, 1949

Dear Dad:

... I got a telegraph today from Collip to say that firm recommendation has been made to the Board of Governors that I be appointed Professor and head of Dept. Also, Robinow is coming and is appointed Associate Professor. I am very pleased – have to do a lot of work to justify such confidence.

I am having a lot of fun and enjoy the people in this department very much. A very stimulating bunch and this trip will prove well worthwhile. The countryside is lovely.

Will write you more later.
   Your son,
    Bob

P.S. Doris arrived safely and seems to be well settled. She was a little tired on arrival but some good nights sleep have mended that. Everyone is well.

              University of Washington
                School of Medicine
             Department of Microbiology
                   Seattle 3
                   21 July 49

My dear Mum and Dad:

Your letters are a great joy to us and we love to hear of your expeditions to the lake and to other places. I look forward to hearing about the expedition to the Mt. Tremblant region. ...

The people here are very energetic souls and very interesting too. I have been learning a great deal of general bacteriology and technique but also have been able to contribute a certain amount myself. I have been working on an organism that Howard Douglas[91] found in mud that is more or less unclassifiable. It certainly belongs in a new genus, outside the *Enbacteriales* though god knows where exactly it should be put. In some ways it resembles *Hyphomicrobium* but is by no means a close relative. It is photosynthetic, and heterotrophic and reproduces by the damndest sort of budding. I have been largely occupied with growing it in slide cultures susceptible of the very best optical resolution and have been fairly successful. I have watched all stages of growth and there is no suspicion of fission.

Many other things interest me too. Cytochemistry is represented by several in the university. Evans is interested in cellular changes produced by animal viruses which ties in with my other work. And Ordal[92] is interested in Myxobacteria and the universe in general! In short I'm learning a lot. Certainly wish you could be here to help enjoy it – time is already half over and how time flies. If you were out here we could have enough room to put you up.

    Your loving son,
    Bob

Godwan River Estate
Ngodwana
E. Transvaal
10th August 1949

My Dear Old Biff,[93]

I have not heard from you in a long time, and hope you are all well. ... We had a spot of bad luck in that coming back from Nelsbruit on the 12th June. In the evening the lights of my car failed going round a bend on the mountain road and it was a case of either going over the edge or trying to keep the curve, which I did, but unfortunately there was a boulder sticking out of the side of the road which collided with my

---

91 Howard C. Douglas was a professor in the Department of Microbiology at the University of Washington, and while teaching there in the summer of 1949 Robert Murray and Douglas collaborated on research on the *Rhodomicrobium vanielli*. Together they published Murray and Douglas, "The Reproductive Mechanism of *Rhodomicrobium vanielli* and the Accompanying Nuclear Changes," *Journal of Bacteriology* 59 (1950): 603-615.

92 Erling J. Ordal was a professor in the Department of Microbiology at the University of Washington from 1937 to 1977. See some of his work on Myxobacteria in Robert L. Anacker and Ordal, "Studies on the Myxobacterium *Chondrococcus Columnaris*," *Journal of Bacteriology* 78, no. 1 (July 1959): 33-40.

93 LAC, E.G.D. Murray Papers, vol. 19.

rear side wheel causing us to fly forward. John and Roger[94] were in the back and they were shot forward into the front seat throwing Nita and I forward. John sustained a broken upper right arm, Nita a broken left fore arm ... and a cut forehead and concussion. Roger got off free and I only bruises. ... Luckily we were traveling slowly or it would have been worse. ...

I hope you, Freda and Robert, Susan, family are well. ... The world seems in a pretty good muddle and god knows what the outcome will be.

John and Roger are working hard on the farm and both want to be farmers, which is lucky for us.

I have drawn a rough sketch of the accident on the back of this page so you will understand how it happened.

  All our love to you all from us all.
    Your loving brother,
      Rory

            September 1st, 1949

Professor E.G.D. Murray
Dept. of Bacteriology
3775 University St.
Montreal, Que.

My dear Dad:

I was surprised to hear in a letter from Sue that you had already returned from your trip in the bush. It is too bad that you were not able to complete the trip you planned, but when the bush is so dry it is probably best to keep out of it. ...

I will be writing very soon to tell you of all the family events and I expect Doris will be writing to Mum.

  Your son,
    Bob

            London, Ontario
            4 September 1949

My dear Mum and Dad:

... The last four weeks in Seattle were lovely – both for weather and for activities – but unfortunately time went so very quickly.

---

[94] The sons of Everitt's brother Rory and his wife Unita (Nita) Murray.

There were several memorable expeditions and some very pleasant times spent with people there. Doris did, I think, write to you of our going to get a close look at Mount Rainier. It was lovely, the clearest day of the year without a cloud for most of the day. It stands in the middle of a very large national forest which is protected ... As you get nearer the views are a light green colour. ...

The Sound is a large body of water when you get out on it and is made quite complicated in structure by the innumerable bays in the enclosing point of land, the Olympic Peninsula. ... It seems to be the choice fishing of the state—except perhaps for the winter run of steelhead. The kings (salmon) run pretty deep and are caught for the most part with deep trolled herring or plug baits. ...

The summer steelhead would have required a great deal of patience to get and time to exert the patience. I was only fishing the part of the Stillaguamish once where they are known to be ... Apparently very different from the winter run where the fish are plentiful. One reason this year was that the snows of last winter were particularly deep and the rivers were high, cold (damn cold) and coloured. ... I wish that we could have stayed a week longer for they were plotting a trip to the mouth of the Columbia for when the Salmon run started when fly and plug fishing sport is hardly to be equalled.

Evans and I made a very pleasant expedition up to a small lake in the Glacier Peak region. Unfortunately we were delayed in starting and so did not get to the jumping off place till just after noon. Anyway you drive along the south fork of the Stillaguamish up to the headwaters over the pass and then down the head of the Sauk river to Elliot Creek. Creek is a misnomer for it is a rushing torrent and quite sizable. Absolute icewater to the touch. It is a good 5 mile walk in to Goat Lake in two tough uphill sections. ... The vertical component is somewhere between 2500 and 3000 feet. We made it in 2 hr. 20 min which I felt was pretty good going. You go through magnificent forest to get there and pretty well the whole way. It is especially fine about the middle of the trail where we saw dozens of western cedar of the order of 10 to 15 feet through the butt and a few monsters a great deal bigger than that. The western Hemlock was also in that region, mostly 5-10 feet through. The lower part of the forest was full of moss and ferns and spruce. These were not so big but they were second growth and very substantial. The trail follows the so-called creek most of the way. Clear and cold, and tumbling down in incredible leaps and falls. It makes a terrific noise. Up near the top of the trail there is a tremendous fall which you don't really see till you get to the forestry cabin at the tail of the lake. The trail makes a great switchback up the last few hundred feet (across windrows of trees knocked down by a snow slide last winter – bad going). Then quite suddenly you come out in the stunted stuff and the cabin is perched on a rocky point about 30 feet above the outlet of the lake. It looks quite a reasonable rapid at this point but you look downstream a further few feet and it disappears into space in a drop which looked well over a hundred feet. As we were watching this we had a beautiful view of a water ousel fishing in the rapids, in fact in the brink of the fall. Diving right into rough water, clinging and creeping on the bottom and then splashing clear completely happy. We were right above it and could see everything in the clear water. The climb was worth it for that alone. The outlet of the lake was one big log jam and we had to bypass that. A beautiful emerald green lake about a third of a mile long and half as wide surrounded by peaks and looking along its length a big mountain with snow fields and impressive rock crags. ...

While we were there a couple of quite hard looking men arrived – no fishing rods but armed with both rifles and 45 automatics. Seemed a lot of weight to pack up a mountain. Considering that most everything was out of season as far as hunting is concerned it seemed a little odd but hardly worth enquiring about. They said that they were doing a bit of prospecting. There is gold in the region (the abandoned mining town of Monte Christo is only a few miles away) and they pointed out high on the mountain ahead just at the side of a big snow field the mine tailings and (through the glass) a two story building. I would not care to make the walk before breakfast everyday. Time was getting on and we rushed down the trail. Even so got home pretty late at night ...

The summer's work in the lab was very satisfying and I think the most worthwhile experience I could have had. The people were so very nice and friendly and we all had a great time – although looking back on it worked pretty hard too. The lecturing was not a great chore and involved none the less a certain amount of reading for I have never known much about some groups. I had an interest in most of the research that was going on and of course had a particular interest in the Rhodomicrobium work. This latter was a lot of fun and provided very useful practice for the photo-micrography and growth studies I intend to do in the future – or am doing now. Howard Douglas was a great fellow to work with. Well, that work is complete for publication purposes and I am now, having written the paper, trying to get the photographs mounted in the right order. They look very good.

Well, we left with regrets. It was such a grand summer. The Friezes whose house we had returned the day we left. Everyone was most kind in getting us off and helping with baggage. Evans drove us down to the train. Everything was in order and our only worry was Peter – who had, as usual!, chosen that day in which to feel not [very well]. As a matter of fact soon after the train started he was quite drastically sick all over everything. However, he began to pick up after that and he was fine for the rest of the trip. The Empire Builder Train is a fine streamline affair and very comfortable. Mother Grace had Roomette and we had connecting bedrooms. We had fine weather again and got a good view of the Cascades and of the Rockies. Unfortunately it got dark just before we crossed the Columbia and I never got a look at that great river. It is amazing to watch the change in the mountains as you cross them from west to east. Forests and greenery on the west side suddenly displacing to brownish range and small trees on the east side – although around Wenatchee there is extensive irrigation and very rich farming. Especially fruit. The kids loved the journey, especially Peter with his passion for trains, and came through it very well even unto sleeping. It was perhaps a little tiring for the adults. We had a six-hour wait in Chicago and took a hotel room to make it more bearable. The most uncomfortable part was the Grand Trunk (which seems to have the roughest roadbed and engineers of the continent). More than that the train got in to London at 5.30 Standard AM.

I forgot to tell you of our trip to Vancouver. Went up by train because we were short of time but I wish that we had made the round trip via Victoria by boat. However, the train follows along the Sound and Sea for quite a long way and, since it was a very fine day we got a nice idea of the beauty of the coast ... the Fraser, which is certainly an impressively big river ... did look a bit dirty when we crossed it at New Westminster. We had quite a time getting a hotel room in Vancouver and eventually settled comfortably in a very old fashioned hotel – almost like a refugee from Baker Street in London. We had a good look at the UBC campus which is in a lovely place for a university. They are building at a great rate and hope to have their medical school going within a couple of years. The only person around the Bacteriology lab was Ranta and he was very kind in showing us around. We had a nice trip round the city and explored Stanley Park. In the evening we went to see friends. Back to Seattle the next day. It would be lovely to have the mountains and the sea and the fishing round you like that. ...

Once back here there was a great deal to do. It was nice to be back among our own possessions again despite the fact that we had a very comfortable nice house in Seattle. In that regard Doris had to unpack all the stuff she had put away before we left because of the decorations and stuff that had never been unpacked after our move. For me there was a lot to do at the lab. Dr. Bittner has gone on holiday and Laz and I have been doing the routine work. Laz has been writing his thesis – completed last Friday and sent in – and so has not been so available. We have also the planning of the course which will be started next week. I have also been writing a paper.

Robinow got here just before we got back and has been house hunting very hard. They have been lucky and found after ten days a stop-gap place that will take them 'till November 1st and just the other day found a very nice duplex with plenty of room which will be available on Nov. 1st. So they are settled and we have no more worries there. Mrs. Robinow and their two boys arrived last Wednesday and we had a hectic day getting them settled. Unfortunately the younger boy got his finger caught in a bight of rope over which the other boy tripped, amputating (not so neatly) the terminal phalanx of his left middle finger. This was rather a freak accident and I would not have believed it could happen. This blow was only partly softened for them by finding their apartment.

I was very shocked to hear of Freddie Smith's very sudden death.[95] He was still a young man and it must have come completely without warning. It will be very hard on Pat with a large and still young family to look after. I suppose that it will also pose quite a large number of problems for you and for the Faculty of Medicine. ...

Robinow is going to be a great asset. We are plotting an attack on the pleuropneumonia group and spent a very pleasant day yesterday working with some material I isolated from a vaginal culture. We want to do some pilot work leading to a real attack on the nature of the beast. Hope to discuss it with you soon.

The kids are very well and the summer seems to have done them a lot of good. Perhaps it is missing the very hot weather that is a good thing – certainly was for me – anyway they are in good health. Alice has started to go to kindergarten at the public school across the road and Peter has improved a lot and is sprouting a face full of teeth – may account for his fretfulness at the moment. He is standing and very active – so should be walking by his first birthday.

Doris is recovering from the rigours of the journey and getting settled into our new abode (which is working out very well) and should have recouped enough energies to write again very soon.

All our love to you both.
Your loving son,
Bob

> The University of Western Ontario
> Faculty of Medicine
> 2 Oct 1949

My dear Dad:

... I have finished the paper on the growth and reproduction of *Rhodomirobium* and have sent it to Howard Douglas (co-author) for him to look over. I would send a copy to you but the arrangement of the illustrations took so long that I would not have time to make them up again (there are 26 photo micrographs!). However, when Douglas returns the copy I will send one with un-mounted photographs. ... The paper is a pretty interesting one, I think, and certainly the work was a lot of fun. ...

... Phage is being a bit pesky at the moment and I have not been able to get the photographs I want of infected cells. We have a paper in rough draft form. In the most

---

[95] Fred Smith had a heart attack when he was welcoming the new class of medical students.

recent J. Path. Bact. there is a note by J.S.K. Boyd[96] on observation of phage infected cells using phase microscopy.[97] So I may send a note there to complement his observations. The paper on rotation should be in this month's issue of the J. Bact. I must write up some of the other more interesting observations on swarming.

Robinow is fitting in very well and is proving to be a good teacher and demonstrator as well as being very active in research. He is trying to finish up his work on spore structure. ...

We have been a little concerned over Peter's health. For at least 6 mos. and probably a year he has been having bouts of fever lasting 2-4 days and going up to 103-104 °F, occurring somewhere around once a month. ... The last time I took him down to the hospital and drew some blood – the odd thing is that everything is apparently normal and *no* elevation of the sedimentation rate. *No* disturbance of white cell count or differential. ...

The other kids are very well. Alice is full of fun and very much her old self. Thomas grows like a weed – is fat and very happy. He is well on the way to the walking stage – hair is still red! ...

I enclose the newspaper cutting re: my promotion and Robinow's appointment.
    All the best,
        Your loving son,
            Bob

                                Montreal
                             4 October 1949

Dear old Son,
  ... Thanks for the newspaper clipping; I'm glad to have it. I had a note from H.R. Dean (from Legard) & he asks me to congratulate you for him.

I have seen the questionnaire you sent & which I return. Some of the British bacteri-

---

96 Sir John Smith Knox Boyd (1891-1981) was a Scottish bacteriologist. After years of hygiene and pathology service in the British army, Boyd became director of the Wellcome Laboratories of Tropical Medicine in 1946 and remained in the post until 1955. Throughout his career he was particularly interested in dysenteric disease. L.G. Goodwin, "John Smith Knox Boyd. 18 September 1891-10 June 1981," *Biographical Memoirs of Fellows of the Royal Society* 28 (November 1982): 26-57; *ODNB* (Online).

97 J.S.K. Boyd, "Morphological changes in bacteriophage-infected organisms as revealed by phase-contrast illumination," *Journal of Pathology and Bacteriology* 61, no. 1 (January 1949): 127-131.

ologists are sniping at Bergey's Manual (C. H. Andrewes[98] & S.T. Cowan[99] are playing a leading part) & I think they are merely displaying an unreasoning prejudice based on an astonishing lack of knowledge of taxonomy.[100] We (the Editorial Board of Bergey's Manual) have prepared answers to each of these questions & are sending them to Cowan. They have stepped off on the wrong foot & are off balance. They are trying to whip up a following. The stupid thing is that there is nothing else to equal Bergey's Manual, with all its acknowledged imperfections which are largely a reflection of the loose & imperfect taxonomic work of the researchers themselves, & we are only too glad to have constructive criticism to improve the manual. I think but few have answered the questionnaire. We are quite prepared to let the Manual stand on its merit. ...

I had a good trip to the Columbus. The fishing was not easy, the weather was much too good & we only had a frost on the night of 30 September 1949. ... I caught fish every day, except one. ... I brought out some nice fish, though I could have done with a few more. Place after place reminded me of you & Mum & so many occasions I wished you were there. I fished alone most of the time. ... There were a number of good fellows there & I enjoyed the trip & feel much better for it but May seems very far off.

98 The British virologist, Sir Christopher Howard Andrewes (1896-1988), worked at the laboratories of St. Bartholomew's Hospital in London, England, and the Rockefeller Institute in New York, NY, in the early 1920s, before joining the staff of the Medical Research Council (UK) in 1927. He spent the remainder of his career working for the Council at the National Institute for Medical Research. He was appointed head of the Bacteriology Division in 1940, and deputy director of the Institute in 1952. His research interests included the spread of viruses, as well as their classification and nomenclature. He is most remembered for helping to discover the influenza virus. *ODNB*; *New York Times*, 4 January 1989.

99 The British bacteriologist Samuel Tertius Cowan (1905-1976) was educated at Manchester University. His most important contribution to the field of microbiology came in his role as head curator of the National Collection of Type Cultures (NCTC). He was appointed to this post in 1947, when the collection was housed at the Lister Institute at Elstree. Following its move to the Central Public Health Laboratory at Colindale, he undertook a major program of reorganization and rationalization which made the NCTC "of the foremost collections of human and animal pathogenic bacteria in the world." He also served as Director of the Public Health Laboratory (1961-1964) and Deputy Director of the Public Health Laboratory Service (1964-1967). *ODNB* (Online).

100 *Bergey's Manual* attempted to reflect the current knowledge about kinds and species of bacteria, and Sam Cowan, who was writing a book on diagnostic bacteriology at this time, was not happy with the way the enteric pathogens were organized. Robert Murray remembers that "even if it did not sit well with EGDM" Cowan made some valid arguments. However, according to Murray his father had his revenge, and Cowan his comeuppance, when the latter was elected a member of the Bergey's Manual Trust and an editor. C.H. Andrewes wanted Bergey's to start a section on viruses, which were not then classified except as names of diseases. He was later a principal in the development of a classification of viruses and got over his discomfiture with the *Manual*. RGEM.

Gertrude & Fred Kalz spent their holiday in Europe. They saw quite a bit of Ted & Sylvia who have enjoyed themselves this past year. I forget what G said Ted's plans are. I'll try to remember to ask her. ...

We hope you can make a trip to Montreal. It would be wonderful. There are lots of things to show you & discuss. I look forward to reading your rhodomicrobium paper. I have been watching for your other papers to appear. Try to let us know so that we don't slip off to the lake.

We are sorry to hear about Peter. Mum reminds me that when you were his age you got bouts of temperature & nausea, sometimes with vomiting & acetonuria. Barley sugar used to cure it. You used to be quite ill at the time & in between you were very well.

It is fine that Robinow is fitting in so well. He can carry a share of the teaching & routine, you will need to rely on some help in taking responsibility because you will have more than enough to do yourself. You must secure time for your own research work.

I expect to have a discussion with the Principal on staff requirements & replacements soon. I would like the opportunity of your visit to talk over possibilities. What about the young fellow you had & who is now interning? There was some talk of him coming here...

 Love to you all,
  Dad

       The University of Western Ontario
          Faculty of Medicine
            29 Oct 49

My dear Mum and Dad:

It seems a little late to be saying "thank you"! Even so I had a lovely weekend with you and enjoyed every bit of it. It was a great joy to see you looking so fit and well. Doris was most anxious to hear all the news and was sadly tortured to hear that I had a lobster at Desjardins! ...

The enrichment cultures of mud from the Columbus have been very negative for the non-sulfur purples except for one sample ... It has developed a nice growth of a Rhodopseudomonus and a few Rhodospirillum. No signs, however, of Rhodomicrobium. Might be worth another try later on. Douglas told me in his letter that he had managed to isolate *Hyhomicrobium* from activated sludge and that he is going to watch it reproduce using the phase microscope. ...

 All our love to you both,
  Your son,
   Bob

Montreal
30 October 1949

Dear old Son,

It was marvelous seeing you here & your demonstration was a great success. It stimulated my people & they were impressed by your talk. I hope you can bring Robinow when your teaching comes to an end; if you could bring or tell me what you want we would like another lesson. Meanwhile Mum & I are planning to go to London after the C.P.H.A. Lab Section in Toronto. We will fix the date & you could reserve a room in the hotel as your house will probably be crowded & it would add to Doris' work. We could still spend the time with you.

Please look in your notes for the name of the man who tied the Cohoe fly, "Buz..." It must be the devil to cast except with a very stiff rod taking a heavy line, such as the old fashioned double-handed salmon rods.

I shall send you some purified Yucatan Elemi in Xylol tomorrow & I expect you will find that many delicate stains will last very well in it. I know the Romanowsky stains (Giemsa especially) last very well in it, I have some dating from when Elemi first came into use about 1921. ...

I am reading Robinow's book of S Africa stories & will return it when I finish it. The photographs interest me most as I knew quite a number of the people & have seen others. ...

There are a couple of interesting papers by N.F. Stanley in The Australian Journal of Exp. Biology & Med. Sci. XXVII Part 2. March 1949 pp 123 & 133 on Listeria monocytogenes.[101] He has shown that the lipid extracted with chloroform, not clearly with ether or alcohol extracts, injected into rabbits causes the monocytosis. The semipurified protein & polysaccharide of the bug do not cause monocytosis. However he was not able to produce antibodies agglutinating sheep cells. I think there is something to be done with this but I have not yet thought out how. Stanley's procedures leave something to be desired. There are a number of complicating circumstances to be sorted out clearly. ...

I was talking to Munroe[102] who used to be a Master at Lower Canada & now is

---

101 In 1946, Neville F. Stanley (1918-1984) was appointed senior research officer at the Institute of Epidemiology and Preventive Medicine and director of bacteriology at Prince Henry Hospital, Sydney, Australia; he became acting director in 1948 and director in 1954. In 1956, Stanley was appointed chair of the Department of Microbiology at the School of Medicine, University of Western Australia, and director of microbiology at the Royal Perth Hospital. He published over 170 original works in his career. Royal Perth Hospital Web site, http://www.rph.wa.gov.au/emeritus/stanley.html.

102 Murray remembers Munroe as one of his teachers in his final year at Lower Canada College (but does not recall his first name). A 1933 school magazine shows that he coached some teams and wrote a history of football at the school and signed it "D.C. Munroe." RGEM.

director of the Education Dept at Macdonald College. He asked me to give you his regards & congratulations. He seems to be quite well these days.

  Love to you all,
    Dad

                    McGill University, Montreal
                          27 November 1949

Dear old Son,

I have not written to you for a long time but you have been in our thoughts constantly for many reasons. I saw your swarming paper in the J. Bact & received the reprints, I return one as you may be short & it is more than I need. I saw, too, your paper on Hyaluronidase in the Can J. Res. I am very glad indeed to see them & to appreciate a clear exposition of good work.[103] I look forward to the others you are preparing.

Ed Hall called on me the other day & I greatly appreciate the compliment. I was glad to show him round the lab, briefly, & to talk over various things which interested him. I think he recognized that a lot was going on & that it required good organization & adequate circumstances. I think you are in a good way to do good work because I am sure Hall & Collip will help you all they can. I gathered they do not want you to be overburdened with administrative worries & are anxious to let you make a reputation by personal research. I thoroughly agree & you will be wise enough, I am sure, to take full advantage of it by devoting all the time you can to your own research. I can assure you a big department is not an advantage & involves a lot of worry & work. Let it come later as the opportunity develops from your own work. In any case the greatest happiness & satisfaction is in research. I often look back to the days when I worked for the N.R.C. & wish I could return to something like it again.

I had to go to Quebec last weekend as a member of a subcommittee to look into the research possibilities of the laboratories there. We visited all the hospitals & all the University of Laval labs which could possibly do research on Public Health problems. Quite a job & rather astonishing to see how a bad organizational design can stultify things. There were only 4 labs in all which seemed to have any reasonable possibility to do research at all. We next have to see the labs in Montreal (Universities, Hospitals, P.H. etc) & then visit any possible outlying labs in the province. It will be a hell of a report to write. …

---

103 R.G.E. Murray and R.H. Elder, "The Predominance of Counter-Clockwise Rotation during Swarming of Bacillus Species," *Journal of Bacteriology* 58 (1949): 351-359; and R.G.E. Murray and R.H. Pearce, "The Detection and Assay of Hyaluronidase by Means of Mucoid Streptococci," *Canadian Journal of Research* 27 (1949): 254-264.

I received two cultures of Listeria monocytogenes via Ottawa from Alberta, one from a canary & one from a fowl; except for one isolated in our lab from an indefinite human case they are the only strains I know of in Canada. There are 2 most interesting papers by Stanley in Australia in which he shows that a $CHCl_3$ soluble lipid produces the monocytic response, but, in his hands it is not antigenic.

We have a pile of snow & fairly cool weather. Quite wintery. Douglas enjoys it with a "shubble" & I expect Alice & Peter would too. Perhaps you have had some too. We are greatly looking forward to seeing you after the Toronto meeting.

Love from us both to all of you,
Dad

Gertrude caught another Nocardia in a sputum of suspected T.B. It seems to be different species to the first one.

Roger Reed[104] caught a Leptospira icterohaemorrhagiae in a case at the M.G.G., almost missed it & would have if we had not told him to hold on to his g-pigs for the proper time. There must be more of them but the clinicians do not cooperate properly.

The University of Western Ontario
Faculty of Medicine
3 Dec 49

My dear old Dad:

I should have written a long time ago and, in fact, started a number of times without any noticeable effect. Somehow there has been a lot to do – especially with the teaching in the latter part of the term – the practical classes are rather exacting. However, the hump is over now but December promises to be bad – not only with Xmas and all but because the NRC stuff has to be in so damn early.

We are looking forward to the meetings and seeing you there. We are giving four papers and no doubt there will be a lot of interesting stuff from other parts. It will be a nice break in the usual daily round. It will be grand to have you come and visit us here. There are a lot of things to show you and Robinow and I are thinking of the things we would like to get out for your visit. There is so much really new in the department – we have been trying to straighten out some old problems but it has not been easy with the teaching as well. All the experiments that I have done in recent weeks seem to have been fizzles for one reason or another; the number of observations have increased but there is nothing new added. ... I have also been following

---

104 At this time, Roger Reed, son of the distinguished Canadian bacteriologist Guilford Reed, was bacteriologist to the Montreal General Hospital. In 1951, he was appointed professor of bacteriology at Dalhousie University, Halifax, N.S. *Science*, 11 May 1951.

up a problem Bob Noble has turned up in rats. When injected with a certain plant extract they *all* come down with a fatal "pyocyaneus" infection. The bug is not in the food, or the plant extract and I have not isolated it from the normal rats. ... Might be interesting.

Bob Elder was very glad to get your letter. It is good that you were able to get such a good salary – he was most pleased. He was writing to you and I hope he has. He is on rather a busy service and may have delayed.

We have got one of the Edwards freeze-drying machines based on the machines that Greaves designed during the war. It looks nice but I have not yet tried it. I hear that you are going to get one.

I saw Ed Hall the other day and he told me of visiting you in the lab and having a very pleasant lunch with you and Mum. He seems to have been very impressed with your set-up and to have enjoyed the visit with you.

I have had to go to a few students' and other functions which have taken up valuable evening time, however they have been quite fun. Doris and I went to the students' show the other evening and enjoyed it a lot. There was a lot of singing that was good and some beautiful dancing. A remarkably well sustained entertainment. Now that it is over the students will have some time to work – they put an awful amount of effort and time into it.

We had a visit from a Frenchman called Klein who is professor of zoology at Strasbourg. A very nice and lively fellow who is a real biologist. He knew a lot of the people that interest us and was able to give news of what is going on and notes on Tulasne and Vandrely; he also knew Boivin well.[105] Spent a very cheerful evening with him.

The winter came with a bang a couple of weeks ago and we had a very nice fall of snow. We have since had some more but there have been the usual messy melts in between. It is marvellous for the kids and they really enjoyed it. It would be better for us if it were dry snow for there is quite a drying problem every night. Alice and Peter are very happy with their schools and it seems to be agreeing well with both of them. Thomas has suddenly decided that it is not necessary to go about on all fours and now walks everywhere. He is very cheerful and apparently well and happy. At the moment we are doing some preparations for Xmas and Alice is having a great time writing her name on cards. She is getting quite good at letters and would learn

---

105 Murray remembers Klein (full name not known) as he told about his experience in a Second World War concentration camp. He was most likely a professor at the University of Strasbourg. R. Tulasne, M. Vendrely, and André Boivin were contributors to biochemical studies of bacteria and part of the effort at the time to discover more about the structure and content of cells, and all of them contributed to the field of bacterial biochemistry. Tulasne and Vendrely worked on bacterial nucleic acids and cellular biochemistry, while André Boivin originated studies of the endotoxin that is an important armament of Gram-negative bacteria. RGEM.

to read very quickly if they would get them going on it in kindergarten. However, the practice seems to be to wait and there is no sense letting her get too far ahead. The system is a bit rigid.

The Robinows have a very comfortable and nice apartment in the south part of the city and I think they have been very lucky. They seem to be happy here and enjoying life. He is certainly a grand fellow to have in the lab and also his experience in other fields has proved useful to other departments. He has a very great biological knowledge.

We will be going to Toronto on the 15th for we want to see what van Rooyen has laid out on the Viruses and we want to have some time with some of his people, particularly Angus Graham. I have reserved a room for you at the Hotel London – but if you would rather sleep here we would love to have you.

All our love,

Your son, Bob

Montreal
26 December 1949

Dear Bob & Doris,

I shall make good use of the dispatch case you thoughtfully sent me. It will help to keep some of my papers in order. Fortunately I do not have to travel now.

We had a quiet & happy day ... [with] Sue, Blake, Douglas & Dr. & Mrs. Robinson. They had most of their presents on their own tree but we kept some little things for them here just as tokens. Douglas had a good time & was tired by the end of the day.

Just before going home Dr. Robinson was not feeling well & on the way home had rigors & a rise in temperature. Later in the evening he was admitted to the Ross & I am told the spot diagnosis is "Virus Pneumonia", though I don't know how it's done. We hear he had a reasonably good night but nothing definite; there were no specimens at the lab last night or this morning. ...

Mum found me a Grenfell Cloth Windbreaker like my old one & Sue, Blake & Doug gave me a Buchley book as well as the spoke-shaves.

It was fine having a word with you on the phone but we should have waited till Sue & Co came. We did not think of it in time. However, we all drank your health at 2pm.

We hope to see you, Bob, with Robinow before very long & will try to make it a worthwhile visit.

Love,

Dad

London, Ont.
29 Dec 49

My dear Mum and Dad:

We have had a lovely Xmas with a feast of both food and presents. As usual, you people were much too good to us – though your kindness and the pleasure of talking to you in the morning did a lot to make the day for us. ...

The dinner was a great feed; even the children consumed their share. We drank your health on schedule and wished you all our love. A relaxed afternoon followed (!) aided by the fine cigars from Fudge (who showed excellent judgement in his choice of presents!). It was a long day for the parents and we were glad for bed! ...

We have had a lovely holiday and many of the pleasures are due to your kindness. We wished we could all have been together but we were fortunate to be able to see you the previous week.

    All our love to you,
        Bob

CHAPTER SEVEN

FAMILY CORRESPONDENCE: 1950-1951

Department of Bacteriology and Immunology
Prof. E.G.D. Murray
January 3rd, 1950

Professor R.G.E. Murray[1]
Department of Bacteriology,
University of Western Ontario,
London, Ontario.

Dear Bob,

I am sending you a culture of *Listeria monocytogenes* today. After some reflection I am sending the Australian strain N.F. Stanley[2] isolated from a human case of meningitis (Med. J. of Australia, August 21, 1948, p. 205) and used for the source of the chloroform soluble lipid producing monocytosis, (Aust. J. of Exp. Biol. & Med. Sci. 27 (1949) 123) and for a study of monocytosis in human cases.

We have separated three types of colony from the culture showing varying degrees of haemolysis. One selectively more haemolytic for human cells, one for rabbit cells and one for sheep cells, but all showing some haemolysis for all three kinds of cells. I hope Anne Masson will do some work on this.

The culture I am sending is the original received from Stanley as we do not know whether there is any partition of the lipid in the three variants. I send this culture as it is the one which provided Stanley with the original active material.

I will be very interested in what you and Rossiter[3] discover about the lipids produced by this organism.

I shall write again soon.

    Love,
        Dad

---

1 Unless otherwise specified the correspondence in this chapter is located in the UWO Archives, R.G.E. Murray Collection.
2 See chapter 6, note 101 for information on the Australian bacteriologist N.F. Stanley.
3 See chapter 4, note 109 for information on the biochemist Roger Rossiter.

>The University of Western Ontario
>Faculty of Medicine
>January 18th, 1950

Dear Dad:

The culture of *Listeria monocytogenes* arrived safely and I am starting it off today. Also arrived was a copy of "Round the Fountain" that I am very glad to have.

We have been very busy getting off N.R.C. reports and also writing papers. You will be glad to hear that the paper on the cytology of infection with T-2 is complete and will be sent off in a few days.

I was very sorry to miss your phone call on Sunday. It was unfortunate that I had rather more than usual to do that day.

I shall be writing again soon.

    Love, Bob

>London,
>Ont.
>22 Jan 1950

My dear old Dad:

This month has been rather a busy one thus far, but things are getting into order now and I think there will be a bit more peace from now on. ... Once everything was in full swing again I had to get down to the N.R.C. annual reports – the NRC seemed to have required a lot of effort this year due to applications just before and reports just after Xmas. ... While we were at it we wrote the stuff up. Fred [Heagy] completed two papers on the DNP work and between us we finished the one on phage cytology. The NRC will have longish reports to read this year but it saved some duplication of effort. I think we have got some pretty interesting work out this last year and I hope we will do better next. ...

I am just getting ready to get started on cytological work on *B. cereus* infected with phage. At the moment I am insolating the phages and noting their general characteristics. I had some good phage and I hope the work will be rewarding. With any luck we should be able to do better work than with *E. coli* B., but the spadework done on the latter will make the work much faster and more defined. ...

Robinow[4] is working hard on spores – especially writing a paper on his work with them. He should be finished that phase in a few weeks, to return to the vegetative cell. Unfortunately we could not get any liaison with the electron microscopes at Toronto (owing largely to the upset in the Physics dept there) and have now collaborated

---

4 See chapter 6, note 40.

successfully with Sweden! By airmail we got a return on the last set in just over 2 weeks from Hedén's Lab in Stockholm![5] ...

Have you started work on *Listeria*? I doubt if we can really get down to work until summer on the nature of the Lipid. As a matter of fact, the papers by Stanley are pretty bad efforts. But his general conclusion that a stimulatory fraction, probably Lipid, is probably true.

All my love,
Bob

London, Ontario
5 Feb 50

My dear old dad:

This last week or so seems to have been very busy – largely because of general duties and visitors, mixed with some social activities and trying to get some work done. I have spent last couple of weeks trying to get the B. cereus phages working properly and have run into the usual snags which are gradually clearing up. Although they produce good lysis they are somewhat slow. Their classification is being held up a bit because it is rather difficult to produce stable resistant strains – however, I am getting them out now. Robin[6] and I have also been collecting non-sporing variants of Bacillus species and have now got one polymyxa and one cereus. They are quite odd looking. Sometime we want to make a study of their structure and see whether they give any clue of the sporing mechanism.

We are still working with the pyocyaneus infection of rats. I have to make up some reasonable method of isolating and identifying possible Pseudomonads that do not produce pigment – in case there is a conversion going on. I am going to try a mineral base medium with benzoate as the sole source of carbon after the fashion of Dooren

---

5 Carl-Göran Hedén (b. 1925) is an internationally respected microbiologist at the Bakteriologiska Institutionen, Karolinska Institutet, Stockholm, Sweden. During his long scientific career he specialized in biotechnical bacteriology, microbial physiology, new approaches in the identification of microorganisms, and defence against biological weapons. He is currently a member of the Royal Swedish Academy of Sciences, and a trustee of the World Academy of Arts and Sciences. Robert Budd, "Biotechnology in the Twentieth Century," *Social Studies of Science* 21, no. 3 (August 1991): 435; World Academy of Art and Science Web site, http://http://www.worldacademy.org/.

6 Although the nickname "Robin" is used by Doris Murray to address Robert Murray in their correspondence, it is also used by Robert Murray when referring to Carl Robinow.

de Jong[7] and Stanier.[8] The batch of extract I have tried did not act as a selective agent for the organism. ...

Don Gillen,[9] who worked on the phage problem last summer, is trying to make up his mind on the training he is to follow. It is encouraging that he is certain that he wants to go into experimental work. What he is thinking of is something in the experimental pathology field but involving bacteriology. He is a damn good man and well worth encouraging. The trouble is to get him started on the right foot. He has no desires to be a tissue pathologist. I sort of feel that he might be best prepared by a couple of years in bacteriology, followed by any pathology training that seemed desirable from the initial experience. What sort of training do you think a fellow like that should follow? If things work out, Don would be an excellent fellow to follow along Bob Elder's footsteps – he would be interning next year and would be available a year from next June.

Kaney Ebisuzaki[10] came to me asking about graduate training in bacteriology. He seems to be serious about it, but deficient in bacteriological training. That being so I recommended that he write to you for he would show his capabilities in the Course III and get essential training. I explained that to him. I never intended that he be considered for graduate work straight off. I have spoken to several of the people who teach him. They think that he is worth-while material. I am going to take him on for three months in the summer and I will find out about him then. He is no great use to us as a graduate student because he needs so much general training in bacteriology – for which we have neither the facilities nor the staff to teach him.

Have you any potential M.Sc. student for us next year? We have discussed this before, but by now you should have some idea of the potential and plans of the course III students and others. If you have none I will write to a lot of departments. I would like to get a capable individual to work with Robin. Robin, Fred Heagy[11] and I will be

---

7 The Dutch microbiologist L.E. den Dooren de Jong worked in the laboratories of Albert Jan Kluyver (1888-1956) at the Technische Hogeschool at Delft, the Netherlands.
8 See Introduction, note 510.
9 While he was a medical student, D.H. Gillen worked with Robert Murray on cytology of phage infection. After interning he moved to E.G.D. Murray's lab and worked under Gertrude Kalz for over a year doing medical bacteriology. He eventually set up a medical practice in St. Catharines, Ontario, with a particular interest in allergic diseases. He died only a few years ago. RGEM. See R.G.E. Murray, D.H. Gillen, and F.C. Heagy, "Cytological Changes in *Escherichia coli* Produced by Infection with Phage T2," *Journal of Bacteriology* 59 (1950): 603-615.
10 Murray remembers that Kaney Ebisuzaki thought about formal training in bacteriology while he was working with Robinow for a number of months on *Metabacterium polyspora*. He eventually completed his doctorate on bacteriophage genetics in Boston and then returned to Western to work in the Cancer Research Laboratory directed by Alex McCarter. When that lab shifted into biochemistry so did Ebisuzaki. Like Murray, he remains active at Western. RGEM.
11 See chapter 6, note 46 for information on Bob Murray's graduate student Fred C. Heagy.

going to the SAB meeting in May. I certainly hope that you will be able to go, for it would be grand to be with you there. Robin and I will probably be going to the histochemistry meeting at Philadelphia at the end of March – where Robin is giving a paper. They are forming a new society and I want to go down as an observer and see whether it is worthwhile for us. It may well be, for it will bring together people from all the biological sciences and should have a broader basis than most of the meetings.

This week I had a busy couple of days with the visit of Robert Chambers (the microdissection man).[12] Collip[13] had to be away and I had to make all the arrangements etc. It was a busy time but a lot of fun. He is a grand old man and is a most interesting and stimulating character. He gave an excellent lecture for the staff and graduate students on permeability of cells, and showed his films in an evening meeting for all comers. I wish he could have been here for an extra day. He is going to be at McGill sometime next week.

Last week we got all dressed up for the annual faculty party. It was a lot of fun, if non-alcoholic, and we had a long but enjoyable evening.

Doris has been plagued with tooth trouble – one wisdom tooth started to act up last Sunday. I fear that there is an abscess there and Frank Kennedy[14] feels that he should take the brute out. That would probably be best since there is every possibility of recurrent trouble if it is left there. The cellulitis around the tooth has calmed down now and she will probably have it out tomorrow.

Young Thomas is progressing fast. He is adding some words into his vocabulary and is making stout if messy efforts to feed himself. He is the same shape as ever, perhaps a little more so. The rest of them are in excellent form.

I suppose that you have been getting the current respiratory infections. We have had one cold run through the family just recently but since it is the only one of the winter we are not too alarmed. It is a fairly hydraulic variety.

We look forward to hearing from you soon.

      Your loving son, Bob

---

12 From 1928 to 1947, Robert Chambers (1881-1957) was a research professor of biology at New York University specializing in cell physiology. M.J. Kopac, "Robert Chambers: Micrurgist and Cell Physiologist," *Annals of the New York Academy of Sciences* 78 (June 1959): 403-404.
13 See Introduction, note 317-318, and chapter 1, note 56.
14 See chapter 4, note 16.

> The University of Western Ontario
> Faculty of Medicine
> 8 Feb 50

Dear Dad:

Here is the M.S. of the paper on phage cytology which may interest you. It is only the draft but it happens to be the only copy I have on hand at the moment so send it back as soon as you are through with it. Some of your people might be interested in this application of cytological methods.

    Your son,
        Bob

> London, Ont.
> 10 Feb 50

My dear old Dad:

... I finally got one of the *Cereus* phages to behave and ran an experiment on him this week – ever since then I have been rushing to get over the material while it is in good condition. The cytoplasmic study was not satisfactory – altogether – but the nuclear changes are very interesting and different to the T2 phage. Does not seem to correspond to the effects noted by Luria for the other T phages. ...

... Robinow has gone down to Providence R.I. to attend a celebration of his father's 70th birthday. Great family do. One of his brothers in the U.S. Army managed to wrangle an air-hitchhike back from Europe for the occasion!

Did you notice on the back of the Abstract form for the SAB meeting that I am put down as an example of how to put your name and address?! ...

    All my love,
        Bob

> McGill University, Montreal
> 11 February 1950

Dear Old Son,

I like your paper on the Cytology of Phage Infection very much. The only serious suggestion I have to make is, on p12, that you delete the last sentence, ending with "The interpretation of these observations is discussed". It seems to me an open question with the probability that the redistribution of chromatin is actually phage chromatin.

A minor suggestion is that Eschemalira (or any Genus name) be spelled in full, or if abbreviated do so to avoid confusion. "E Coli" could be Escherichira or Entamoeba.

So I suggest using, as I do, "Esch." & "Ent." "Sal.", "Staph.", "Strep" etc. Where are you publishing this? ...

Love,

Dad

P.S. Did you receive the thesis (L. Loeb)[15] sent by registered mail.

Montreal
12 February 1950

Dear old Son,

For the past ten days we have been laid low with a curious form of cold. Sue & Blake were knocked out at the same time. Mum & I are not quite over it yet & it is difficult to do what has to be done of necessity.

Anthrax is a nasty thing to have about & in a department doing hospital work you cannot allow it, but its sporulation can be controlled by O_2 & by temperature of growth & would perhaps be more instructive than non-sporing variants could be. ...

If Gillen wishes I could reserve a place for him in a year's time & he could get all there is to offer in this lab. Duff[16] seems to be developing experimental pathology in his department. Some of his people work with bacteria or immunological sensitivity to produce lesions without quite good enough bacteriological training. However it is a step in the right direction & it might be a good place for Gillen later on.

What is Kaney Ebisuzaki? It sounds like Japanese. I wrote to him, as you know, & it can develop from there. If you think he is good enough he could do Course III next year & if he proves good enough proceed further.

I have not looked into the potential students in Course 3 as I was hoping you & Robinow would be coming. There is a good girl who got 1st class in Course 3 & has done a year under Gertrude in the Clin Div & was a year with Fred Smith[17] to no great good & has also done the course in medical mycology at Duke Univ. She is a clever girl, presentable & hard worker & would like to work for M.Sc. at least & might be tempted to a research job. Elizabeth Seale[18] is her name.

The histo-chemistry meeting seems a good idea & might be just what you want. Well worth looking into. Chambers seems to have given you some fun. I have not

---

15 See chapter 6, note 48 for information on Bob Murray's graduate student and later allergist Lazarus J. Loeb.
16 See chapter 2, note 84 for information on Lyman Duff, dean of the McGill Medical School.
17 See chapter 1, note 87; chapter 4, note 43; and chapter 6, note 95 for information on Fred Smith.
18 See Elizabeth E. Seale, "New Fungistatic Drugs," *Canadian Medical Association Journal* 65, no. 6 (December 1951): 582–584. The article places her with the Department of Bacteriology and Immunology, McGill University.

heard he was coming here & have not been invited to meet him. I would be sorry to miss him.

I think I told you we found that the $CHCl_3$ extract of Listeria does cause monocytosis. Stanley wrote that he is coming to the USA & he hopes to come to see me in March. I hope he does not come while the City Council Budget meeting is on. ...

I'm sorry Doris is having trouble with a tooth. Those apical abscesses have not been properly looked into. I have found a couple which were pure bacterioides & there might well be some of other causes.

The Grosbeaks arrived yesterday at lunchtime. Quite a large flock settled on the box-elder tree & looked very fine. It seems early for them to come, but we have not kept a record of the dates. I should start tying some flies.

 Love to you all,
  Dad
PS. I must buy a new pen.

London, Ont.
20 Feb 50.

My dear old Dad:

I am glad you like the paper on the cytology of phage infection. I sent it to the J. Bact. nearly a month ago and, bar knowing it was received, have heard no more. I fear some referee may be sitting on it. I agree about generic names and their abbreviations but Editors have their policies – the generic name being spelled out in full the first time used in the text. Your comment on the redistributed chromatin is probably not justified – redistribution occurs within two or three minutes of infection even when the cell is infected with only a single phage particle (as is probably true when only 20% cells are infected). After the 7-10 minute period some of the visible chromatin *may* be phage chromatin; before that it is practically impossible. ...

I have not seen Gillen again to talk too seriously about prospects since I got your letter. I will be seeing him this week. He is a good man, stable, and interested in experimental work. He can't make up his mind which side (Path or Bact) he wants to concentrate upon but is leaning heavily towards bacteriology. I feel that only a little more experience will decide him. My advice is to plan on spending at least two years in bacteriology and not to go making decisions until at least one of those years has gone. He will intern for a year first – I expect he will want to come to you and will let you know. I think he is a good man. ...

Collip tells me that Sir Henry Dale[19] is going to be visiting here the 16-18th May. This is a most unfortunate date being the time of the SAB meeting. He may be in Montreal soon after. ...

    All the best to you both,
        Your son,
            Bob

London, Ont.
26 Feb 50

My dear old Dad:

... On Thursday I went to Toronto with Roger Rossiter. He was going to see Angus Graham[20] at the Connaught about the set-up of a radioactive-tracer lab so I thought I would go along too. Saw quite a lot of the people. Also heard a seminar by R.A. Fisher on the genetics of the Rh factor – very interesting. Donald Fraser[21] looks very well and says that he is going to the Laurentians to ski – coronary or not! Saw van Rooyen, Rhodes and a number of others for short time.[22] Saw Ted Roy[23] in the distance but did not get a chance to speak to him. The Sick Kids Hosp new building looks pretty good from the outside and they seem to be well along. ...

Saw Don Gillen yesterday and talked with him about possible plans for the future. He seems very attracted to the possibility of working in your lab starting in June 1951. He would then be following a year behind (Bob) Elder.[24] He is definite that he wants to do experimental work more on the applied than fundamental side. I think he would make a good man. ...

    All the best to you all.
        Your son,
            Bob

---

19 See chapter 5, note 110 for information on Henry Hallett Dale (1875-1968), British physiologist and pharmacologist.
20 Angus Graham worked with van Rooyen at the Dufferin Division of the Connaught Laboratories.
21 See chapter 5, note 9 for information on the bacteriologist Donald Thomas Fraser (1888-1954).
22 See chapter 6, note 61 for information on van Rooyen and chapter 5, note 109 for information on Rhodes.
23 See chapter 1, note 89 for information on Ted Roy, a close Murray family friend.
24 See chapter 6, note 55 for information on the bacteriologist R.H. Elder, who studied under both Bob and Joburg Murray.

University of Western Ontario
375 South Street,
London, Ontario
March 9th, 1950

Professor E.G.D. Murray
Dept. of Bacteriology
3775 University St.
Montreal, Que.

My Dear Dad:

I wrote to James Gibbard[25] recently to find out what, if anything, had been done about the formation of a Microbiological Council. He says he has not done anything and it looks as if he won't be able to do anything in the near future due to other jobs. He indicates that I should go ahead if I have any ideas.

I think some useful spadework could be done – even remembering that letters rarely get good response, which was your experience. The thing is to find the key people in the various fields, even where they are not represented by societies. The Phytopathologists seem to be the only ones with a definite organisation and they are definitely interested, because Professor Walker[26] asked me recently if anything had been done yet. Tom Cameron could give some idea of the parasitologist population in Canada. In fact, I think the only way to get the thing moving is through selected individuals. I would like to know what specific individuals you think would fill the bill. Also, what specific points do you think should be made to these individuals?

I have just been speaking to a friend who is in charge of the fermentation side in a local winery. He says most of the people on the industrial side would be represented in the nutritional and industrial sections of the Chemical Institute of Canada.[27] There is a Brewmasters Association (affiliated to a U.S. organisation) and a Winemakers Association, but I gather they are fond of golf and other pursuits![28] He is going to give me the names of three or four people who would be most interested in such a Council.

I look forward to hearing from you.

    Your son,

        Bob

25 See chapter 6, note 36.
26 A. R. Walker was a senior member of Western's Botany Department at that time. He was a respected phytopathologist and was active with Robert Murray and Norman Gibbons in the initiation of the Canadian Society of Microbiologists. RGEM.
27 The Chemical Institute of Canada is a professional association of chemists, chemical engineers, and chemical technologists working in industry, government, and academe.
28 At the time Bob Murray was acquainted with Henrik Schoenfeld, the winemaker for the now-defunct London Winery. RGEM.

Department of Bacteriology and Immunology
Prof. E.G.D. Murray
March 11th, 1950

Dear Son,

I am sending you my file on the proposed Canadian Microbiological Association. I am encouraged by you interesting yourself in getting such an organization moving and I think you might well succeed. The attitude of various opponents has changed and I think you will get more support than I did. However, you must expect that many, even of those who are sympathetic, will not take the trouble to do anything or even answer letters.

I owe you several letters and will try to write this week-end.

Best of luck.

375 South Street,
London, Ontario
March 16th, 1950

My Dear Dad:

Thanks very much for sending the file of correspondence concerning the Microbiological Council. It is a lot of help. I am just trying to find out who would be the key people interested enough to push the idea within their own group. Once a few of these are identified the job should gain momentum. Please let me know of any organised groups you may hear about.

Some time ago you said that a Miss Seale might be interested in Graduate work, perhaps at Ph.D. level. From what you said about her, I suppose she is a pretty competent girl. Is she still interested? It is too late in the year to get a fellowship or to get money from the University so that she would have to pay her own way. ...

Hope to hear from you soon,

Your son,
Bob

Montreal
19 March 1950

Dear old Son,

I have a pile of letters from you which I have not answered. I have so much to answer that some will, no doubt, be missed out.

I enclose a snapshot that was clicked by Kate Waugh[29] while I was unsuspectingly

---

29 Kate Waugh worked in E.G.D. Murray's department. RGEM.

signing papers. I have also located the snap of Fred Smith & me taken by Max Sauer years ago. ...[30]

I see your paper on Rhodomicrobium in the J. Bact. & it has come out very well.[31] I have not read it carefully for want of time. I suppose the paper you mention Sherman accepting unaltered is another one, probably the phage one. You have some very nice work out & coming out & I am delighted. I probably misread the point you corrected me on concerning the rearrangement of chromatin.

The problems of fixation should prove most interesting. Apart from the question of selecting methods which give the minimum of distortions from the natural arrangement (which is not properly avoidable) there is merit in making true comparisons between various methods, each of which may reveal something worthwhile.

I look forward to learning about the nuclear membrane. I would think one of the acid blues (water-blue, china-blue, methyl-blue) would be valuable, intensified by acetic acid. Remember that though these blues differ little chemically they differ greatly in their selectiveness as histological stains & each has uses. Very beautiful differentiation of achromatic structure can be shown with them in protozoa & other animal nuclei. ...

We are still waiting for our freeze drying machine & could use it to advantage. Everything ordered from Britain seems to be snagged in some way these days. Even filters from Gallenkamp. I got quicker service from Germany on some good filter membranes which are proving to be what we wanted. ...

Are the plans about rearranging the lab still to the fore? It is essential & will take up a lot of your time while being carried out. The larger & more active your department becomes the more of your time it will drain. Some you cannot avoid as Head but don't make my mistake & let it dominate your own interests. Not only is it important to do your own research but it is essential to happiness, for the other way is just slavery. I believe it is possible to insist on providing a definite proportion of time at a fixed part of the day sacred to your own work & do what you can with suitable help with the rest. Universities & hospitals these days have become bloodsuckers. They must pay for service & benefits just as others have to & knowledge & experience must become price-able property. Scientists have given far too much for nothing.

I ordered the symposium on Bacterial Surface a while ago. ... I will try to get Science News 14.

You & Rossiter seem to have had a good trip to Toronto. Butler[32] is a very good man indeed & a clever protein chemist. A good man to know & a nice fellow.

30 Max Sauer, the Montreal-based photographer.
31 R.G.E. Murray and H.C. Douglas, "The Reproductive Mechanism of *Rhodomicrobium vanniellii* and the Accompanying Nuclear Changes," *Journal of Bacteriology* 59 (1950): 157-167.
32 "Butler" most likely was someone who worked at the National Research Council of Canada, which according to Murray is "probably why my father knew about him." RGEM.

You have not said any more about going to the histo-chemistry meeting in Philadelphia this month. I hope you can do so. We (Mum & I) look forward to seeing you at the S.A.B. in May. I don't know if Chambers came here. If he did I did not hear of it.

I suppose you received Laz Loeb's Thesis; I sent it registered & Paterson keeps waving the receipt at me to know if he must keep it. You have not mentioned it in your letters. The papers on it will be most interesting.

Don Fleming[33] has resigned from my department & is going over to Public Health, with Vivian[34] ("Health & Social Medicine" or some such name) as Associate Professor. He holds that rank with us. He is interested in continuing as Secretary of the Faculty of Medicine & feels he can do more good with part time in Public Health than he can in Bacteriology. I think he is correct.

I don't quite know what scheme to put up to the University to provide a replacement & I have not tried out what they would consider. I fear there will be some difficulty & not the least will be to find a man. I wish it had been delayed a year as Elder might be good for the teaching job when Gillen comes for the Clinical job. I have an inkling the Principal will want to do it on the cheap & I shall have an argument on my hands.

I think I wrote to you about Gillen. I shall reserve the place for him in 1951; a young pathologist is coming in July for one year. I have not heard any more about Kaney Ebisuzaki but probably will not until later in the summer.

N.F. Stanley of Sydney Australia, who extracted the monocyte stimulating lipid from Listeria, visited us last week. He is a good young fellow & we enjoyed his visit. He hopes to get back to some more work on the lipid but is now having to take an interest in viruses. ...

(Charles) Evans[35] wrote to me offering me a similar job to that you had last summer. Very nice of him, but I feel too tired to take on summer teaching & I don't feel I have much to offer them. I wouldn't like to disappoint them.

We had a troublesome respiratory infection run through the lab a couple of times during the winter. It was a curse & left people feeling rotten. All seems well again now. Mum & I each had a couple of goes of it & it left us washed up.

Douglas spent a happy 3rd Birthday on 17th March. We went to see him & had an amusing hour. He is very well & very bright.

I am glad you still play Badminton as it is very good for you & you must keep fit, even if it does take up a little time.

---

33 See chapter 3, note 123 for information on the medically trained bacteriologist Donald S. Fleming.
34 See chapter 5, note 18.
35 See chapter 6, note 76 for Charles Evans, chair of the Department of Microbiology at the University of Washington, Seattle.

I have not tied any flies this winter & I did intend to replace most of my stock. I also meant to introduce some neglected old patterns. I must organize my time rather better.

Mum & I went to see "A Tight Little Island". It is a marvelous & hilarious film & if you have not seen it do not miss the first opportunity.

Love to you all,
Dad

The University of Western Ontario
Faculty of Medicine
22 March 1950

My dear Dad:

We were most pleased to get your long letter yesterday. I am writing a hurried note for I will be setting off for N.Y. and Philadelphia tomorrow and have a great deal to do between now and train time.

I was most pleased to hear that Evans had offered to have you out there for the summer. I think that you are wrong to argue that you would find it tiring or that you would have little to offer them. On the contrary. I found that it was a most restful summer because it gave us a change – I am sure you would find it the same. It gives you a real opportunity to see the country and have a close look at the fishing – N.B. it is only a short trip to Vancouver Island and to other places of real fishing interest.

They do not require much of you and you can suit things to the energies you have. I am sure that Evans would really like to have you there for a while – not to make you work for him but to get your advice on various matters. I know that he is very concerned with getting started on plans for a first class clinical lab and the advice you could give him on the spot would be of real value to him. He has a most interesting department with really good men (young) who would appreciate having a senior man come in and talk things over with them. They do not make offers lightly and they are all discussed very carefully by the whole group before they are made. ... You and Evans and others would be off fishing at odd but frequent times. Mum would love the place – and the sight of real green grass, mountains and trees would be like a transfusion to her and for you too. Robinow agrees with me and insists that I tell you that you have a lot to offer. They are grand people – it is a pleasant city and you can get to a lot of interesting places from it. Please Go!!!!

During the past week I have been playing with blood and have found a very nice correlation with our ribonuclease. We will probably be writing it up. It is nice to have the parallel chemical and cytological data. ...

We are all well. Will write again when I get back from the trip.

All my love, Bob

Montreal
23 March 1950

Dear old Son,

A very beautiful book has arrived from you and we are enjoying it. "An Album of the Chalk Streams" by E.A. Barton. The photographs are splendid & the letterpress most attractive.

In "The Woman's Home Companion" for April 1950, on p 32, there is an article, "Menace in the Medical Labs", you should read. It is written sensationally to stir up mud & argues from the particular to the general in journalese, but there is merit in places. It is something you can use to emphasize that the hospital must pay to provide well trained personnel & proper space & equipment. We cannot get the right people because we cannot pay properly & also overwork them. ...

In haste, Love to you all,
Dad

Montreal
9 April 1950

Dear Old Son,

I sent you an old polyvalent dysentery "phage" which may not be alive. The phage classes in Course 3 have been dropped & I have not yet reconstituted them as they must be. Meanwhile the Sonne phage cannot be found; Freddie [Smith] used to look after those classes & when he altered the course while I was away he may not have kept the cultures. I have a collection of phages & will find them when I get my things straightened out a bit more.

I gather your trip to Philadelphia was quite a success & that you are glad to have been to the places you visited. Gertrude[36] is looking forward to visiting you & I am glad she is going.

Don Fleming resigned in order to join Vivian's Department ("Health & Welfare"); it seems to be what he wants to do. He ... realizes that with only 1/3 of his time available to the lab he cannot get very far in bacteriology & it is not of great value to the lab. Now Joan de Vries has resigned too. She wants promotion & more money, neither of which can I promise & neither of which is she quite worth out of proportion to others. However, I am in a difficulty & at present see no solution to it. I am not hopeful of a satisfactory solution as the Dean has been to ask for cuts to be made in the budget & now the Principal has called a meeting of all heads (Chairmen) of departments to discuss the budget. An unprecedented & bad sign.

36 See chapter 3, note 101 for Joburg's colleague Gertrude Kalz.

University salaries are out of line in relation to cost of living by comparison with industry, street-cleaners, etc. There seems to be little realization of what is likely to happen & how important it is to get good men & keep them in the Universities. I don't see how to keep the department up to maintain such position as it has gained during 21 years of very hard struggle.

I hear the Cambridge man who went to Saskatoon is dissatisfied there & wants a job. Do you know anything about him? I am told Collip was interested in getting him placed. You might enquire & let me know. I might be able to offer him something here.

It is going to be the devil to have the lab in working order next September. ...

We are greatly enjoying the book you sent me, "An Album of the Chalk Streams" by Dr. E.A. Barton. It is delightful. The photographs are magnificent & the letter-press intriguing. I wonder what will happen to "The Chalk Streams" where so much development in British fly-fishing was effected. Some famous part, I believe, has been flooded under a reservoir.

I bought Mum a 3 piece Hardy "De Luxe" rod. She is very pleased and itching to use it. We are always afraid of damaging her "Grandpa Dunne" rod traveling with it. This is 4-3/4oz & balances very nicely & feels good in the hand. ...

W.H. Cook[37] wants me to go to Ottawa for a couple of days to discuss some of the problems of his division in the Nat. Res. Council Labs. I have offered to go between 18th & 22nd. I don't know quite what it involves & it may well be more than I can do. Mum will go with me for fun & a change.

This coming week is pretty full up & ends with a City Council meeting, which I hope will not last more than one day, but there is no knowing as it involves the recent "Montreal Bill" in the Quebec Parliament as well as a lot of routine resolutions.

Gertrude is writing a chapter on bacteriology for a book on surgery the surgeons here are producing. She makes me criticize it, which takes time, but I am glad to help so that it may be reasonably good. They want her to cover ground that should rightly be in their chapters but if it is read it may well do some good. The surgeons we find are least informed on bacteriology & make no effort to improve. The book will have a local sale, like Oertel's "Pathology" & for the same reason, but I don't know whether it will be good enough to attract a wider field. Anyway Gertrude is putting a lot of experience into it.

---

[37] The food technologist and biochemist William Harrison Cook (1903-1993) worked in the Applied Biology Division of the National Research Council, and was appointed director of the division in 1941. He recorded his experiences at the NRC in his book, *My Fifty Years with NRC, 1924-1974* (Ottawa: National Research Council of Canada, 1977). *The Canadian Encyclopedia*; Institute for Biological Sciences Web site, National Research Council (http://ibs-isb.nrc-cnrc.gc.ca/events/75th_history_e.html).

Do you know "The Cytology & Life-history of Bacteria" by K.A. Bisset (Birmingham) 1950. I would like your opinion of it.

What are you planning to do for a holiday? If you are coming our way we would like to arrange things suitably.

  Love to you all,
    Dad

The University of Western Ontario
Faculty of Medicine
14 April 1950.
17 April

My dear old Dad:

... The staff troubles you are having must be very discouraging and give you a lot of work to do. It is too bad that the University will not allow replacement at proper salaries. Probably the worst part will be to find suitable people even if the jobs are there to be filled. You will be having considerable difficulty keeping up with the teaching if you don't get some help. ...

Talk about both rearrangement and extension of the lab is very much to the fore. Unfortunately it is hard to get it beyond the phase of talk, and concrete plans are noticeably absent. Despite this I have been putting quite a bit of time into it and most of the right people are persuaded ...

We will have quite a lot of work to get done this summer but much of it should run smoothly for a lot of the pilot work is there and it will be a matter of following some well defined leads. We will have 3 students (paid by N.R.C.) – one for each of us. I also expect an M.Sc. student and, later in the summer, a Ph.D. student. It will require a bit of energy at first to get the ball rolling but it should be a considerable help in the work. We look forward to seeing Gertrude and she should get some interesting material to think about. We have lots of ideas and a very few facts but it should be a stimulating summer. ...

Bisset's book is probably a bad one but I have not seen it yet. I have had it on order ever since I first heard of it but it hasn't come yet. I am prepared to bet it is most misleading and probably will be a bit upsetting to Robinow (to say nothing of myself!). If you read his papers it is quite evident that he has not been sufficiently critical in his evaluation of observations and has made some (if not many) unjustifiable conclusions from poor material. The list of chapter headings sound like a program of work for the next century. ...

I saw the article in the C.M.J. and it reads very well.[38] Please ask Gertrude to send me a couple of reprints when they come. It is a pity that Foley didn't get going quickly on his part of it, for it should have been out long ago. I am still writing papers and have the micrococcal antibiotics stuff in final draft and pretty well ready to go off. I will be glad but then there will be others to worry about. Rossiter and I have to get to writing the blood work soon if not sooner. ...

Doris had a very nice birthday and enjoyed herself a lot. We took an easy weekend and had a good time. The kids had a great time getting her a little present each. Your presents were very much appreciated. We went out in the evening and had a Smorgasbrod and some drinks. ...

I took Alice to see Cinderella last week. She loved it and talks about it. Sat on the edge of her seat and laughed and clapped – it is an excellent film and very funny. I think it is one of the best Disney films. It is doing a great trade in kids young and old.

I will write very soon and try to keep up with all the things I would like to tell you both. ...

  Your son,
    Bob

                Montreal
               16 April 1950

Dear old Son,

I hoped to hear from you & I hope it is only that you are very busy & chiefly with research. I see Heagy's is out in the last J. of Bact, though I have not read it yet.[39] It is hard to find more than moments & at night I find myself rather weary. This past week I have had late meetings every night.

We are going to Ottawa on Wed 19 until Sat 22. I have promised to view the Biological Division of the N.R.C., discuss their work & comment. I only hope I can give them some useful suggestions. It is quite likely their work is out of my line.

I would like to know about [Guy] Richards who is at Saskatoon.[40] I have an idea he was at Cambridge about your time but I think you told me he did not take the trip. On the other hand you might be thinking of him for your clinical lab job.

I have not worked out my problems yet & so do not know what "the authorities" might consent to. I shall have to make an issue of the situation soon.

---

38 G.D. Denton, G. Kalz, and A.R. Foley, "*Staphylococcus folliculitis* (Pemphigus Neonatorum)," *Canadian Medical Association Journal* 62, no. 3 (March 1950): 219-228.

39 Fred C. Heagy, "The Effect of 2,4-Dinitrophenal and Phage T2 on *Escherichia coli B*," *Journal of Bacteriology* 59, no. 3 (March 1950): 367-373.

40 See chapter 6, note 51.

We have planned to have a septic tank installed at the lake towards the end of May. It should put the privy business in order permanently & safely. ...

There is a Dr. Giuseppe Penso,[41] Professor of Bacteriology in Rome, who has been giving a series of lectures in l'Universite de Montreal which I am sorry I have not been able to go to. They tell me he is very good. I met him a couple of weeks ago & he is a nice fellow & I would judge very able. He speaks French perfectly without any Italian accent, but he is an Italian. He tells me Chain[42] left Oxford & is now Professor of Biochemistry in Rome; quite extraordinary. ...

What is your housing situation? I thought you had the present house only for a year.

We expect to hear any time that our rent has been put up & if they go to the limit allowed we will have to move if we can find a place; 18% rise will be too much for us.

The weather is fine & the maple tree opposite the house is breaking into bloom. Various birds have arrived & it will be marvelous to get away to the country. I feel I could do with a rest.

  Love to you all,
    Dad

        The University of Western Ontario
          Faculty of Medicine
           27 April 1950

My dearest Mum:

You must admit it is a rather rare thing for me to remember birthdays etc., but, with a little conscientious help from Doris, I have managed at last. We hope that you will have a nice birthday (weather and all) and that Dad will give you a drink and a nice feed. We have sent you a little something and I hope they don't fall in the class of useless duplicates! Keep a nice time! I wish we could see you – but anyway I look forward to seeing you at Baltimore. (Don't forget to let me know *when* you expect to arrive there and *where* you will be staying – I will be there on Sunday and at the Lord Baltimore (headquarters) Hotel).

I am having a very busy life – not only is it busy now but promises to get worse with

---

41 Giuseppe Penso was a member of the Department of Microbiology of the Instituto Superiore di Sanità in Rome, Italy.

42 The German-born biochemist Sir Ernst Boris Chain (1906-1979) immigrated to Britain in 1933. That same year he joined the Department of Biochemistry at Cambridge, and in 1936 he accepted an offer by Howard Florey to join the Sir William Dunn School of Pathology, Oxford University, where he was appointed university lecturer and demonstrator in chemical pathology. Chain, H.W. Florey, and Alexander Fleming shared the 1945 Nobel Prize for Physiology or Medicine for their discovery and development of penicillin. *ODNB* (Online).

sticky uninteresting interludes of examinations, meetings etc all around the time of the SAB meeting. Doris, though busy, is well and happy – so are the kids. We all look forward to a little fresh air and sunshine when the weather stabilizes and warms up – I hope this will be soon. I am a bit sick of this semi-winter. ...

Must rush.

 All our love,
  Bob

London, Ont.
10 May 1950

My dear Mum and Dad:

 Well, it was quite a shock to hear that you are going to be moving at the end of the month and I sort of wish that I could pay a quick visit to Montreal to see the last of the old house and pay my respects. The University seems to have become singularly rapacious of late years – I agree with Collip who thinks it is the sort of action that can be expected of them. On the other hand I think perhaps, if the new place is as nice as Mum's description, that you may be pretty comfortable and that Mum will certainly have a lot less arduous housework to do. Since I will be seeing you next weekend I don't really need to go on.

 I seem to have an awful amount of paper work to do at the moment and with exams coming up it will get worse instead of better.

 Unfortunately I will have to cut my visit to Baltimore a little short since I have to [be] back here to conduct some examinations and get them out of the way. The following week Gertrude is coming and I want a little time with her before I go away again for a week. Roger Rossiter and I have decided that we have to have a little peace and relaxation before long and we are going to take the canoe and go off for a week. It should be fun as well as a much needed relaxation for both of us. We will probably go to Algonquin Park again and will leave directly after the faculty meeting and convocation on June 3rd.

 We have booked through to Baltimore and will be arriving on the afternoon of the Sunday. We have a room in the Lord Baltimore Hotel and you should be able to track us down there alright. I have no idea when to expect you or where you are staying. We will talk about plans and news when we meet.

  Your son, Bob

Montreal
19 May 1950

Dear Old Son,

This is your birthday & Mum & I are so glad to have seen you so recently. The traveling interfered with this little present & good wishes reaching you on time but you know our thoughts & love are with you.

It was a great pleasure to us to be at the meeting with you & the nice things said about you by so many of our old friends gave us great happiness & pride. However, we enjoyed your paper & discussions on their own merit. May you have many years of happy success.

It is good news that you are going on a short trip in Algonquin Park & you will be glad of it & be much the better for it. We hope you will not allow anything to put it off.

Doris & the children were no doubt glad of having you home today. We hope you found them flourishing. To us this is always an important day & one which floods with happy memories, to which every year adds more to be proud and glad over.

With our dearest love & best of wishes,
Mum & Dad

The University of Western Ontario
Faculty of Medicine
1 June 1950

My dear Mum and Dad:

It has been a devil of a busy two weeks and I have only that excuse for not writing long ago. I certainly should have written to thank you for your very kind and flattering birthday greetings. I was very pleased and touched. It was grand also to talk to you on the phone. Best of all was the change to see you both and spend a little time with you at the meetings at Baltimore. It was a busy time but I was very glad of the opportunity to be able to see and talk to you both.

When I returned from the meetings I had to go straight to marking papers and the like. I ... got it all finished and turned in very soon after Gertrude arrived. About that time both the University and the hospital got in a hell of a dither about the necessity of giving more space to the labs. ... So it was that I had to (*very* regretfully) cancel our projected trip to the Algonquin. This was most disappointing. But it is a good thing that I did not go away for we have had three very important meetings since then and decisions are being made. We will probably be getting about 1000 square feet for ourselves next to our present quarters. ... It has been a most irritating series of meetings, although somewhat successful, because of diplomatic necessities. However, some important concessions have been wrung out. I think my neighbour to the east is getting very much more than he should but as long as I have what I need I don't care too much.

It was very nice to have Gertrude here. I think that she enjoyed herself and she was certainly stimulating to have. She and Robinow worked long hours and we had some good discussions. She will I think be a great help to you in the course 3 teaching and may be able to interest some of the staff as well. She is a sterling character. The kids loved her and I wish she could have stayed a little longer.

We have been very well and the weather being kind to us, the kids have been in the open most of the day and seem to be thriving. The garden has been a great pleasure and has been very beautiful during the last week or so with apple, pear, peach and lilac blossom out in every direction. Alice has planted some radish, lettuce and carrot which seem to be coming up in spite of onslaughts by Tommy and the birds. ...

The cheque was most useful and you should not be so generous!

Love from us all, Bob

The University of Western Ontario
Faculty of Medicine
15 June 1950

My dear Mum and Dad:

... Both Peter and Thomas had been a little off their oats during the previous week and had run fevers of no very great significance. Well Alice looked to be blowing up much the same on Sunday when we went out to the Nobles. By Monday she was coughing, had a conjunctivitis and fever looking more and more like measles – which was fully confirmed yesterday by a fine rash. She was rather proud to get them at first but now she is quite sure that she does not like the measles at all. She is a rather sick poor thing. The sad thing is that we took her to that party on Sunday and I fear that she must have exposed several small children – however, we have let them all know and they will be filling them up with globulin. I was the other victim – woke up on Monday morning aching in every joint and a suspicion of a sore throat. By noon I was feeling terrible and spent an afternoon of rigors and acute misery, especially of the joints and muscles. We got started on penicillin and salicylates which didn't seem to help much till Wednesday night. Oddly enough my throat got worse as my fever dropped. Today I feel well again, and a good thing too since other complications have arisen. ...

This was not the only problem of the week (additional) for Doris, who is in the second month of incubating our fourth, started to have intermittent bleeding and it looks more than likely that the pregnancy is terminating itself. In that case she will have to go into hospital for three days or so (may be as soon as this weekend) for a D&C. Though she is not actually ill at the moment she is not feeling too happy and needs moral support and not a house full of crocks! Last night her brother John, who was in Detroit on a consultation, dropped in for two hours on his way back to New York. This was very nice for Doris since she has not seen him for some eight years. However, he could hardly have picked a more confused time and things would have been more organised if he had come for the birth of any of the youngsters! ...

If Doris has to go to the hospital we will get in a Red Cross nurse who does everything for the children from 8am to 7pm. In addition Mother Grace is probably coming here next weekend or soon after. She will be staying in the house next door so will not have the problems of former years. She is no longer active enough to be able to look after things unaided but could be of great help in taking out one or two of the children for an hour or two.

Things have been going on at the lab while I have been laid up and everything is alright. The only problem is that I want to get the boys who are here for the summer properly established and working for the time that I may be away on holiday. I am getting very behind in correspondence and other rather routine things as well.

This is a sort of bulletin of the moment. There are really many other more cheerful things to talk about but I must rush to other things and will write again as soon as I can.

    Your son, Bob

The University of Western Ontario
Faculty of Medicine
17 June 1950

My dear Mum and Dad:

The crisis is pretty well past. Doris went into hospital last night and was in the O.R. this morning. Although I have not seen her Helen Rossiter,[43] who gave the anaesthetic, phoned me up this morning after the event and told me that she was fine and that all was well. It has been a bit of a blow to Doris but I am sure she will be well very soon. Alice has got better – as is usual with measles she was flat as a pancake one night and the next morning was sitting up and demanding attention. The others have been given globulin concentrate and we hope will get a mild attack so that we may be free of the measles in the future at any rate.

---

43 Helen Rossiter was trained in medicine and anaesthesia at Oxford University, which is where she met and married Roger Rossiter. They came to Canada in 1947 when he was appointed professor of biochemistry at Western. She gained Canadian licensure and practiced anaesthesia for many years until she became the University Physician looking after student health until her retirement. RGEM.

I have been holding the fort for the time being. Doris' mother will be arriving tonight which will allow me to get out a bit. Things will work out alright. It has been hectic but the worst is over. I am all better myself and seem to have my streps under control for a bit.

Will write again soon.

 Your son, Bob

The University of Western Ontario
Faculty of Medicine
28 June 1950

My dear Mum and Dad:

... Since Thomas and Peter seem to be recovering from their *very mild* attacks of measles – all should be well by the weekend. We have therefore come to a decision about traveling and will be leaving here on Wednesday 5th July. ... This time we should not have too much baggage but *please* let us know *immediately* if there is anything we should bring up with us – e.g. blankets etc.

Considering everything we have come through our siege very well. ...

Glad to have a long letter from Sue. They sound well and happy. She said you had accomplished the move without too many hitches and that it was about time you caught your breath and had a rest. I can quite believe it.

 All our love,
  Bob

The University of Western Ontario
Faculty of Medicine
Sunday August 7th 1950

My dear Mum:

Although this should have been written sooner this comes with all our love and thanks for a fine vacation. It must have been a trial to have so many mouths to feed: a task well done for every one of us, at any rate, have put on pounds. ...[44]

Since getting back it seems to have been hectic for me – a lot of work, a return of the sore throat now being treated heroically even if I am still ambulatory. I am feeling fine today and hope to keep so. ...

Not many startling things have happened. It is a little sad that Jim and Peggy

---

44 This refers to a summer holiday at the Murray family cottage at Bark Lake.

Thompson[45] will be leaving in a few weeks to go to the Univ. of Alberta at Edmonton. We will miss them.

I hope that the mucous membranes are behaving themselves now and that you and the whole family will be free of plagues.

    All my love,
        Bob

                The University of Western Ontario
                        Faculty of Medicine
                            7 Aug 1950

My dear old Dad:

Having returned to the grind again it is a little hard to work out the fine vacation you gave us with more than a detached view. I am brown and fit, which is a great deal more than I was before. The trips and expeditions with you are always of the greatest pleasure to me and I hope you enjoyed them as much. We must invent a simpler way of getting out those strong (if thin) worms. ...

I was very glad to see you looking so much better when I left. You were so much in need of change, rest and exercise. I hope you can plan things so that you and Mum can get away on a proper trip. ...

I will write again in a few days, for I have a few other things I have to tell you about. The work seems to have gone well but there are some highly confusing things.

    All the best of love ...,
        Bob

                      1569 Pine Avenue West
                                Montreal
                      (Phone Wilbank 2279)
                            8 August 1950

Dear old Son,

I meant to write yesterday & your letter of 7th arrived this morning. The cheque seems too much & Mum will have to decide about it. I don't know what her figure was. I know you thought it was low but you must remember that living is much cheaper as

---

45 Robert Murray and his family knew Jim and Margaret Thompson very well while Thompson was an assistant professor of anatomy at Western. Margaret was trained as a geneticist and Murray recalls her as "a good one too." He obtained a professorship at the University of Toronto and Murray believes she went to the Hospital for Sick Children in Toronto where she also had a distinguished career. RGEM.

we do it at the lake than it is in town.

I am very distressed you had another strep throat. You had better test its susceptibility to various antibiotics & then take a thorough going course of treatment. It might be well to have a course of vaccine. I would do it this way: Kocto-vaccine 1000 x 10^6 per cc boiled 15 minutes; make dilutions & do intradermal tests & start subcutaneous dose at the quantity just giving a definite reaction; doses doubled at first & then increasing more slowly at 48 hour intervals for 6 doses to a maximum of 1cc. Then formalized vaccine subcutaneous doses at intervals of 5 days say 200 x 10^6 ranging up to 2000 x 10^6 depending on reaction. This has been successful sometimes. ...

I think you all looked ever so much better at the end of your holiday; particularly you & Peter most noticeably. We are very happy to have had you with us. I loved our trip to the Columbus & it did much for both of us. I wish we could do it when fishing is better than in July. We must try to think out some scheme for it.

The only thing we found which seems to be yours is a pair of socks ... I am enjoying "Trout Problems" & I am very glad of the other useful birthday gifts.

Mum & I went to Pike Lake to try out her new rod. She loves it. It is a beautiful little rod with perfect action & amazing power. Its weight & balance suits her to perfection. I must change the snake-rings for good bridges & get her a better line than that damned nylon one. The fish ran small & we kept only one – very strange, but we had a good time. I hope we can manage a short trip together.

I am clearing up tag-ends of things & there is a City Council meeting next Monday which makes a difficulty & September is very near at hand. This summer seems to have gone in a rush with little time for anything – the moving messed us up.

 Love to you all,
  Dad

London, Ont.
21 Aug 50

My dear Mum and Dad:

I have taken so long to write after my short note soon after we returned that you must wonder if we are still around. At any rate I have not had any more trouble from my throat and the streps are no longer to be found. A combination of penicillin and sulfadiazine seems to have done them in. ...

We have been very busy in the lab and there are many more problems than solutions. The phage work, although extremely interesting, is extremely frustrating. ...

We have not looked into the housing situation yet but will start very soon. I want to learn a bit about the tricks of the trade first because I don't understand how many of these things work. It is most awfully kind of you to offer to back us up – I think that, despite the present level of real estate we would be wise to have our own home and put money into it as an investment. What I would like to know is how much you are *sure* you could lend us and over what period of time. I am thinking in terms of perhaps 2, 3 thousand dollars. I think the payments on mortgages run around $8 per thousand per month and we can't really go beyond about $100 a month for rental or for payment of mortgages and loans. ...

Rosy[46] leaves tonight for N.Y. on her way to England. She will be coming back via Montreal and hopes to look up you people. She has your address and says she will write from England.

All our love to you all.

Your son,

Bob

29 August 1950
The University of Western Ontario
Faculty of Medicine

My Dear Dad:

I don't seem to have heard from you for a very long time but maybe it is partly my fault for not writing too much myself. We have been somewhat busy. There seem to have been many small dislocations and at the moment I am finishing up work in preparation for the teaching term, which seems very close. On Saturday I am taking off for New Haven and the Congress for Cell Biology which should be a lot of fun. I shall be taking various photographs of our work down with me for there will be a number of interested people there. Our main preoccupation, other than teaching, has been to get the research work into good shape for less intense but more critical experiments between now and Xmas.

The cereus phage work is becoming a little more intelligible and I think that it should be in fair shape soon. There are some knotty problems to clear up but they only require well designed experiments and some of these have already been piloted. It looks fairly certain that all the cereus phage we have obtained (5) do much the same things to the nuclear structure. But one pair of phages, which are very closely related and apparently will reproduce concomitantly in the same cell, produce a different set of changes when the cell is infected with both. This confused us for quite a while.

We have failed to grow Metabacterium but there are still a number of possibilities to be tried. I am sure that it is a strict anaerobe but it has some requirement that we

---

46 Rosy Robinow, the wife of C.F. Robinow.

don't understand. There has been no sign of germination of these enormous spores in the medium.

The Japanese boy who wrote to you about graduate work, Kaney Ebisuzaki, has decided to go to Wisconsin for his MSc work. Largely because the biochemical work he is interested in is going on there. He may come back for a PhD later.

This rail strike is a most damaging and annoying happening. I certainly hope that something is done soon but I fear that the govt has pussyfooted around so much that a proper solution is not too likely.

We have not got our 60 cycle power yet – even in the lab and we were supposed to be changed over in June. This is a nuisance for I have got one piece of 60 cycle equipment and would like more. I expect that the rail strike has held this up too.

I enclose the report of the Department of Microbiology at Seattle that Evans sent to me. I thought that you might be interested in it. They are a pretty active group. Please return it when you are finished. ...

All my love,

Your son, Bob

Department of Bacteriology and Immunology
Prof. E.G.D. Murray
August 29, 1950

Dear Bob,

I think your draft memorandum on a Canadian Association of Microbiologists covers all the important points. I don't think it would be wise to make it longer because it might not then be read, but you might add that comments and questions would be welcome. ...

When the Memorandum is circulated, I shall canvas everyone here. Don't forget to send it to Armand Frappier (U. of Montreal),[47] J. Edouard Morin (U. of Laval), and H.M. McCrady (Provincial Labs, Que.).

I think any suggestion to expand the Laboratory Section into a representative society is out of the question. It would be opposed by both Public Health people and by microbiologists other than Medical.

Once it is started, it will gather in the lethargics and also the opposition, such as now remains, will die down.

Best of luck,

Dad

---

47 See chapter 5, note 48 for the Quebec physician and microbiologist Armand Frappier.

McGill University,
Montreal
30 August 1950

Dear old Son,

I should have written to you before & we have been glad of your letters. I answered the one about the Can. Microbial Association quickly & I shall answer the one I received today later when I have read Evans' Report. I am glad you are going to the meeting on cell biology. Now to answer your letter of 21 Aug 1950.

It was a delight to us to have all of you with us, as you know. I only wish I could have given you some worthwhile fishing; however you had a good rest which I am sure did you great good.

I am glad your throat is not troubling you but it would be well to do repeated cultures at intervals to watch events.

Look into the house question immediately as there is no sense in delaying & it only makes the paying off more troublesome. I have talked it over with Mum & we are anxious to help as much as we can. I am sure we can lend you $3000 for down payment without any interest. The time we shall need money is in 5 years when I have to retire & by then you ought to be receiving more salary. You might enquire from the University about a mortgage. McGill takes a mortgage for staff at a much lower rate than they can get outside & I am told of several people who have availed themselves of it. Perhaps Western may do something like it too. I think you must act at once because 5 years pass very quickly. I have your life insurance I took out for you when you were a baby so that it has a low premium of 10L a year for 1000L & it has its paid up bonuses attached. I could have it made over to you & you could cash the bonuses & perhaps borrow favourably on it. You could get a sound financial friend to advise you how to arrange the whole thing & failing that you might well find your bank manager a source of useful information & advice. There is always Hilary Robertson[48] too who would tell you who to ask for advice on a scheme.

I am sure you will find it can be done cheaper or no dearer than paying rent & you have your property in the end. There is also the government housing scheme which helps advantageously & is worth enquiring about as it offers a very low rate of interest I am told. Something reasonable must be possible.

We shall be delighted to see Rosy [Robinow] on her way through Montreal. Mum likes her very much indeed. She could stay with us.

Your observations on the B. cereus phage are most interesting & the combined effect is extraordinary & an achievement in discovery. I have not tried anything with metabacterium; I shall try later. I don't know what your bacteroids-like organism

---

48 See chapter 3, note 2.

may be. Strange things are found when looked for & I think many are overlooked because most people ignore everything unknown. A lot depends on methods too.

Mum hurt her back badly last week & was crippled. Some displacement went back with a snap (audible) last weekend & she was better but still in much pain & discomfort. Sue with Ruth Elder caught a 15-3/4lb lake trout, to their delight & huge excitement. It was as big as Dougie ... Douglas caught his first fish, a small Bass, last weekend & he was very delighted. He ate it with relish.

I was kept in Montreal by City Council & we have another meeting called for Sept 1st which will be another long one I fear. Then teaching starts on Sept 8th & other things are looming up. I don't yet know how to get a few more days holiday. We had hoped to get away, above Lake Munroe, for 10 days or 2 weeks before Sept 1st but the special meeting of the council scotched that scheme.

Gertrude is just back & had a fine holiday in California but a difficult journey back from Chicago (via Malone!!)

Love to you all,
Dad

10 September 1950
Montreal

Dear old Son,

Your P.S. from Yale is good news & I am sure the meeting has proved valuable to you. It is splendid that you are able to go there.

I have not got round to the things of interest to you raised in your letters. Three of our technicians are leaving for more money & just when we are becoming most busy & their training is most required they have to be replaced by untrained people. It makes it difficult to keep things going. A young Englishman walked in the other day, name of Jacobs, who has three years training in British army bacteriology labs & has the Intermediate Lab. Assistants Certificate, he has also worked 2 yrs with Klotz at the Ottawa Civic Hosp. and a year at Hamilton. I think he would exactly fill the vacancy of a Senior Technician ... but I am sure he won't accept because the wage is too low. He has been getting $180 a month & my job is only allowed $100. I fear I may lose others yet. It is hard to run a good place on finances below economic standards & the "Administrators" do not & will not understand the situation. I am weary of it. Is this man of any interest to you? ...

Mum was to have come back yesterday for a couple of days but some-how was not able to make the train. She was to have gone with me to the Canadian Veterinary Medical Association Banquet. It was a friendly & quite pleasant dinner & I have quite a number of friends among them. ...

Burnet[49] is going to come here to give a lecture on [the] 23rd on the Epidemiology of Poliomyetitis & the story of paralysis following vaccination ... as experienced in Australia & England. Can you make it an excuse to come for a couple of days. Our spare room is in good order & Burnet is worth hearing & the vaccine story is important.

A few more names of people you should approach regarding the Can. Microbiol Assn: Dr. C.A. Mitchell (Animal Diseases Research Inst., Mountain Rd. Hull P.Q.),[50] Dr. Genest (Veterinary Research Laboratory, St. Hyacinthe, P.Q.) Dr. A. Bertrand (Notre Dame Hospital, Montreal). I am only thinking of people in this province & it is important to work on Quebec. I think I already mentioned [H.M.] McCrady & T.W.M. Cameron.[51]

Robert Breed & N.R. Smith[52] are coming here Tuesday on Bergey Manual business. Breed has just returned from the Rio Congress & there seem to be some important matters of policy to discuss.

Cecil McDougall has given me his light specially made Chestnut Canoe. He has had it a number of years & has given up portaging & thinks it would be useful to me. I have no doubt it will ... It is at his place at Lake Manitou & we are arranging that I go up with him & then he drops me with the Canoe at Caribou & I take it through Caribou, & Green to Bark. We are waiting to get a mutually suitable week-end. At present my lectures on Fri, Sat & Mon. are in the way but they will soon be over. ...

V.C. Wynne Edward[53] phoned. ... He has just arrived back from Clyde River & says he had a wonderful time there & the expedition was a tremendous success in every way. I expect we shall hear about it at one of the meetings of the Arctic Institute later on. He is going back to Aberdeen tomorrow by air.

I have had the other two ... pipes repaired. The better one is the Negro Head & I am sure you would like to have it. The colouring & the carving are good & it is much admired by Blatter the pipe maker. ... I have had it since 1908 & it would be hard to get one like it now.

    Love to you all,
      Dad

---

49 See chapter 4, note 10 for the distinguished Australian microbiologist Frank Macfarlane Burnet.
50 See chapter 2, note 36.
51 Institute of Parasitology, MacDonald College.
52 See chapter 3, note 138 for information on Robert Breed and chapter 5, note 97 for information on N.R. Smith.
53 See chapter 4, note 59.

Montreal
17 September 1950

Dear old Son,

Our medical classes are well under weigh & I have given the first 5 lectures. There are troubles which are quite serious because technicians are leaving as their salaries are too low & they are right. I can't do anything & will have no trained staff other than Turko Salo by the end of the month. I am sick of the struggle. ...

Vollum of Oxford should be here this week (about Wednesday) & it will be interesting to talk to him. There is a splendid paper by Kairno, Smith & Vollum in the last number of the J.A.M.A. on Tuberculosis Meningitis, which you should read.[54] Kairno was here recently & lectured on it very impressively & very modestly. There is a move to do the same sort of work at the Alex & I hope to reopen the lab there very soon. I have a bacteriologist but I don't know where to get technicians.

Mum is still at the lake & I hope to be there for the last week of Sept. I do not intend to go to the Columbus. We will get some fishing round & about. Mum writes that there are still large numbers of blackberries; we have had great amounts of them this year & Mum has made some wonderful jam. She is annoyed at two boat-loads of Jews who come day after day to fish just off our wharf. She has seen them get one or two good fish as well as small ones. Anyway their continuous presence annoys.

The RCAF put on quite a show with jet fighters & bombers today. The prize effort was the "F.86" Jet which I saw do two runs over the city, quite low, at what the radio said was just under 700 MPH & then turn up at an astonishing sharp angle & fly up out of sight in a very few seconds. Amazing & I had a fine view of it.

The lab is getting busy & will be very much so soon. ...

  Love to you all,
    Dad

The University of Western Ontario
Faculty of Medicine
24 Sept 50

My dear Mum and Dad:

... The house hunting has gone slowly because I have been so busy getting class started although we are started on a "looking" program. The sort of house we will really want in the future seems to be a bit out of reach but with patience we should

---

54 R.L. Vollum was professor of bacteriology at Oxford University. He supplied the anthrax strain 14578 used in the Gruinard Island anthrax tests in Britain during the Second World War. Geoffrey Holland, "United States exports of biological materials to Iraq: Compromising the credibility of international law," http:// deepblade.net/journal/Holland_JUNE2005.pdf.

find something approaching it in our price bracket. The trouble is that we would like four bedrooms which splits one between the old house and the newer expensive models. As a matter of fact they are *all* expensive. 3 bedrooms are easier to find and we would settle for that if there were attic or other space that could be converted into usable room. We shall see.

I have not had time for much real work myself – although I am in the process of back-checking some of the summer's results. The most interesting thing occurred in the last week. I have Phil Fitz-James[55] working on the biochem. of spores. The first problem to solve was how to break them up for extraction and analysis. Grinding in usual ways proved relatively ineffective. He found that you can do a complete job if you grind them *dry* with quite coarse alundum. ... The job is complete in 10 minutes in a cold mortar with an ordinary pestle. ... This should be useful for any non-pathogenic organism ...

I have been doing some general reading that is proving useful to me for its stimulation value. In particular the new (2nd) edition of Burnet's "Production of Antibodies". This is a most refreshing slant on immunity and gives the subject some biological perspective. ...

I hope you will be able to give your Course 3 some phage work. It is important stuff to understand nowadays and is well adapted to class work. Useful for illustration of principles.

All the best of love.

Your son,

Bob

London, Ont.
October 6th, 1950

My dear Dad:

... How do you mark your glassware? You have always had a stamped "Bacteriology" on glassware and I want to do the same. Too much expensive stuff seems to disappear and I want to minimize the loss – or, at last, be able to spot the glassware.

About Burnet – the date you gave us was, I think, the 23rd of October. This seems to be the week after the Med-Chi meetings so I wondered if I am right. I hope I may be able to come along.

Yours aye, Bob

---

55 P.C. Fitz-James (1920-2006) was one of R.G.E. Murray's early PhD students. He obtained his degree in 1953 and subsequently became a distinguished microbiologist and biochemist. He held appointments in both the Department of Microbiology and Immunology and the Department of Biochemistry in the years 1953-1987. University of Western Ontario, Department of Microbiology and Immunology, http://www.uwo.ca/mni/faculty/mourns.html.

McGill University, Montreal
10 October 1950

Dear old Son,

I have owed you a letter for a long time. We were delighted to get your long letter of Sept 24 & got one of Oct 6 today. The former arrived just when I was off to the lake to join Mum for the last week of the fishing. I decided not to go to the Columbus though both Bill Newman & Hilary Robertson each offered me a drive up.[56] I have heard since that the fishing was found to be not very good.

Sue & Doug were still at the lake as Blake was away east on a month long inspection trip & they returned to Montreal on Sept 26th. Mum & I spent the remaining days fishing. Our first trip was to Twin L. for the day. The trout were not grouping & not taking freely but we got 20 by chasing rises about the lake. A nice day & good fun. The fish ran small but nice to eat, we kept seven. ...

We then returned to Montreal. There were a host of troubles at the lab as we cannot keep technicians at the wages paid by McGill. Graduate students had to be started & registered. A number of other things had to be arranged & set up.

This past weekend (Thanksgiving) was wonderful weather, like late summer. On Saturday we hired Edward Miller & his truck to fetch the canoe Cecil McDougall has given me from L. Manitou. ... Yesterday we went to the end of the lake for lunch & caught one bass on a plug over the reef, the only touch we got, about 1-1/2 lb. We had a tiresome journey back to Montreal on a very crowded train. The bass season closes next weekend but we cannot go up for it. So, the fishing is over.

I hope we can get in a couple of weekends before freeze-up. We must get in one to close the cottage.

The phage filtrates are very old & came from l'Institut Pasteur. The host strains have been lost. I thought you might pool them & try them wild on anything you are interested in on the chance of picking up something useful.

We mark the glassware with fluoric acid & a rubber stamp. It was necessary at one time & would be useful now but everyone is too busy to do it. It was quite easy to do.

Burnet's lecture is a Special Meeting on Oct 23rd. It will be fine if you can come, but let me know a day or two ahead. He will talk on Polio & on the Australian & British experience with whooping-cough vaccine & Diphtheria toxoid predisposing to Polio. ...

I am glad Peter's T&A went off well & I hope it will put things right for him. It is good that the other children are well. I hope you are free of your strep.

The dry grinding with alundum is extraordinary. We will try it with the lactobacillus from which we are trying to get cholinesterase.

---

56 See chapter 3, note 2.

I have the 1st Ed. of Burnet's "Production of Antibodies", I don't know if I have the 2nd, but if not will get it. I forgot to send you "The Biology of Infectious Disease", but I have made a note to do so.

I do hope you can come here for a couple of days.

  Love to you all,
    Dad

London, Ont.
16 Oct 1950

My dear Mum and Dad:

Time has certainly been both hectic and busy – with little time to do the many things I should do such as letter writing. It was very nice to talk to you on the phone and most kind and generous of you to help us so much in the matter of house-buying. It will undoubtedly be a turning point for our finances – the continual drain into rents without possibility of return was a rather alarming prospect for some years to come. As it is we are going to be in the position to have some assets (in the form of a negotiable property) and can repay you at the same time.

The cheque arrived safely today and we will be ready to do the main financial deal at the end of this week. Meanwhile I have a lawyer working on the mortgage, title searching and other tricky parts of house buying. ... I would like to ask what method you would think best for managing your loan. 1. I can send you a cheque every month 2. I can set up a savings account here and add to it all the time, or 3. I can buy savings bonds or do any thing you direct. Please let me know. ...

The house is the normal yellow brick rather than boxy house – improved a little in appearance by pinned-back-shutters.[57] It has enough space for us but not much spare (which is another financial restriction). It is light inside and fairly well kept up. A little carpentry in the kitchen will improve it but Doris feels it will be alright. The only thing that will need serious attention is the back shed (or "summer kitchen" as they call them here!) which will have to be demolished next spring. There should be enough wood in it to build a smaller annex on the back. The garden is quite large with an apple tree and a walnut ...

I think we will be happy there and will have pleasant neighbours. It is still handy to Peter's nursery school and to all their little friends. One block north is that large old gravel pit which is being turned into a playground – it is very handy. There are quite a number of university people that I know within half a block. ...

Must go to bed.
  All our love,
    Bob

---

57 The new house was located at 875 Helmuth Street, London, Ontario.

The University of Western Ontario
Faculty of Medicine
22 Oct 50

My dear Dad:

There have been a lot of things happening lately and it's hard to know where to begin.

We will be moving on Nov. 4th into the new abode. I think we will be comfortable there. The final stages of purchase are going through. We will be able to get the mortgage (15 yr) at 5% which is a help. The things we have will fit in there very well but we will have a few things to get in the course of time. …

We had the fall convocation on Friday night and it was quite a good dignified show. Fred Heagy got his Ph.D. (*in absentia*). The best thing was the speech by Ralph Bunche[58] – a very good address and most impressively delivered. Such a speech is most important to the realisation of the efforts of the U.N. and I wish more people could hear it.

I have had a couple of letters from Fred Heagy. He and his family have found a place to live in Glasgow and are adjusting to the rather persistent intrusion of the elements – soot, wet and cold – of the west of Scotland. He is having a good time in Davidson's[59] department and seems to be doing a lot of active work already. He should go far. …

The two papers on the antibodies are now out in the C.J.R. – apart from one error they seem to be alright. I will send you reprints when they arrive. We are just getting ready to send off a paper on reticulocytes – work done with Roger Rossiter. Nothing like a varied fare! …

Yours,

Bob

---

58 The political scientist and diplomat Ralph Bunche (1904-1971) was Chair of the Department of Political Science of Howard University from 1928 to 1950 and taught at Harvard University from 1950 to 1952. He is most remembered for his work in the United Nations Organization. In the years 1947-1949 he held several posts, including assistant to the UN Special Committee on Palestine, Principle Secretary of the UN Palestine Commission (the task of which was carrying out the UN-approved partition), and most significantly, acting UN Mediator on Palestine. In this last position he convinced Israel and the Arab states to sign cease-fire agreements. He was awarded the Nobel Prize in 1950. Nobel Foundation Web site, http://nobelprize.org/.

59 James Norman Davidson (1911-1972) held the Gardiner Chair of Physiological Chemistry at the University of Glasgow. Fred Heagy spent some time there before returning to the UWO Cancer Clinic. A. Neuberger, "James Norman Davidson, 1911-1972," *Biographical Memoirs of Fellows of the Royal Society* 19 (December 1973): 280-303; RGEM.

McGill University, Montreal
29 October 1950

Dear old Son,

I suppose you have concluded the house buying & are preparing to move in next week. May it all bring you great happiness. We look forward to seeing you there when you have settled in.

Burnet's visit & lecture was completely successful & we all enjoyed it. He put forward a very interesting & reasonable conception of the epidemiology of Polio. However I cannot see what interpretation is to be made of the British & Australian observations on what seems to be an augmentation of Polio by immunization against Dipth & Pertussis. There seems not to be evidence of it in the US & in Montreal. There must be something more to it than appears at present.

Wyckoff[60] was here & gave a couple of lectures I was not able to go to. The outstanding feature of his visit was his very fine electron microscope photographs of bacteriophage infection of Esch. Coli. He is a very charming fellow & well worth meeting.

Much time was taken up by Med Board & Bd of Governors of the Alexandra Hospital & the laboratory. This business is not going as it should & I shall have to take a very firm stand on it at once. There is also the proper treatment of T.B. meningitis involved.

Old Leopold Negre[61] gave some lectures (one I went to) on tuberculosis & also a reception. It was nice to see an old friend & he has some good material on tuberculosis.

The week started with high ceremony on Sunday 22nd Oct with the opening (& blessing) of "l'Hopital St. Joseph et L'Institut Lavoisier" which we had to go to. Quite a place & a very much bespeeched & besprinkled business.

Yesterday was fully occupied by receptions & scientific papers & descriptions of the functions of "l'Institut de Microbiologie et d'Hygiene" at the U of M. in a "Journee de Recherche". It was followed by a banquet & as the Mayor had another

---

60 Over the course of a varied and productive career, Ralph W.G. Wyckoff (1897-1994) published over 400 papers and worked in eleven laboratories, including those of the Rockefeller Institute for Medical Research (1927-1937) and the National Institutes of Heath in Bethesda (1946-1952). He was renowned for his recognition and use of new techniques and instruments, and his most important work came in the areas of X-ray diffraction, electron microscopy, and crystallography. R.G.E. Murray collaborated with Wyckoff on a paper examining electron microscopy. See Murray and Wyckoff, "A Method for Obtaining Comparable Specimens for Both Light and Electron Microscopy," *Journal of Bacteriology* 66 (1953): 242-244; "Ralph W.G. Wyckoff (1897-1994) – Obituary," *Acta Crystallographica* A51 (1995): 649-650; New York Times, 9 November 1994.

61 Léopold Nègre (1879-1961) was a French physician and was educated at the Pasteur Institute as a microbiologist. He held numerous appointments during his long career at the Pasteur Institute, including Chef de laboratoire à l'Institut Pasteur d'Alger (1910-1918), Chef de laboratoire (1919-1931), and Chef de service (l'Institut Pasteur, Paris).

engagement I was suddenly requested by him to represent him at the banquet & unexpectedly was called on to speak (I should have known it would happen) but I was caught unprepared.

So with lectures & numerous other things it has been a busy week. Next week is building up; starting with a dinner of local members of the Royal Society of Canada; J.E. Morin of Quebec coming to discuss the development of a society Path & Bact (provincial) & the City Council with a formidable agenda including the membership of the Transportation Commission; there are also lectures & other business.

I enclose a photograph, which, please, return to me. The students gave it to me. It was taken at a party they gave & in the midst of a most amusing skit the bearded fellow in the foreground pointed at me & said "I see an imposter" & then called me up onto the stage to join in a song, during which the photo was taken. A jolly evening & good hearted rap. ...

I am glad you are taking up squash & badminton again. I am sure it is most important for you to do so. Remember me to Tony Brown – I like him.[62] Is Heagy working with my old friend Stanley Davidson?

I liked your papers in the C.J.R. & look forward to the one with Rossiter.

I hope the kids have got over their colds & that you have no more of them. It is good that Peter's treatment is going satisfactorily. He will no doubt show the benefits of it presently.

I must look over some notes for 9AM tomorrow.

    Love to you all,
        Dad

                29 October 1950
                      London

My dear Dad:

Another week gone and not a hell of a lot to show for it. Sundry trials and troubles continue to waste time or make things difficult. Not the least of problems these days is to provide good material for teaching labs. This year seems to be worse than ever and such things as pus (of any sort in reasonable quantity) seem to be unattainable. ...

The hierarchy are still talking about building a medical school on the main campus – which might be fine if the money is forthcoming. If they do this they are going to be far removed from hospitals until more money is forthcoming for a University Hospital. I am not sure how good the scheme is, but it is in the cards. I have to put in some tall thinking about how I would like things to be if it occurs. If we did move

---

62 See chapter 6, note 29.

up the pressure would be upon us to provide classes for Science students in addition to the Medical students. This would not be too difficult if staff and space is allowed for. But the problem for me would be the clinical lab – whether it should remain under direction of the department; or, should be given over to a hypothetical Clinical Pathology department. Really, it is a decision between a department doing a bit of everything (but of necessity largely medical) and an academic department (say microbiology generally). The clinical lab of course is very valuable for maintaining touch with the medical field and with the men, and more importantly is a source of research material – but I wonder how easily one can dispense with it? Collip has asked us to think about the requirements should such a plan go through.[63]

Was in Toronto last Monday about the program of the lab. section meeting. It will be quite a full one – to make it workable we refused three papers and it will probably be fairly interesting. I was glad to see several papers from your lab. That Toronto bunch is very wishy-washy and insular! Notable exceptions of course such as Don Fraser (who is looking extremely well & sends you his best).

Have you ever run into a fellow called Bergold at the Dom. Insect Lab at Sault Ste Marie? He is an expert on insect viruses and has written some interesting papers – several people from Wisconsin and other places in the U.S. told me that he is a damn good man.

I am reading a fascinating and scholarly book called "The Advance of the Fungi" by E.C. Large (Jonathon Cape, 1940). It is a fine history of fungus diseases of plants and is most stimulating to read not only in terms of microbial diseases but also the peculiarities inherent in any investigation. I think you would enjoy it. We have it in our library – you may be able to get a copy to look at.

We have nothing particularly new to report about our move which is due next Saturday (Nov 4th). The address is 875 Hellmuth Ave. The other vital statistics will follow when we have them.

Alice and Thomas (particularly Thomas) have had vile colds, nasty coughs etc. They are now much better. For a few days they were miserable and extremely hydraulic! Peter is getting X-ray therapy twice weekly and we will have to wait awhile to see whether it has had any favourable effect. We certainly hope so. He is very good about having these large frightening machines pointed at him. ... He seems very happy at his nursery school – he needs it very much more than Alice did. Alice is very thrilled with her school work and seems to thrive on it.

It is most discouraging for you to have to put up with so much trouble in the lab at this stage. The problem of paying technical assistants has got to be solved somehow. I was astonished to find out that many of the big U.S. universities (but probably not

---

63 This refers to plans to build University Hospital at UWO.

State universities so much) do not do much more than house their big departments – and most of the people, professional and technical, are paid by outside granting institutions. I don't know what the solution may be but it must be found. It would be better to have fewer but adequately paid and able technicians.

How is Bob Elder working out? You have not mentioned him and I would like to know.

    All the best of love.
        Your son,
            Bob

                          The University of Western Ontario
                                  Faculty of Medicine
                                      12 November 1950

My dear Mum and Dad:

What with one thing and another it has been a rather hectic week and I have not really had time to do more than drop you a post-card to indicate that we had moved. ...

... We spent a pretty busy day getting the essential services going; getting a few curtains up etc. I must say the place was astonishingly neat by late evening. However, on Sunday we started on the large number of packed cartons, and we seem to have kept going on them ever since. There are still a few odds and ends in both attic and back shed. At any rate, we have got everything in without any serious hitches and are settling very comfortably, thank you. ...

    Will write again very soon.
        Love, Bob

                            The University of Western Ontario
                                  Faculty of Medicine
                                      1 December 1950

My dear Dad:

This has been a bad week. Since Saturday I have developed a real brute of a cold ... To make matters worse we have been doing immunological practicals in the class this week and things have gone wrong consistently. Nothing wrong with reagents but individual errors have conspired to put everything wrong. First, someone diluted the haemolysin with distilled water (probably old and well carbonated!) and this inactivated the complement in the first series of titrations. Then, last night one of our people was clearing up the mess in the lab and chucked out all the tests that were carrying on to today. I am thinking of having all the lab coats dyed black! As Pooh would say – Bother.

We are getting along with the paper on the localisation of Gram staining but have some pictures to take. There is a good bit of polishing to be done yet.

I am sending on a copy of a letter from a Mr. C.H. Unwin who is looking for a post as a senior technician. He is at present with ICI[64] but has done a lot of varied work in the technical line but most recently on antibiotics. A Dr. Taylor from the ICI research labs wrote to say that he is a good man.

I was reading a paper by Prevot last night and he persists in referring to *Cl. welchii* as *Welchia perfringens* – has he been doing that for long?[65]

You are being quite severe about genetical work on bugs. In some cases this is warranted. In the case of Lederberg's[66] work I think it has to be accepted and I am pretty sure that his work is very carefully controlled. Unfortunately, it is only one strain of *E. coli* that has shown the segregation phenomena but the experiments can be, and have been, repeated by several people. I see no reason to exclude genetical principle from bacteria, or the converse. As to cytoplasmic systems – these are probably of great importance but the interrelations of nucleus and cytoplasm are still damn hard to study. Danialli is doing some damn interesting work on amoebae by putting the nucleus of one species into the cytoplasm of another "species" and seeing what happens to characters.

Hammerling's[67] work on the Acetabularia is equally interesting (grafts). The best work of the moment and the most complete is that of Tracy Sonneborn[68] on Paramecia – he had an extensive article in the American Scientist (Sigma XI) and a more popular version in the last number of the Scientific American. Of course most of the work cannot yet be translated directly to bacterial genetics – the important thing is that systems susceptible to such study on bacteria be recognised and collected. There is quite a lot being done with regard to characteristics such as phage susceptibility and

---

64 Imperial Chemical Industries.
65 André-Romain Prévot (1894-1982) of the Pasteur Institute, Paris, France.
66 Joshua Lederberg (b. 1925) is an American geneticist and microbiologist. He obtained his PhD at the Department of Microbiology and Botany at Yale University and was appointed assistant professor of genetics at the University of Wisconsin in 1947, with promotion to professor in 1954. He then moved on to organize the Department of Genetics at the Stanford University Medical School, of which he was appointed professor and executive head in 1959. He was awarded the Nobel Prize in Medicine or Physiology in 1958 for his work on bacterial genetics.
67 See J. Hammerling, "Nucleo-cytoplasmic Relationships in the Development of Acetabularia," *International Review of Cytology* 2 (1953): 475-498.
68 Tracy Morton Sonneborn (1905-1981) worked with species of the *Paramecium aurelia* group and studied the basic biology of protozoa. He joined the faculty of the Department of Biology at Indiana University in 1939 and remained there for the remainder of his career.

radiation sensitivity, and an increasing number are working in biochemical systems following Lederberg. However, the characters being followed are not simple and the problems of how and what to study are very great. As usual with such branches the beginnings are hazy and complicated. I fully expect that someone will find a more simple set of characters to follow. We are keeping our eyes out – for example we have a non-sporing variant of *Bacillus cereus* which might be helpful: I will hunt up some good references and send them to you.

Xmas is almost upon us with all its attendant social activity. We are going to have to go somewhat carefully this year with all the extra expenses we have had in the last month or two, and a few more to come. With lawyer's fees, coal bills and such like I have had to mobilise as much as I can safely spare of the little cash we have – so the only thing to do is to be careful for a month or two until we are stable again. That looks worse than the situation really is.

I don't like all the international bickering and profusion of silly and undiplomatic statements that are coming out right left and centre. It is so damn silly that the majority of the people of the world are of peaceful intent, yet national leaders seem to be unable to cope with each other.

I hope that by next week I will be fully recovered from this beastly cold and better able to get out all the work that is piling up. I am going to go to Ottawa a day or so early. I want to see Eugene Munroe and I want also to talk to some of the people there about the Microbiology Society business. I will probably arrive there on the Saturday morning. Unfortunately I have to be back right after the meeting so I cannot make any extended visit. I have had a letter back from Morin but not from Frappier. Morin is quite enthusiastic and from what he says I gather there has been a bit of a damper on the possibility of a Quebec society – this is only between the lines and not said directly. Anyway, I want to get a few people from various fields of microbiology together before (not at) the lab section meeting so that I can get personal estimates of the probability of success. Since we will have to make some sort of committee report at the meeting I want to be well armed with fact rather than fancy. We will have to decide when and where to hold an organisation meeting. It might be possible to hold it in either Ottawa or Kingston – have you any ideas?

Well, I must away to work.

    All my love, Bob

The University of Western Ontario
Faculty of Medicine
30 Dec 50

My dear Mum and Dad:

All your good wishes and the many presents that you sent were the making of our Xmas. As usual you were very generous but all the presents were and are much enjoyed. Peter and Tommy look very smart in their new suits with cowboy shirts. Thomas was particularly impressed with his handsomeness and examined himself in the looking glass at frequent intervals. I am very handsome and well-dressed in my nylon shirts – they are the first I have had and will be most useful. With the help of your cheque, Dad, Doris has got herself a very nice suit and can wear the new blouse with it. She looks very handsome all round, and has adopted a new hairdo! Alice got lots of nice things and had herself a wonderful time. ...

Since returning from Ottawa the financial situation has eased a bit – enough for us to recover from all the fees etc of the last two months. The university gave a salary supplement that made all the difference, though they have not said if they are permanently increasing salaries. Your cheque, and a *very* large one, does not therefore have the necessity of current use and I will be able to use it to advantage. Another future help, we hope, is the rebate from the U.S. income tax which is long overdue – they have at long last acknowledged receipt and promise some action. It is comforting to be caught up again. We have also completed all business regarding the house, so that we feel quite settled. ...

Lots of love to you both from all of us. Doris is up to her ears in bits of housework but promises she will be writing very soon – she has lots to tell you.

All my love,
Bob.

P.S. – Give Fudge a loving pat from me for the ... *cigars*

The University of Western Ontario
Faculty of Medicine
19 Jan 51

My dear Mum and Dad:

So nice to get your letters but I fear you have been sadly neglected in return. This has been a mighty busy month so far and more to come. ... Big jobs loom ahead and greatest among them the paper work to get the Microbiological Society going. I am accumulating lists of names. There is some index that the spadework at the Ottawa meeting got some people interested enough to write in. It will be a lot of work but worth it in the end – I may need a little help later on and may have to shout. Robin and I are trying to get out of the talking stage on the writing of the projected book and have got as far as roughing out headings for sections – which, I suppose, portends a lot of rather discouraging attempts to define and express observations without falling into the trap laid by Knaysi, Bisset and Co! ...[69] Apart from all this I have a lot of other stuff I want to get after – notably the phage work, which is now getting really interesting. ...

During this week the lab was changed over to 60 cycle current. Quite an event. People descended on us like locusts and got all the motors, etc changed in a morning. Our only residual troubles are with our centrifuge and a spectrophotometer – as usual these are things much in use, damn it – but the troubles should be over next week. It will be nice not to have to worry about cycles any more for special equipment. ...

I look forward to seeing your review on listeriosis. I would particularly like to see your reports on the Lake Monroe projects because I have little idea what you have been doing there. ...

I will be writing separately in the next day or two about special matters. By the way did you get some of Robin's reprints on Bartonella?[70] Send them back when you have read them. Also, many thanks for the oil bottles – I am very glad to have them.

All our love to all of you.

Your son, Bob

The University of Western Ontario
Faculty of Medicine
23 January 1951

My Dear old Dad:

I wrote a long and, probably, semi-unintelligible letter the other day and now I have yet another letter for you. Your long and newsy letters are most welcome to us both and the only reason I have written so little is that surplus energy has not been at all abundant. ...

Sorry to hear that both Doug and Sue have acute respiratory infections – as I know well, they are the devil for making you sick. I am interested to hear of this Australian work (*Please send me the reference*). I have a feeling that I have read it but that I must have misinterpreted it or missed the significance. The antibiotic business is getting

---

69 Georges A. Knaysi worked at the Laboratory of Bacteriology, New York State College of Agriculture, Cornell University, Ithaca, NY. K.A. Bisset was with the Department of Bacteriology at the University of Birmingham.

70 *Bartonella* is a genus of Gram-negative bacteria that causes the disease bartonellosis. The bacteria is transmitted by insects including ticks, fleas, and mosquitoes, often to cats, and to humans, often through cat scratches.

infernally complicated and hard to rationalize. I will have to give a talk on the general subject to the Hospital staff in about a month and will have to do some general thinking. ...

Long letter from Fred Heagy. He seems to be enjoying his time in Glasgow – and they seem to appreciate him. In fact, they have given him a University title even though it was not a part of the bargain and the Davidsons had them out to share their Xmas. He is also speaking for them at the joint colloquium of the biochem departments at Edinburgh.

As part of my Xmas present I went out and bought some carpentry tools. My, they have gone up in price! Hope to start and get a bench going soon.

  All our love,
    Your son,
      Bob

                 The University of Western Ontario
                     Faculty of Medicine
                        January 31, 1951

My dear Dad:

Thank you for the reference to Hayes' work. I think that was the paper I read but I must have misinterpreted the conclusions. I will look it up.

The bacillus from milk that has very large spores sounds very interesting and we would like to have a subculture. This organism may have some relation to an organism found in tainted milk. M.I. Christian, "A contribution to the bacteriology of commercial sterilized milk. Part II: The coconut or carbolic taint ..." J. Dairy Research 3: 113-132, 1931. ...

I got a copy of Knaysi's book the other day.[71] It has to be in every library, unfortunately, but it is a foul piece of work and has a most horrible veneer of authority and scientific integrity that will not stand critical evaluation. I fear that most people will take it as authoritative and not question the scientific crudity. Again the electronmicroscope is taken as a technical god and there are lavish, uninformative illustrations that would horrify any reasonable cytologist whatever his special field. This is too bad.

I am trying to make a card index of potential members of our projected Microbiological Society. Could you send me as complete lists as possible – concentrating on Montreal area but including everything else if possible? It is badly needed for completeness. Industrial labs are hard to check up on and any help in listing these would be appreciated.

---

71 Georges Abdallah Knaysi, *Elements of Bacterial Cytology*, 2nd ed. (Ithaca, NY: Comstock, 1951).

We are very sorry to hear that Ruth Elder is having trouble. I trust that it will respond to treatment. Please let me know what the verdict is.

Your son, Bob

Lionel Wolman[72] spent the weekend with us and seemed to enjoy 2 days talking to various people around here. He is a very nice fellow. He asked me to send you his regards.

Montreal
4 February 1951

Dear old Son,

Mum has been none too well this past week. She has had, & still feels poorly, a trachaeitis with headaches & vague pains. She looks pale & drawn & fatigues easily, lacking her usual drive. There is nothing definite to go on & I don't know what to do about it. In addition she slipped on the icy steps outside & again within a little time on the stairs, in this second slip she ticked a muscle in her back which caused her a lot of pain. She is better from that now.

Douglas is still staying with us & is much better & evidently mending. He only got up three days ago & quite evidently the long siege took a lot out of him. He is still rather pale & is glad to rest from time to time. The cultures 3 days ago showed the staphs & the streps gone, but a marked preponderance of a yellow neisseria which is probably of no significance & taking advantage of the reduction of gram +ve organisms. I'll take another culture tomorrow to see what goes on.

There is much to learn about the balances disturbed by antibiotics. In the pre-antibiotic sulphonamide days the throat cultures on nurses & students which the mat-wives used to insist on showed 10-20% had some strep pyogenes, but for the past year they are hardly ever seen. I don't know the reason; whether pre-culture troche or prescribed treatment is a practice we are not told of. I was interested in a report, I think from Boston & I must find it again, in which the finding was that pneumonia is just as common as in pre-sulpha-antibiotic days but the age incidence is shifted & the mortality reduced. I don't think we could get information on the point as the clinical data is not reliable.

I got your letter & will try to make a list for you of P.Q. microbiologists & as complete as I can. I shall add a miscellaneous list, but I cannot pretend to making a complete list.

I have a Swiss mycologist starting work in the lab, Fritz Blank, who has had a lot of experience in Switzerland, Holland & France under good people. He seems a good

---

72 Lionel Wolman (1921-1969) was a faculty member of the Department of Neuropathology, Royal Infirmary, Sheffield, and a member of the Royal College of Physicians.

fellow & seems to know his stuff. After years of trying to get a mycologist lab going & suffering every obstruction & petty jealousies, I gave it up & now out of the blue (we) found the money, to try get a start; at least for a year. ... I could not refuse to do anything with it, so here goes & we shall see what comes of it. There is much to be done.[73]

I am trying to get the Lab at the Alexandra [Hospital] on its feet. The bacteriologist is appointed & one technician of sorts. Other technicians have to be found & equipment too. I have checked & approved lists of things needed, but what the delivery time will be is beyond guessing. It will probably prove hard to get the shift over effected & the work is still being done in my lab.

We had a fair snowfall & for some reason it has not been so well managed as usual; probably because it was the first of any size this winter. It was cold as well. This street was cleared today so my walk to the lab tomorrow will be easy going for a change. All this week I felt that it would be easier with snowshoes. I have been very glad of my Otterskin cap & my warm coat. Mum has not been out of the house.

Sue & Blake have their things moved in but have not moved in themselves. They are staying with his parents & go out to the house of an evening to do some work on it. Sue cannot do too much. ... Blake insists on doing the painting & I don't know when it will get done, not for months with only evenings for work. Sue is looking better again; the sore throat pulled her down.

I shall have to start hunting down information & ideas to give to the Public Health meeting in May. What is wanted is to bring home the need for research & its direction along proper lines. I must think out a theme in order to give them something worthwhile.

I have not yet dug out my fly-tying materials, nor have I given a thought to tackle repairs. ... I hope you will be able to arrange a spring trip, you must do your best to do so.

   Love to you all,
    Dad

The University of Western Ontario
Faculty of Medicine
7 Feb 51

My dear old Dad:

... I had an interesting and surprising communication in the mail the other day from Nungester[74] (this years president of the S.A.B.) asking me to be a member of the

---

[73] See discussion of the McGill mycology lab in Appendix E.

[74] Walter J. Nungester (1901-1985) worked in the Department of Bacteriology, University of Michigan Medical School. He served as president of the Society of American Bacteriologists in 1951.

editorial board of the Journal of Bacteriology. This will involve a little work – 2 or 3 papers a month to review and some say in decisions on policy. But it is a very flattering offer and brings a little kudos I suppose. Very nice of them to ask me.[75]

I had a very enthusiastic letter about your student Whitfield[76] from Stevenson.[77] He has an excellent record and just the sort of background training that we can build upon. I said I would take him and recommended that he apply immediately for an N.R.C. bursary. If he acted post-haste he probably got it in in time. That fills me up for the year and I have no more room for anyone – whoever they be! ...

In between experiments I seem to have an ocean of other work to do but most of it is under control. We have a tremendous bit of people building up for the Microbiological Soc. business. Putting them all in a card index. There are about 400 so far and a good many yet to come! Keeps my secretary good and busy. ...

All my love to you both. Will write again soon

Your son,
Bob

The University of Western Ontario
Faculty of Medicine
12 Feb 51

My dear Dad:

The enclosed specimen will remind you of the search for bits of bass in my finger last July! I have had a bump on that finger ever since and it migrated across the finger. For the last couple of months it has been painful and looked as if it were about to disgorge. ... A game of squash and some carpentry persuaded the point to go out far enough for me to grab it with forceps. ...

Work is confusing but interesting. We seem to be getting close to the structure of the nucleus in *B. cereus*. ...

All my love,
Bob

---

75 Bob Murray served on the Editorial Board of the *Journal of Bacteriology* from 1951 to 1954 and 1980 to 1986.
76 James F. Whitfield (Institute for Biological Sciences, National Research Council of Canada) studied under both Murrays at McGill and Western. He obtained his PhD in 1955 under R.G.E. Murray's supervision. He subsequently developed cell biology laboratories at Chalk River (1955-1962), the European Atomic Energy Community Center in Ispra, Italy (1962-1965), and the National Research Council of Canada in Ottawa.
77 This is a reference to J.W. Stevenson from E.G.D. Murray's McGill department; see chapter 4, note 74.

Montreal
17 February 1951

Dear old Son,

On Thursday, 8th, evening I was suddenly stricken with flu & had to spend a week in bed. I still feel rather washed out & tire very easily. There are also sudden severe pains in any muscle, bone or joint without reason or constancy. It will blow over in time. Mum is still troubled by the same sort of after effects as I am, though she got flu a week or more ahead of me. She is still pale & drawn & tired.

Douglas is well again & active with his laugh back again, however, he tires too easily.

Sue is just recovering, Blake had it the same time as I did, & Dr. & Mrs. Robinson have it now. ...

Eight people in the lab seem to be down at a time; as fast as any return others go down. It makes things difficult. There is a great deal of it in Montreal; variable though not high temperatures, headache & variable aches & pains, variable coughs & upper resp. tract effects & considerable residual weakness. No particular complications I have heard of. In all mild but disorganizing.

I enclose a reference to a paper on gram-staining and nuclear structure; if you have not the journal available I will send you my copy.

Many thanks for the information about Corke, it is very helpful to me. I don't feel I can just take a man just because he is a good undergraduate student in mycology. If he is anxious to go further in microbiology & broaden his knowledge of bacteriology & immunology he could spend one session taking Course 3 & some biochemistry. Then he could go on to a Ph.D. if he wished to on some problem we could cope with. I have quite enough mould work with Fritz Blank starting & I have limited space. If Corke wants to do some bacteriology that is one thing but if he wants to do mycology he would do better in some other department & place, such as Duke Univ. However, there is a difficulty for Canadians going to U.S. universities at present as they have to sign that they are prepared to be called up for military service & they are called up at once!

Have you read any of the work on inorganic ions by MacLeod (Biochem Queen's). Some of it must be quite important to you & the synthesis of nucleic acids. It is worth a look.

I don't know whether Cliff[78] sent you the sporebearer from milk. I have reminded him several times ...

---

78 See chapter 4, note 95 for information on the general bacteriologist Cliff Kelly.

22 February 1951

This is as far as I got. I quite suddenly had a pain in my right ear, which very quickly got worse. After a sleepless night McNally[79] came the next day & found the drum bulging. He did not pierce it but put me on antibiotics & ephedrine, after 3 days the pain started to subside & has not been bad yesterday & today. I am tired of the continuous singing note which never lets up & the full feeling & the deafness. I am just going to McNally's office & I hope he will at least catheterize the Eustachian tube. I have done no work since Saturday 17th.

    Love to you all,
        Dad

The University of Western Ontario
Faculty of Medicine
26 Feb 1951

My dear Mum and Dad:

    Very sorry to read in your letter that you have both been severely afflicted by the current respiratory epidemic. You seem to have had quite a severe double dose and to have a hot ear on the top of it all is rather bad luck. I hope that you are both feeling very much better by now and that the bugs will leave you alone for a while.

    We have had quite a hectic couple of weeks too. Doris has written to you and she has told you some of the story. I seem to have been the only one to escape any severe respiratory infection thus far although for a couple of days I have not felt too full of pep. All four of them have had what I think is a severe cold – at any rate it is not the same symptomatically as the currently described influenza. They have had fevers going to 104 with a severe dry cough and malaise. None of the bone and muscle pains have appeared and the headache has not been evident. At any rate it made them sick enough and required about a week of bed to get them well. …

    The budget meeting has come and went! I can't say that I will be either happy or sure of the arrangements discussed until they send us the decisions in a couple of months. The financial basis of the University is peculiar in many ways and leads to many troubles; not only for us but for the administration as well. I hope that the provincial government can be made to see the light and give some real support to universities other than Toronto. …

    Rossiter and I wanted one fairly ordinary journal for our paper on reticulocytes and it took five weeks! Incidentally that paper has been accepted by the journal BLOOD. Other writing is not complete but is getting along nicely … a lot of interruptions to

---

[79] Unidentified.

that sort of work at the moment.

I hope you will send me your lists of qualified microbiologists in Montreal and PQ as soon as possible. I want to get the preliminary notices out as soon as possible. Norm Gibbons[80] has done a lot of spadework concerning the meeting itself. It seems to be best (and also cheapest) to have the meeting at the University of Ottawa. Other plans are shaping up. Wish we could get together more easily to discuss these matters. We have a tremendous list of people in the card index but there are many gaps – some of these are inevitable and will have to be filled in later on.

Horrible changing weather these days. Very hard to get the heating under control with coal and we are usually much too hot. Today it is raining hard too. At last I have had my raincoat cleaned and waterproofed again. I have been going around for ages with a coat no better than a piece of blotting paper.

I must get back to the lab again and will post this letter on the way. It is hard to get the marketing done these days with all the kids CB [confined to barracks].

All my love, Bob

P.S. Where are Sue and Blake staying? Are they in the house now? What is their address?

The University of Western Ontario
Faculty of Medicine
5 Mar 1951

My dear old Dad:

... There is no lack of work to do. The J. Bact. job was heard from and I had a sheaf of questions to go through concerning Journal policy. There is the microbiology business (the meeting is to be immediately after the R.S. meeting [June 7-8] I think) and is going to be held in the University of Ottawa. ...

I spoke to Don Gillen on the phone tonight. He wrote to you and posted it yesterday. He is definitely ready to come to your lab in June. I thought it was all fixed up and you had planned on him being with you. I would have certainly let you know if there were any hitches in the deal. He is a good man and a hell of a hard worker. He is more than happy about the salary offered – he would hardly do better anywhere else at that stage – anyway it's a good deal more than you can get on an NRC graduate medical Research Fellowship such as Phil Fitz-James has. ...

All the best,
Bob

---

80 See chapter 5, note 80 for information on the Ottawa-based microbiologist Norman E. Gibbons.

> The University of Western Ontario
> Faculty of Medicine
> 20 March 1951

My dear Dad:

It was kind of you to send me the box of books so promptly. They arrived in good time and in good order. We have been pretty busy and I wish I could have been able to have more time to read some of them fully before finishing my talk. Most I have read before and my memory was sufficiently good to give me good chance to select. The talk was last night and it really went over very well. I made no attempt to make a technical talk out of it and really designed it as (I hope) an interesting and informative discussion of recreations in their ramifications with Angling as the source of examples of art, craft and literature. As a side dish I put out the books, plus some of mine and others I got from the library.

Since the Chirons are a gathering of scholars of all sorts this was much appreciated.[81] I was really very pleased. I gave the opening talk of the Chirons three years ago and was very glad to have this one a reasonable success. We have a very varied lot of talks and much good discussion. I think it is one of the best things around here. I will send your books back the beginning of next week – there is a little reading I want to do.

We have had a Dr. Ian McDonald[82] (from N.E. Gibbons' lab at the NRC) with us for the past 10 days. He is here to learn all we can give him of cytological techniques. He is a nice and very competent chap – originally UBC and a fairly recent Ph.D. from Wisconsin. It provides extra work for both Robin and myself but it is fun and makes us think hard about what we *think* we know! I think it is a good project for us to do this sort of thing occasionally. ...

Don Gillen was in yesterday and very pleased and happy with your letter and the information it gave him. He is looking forward to being with you. He is a good chap.
...

I have a lot of work and interesting too. Still a lot of less interesting but nevertheless important general business as well that takes time. I have now to serve on a small committee to survey the general business of running the medical school and make recommendations. It will be a biggish job. We are about to send out notices of the meeting for the organisation of a Canadian Microbiological Assn. If I haven't got the

---

81 See chapter 5, note 119 for details on the Chiron Club at Western.
82 Ian J. McDonald worked in the laboratories of the Division of Biological Sciences at the National Research Council of Canada in Ottawa. See Ian J. McDonald, Teena Walker, Byron F. Johnson, Antonio J. Aveledo, and C. Stan Tsai, "Effects of Ethanol and Acetate on Glucose-limited Chemostat Cultures of *Schizosaccharomyces pombe*, a Fission Yeast," *Canadian Journal of Microbiology* 33, no. 7 (July 1987): 598-601.

Quebec and Montreal names by then I shall send [illegible] them to you, Morin and Frappier. ...

>Your son,
>>Bob

>>>The University of Western Ontario
>>>Faculty of Medicine
>>>8 April 51

My dear Mum and Dad:

... Not so long from now I have to go to the SAB meetings. Will you be going? I have to go for a meeting of the J. Bact. Editorial Board. I will probably only be there for a couple of days. Soon after that is the meeting in Ottawa.

... *11 April*: I had a letter from Robert Breed the other day full of plans for the Bergey dinner. He wants me to give an opinion on the cytology of the *corynebacterieae* and how it may affect taxonomic considerations. This I will do but I shall have to do some work on it between now and then. ... He has also asked me to take part in deciding the cytological work that should be included in the next edition of the manual. From the comparative morphology point of view the work has hardly started and most of our work is directed to the detailed study of a very few "model" organisms so that a basis for general study is laid. Breed sent me a copy of Cowan's[83] submission on taxonomy to the Internat. Congress. It is an interesting document and is, perhaps, sensible in a few places – the whole, however, smacks a little of the "young man in a hurry" syndrome! It will be hard for me to look over all the various kinds of organism that have been or are allied to *Corynebacterium* and its family. If you have an authentic collection of types that you think I should look at send them along. ...

You people had better get some holiday. Why don't you cash a few pounds and go to Jamaica or Bermuda?

>Your son,
>>Bob

---

83 See chapter 6, note 99 for information on the British bacteriologist Samuel Tertius Cowan (1905-1976).

>Department of Bacteriology and Immunology
>Prof. E.G.D. Murray
>April 16, 1951

Dear Bob,

I'm sorry I have not been helpful in the Canadian Microbiology effort. You seem to have done very well and I shall distribute the circulars where it seems necessary. I enclose my form completed. I don't think there are very many Microbiologists in Quebec Province you will have missed through those you have contacted. However, the information may not have been passed on.

I suppose you got in touch with the Chalk River people. H.B. Newcombe[84] might be the one to help there. I suppose too you contacted T.W.M. Cameron, Institute of Parasitology, Macdonald College, Ste Anne de Bellevue, Quebec.

>Dad

>Faculty of Medicine
>The University of Western Ontario
>20 April 1951

My dear Dad:

... You sent me a culture of *Listeria* some time ago but I fear it was not alive and I would appreciate another one to look at.

... I have written to H.B. Newcombe – he was sent the circular before but has not replied (we have had 120 replies and more coming in). I asked him if he would give a paper. Unfortunately, the plant people and many of the mycologists are going to their usual meeting at the end of June and are probably unwilling to give their papers at this new meeting. Don't blame them but wish they had more strings to their bows! Who is coming other than yourself? How about a general paper?

Are you going to the SAB meeting?

Must get on with the work – I have a pile of slides to look at before I can get home.

>Your son, Bob

---

84 Howard B. Newcombe was with Atomic Energy of Canada Ltd., Chalk River, Ontario, from 1947 to 1983. He rose to head of the Biology and Physics Division.

Department of Bacteriology and Immunology
Prof. E.G.D. Murray
April 24, 1951

Dear Bob,

I am sending you some proved Listeria, including the type Species, and some Listeria like cultures. I shall have to hunt for an old culture of Erysipilothrix if you can't get one from Charles Mitchell; at present I don't know where it is.

... I can't go to the S.A.B. as it conflicts with a meeting I can't avoid here. Breed will be annoyed but I can't help it. I have too much piled up here and am getting near difficulties. I would like to get to the lake for a week to get properly fit but can't get a clear time yet.

Love, Dad

The University of Western Ontario
Faculty of Medicine
29 April 51

My dear Mum and Dad:

... The end of term is almost upon us and there is a small spot of examining coming up – however, this is no worry, and only takes up a little time. I have a fantastic amount of paper work to do and also a few small jobs such as the stuff for Breed. (The *Listeria* arrived safely). Robin and I have completely changed the attack on a couple of papers (which were not working up nicely) and the writing up, though, is looking better. New graduate students will be here soon and have to be started on the right foot. I will need all my energies.

I hope to hear that you are both of you planning a short healthful holiday as a preliminary to the summer. Do try to do it.

... Best love to you both from both of us. Take care of yourselves and find some sun to sit in.

Your son,
Bob

The University of Western Ontario
Faculty of Medicine
22 May 1951

My dear Mum and Dad:

... We have been more than busy. The examinations of medical students and others have taken up several days – orals are now over and we still have papers to mark.

This has to be completed by the end of the week. I have also been working very hard on the *corynebacterieae* for the Bergey meeting – it is quite a lot of work and has taken more time than I bargained. I wanted to get some representative photomicrographs to show and getting these has been the problem. ...

... I am telling Robinow to look out for Gertrude at the SAB for he will be there on Sunday. He knows lots of people and can help her. I look forward to seeing her again.

Well, many thanks for a happy birthday, your greetings and your present. All are very much appreciated and enjoyed.

All my love,
Your son,
Bob

Bob Murray's attack of Meningitis following the Canadian Society of Microbiology Meetings in Ottawa

10 June 1951
Ottawa Civic Hosp.

Dearest Doris,[85]

This is a hell of a thing to happen when away from home but I suppose it is always possible. At any rate, there is nothing to worry about now – though I confess to being a little worried myself Friday and Saturday.

No one is quite sure what was wrong – but anyway it was an infection, probably virus, with some tendency to involve the C.N.S. and gave me the world's worst headache. I hope I never have one like it again. This morning I woke from my (drugged) sleep to find the headache practically gone and I can at least read again. The doctors are cautious but think that with some care and caution I might fly home in 48 hours. We will hope so and I will telegraph my plans. It was very lucky that Mum and Dad were here and Robin to help with advice – I much needed it for I was in no state to make decisions. However, it is mostly past history now so don't do *any* worrying.

I am concerned to miss Peter's birthday – please give him my best love and a hug.

...

The meetings went very well and I think we are a success! I am sure that Mum and Dad did all they could to keep you informed of happenings and progress. The care I have been given is *very* good and I am sure you would not think otherwise.

All my love darling,
Your Robin

---

85 Correspondence between Robert Murray and Doris Murray, private collection, Robert Murray.

Chateau Laurier
Ottawa,
Ontario
June 1951

Dear Doris:[86]

We have just come back from seeing Bob at the Hospital. He is vastly improved & we can see a difference since our visit this morning. In addition to looking so much better, he is – except for a bit of a backache – quite comfortable & cheerful. We saw him eat a good lunch at 10 o'clock & a good supper at 5:15 & we really feel very happy to know that he is so obviously better and on the mend. ...

We leave for Montreal tonight – but Dr. Plunkett, in whom we have deepest confidence, says there is no reason why we should stay & that we can feel quite at ease in our minds. Bob has lots of friends here who will be going in to see him now that he is feeling up to having visitors & I believe he will have lots of company to cheer him through the few days that will elapse before he can travel home to you. He will have to 'go easy' for a while to get his strength back again ...

Everyone has been extremely kind & helpful & we are so glad that when it happened Dad was here to get the right people to look after him. ... But Dad and I will keep in touch with Dr. Plunkett & will hear how his convalescence progresses. ...

    Love,
        Mum

15 June 1951

My darling:[87]

Well here I am again and trying hard to get well and not *yet* allowed to get up but have *hopes* that they will relent tomorrow. Have just written to Robin – he has sent me two long cheering letters but my effort in return pretty poor. I find writing a letter exhausting probably because of the ungainly positions I must assume.

I am improving I think and feel fairly well – not much appetite or ambition yet so do little but lounge in my bed and look forward to your daily letters. Wish I wrote as well but somehow my brains are addled and my fingers are inept.

Went down for EEG this morning without breakfast. He fastened a tight cap made of rubber tubes to my head and many (16, I think) pressure electrodes to my skull. Gave me a beastly headache at the time – made me do mental arithmetic, over breathing, etc. Rather an ordeal I fear. No new information out of it. ...

86 UWO Archives, Freda Murray Papers.
87 Correspondence between Robert Murray and Doris Murray, private collection, Robert Murray.

I wish you could be here but don't go trying to make the trip. It is expensive and unnecessary for I should be home early next week. I long to see you and the kids.

Your loving,
Robin

McGill University, Montreal
25 June 1951

Dear old Son,

We are glad to think of you home. You must take things very quietly for a while and avoid getting tired in any way. No doubt you will soon get on your pins again & quickly regain strength so that you can be out & about freely. Once that is so, it would be well to have a good spell in the country. When you have a scheme to do that let me know, for we can easily help you to realize it without any embarrassment to ourselves. At the moment, the most important thing is what is best for you.

As soon as we get out of this place we are going to the lake. The date set for moving our things into storage is 5th July & we are packing up for that purpose.

Meanwhile we have taken a room at 1570 Pine Ave. West for July, August & September, to have a pied a terre. It is all very inconvenient & has put us out of our stride, but we will soon settle down again. I propose to take a good holiday once I can get away.

I have got most of the difficulties at the lab straightened away as well as may be. I have had to take on some D.P. types who have not the basic qualifications our Course 3 people have & they will take quite a bit of training & watching. Gillen will learn the ropes quickly & will have to give all the help he can [while] Anne-Marie is looking after the training & supervision of the others. Bob Elder staying on to help over the most difficult time is of the greatest assistance. The only old hands left in the Clin Lab are Anne-Marie, Wardie & Salmi Nomik; that means seven new people. There are changes in the general technicians too which are disturbing but I was able to avoid worse happening. ...

Among other changes that of secretary is troublesome & it impedes a lot of business. The salary we can offer does not compete for efficient & reliable people, so I must put up with what I can get.

I must let a number of things drift & do what I can with them as I find means & ways to handle them. I am only struggling with the most urgent & essential problems as they crop up. My illness set me back quite a bit but after a holiday the difficulties will not loom so large. I grudge the time taken up by many outside affairs. I have too many committees, Boards, etc, and appeals by other institutions which should do better for themselves.

I have not seen Sue & family for quite a while. They seem quite a way off at Lakeside.

Sue & the kids will be going to the lake next month sometime, but I have not heard their plans. They are not so oppressed there as they were in town & when it is stuffy here it is quite fresh at Lakeside. ...

I expect you had quite a reception on your homecoming. The youngsters will have given you a great welcome & Doris must have been very happy. I hope the journey was not too tiring & difficult. The train would have been worse ...

Don't forget to let me know about the hospital bills & let us help a bit. We don't want you to be worried by any difficulties. We will remember what a difference a little help would have made to us, years ago, when times were hard for unexpected reasons.

With all our love,
Dad

McGill University, Montreal
28 June 1951

Dear old Son,

We are very glad of Doris' letter. It is a pity you had so much bother with the journey home. There is nothing wrong with the people who fly the Canadian planes but the office work of the air lines is abominable. The journey must have tired you quite a bit. You can take it very quietly in your own familiar surroundings but don't let problems be brought to you. Robin can look after things for the little while you must rest. You can get some fun watching the garden grow & with some quiet reading for fun. Let me know if you want any of our books, before they are packed for storage on 4th July. You have most of the best of the fishing books, except Radcliffe's "Fishing from the Earliest Times". ...

We propose to spend a good time at the Lake & that will not harm either of us. We propose to go to Lake Munroe for a while, though I have no intention of working there. There is some nice country to ramble in there. It is wilder than round Bark & so there are more wild things to see & very handsome scenery. First we must spend a while getting into condition.

I think things should keep in reasonable order at the lab as I have arranged for most of the essentials I can foresee. Dolman[88] is looking for people. I sent him the name of the young fellow you mentioned ... in case he can use a junior M.D. of

---

88 Claude Ernest Dolman (1883-1976) came to Canada from England in 1931 and joined the Connaught Laboratories at the University of Toronto. He moved to British Columbia in 1935 where at the University of British Columbia he was head of the Department of Bacteriology and Preventive Medicine (1936-51), and its successor, the Department of Bacteriology and Immunology (1951-65). University of British Columbia Archives, Claude Dolman fonds description.

promise. It would not be a bad opportunity. ... Otherwise, if he wishes to wait a year I shall have a vacancy in the Clin Lab then. ...

I have acquired some more unavoidable committees related to the Montreal Board of Health.[89] They will take up time & some trouble. These things become hard to refuse & I lose count of how many there are. I got rid of some last winter but more seem to have come than I rid myself of. I must try to slough off some more.

... Tell Doris we are extremely happy to get her letters, but, as she has more than enough to do, we will take it that "no news is good news" & not to put herself out. A postcard will do.

... Goodnight, old Son, & take care of yourself.

    Love,

        Dad

                9 July 1951

       The University of Western Ontario

             Faculty of Medicine

My dear Mum and Dad:[90]

... I am doing very well and I feel quite well now although I am taking things very easily for the moment. I can now begin to take some real exercise and I think – so far it has been mainly sun, air and gardening. All of which are very good for me ... We are certainly very happy with our garden and our abode – it makes a lot of difference to us after the trials of renting. There are still lots of things to be done and will keep me busy.

Robin seems to be keeping things going very well at the lab and there is not much to worry about. I manage to restrain myself from wanting to get down to work and I hope that I can keep it up. There are many interesting things going along and everyone seems to be happy.

The kids are growing just about as well as the garden – appetites and activity seem to be flourishing and they are good and brown. Peter is still thrilled with his wagon and plays with it constantly. Thomas refuses to be outdone and can handle it very well too. In fact, he seems to be increasing his vocabulary and skills at a fantastic rate, and his shadow (all 36 lbs of it) grows no less. (He weighs the same as Peter!). Alice has tons of friends including a boy-friend who rejoices in the name of John Yocum and who is very nice.

Doris is very well but plagued with a sore foot that I can make nothing of except that it seems to originate from wearing a particular pair of sandals. I am very glad

---

89 Joburg participated on several Montreal Board of Health committees during his career, including those related to problems of tuberculosis in the city.
90 LAC, E.G.D. Murray Papers, vol. 19.

that she has now got a cleaning woman again coming once a week. This makes a big if expensive difference for her and she does not have to worry about the heavy polishing and cleaning. As a sort of bonus the cleaning woman brings along her boy, who is Peter's age and they get along very well indeed.

I look forward to hearing from you about yourselves, the lake and the garden.

 Lots of love,
  Bob

             Bark Lake
            13th July, Friday
               1951

Dear old Son,

... We were very glad to get your letter & to know you are so much better. It is good you feel you can do more & are resolved not to work yet. I am sure Robin can hold the job down for a while & will gladly do so. ...

The water is perfect for bathing. The bush is astonishingly thick with new growth & the birds are active & sing marvelously.

Sue, Blake, Doug, Joan, Ruth & Bob Elder arrived by car in the middle of the night. Blake & Bob returned at 6AM the next day. The others look better even after two days ...

It is good to know Doris & the young uns are so well. I hope you can get away for a bit before the summer ends. It would be wise to do so as soon as you can safely do so.

 Love to you all,
  Dad

We are visited by a family of 5 barred owls. They have worn the back off one of the cedar arches by the kitchen door.

             Montreal
             18 July 1951

Dear old Son,

Mum & I came back here yesterday & I have spent all today in the City Council. Mum returned to Bark Lake this afternoon & I propose to join her there on Saturday. We plan to go off to Lac Escalier the end of next week if possible, but we will let you know when we leave.

On Sunday we went into Pike & got 5 nice bass. These are (2 of them) the first fish I have caught this year. They were very good to eat. ...

Mum's garden looks fine & there promises to be a big crop of raspberries in the bush. The big wild strawberries Mum transplanted last year have bred true & gave enormous berries this year. She is going to propagate them.

I intend to fix up a few urgent things at the lab tomorrow & Friday. Gillen is settling in nicely & seems very well pleased with the lab. Bob Elder is lending me a great good hand & I am very grateful. Ruth seems to be very well. I wonder a bit if the clinical diagnosis is correct. We have not been able to catch any T.B. from her with quite extensive trials.

Robinow very kindly wrote us a very nice note, telling us of seeing you. It pleased us greatly & we are grateful for his very kind thought. In a recent letter Collip also mentioned your progress. We have a number of fine friends.

Montreal is in the hands of Tourists & I shall be glad to get out of it. The club is closed for three weeks, Damn it!, & the competition in restaurants for meals is the devil & all.

Our room is quite comfortable & we will manage all right.

You have not let me know of your hospital bill. We would like to help out. Please let us do so.

      Love to you all,
          Dad

                                                Bark Lake
                                               21 July 1951

Dear old Son,

I have just arrived at the lake & the train was very late, due to funny doings. It is good to be starting a holiday. ...

I sent you 2 Norwegian spoons that Bill Gulline[91] calls "Smorgasbord" & he & others swear they are marvelous catchers of Bass & Trout too. I shall try them.

Everyone here is very well & send their love.

Mum & I will start our trip from Monroe Lake on Thursday & will be away 3 weeks, we hope. If there is any reason to get in touch with me A.J. Paterson at the lab knows how to do it. Otherwise if you write to Bernard Cooper, Mount Tremblant Biological Station, Lac Superieure, P.Q. telling him to get the fire warden to get after me will be effective. We will not be able to write till we get back.

      Love to you all,
          Dad

---

91 See chapter 4, note 50.

21 Aug 1951
The University of Western Ontario
Faculty of Medicine

My dear Mum & Dad:[92]

I should have been writing to you at regular intervals all the time you were away but somehow things got away from me and I kept putting it off. ...

I am very fit now and very impatient to get back to work again. Rest is all very fine but it becomes a bit irksome and stagnating after a while. Puttering in the garden with as few clothes as possible tanned me up and finally a week's trip with Roger Rossiter has given the final touches so that now I am toughened-up a bit. Although they have not been away the kids have had a good summer and are very well. They had a slight set-back when each of them were stricken by an acute pharyngitis[93] – rather a severe infection that responded well to treatment in the cases of Peter & Alice. Thomas, who was the last to fall, was more severely hit and was quite a sick boy for a week with temperatures up to 106. ... Luckily, neither Doris nor I were afflicted so that we were not dragged down by it. They are now all well and have caught up their usual good spirits and toughness. ... They have had a couple of short day trips down to the beach at Port Stanley (Thomas insists that it is "Airport Stanley"!) and they have a good time despite the number of people and general surroundings. We were down there yesterday and had a good hot sunny day – they dug sand and got splashed by the waves, had ice cream cones and a picnic. ...

Roger Rossiter and I managed to carry out our previous plan to go on a trip. We, however, did compromise a little on where we would go and decided to go where we would not [have] too much effort (but enough). We decided to go and have a canoe-eye view of the Georgian Bay region – since we had only a week we had to choose a rather short trip. We drove up to Bala which is on the road to Parry Sound and which is at the exit of Lake Muskoka. We decided to go from there through the Muskoka (called Muskosh locally) River to Go Home Lake, out into Go Home Bay and there visit Bob Noble at his cottage, then go south in Georgian Bay and back to Go Home Lake via the main exit of the Muskoka River. Not a long trip and could be taken in any number of easy stages and replanned if necessary at any point. We had little or no idea of where and how the jump off place near Bala would be situated. When we got there we cornered a provincial policeman who told us how to get to the power dam by road but he couldn't say whether we could leave the car there. His directions were accurate but neither of us knew that the maps were inaccurate because of the building of new dams. ...

---

92 LAC, E.G.D. Murray Papers, vol. 19.
93 Painful inflammation of the back of the throat.

Big [illegible] power boats abound and some houses had a regular fleet ..., a couple or three sailing vessels and outboards and maybe even a big sea-going boat. We had arranged to call on Bob Noble who was up here at his family place on one of the outer islands of Go Home Bay (Bushley Inlet). We timed it very well (or Roger did) and we rolled up just as Bob was sharpening a knife for a big roast. Very good too. Bob, Eileen and the two elder children were there. We stayed over Sunday and had a very good time. We fished around quite a bit but not very successful – this is not the usual in those parts. It couldn't have been a more perfect day: light west wind and no clouds in the sky. We paddled down through the rocky islands over innumerable reefs and got a good sunning.

There are lots of houses there really but many of the inner places and recesses are completely free of them. ...

Perhaps the prettiest section was from the mouth of the Muskoka River to the camp place for the night. (MAP) Moderate current and high water. No particular rapid at the place marked but a considerable one at the site of an old dam just before the shallow lake-like widening of the river. Above that a fairly deep gorge and lots of foam. Very rocky territory and Oaks about 80% of the trees! At the head of the gorge a brisk rapid and a couple of hundred yards above that a hell of a cataract and complicated fast flowing water and high rocky ledges with grassy flats – we camped quite high above the falls. The portaging was complicated by all the rock faces and we had to pass the rapid on one side and the falls on the other. Nice place but fishing disappointing; got about 8 ½lb bass below the falls which were welcome for breakfast. Not far above the Muskoka is joined by the Gibson (or ? vice versa). The Muskoka drops into the Gibson in another unmarked cataract (about 10 foot drop) ...

Getting up the river was a [illegible] formidable business for the hydro dams are apparently opened in the morning and we had the swiftest current to deal with. It took us 2 ¾ hours to get to the dam. Some places were very hard work and we had to "give her 10" – as Roger ... We got there hungry as the devil and got packed away quick, lunch in Bala and long drive home arriving there 8pm all well. It was a good trip, not too strenuous but with a few good bits of exercise as there should be. As it went on I got good and fit and now feel ready for most anything. We ate well and slept well and got lots of sun. ...

Look forward to hearing from you.

    Your son,

        Bob

Montreal
31 August 1951

Dear Bob & Doris,

When we get back to Bark Lake we found Doris' letter & got Bob's the next morning. We were delighted to get both of them. We were so glad to learn that you had got over all the pharyngitis in the family without residual effects & that Bob was so much better & had gone on a trip. It was a great relief to us as we had been away without news for a month.

We thoroughly enjoyed your description of your trip, Bob & it is good to know you are so well as to be able to do it & enjoy it. You went through most interesting country with nice variation. Surprising you did not get more fish as I supposed those to be good waters for bass. The map shows a lot of varied possibilities for trips in that region. The falls must have been quite impressive & reasonably fast water is quite a lot of fun. Roger Rossiter must be a real good traveling partner & that adds greatly to a trip. ...

I'll send back the maps as soon as Sue has read your letter. We came back to Montreal this morning. I expected a meeting with Robert Breed & N.R. Smith tomorrow, but I find they are now not coming. Had I known we would not have returned until Monday. ...

We had a very good trip & feel much the better for it. We were not fit enough to do anything very strenuous. ... We made our camp more comfortable on the island in Lac Escalier & did some day trips from there. We stayed there 32 days & ate the last of our food for the breakfast of the 33rd when we broke camp & returned to Bark Lake. We lived largely on pike, we must have eaten nearly 70, & found them good, satisfying & not tiresome ... We also had lots of raspberries & blueberries; we had fresh berries every day & the best way was Mum's way of making berry pancakes. ...

We are both fit, rested & dark brown. The only trouble is the persistent joint pains dating from our flu in February.

Glad you feel well enough & incline for a little work in the lab. Take it easy & carefully. Just enough wisely chosen is probably better for you than boredom & irritating inaction.

      Love to you both & the youngsters,
         Dad

9 Sept. 51
The University of Western Ontario
Faculty of Medicine

My dear Mum and Dad,[94]

Very nice to get both your letters and to hear that you are now very bronzed and fit. ... We are still going along quietly and not being too energetic. I have been going down to the lab for a few hours each day and have not been doing too much. The children have started school – rather a relief for Doris, since they were beginning to be a bit bored with too many play hours.

We had a very pleasant visitor in the lab this week in C.G. Hedén[95] from Stockholm. An old friend of Robin's. He was most charming and we had a grand talk and one most enjoyable evening with him. Lyman Duff was in town for the Dean's conference. Had a good talk with him at a party Collip gave the other night. He had not seen you since your return but he will be able to give you a good report that I look well!

Teaching starts immediately and I fear that Robin will have his hands full for a while. I will be able to give some help but he will be taking the main burden. We are better off than usual with demonstrations.

I am going to Ottawa for the Council meeting of the new Society (Canadian Society of Microbiology) – the meeting is on the 24th. The date was set by the visit, on other business, of a couple of councillors from a distance. Otherwise it is a bad date for Ottawa will be full to the gills due to the Atlantic Pact conference. I have not made exact plans of the trip yet but will probably go down a day or so ahead and will be staying with Norm Gibbons. I wondered whether I might come down to Montreal but it depends on other circumstances and also you will be all tied up with a move ahead.

Saw Bob and Ruth Elder a week ago. They were here for the weekend after checking in at Ottawa. They have found a place to live and will be moving in soon. Nice to see them again.

The canoe stood up to the trip very well. Being rocky country there is plenty of scratching and scraping however careful you are. ...

All the best,
Love,
Bob

---

94 LAC, E.G.D. Murray Papers, vol. 19.
95 Carl-Göran Hedén from the Bakteriologiska Institutionen, Karolinska Institutet, Stockholm, Sweden. See note 5 above.

Montreal
20 September 1951

Dear old Son,

I am glad to have had good news of you from both (Lyman) Duff & Dowson who have seen you fairly recently. I think it is well for you to be doing enough to prevent you getting bored but not enough to cause you any fatigue. ...

Yesterday I signed a lease for 1559A Pine Ave West from 1st Oct. We will not be able to move in until somewhere after Oct 7th because the movers won't do it earlier. It is encouraging to think we will be with our own things again soon. It seems a long time since last June; much longer than it is in fact. I am weary of this living in a lodging house.

I am preparing to go to the Columbus on Saturday 22nd Sept for a week's fishing. I wish you were coming with me. We are leaving at 7AM & I must get something I can eat before going. Probably I should not go. I cannot possibly deal with all there is on my desk & there are some clouds on the horizon; however, I shall do what I can with the most important matters & let the rest wait & ripen.

... The kettle you & Doris gave us is a blessing in my lodging. I make a cup of tea in a few minutes & it is especially welcome early in the morning before I go to St. Catherine St to get some breakfast. It is wonderfully efficient & we think of you every time we use it.

  Love to you all,
    Dad

Just got a letter from Ruth Elder giving good news of all of you. They seem happy in the new job.

London, Ont.
30 Sept 51

My Dear Dad:[96]

It was very nice to get your letter and hear that you have been to the Columbus for a week. It would be very nice to have a go at the trout and with any luck you will have had good sport. At any rate, the bush is beautiful at this season and the extra relaxation will help you through the winter.

When I was in Ottawa at the beginning of the week I rang up the lab and heard you were away for the week. Had you been there I would have come down for a day. I was sorry, but yet glad that you were fishing. ...

I had a pleasant trip to Ottawa and stayed with N.E. Gibbons. They were most kind and arranged things so that I could rest when I needed. The meeting went off

---

96 LAC, E.G.D. Murray Papers, vol. 19.

very well and we got a lot of work done. The membership drive will be getting under way immediately and we are planning on a meeting next June. It was surprising (and gratifying) that all the Council were there and they were very helpful. I spent a little time in labs there but was not able to see everyone I would have liked to see. I had a very interesting morning with Lochhead.[97] The return journey was not as good as the going – too many damn bumps and squeaks. I was a bit tired when I got home.

I have been working part time: going down to the lab in the morning and returning about 3 pm: it is as much as I can safely do without getting worn out. ... I have 5 semi-written papers and they don't progress very fast either. We are getting a lot of work done with 3 graduate students and all that is going well. ...

We look forward to seeing Mum's journal of your trip and also your account of the Columbus.

All our love to you both.

Your son,

Bob

P.S. Doris just wrote a long letter to Mum with most of the family news.

London, Ont.
October 8th, 1951

My dear Mum and Dad:[98]

... There are lots of jobs about the house and not all of them too pleasant. We got the furnace ready for winter and are glad of it; also 6 tons of coal at the current high prices! I am trying to de-dust the basement – a hell of a job – all the "crooks and nannies" being full of the dust of ages. Soon we will have to get the eaves cleared of leaves and the storm windows as we are just beginning to notice the draught around the windows.

We finally got the accounting on the Ottawa Civic Hospital bill and it was quite a shock! The total for 15 days was 340 odd dollars of which the insurance paid 175. The main cost was the room at 13.50 a day – I can't imagine what they charge for other rooms for mine was no great hell and no hotel could get away with it. The Pathology charges were $36 (!) and the drug bill was 84!!! No professional reduction was made ...

The local enthusiasm for football has started again. I suppose that not long from now we will have to get Peter and Thomas a football so that they can compete with the neighbours. Now that the old standby medical students have graduated I suppose

---

97 The Canadian bacteriologist Allan Grant Lochhead (1890-1980), son of the renowned Canadian biologist William Lochhead (1864-1927), was appointed Dominion Agricultural Bacteriologist in 1923. He devoted much of his time at the Department of Agriculture to the study of soil bacteria. A Fellow of the Royal Society of Canada, he was awarded the Society's Flavelle Medal in 1958.

98 LAC, E.G.D. Murray Papers, vol. 19.

that the team will have to struggle for talent. However, they didn't seem to have much trouble beating Queen's yesterday.

I am still going along at the same gait. Working about a half-day and rather glad to rest after it. Things are going reasonably well and research keeps going thanks to graduate students. I wish that I could do more but I have to be patient and wait my time.

We have been wondering if you are planning to come down here before or after the CPHA meeting in Toronto? We would love to see you and so would the kids.

Everyone is well and cheerful. We all send our love and hope that the moving goes smoothly.

    Your son,
      Bob

> 1559 A Pine Avenue West
> Montreal
> (Phone Glenview 4740)
> 21 October 1951

Dear Old Son,

... The 30th of this month will see the Royal visit to Montreal. We have the University Ceremony in the morning, then the Civic Reception in the afternoon & the official Civic Banquet that evening. I expect the Royal Couple will be glad when this very strenuous tour is over. They are getting no rest & it must be exhausting.

The preparations you sent are good & the method offers possibilities. I have not discussed them with Blank yet as he is on a holiday. He will return next week; however, he is not a histologist & knows no pathology. He is rather limited to the fungi themselves, but at that he is very good. I hope I can get that division (mycology) firmly established. I have given up any idea of getting a Virology Division. The possibility of that was frustrated by previous Deans (Meakins[99] & F. Smith[100]) when there was a real chance to do it.

The Columbus trip was not spectacular other than the weather & the fall colours were wonderful ... On the whole the weather was too good to get the best fall fishing. ... I drove up with van Patter & then drove back with Lytell, which made the journeys both ways very pleasant. The road in from Nominingue is thoroughly bad; it is cut up with heavy trucks loaded with logs when it was soaking wet during a rather wet summer. The lakes are all very high, surprisingly so. ...

Having moved our things in we went to the lake for the weekend. It is probably the last I shall have as I have a string of lectures on Sat mornings & Monday mornings,

---

99 See chapter 1, note 26.
100 See chapter 1, note 87; chapter 4, note 43; and chapter 6, note 95 for information on Fred Smith.

which spoil the chance. We had a paddle or two each day for fun, otherwise we did jobs round the cottage, such as oiling tools, putting things away, piling wood, dropping kindling etc. It was a weekend of marvelous weather & we enjoyed it to the full. The colour was mainly over & the leaves were largely down, but there were many beauties by day & moonlight. The "hunters" were out in force so we did not go in the bush at all. We had a few casts in Bark Lake for bass or a pike & got nothing worthwhile & very few not worth it. The fishing in that lake has deteriorated remarkably.

Let me know if there is any modification of the bill from the Civic Hospital. One thing I am certain of is that it would have been a much heavier bill had it been the R.V.H. The cost of sickness has reached such absurdity that the hospitals are forcing state medicine to the forefront. The enclosed cheque will help you & if they make some reduction so much the better for you.

I used to withhold all returns of lab work for doctors & their families, nurses & all students. This held ever since I have been here until the past year when the R.V.H. insisted on the returns being made to them saying they would make proper allowances. I fought against it but had to comply with strong protest. I don't suppose they make any allowance. I still have no charges if specimens are sent from outside & only make returns when R.V.H. requisitions are used.

I don't yet know how things will work out, but we will do our best to visit you about the time of the C.P.H.A. Lab Section meeting. I have to be in Toronto on 19th January & may be able to go to London then too. I'll let you know as soon as I can see clearly how things stand.

Mum did not quite finish the journal of our trip & up to now has not had time to finish from her notes. I notice she was doing some of it last night & tonight. Possibly she will finish it & send it along soon. I expect it will be quite good reading.

  Love to all of you,
    Dad

London, Ont.
21 Oct 1951

My dear Mum and Dad,[101]

We are very glad to hear that you have completed the moving job though I suppose that you have a lot of settling yet to do. I was very sorry to miss your phone call the other day and feel that I have missed some of your news. I had to be in Toronto for a meeting concerning the Xmas meeting and took the opportunity to spend a day with van Rooyen and his people. I stayed overnight with van Rooyen and we had very

---

101 LAC, E.G.D. Murray Papers, vol. 19.

interesting and amusing conversations, as usual. They have a lot of most interesting work going on and of particular interest to me, the phage work being carried on by (Angus) Graham, French and Lesky.[102]

Had a pleasant talk with Don Fraser but he has a rather sad story to tell. He had noticed an asymmetrical lump in his thyroid but the medical types didn't think too much of it because it was so slight and he had a few nodules for years. However, he had a slightly increased Iodine uptake and they finally decided to do a partial resection. This was done during the summer – they found some odd lumps that, on a lot of opinions, were thought to be malignant. So he has since been treated with radioactive Iodine and with deep Xray. He had a good summer but after a while developed signs of myxoedema and he does not look well now. He is very cheery withal and, and apart from his face and a very husky voice, he is much as usual. It is a very sad story – but he is a tough old bird and may keep going for a few years yet.

Van Rooyen is working hard at the pox and CNS viruses for the Manual (Bergey's). He seems to have got in a fair amount of good work and I think it may be very helpful. He said he is going down to N.Y. for a meeting with Breed sometime soon – I suppose that you will be going too.

Tommy has had a successful birthday. He was up early to see what he got! The most successful presents are a wagon (largely your contribution) and a very nice train (from G.G.). Doris made him a cake that pleased his little heart – shaped as an engine with peppermint creams as driving wheels and a chocolate cow-catcher. Peter got a great kick out of it too! Tommy loved his cards – especially the one from Sue and Blake showing the active worm in the apple. Worms are definitely his favourite animal!

I suppose that you will soon be tied up with the Royal visit. I don't know what sort of fandango Montreal has in mind but I expect it will be fast and furious. We didn't have much of a chance here and it was late in the evening. Alice would have loved to see the princess but we thought it best not to go down. The crowds were packed like sardines in that small space opposite the CNR station and I was surprised how they managed to get them all in. The unfortunate couple seem to have been rushed off their feet and have had little chance to enjoy Canada.

I had a letter from Claude Dolman the other day. He tells me that he is taking on John Stock – who should be a good man for him. He is also hoping to get Derrol Pennington as Assoc. Professor.[103] That would be excellent if it goes through, for Derrol is a very good man.

Robin has had a tough term with more than his usual share of teaching. He is

---

102 Connaught Laboratories, University of Toronto.
103 Derrol Pennington was a biochemist working in the Department of Microbiology at the University of Washington at this time. He published numerous articles in the *Journal of Bacteriology* in the late 1940s and early 1950s. RGEM.

away for a few days now – has been invited to give a lecture to the Northern NY State branch SAB in Rochester. He will give them a good show. He is going to put forward some of our recent work on nuclei.

I am very well really and feel better than I did a month ago – at any rate my tolerance for work and difficulties seems to be greater. I am starting a few experiments and have got the graduate students working hard. Experiments are rather hard to do in a half day! We have got Whitfield properly settled in a problem and he is working quite well on the *Sh. Sonne* phages. This is an interesting phage-system and we are going to have to do some nucleic acid chemistry to keep our cytological work on an even keel. ...

We look forward to your letters.

    Lots of love,

        Your son,

            Bob

London, Ont.
5 Nov. 51

My dear Mum and Dad,[104]

... Things haven't been too bad for me; I have had a few bad days when nothing seemed to go right and I just have to give up. Most of the time I can get some work done and not get too worn out. Doris has much the toughest time: I am very glad that this opportunity to go to New York has come about – it will do her a world of good to get away for a few days and she hasn't seen New York for over seven years.

The chance is fairly incidental and is a N.Y. Academy of Science confab. On "Viruses as a Cause of Cancer". Most of it is biology and not medicine and it looks as if it may be quite interesting. The first session is most interesting to me and I will be glad to hear it. It will give me a chance also to go and see Mark Adams[105] at N.Y.U. who is working on T5 and serologically related phages and which overlaps into our work on the same system. May be able to save ourselves some labour. Will also see Ash, and, I hope, some others. We are planning to leave here on the night of the 13th and come back on the night of the 17th. That will give us two clear days uncluttered by meetings. It will not be too expensive, since my side is looked after, and it will be well worthwhile and give Doris a nice bit of a holiday. ...

---

104 LAC, E.G.D. Murray Papers, vol. 19.
105 Mark H. Adams (1912-1956) was an associate professor in the Department of Microbiology, New York University College of Medicine. He joined the department in 1942 after spending three years as an assistant at the Rockefeller Institute for Medical Research. He was a fellow of the New York Academy of Science and the New York Academy of Medicine. *New York Times*, 19 October 1956.

The correspondence between the superintendents of hospitals concerning my bill at Ottawa came to nothing. The Ottawa fellow said that as a civic hospital they had no policy of professional reductions; he had taken it up to their financial committee and they refused to do anything. (Our hospital is a civic hospital and they have a policy; also most expensive room is $11.50). So we have the full bill to pay. Kirk (our Superintendent) said "I wouldn't pay it in too much of a hurry!" Anyway your cheque is most welcome at this time with our various expenses mounting. But I feel that we should not be imposing on you like that.

Winter has come on us with a huge bang. The last three days have been very cold and we have a few inches of snow lying around. The kids were very excited to see the snow on Saturday morning and demanded breakfast, snowsuits and sleds all in one breath and were out by 8 o'clock! This means, also, that the furnace must be attended with usual precision; and I am getting practice for later on. We are lucky that the house heats easily.

Doris is very eager to hear of the Royal visit and the affairs that you both and Sue were to go to. It must have been rather nice as a gala show, even if a little exhausting. Doris has been following the tour with great interest and so has Alice (she has been making a scrapbook).

Even though I have been cutting down on affairs social and otherwise, we have been out a few times. Had a monstrously dull fall convocation a couple of weeks back with a most fatuous speech from the provincial attorney-general.

Spent a pleasant morning with my friend [Henrik] Schoenfeld who is Winemaker at the London Winery. Rather fun, looking into the great 20,000 gallon vats (made of B.C. fir or Redwood). We have done a little for him on preserving wine yeasts. He gave us a case of wine which is very nice to have. ...

Your son,
Bob

1559 Pine Ave. West
Montreal, Quebec
November 8, 1951

Dear Family,[106]

This has been a very busy three weeks for us. ... The apartment now looks like home & we are very pleased with it. It is amazingly, much more spacious than 1569 ... & and we are reveling in lots of hot water & hot radiators & last but not least a very nice & considerate landlady [who] makes life very peaceful & quiet!

106 This letter recounts the 1951 visit to Montreal of Princess Elizabeth and Prince Phillip. UWO Archives, Freda Murray Papers.

This has certainly been a tumultuous month. ... Sue & Douglas had a series of dentist appointments & Sue had to have her first injection so we were a family party for three days of the week following. In between time I had to get in fittings for two dresses, then the Royal Visit and the excitement attending that ...

Dad has been cursing the movers for damage done to books – but he doesn't realize that lots of the damage is cumulative & that books don't much like being packed & re-packed as often as ours have been. ...

Now, for the Royal visit which of course was a great & exhilarating time & we seem to have made a great affair of it! ... Lots of flags & the Arts building was handsomely draped in red, the bands playing (the McGill Band almost next to us) & then the Air Force Band at the upper level & students singing in between. The time went very quickly – no one was bored by the long wait. It was 10:30 when the Procession arrived ... the Provost Marshall in his white jeep got out wondrous cheers ... & there was much appreciation of a late & portly major who marched up & down the Centre drive innumerable times. ... When at last the Royal car arrived a terrific roar of cheers went up. We were in an excellent spot & had a lovely view of the Princess & the Duke. In fact we were just below the photographers who took a marvelous picture from the top ... of the gates & we saw just what he photographed (I have marked this picture in those I am sending Alice for her scrapbook).

The car crawled very slowly up the avenue & we could see it all the way. Dad was with all the elite at the Arts building & says that it was a very pleasant ceremony there & that the Princess looked lovely. It was over very quickly – 15 minutes – but Douglas & I got out long before that as we were near the gates & we were fortunate enough to catch the first bus allowed along Sherbrooke so we were home again by 11:00. ...

I had just time to change into my nice new dress while Dad climbed into his rented morning coat, striped trousers, grey waistcoat & tie and grey gloves, of course. He looked very distinguished & handsome. ...

... at lunchtime Dad had phoned to say that it might be impossible for us to get through the crowds in a taxi from here for the banquet at the Windsor (Hotel) & that he had taken a room at the Dominion Sq. Hotel where we could take our evening clothes in suitcases as we went to the City Hall in the afternoon – when after the reception we could dress for dinner at the Windsor ...

... Crowds were beginning to gather around the City Hall. ... We had to show our cards innumerable times & were passed from one doorman to another and eventually arrived at the steps up to the staircase ... The decorations were magnificent & the Hall, always handsome, looked really beautiful. The silk banners of flags were lovely against the white marble walls ... The table holding the City's famous Golden Book was covered with gold brocade & the 4 chairs were of dark brown wood with crimson cushions. Flowers & ferns were around the bases of the columns & dark red carpets covered the aisles between the sections of chairs.

There was no clatter or noise & everything was quiet & very dignified. ... Everybody had to be in their places by 4:30 – thank goodness the chairs were comfortable for we had a long wait. It was most amusing to watch all the people arriving – the men very correct in morning dress: the ladies equally correct in cocktail dresses but exhibiting every variety of taste, from the opulent to the severe. ...

... The balconies were filled with reporters & photographers & I hope you heard the broadcast – by a B.B.C. man who was immediately opposite us. We had quite nice seats, next to the main aisle & the Mayor's office – about 10 rows from the dais: so we could see well. Fortunately, were sitting next to friends so we could enjoy quiet conversation during the time of waiting. A Mr. Bogert sat in front of us – a delightful elderly councillor who Dad likes very much indeed, though I had not met him before. ...

... At 5:00 ... the excitement mounted but at 5:15 the Mayor [Houde] hustled in & made an announcement that the lengthy (75 mile) tour of the City had made the Royal visitors late & that they would not be here for another hour! ... Everyone settled down for more talk a little more slowly now.

But at last we heard the roar of the motor cycle escort & then the fanfare blown by R.C.M.P. buglers & then the Royal Party entered the main doors escorted by the mayor, paused a moment for the national anthem & then walked closely past us & to the Mayor's Office. Dead silence!!! You could heard a pin drop & Dad said 'Good God! They must think we are all dumbstruck! I'm going to start a clapping of hands when they come out again.'

We hoped the Mayor would give the Princess a glass of sherry, for she looked quite weary – who would not after a 75 mile tour of the City – McGill reception, Canadair reception, Vickers reception, Chalet reception ... an exhausting day! After ten minutes they came out ... and I am sure the [Mayor] had insisted on that glass of sherry for the princess had her lovely colour again & looked most charming in her green velvet dress & very pretty hat. She walks beautifully with perfect balance & poise.

The moment they appeared Dad started the applause, which continued till they were seated on the dais. The Mayor, representing his chair of office, read the address of welcome in French & English: the Princess replied with a graceful little speech. The Royal signatures were written in the Golden Book & the Princess replied that everyone in the hall should be presented to her (all 475 of them). The first, of course, were Archbishop Leger & Archbishop Roy (of Quebec) resplendent in their magnificent robes ... Then the Executive Committee & wives, Councillors & wives & so on. It was well organized ... no confusion at all. The Mayor exhibited his marvelous memory for names, introducing most of the people without ever glancing at the personal card each had ready. Dad was introduced as Councillor Dr. Murray, Representative of McGill University & the Princess shook hands with him & then stopped him & had a short chat about the University, the number of students, his Department & so on & then he passed on to shake hands with the Duke, who also had a little talk about his rather unique position as Professor Councillor whilst I made my curtsey & shook hands & received a very charming smile from the Princess and then the Duke. Then we passed around via the Council Chamber & came back to our seats.

When everyone had been presented the Royal Visitors left, escorted to their car by the Mayor & as they walked slowly through the Great Hall, a perfect wave of applause – cheers arose from all those staid & correctly attired persons present – a very nice tribute! We very quickly got our coats from the Councillors room whilst others struggled with the 'Vestaire' – Mr. Bogert & his daughter gave us a lift to the hotel in their car. By this time it was 7:15 – the deadline for the Banquet was 7:45! ...

... I do not believe I have ever dressed so quickly for any function. In exactly 20 minutes Dad & I were ready & we had only to put on our coats & walk over the Square. You can imagine the milling crowd ... but we managed to squeeze through all right. It was an icy cold but clear day ... and we made a grand entrée through surging crowd of sightseers. ... There were three dining rooms full of guests ... we were in an excellent position to see the High Table. ... The buglers of the R.C.M.P. ... announced the arrival of the Princess & Duke & Mayor & Mayoress. ...

... Dad looked very handsome & distinguished & we were sitting with Colonel & Mrs. Vautelet who were both interested & amusing – keen, informed & very conversant with the North Country. She is a personage in the Liberal Party, I believe, & writes & lectures a good deal but she is not at all oppressive & we had an interesting evening with them. Our other neighbors at the table were a most objectionable youngish man & his much opinionated wife. He was obsessed with his ideas on "The New Order" and "Down with Tradition" was his watchword. I could not help feeling that he should have been strong-minded enough to refuse to attend this Banquet – but he had a reputation for keeping with the right people!

We spent about ½ an hour making our way through the crush to chat with spectators here and there. ... At last we managed to get our coats & ... squeezed out through the East door into the frosty night. ... We walked around to the Hotel & I changed into my afternoon dress – packed my gown into my suitcase, while Dad packed his 'morning dress' into this case ... we paid our bill, got a taxi & left for home. We didn't have to worry about the dog ... for I left him with Dr. Baker for 24 hours.

It was 1:30 when we went to bed after this most interesting day & Dad & I agreed that we both enjoyed ourselves tremendously throughout & that although we were not exhausted we were jolly glad to get to bed after all. ...

... We are ever so happy & comfortable here & and I am very happy that Dad likes it

so much. ... It really looks like home now & we wish you could come & see it. ...
>
> Lovingly,
> Mum

London, Ont.
11 November 1951

My dear Mum and Dad,[107]

Doris is scurrying around getting things wrapped for Alice's birthday – tomorrow as ever is. It is hard to believe that she is seven – even though she is beginning to seem quite grown up. She is quite the schoolgirl and enjoys things very much. She reads very well and has lots of resources.

Doris is also very excited about going to N.Y. as you might expect. She has not been back since before Alice was born, and has made a few (!) plans for our four-day stay. I will be badly tied up one day and she plans a flying visit to Princeton. She needs this short holiday badly, having had no rest this summer and with winter yet to go through. It will be fun to go away together for a change. ...

Doris' mother has arrived to be sitter-in-chief for the week. We have things worked out fairly well and it should not be too hard for her.

Look forward to your letters.

Your son,
Bob

London, Ont.
1 December 1951

My dear Mum and Dad:[108]

... We look forward to discussing so many things when you come down for a visit. This is an event we have been awaiting. I have made a reservation for you at the Hotel London – you will be much more comfortable there really and you would not be exposed to the early morning play of the kids (that can only be described as a decibel attack).

Plans for the meeting in Toronto are all set and it looks like an interesting program even if it does follow the usual lines. I have not had the energy to organise anything special but have managed to keep the number of papers down a bit so that there should be enough time for discussion.

---

107 LAC, E.G.D. Murray Papers, vol. 19.
108 Ibid.

We had a wonderful trip to N.Y. and wish that it could have been a bit longer. We had a few people we had to see and budgeted our time for them. Had a pleasant evening each with Ash,[109] John Jewell[110] and Doris' brother John.[111] Saw a great many friends at the meetings and derived quite a bit of useful information. The meetings were very good and the discussions were unusually lively. Many old friends of yours were there and sent their greetings. Among them C.P. Rhoads,[112] Peyton Rous,[113] Stuart Mudd,[114] and many others. Doris saw many of her bosses of Rockefeller days and had a great time seeing them all. Frank Horsfall,[115] by the way, had a severe viral meningitis a year ago ... and ... it took him 8 months to get back to normal. I had some very valuable, to me, talks with phage workers, especially Seymour Cohen,[116] Mark Adams and Ash – they were able to give me some interesting and helpful advice; they were quite interested in our stuff too.

There were lovely things in the shops. Many things expensive but many things equally cheap and of excellent quality. I wish we could have had more spending money – especially for materials. I went, briefly, to the "Anglers Roost" which is a very small shop in the basement arcade of the Chrysler Building ...

We had comfortable journeys both ways and in general the trip was not too tiring. The meetings were a little concentrated and I had to miss one session. It was a

---

109 See Introduction, note 377 for information on Igor N. Asheshov.
110 John Jewell was a scientist at the Ayerst, McKenna and Harrison Ltd. Laboratories.
111 John Marchand was a physician. See chapter 3, note 94.
112 C.P. Rhoads was director of the Sloan-Kettering Institute, the cancer research centre that was established as the New York Cancer Hospital in 1884 and today is known as the Memorial Sloan-Kettering Cancer Center, from 1939 until his death in 1959.
113 The American pathologist Peyton Rous (1879-1970), known commonly as the "father of the tumor virus," was a world-renowned cancer researcher who was awarded the Nobel Prize in Physiology or Medicine in 1966 for his discovery of cancer-causing viruses.
114 The bacteriologist Stuart Mudd (1893-1975) was most renowned for his work on the freeze-drying of blood plasma and combating patient infections. He joined the Department of Pathology, School of Medicine, University of Pennsylvania in the 1930s and eventually became head of the school's Department of Bacteriology when it was established. His wife, Emily Hartshorne Mudd, was the consulting editor of the Kinsey Report, *Sexual Behavior in the Human Female*. Luigi Mastroianni, "Emily Hartshorne Mudd, 6 September 1898 – 2 May 1998," *Proceedings of the American Philosophical Society* 144, no. 1: 99-104; University of Pennsylvania, University Archives and Records Centre, "A Guide to the Mudd Family Papers, 1919-1980."
115 See chapter 4, note 120.
116 The biochemist Seymour S. Cohen was assistant professor of biochemistry and pediatrics and subsequently professor of biochemistry and of pediatrics at the University of Pennsylvania. Later in his career he held posts at the University of Colorado, the National Cancer Institute, the University of Tokyo, the State University of New York at Stony Brook, and the University of California in San Francisco. In the 1940s he "offered the first systematic exploration of the biochemistry of virus-infected cells and of how viruses multiply." Seymour S. Cohen Papers, American Philosophical Association, http:// www.amphilsoc.org.

worthwhile trip for both of us; it was especially good for Doris.

Doris' mother did not have too much trouble looking after the kids. Various people lent a hand and took a child for a while to give her a rest. The O'Neills, next door, are grand neighbours and helped enormously. Clint kept the furnace going and kept an eye on things generally.

When we came back there was about 3 inches of snow at St. Thomas, only ten miles away, but when we got into London there was about 14 inches. They had a very brisk snowfall, followed by a week of winter. It has been mild since.

... Sir Edward and Lady Mellanby[117] were here and we went to a pleasant (if prolonged) dinner party that the Collips gave for them. Mellanby gave us a very good seminar on Vitamin A and lead up to the work they are now doing on tissue cultures of bone with Honor Fell[118] at the Strangeways. It was very nice to see a piece of work that he has followed, discontinuously, most of his active life, bear much excellent fruit now that he is officially retired. I was very impressed. Apparently Honor Fell is very excited about it and is putting full time on the problem. I expect that you have seen the Mellanbys since then for they were expecting to see you in Montreal.

I have been in Toronto a couple of times lately to get the program cleared up. The last time I saw Van Rooyen he was worrying about publishing the work he has done on his section of viruses. Apparently the NY Academy has a meeting on virus classification sometime in the New Year (I wish I could go to that too but have to draw the line somewhere and we haven't the funds for more such trips at the moment). I gather you have been in contact with Van about his stuff; I think he has done a good job.

Work has had its ups and downs. I hope that with the New Year I will be able to get more done. For the time being we have to get out applications and reports; irritating to have to do these things such a long time before the ... Year is out.

Peter has been under the weather today with some vomiting and diarrhoea; he seems better already. I fear this will mean a little family epidemic. Goes the rounds here every year. Everyone else has been very well so far.

The next visitor is Roger Stanier who is coming on the 11th or 12th. It will be nice to see him.

We look forward to seeing you both and the children talk about it a lot.

    All our love,

        Your son,

            Bob

---

117 The British medical scientist and administrator Sir Edward Mellanby (1884-1955) was a professor in physiology and secretary of the British Medical Research Council (1933-1949). Some of his most important work involved the study of rickets. Lady Mellanby (1882-1978) collaborated with her husband on many nutritional studies, and particularly the impact of dental health on nutrition. *ODNB* (Online).

118 Dame Honor Bridget Fell (1900-1986) was a distinguished British cell biologist. *ODNB* (Online).

>Montreal
>28 December 1951

Dear old Son,

We loved the few days spent with you & it was delightful to see the family so well and happy. We found the children grown & splendid and great fun. It was good to see Doris so well & happy in your own house. We came away very happy.

We spent Xmas day with Sue & family. They are all well, except Blake who has a cold similar to Mum's. The presents were under the tree. Thanks to all of you for my nice gifts. Mum is afraid I shall be undressing to show people my handsome underpants; they are splendid trout & I shall have heaps of fun with them. "The Pascoe" will be jealous. "The Fly & the Fish" by John Atherton looks interesting & I shall read it soon. ...

The staff gave me a most elegant umbrella. I may learn to use it. I suspect Mum had something to do with it. They seem to have had a good party at the lab & a lot of fun. One for the senior staff & one for the technician staff. Everyone draws a caricature of one whose name they drew & some were very good.

I have to give the New Years Day address to the luncheon of the University Club. I don't feel much like it but could not well get out of it.

The train back to Montreal was 2 hours late. Montreal is piled high with snow & though the streets are cleared there are high piles on both sides & the sidewalks are difficult. The last few days have been cold, today is 15 below & colder outside the city, but it is bright & clear. I am glad of my warm coat & fur cap.

I must get going.

>Love to all of you,
>>Dad

>29 December 1951
>The University of Western Ontario
>Faculty of Medicine

My Dear Mum and Dad:[119]

It was so nice to see you for the week before Xmas that the visit in itself would have been more than enough of a present for us. We were not the only ones that enjoyed the visit. The children loved it and are still talking about it even through the haze of Xmas itself! It concerned us a little that it must have been rather expensive for you – the main thing is that we had a real visit together and enough time to enjoy ourselves without being rushed. We wished, so much, that you could have stayed and enjoyed and shared the happiness of the children and ourselves on Xmas day but we know you would have stayed if you could.

---

119 LAC, E.G.D. Murray Papers, vol. 19.

It was a long day and voices were heard before 6 o'clock! ... we had things under fair control and insisted on a proper breakfast that started soon after 6. Then the hurly burly began. The kids got such a galaxy of presents that the list boggles my eyes but certainly yours were among the most successful – especially the fire engine for Tommy: he has been raising and lowering ladders all week. For Peter the train and the accessories were a "succés fou". He has used them for practically all his waking hours and could only be persuaded to eat his Xmas dinner on the plea that engineers have to eat too! The engine works perfectly and certainly the freedom from winding problems is a great help to real enjoyment. He used 2 sets of batteries in 24 hours! Alice had a very happy time and has a tremendous set of good toys and books.

The turkey came out excellently. It was treated according to prescription and did not give us any bad moments! A great dinner with nice wine and all the trimmings. We were all pleasantly full.

It was grand to talk to you on the phone and give our greetings. Certainly, what we said then could not have done justice to all your thoughtfulness and kindness and presents. We were very touched by the notes of love and happiness that accompanied the cheques – the latter were much ... but will be put to good use.

The rest of our Xmas day was a little more hectic for we went for tea with the Hannays[120] and Robinows. A little noisy! though fun. The going underfoot was heavy for we got a good six inches of snow on Xmas morning on top of more fresh stuff. The weather couldn't have looked more like a Xmas card.

Robin gave me a beautiful book by C.M. Yonge called "The Sea Shore" – really on collecting in the intertidal zone around the U.K. with the finest colour photographs of natural materials I have ever seen. Really lovely and excellent reading.

Our post Christmas state has taken a new turn for I suspect that Peter has a very mild go of the mumps. He has tender and slightly swollen parotids without anything else to account for them. I fear the final decision rests on what happens to A & T! Otherwise we are very well and full of ambition even if still suffering from Xmas lethargy. For myself, I really think that I feel well again without any of the vague but unpleasant tiredness that has plagued me for so long. Also unusual, is that I really look forward to getting back to work again. Don't be alarmed, I will go carefully and not invite trouble.

We send you both our fondest love and best wishes for every good thing in the New Year, from fortune to fishes!

    Your loving son,
        Bob

---

120 Christopher Hannay was a researcher at the Agricultural Research Institute, Canada Agriculture, at the University of Western Ontario from 1950 onwards. He, his wife Hilda, and their children became close family friends to the Murrays. RGEM.

CHAPTER EIGHT

FAMILY CORRESPONDENCE: 1952-1955

McGill University, Montreal
8 January 1952

Dear old Son,[1]

I do not want the antibiotic resistant strains myself. Dr. Frieda Fraser,[2] Connaught Labs, wanted strains of any bacteria resistant to 2 or more antibiotics for an investigation she was doing. I suggested you might send any you found to her, if she still wants them.

We had a letter from Bill Farren (now Sir William)[3] saying he hoped to see us here early in February. We look forward to it greatly. His father is 87, less active physically but unchanged mentally.

Ashley Miles[4] sent us a lovely little book of Ackerman prints of Cambridge in the "King Penguin Series". Splendidly done.

      Love to all of you,
          Dad

---

1 Unless otherwise specified the correspondence in this chapter is located in the UWO Archives, R.G.E. Murray Collection.
2 Freda Helen Fraser (1899-1994) was a prominent lecturer in microbiology and researcher at the Connaught Laboratories at the University of Toronto. She was appointed professor of microbiology in 1955. Katherine Perdue, "Passion and Profession, Doctors in Skirts: The Letters of Doctors Freda Fraser and Edith Bickerton Williams," *Canadian Bulletin of Medical History* 22, no. 2 (2005): 271-280.
3 Sir William Scott Farren (1892-1970) was a Cambridge-educated aeronautical engineer. He held many high posts in his career, including director of the Royal Aircraft Establishment (1941-1946), technical director of Blackburn Aircraft Company (1946-1947) and A.V. Roe and Co. (1947-1961), and director of Hawker Siddeley Aviation. He was elected a Fellow of the Royal Society in 1945 and was knighted in 1952. *ODNB* (Online).
4 Sir Ashley Miles (1904-1988) was a British microbiologist whose early interests in pathology and bacteriology were encouraged at Cambridge by E.G.D. Murray and H.R. Dean. He quickly rose to prominence, becoming chair of bacteriology at University College Hospital (London) at the age of thirty-three. He later became deputy director of the National Institute for Medical Research and, in 1946, head of its biological standards department. In 1952 he assumed the directorship of the Lister Institute of Preventive Medicine and was named professor of experimental pathology at the University of London. Significantly he co-edited five editions of *Topley and Wilson's Principles of Bacteriology and Immunity* with Sir Graham Wilson. A. Neuberger, "Arnold Ashley Miles, 20 March 1904-11 February 1988," *Biographical Memoirs of Fellows of the Royal Society* 35 (March 1990): 304-326; *ODNB* (Online).

1559 Pine Ave. West
Montreal, Quebec
Jan. 12, 1952

Dearest Bob,[5]

Time has a habit of sneaking by when one is busy & having been seriously occupied with a number of things such as house-painting, a bit of sketching, a sudden spate of photographs at the Arctic Inst., our season colds, & so on. ... Dad is still miserable with sniffles & general malaise. ... He is very depressed & feels the burden of piled up work at the Lab plus the annual budget & other troubles. ...

... The New Years came in with a few pleasant parties of which the most enjoyable was given by (the Starkeys)[6] – hot mulled wine ... that went down very agreeably – "straight to the cockles of your heart". Their delightful house was crammed with their innumerable friends – about 50 arrived in 10 minutes & others at diverse times & we all had a good time. ... After the party we were easily persuaded to stay on & have a buffet supper & we spent a most pleasant few hours eating & talking 'in the bosom of the family', as Hugh said. The Duffs[7] & the Howletts turned up later in the evening; and one way & another time sped by & we walked home over the hill at about 1:30! ...

Your loving Mother

McGill University, Montreal
15 January 1952

Dear old Son,

This cold I have is a stinker. It hangs on making me feel very poorly & disinclined to do anything.

Douglas suddenly started a swelling at the angle of the jaw both sides, painful; with a temperature of 100 & a little over; no injection of the duct. They thought first it might be mumps but it lasted only 2 days & they now think it to have been lymphatic glands due to his tonsils ... The test will be the other kids he plays with. However, there have not been any cases in the vicinity & it is difficult to guess where he might have caught mumps.

Today is your Wedding Day & we wish you long years of happiness & success.

... I was talking to Karl Stern[8] yesterday & he asked me to remember him to you. He is a very nice & able fellow. ...

---

5 UWO Archives, Freda Murray Papers.
6 See chapter 3, note 102 for information on Hugh Starkey of Queen Mary Hospital, Montreal.
7 See chapter 2, note 84 for information on Lyman Duff, dean of the McGill Medical School.
8 The German-Canadian psychiatrist Karl Stern (1906-1975) taught in the McGill Medical School.

Mum is painting the apartment. It makes quite a difference & she is pleased with it. It will take a while before it is all done & may be like the Forth Bridge. ...

We had tea with Bill & Mrs. Newman[9] on Sunday. He sends you his best. They seem well & are very keen on their place on Dorval Island. They spend the summer there & he drives in & out to work. I would think it requires considerable means.

Love to you all,
Dad

17 Jan 1952
The University of Western Ontario
Faculty of Medicine

My dear Dad:[10]

I am so sorry that you have such a beastly cold. They can be so miserable and incapacitating. We have been lucky so far and have not had any colds to speak of. ... We have had a few other bits and pieces, mainly mumps. This latter was very mild and was only decided by exclusion. Very little swelling but confined to parotids; slight fever and little else. Tommy and Alice got exactly the same thing two weeks after Peter came down with it. Duration of swelling only 4 days. Funny business.

Thanks for the greetings on our Wedding Anniversary. We spent an ordinary day because I was very busy preparing a seminar for yesterday. We had a nice quiet evening last night listening to the "Marriage of Figaro" and had a drink or two.

The seminar was one of those complicated and confusing things on the current status of phage work; fun to look back upon but hell to prepare. It went well and staff and students seemed to get a lot of stimulus from it – so the effort was worthwhile. ...

About the Toronto trip: I have one or two things I have to do but would enjoy seeing you if we can fit our times together very well. ... I think I will come down on the Monday morning train that gets in about 10:45 am and go back by the evening train on Tuesday. Have you booked a hotel room? If so is it a twin bed room so we could share it? ... While there I want to see the Sick Kids hosp and Ted Roy's lab.[11] I want to go out to van Rooyen's Lab.[12] and I want to see a couple of people in the Botany Department. We can probably combine our visits to some extent.

I have become very busy again and seem to thrive on it! It is a really nice feeling – no more of that constant lassitude and tiredness. Some of the work was badly in need

---

9 See the beginning of chapter 3 for information on the Bark Lake group.
10 LAC, E.G.D. Murray Papers, vol. 19.
11 See chapter 1, note 89 for information on Ted Roy.
12 See chapter 6, note 61 for information on the British-trained bacteriologist C.E. van Rooyen.

of attention and it will take several weeks of slogging to get it moving smoothly again. However, we are getting some very interesting and worthwhile results.

The weather is lousy and would be a credit to Glasgow! Nothing but fog and rain. Doris has had her hands full with all three kids home most of this month. Will be glad to get them all back to school this week.

Glad to hear news of Bill Newman and of Karl Stern – wish I may see them again.
 Love to you both,
  Bob

London, Ont.
21 January 1952

My dear old Dad:[13]

... Your greetings on our anniversary were much appreciated. We have much to remember and much to look forward to. You do us proud to congratulate us with our tenth – but by all methods of arithmetic we come to but nine! We had us a dinner and a quiet evening – enjoying ourselves in seemly fashion.

I had a flattering and most pleasant letter from C.A. Evans (Seattle)[14] saying that he would have an opening in his department – and that if there was any chance of my being interested, he would fight for a good and sufficient position. Remembering the people, the surroundings and the pleasure we had there I had some deep thoughts. However, I feel that there must be much deeper attractions before I submit to the disruptions, loss of valuable working time and changes in the freedoms I enjoy. I must say that if I was offered a research position with some interesting and not too burdensome teaching I would be deeply tempted!

There have been some pleasant and stimulating evenings with friends; and one excellent lecture by Lloyd Stevenson[15] (medical history) on the development and origins of ideas concerning vivisection and anti-vivisection. This seemingly dull subject was decked in such scholarly fashion and thoughtfully presented that it was fascinating. ...

We have a lot to do these days – both bench work and writing – and the days hardly

---

13 LAC, E.G.D. Murray Papers, vol. 19.
14 See chapter 6, note 76.
15 Lloyd Stevenson (1918-1988) was known for his work in medical history. He was the author of *Sir Frederick Banting*, an early biography of Sir Frederick Banting, and of *The Meaning of Poison* (Lawrence: University of Kansas Press, 1959). Leonard G. Wilson, "Lloyd Grenfell Stevenson (1918–1988): Medical Historian and Man of Letters," *Journal of Medical History and Allied Sciences* 43 (1988): 377-385.

seem long enough. It has been more difficulty with staff troubles – soon to be rectified I hope. ...

All our love to you both.

Your son,

Bob

10 February 1952
The University of Western Ontario

My dear Mum and Dad:[16]

... The trip to Toronto was grand and I very much enjoyed the chance to see you, Dad. I was there again yesterday for DRB committee meeting but did not stay on ... I saw van Rooyen at the meeting and he was very sorry not to have had the chance to see you. Armand Frappier[17] was there and I had a long talk with him about the local arrangements for the C.S.M. meeting, the provincial Soc. and many other things. I was very glad of the chat because he never seems to write or answer letters and so I found out a lot of things that I wanted to know. He is a very pleasant fellow to talk to.

I have had a busy time in the lab and time has gone very quickly. Robin [Robinow] and I have been concentrating on a paper on "plasma membrane" and cell wall but we have done quite a few experiments. The work on the odd organism "1297" has suddenly turned up as a useful clue or two. We did some comparative work with the 4 sp. of *Azotobacter*. They are astoundingly similar in morphological features: capsule size, "vacuoles", and peculiarities (and difficulties) of cytoplasm and nuclei. No flagella, however. A couple of them have a need for iron and grow well on Blood agar! ...

We have had various social do's and some pleasant friends in for evenings. Among them is an old friend of Eugene's,[18] Dave Scott by name, who is now in Tony Brown's department.[19]

The sudden death of the King was a great blow. I found it very hard to believe and most surprising without warning. Very hard on the new Queen to have been away at the time. I am sure that Elizabeth will be a fine monarch; and with Philip's obvious good sense and cheerfulness might even be a great one. Churchill made a fine oration, I thought, and as impressive as I have ever heard him.

Was visited in the lab by Robert Cruickshank (Mary's Hosp) the other day. He came to visit some relatives (eat Haggis and drink whiskey, he said!) and spent a

16 LAC, E.G.D. Murray Papers, vol. 19.
17 See chapter 5, note 48 for the Quebec physician and microbiologist Armand Frappier (1904-1991).
18 See chapter 4, note 11 for information on the entomologist Eugene Munroe.
19 Tony Brown was head of the Department of Zoology at the University of Western Ontario.

couple of hours in the lab. He sends you his regards and regrets that he cannot visit Montreal. He is on a commission job looking into general problems in medical education in the U.S. He seems to be very well and quite lively.[20]

All our love to you both.

    Your son,

        Bob

17 February 1952
Montreal, Quebec

Dear old Son,

I have not written you since my return from Toronto. I loved the time spent with you & enjoyed our visits to various friends, but most the time we were together.

The journey back went well. I met with some friends, old pupils & talking with one & another took up the time. One particularly was Crabtree, an old schoolfellow of yours & he asked particularly to be remembered to you. He is a good fellow & someone of authority in the lumber or paper company which owns the rights in the Mattawa country you wish to go to. This came out in conversation & I thought you might find it useful to know it.

I saw Norm Gibbons[21] here on my return. We had a diffuse kind of conference ... which I hope Norm got something out of. I fear it left me rather on one leg. If Norm lets me know just what the Council of the Microbiol Soc. wants particularly fixed, I will do my best to have it arranged just so. Your talk with Frappier will have done great good. Armand F. is a very nice fellow & very competent. I get on with him extremely well.

I am sorry not to see Robert Cruickshank.

Burkholder[22] was here as a (guest) lecturer & spent the morning in the lab. He seemed to enjoy it. In the evening he gave an exemplary talk on antagonism & cooperation amongst microorganisms; mostly antibiotics & some vitamins. Very well put together & delivered most attractively & modestly. He was surprised at all that goes on in our lab, particularly as he had been invited to look over the Botany situation a little while back & had refused the offer. He was surprised that he was not shown our

---

20 Robert Cruickshank was professor of bacteriology at St. Mary's Hospital, and he later became professor of bacteriology at the University of Edinburgh. He was a major figure in British bacteriology and he visited UWO several times. In 1952, he was on a tour of sixteen American and Canadian medical centres. *Science* 115, no. 2989 (April 1952): 390.
21 See chapter 5, note 80 for information on the Ottawa-based microbiologist Norman E. Gibbons.
22 W.H. Burkholder, an American bacteriologist at the Department of Plant Pathology, College of Agriculture, Cornell University. He was a contributor to *Bergey's Manual*.

lab & said it would have made a difference to his opinion. The "authorities" at McGill have no idea how to handle the situation.

I can see it would be hard for you to evade involvement in any "Civil Defence Scheme" that is developing in the region. I am glad I resigned from everything to do with D.R.B. [Defence Research Board]. I saw recently in the paper that Otto Maass[23] has again been appointed to authority in National Defence over B.W. & also another chemist, Hugh Barrett, at Suffield. It surprises me that self respecting bacteriologists tolerate the situation. I read a report of a meeting in the U.S. & a talk by a Canadian entomologist. I was annoyed exceedingly by the brag that they had established cooperation with two American immunologists, whose names mean little to me, to identify blood sucking insects from the NW Territories. The D.R.B. bacteriology committee cannot be doing anything. The situation seems ridiculous. I notice various bits & pieces that confirm that I am right.

I, too, am sorry not to have seen van Rooyen.

It is good news that you & Robin are getting back into your stride. You have a lot of good stuff on the way.

The King's death was a shock for the lack of warning. He did magnificently throughout his Reign. Churchill's tribute to him was one of his very finest accomplishments. Queen Elizabeth will be splendid too & I am confident that it is great good fortune that the Old War Horse, Winston, is there to advise and encourage her. It would have been a sad & sorry situation if She had been forced to rely on Attlee & Co.

I was greatly surprised the other day to see in the papers a statement by the Hon. Paul Martin about our work on the isolation of T.B. I was not consulted. Fortunately it did not say anything outrageous. I wish I could get that problem properly cleared up, either by having time to do it myself or find a suitable assistant to carry it through with guidance.

Mum has done a beautiful pencil drawing looking N. over Lac Escalier to the entry of the Devil Run & up to the Devil valley. It is the best she has ever done & has wonderful character. I'm keeping this one, for a change. The more you know the place the more you see in it.

We thawed out [the] pheasants we were given at Xmas & I skinned them this afternoon. Mum is cooking them in the French Canadian way for partridge "Au Choux" & the place is pervaded by a wondrous smell. I wish you were here as it is really "very spechul". ...

   Love to you all,
    Dad

---

23 See chapter 2, note 30.

20 February 1952
The University of Western Ontario

My dear Dad:[24]

I am glad that you have had a good journey home and had some good companions to talk with – it can be a weary seven hours sometimes. The last time I saw Crabtree he was doing a PhD (I think) in cellulose chemistry. He must be getting up in the firm by now – as I remember it he was being supported by a very good scholarship provided by the Paper companies.

Norm (Gibbons) seemed to extract some general information from the "diffuse" meeting and may have helped to get some of the objectives straight. I think my talk with Frappier may have been more helpful for he is most anxious that the whole thing is a success. It was the first time that I have had a real talk with him and I was very impressed by the breadth of his understanding of many and various problems. ...

24 Feb

We have had a hectic time lately with the social activities of one sort or another. Since writing this things have become more hectic because Doris has been ill. We had been out to a Faculty party on Wednesday night and had a bad night because Alice's tummy was very upset. Doris was very tired the next day, and by evening complained of chilliness and had an obvious fever. She must have had quite a fever during the night and it was 102 on Friday morning. During Friday it was very alarming with four spikes up to 104-105 and rigors. Frank Kennedy[25] came in to see her and could find little or nothing to account for it and could only suggest that it might be a virus pneumonia. ... She is now cheerful but still abed. So I have been chief cook and bottle washer. Luckily I have been able to leave the children for short intervals with friends to get essential business done.

Robin is in hospital for a short time getting an inguinal hernia repaired so that I have to get down to the lab. He is doing well and will be back soon.

We have been tearing the guts out of our refrigerator room as part of our campaign to reduce the mould menace! We will then treat all surfaces with these chemical compounds. Talking of refrigerators we have done a deal with the old GE refrigerator. I was beginning to be a bit alarmed by some clunks and other mysterious noises in the mechanism – also by the hinges wearing out and in imminent need of replacement. ...

I must turn to the cooking again for lunch time approaches. The kids seem to treat my offerings with great fairness and lots of advice. The roast seems to have turned

---

24 LAC, E.G.D. Murray Papers, vol. 19.
25 See chapter 4, note 16.

out alright, but I am not a great hand at desserts and I think I will have to buy a brick of ice cream!

    All the best,

        Bob

Department of Bacteriology and Immunology
The University of Western Ontario
2nd March 1952

My dear Dad:[26]

... Another excitement, especially for the kids, was that the cat produced a litter of 5 kittens yesterday (Saturday) morning. They are very thrilled! It was a well timed arrival and gives the kids something to talk about!

Robin is back at work again after an uncomplicated hernial repair. Although he says that he has to try to avoid laughing for the time being, he is as full of jokes as ever! ...

We are getting some good work done in the lab. The necessary details on the cytology of T5 ... are filling in nicely.[27] We are having trouble, however, with the chemical work on the nucleic acids during infection. The methods we have been using are good enough in theory but in practice are subject to a lot of interference from other extracted substances, especially some proteins. The cytological work, however, fills in the times of difficulty and it is coming along well. I have been corresponding with [Ralph W.G.] Wyckoff[28] on some possibilities of collaboration to get E.M. [Electron Microscopy] and straight cytological work into some sort of working agreement. He has offered to finance a trip for me to go down to Bethesda for a day or two at the end of the month. He had done some work on T5 infection and has a man ... doing very good work on phosphorus uptake by infected cells.

The work on our comic bug (1297) has taken a turn for the better since we have found that there are striking morphological similarities to *Azotobacter* – it differs physiologically from the four representative species that we have for comparison. The problem now is to determine whether it fixes nitrogen. This is not so easy to determine because it requires complex nitrogen anyway – it may well fix nitrogen as

---

26 LAC, E.G.D. Murray Papers, vol. 19.
27 This is a reference to T5 bacteriophage. Robert Murray had been working on two of the six T-phages at this time, the T2 and the T5. The T-phages were and remain useful for research purposes because they exhibit differences for study of the infection processes. RGEM. See R.G.E. Murray and J.F. Whitfield, "Cytological Effects of Infection with T5 and Some Related Phages," *Journal of Bacteriology* 65 (1953): 712-726.
28 See chapter 7, note 60 for information on the American microbiologist Ralph W.G. Wyckoff.

well. About the only certain method would be to use N15 (the heavy isotope) which would require equipment and techniques that are not yet available in Canada as far as I know. ...

All the best to you both.

Your son,

Bob

16 March 1952
The University of Western Ontario
Faculty of Medicine

My dear Mum and Dad:[29]

So nice to get long letters from you both. You seem to be keeping more than busy with Councils, institutes and paint! ... We both seem to have some trouble finding time and energy to get started. ... We have to make up a priority list. ...

We seem to have been semi sickly for quite a while. ... As a result we have had a quiet time which is, perhaps, fortunate at this time of year – heavily overloaded with social events, etc. We do regret missing the 3rd/4th year party which is easily the best party, student or otherwise, of the year.

We had a few days of cheerful spring-like weather with bright sunny skies and a pleasant feel to the air. However, last night it clouded up and this morning we have several inches of snow. ... I am afraid we are going to have a lot of false starts before we ... have more than a promise of spring. It must be "around the corner" because we have already got a seed catalogue with the usual pictures of unattainable blooms and vegetables!

I hope to go down to Bethesda at the end of the month for a visit with Wyckoff. It would be worthwhile and a help to both of us. They are going to pay my way. It is not too long now before the SAB meeting and I hope that I will see you there. We are giving two papers (one on the spore biochemistry and one on the phage work); I also have an Editorial Board meeting. I have had a letter from Robert Breed[30] about the Bergey dinner and plans for the meeting to follow. I hope that this year they will not fill the time up with statements but provide time for discussions of problems. Last year there was no time to discuss anything! He has also given me a vague idea of what he wants me to do: to examine morphological definitions in higher ranks from Class to Families. I haven't yet determined how much use this can be due to the lacunae in work.

I saw that the CMAJ had republished your talk to the CPHA. I am very glad for it

---

29 LAC, E.G.D. Murray Papers, vol. 19.
30 See chapter 3, note 138.

contains some excellent points that really need further hammering. I read it again with great pleasure.

Our note on cell wall was in the J. Bact last month – you probably saw it. It ... has a useful purpose. We have another paper finished and in stages of gluing the illustrations together as plates. We should get it away soon. Others are coming along slowly. It is hard to work and write too but we manage.

I am starting (have started) some work on the development of phage symbiosis using the cytological work as an indicator of what is going on. I have been swearing at my lysogenic *B. cereus* phage system[31] ("hazy") for several weeks because it has been impossible to get sufficient titres of the phage that does it. A lytic mutant keeps cropping up at the most awkward times ...

All the best,
Bob

30 March 1952
The University of Western Ontario

My dear Mum and Dad:[32]

... We had a long and cheerful letter from Sue – she seems to be enjoying her life very much. Dougie and Joan sound to be in grand form (barring colds and tonsils). Doris had a great time last week making a dress for Joan and has been a constant user of the old sewing machine.

There seems to have been lots of things going on but we have done little ourselves – mostly staying home and saving the pennies. A spate of lectures by Selye,[33] Evelyn[34] (whom I did not hear) and the other night David Thomson.[35] The latter was in great form and gave an excellent lecture; we had an amusing little party at Rossiter's after-

---

31 Robert Murray explains this passage: "Some phages infect cells but a proportion of the cells do not produce a new lot of phages. Instead they incorporate the phage genes in the chromosome of the host and it stays there until, at some later time or under a specific cellular stimulus, the gene is transcribed and a number of phage are produced. True of many phage systems and this one was found for a strain of Bacillus cereus and its phage." RGEM.
32 LAC, E.G.D. Murray Papers, vol. 19.
33 Endocrinologist Hans Selye was a member of the McGill medical faculty until 1945, at which point he accepted the opportunity to found and direct the Institute of Experimental Medicine and Surgery at the University of Montreal. See his autobiography, *The Stress of My Life: A Scientist's Memoirs* (Toronto: McClelland and Stewart, 1977).
34 See chapter 4, note 85 for information on Kenneth E. Evelyn of the Donner Research Institute at McGill.
35 David L. Thomson was chair of the Department of Biochemistry at McGill University (1941-1958).

wards.³⁶ We had a pleasant visitor yesterday and today from Yale: David Bonner.³⁷ A very charming and able fellow who is an excellent worker in biochemical studies of mutation and especially well known for his work on *Neurospora*.

Robin and I had some worthwhile discussion with him. I must say that I am alarmed by all the stories I hear, from people who have experienced it, of the way the "security" program and the investigations by congressional, Senate and other bodies are being handled in the U.S. Bonner has just had a most troubling experience of this having being asked to take over some important (but *unclassified*) work at Oak Ridge – after it was all settled he was refused security clearance because, on the basis of much malicious hearsay, he was considered not disloyal but a bad security risk in the future; i.e. it seems to be a "crime" to be at all independent in thought even scientifically. With all this he had to attend "hearings" (a number of them) *at his own expense*. Apparently this is the rule, that once the suspicion is officially docketed, even though false, the unfortunate is immediately deserted by his own government and has to pay his own way! This fact cost Bonner about $2000! I am astounded and very upset by the effects and implications of these happenings, and can only hope that some good sense, fair play and integrity will prevail in Canada.

I have had to put off my Washington trip for a week. This is a nuisance because it brings it nearer than I had wished to the SAB meeting. It could not be helped – Wyckoff had some sudden business come up this week.

Several of us are going to the SAB. We have space booked for ourselves and will be arriving the evening of April 27th. If you cannot get a booking, we could get an extra bed in one of our rooms. I have done this before without difficulty. Let me know. I think that you should go down if you can – Robert Breed would probably be disappointed if you didn't and many old friends would miss you.

Norm Gibbons *wishes* that he had heard from a number of people going to the C.S.M. in Montreal but he was a bit alarmed at the paucity of papers. I hope that there will be a reasonable number from your lab and some other Quebec places. I told him to expect the papers to roll in slowly! If there is a shortage of papers a symposium of invited speakers is worthwhile and could be arranged.

I have been working hard at the "symbiotic" phages described by Boyd³⁸ for *S. typhimurium*. They are very interesting but I am having difficulty getting quite what Boyd shows in his papers in his Optical Density vs. Time curves. Otherwise it is working out well.

36 See chapter 4, note 109 for information on the biochemist Roger Rossiter.
37 David Mahlon Bonner (1916-1964) was well known for his work on *Neurospora* biochemical genetics. Maarten J. Chrispeels, "David Mahlon Bonner, May 15, 1916-May 2, 1964," *Biographical Memoirs* (National Academy of Sciences) 88 (2006).
38 See chapter 6, note 96 for information on the Scottish bacteriologist J.S.K. Boyd.

Are you making plans for the summer yet? We have not made any decisions ourselves and I find it hard to plan at the moment with a lot of unfinished business ahead of me.

 Lots of love from us all.
  Your son,
   Bob

                Montreal
                6 April 1952

Dear Old Son,

... The snow has disappeared astonishingly & the robins, grackles & a song sparrow have been round for some days. We have not been on the mountain, but I suppose other birds have been through. We saw a small hawk the other day. Elsie says the ice is 30" thick with 14" of blue ice, so it won't be out for a while.

All the family at Lakeside have had "flu-like" colds but seem alright. Dougie will have to have his tonsils out soon, they are glazed over with scar tissue & have been enlarged & troublesome for too long. They yield Staph pyogenes every time. They show improvement since he was immunized with toxoid but it was done too late to save the situation. It ought to have been done long ago. Joan was a sick girlie on her birthday. Mum was out there & the little one had good fun with her one candle, trying to blow it out. Sue is delighted with the little dress Doris made for her.

I am much better but I do not feel very energetic & there is a lot to do. There is the usual pressure of graduate student theses to be ready on time and some being held over to fall. It is always a struggle. My biggest trouble is the mycology lab as the money for it is just running out. I have spent a lot of time trying to get support for it without luck & so I put in an application for a Public Health Research Grant as a survey of mycotic diseases here. The committee turned it down as a diagnostic service, though I argued that diagnostic work has to be offered to get material & cases for a survey & that it involves a lot of research. I have appealed to both the University & the Provincial Ministry of Health & I am waiting to hear what either (or both) will do about it. Anyway it will keep me tied up this month. If I cannot get support I must close down the lab & must try to find jobs for my two people. I have to read some of our own theses & some from other departments & there are examinations for four courses. I don't see how I can get away.

David Thomson said he saw you in London & that you were well. I don't think I know Bonner. What he tells seems familiar with added features & exaggeration of so-called "Security" investigations. The authority is in the hands of the wrong kind of people. The implications are bad & unjust without any possibility of redress for malicious actions. Science is being used as a political tool, both nationally & internationally, to the great disadvantage of science & to the danger of scientists. The accusations about bacteriological warfare are ridiculous but even the statements reported in the press by the US "authorities" are inaccurate & misleading. They might at least take care that what they say is true. It is best not to have anything to do with any of the classified work.

I shall try to have something offered to the C.S.M. Fritz Blank[39] has offered a paper & a demonstration. None of my senior people have anything ready this year. A couple of graduate students getting their Ph.D. this year could do something. ...

There is the Royal Society 1st to 4th June, then the C.S.M. June 5 to 7th. A.A. Miles says he is coming here about 16th May for a couple of days on his way back to England after visiting the US.

I hope you are planning a fishing trip with Rossiter. I suppose you are planning the family holiday too. Let us know if we can help any way. You have to take some time off to get physically fit.

>   Love to you all,
>   Dad

6 May 1952
The University of Western Ontario

My dear Mum and Dad:[40]

... Last week I spent in Boston at the SAB meetings. Four of us went down from our lab and Chris Hannay (from the Science service lab)[41] joined us – so we had quite a local representation. Hannay, Truant[42] and I went down by car. It is quite a long way and is about as long a drive as one could wish to take in one day – 620 miles. It was a pleasant trip though and the weather as far as Springfield was beautiful. The scenery in the Finger Lakes region and as far as Albany where we had lunch was very pleasant. But the real beauties are in the Berkshire Hills between Albany and Springfield. Fine wooded hills with a delicate tracery of green and a great mixture of dark colours to back them up. On the eastern slopes going down towards the Connecticut River valley we went along a fine tumbling stream (the Westfield River) that was well decorated with anglers – a most tantalising sight. Soon after we ran into the clouds and streams of rain and the going was slow with lots of Sunday drivers clogging the

---

39 E.G.D. Murray established a program in medical mycology in his department in 1952, under Fritz Blank.
40 LAC, E.G.D. Murray Papers, vol. 19.
41 See chapter 7, note 120.
42 J.P. Truant obtained his PhD under Robert Murray's supervision in 1954.

roads. We started from here at 2am and we were in the hotel in Boston at 7pm. The place was crowded with bacteriologists – some 1900 were said to have registered and I can well believe it. Some of the meeting rooms were a little small, hot and hard to get into. However, things worked out very well and, instead of getting snowed under with too many papers I was able to talk to most of the people that I really wanted to see in relative peace and quiet. Altogether it was a good meeting and some interesting stuff was presented.

My paper came early on in the first day and it went very well. The slides were good and showed to advantage even in the huge ballroom. It would have been best in the phage session but altogether discussion was limited (because [Georges A.] Knaysi[43] talked for half an hour and would not be shut up). I got lots of favourable comments and a number of the really interested people got hold of me afterwards and we arranged for our own discussions. The cytological papers were, for the most part, disappointing and I think that the very best and most exciting current work was reported in the phage, virus and genetical sessions. This is too complicated to talk about here and I will save it up until I see you for I think that you will find it interesting. Aside from these rather interrelated observations (which were worth the price of admission) I did not hear of anything too world shattering. Our other paper on the spore work, given by Phil Fitz-James,[44] came in the last session of the meeting, unfortunately. He gave it very well indeed and by good fortune there were some interested people there and he got a lively and encouraging discussion. So much so that Foster[45] (Austin, Texas – who has and is doing excellent work on spores) sought him out afterwards.

The first afternoon I had to spend in the meeting of the editorial board of the J. Bact. A long but most interesting discussion. That evening we spent (Robin and I) with (Salvador) Luria[46] and Schlessinger[47] and had a most profitable time. On the Tuesday noon I had to go to a luncheon for the presidents and Secretaries of the local

---

43 See chapter 7, note 69.
44 See chapter 7, note 55 for information on the microbiologist P.C. Fitz-James (1920-2006).
45 J.W. Foster, Department of Microbiology, University of Texas.
46 After spending several years at Indiana University (1943-1950) as an instructor, assistant professor, and associate professor, Salvador E. Luria (1912-1991) became professor of microbiology at the University of Illinois, where he would remain until 1959. At that point he left for the Massachusetts Institute of Technology, where he was professor of microbiology (1959-1964) and Sedgwick Professor of Biology (1964-1965). In 1970, he was appointed the Department of Biology's Institute Professor. He shared the 1969 Nobel Prize for Physiology or Medicine with Max Delbrück and Alfred Hershey for their work on the replication mechanism and the genetic structure of viruses.
47 R. Walter Schlesinger from the Department of Microbiology, Rutgers Medical School, New Jersey.

branches as a "honoured guest" (!) – unexpectedly, for me, I had to give a short speech in my Canadian capacity in company with Dubos[48] and Scherp.[49] It turned out to be fun despite the threat of speaking – incidentally the "speech" was a very informal affair and went down fairly well as far as I could tell. Anyway it was an unexpected but important bit of public relations work.

The formal functions of banquet and Lilly award address were not too outstanding.[50] (Ashley) Miles was there and had to say a few words, which he did very gracefully. Dubos made a short historical speech delving into the "bacteriology" in ancient literature – I have an unhappy feeling that this has been done better by others in the past. After the banquet Miles was looking weary and I grabbed him and fed him a drink. We had a grand chat, and stayed up much too late. He is a most pleasant fellow. It is too bad that he cannot stop in here on his way to Montreal but he has a very tight schedule.

The Bergey dinner was on the Wednesday night. Dinner was fine and we had some amusing chat but I fear, and this is a private opinion, that the subsequent discussion was not much help to anyone, even the editors and trustees. Breed was not there and is apparently badly crippled by a neuritis; N.R. Smith[51] arrived only that day and was not very au fait with what was going on; Conn[52] and (R.E.) Buchanan[53] were there and the latter contributed, off the cuff, about the only really useful guiding and informative remarks of the session. As usual there was more than too much on the agenda and hardly anyone seemed ready or anxious to do more than listen to a few set pieces. Breed's outline of a possible classification for the next addition got only token discussion when it could have been heated and informative – the pundit, Buchanan, was obviously against many of its basic provisions and I can hardly blame him – I am too. It seemed to me both undigested and indigestible. The arrangement of the pseudomonads seemed acceptable, the rest was *not* discussed. Much time was taken over a discussion of the decision of the judiciary committee on the status of

---

48 See chapter 4, note 66 for information on the renowned microbiologist René J. Dubos.
49 Henry W. Scherp specialized in oral bacteriology in the Department of Bacteriology, University of Rochester School of Medicine and Dentistry, and later at the National Institute of Dental Research, Bethesda, Maryland. He served as secretary of the Society of American Bacteriologists in 1950.
50 The US-based pharmaceutical company Eli Lilly provides an annual award for the most outstanding researcher in the field of microbiology. The recipient presents a public lecture at the annual banquet of the American Society of Microbiology (name changed from Society of American Bacteriologists in 1960).
51 See chapter 5, note 97.
52 H.J. Conn was chair of the International Stain Commission and was at the Agricultural College in Geneva, NY.
53 The American bacteriologist Robert E. Buchanan (1883-1973). See chapter 3, note 140.

Bacterium – it ended on a general assent that no inflammatory condemnation of the decision be sent in the form of a resolution (adequate attention having been drawn to shortcomings already) but that individuals with good reasons could write to the general secretaries about the matter if they wished. I had breakfast with Conn, Buchanan and Smith; the financial status of the Manual seems to be good, as usual, and there is nothing much to add to the data that Breed sent to you.

I think that people were not in very great accord with the idea of printing the new edition with a first volume consisting only of descriptions *without* synonyms. However, this is a publication problem. I will tell you more about the meeting when I see you.

We started back at 2am on Friday and we were back home by 6pm in time for a much needed supper. We were lucky in having another beautiful day to travel in. I wish that we could have had the time to explore around a bit as we had hoped but it was not possible.

Of course, there were a large number of your old friends there and all of them asked to be remembered to you. I would find it very hard to recall all of them. On one of the bright clear sunny afternoons I went up to the Hancock Tower with Miles to look at the view – very fine it was too – and there met Lochhead[54] and his wife. Mrs. Lochhead asked after you Mum and said that she was very sorry not to see you there to explore Boston again. I have no doubt that it is a nice city but I did not see very much of it!

Next year's meeting is to be held in San Francisco! This is going to make quite a problem for us to attend, but as one of the westerners put it: "you will find out what we have to contend with every year". Should be a nice trip for someone.

During April we had a visit from Brieger[55] (the man from Papworth and the Strangeways) and we saw him again at the SAB. He was hoping that he would see you in Montreal on his way back home. He is very interested in your work trying to grow tubercle after treatment with enzyme etc. He has some most extraordinarily interesting observations on the behaviour of virulent and avirulent strains of tubercle, also avian strains, during their early growth phases that may be part of the explanation of the difficulty of growing the brute. He is a very charming and able man.

The other exciting event that I have to tell you about was my visit to Bethesda and a very pleasant and worthwhile 4 day's work with Wyckoff. As I told you on the phone they (Dr. and Mrs. Wyckoff) send you both their best regards and hope that they may visit you again and that you will do likewise. Most charming people, and they made my stay very pleasant. The work went very well and we worked hard at it. We got some

---

54 See chapter 7, note 97 for information on A.G. Lochhead.
55 This is a reference to E.M. Brieger of the Papworth Hospital and Strangeways Laboratory, Department of Pathology, Cambridge University.

useful information on the phage infection series of T5 and its relatives but we were not able to get at a couple of the problems at the end of the latent period – maybe we can do that later. However, the real progress was made on the normal host cells. It turned out that *Sh sonnei* is not only a good straight cytological object but is also very transparent to electrons and reveals much more internal detail than most of the enterics – at least our strain does. More important, we found that we could transfer the fixed, treated or untreated cells, from our coverslip preparations to collodion film and examine them in the EM. That means that we can fix in any way we wish, treat in any way we wish etc. and gain the advantage of looking at exactly comparable material by the two main methods. This is most encouraging and is, I feel, a technical advance that can be very useful. There should not be any excuse for blind electron microscopy and we can utilize our preparations to the fullest advantage.

Doris had a very cheerful birthday, despite her feelings of advancing years! She has written to you all about it and I will say little about it. Alice decorated the cake, very tastefully I must say, and asked Doris most embarrassing questions about the number of candles. We had a joyful party. We went out that night to a dance that Collip[56] gives every year for the final year. A very nice do.

We have been having very fine weather – almost too good – and I have got much of the garden planted. Some things are well sprouted already and I can only hope that there will not be a frost for a while if at all. I expect that you will be getting up to the lake very soon for the digging and planting session, now that the ice is out (isn't it very early this year?) ...

Roger Rossiter and I hope to set off fishing for a short week starting on the 20th – we will go to the Algonquin Park I think because of the restricted time that we both have available. I trust that nothing will get in its way.

About the Montreal meeting. I am thinking of coming down on the day train on the 4th – this does not get in until late at night – about 11.45 day light time I think – are you going to be back from Quebec by then? If not, you had better send me a key in case I am earlier than you.

I talked to Saint Martin in Boston and it seemed to me that the program is well in hand. I am sorry that they have booked you for a talk, but I am glad of your doing it and I think it should be fun. I hope that it will not overburden you at this time. I very much look forward to the meetings and especially to seeing you again. I wish that we could make these visits a little more frequently. ...

All the best of love to you both.

Your son,

Bob

---

[56] See chapter 1, note 56 for information on J.B. Collip, the dean of medicine at Western.

13 Joburg leaves Biological Warfare Meeting (1944)
LAC, Photographic Division, E.G.D. Murray Collection, E008406493

14 Joburg captivates his audience at the Suffield BW Station (1947)
LAC, Photographic Division, E.G.D. Murray Collection, E008406495

15 "Extended honeymoon": Doris and Robert visit Toronto's Royal Ontario Museum (1944)
Peter Murray Collection

16 Robert Murray in his UWO medical laboratory in Victoria Hospital
Peter Murray Collection

17 Two outstanding microbiologists: Robert Murray and Carl Robinow (1954)
LAC, Photographic Division, E.G.D. Murray Collection, E008406492

18 Bob Murray's lecture at the 1970 Congress of Microbiology in Mexico City
Peter Murray Collection

19 Joburg receives honorary degree (D.Sc.) from McGill University (1957)
Peter Murray Collection (Canada Wide Photo)

20 Joburg greets Soviet Virologist Zhdanov at the IAMS 1962 Montreal Congress
Peter Murray Collection

21 Former prime minister Louis St. Laurent and Joburg exchange stories at a Quebec City reception (1962)
Peter Murray Collection

22 The Murray family and friends enjoy the 1963 CSM Conference in Guelph, Ontario
Peter Murray Collection

23 The Murray children pose for a picture in the early 1950's (Tom, Peter and Alice)
Peter Murray Collection

24 Tom, Peter and Alice celebrate their father's D.Sc. (Hon) award at the May 2007 McGill Convocation
Peter Murray Collection

McGill University, Montreal
9 May 1952

Dear Old Son,

We are just off for our first trip to the Lake & this is to tell you your cheering letter arrived this afternoon. I won't answer it now but will on our return in 5 days with the news of the lake.

It is good to know you had a good time in Boston with many friends. Steve [J.W. Stevenson] told me you looked well.

We have had various visitors. Some old friends like Sir Charles Hercus[57] who was sorry not to meet you while he was in London Ont. A friend of yours, Henry from Seattle, dropped in yesterday & we all like him very much. He came to see what we could do to help him in starting a Clinical Bact. Lab in Seattle. A very amusing little Persian, the Director of Public Health for Iran, a Dr. Hussein Marchoux, has been in several times. He knows a lot & is very friendly. Some others too.

I'm glad you saw Miles; he is a very nice fellow & will do well as Director of the Lister Institute. He is coming here on 17th & I am looking forward to it.

I'll do my best with the paper for the C. M. Soc. I think it can be a mixture of thoughtful stuff and some fun. ...

 Love to all of you,
  Dad

London, Ont.
20 May 52

My dear Mum and Dad:[58]

Just a note written before we take off on the trip into trout country!

Your greetings on my birthday were very nice – both telephonic and by letter – and thank you for all your good wishes. Your cheque (much too big and too handsome as usual) has come in useful already! ...

It has been a hectic week or two, especially the last few days with Orals, papers to mark etc. but that is finished and we can go off on our trip with a fairly clear conscience. I have got it all packed and waiting for Roger to call in a few minutes. Personally I think we have too much gear but it promises to be wet and cold and we may be thankful for the extras.

All your love and confidence is only equaled by my love and confidence in both of

---

57 The New Zealand-born Sir Charles Ernest Hercus (1888-1971) was a professor of bacteriology at Otago University from 1922 to 1959 and dean of the university's medical school from 1937 to 1959.
58 LAC, E.G.D. Murray Papers, vol. 19.

you. I look forward to seeing you in June. Be sure and write me of your *exact* plans so that I can arrive at the right time!

  All my love,
    Bob

26 June 1952
The University of Western Ontario

My dear Mum and Dad:[59]

... The visit with you was grand and I very much enjoyed the chance to see you all. You were too good for me – I have not eaten so grandly for a long time. Doris got quite hungry listening to me and had to go out and buy two lobsters to get into the spirit.

It was particularly good to find that both of you are so well and so much better than last year. I am sure that with a summer vacation of the usual type that you will be in top condition before long.

After a fishing trip and a couple of hours each day in the garden I am feeling pretty chipper myself and fairly well able to cope with weather and conditions. ...

Don Gillen[60] was here for a little over a week and worked quite hard at developing labs for the course III cytology – I think that he will be able to provide them with a fair set of labs. The most important thing for him was to find out how much was *not* known and to scale down his labs to a more simple level. I feel that, although there are many things that can be done as experiments in many groups of bugs or biological situations, most of them come under the heading of research projects and are not suitable for students. However, he is armed with a number of alternatives and he will be able to keep them more than busy.

Since he left I have been able to get some really good phase preparations of living cells. I wish that I had concentrated on this a long time ago instead of leaving it as I did. Clifton has some quite good pictures in the April J. Bact.

The family is very well and very happy. Peter came home yesterday very pleased and with a little card that he had made himself saying "promoted"! Alice has also done well. Tommy has finished and the others are on the last day. They have had a marvelous time with the wading pool that I got in Montreal. It was an immediate success and has been much used. ...

  Your son,
    Bob

---

59 Ibid.
60 See chapter 7, note 9 for information on D.H. Gillen.

> Hotel Palliser
> Canadian Pacific Hotels
> Calgary, Alberta
> July 3rd, 1952

Dearest Bob & Doris,[61]

Life has been full of surprises for us in the last couple of weeks. You possibly know that Dad's name was drawn from the hat to come to Calgary as one of the Municipal Representatives for the Canadian Federation of Mayors & Municipalities & we had an idea we'd a chance to go together if finances permitted & we thought to be leaving about the 10th July, or thereabouts.

I was at the Lake with Doug for a week – very busy effecting weed control in the garden & Dad had an invitation to go the Marcotte ... for a couple of days fishing on the evening of 25th. Doug & I left a.m. on 26th – being hot & steamy & a terrific thunderstorm all the way down – & arrived home to find Sue & Joan there & Sue said 'you don't know what you are in for today, do you? Well, you leave for Calgary at 8:15 *tonight!* Apparently Dad had arrived home at 7 previous evening & within ten minutes had been phoned by his Sec. that arrangements had been made & the Delegates were leaving by train instead by plane & that the Convention was starting Monday 30th at Calgary. ...

At 7:00 we were all packed & Blake took us down in the car to check our two backs & then we had a quick dinner at the [Faculty] Club & went down for our train at 8. There were quite a few Delegates on the train, from Montreal & points east, & we found many interesting people to talk to. The journey was comfortable and most interesting tho I must say I was very glad to get to Calgary at least late on Sunday a.m. – so much sitting is rather wearing! ... The scenery along the N Shore of Lake Superior is magnificent – lovely purple islands & tall headlands & forested mountains. ...

[Mum]

> 8 July 1952
> The University of Western Ontario

My dear Mum and Dad:[62]

It was quite a surprise to hear that you had gone out to Calgary. The first I heard of it was a clipping ... that Joe Truant found in the Windsor Star. Then Peter got Mum's letter (and was very thrilled with it). Very nice that you could both go. I hope that you

---

61 UWO Archives, Freda Murray Papers.
62 LAC, E.G.D. Murray Papers, vol. 19.

have taken the opportunity to see a bit of the west, and I long to hear where you have been and what you have done.

We have had a fairly humdrum existence lately and have not done much exciting. Doris took the kids down to Port Stanley one sunny morning. ... She has also arranged for Alice to have swimming lessons at Gibbon's Park and we hope that Peter may be able to do it later in the summer. They teach them well from what I hear. Otherwise it has been mostly hot and sticky.

... I saw in Science the other day that R.A. Kelso[63] died in the States – he was much older than I thought.

We all went out to see Bob and Eileen Noble at Delaware last Sunday. Had a very nice time and enjoyed the change.

All the best of love to you all,
Bob

London, Ont.
11 July 1952

Dearest Mum and Dad:[64]

It was grand to get your long descriptive letter, Mum, and to realize how much you must be enjoying your western voyages. It is certainly the most lovely country. ...

When you are in Victoria I expect that you will be seeing Ted and Sylvia (Rose).[65] Please give them our best love. ...

Doris' mother arrived from California the other morning. She seems to be extremely well and must have enjoyed herself a lot during her stay in Pasadena. ...

The kids are well – though Alice has been abed for a few days with a fever of no obvious origin. Doris is still not smoking and is still finding it hard going.

All our love, and enjoy all your trips.
Your son,
Bob

22 July 1952
The University of Western Ontario

My dear Mum and Dad:[66]

It has been very nice to get your cards and letters all along the route of your trip. I would guess from your last card that you must be almost home by now and I had

---

63 Robert Murray does not recall this individual. An American archaeologist named James A. Kelso appears in the obituary notices section of *Science* 114, no. 2972 (December 1951).
64 LAC, E.G.D. Murray Papers, vol. 19.
65 See chapter 3, note 50 for information on Ted and Sylvia Rose.
66 LAC, E.G.D. Murray Papers, vol. 19.

better get this off quickly before you get off to the lake. ... We have envied you all along the way and I wish that we could make a similar trip someday. Maybe I will get some sightseeing done when I go to San Francisco next year ... as I hope to.

We have had the hottest and stickiest weather that we have suffered in many a year and we are likely to have it for a day or two yet. It was all but impossible in the lab and far from comfortable anywhere else. We have had storms every day but they seem to make the situation worse rather than better. I am now taking a couple of weeks away from the lab. Doris is taking a few days off in Rochester N.Y. with Alice and GG. So the boys are all together making various kinds of hay! ...

Everyone is very well and happy.

Love to you both.

Love, Bob

Department of Bacteriology and Immunology
The University of Western Ontario
375 South Street, London, Canada
29 August 1952

My dear Mum and Dad:

... Work at the lab has been fairly hectic with all the usual end-of-summer problems – and, additionally, the clinical work, since Dr. Bittner[67] is on vacation. We share these things out but even so they can be time consuming. I am rather sick of reading drafts of theses; and the drafts of my own papers are equally annoying!

The kids are very well and enjoying themselves. Alice has her swimming and many friends. Peter is a great helper; and Tommy is around! They had a great time yesterday with Doris when they all went down to Port Stanley for the day. They seem to enjoy these trips very much and have lots to tell you when they get back. ...

All the best to you.

Your son,
Bob

The University of Western Ontario
3 Sep 52

Confidential

My dear Dad:

I was glad to get your letter and the enclosures yesterday – thanks very much. I am glad to have copies of the introduction to your speech by Frappier. ...

---

67 See chapter 3, note 90.

Sorry to hear that the possibilities of a virus lab are completely through. I thought I had better write you details, in confidence, of the person I had in mind because, although he has experience with viruses, he is so broadly trained that he could be considered and would be interested in other aspects. He came to see me yesterday and he is such a nice, able and competent chap that there must be a good job for him somewhere.

He is one of Van Rooyen's chaps, George Dempster by name, and his life and work has been made miserable for him by the blighting efforts of Defries[68] and Orr.[69] I have heard the whole tale from Dempster and Van and I am sure that that is what has happened. (Van is very upset at the possibility of losing him but in that odd organisation his hands are tied!) Anyway he is now fed up with what they have and are doing to his work and has made up his mind that he will have to break away from that environment. However, this has not gone up to the Director etc. for he wants to help finish his present work with Van and he is prepared to wait until next spring, if necessary, before making the break.

Dempster was trained in Mackie's[70] lab at Edinburgh and has a B.Sc. with first class honours in bacteriology and pathology. He has the M.B. and is going to submit his MD thesis this year. He was then Assistant in the Public Health lab and subsequently Lecturer in Bacteriology. He then became an army pathologist and spent a lot of time in West Africa and got a lot of parasitological experience among other things. On his return he was Lecturer in Bact. at Edinburgh again and taught general as well as medical bacteriology. Two years ago he came out to work with Van and has been doing some good work on `flu and other such things. He is, therefore, a real general bacteriologist with good general experience. He wants an academic job and wants to see if there is anything worthwhile here for him (he would rather stay in Canada) before he explores his alternative of returning to the U.K.

He is a damned good man and I would not like to see him leave Canada. I am told he is a good teacher and I can quite believe it.

Keep this under your hat for I would hate to have this creep back to Defries et al.

Have you heard of any other possibilities for such as he? I wish I had room and could take him on, for he is genuinely interested in Clinical Bacteriology.

I am also looking round to see if there is a niche for Phil Fitz-James who will be

---

68 Robert Davies Defries (1889-1975) of the Connaught Laboratories.
69 Most likely a reference to Mort Orr, who was active in the Connaught Laboratories at the University of Toronto at this time. Henry B.M. Best, *Margaret and Charley: The Personal Story of Dr. Charles Best, the Co-Discoverer of Insulin* (Toronto: Dundurn Press, 2003), 432.
70 Thomas Jones Mackie (1888-1955) was professor of bacteriology at the University of Edinburgh from 1923 to 1953, and dean of the Faculty of Medicine from 1953 to 1955. "Thomas Jones Mackie, 5th June 1888-6th October 1955," *Journal of Pathology and Bacteriology* 76, no. 2 (October 1958): 605-620.

taking his Ph.D. next summer (also has M.D. and M.Sc.) and is a very able biochemist who wants to continue working in the microbial side. He was originally trained in bacteriology at UBC by Blythe Eagles[71] et al. He is a good man too.

All the best,
Bob

London, Ont.
10 Sept 1952

My dear Dad:[72]

Well, the grind has started again and teaching has started. In a couple of weeks we will be well enough into it and will find it's running smoothly. We still have some thesis problems ahead of us but they will be finished by this weekend. ...

21 Sept.

... I look forward to hearing from you about your Geneva trip. Is Breed still going ahead on that "tentative" rearrangement of Kingdoms, Orders and Families? I have not heard any more from him and I fear that I have not yet done what he asked. How about the Rome expenses?

Concerning Rome: it is likely that I will be going too and it would be fine if we could get together on the trip. Let me know what your plans are. ...

It is a pity that you were not able to get in a toughening canoe trip this summer. I hope that you managed to get in even a short fishing trip.

All our love to you both.
Your son,
Bob

---

71 Blythe Eagles (1902-1990) was born in British Columbia. He received his PhD from the University of Toronto in 1926 and subsequently held numerous posts at the University of British Columbia, including head of the Department of Dairying (1936-1955), chairman of the Division of Animal Science (1955-1967), and dean of the Faculty of Agriculture (from 1949 until his retirement in 1967). Biography of Dr. Blythe Alfred Eagles, City of Burnaby Web site, http://www.city.burnaby.bc.ca/.

72 LAC, E.G.D. Murray Papers, vol. 19.

Montreal
18 September 1952
Confidential

Dear Old Son,

About my address to the C.S. Microbiol. After I gave it, Panisset[73] asked me for it, wishing to translate it into French with the purpose of it being published simultaneously in both languages. He left it at the lab one day with his effort as far as it had gone, for me to see. With it I found Frappier's introduction in both languages & I had it copied for you. I don't know any more about it.

I took an opportunity yesterday to talk to Dean Duff & he sees no chance of getting on with a virus lab & offers no encouragement. It seems impossible to get them to realize the importance of it and nothing short of a major epidemic will do it. Meakins[74] & later Fred Smith[75] ruined my efforts repeatedly & Duff is cold to any present move. For my part, I am weary of major battles & with only three years to go I am not embarking on a project I see no hope of finishing. At present I have a brief respite for a good thing like the mycology lab & no permanent assurance for it; it will have to be battled for again before next spring. Therefore I say there is no hope for a Virus Division, much as I want it & have for many years.

This situation does not preclude any hope for a good stepping stone for George Dempster here next year. A job as "Lecturer" in the Clinical Lab will be vacant when Dr. Gordon, who holds it now, returns to Ottawa. Confidentially I am not too easy about (Don) Gillen; he leaves me with a sense of being uncertain in his own mind & I will be surprised if he settles down to serious work. He will do what is required of him honestly & carefully, without any doubt, but I think would prefer something half clinical with a pretense of lab work in it. If at the end of the year (teaching year) something he liked turned up I would not urge him to stay, though I would not turn him out. If he goes, then a better job than the clinical job would be available. One job is certain & it is not impossible that there may be two. When the time comes we can review the situation with both Dempster & Fitz-James in mind. ...

I was dining in the University Club & then talking to a couple of British doctors with Archie Campbell.[76] Very nice fellows & we have many friends in common in

---

[73] This is a reference to Maurice Panisset (1906-1981), a French-Canadian bacteriologist who worked at the University of Montreal where he specialized in veterinary science. In 1951, he became chair of the Department of Immunology and Epidemiology and was named head professor of the School of Hygiene. From 1970 he headed the School of Veterinary Medicine in the Department of Microbiology and Pathology.

[74] Jonathan Campbell Meakins (1882-1959), the Canadian surgeon and medical administrator. See chapter 1, note 26.

[75] See chapter 1, note 87; chapter 4, note 43; and chapter 6, note 95 for information on Fred Smith.

[76] Obstetrician, gynecologist, and surgeon at Royal Victoria Hospital.

the UK. I was called to the phone & Blake gave me a rather alarming story about Sue who was suddenly taken ill this afternoon. Started with nausea & severe diarrhea & vomiting, followed by rigors, acute muscle pains & he said stiffness of the neck; temperature rose to 101.6 and pulse 120. I told him to get hold of John Hay, who ... says there has been a lot of that sort of thing out there & it is usually over & done with in 48 hours & does not respond to Sulphonamides or any antibiotics. No indication of what it is due to ...

What a sorry mess Defries makes of a magnificent opportunity. He has impeded many things outside his own domain too.

 Love,
  Dad

McGill University, Montreal
18 September 1952

Dear Bob,

I have returned from a very strenuous meeting of the Bergey Board. We worked not less than 10 hours a day for the 3 days on difficult problems.

This is to let you know that it is agreed that I shall go to Rome Sept '53 & the cost of the fare will be met by the Bergey Fund. You can therefore distribute your money as is thought best for other representatives & I can go as designated by the Canadian Soc. of Microbiologists as Canadian Representative & can do the work in the Committees & Commissions to which I have long been appointed & could seldom attend.

There is a City Council meeting tomorrow but I shall join Mum at the lake on Saturday to stay a week. I am not going to the Columbus for the fall fishing but Mum & I will fish in the lakes round about Bark Lake.

Doris' cheerful & informative letter, with snapshot, gave us a lot of pleasure. It is good to know that all is well & that you & the kiddies are fit. The kiddies look fine in the photographs.

I still am without a secretary & it makes many things difficult & laborious.

Breed & N.R. Smith send you their regards. The more you get to know N.R. the nicer he seems to be. I enjoyed seeing much of both him & his wife, who is very nice indeed. Breed is far from well & I am worried about him, though mentally he is just as good as ever & his memory is magnificent. He is wonderful for 77.

 With love to all of you,
  Dad

> Montreal
> 5 October 1952

Dear Old Son,

Our letters must have crossed as I seem to have answered some of the points in your last letter.

It is good you are all well & busily happy. Sue is still feeling the effects of her 48 hour bout of illness; she was followed by Blake ... Three days ago I got it badly enough to stay home for the day & it has left me rather washed out & Mum got it more mildly the next day. We are alright but lack energy & get tired. I have to go to Ottawa to the N.R.C. Med. Advisory Committee meeting & leave this evening to return Tuesday evening. ...

The construction job (Bob's shed) is quite an undertaking. Peter seems to have a good idea of purpose & it is very good for him to take some part in the work as it gives him the right idea and confidence. The price of lumber is outrageous & I gather there is more available than can be sold. The lumber people in B.C. regarded the loggers' strike as helpful because they were stocked up, for one thing, and also the bush was too dry for safety. Yet this price is maintained. We have to rebuild the wharf this fall & it will cost a bit. Rosie is doing it & I hope it will be finished when next I can go up. Unfortunately lectures fall on Saturday & Monday which wrecks many weekends & this year those I have free of lectures are somewhat taken up by having to go to meetings of some unavoidable kind.

I took off the last week of September, as is my custom, but I did not go to the Columbus, where I hear they had the best fishing there has been in many years. Everyone caught many & good fish – running in on the beds ... I went to Bark Lake with Mum instead. ... We went up to Twin Lakes on three days. On one we got 15 small trout in Little Twin, 3 on another day & on one day 6 in Big Twin. The trout are small & not too plentiful.

We did some chores round the camp; cutting trees, piling wood etc, most of it fairly strenuous. It did me good & I sweated off any excess fat. My wind seems to be good but my legs are not what they used to be & they seem to tire first.

I gather you found something wrong in Gerard's thesis. I think David said in the graphs. Please let me know what it is. I am glad (J.F.) Whitfield[77] did well enough – he needs to mature but I think he has good makings.

I think I told you that my expenses will come out of the Bergey fund which allows you a better chance to cover the other delegates' expenses. I am delighted to hear you will go to Rome. We had better plan things out & make definite bookings. I am told the best way to go is B.O.A.C. via England. I have not made enquiries yet & I have a

---

77 See chapter 7, note 76 for information on the cell biologist James F. Whitfield.

lot to do besides that. There are meetings & Royal College Certification Exams. I go to Kingston 14 & 15 Oct. There are some papers I have to edit carefully & I have to stimulate the writing of other papers & some reports. The resistance & inertia is hard to overcome & the others seem to expect me to think of everything & initiate every move.

We will go to the Lab Section meeting but we cannot offer many papers. I have asked some to send in titles. Some of those who could have given something have now left.

I will send you Potel's papers to see.[78] The condition should be watched for, though it may be a local thing. However, it is only recently that Listeria has been found in Canada: Alberta & Ontario in cattle & birds, and in Lemmings brought from the Arctic by Tom Manning.[79] They must be being missed & I think there is more than one species but I am not sure that Potel's strains are a new species.

The copy of the article by Terence McLaughlin, in the Statesman & Nation, you sent me is amusing and at the same time very suggestive.[80] I drew Breed's attention to the other article you sent me on taxonomic problems; it has a number of important points & we would do well to emphasize them in the Manual.

There are some interesting articles in the last W.H.O. Bulletin. One by (Ernest B.) Chain[81] is very surprising. He shows that aeration is much more effective by creating a vortex about ½ way down a volume of liquid than by any sparger or other method. As we are interested in getting massive growth & in effective aeration in three of our "projects" we will have to try it out.

I have been inveigled into giving an address to the St. James' Literary Society. I have no idea yet on what to build it & have to get busy on it soon. One cannot refuse all these invitations but I don't enjoy them.

I still am without a secretary. The salary is not enough & five suitable people have refused. I am preparing a battle to improve it but the "Administrators" are against raising the salary from any source of money because they might have to pay others more too. I am weary of struggling to keep key people on less than a living wage.

---

78 J. Potel, a bacteriologist at the Institute of Pathology, University of Halle, and later the Institute for Medical Microbiology, Medical Academy, Hannover, Germany, was known for his work during an epidemic of listeriosis in newborns in Germany in 1949. Herbert Hof, "History and Epidemiology of Listeriosis," *FEMS Immunology and Medical Microbiology* 35 (April 2003): 199.

79 T.H. Manning, "Bird and Small Mammal Notes from the East Side of Hudson Bay," *Canadian Field-Naturalist* 60, no. 4 (1946): 71-85; T.H. Manning, "Birds of the West James Bay and Southern Hudson Bay Coasts," *National Museum of Canada Bulletin* 125 (1952).

80 Terence McLaughlin, "The Laws of Science," *New Statesman and Nation*, 6 September 1952, 261.

81 The German-born biochemist Sir Ernst Boris Chain (1906-1979). See chapter 7, note 42.

I see the applications for Public Health Research Grants have to be in by Nov 1st!! to become effective April 1st.
 Love to all of you,
  Dad

1559 Pine Ave. W
Montreal, Quebec
October 8th, 1952

Dearest family,[82]

... I had a wonderful two weeks up at the lake. It is quite the nicest time up there – late in the fall – practically no one other than myself to disturb the peace and quiet; some lovely hot and shinning days; some hazy and calm; some wet and furiously windy ... We spent a couple of days on various work around the [property] – things that had to be done and the rest of the time were out on 'Expositions' of various kinds. ... It did us both a world of good and we came back feeling toughed up and healthy.

Dad will have it that his legs are not what they used to be but his wind is as good as ever and I did not find mine failing me either. The poor old pup was heartbroken that we did not take him too ... and would sit there sadly nodding his head till we were out of sight. ...

Rob, I hope we shall see you at the Quebec meeting in December. ... Is there any chance of you coming too, Doris? Our spare bed is rather small but two can just manage to fit into it.

Now I must simply fly off to do my shopping; Dad will be back early to lunch – I hope – as he has a meeting.
 A hug to each of you.
  Lovingly,
   Mum

19 October 1952
The University of Western Ontario

My dear Mum and Dad:[83]

Times have been very busy and it has been hard to fit in letter writing – your long letter about general doings was a lot of enjoyment to us. We envied you going up to the lake for the end of September – it is such a lovely time of year. We have had a very

---

82 UWO Archives, Freda Murray Papers.
83 LAC, E.G.D. Murray Papers, vol. 19.

pretty fall here, too. Some quite good colour and many very beautiful days ...

... I have been working on the "back parts" on all spare daylight hours and there have not been too many of these. It is now in a shape ... The roof is on and looks neat and true – but I have to put in the flashing. I have the sides on and they will have to be shingled. And finally the two doors will have to be set in and hung ... Plenty of errors have been made and it could not be called a "professional" job – one friend asked me why in hell I struggle to remain a bacteriologist!

We had a very pleasant visit from Ash[84] and Elizabeth[85] a week ago. He is very well, looks the same as ever. He sends his best regards. The work he has been and is doing is most interesting and worthwhile; some activity against animal viruses has been shown for three of his antiphage agents but there is no pattern yet visible to indicate how one might correlate anti-phage and anti-viral activity. Despite his good work I hear that the Nat. Found. Infantile paralysis is not likely to maintain his grants – they are more interested in the possible quick rewards of the tissue culture and ... globulin work. So the fundamentals perish.

I had to go up to Toronto recently about the program for the Lab. sect. meeting. A fair number of reasonable papers are submitted and it should be rather fun. The Connaught was in its usual form! Saw Charlie Hanes[86] for a moment – he sends his regards – and hope to have dinner with him this weekend when I have to go again.

About the jaunt to Rome. Have you any definite plans? I would like to see a *few* people in England, also in Paris. These trips are so damn expensive these days that one must plan rather carefully. It is a poor time to try and see people in labs in England, at any rate, during the last part of August. But with courses and other responsibilities here we (or I) cannot prolong the trip beyond September. I would rather go by boat, if anything, but the time almost militates using air transport even if it eats into the money. We should get bookings held soon because it is a Coronation year and everything is liable to be booked way ahead and we may have to wait for cancellations even now. ...

   Your son,
   Bob

---

84 See Introduction, note 377 for information on Igor N. Asheshov.
85 See chapter 4, note 122 for information on Asheshov's wife, Elizabeth Hall.
86 The Canadian-born Charles Samuel Hanes (1903-1990) received his PhD at Cambridge University and spent much of his professional career at the institution's Low Temperature Station for Research in Biochemistry and Physics. He was elected Fellow of the Royal Society in 1942 and the Royal Society of Canada in 1956, and was the recipient of the 1955 Flavelle Award. J.T.F. Wong and W. Thompson, "Charles Samuel Hanes, 21 May 1903-6 July 1990," *Biographical Memoirs of Fellows of the Royal Society* 39 (February 1993): 148-155.

The University of Western Ontario
18 November 1952

My dear Dad:[87]

... I had a very interesting and stimulating trip to Cleveland a couple of weeks ago (with Robin and Chris Hannay). We went to the meeting of the Electron Microscope Society. Really, they have come a long way in the last couple or three years. The sections of tissues, and especially those by Sjöstrand (Karolinska Institutet) were really magnificent. The detail and clarity and resolution of cytoplasmic structure was magnificent. It is beginning to show what can be done – it also showed that there are special problems with bacteria but that similar techniques will give necessary and essential information. I fear that we (and I mean *we*) are getting to the point that we need an E.M. The problem is that there are a sufficient number of technical problems to solve while we go through our material at the E.M. that it will soon be uneconomic to have to go away to another laboratory each time. We had some very valuable discussions with people there and I think that it was one of the most stimulating days I have spent in a long time. Stuart Mudd[88] was there and sends you his best regards. ...

We are all very well – astonishingly so – and the kids are enjoying themselves at school. Alice had a very nice birthday and got the sorts of presents she really wanted.

All our love,
Bob

McGill University, Montreal
1 December 1952

Dear Old Son,

... I have started enquiries & I find there will not be any particular difficulty in getting reservations to Rome & back with stopovers where you like. The return fare (BOAC or TCA) to Rome is $634. Tourist class in a ship both ways saves some $300 but takes up between 2 & 3 weeks. Ship (tourist) one way & plane back costs $420.

I haven't fixed details yet. I am thinking of leaving here 1st or 2nd Sept & having a couple of days in Rome before the meeting. Dumas of l'Institut Pasteur has asked me to stop over in Paris & would like me to give them a lecture but I have not decided whether or not to do so. Mum may go to England at the same time & I would join her there after the congress for a couple of weeks to see old friends.

We can discuss these things when you come, I hope Thursday 11th.

Love to you all,
Dad

---

87 LAC, E.G.D. Murray Papers, vol. 19.
88 The American bacteriologist Stuart Mudd (1893-1975). See chapter 7, note 114.

McGill University, Montreal
3 January 1953

Dear old Son,

We had a grand time with you over Xmas. It was splendid to find you all well & see the kiddies coming along so well. They are remarkably well advanced. Our journey home was uneventful. ... We spent New Years Day with Sue, Blake & the kids. Doug spent much of the time with his skis & toboggan; Joan does not think much of the snow, it is uncomfortable stuff best left alone. ...

I hope our visit did not tire you & Doris too much. It was a busy time for you & exciting for the little ones & it can be a bit of a strain. We had a cheerful letter from Doris which pleased us & she told us she was not tired. It must have been fun to see Alice & Peter enjoy "Peter Pan". I remember so well you going to the film version of it in Cambridge & to Treasure Island & to Maskeline & Devant's & the fun it was. ...

I'm sorry I forgot to try to see Tony Brown. I meant to do so & remembered only at moments when I could do nothing about it. Work he is interested [in] is on insects – developing a breed resistant to the various insecticides seems to me much like bacteria & antibiotics – also mice to infections & poisons. There is some important principle behind it. Possibly it would be found with phage too if you could recover uninfected cells to grow phage free cultures from infected ones. The curious thing is the resistance is not absolutely specific.

Mum & I drank a toast to you when New Year came in. We were our own little party for it & except for the sirens & hooters little seemed to be doing.

    Love to you all,
        Dad

McGill University, Montreal
27 January 1953
Midnight

Dear old Son,

Sue had a very tiny baby tonight (about 10pm) which will probably not survive; about 5th month. ... It's a nuisance & has been a worry. Sue is well & takes it reasonably. It's a pity it did not run a normal course. ...

We had a meeting of the P.Q. Soc of Microbiology this evening & it went off quite well; 60 turned up out of the 100 members & a couple came from Quebec. We had it at my lab. Stevenson gave a splendid talk on botulinus toxin & one of Frappier's men talked on T.B. cultures. I'll send you the programme. ...

We all sent Norm [Gibbons] our ideas on a Journal of Microbiology & you no doubt had a useful meeting. I wish you could have come on here from Ottawa. I expect the journal question will be reported at Guelph. I can't see Microbiology & Botany; it could fit in some ways with biochemistry at the present time. I feel there is room & place for a straight journal of microbiology.

Panisset says my "Why be a Microbiologist" will be in the next number of "Revue de Biologie" or whatever it is called. That journal appeared quite regularly but something seems to have held it up.

I think it probable that you will have to consider offers from time to time. Some of them will be hard to decide & some will be hard to assess correctly. It is sometimes hard to be sure of the security & political influences have an awkward prominence in some places in the USA. Happiness & opportunity can outweigh glitter. I am sure of your sound judgment. It is well to tell Collip of the offer; that they may realize the situation; the better they make your position the better offers have to be made by others. This is very important, not merely that it strengthens your own position but it raises the standards & values of Chairs in other places. My fights to raise the value and standing of appointments here has done something for other places. We must not undervalue ourselves if we hope to have others appreciate us & it is important to get appointments raised to a standard to attract good men. Bacteriology has suffered a lot from being represented by inferior people.

It is good to know the youngsters are well and actively happy. Give them my love. We had a very happy Xmas with you. …
    Dad

London, Ont.
3 February 1953

My dear Mum and Dad:[89]

We were very sorry to hear of Sue's troubles. It was a great relief to hear that she is doing well and will soon be back with her family again. …

The Canadian National Ballet came here for a week and we went twice (between us) and enjoyed it very much. Doris took Alice one evening and she was most thrilled and impressed. …

It was nice to have your comments both in your letter and on the telephone about this business of being offered jobs. Judgement and decision is all very fine but it is nice to have some reassurance in some such situations. It is nice to know that people still think about me as a possibility. …

I am very glad to hear of a successful P.Q. Soc. Of Microbiol. Meeting – it is good to know that it is flourishing. Several people have mentioned it in letters in the past few days and they seemed to enjoy it.

---

89 LAC, E.G.D. Murray Papers, vol. 19.

We have been very busy just lately. I have one of [Salvador] Luria's people, [Giuseppe] Bertani, visiting for a couple of weeks.[90] We are working together on a part of the problem of lysogenic (symbiotic) infections with phage. So far we have found out some very interesting things that we were *not* looking for (serendipity!) – however, the difficult problem is still there. One trouble is that Shiga's dysentery bacillus has a very odd structure. I don't know if it is a peculiarity of this strain or is common to all. Could you send me a couple of strains to try?

Bertani is a very competent, quiet and able fellow and we are having a lot of good fun working together. With luck we will get something useful, aside from the interesting side lines. ...

    Best love to you both
      Your son,
        Bob

London, Ont.
22 Feb 1953

My dear Mum and Dad:[91]

... We are enjoying an enforced quiet life – in part, Doris is beginning to find it difficult to get around at her usual pace – we are also well poxed since Alice and Tommy grew a grand crop last Friday. ...

Bertani's visit here was a great success. He is a very pleasant, quiet and capable fellow doing excellent work in Luria's lab on the less dramatic but most interesting lysogenic (or "symbiotic") phages. We got together to try to find out details of what happens when you establish a symbiotic infection. We got some fascinating morphological data that sets up excellently for the next years work along that line. We used a Shiga phage which is temperate in the sense that a large proportion of cells infected recover and thereafter *carry* the potential to produce phage at some later time. A proportion type and produce more phage with the same characteristics except for an occasional "mutant" particle which may show a purely lytic or virulent character. We followed the lytic series of events and then the lysogenizing series. ...

We also have some very interesting data on super-infection of carrying cells but that is too much for the moment.

A byproduct of the work is interesting – the normal nuclear structure of Shiga

---

90 In 1949, Giuseppe Bertani joined the laboratories of Salvador E. Luria at Indiana University in Bloomington. He is currently with the Biology Division, California Institute of Technology, Pasadena, California. See Bertani, "Lysogeny at Mid-Twentieth Century: P1, P2, and Other Experimental Systems," *Journal of Bacteriology* 186, no. 3 (February 2004): 595-600.
91 LAC, E.G.D. Murray Papers, vol. 19.

is *very* odd and each nucleus seems to contain a central or excentric basophil mass. Since this is removed by hot acids *and* by ribonuclease it is possibly a ribonucleic acid containing structure ... It seems to divide with the nucleus which is interesting for in most fungi the old one is discarded and two new ones formed. I am trying to find out more about the thing now. ...

I saw a report in the paper today that McGill is not getting the "Federal Aid" this year. Is this true? If so it sounds like one of Duplessis' more scoundrely tricks. I hope that does not mean trouble with salaries.

They are making a relatively enormous fuss around here about the 75th anniversary of the founding of the university. Formal dinners, etc., the Governor-General coming (which is fine) – but it seems odd not to wait 'til the 100th. ...

Much fuss over the Red Dean![92] A whole lot of damn fools shouted him down the other night instead of listening and raising their hell afterwards. Ed Hall[93] made a tactical error in talking to the newspapermen about it and saying that people had a right to heckle! The result is that the churches, editorials, etc. fall on Ed's neck as a supporter of rowdyism. I'm sure he didn't mean it – but there it is. The local paper has a good cartoon of the aforementioned Red Dean with an enormous great cork in his mouth and captioned "A smaller cork might have done just as well". ...

Best love to you both.
   Your son,
    Bob

            McGill University, Montreal
                 27 February 1953

Dear Old Son,

Everything is rather a mad rush of too many things to do at the same time. Salary difficulties are prominent again & part of it is due to "administrators" stupidity, but the situation is accentuated by Mr. Duplessis's refusal of Federal Grants to the Universities in P.Q.; & that is the present furor which is getting out of proportion.

Sue improved a lot though slowly but has been set back by getting chickenpox from Doug who is well & back at school. Joan & Mum have chickenpox too, so there

---

92 This is a reference to the Very Reverend Hewlett Johnson (1874–1966), the English clergyman, Dean of Manchester, and subsequently Dean of Canterbury. He was a campaigner for the rights of the poor and a committed socialist, and on his travels to the Soviet Union he became convinced that the salvation of humanity rested on the reforms being made in that country. He earned the nickname "Red Dean" as a result of his public pro-communist and pro-Soviet positions. *ODNB* (Online).

93 See chapter 3, note 73 for information on Western's president, G. Edward Hall.

are three generations of pockers in the pest house. ... All told the last month or more has been depressing.

I went over the proofs of "Why Be a Microbiologist" & it will appear soon. I don't care for their paragraphing but I suppose it is the Editor's idea & they did not follow *the HPS* in that. It does not read too badly. The French version only covers the first part.

I sent "Nature's place in man's world" to the American Scientist & will hear in time if they accept it. Our papers on preoperative antibiotics in *gut* surgery will be out in the Am J of Med Sciences shortly & a paper on the Middlebrook-Dubos test in TB has been sent to the Can Med Ass J.[94] I have to re-write one or two other M.S.S. & edit a couple more. I find it hard to find the time between lab troubles & interruptions, numerous committees & City Council.

Listeriosis in Canada is looking interesting & I must find time to tell you, but I am in a hurry just now. However, a case like Potel's cases has been unearthed in Toronto from 1951 when it was limboed. It was awakened by Allen getting L. monocytogenes from a blood culture in Fort William.

I hope you are all well & that things are going well.

Love,

Dad

McGill University, Montreal
1 March 1953

Dear old Son,

... Soon after I wrote to you on Friday (27 February '53) I was cleaning my teeth, just before going to the U of Montreal to hear a lecture by Honor B. Fell (Cambridge)[95] & I pricked my left thumb very slightly with a clasp on my lower bridge. It was a very slight prick, not sufficient to draw blood, but within 5 hours the thumb was swollen, painful & throbbing. I then took sulphadiazine steadily & am still on it ... I guess it to be a strep pyogenes by the behaviour, but no culture was possible without incision which did not seem wise. Years ago that slight accident might have been dangerous; it was no more than that killed M.B.R. Swann[96] in about four days & I have known other cases.

Tomorrow starts the City Council Budget meeting. It will continue throughout the next two weeks to end midnight on the last statutory day. I have to go to Ottawa

---

94 Salme Nommik and J.F. Meakins, "Hæmagglutination Reaction in Tuberculosis: The Middlebrook-Dubos Hæmagglutination Reaction in the Diagnosis of Tuberculosis," *Canadian Medical Association Journal* 69, no. 2 (1953): 140-146.
95 See chapter 7, note 118.
96 In 1926, E.G.D. Murray, R.A. Webb, and M.B.R. Swann isolated *Bacterium monocytogenes* (later renamed *Listeria monocytogenes*) from rabbits.

in the middle of it for a couple of days & there are several other things piling into the same time. There are too many outside things to do. ...

I have a letter from (S.T.) Cowan,[97] with a timetable for the Nomenclature Committee, the Judicial Commission & subcommittees, saying there is much to be done & that it is desirable to have certain meetings the week before the congress. It looks as if I shall have to be there by Sept 3rd, 9AM. Some of the committees I would like to watch clash. It looks very like work. I haven't got the Canadian list here, so I find I cannot suggest names as you request, but will do so tomorrow from the lab. As far as I am concerned personally I would be interested in any one of Sections VI, IV, V and XIII & even XVI; probably the most suitable would be IV or VI.

I had to represent the Mayor & the City at the Annual Banquet of the S. African War Veterans Association last night & make a speech. As often happens I was lucky & what I said appealed to them. It was easy because I could speak with knowledge of things of that time, but I hit just the right note. After it they cheered me & sang "He's a jolly good fellow" & afterward took me to their Club very kindly. The Commissioner for N. Zealand & Arch Deacon Gower-Rees[98] each made good speeches; there were eight speeches, songs, The Last Post (Bugle) & a Lament on the Pipes. It was a pleasant evening and very dignified in proper places, with very remarkable good comradeship. Much more pleasant occasion than some of the Mayor's assignments.

It is good that Doris has not caught the chicken-pox from the kiddies. I hope she is immune as it would be a bad nuisance. I suppose all the children will get it & I hope lightly. ...

I am looking forward to the meeting of the Royal Society in London, though I am not so pleased at having to take part in the Symposium as the subject is hard to handle. I must get the paper prepared, it is difficult as I have so much on hand and the illnesses in Sue's family have been upsetting.

      Love to all of you,
          Dad

                                    London, Ont.
                                    3 March 1953

My dear Dad:[99]

... I expect that the symposium (at Rome) will be full of contention but it will be a good thing to get the stuff out into some critical air. They are going to publish the text in extenso, so Robin and I have both to get MSS ready by the middle of April. There

---

97 British bacteriologist Samuel Tertius Cowan (1905-1976). See chapter 6, note 99.
98 Archdeacon of Montreal and Rector of St. George's Anglican Church.
99 LAC, E.G.D. Murray Papers, vol. 19.

seems to be too much writing to do at the moment and hardly enough time to do the necessary reading. Our "priorities" have to be changed all the time!

I have not heard any of the plans for the Royal Society meeting. They have not asked us to play any part but I expect that we will hear something about it soon. What symposium are you involved in and what are you talking about? We look forward to seeing you here again.

The South African War Veterans Association banquet must have been rather fun for you and you must certainly have been able to give them a talk to their taste. They must be getting a fairly aged lot by now. ...

Best love from us all and please give Mum and Sue our best wishes for a speedy return to well-being, even if they do remain somewhat scabrous for a while.[100]

Your son,
Bob

McGill University, Montreal
8 March 1953

Dear old Son,

... I am not convinced that there is any relation between chicken-pox and shingles. As far as I can gather it is merely a clinical impression argued rather vaguely from that particular to the general. I had chicken pox about 1903 & have seen many cases without catching it again, but I had shingles quite badly in 1932 without any of you getting C-pox. I am sure I caught it from a barber who cut my hair & who part way through I noticed had perfect herpes. ...

It is strange enough that Spekk-finger[101] has been so very poorly investigated. The paper in Arctic is a good example of the need of a bacteriologist of some experience to do the work. I tried long ago to get some material & I have recently tried again. It is interesting that some have found what they think is a corynebacterium & others something like Erysipelothrix. I have little confidence in the indefinite cocci found. But these things could be properly examined. There is no reason against a possible severe localized & destructive infection by a corynebacterium of particular character; after all *Corulcerogenis* does a very destructive job to the skin in the three cases I have seen.

I don't think I have Winogradsky's collected work. I forgot to look yesterday to make sure. I also had not time to see what I could do about the Shigas you asked for; I don't think I can do much for Shiga as it is difficult to keep easily & mine have been neglected, the other dysenteries are easier.

You can help a lot to get some things straightened out & it will help to have the

---

100 A reference to the chicken-pox outbreak.
101 A severe finger infection.

texts published in fall. I hope they will do better than the Rio de Janeiro people as that Congress has not been reported yet. This lack of the V-th Congress Proceedings is a nuisance in many ways.

I have not heard from the "American Scientist" yet. They may find merit in the M.S. & if they do I shall send you the copies you wish. I was amused at someone wishing to alter the spelling in my quotations from Middle-English literature in "Why be a microbiologist", but the latest development is astonishing. It seems the "Revue Canadien de Biologie" is now the property & responsibility of the U of Montreal; the editors are afraid to include my story of Wyatt G. Johnston and "the immaculate infection". It seems they have a very mild religious qualm over it but are much more afraid I might annoy Mr. Duplessis; the other horn of the dilemma I have pointed out to them is that across Canada the story would be expected to be there & its absence will be recognized as bigotry, which will do no good to the representation of this province. Perhaps it is even more surprising that the solution they intend is to leave it out in the journal but put it in the reprints! ...

Your work with Bertani seems to have been happy & good. I don't know the characters & destructions of the phages & find papers difficult to read because of the terminology used. However, you seem to have found some good approaches to important questions. The separation of lysis & reproduction is quite exciting & the "process of recovery" in development of lysogenic strains is most interesting. I wonder if they might produce antibodies.

The report [is] that Quebec will not allow Universities to get the Massey Report subsidy offered by Ottawa. It means a loss of something over $600,000 for McGill. No one seems to understand the politics of the situation. The upshot is that projected improvements are completely out. In consequence I have had to revise my budget application & was instructed to reduce the amount by $2000 below last year's budget. It makes a number of difficulties. I have lost a highly trained technician for salary reasons & cannot now raise the salary to be able to get an equivalent replacement. There is danger, too, of losing others who are essential. Duff is a good fellow & has allowed me to do the readjustments to try to avoid serious trouble as much as possible, he argued for this rather than that the principal et al. should just cut off the money wherever they [wished] & they agreed.

The Red Dean is a curse & so are others of the cloth, & you seem to have quite a sample of them in London, Ont. Ed Hall could have kept out of it but I think too many are shirking situations & the hot-heads are getting away with too much. Then panics & witch-hunts are developed by other hot-heads for political purposes. It would help a lot to have some sense & judgment exercised by those in high positions. ...

I have to go to Ottawa to the N.R.C. Med. Advisory Committee & return on Wednesday. I leave here in about an hour. It looks like being a busy time. The Budget

Meeting of the City Council is on; it started last Monday, 2nd, & will go on till 14th. Until 11th the meetings are afternoon and night, but after that anything may happen. It makes it difficult to get some things done in the lab. There is quite a lot to do & I feel a bit tired at times. I shall be glad when Mum gets home.

I must get moving now.
>Love to all of you,
>>Dad

>>McGill University, Montreal
>>12 March 1953

Dear old Son,

I got home from Ottawa 7:30 last night & had to go straight to the City Council. On getting home I found a still spotty Mum here, though she feels much better & the spots are healing. Sue is better too.

This morning I received a letter from the Royal Society of Canada saying that I have been chosen for the 1953 award of the Flavelle Medal. I was quite unaware I was being considered & it is a complete surprise. Mum is very pleased & I know you will be too. It will be an added pleasure to me that you will be able to be present when the Medal is presented. This recognition is a great encouragement. ...

I have not looked for the things I promised (Winogradsky's book, Shigas etc) but will try to do so on Saturday. There are Council meetings morning, afternoon & night until midnight tomorrow (Friday) & some other troubles, of which the department budget is not the least.

>Love to you all,
>>Dad

>>THE ROYAL SOCIETY OF CANADA
>>Regulations for the Award of the Flavelle Medal[102]

The following regulations were adopted by the Council of the Royal Society on February 28th, 1942 on the recommendation of a committee appointed by Section V for that purpose:
1. The Flavelle Medal shall be awarded for original research of special and conspicuous merit in the biological sciences. The nominee for the award shall have made an outstanding contribution to biological science during the preceding

---

102 LAC, E.G.D. Murray Papers, vol. 8. For more information on the Flavelle Medal, visit the Royal Society of Canada Web site, "RSC: The Academies of Arts, Humanities and Sciences of Canada," http://www.rsc.ca/index.php?page_id=61&award_id=12&lang_id=1.

ten years or, during that period, he shall have made significant additions to a previously made outstanding contribution to biological science.
2. The selection of the person to be recommended for the award of the Flavelle Medal shall be made by a committee appointed by Section V. This committee shall consist of not less than six members, and Botany, Zoology and Medical Science shall each be represented by two members.

Department of Bacteriology and Immunology
The University of Western Ontario
375 South Street,
London, Canada
14 March 1953

My dear Dad:[103]

I was most pleased to hear that you are to receive the Flavelle Medal for 1953. It is a great surprise! At last you are getting some recognition in Canada for all the things you have done. So we shall look forward to the Royal Society meeting and your visit even more than before. ...

The work goes on fairly well and at the moment I am trying to get my mind clear on the problems of normal nuclear structure in our various bugs. I think that, between us, we are beginning to get at a sound basis for building some concepts. These detailed studies cannot be done in a hurry and I now see some of the "gyrations" in my own thinking over the past two years and a bit. It will be, for the time being, the definition of the problems – a type of study that is too often neglected because it is so discouraging. Otherwise I am immersed in writing and preparing (reading) for writing. It is sometimes rather hard to get going both at the lab and at home. Peace is a scarce commodity – in all senses! ...

Well, love from us all and congratulations on the award.

Your son,
Bob

McGill University, Montreal
29 March 1953

Dear old Son,

The enclosed cutting makes the award of the Flavelle Medal public. Various friends have shown their pleasure. I have to write an address to give when I receive it, to Section

---

103 LAC, E.G.D. Murray Papers, vol. 19.

V, & I thought to tell the story of *Listeria monocytogenes*. It should be about something in the work of the recipient & this subject seems to offer something that could be of fairly general interest. Most Fellows of the section are not medicals & only a few are bacteriologists & it should, I think, be of interest to other biologists. I'll see how it works out.

I did not know the Medal was in the offing so I had agreed to take part in a Symposium; they wanted me to speak on recent advances in Medical Bacteriology. I have written that one & I have tried to make it of general interest by dealing with the principles & the problems raised by recent work, rather than the medical applications. I hope it works out alright.

I am glad Fitz-James is interested. I have no intention of interfering with his finishing up his present work & that is why I said we would arrange the time he would take up the job if he accepted. I hope he can come to see me & the place. I know he is interested in the biochemical aspects of bacteriology & that suits me very well; but he can learn some other sides without disadvantage & to take his share in the teaching. His own research can be whatever he likes. ...

Today is Joan's second birthday. We saw her yesterday & she is very well & a dear little girlie. Sue is much better but pale & washed out & still troubled a bit with a cough. She has had a bad time. Douglas is fit again but he's lost some weight & needs some outdoor exercise. Mum is still a bit played out but is picking up.

I was invited to a small luncheon this morning given by the High Commissioner of India in honour of Mrs. Pandit. She is a charming and very remarkable woman. I had the honour of sitting next to her. The Mayor [Camillien Houde], Sir Frederick (Head of ICAO) & some others were there (about 10) & it was a pleasant occasion.[104] Mum did not feel well enough to go, unfortunately, as she would have enjoyed it.

I have not got the Winogradsky book you want. You might write to Robert Breed, who might well have it & would lend it to you. My Shigas are hard to get at & will take time & I don't feel sure they are alive. If you write to Jim Gibbard[105] I am sure he can send you several & they would be "recently" isolated. I enquired & Bailey tells me they have them.

The snow has gone from here & the ice is nearly all out of the river. It looks as if we might get up to the lake soon. That would do Mum a world of good & I would not mind a bit of a rest. I get tired rather more easily than I used to.

I hope the kiddies are all well again & that Doris is not finding it too difficult. Her recent letter was cheerful.

  Love to all of you,
    Dad

---

104 Reference to the civil aviation administrator Sir Frederick Tymms (1889-1987), who at this point headed the International Civil Aviation Organization.
105 See chapter 6, note 36.

London, Ont.
7 April 1953

My dear Mum and Dad:[106]

... How are plans coming for the jaunt to the U.K. etc.? I have no idea how you are arranging things. A cheerful letter came from Nan the other day. I am going up to see her, since I will touch down at Prestwick. She asks me to tell you that she and Jimmy intend to be away for the first three weeks of September but that they hope for a visit either before or after.[107] I hoped that we might be able to visit Paris together for a couple of days after the Congress, but maybe that is impossible.

Work is confusing at the moment and I am trying to get some difficult points on the road to settlement before I send off my MS to Rome. Lots of other things are boiling in the pot. We have some papers in the press and most everyone in the lab is represented – some good stuff too. Joe Truant and I are struggling with how to present the stuff on *Moraxella* and *Azobacter* (each separately) and the first of these will be ready soon. ...

The tentative program for the Guelph meeting looks very good and I think it is certain to be a lot of fun.

We look forward to the R.S.C. meeting here and seeing you. It will be a great pleasure to us: both seeing you and knowing of the honour and respect paid to you. ...

Best love to you all,
Your son,
Bob.

McGill University, Montreal
12 April 1953

Dear old Son,

... I'm sorry you did not take up Burton's offer as it would be a good thing to do & easy as the meeting is in London. Perhaps you can reconsider. I would be happy if you did, but don't want to press it. The society likes to give opportunity to selected workers.

We are looking forward to the visit to London & to Guelph & seeing you. Andy Rhodes gave a good talk here on Friday & spent part of the day with us. Very nice.

I have not yet & must at once work out the plans for Rome. I have so many things to do I can't remember them all & every day is full, to leave me rather weary. I hope the ice will be out & we can open up the lake & get a whiff of wood smoke & a paddle.

---

106 LAC, E.G.D. Murray Papers, vol. 19.
107 See chapter 3, note 29 and chapter 4, notes 79 and 90 for information on Robert Murray's cousin, Nancy Heard, and her husband, James "Jimmie" Stuart Abercrombie Pearson.

...I have to go to Kingston on 24th to examine for a Ph.D. I would like to see Guilford[108]

It's good to know you are all well & that the gardening is starting. I should tie a few necessary flies – we're nearly out.

I'm glad you like Frances Russick's Review & she will be pleased to know you do. Reprints are not in yet.

    Love to you all,
        Dad

                McGill University, Montreal
                26 April 1953

Dear old Son,

We hope Doris is home & well & content again. We are impatiently looking forward to seeing all of you on 30th May. I have arranged to be in London a couple of days before the meeting to spend the time with you before the turmoil, as the C.S.M. meeting is immediately afterwards. A few days after the C.S.M. I have to be in Ottawa for a week because the Gov't has appointed me a member of the National Research Council. It is quite an honour & will involve some more work & I don't yet know how much. It is sure to be quite interesting.

I have just returned from a couple of days in Kingston, examining for a Ph.D. They get their External Examiner to go there for the Oral. I saw over the just completed new D.R.B. bacteriology laboratory. It is quite a large three story building & very well arranged, under the direction of Guildford Reed & it will keep him busy.

My effort to get an Honorary Degree (D.Sc.) for H.M. McCrady [Provincial Labs, Quebec] has been successful & I have to present him for it at the Convocation on 27th May. I am sure it will please all the Public Health people. I suppose he is a member of the C.S.M. & it should be taken notice of in the Business Meeting at Guelph as an honour to a member.

The Flavelle medalist address I have to make is written & I hope it will prove interesting to the section as a whole as I have introduced a variety of biological problems into "The Story of Listeria". I have also completed the "Implications of Recent Advances in Medical Bacteriology" for the Symposium Lochhead arranged for Section V of the Royal Society. I hope this, too, will prove to be of general interest as it involves a variety of problems. ...

Sue has ups & downs ... Sue's kiddies are very fit. Blake was offered quite a good job at Ottawa, but has not taken it because the offer made his present bosses boost his

---

108 The renowned Queen's University bacteriologist. See chapter 1, note 115.

position so that the difference financially was hardly worth the expense & bother of moving. I gather the response was that the head of the Department in Ottawa came to see him here & I don't know yet what transpired. Anyway it has done him quite a bit of good.

We are in the midst of examinations. I am leaving it largely to Steve (J.W. Stevenson), Gertrude,[109] Cliff [Kelly][110] & Don [Fleming],[111] without compunction.

Phil Fitz-James seemed to enjoy his visit & I look forward to learning his decision. I think the position would afford him a good opportunity to establish himself & work up a position to demand the kind of job he wants eventually. It is better for him to have a full fledged University Appointment than to carry on with fellowships & grants, which can only advance him in a limited way. I have seen people keep on too long with them & spoil their candidature for jobs because people ask, if he is good why did he not get an established job before this.

I still have not got a candidate for the Clinical Bact job. I can fill it with a D.P. of course, but I have enough of those to have some say there is no one in the lab that can speak English. However, I have not advertised it & don't know what may be possible. ...

 With love to all of you,
  Dad

Godwan River Estate
Ngodwana
E. Transvaal
1st May 1953

Dear Ever,[112]

You will have doubtless read in your press the results of the General Election in the Union of S. Africa. It was a severe blow to us and to many others, actually the United Party polled a much higher percentage of votes but the Nationalist Party gained the majority of seats. For many days, I felt quite stunned, could not settle down to work. Poor S. Africa, what of her destiny? There is no doubt the election was won by fear. Fear of the Black People, fear of the English, their traditions, language, etc. What a tragedy it all is. There are thousands of Afrikaans people who are with us, but I am afraid in five years we will have lost many of them, those in Government Positions, i.e., Civil Service, Railways, Defence, Police, Schools, etc., etc., will in time join the Nats, as they will find promotion etc. will be jeopardized. The Nats will of course

---

109 See chapter 3, note 101.
110 See chapter 4, note 95.
111 See chapter 3, note 123.
112 LAC, E.G.D. Murray Papers, vol. 19.

deny this is their aim, but there is no doubt any Afrikaner who is known to be a supporter of the U.P., and holds any Govt. Position, can and will be victimized. Then the young people are undoubtedly "fed" with Nat. propaganda in the Government schools, and quite a few of the Universities. There are of course the many private schools (similar to English Public Schools) ... and of course many Convents and other Girls Private Schools, while they are obliged to teach according to the Govt. curriculum, the method of teaching has been left to the schools, but how long will this be allowed? Further, it is not possible for everyone to afford the fees, and as they are generally Boarding Schools fees are high. Is S. Africa going to be like a second Hitler Germany? If one wishes to holiday "overseas" there are many questions put to one and one of the things no S. African is permitted to do is to criticize the Govt. (otherwise a passport will not be issued). I suppose I could be "hauled over the coals" for this very letter if it became public.

One of their great aims is to have a Republic. God save us all, I cannot see Rory living under an Afrikaans Republic, he has an inherent loyalty to England and the Queen, nobody would ever be able to "muzzle" him, he would become a "rebel" in S. Africa. Senator Heaton Nichols of Natal has now stirred up a hornet's nest – he wants Natal to break away from the Union and become a Monarchy, what an unholy mess we are in.

I love S. Africa, it is truly a grand country, but I am so worried and grieved about the future, so many people will be getting out. Some folk say "Rats leaving a sinking ship", that is all very well, but when one sees everything one has strived and fought for years gradually being taken away or destroyed, it makes one think. It might eventually be made impossible for anyone to leave, or at least to be permitted to take their worldly goods with them. I know this sounds very pessimistic and rather a defiant attitude, but I personally can see nothing but "black clouds" ahead. In my opinion the United Party will break up into splinter parties and eventually it will be Afrikaans versus English, what future will there be for my boys and their possible children? The United Party have more or less admitted the Nats are "in" for many years, and if their past five years is any criterion what will the next five and those following bring us? ... It makes one wonder "Did England adopt the right attitude after the Boer War," she wanted to be fair and just, gave them their language and made it an official tongue together with English. Since 1910 (the date of Union) we have been led by Afrikanns, Botha, Smuts, Hertzog, Malan (and even had the U.P. been successful this time, it would have been Strauss). Botha and Smuts were two fine men, and Hertzog too in his way was certainly not without greatness, and he would never have consented to the messing about of the Entrenched Clauses (i.e. the two language rights and the Coloured Voters) but these people are determined to go ahead, in spite of being "blown out" in the highest court in the land over the Coloured Voters, they will now find ways to overcome it. The natives and Indians are combining – they never would have done so before – what will it all lead to? I honestly feel it would be wiser to leave and start afresh in Rhodesia where one would at least not be branded as a "Jingo" just because one is an admirer of anything English. Dear Godwan, how I love the place, it breaks my heart. Do forgive me for this depressing epistle, but I just felt I had to write you.

May 2nd

I was disturbed yesterday, so could not continue and this morning Rory received your letter 3rd April. We were delighted to hear you had been awarded the Royal Society Medal, the Flavelle, our heartiest congratulations (Rory will write you). I know you richly deserve this great honour, Freda and your family must be immensely proud of you. ...

 Heaps of love to you all,
  Nita

Godwan River Estate
Ngodwana
E. Transvaal
4th May 1953

My dear Old Biff,[113]

 Many thanks for your letter of the 3rd (April). We were all most gratified to hear of your award of the Royal Society Medal which you richly deserve – far more. ...

 The political situation in this country is shocking. The Nationalists have no intention of respecting the Constitution, language rights, or any other rights which obstruct them. In fact, Strydom (Strijdom) has said they want only one language in the country and a Republic outside the Commonwealth and no compromise. ... I see big trouble ahead. ...

 With love to you all.
  Your loving brother,
   Rory

---

113 Ibid.

London, Ont.
4 May 1953

My dear Dad:

... I was most pleased to hear of the honour of appointment as a member of the National Research Council – really a grand thing, even if it will add a little to you in more work. Several people have spoken to me and said how glad they were to hear of the appointment.

It is good that McCrady is to get an honorary degree from McGill after many years of distinguished service. I saw in the paper, and enclose a cutting, that Guilford Reed is to get an honorary degree from Saskatchewan on May 8th.

I am much relieved to hear that the family health is improving and that the "alarums" have more or less ceased. ... Blake must be happy to have got some increase in salary [which] will be a great help to them. We are all quite pleased too because the University has done a lot to increase salaries, both upper and lower levels. Come July 1, we all get a good increase. In my case they have added $800, which is quite handsome considering all (brings us up to $7,800). The damn dollar is not any too valuable these days but it seems to be fairly stable over the past year. ...

"Medical politics" around here is in a parlous state and a group of self-opinionated blighters have been making things very uncomfortable. Mostly petty jealousy of the success and prestige of McLachlin, Brien[114] etc and a long smoldering resentment of the way Ed Hall cleaned things up around here 8-9 years ago. They have caused much anguish to Bert Collip, who is extraordinarily sensitive, and I have been a little worried about him lately and so has Bob Noble and Roger [Rossiter]. However, he is away for a week and I hope to see his spirits improved when he returns. ...

Phil Fitz-James was very impressed with your lab. If you have not heard from him yet it is because he is trying to reconcile his financial difficulties and Montreal. He is *not* at all well-off, has three kids, and he does not want to perpetuate his difficulties in a high cost city. I have not badgered him about it, because I know how difficult these decisions can be to the individual. Your comments are very correct and he realizes it.

I must back to work now – will try and write again soon.

Your son,
Bob.

---

114 A.D. McLachlin, professor of surgery, and F.S. Brien, professor of medicine. Both were appointed in 1945 and became important figures in the university as well as in Robert Murray's life.

Montreal
5 May 1953

Dear old Son,

Sue is continuing to have a hard time. Dougie was taken with others to see the school they will be going to next year. A week later he developed a sore throat and a rash. The description I have from John Charters is very like Scarlet Fever. His throat was full of strep pyogenes but is clear now after a course of penicillin & his rash has gone after being intermittent for a few days. ...

I have been appointed a member of the National Research Council for 1953-56. It is an honour with work attached but will be interesting. David Thomson[115] & Ray Farquharson (Toronto)[116] are the other two new members. By the way, Guilford Reed is about to receive an honorary degree from the University of Saskatchewan, this with McCrady's degree from McGill should be recorded as honours to members of the C.S.M. & go in the archives. By the way, the publication of the Proceedings in the Revue Canadienne de Biologie is very slow. The journal belongs to the University of Montreal now & is no doubt delayed by prayers & expurgations. ...

Besides two lectureships vacant there is present & impending worse technician trouble; a matter of salary. I still rather hope Phil Fitz-James will come here. I have no one in view yet for the Clinical Lab job ($3500.00 seems good enough for it considering other jobs & things) which offers good training.

There is a spate of retirements; among those in the Faculty of Medicine are Gavin Miller (can't be age), Goldbloom[117] & Stekle.[118] Don Webster has been appointed Prof of Surgery & Surg in Chief to CMH.[119] I imagine the successor to Stekle is a problem & Melville is hard to pass over in that subject; I would abolish the department & make medicine teach a modernized combined pharmacology & therapeutics. We have to try to undo what they teach on sulphonamides & antibiotics as it is since they do not know any bacteriology. We cannot get any time for medical mycology. ...

Our exams are almost over & there remains only the dentists. There were only 4 in Course 3 & no firsts this year. A pretty good med year & a good course 4. Only one Ph.D. (Whitehead) & he was excellent & 1 quite good M.Sc; other graduate students will put in their theses in the fall. Space is a problem.

---

115 Chair of the biochemistry department at McGill.
116 Ray Farquharson was head of the Department of Therapeutics at the University of Toronto. From 1960 to 1965, he served as the first president of the Medical Research Council of Canada, which had been established largely as a consequence of his 1958 report to the Medical Committee of the National Research Council.
117 Most likely a reference to Alton Goldbloom, who joined the Faculty of Medicine at McGill in 1922 as an assistant demonstrator in pediatrics and later became chair of the department.
118 Unidentified.
119 Children's Memorial Hospital, Montreal.

Not having heard to the contrary we suppose you are all well. We are greatly looking forward to the end of this month when we shall see you.

 Love to all of you,
  Dad

London, Ont.
12-14 June 1953

My dear Mum and Dad:[120]

 Peter has had a very happy birthday – he not only says so, he looks it. We made the mistake of putting his presents in his room last night – we awoke at 4:40 a.m. to realize that all three kids were up and sharing in the fun! Luckily they were amenable to persuasion and did return to sleep – to the extent that we had a terrible rush to get to school on time! ...

 We all went to see the film of the Coronation (J. Arthur Rank) at 5 p.m. the other day. It was a magnificent show and the colours were beautiful. It is amazing how well they did in the outside scenes on what must have been a rather dim day. Alice, of course, followed every detail and enjoyed it tremendously. Peter and Tommy were not so impressed with the Abbey but took great note of the parade back to the palace. I think they enjoyed it all, really, but it involved a lot of detailed attention and they haven't the staying power as yet. I think that they will remember it.

 It was grand to see so much of you both while you were here and at Guelph. The children were thrilled to see you and to tell you lots of their special interests. We were most proud of you, Dad, and so pleased at the medal; and we enjoyed your papers very much indeed. Wish that you could have stayed longer. ...

 By now you will have spent an exhausting week in Ottawa, luckily it has not been hot, and you will be glad of a change. Mum says she is going to try and persuade you to take a few days at the lake, I hope that it can be done.

 Your son,
  Bob

London, Ont.
22 July 1953

My dear Mum and Dad:[121]

 ... At last I have finished with most of the Rome stuff and have only the usual last minute doubts and editing to do! Have got one more paper off to the journal and it is

---

120 LAC, E.G.D. Murray Papers, vol. 19.
121 Ibid.

accepted (*Moraxella*) but there are still at least two to go and I must try to have them finished before I leave.

Still no word from the people at Rome about where and how we are to be accommodated. Have written again (a couple of weeks ago) indicating firmly that it was nice to know about such things when arriving at a strange city. Have also written to the Hôtel de Brésil, Rue le Goff, Paris V for a reservation. It is a small hotel near the Sorbonne and not too far from the Inst. Pasteur. Roger Rossiter stayed there last year and says that it is O.K., reasonable, and surrounded by good restaurants. No sense spending a fortune.

Plans are made for Doris and the kids to leave here the 31st arriving in Mtl morning of August 1st. Blake is going to meet them and take them up to the lake. ... I shall try and leave here on the 14th to spend a few days there before flying off. ...

Chris Hannay is having great fun (and frustration) with his "crystals". All the insect pathogens that we have examined have the things! He is trying to see if he can induce crystals in non-crystal strains A. cereus. Have you any idea if anthrax has them??

We look forward to hearing about your trip and of course very much look forward to seeing you.

All our love.

Your son,

Bob

London, Ont.
2 August 1953

My dear Dad:[122]

Doris and the kids got off safely from here and I hope arrived in good order. The kids were so much looking forward to the lake, and all that goes with it, that I thought they would bust with suppressed excitement.

I am planning to fly to Montreal in the afternoon of the 14th and jump to the lake with Blake. Weather permitting, of course, and if not I shall take the train and arrive the Sat. a.m. ...

It is hard for you to have to write so many speeches this year – always an exhausting business and hard on the time. I am sure a good topic will turn up and fit nicely.

I was sorry to read about Leonard Huskins, who was always most kind and considerate to me.[123] I knew that he had suffered a coronary but hoped it was not too serious. It is a fine and appreciative leading article in the "Star".

There is a most interesting lead article in the current number of Science which is

---

122 Ibid.
123 See chapter 5, note 107. Huskins was a professor of genetics at McGill.

worth reading. By a fellow called Richter, on the character and implications in the mechanisms supporting research. It is excellently written and carries a punch. You would enjoy it. ...[124]

Am up to the ears with Phil's thesis and my own pieces of paper.

>All the best.
>>Your son,
>>>Bob

>>>>>London, Ont.
>>>>>Aug 3, 1953

Barkmere, Quebec[125]

Dearest:

You don't have to worry about me! I am well fed by my own hand ... and am still living on beef. Have not yet had the necessity of shouting for help or asking someone to ask me to dinner. However, to allay any other fears you may have, I miss you. ...

It was a bit of a wrench seeing you all go off – but the kids, and even you, looked so happy that I was glad for you.

I did (and do) feel badly that the kids are not provided with *some* fishing equipment. Maybe you can let me know what is "de rigeur" with the small fry (meaning no pun). Maybe Doug will feel important teaching Peter how to fish! ...

Yesterday Robin found a tall Frenchman wandering along the corridor looking for an "Urologiste" (I thought he said "Neurologist") – out of the kindness of his heart he took him down to the outdoor Dept – came back bearing a specimen that was full of the finest and fattest gonococci we've seen in a coon's age. So the fellow came to the right spot in the first place!

Look forward to, expect and hope for an interim letter from you, and send you my best love.

>Your loving,
>>Robin

(P.S.) Hope that A, P & T are settling down and having a good time. Wish them good swimming, ... fishing, carpentering and exploring from me!

---

124 Curt P. Richter, "Free Research versus Design Research," *Science* 118, no. 3056 (July 1953): 91-93.
125 Correspondence between Robert Murray and Doris Murray, private collection, Robert Murray.

London, Ont.
Aug 5, 1953

Barkmere, Quebec[126]

Well? – What do you think you are up to? I do not find any letter from you today and only things for you, such as the enclosed P.C. and the odd bill (Dentist - $10.00!). ...

Very funny (true?) tale on the radio roundup last night. French professor was annoyed by rabbits eating his garden, so he imports Myxoma virus and starts a fine little epidemic that wipes out the pests. Writes paper for Acad. Des Sciences; sits back awaiting Légion d'Honneur. Rabbits start dying all over France – roads are littered with corpses. Hunting season opens, but no rabbits. Sporting goods people furious; hunters furious; price of skunk, mink, lapin coats goes up. No rabbit pies. Outcry against professor. Suggestion to import rabbits from USA which are resistant to Myxoma: outcry from communists against such a move; worried farmers point out U.S. rabbit eats more. Professor worried everyone worried. Even court action for "restraint of trade" – epizootic said to be spreading fast!

Going to have supper with Hilda and Chris tomorrow night.

 All my love, dearest,
  Robin

(Hellensburgh, Dumbartonshire)
Sunday, 23rd Aug [1953]

Barkmere, Quebec[127]

Dearest:

This is a most lovely house[128] and I (to say nothing of Nan and Jimmy) wish that you could be here with us. It is right on the shore of Gare Loch, where it enters the Clyde and is, maybe, 20 miles from Glasgow. It is in just a little village but is hard by the pretty little town of Helensburgh. The road runs along the shore and separates the house from it – but that is not noticeable. The view is magnificent and as I write I look across the front lawn (say twice as big as our whole lot), over the small yacht anchorage. ... The moors running over the hill are purple with heather. ...

Miss you terribly, I might say and resolve that we *must* find some way of doing this sort of trip together in the future.

126 Ibid.
127 Ibid.
128 Cousin Nan's (Nancy Heard's) house near Gare Loch, Scotland.

Both Nan and Jimmy send their love to you and to Sue.
All my love darling and to the kids,
Yours,
Robin

Stockholm
19 Sept. 1953

My dear Mum and Dad:[129]

I have a while to wait while Hedén[130] is busy and it will probably be about the only chance for me to write before I leave on Monday morning.

It was grand to have such a long visit with you, Dad, both in Rome and in Paris. I enjoyed the chance very much and have accumulated some valuable and indelible experiences. Our visit in Paris, being more restful and less rushed, was a welcome oasis. However, I really wish that both Mum and Doris could have been with us. Mum should have come for she could have had a real chance to see the environs of Rome – and I wish that we had had the chance to see more.

... The plane to Copenhagen was a DC 4, quite comfortable and the service and food really excellent. The Scandinavian airlines do you very well. The weather cleared a bit and we had a good view of the ground as far as Brussels but from there until just north of the Zuider Zee it was both cloudy and bumpy – had to take good care of the coffee so that it stayed in the cup! I was fortunate in sitting next to a *very* nice USAF colonel who had been stationed in these parts (as an attaché and now in SHAEF) for 7 years and who knew the region perfectly from both air and ground. A soft-spoken and well-educated man. ...

We flew around Copenhagen and it looked beautiful and gay, all lit up, and landed at the airport right next to the *water* and over the water you could see the lights of Malmöe in Sweden. Bright moonlight helped to make the scene even more beautiful. Nice airport buildings and comfortable waiting room done in modern Danish style. Had to wait about 40 mins there and transfer to another plane (DC 6 this time) for the last leg to Stockholm (As usual had to fill in a fistful of papers!). Very comfortable flight but nothing to see except the *occasional* town and then Stockholm itself, built on a number of islands with the sweeping curves of bridge lights to join them. No difficulties; Hedén met me with his car and we went straight to his home. This is a nice red brick modern and very comfortable house in the north suburbs and very close (2 ½ mile) from the Karolinska Inst. The good taste of these people is remarkable – they

---

129 LAC, E.G.D. Murray Papers, vol. 19.
130 See chapter 7, note 5 for information about Carl-Göran Hedén of the Bakteriologiska Institutionen, Karolinska Institutet, Stockholm, Sweden.

furnish their rooms sparsely but very well in the semi-modern style and everything fits together harmoniously. ...

This is a most extraordinarily beautiful city, with an old part that has retained all of its medieval character except for the modern interiors of many shops. It is slightly hilly and is divided by many waterways and canals, clean and fresh looking, and the harbour full of the White and Blue Swedish Line ships. Hedén took me sightseeing the first day and we not only looked about the town from the best vantage points but also went for an hour's boat trip. Hedén used to have a seagoing boat (and holds a North Sea skipper's licence too!) but this was one of the usual sightseeing launches. The leaves are turning and the rocky country looks very like the fringes of the Laurentians. You people would *love* it here.

The Karolinska Inst.[131] is an enormous place – still being built – and is quite the nicest and best equipped (both men and materials) that I have ever seen. Everyone is friendly and cheerful but they are obviously very hard workers and serious too. The microbiologists only have a corner of a building at the moment and their new inst. is just about to start construction. I have also visited the Karol. Hospital and its lab. Most impressive and they do careful good work and some very practical research.

Yesterday morning I gave a lecture – a lot of people there – and it went very well. Kind things were said and attention must have been good for people keep bringing up possible ways of attacking the problems. ...

Well, ... I must away, but lots of love to you both. You should visit Scandinavia!

    Your son,
        Bob

London, Ontario
4 October 1953

Professor and Mrs. E.G.D. Murray
c/o Bank of Montreal
Waterloo Place
London, ENGLAND

My dear Mum and Dad:

I have been hectically busy since I got back and it has been hard to find time to write. I hope that you got my letter from Stockholm which will have told you some of the nice impressions that I had of the place. It was certainly lovely and hospitable – and if ever the opportunity came I would go again. I think that I wrote on the

---

131 The Karolinska Institutet, Stockholm, Sweden.

Saturday morning – after that interlude Hedén and I went over to the Central labs of the Stockholm hospitals. This is the lab of old Davide, who is not at all well and is to retire soon. A lovely new lab with some very good ideas in design. They had lots of equipment and first class too. After lunch with them we went and did a bit of shopping – the stores are *very* fine and very tempting. ...

That evening, after dinner, we went into the old town of narrow streets and mediaeval fronts – walked around for a bit and looked into the windows. Old shop fronts (they are not allowed to change them) but modern interiors that blend remarkably. Then went to an old cellar restaurant that has been in the business for four centuries. Rather fun and very cheerful; everyone joining in the folksongs. ...

Well more in the next ...

    Love,

        Bob

                London, Ont.
                24 October 1953

My dear Mum and Dad:[132]

It was nice to hear your voices the other night, and quite a surprise too because I was certain that I had been told you would not be back until the 22nd. ...

I was in Ted Roy's lab a couple of weeks ago just after they isolated the *Listeria*. Most interesting but we have not had any material yet! It has been a rushed time lately and when the paper work subsides the teaching looms up, and so I see-saw through the fall. Hardly any worthwhile experiments done and I am getting the itch to get back to the bench.

Have to go to Ottawa this weekend for a meeting with Marion[133] on Monday to discuss the journal business. Wish that I had time to come on to Montreal but I think that I had better stick to the business.

Bert Collip is in good form these days and seems to have got himself into the right frame of mind to slough off the pinpricks of local medical politics. Saw both of them yesterday at convocation and they send their best.

Must go off for supper but will write at length next week.

    Your son,

        Bob

132 LAC, E.G.D. Murray Papers, vol. 19.
133 Leo Marion, a distinguished chemist from the National Research Council laboratories, was editor-in-chief of the Canadian Journals of Research, which included a number of different titles, among them the *Canadian Journal of Microbiology*. This meeting involved issues related to the establishment of the *CJM*. The first volume came out in 1955. RGEM.

London, Ont.
7 November 53

My dear Mum and Dad,[134]

. ... Doris has started drawing again at the Library on Tuesday evenings and it is very good for her. They have a class and some instruction – someday she will have time for more of it. I am going to have to find a way of getting some exercise for I am getting terribly soft (and quite fat, for me!), but it has been hard to find both time and energy so far this term (life has been very busy).

We do not seem to have done anything very exciting and have only been out to a few evening do's, such as the Little Theatre. There have been some good movies and some more are coming.

Both Robin and Chris are off on their travels this week. Robin has gone down to Rutgers to give a lecture at Waksman's Inst. And will take the opportunity to talk over some problems (particularly macronuclei) with people in New York etc. He needs a bit of a change and it will do him good. I fear that Waksman is going to offer him a job and I hope that it will not be too attractive. Chris has gone off to the Sault to confer with Bergold et al, do some EM work on his crystals and plan the battle on toxicity etc. Very exciting that the crude crystal traction has proved to be toxic to insect larvae and opens up a lot of possibilities that insect people will have to explore. ...

We had an excellent seminar this last week from W. Feldberg F.R.S. who used to be in Physiology at Cambridge. A magnificent and most interesting talk about a complex subject. He has got some most interesting pharmacological observations on drugs acting on the C.N.S. ...[135]

All our love to you both.
Your son,
Bob

The University of Western Ontario
November 12, 1953

Professor E.G.D. Murray[136]
Department of Bacteriology, McGill University,
3775 University Street, MONTREAL, P. Que.

Dear Dad:

Thanks for the encouraging letter telling of the discussion of the new Journal by the N.R.C. I am glad that they and Marion feel that it is a good step; it is up to us to

---

134 LAC, E.G.D. Murray Papers, vol. 19.
135 Central nervous system.
136 LAC, E.G.D. Murray Papers, vol. 19.

get it started on a good basis. I wish that you could have a discussion with some of your people and others in Montreal to see if there are any general suggestions that we might consider in setting up policy.

The Editorial Board will have to be considered with care and must be reasonably representative without being too large. Marion seems to have the idea that an Editorial Board should be used for arbitration only and that the bulk of the reading be done by anonymous referees. I would rather have a working Editorial Board. We must also consider the possibility of appointing a member from the U.K. and from the U.S. This would be of help if we can expect any papers from overseas, and would have some value [in] prestige if the individuals are well chosen. ...

 Yours ever,
 Bob

            University of Western Ontario
            November 18, 1953

My dear Mum and Dad:[137]

... Another cause of depression last week was that Robin went down to Rutgers over an offer of a new job at Waksman's new marble palace – 'the house that Streptomycin built'. Was much relieved, on his return last Sunday, to hear that he is *not* attracted. I must say that it does not sound too attractive to me either, even though there are a few good people there such as Nickerson[138] and Ruth Gordon,[139] and Harry Eagle is to be asst. director.[140] He brought back a marvelous ... picture of Waksman warming to his favourite subject, the Actinomycetes, rising to his feet and gripping the edge of the table, saying "They're a gold mine! A gold mine!" in a strong and rising voice!! That about sums it up. Robin, although taken out to lunch and shepherded about a bit was left to his own devices after his seminar, and had to find his own way back to a hotel and fend for himself, in the somewhat uninteresting town of New Brunswick, for the rest of the evening. A bit shabby and unnecessary, I think. However, he took the opportunity to see various good people in New York (and even Boston, for a day); Schräder and Ryan and Porter[141] treated him well. The people in Columbia snatched

---

137 Ibid.
138 Walter John Nickerson, professor of microbiology at the Waksman Institute of Microbiology, Rutgers University.
139 Ruth E. Gordon, professor of microbiology at the Waksman Institute of Microbiology, Rutgers University.
140 The American pathologist and cell biologist Harry Eagle was a long-time trustee of the Waksman Foundation for Microbiology (1951-1981). He also served as president of the Society of American Bacteriologists in 1958.
141 At this time, the Canadian cell biologist Keith R. Porter (1912-1997) was doing pioneering work

him for a lecture and in their thanks expressed their gratitude to Rutgers for the opportunity! He had a grand scientific time with them and was very much heartened by their interest and encouragement concerning ... the nuclei of fungal hyphae.

The other excitement is that Chris [Hannay] went to Sault Ste. Marie to do some E.M. work on the crystals and the chemical fractions that he and Phil Fitz-James have worked so hard on for 2 ½ months. Very exciting ...

We look forward to seeing you and enjoy all your letters. Your long letter about the lake came today Mum – it must have been grand up there.

All our love.

Your son,

Bob

> The University of Western Ontario
> November 29, 1953

My dear Dad:[142]

Things have been quite exciting for the past few days because we have now got the money for an Electron Microscope and have found (practically bulldozed) a room for it! Admittedly we have been agitating, in an active way, for an E.M. but it had looked rather a slow business from the outside. The final stimuli proved to be Chris and his crystals, and Robin's invitation from Waksman! (Incidentally, Waksman did not have any provision for an E.M. and we shall beat him to it). But Chris really sealed it by convincing the Science Service Advisory Committee in his lab and giving them a real talk – the result was that it was agreed that there should be an E.M. in London for the common good and that means must be found. On Friday Ed Hall went to Toronto well armed with good arguments and talked to the Atkinson Foundation – they provided us with $25,000 which will get us well started. I must say that I never liked the Toronto Star as a paper but the Foundation seems to have some uses! So we are planning madly and we should be ready to work in a few months. It will be very welcome because we have a number of problems on which we have enough background understanding to be able to get somewhere with electron microscopy.

With all the burgeoning problems of our own and Chris' work we have to make a number of plans for other work than the electron microscopy. I have been worrying about Phil Fitz-James for a long time now. I feel that we must find some way of retaining Phil for some time not only because his work is good but because he is needed badly for the progress of some things. And especially for the crystal work, which is

---

with electron microscopy at the Rockefeller Institute for Medical Research in New York. He later moved on to Harvard and the University of Colorado.

142 LAC, E.G.D. Murray Papers, vol. 19.

linked in formation to sporulation in some way. Chris would like to have him but Science Service establishments and the Civil Service Commission are slow movers. Collip brought it up to me himself and felt that he should be held on an NRC Post-Doctoral Fellowship. We shall have to work on this, and quickly. ...

I enclose a reprint of Billingham's paper in case you have not seen it – I think that it is rather exciting and bears some thinking about.[143] I had things like that in mind some years ago but never really formulated an experimental attack. I am very glad they have done it. Experts like that may help force immunologists to think about the fundamentals rather than peripheral phenomena.

All our love.
Your son,
Bob

The University of Western Ontario
1 Dec 53

My dear Dad:

Nice to get your letter yesterday – since then, and my letter of a few days ago, quite a lot has been happening and some of it surprising. I doubt if I shall ever be in the position of having $25,000 and nothing to spend it on! Today spent in talking to representatives of both RCA and Philips. The RCA people are discontinuing their EMU series and are starting up next year with a newer, bigger and "better" model. Result is that EMU not available and the new "monster" will be too expensive (and will not be "out" till March). A smaller version of the monster is planned for next summer (June, say they) at about present price. BUT NO firm ... delivery dates *or even* data on characteristics etc. Suspect with *new* designs that there are "bugs" in them – and there usually are. Result is that Philips looks more attractive; especially with new lens design and the hope that the photographic difficulties are eased. Result is a bit of a turmoil in the midst of a lot of business anyway! Very provoking. ...

Sorry to hear that Robert Breed is under the weather. He will have to be careful of himself, I fear, for he seems to have had a lot of trouble. I fear that I can give [little] help to the Manual yet, and *truthfully*, I don't think that I want to be embroiled too much. On the other hand I shall be glad to help where and when I can but the

---

143 In the 1940s and 1950s, the British-born American immunologist and transplant researcher Rupert Everitt Billingham worked with his mentor, zoologist Peter B. Medawar, on experiments that provided the basis for successful tissue and organ transplantation in humans. Medawar was awarded the Nobel Prize for Physiology or Medicine in 1960 and shared the award money with Billingham. From 1965 to 1971, Billingham chaired the Department of Medical Genetics at the University of Pennsylvania School of Medicine.

structure of bacteria needs a lot of systematic work *after* a solid groundwork has been laid. So we have to concentrate on the groundwork for the time being. ...

 Lots of love,
  Bob

The University of Western Ontario
28 Dec 1953

My dear Mum and Dad:

We missed you at Xmas time this year and wished you could have shared the joys and excitements of the day. Still, we recognized that you were having a grand time with Sue and Blake and their kiddies. We drank your health at two p.m. amid the wreckage of an enormous bird (which is still with us!) – Peter, the only one who likes wine, took a great swig and in the right spirit too.

We had a grand day, all of us, and none were too tired or crotchety to feel overcome by it all. We had been out, but not too late, on Xmas eve on a round of deliveries ending up at Rossiter's for a bite and a drink. ...

We wish you all the best of luck and good fortune in '54 and send all our love and thanks for your kind thoughts that mean so much to us.

 Your loving son,
  Bob

Department of Bacteriology and Immunology
The University of Western Ontario
375 South Street,
London, Canada
Monday 1st March 1954

My dear Dad:[144]

I seem to have been snowed under with paper work of all sorts these days and the consequent mental indigestion has not been at all conducive to letter writing. The delay has been too long and I am sorry that there is so much that I will forget to tell you about. We have budget meetings this week, news releases on the E.M. grant, problems of the future moves in the lab, and maintenance problems all on the cards this week – to say nothing of a spate of committee meetings and two social events. This sort of thing seems to have been my lot for the past month. I also have this new Journal and some worries about it hanging over my head – I want to discuss it with you later in this letter. ...

Robin has been away in N.Y. for the past week looking into E.M. techniques again in preparation for our own equipment which should be installed about the middle

---

144 LAC, E.G.D. Murray Papers, vol. 19.

of the month. He has had a fine time by all reports and has learnt a lot – he should come back today if he is not delayed by last night's storm (which tied the roads up in this region). ...

The money for the E.M. was obtained from the Atkinson Foundation which is rather closely connected with the Toronto Star. We shall have to put up with a bit of publicity but no doubt we can live it down! They were up here with photographers just over a week ago and I believe they are "releasing the story" tomorrow. They are too damn impatient to wait until we have the equipment here and in working order. The Press is rather difficult to handle and many things that they do are just plain irritating. ...

The new Canadian Journal is a little worrying betimes and, although I have confidence in its future, I sometimes wonder why I took on the job! We have had a few MSS as you know but so far none of them are of very high caliber and could be classed as honest and acceptable work but not really the sort of thing that I would like to see in the first number. I hope that more and better stuff is in the offing. Have you anything?

The matter of an Editorial Board is very much on my mind and you will have to deal with it at the appropriate subcommittee meeting in Ottawa. ...

I have approached the following and have their consent (subject to approval by the NRC):

A.G. Lochhead (General Bact.)[145]   T.E. Roy (Medical Bact.)[146]
C.E. Van Rooyen (Viruses, etc)[147]   J.J.R. Campbell (Physiol & Bioch.)[148]
T.W.M. Cameron (Parasitol.)[149]   R.W. Watson (Industrial & Bioch.)[150]
E. Silver Keeping (Mycology)[151]

There are two problems, *Plant Pathology* and *Veterinary*. I consulted the plant path people and their opinion was R.O. Lachance[152] – who is not willing (though I have written him again – no answer yet!) because he said he felt linguistically and otherwise inadequate. Another problem is Veterinary: Savage is not willing because he

---

145 See chapter 7, note 97.
146 See chapter 1, note 89.
147 See chapter 6, note 61.
148 Professor of Microbiology at the University of British Columbia.
149 Institute of Parasitology, MacDonald College.
150 R.W. Watson was a microbiologist at the Division of Applied Biology, National Research Council, Canada.
151 The British-born botanist Eleanor Silver (Dowding) Keeping (1901-1991), who taught at the University of Alberta, was one of the first women to teach science at a Canadian university.
152 From the Research Station of the Department of Agriculture (Canada) at Ste-Anne-de-la-Pocatière, Quebec.

retires this year and is uncertain of his future. I rang up Chas. Mitchell[153] and his suggestions were Swatkin or Moynihan (who is in the Div. of An. Path. at B.C. and attached to U.B.C.). Don't know the latter and not happy about sight unseen. Do you know him? Any suggestions?

About the overseas representatives things are better. Have contacted and had favourable replies from ... Miles[154] for U.K., from Lwoff[155] for France and Frank Johnson[156] for U.S.A. – good men all. Lwoff will be a bit tied up for the rest of the year but wrote the most helpful and encouraging letter of the whole bunch. ...

Saw your paper (lecture) in the Lancet and so did many others around here. Nice comments from Rossiter, Lloyd Stevenson and others. ...

I have an awful lot of writing to do and also hanging over me is [Joe] Truant's thesis – the poor fellow is not too good at digesting his own work. We have a number of things that are too good to lay aside and should be written now – but time is hard to find. ... Damn! Good writing demands peace and there is precious little of that.

Doris is taking more time for her art work and is working very seriously at water colours. She is an excellent draughtsman with the pencil and shows some promise with colour. I am glad of it for it gives her a good outlet for her talents.

I must get this posted and off to you. I will write as soon as I have time.

    Your son,
        Bob

                          The University of Western Ontario
                                          7 Mar 54

My dear Dad:[157]

The clouds are clearing a bit but life continues to be rather full of trials and difficulties.

The budget meeting was fairly clear sailing and, I think, they do their best for us. As usual, I discussed the clinical bacteriology situation and the future necessity of

---

153 Charles "Chas" Alexander Mitchell served as chief of the Animal Pathology Division, Department of Agriculture, Canada. See chapter 2, note 36.
154 See note 4 above.
155 The French microbiologist André Michel Lwoff (1902-1994) was appointed head of the Service de Physiologie Microbienne at the Pasteur Institute in 1938. F. Jacob, M. Girard, "André Michel Lwoff, 8 May 1902-30 September 1994," *Biographical Memoirs of Fellows of the Royal Society* 44 (November 1998): 254-263.
156 Frank H. Johnson (1908-1990) was a professor of biology at Princeton University and was involved in the study of light production by bacteria and other organisms such as jellyfish. RGEM; *New York Times*, 26 September 1990.
157 LAC, E.G.D. Murray Papers, vol. 19.

providing a really first class man for the job who can do something with the good material that goes through. At last they have said something definitive and would be committed to providing for this as of July 1 of 1955. He would have to be someone with the ability and energy to set up a clinical division because the future looks as if it will bring a considerable change in the organization of the university. So I shall be on the hunt for someone and I would like to collect some possibilities in front of me for thought. Bob Elder is one distinct possibility if he can be lured away from the machinations of Ottawa. ...

A most interesting Dane (Asboe-Hansen by name)[158] gave a very long seminar but a most stimulating one on effects of various hormones on cells and tissues. Had a chance to talk with him because Watson gave a nice dinner party on Friday night.

Robin was most impressed with some work being done by Councilman Morgan[159] at Columbia on the development of the pox viruses. ... So far this is not published but I look forward to seeing them for R. says that they really take on where he and Bland left off in 1938. ...

        All the best
            Your son,
                Bob

                                        The University of Western Ontario
                                                    14 March 1954

My dear Dad:

You have been having a hell of a time with meetings and illnesses; the whole thing must be very exhausting. I hope that this reaches you in Ottawa, and I hope that these particular meetings are not too exhausting or exasperating. Anyway you will have some good company and some old friends to see.

Your long letter was grand and cheered me up quite a bit. I hope that some of the more enthusiastic supporters of the Journal will do their best to see that it is supported by others and that encouragement is given to sending papers to us. I agree that there must be some more effective notice of the Journal – so far I think that this has been largely directed to the subscription list. This is the wrong way round and it should be directed towards getting papers. The subscription will follow the quality of the Journal. I have written to Marion twice about this but have had no answer. I

---

158 The Danish microbiologist G. Asboe-Hansen from the Connective Tissue Research Laboratory, University Institute of Medical Anatomy, Copenhagen.

159 The American microbiologist Councilman Morgan (1920-1990) taught at Columbia University and was associate dean of students and curriculum affairs at the school (1970-1978). *New York Times*, 12 October 1990.

feel that ¼ page advertisements in the J. Bact and in J. Gen Microbiol would be infinitely more important than the *free* notices in Science and in Nature that have been published (once only). However, the most effective tactic in *Canada* is personal and requires that we *encourage* people to send in their best work. ...

What is Bergey's Manual going to do about the broad outline classification? Is Breed sticking to the outline he had outlined to us two years ago, new kingdom and all?

Jim Whitfield and I are doing some studies on the "nuclei" of enterics and have picked three that show the range of conformations. Rather interesting and emphasizes a few good basic structural elements. He is writing up the Shiga phage work and it should be good – we fit the other work in as a stimulating recreation!

All the best.

Your son,

Bob

Godwan River Estate
Ngodwana
E. Transvaal
17th March 1954

My dear Old Biff,[160]

Many thanks for the reprints from the American Scientist "The Peace of Nature in Man's World", "Why be a Microbiologist". Armand Frappier has given you some of the praise due you for a number of years. And I am pleased to know that the true scientists know the worth of a great scholar. Which you are. ...

Nita, John, and Roger are all well. I am not so fit at the moment as I am just recovering from a motor accident. A car ran into my car at rt. angles turning my car over 3 times. I sustained concussion and ... knee lacerations, cuts, and after 10 days in hospital ... it is now a month since the accident and I am still stiff all over and none too fit. But was lucky to get out of it alive. But I hope to be quite fit again soon. ...

I hope Freda, yourself, Bob and Susan are all well? ...

All the best to you all from us all.

Your loving brother,

Rory

---

160 LAC, E.G.D. Murray Papers, vol. 19.

London, Ont.
28 March 1954

My dear Mum and Dad:

... We have not been doing too much of interest lately partly because I have had an awful lot to do and have had to work most evenings. Had a pleasant visit during the week from Charles Chaplin who was returning to Ottawa and Lochhead's lab after two frustrating years at Suffield.

We are still waiting for the EM and I have a nasty suspicion that it is sitting in a ship in N.Y. harbour. It seems to have been a long time on the way but maybe it is just as well and will allow us to get some of our chores done before the problems of breaking in the machinery are upon us.

Collip got back from Ottawa looking quite benign and happy, and he told me that Dad was looking well despite the siege of streps. etc. I gather that our grants, fellowships etc have been approved alright. Concerning the Journal I have only had a short note from (Leo) Marion *yesterday* to say that the appointments were approved and that (the NRC financial officer) had agreed to the expense of advertisements in the J.G.M. (Journal of General Microbiology) and J. Bact. ...

All the best.
    Your son,
        Bob

The University of Western Ontario
10 April 1954

My dear Mum and Dad:

... I am glad that you met Lloyd Stevenson. He very much enjoyed his lunch with you people and said he had not had so much fun in ages. You may well see more of him since he has decided to go to McGill and eventually take up the long vacant Professorship in Medical History. ...

I had to give a talk to the Chirons last Monday and had to spend last weekend reading widely in preparation for it. The talk went very well and provoked a good discussion on patterns of life and living rather than careers. During all this reading I read Franklin's biography of Joseph Barcroft.[161] A most interesting "life" but very dull writing and quite in character with Franklin who is a bit of a dull dog. Well, that may be a bit hard on him and looking back through it I should say that "parts of it are excellent".

Aside from other things we have had the engineers with us for a week getting the

---

161 See chapter 5, note 70 for information on the British physiologist Sir Joseph Barcroft. The text referred to is K.J. Franklin, *Joseph Barcroft, 1872-1947* (Oxford: Blackwell Scientific Publications, 1953).

E.M. put together and now there are a few more days work of alignment, testing etc. It looks very nice but I won't commit myself until I see what sort of operating troubles we will have. ...

All are well here and enjoying life even if it is busy. It was grand to get your long letter, Mum, and to catch up on some of the doings.

Best love,
Bob

McGill University, Montreal
11 April 1954

Dear old Son,

I don't remember where our correspondence stands & what news I have or have not sent you.

I still get letters about the generalizations I embarked on in the 4 lectures & they seem to be most kindly received. I am still worried about the talk Lloyd Stevenson induced me to promise; I wish I had it off my hands as I had some papers to rewrite on work we have done. I wish people could do a half decent job of writing up their work.

The monocyte work is trickling along & we have a crucial experiment on hand. We are giving the monocytes horse serum staph antitoxin to work on in vitro & by titration of the x antitoxin can determine their take up & will look for the production of anti-horse precipitin in the later stages. I have some hope of resolving the difficulties of the oil-partitioning of T.B. by adapting R.W. Reed's[162] technique to it; results at present are promising. The chloromycetin & chloraphenical treatment of typhoid bacilli is giving most exciting indications of possibilities. ...

There are eight other projects trickling along at different rates, and two outside ones I am interested in (Dr. Mankiewicz on a T.B. observation & Dr. Morganti[163] on T.B. meningitis) are good, especially Dr. Mankiewicz.[164] The lab is thus very busy & I shall have it all with some regrets next year.

I have sent Norm Gibbons a letter from (Giuseppe Penso)[165] & my reply. He asked that the VII Congress of I.A.M.S. be invited to Montreal in Sept 1956. This question must be seriously considered at Kingston.

All this past week has been taken up largely by the City Council (Budget Meeting) & it will continue until midnight on Thursday next week. Most days the meetings have gone on to midnight & will continue that way. There is a threat of another

---

162 See chapter 6, note 104 for information on Roger Reed.
163 Odesca Morganti.
164 Edith Mankiewicz of the laboratories of the Royal Edward Laurentian Hospital, Montreal.
165 See chapter 7, note 41.

session being called a few days after this one ends. These prolonged & argumentative meetings are quite tiring.

What do you think of "Bacterial Genetics" by W. Braun?[166]

I had forgotten, in the rush of things, your enquiry about diploc. mucosus. I don't believe I have a culture & it would be difficult to check as my things are rather badly disarranged. (S.T.) Cowan was interested in it at one time and may have cultures & it is possible Dr. Sara Branham[167] has too.

I don't think the character of the General Outline Classification in Bergey will be changed except in detail. However, I believe a day will come when it will be necessary to have a Bacterial Kingdom. At present there are too many diversified forms being squeezed into too few divisions. There seems to me as much difference between a bacillus, a spirochete, an actinomyces & others as there is between a crustacean, an insect, & a mammal. I also recognize the general consensus of bacteria as different from plants & animals as these are from one another. This might make a good essay subject one day!

I have been invited to give the annual Foundation Lecture to the Academy of Medicine of Ottawa in March 1955. I shall have to find a general topic to interest all kinds of medics. Not too easy to do. I also have to think of a more or less general biological topic for my Presidential Address to Section V of the Royal Society of Canada in 1955.

On Thursday night we had a meeting of the Quebec Div of the CSM all on different aspects of the staphylococcus problem. There were five papers; some quite good & some rather superficial. No depth of penetration to any of them though a few interesting points were raised, with little discussion with proper understanding though much talk. It is probably good for them to get together from time to time.

Yes! ... Elder[168] would be a good man for the Clin. Bact Lab if you could get him; I think it would be a question of salary only as he is quite well paid (for a bacteriologist as things are) & is building a house in Ottawa. I would imagine he would be in the running for Gibbard's job when the time comes, but I don't know when that will be. There is a very good man in Edmonton (it seemed to me when I was there & I wondered why they had fetched Stuart out) & I think his name is Sterrit. He is another Scot; in his thirties; vastly interested in pathogenic bacteriology & with experience in Britain & I think in India & Middle-East; well founded in pathology too, so that

---

166 Werner Braun, *Bacterial Genetics* (Philadelphia and London: W.B. Saunders, 1953).
167 Sarah Elizabeth Branham was a senior bacteriologist in the Hygiene Laboratories of the National Institutes of Health, Bethesda, Maryland. Her research focussed on influenza and meningococcal infections, and she collaborated with E.G.D. Murray on the latter and extensively on the Neisseria Subcommittee of the Nomenclature Committee of the International Association of Microbiologists. See also Introduction, notes 464-465.
168 The Canadian microbiologist R.H. Elder. See chapter 6, note 55.

not being able to get along with Stuart (who was a student with him in Scotland) he went over to the Pathology Department as a pathologist. He is worth having a look at & perhaps more in reach than Elder in salary attraction. I liked both him & his wife personally very much indeed, I liked his quick understanding, his manner, his considerate independence of spirit, & his apparently good range of knowledge; from what we saw in a visit to his home, I liked his way of living.

Mum had a nasty fall a couple of days ago; she wrenched herself and bruised herself & has been very uncomfortable. She feels better today & otherwise she is very well. She is very busy with the Arctic Institute job she does with much pleasure. She clicks away on her typewriter with great concentration from time to time, but she has not said anything about what she is doing. ...

It is good to hear Doris is getting so much fun out of her painting. It is very well worthwhile & it is good to have an interest outside the daily chores. A number of Mum's drawings make interesting records besides being very attractive.

Lochhead told me Charlie Chaplin[169] was returning to him. I am not surprised as I did not think D.R.B. would suit him. I have not seen him since he returned from Suffield.

We are going to supper with the Norman Shaws[170] presently & David & Mrs. Keys & other old friends are to be there. It will be fun seeing them. Keys[171] is giving a talk tomorrow to the Canadian Club on Atomic Energy. The panic is being played up beyond reason & in the end will do more harm than good for lack of a proper perspective.

I wish I could have a yarn with you one day soon. There are several things I would like to talk over.

All our love,
Dad

The University of Western Ontario
21 April 1954

My dear Dad:

It was nice to get your long and cheerful letter and to hear all your news. It is grand that you have so much exciting work going on in the lab – it is a most impressive list and there are some things of fundamental importance. May they flourish. ...

---

169 Charles Chaplin was a bacteriologist who worked as a researcher at the Experimental Farm (Agriculture Canada) in Ottawa. RGEM.
170 See chapter 5, note 28. Shaw succeeded David Keys as professor and head of the Department of Physics at McGill.
171 David Norman Keys (1890-1977), vice-president of the atomic laboratories at Chalk River, Ontario. See chapter 2, note 4.

The EM is installed and we have done some testing. Got some respectable pictures of flagella and have tried out our new enlarger etc. ... No vacuum or electrical difficulties; mainly small mechanical difficulties. ... The machine is nice to handle and let us hope that it continues to behave itself.

Roger and I both have to go to Winnipeg, which is too bad in a way because we had a fishing trip planned at the time. However, we shall put the canoe on the car and fish a couple of days each way and see some new territory. So we shall see you there for a day and have a chance for a chat.

I have not read Werner Braun's book through. Jim Whitfield has and says that it should be useful for senior students and grad students. It is the only book at the moment. ... There are some oddities – but few books are free of these.

 All the best.
  Your son,
   Bob

McGill University
Montreal
April 27, 1954

Professor E.G.D. Murray,[172]
Chairman,
Department of Bacteriology.

Dear Professor Murray,

I still have not recovered from the shock you gave me when you told me some time ago that your date for retirement would be coming up August 31, 1955. That is still a year and a half away, but we must face the situation. It is not too early to start thinking about possible successors. For this reason I wonder if you would be good enough to do as I asked you in a recent conversation, namely, to prepare a list of candidates that you would consider suitable for appointment, including the possible candidates in your own department and certainly not omitting the name of your son, Bob.

I would very much appreciate it if you could, in each case, add to the name whatever biographical data you can collect without going to an undue amount of trouble.

This note is intended really just as a reminder following our conversation.

 With kindest regards,
  Sincerely yours,
   G. Lyman Duff, M.D.,
   Dean, Faculty of Medicine

---

172 LAC, E.G.D. Murray Papers, vol. 19.

London, Ont.
19 May 1954

My dear Mum and Dad:

Your nice birthday greetings by phone yesterday morning were most pleasant and your letter (and more than handsome present) arrived today. ...

Have been doing quite a bit of work with the E.M. – which is working very well although there is still one very technical fault to be dealt with. I have got some quite reasonable sections of young *B. cereus* and hope to go after some more tomorrow. Chris has been looking into crystals and has remarkably interesting pictures of a regular lattice structure in the Xtal surfaces, and he is trying to find out if the particles in the dissolving Xtal are artifacts due to alkali or not. ...

Glad to hear that the Boston trip was a success and generally interesting. Pittsburgh was the usual set of affairs. Nice to meet Kluyver[173] again and to have a chance for good chats with many interesting people. We raised a certain amount of dust in the cytological sessions and there was some pretty plain speaking. It may not have done much good as far as the Philadelphia crowd is concerned (they are impervious) but at least it made the position more clear for others. Will tell you about it when we meet. You were missed by many people and many of your old friends were there.

All my best love.
Your son, Bob

The University of Western Ontario
4 August 1954

My dear Mum and Dad:

... Several cheery communications from Robin who seems to be having a good business holiday, oscillating between various labs and the British Museum. He says that the EM conference was most worthwhile and it was important that at least one of us were there. He should be back next week sometime and I am looking forward to seeing him and hearing about it.

I have the routine in front of me for another 3 weeks while Dr. Bittner is away – fortunately light work so far but the lid could pop off. I hope not, because there is much that I would like to do before the necessity of teaching is upon us.

The new Journal is out and looks pretty fair with a few errors that are not too glaring. Wish we had a better 2nd issue! Hope that more papers will flow in the fall. We need them – can't maintain standards on too little material. ...

Your son,
Love, Bob

---

173 Albert Jan Kluyver (1888-1956) of the Technische Hogeschool at Delft, the Netherlands.

Montreal
17 August 1954

Dean, G. Lyman Duff[174]

Dear Lyman,

This is an unofficial commentary for your own use in devising a means of selecting my successor, and of necessity it is incomplete for lack of *curriculum vitae* for each person. I cannot well ask for such information as it would immediately create a situation which might embarrass the deliberations of the selection committee. Such information you can get officially without stirring comment through Deans and immediate superiors, who can seem to collect it for their own purposes.

In general, I think it is desirable that the selection be based on a wide interest in bacteriology with research, teaching and organizing ability. Scientific achievement and standing in the opinion of other bacteriologists (especially Canadian bacteriologists) and personal attributes should be evaluated largely by the respect they would collectively command, to retain the present desirable members of the staff and attract suitable replacements or additions. Medical Qualification seems to me essential, unless association with the hospitals is to be abandoned. Professed virologists seem to me unsuitable because of their restricted interest and the unlikelihood of their requirements being made available at McGill.

Most of this you no doubt have in mind. There may also be new policies, conditions and existing discontent of which I am not aware, that will have to be considered, which I cannot take into account.

You asked me to give you a list of possible candidates and I find it difficult without making specific enquiries. I can only give you my impressions. I would arrange the names of candidates in the following order:

A. *Canadians*
1. *R.G.E. Murray* (without personal special interest on my part) is most eminently suitable and particularly qualified. I know my present staff would accept him without question and he commands respect not in Canada only.
2. *T.E. Roy* Very desirable on the grounds of knowledge and ability as a bacteriologist and an administrator. He lacks high international reputation, though he is greatly respected in Canada. Very experienced in clinical bacteriology and extremely sound. Would be well accepted by present staff. His only fault is that he tends to underrate himself. His department I think is the best of its kind in Canada and he has a happy and admiring staff. He commands the respect of his Clinical colleagues.

---

174 LAC, E.G.D. Murray Papers, vol. 19.

3. Bracket together *J.W. Stevenson*[175] and *Roger W. Reed* (Dalhousie University) but there are reasons to rate Reed somewhat higher. Both have decided research and teaching ability. Stevenson would be accepted by the present staff but he might not be content to work under Reed; they worked together during the war and are great friends but of equal general standing. Neither have other than Canadian reputation at present but both will achieve much more. I believe Stevenson would tend to try to keep things much as they are and I am not sure he would stimulate wide fields of research but what he undertook would be good.

   Reed has much more vision and searching curiosity than Stevenson and more progressive organizing power and command. Reed will, I think, gain an important reputation and rank high in Canadian bacteriology. As a long range bet, Reed may well go further than Roy in time. There is no doubt Reed is a man coming on rapidly.

4. *D.H. Starkey*[176] has remarkable organizing ability, common sense and a wide range of knowledge. He would not inspire wide fields of research, but would not impede research. He has close experience in official organizations, in Government Committees and has a flair in political matters. I am not sure he would give inspired teaching. He is well known in official circles but not because of his published work.

5. *C. Robinow*[177] is a most distinguished bacteriologist of world wide reputation. An excellent teacher and inspiring research worker. I believe he would have great difficulty with administration and prefers to avoid that kind of responsibility; except for this he should be rated much higher. He is a great ornament to any department (or Institution as a whole) wherever he may be and commands everyone's respect, admiration and friendship.

6. *G.G. Kalz* has wide knowledge, experience, great teaching ability, sound sense and commands respect. Would run things well but might not develop progressively. Might not inspire wide ranges of research but would encourage them.

Other names must be mentioned:

   C.E. van Rooyen and *A.J. Rhodes*[178] are very distinguished virologists of extensive reputation with general bacteriological knowledge and training but would not now abandon their very specialized interest.

---

175 J.W. Stevenson was a bacteriologist and an associate professor in the department at McGill. He served as dean of medicine at the school in the 1950s. See chapter 4, note 74.
176 See chapter 3, note 102.
177 See chapter 6, note 40.
178 See chapter 5, note 109.

C.E. Dolman[179] has personality difficulty and has not achieved what his marked ability promised even with great opportunity.

N.E. Gibbons (Ottawa University), *F.O. Wishart*[180] and *R.J. Wilson* have not achieved required standing, nor shown positive powers.

Marion Ross is an experienced bacteriologist of wide knowledge in clinical bacteriology. Very well thought of and a very nice person. I do not know her well. I do not know much about her work and have not seen anything sterling by her in the literature.

B. *New-comers to Canada*

R.W. Stuart (Alberta) seems to me chiefly a public health laboratory kind and my impression is that his staff find him difficult to get on with. I am not sure he has the kind of knowledge and interest desirable for the position here. I don't know him well.

C.R. Amies (Alberta) seems to me a man worth enquiring about. He has a very wide experience and a great range of knowledge. He is said to be an excellent teacher. I know him very little but like him very well. I would not ask for information about Amies from Stewart, who is his chief.

J.C. Colbeck (Shaughnessy Hospital, Vancouver) seems to me a very well trained and experienced bacteriologist. Attractive character and good worker. I know little about him but enquiry might be made.

C. *Recommendations by Professor A.A. Miles* [E.G.D. Murray quoting Miles]

"Of the three *S.T. Cowen* seems to me far and away the best. He is very widely known and is one of the two Permanent Secretaries of the Nomenclature Committee of the Judicial Commission of the International Association of Microbiological Societies. He is a very able bacteriologist, at present chiefly interested in taxonomy and classification, but with a good deal of other experience. Has very sound judgement. He is widely and diversely liked personally. He has done a great work as Director of the U.K. National Collection of Type Cultures, as well as in various Commonwealth and international organizations. He should be very seriously considered and I would only rate him second to R.G.E. Murray. He seems very much liked by his staff and his department is splendidly organized. *MacCallum* is a virologist and I do not know much about him. *Lacey* is a well thought of man but I do not know him."

I hope you find this helpful. It goes as far as I feel justified.

There is an urgency to the announcement of my successor. It would allay an element of unrest in the staff of a personal nature. But even more important is the provision for continuation of projects which depend on Grants in Aid. These have to be

---

179 See chapter 7, note 88 for information on Claude Ernest Dolman (1883-1976).
180 See chapter 5, note 114.

applied for in November and December and the granting bodies might view them favourably if I can name and secure the concurrence of a successor of whom they approve. Otherwise they would hardly support a contract I am not able to complete.

Yours sincerely,
EGDM

McGill University, Montreal
12 September 1954

Dear Old Son,

It was marvelous to see you, even for so short a time, and all the more happy that the brief change & rest did you much needed good. I was shocked to see you so tired. You must make better arrangement for an essential & sufficient holiday. I wish I had prepared my paper earlier so that I could have gone with you to the lake. I am glad the journey home was not protracted.

The Chicago Committee seems to have been of little purpose, as is often so, & French Lick will be a troublesome place to reach for the main meeting. I hope it will be worthwhile. I was not favourably impressed with the questionnaires I had to fill out & some questions were meaningless. I was in the Chicago University Club years ago: I forget details, but it seemed to lack what I find necessary in a club. ...

The Duchess of Kent & Princess Alexandra have spent a couple of days here & visited the Nero & some other set pieces. They are spending the weekend at some "private estate" in the Laurentians (probably near Ste. Agathe) & it is a good thing that at least today is clear & fine. Tomorrow evening there is a Civic Banquet, with full dress, to which Mum & I are going & on Tuesday a McGill luncheon in the newly furbished "Redpath Hall" (this we saw with L. Stevenson). We have evaded various other functions.

The Saskatoon trip was good. There were a number of interesting people (especially from the Rocky Mountain Laboratory, Hamilton, Montana) & some good discussions. Most of the talk was on viruses, rickettsias, ticks, lice, & other vectors. All very friendly, informal & frank. Our talks (Fritz Blank & mine) were well received & evoked a number of questions. There was a lunch by the University & a dinner by the Provincial Government without introducing formality except for a couple of speeches.

I spent some time with Wendell MacLeod (Dean of Medicine)[181] & some of his staff. Their new 500 bed University Hospital, on the campus, is very good indeed.

---

181 J. Wendell MacLeod (1905-2001), known popularly as the "Red Dean" of Saskatchewan because of his strong progressive views and support of universal medicine, was the dean of medicine at the University of Saskatchewan from 1952 to 1962. See Louis Horlick, *J. Wendell MacLeod: Saskatchewan's Red Dean* (Montreal and Kingston: McGill-Queen's University Press, 2007).

The laboratories in it are very well provided for & Don Moore, the pathologist, has a very good appreciation of bacteriology & mycology of which he has charge for the hospital. The University Department of Bacteriology (under Rae) is completely dead, and, although it worries him, MacLeod cannot yet do much about it, but he hopes to appoint an assistant professor (when he can find one suitable) who will introduce some life into it by way of research. Rae could not contribute to it but I believe … a proper man could succeed him & develop it with every encouragement. The physiology & biochemistry are very good & active. The new fulltime Professor of Medicine is excellent & far seeing. I did not meet the Professor of Surgery but they are very happy about him. MacLeod is doing an excellent constructive job & there is every sign of real possibilities.

The serum from the typhoid case … promises to be interesting. … I shall send you details later. I hope we can get more such sera & would like a later sample from this case, especially if chloromycetin has been continued.

Thanks for the cheque; I had forgotten about it.

I hope to get away for about a week starting 19th or 20th to the Columbus. I have to be back by 27th as Neville Stanley[182] from Australia should be coming then & I would like to see him. He demonstrated the monocyte producing lipid.

  Love to all of you,
  Dad

Chateau Laurier, Ottawa Ontario
6 November 1954

Dear old Son,

I have not written a private letter to you for a long time. We did little this summer past & the time I had at the lake was wet. I did not fish but took Mum & Doug out a little & they caught a few in Pike Lake, not quite as good as usual, but in Bark Lake the bass fishing was hopeless. Very few fish and those small; it goes from bad to worse. …

The time was more usefully spent in making a few things including a couple of paddles. I got down a bit of unnecessary fat by sawing & splitting some stove wood which Mum thought was required. I was not quite feeling up to hard trips; pain in my right acromio-clavicular joint and in both knees troubled me.

The end of Sept I had a week at the Columbus. It was very wet & very windy; the fish were not bunching atall & had to be hunted here & there. … Charlie Pascoe was up & we fished together all the time & it was like old days. … A couple of new members, up for the first time, were very disappointed with the fishing. They had nothing & were astonished at what we had. Their method was to run their outboard at a fast trolling

---

182 The Australian bacteriologist Neville F. Stanley (1918-1984). See chapter 6, note 101.

speed down the middle of Pie IX casting about 20ft of line out at each side. They were told the end of the little bay at the old Trudel Portage was a good spot so they ran their motor all over it & found it no good. It seems very wrong that such types are elected.

We have had a couple of weekends at the lake this fall, after Oct 1st, & did little but small jobs round & about, mostly cutting wood, & a few paddles. There were too many "hunters" in the bush for comfort or safety. Mum is up at the lake now while I am here (Ottawa) until the afternoon of Friday 12th on N.R.C. business. ...

Amongst offers was one from Yale, but I cannot believe I could live in the States. However, it is flattering & nice of them to think of me.

Four people who were at French Lick have told me of your triumphant success there. It is a great compliment to you that you were chosen the spokesman for the bacteriologists and the way you did it is said to have impressed and captured everyone. The Canadians are proud of you & happy for it. Mum & I are elated and proud, but most happy for the recognition it brings you. If you have a copy of your talk we would love to see it. Duff was one of the people full of your praise.

I was surprised to get a S.A.B. ballot with my name on it for Vice-President. I had not been asked to stand & that was the first I knew of it. Something wrong about it & I suspect they got us mixed up. I won't get many votes as I don't have much to do with them & I don't think my things have been read much in the U.S. I was up for it many years ago, but Ira Baldwin was the other and he had just retired from many years as a splendid Secretary of the Society, so there was no chance at all for me.

I expect Collip will tell you, your letter about visiting research workers of high merit on consultation & cooperation was well received. Something may yet come of it. It is a sound suggestion.

I am sending you a M.S. on the isolation of TB from CSF by filtration. The merit of it is the much greater chance, almost certainty of finding them and the appreciation of their persistence during treatment. It will be of considerable interest to those concerned with TB meningitis. Odesca Morganti is doing very good work and she will have some more to publish later on TB meningitis.

The clinical people at the Alexandra have no conception of what is meant by research. They neglect completely the laboratory findings and modify their treatment on hunches and a spurious theory of the pathology of the condition that they have invented without reference to any direct observation. They do not read and have no idea whatsoever of the effects & nature of the drugs they use. Consequently they are missing a splendid opportunity. I wonder the P.H. committee awarding the research grant does not criticize the clinical report severely. I have been severely outspoken on it without effect, but they can withhold the money and give their reasons and make stipulations effectively.

Both the deflection of specificity by chloramphenicol and the production of antibodies by monocytes promise well. I hope to get them far enough to publish something by the end of my time at McGill. I think, too, we should get the TB isolation by oil partition a step further. It seems we have got a means of overcoming the inhibition of growth by the oil extraction.

I cannot get the reprints of the paper by Hawirko & myself from the C.J. Public Health.[183] They were extremely slow in publishing the paper & postponed it from number to number. Then, at last, it appeared in May & in spite of numerous letters we have not got the reprints yet, though they promise them by the week.

Fritz Blank has quite an investigation going with the City Health Department & this has revealed quite an epidemic of ringworm due to microsporon canis, which is strange. The epidemiology will take some tracing. Some infected cats & dogs have been found but many cases must be person to person spread. The extraordinary thing is the ineffectiveness and lack of understanding of dermatologists.

The CBC started on Thursday weekly interviews with pairs of members of the Montreal Medico chirurgical society; this first one was a physician & the head of the Dermatology of the MGH and I could not have believed such a display of ignorance was possible. There was no indication of any understanding of the principles of etiology nor of treatment & practically everything they said was nebulous confusion. Their teaching of students must be on the same plane. There seems to be an intensive enquiry into programs, systems and circumstances of medical teaching but I have seen little to indicate a realization of the importance of selection of the teachers.

No news to the contrary I suppose means the family are well and happy. People who saw you at French Lick say you looked well & I was glad to hear it. I was disturbed that you did not have a holiday last summer & to see you so tired when you quickly came to Montreal. You must give some attention to your own needs.

  Love to all of you,
    Dad

                The University of Western Ontario
                     10 November 1954

My dear Dad:[184]

... Times have been busy and in some ways eventful – and in many cases days have either been too short or too full depending on stamina! Teaching is, of course, in full swing and, probably because of the events around us, is not nearly so neatly organized as I could wish. But it takes time that has been precious short. Despite all

---

183 Roma Z. Hawirko and E.G.D. Murray, "Oil Partition for the Collection of Small Numbers of Tubercle Bacilli from Aqueous Suspensions," *Canadian Journal of Public Health* 45 (May 1954): 208-215.
184 LAC, E.G.D. Murray Papers, vol. 19.

this I have managed to get a few experiments going and hope to get some auxiliary information on nuclear structure from the condensed nuclei in the cells exposed to increased salt in their environment. I have just got a whole batch of sections ready and, damn it we have some circuit trouble (the first) with the E.M. When next term comes along we should get along famously.

Doris and the kids are all in good form and have all been very well. School seems to be suiting them all – Doris gets a bit of time for her needs – and they are doing quite well at all their pursuits. ...

Your son,
Bob

London, Ont.
17 Nov. 54

... I have no good prospect for the clin. lab. yet. ... Bob Elder seems to be properly entangled in Ottawa and there is little chance of disentangling him. So I must go on looking. ...

We were out at the Collips just before he saw you in Ottawa and I had a word or two with him on Monday. He seems to be planning definitely on your coming and is worrying about grants and things like that. ...

There are too many things packed into these fall months and we only appreciate it when Xmas is past and vistas of peace stretch out a bit! ...

Your son,
Bob

McGill University, Montreal
16 November 1954

Dear old Son,

I return your copy of the talk you gave at French Lick. I like it very much & quite understand why people were enthusiastic about it. You covered a great deal clearly & thoughtfully, with a light though firm touch on the most serious problems. I have made a copy for ourselves. ...

There was quite enough work at Ottawa & I also saw several friends. I went to the dinner of the Ottawa branch of the CSM (I could not go to the meeting) & Grant Lochhead gave a talk on the memories of a microbiologist. I was made to thank him. It was all very friendly & a good turnout.

Mum is home & well. The weather was cold; down to 11°F but she enjoyed it. The ground was too frozen to dig the garden. ...

Love, Dad

The University of Western Ontario
28 November 1954

My dear Dad:[185]

... I have had a letter from Ettinger[186] saying that the proposal about bringing in senior investigators for short periods into the laboratories was well received. However, the thing stops a bit short of usefulness. We could ask without any great trouble to bring in someone from anywhere in N. America. The sticking point is in getting someone from G.B. or the Continent, and they are the people that we need to provide for. As far as I can make out the NRC labs could bring in anyone from anywhere if they could give good reason, and that is what we need.

Birch-Andersen,[187] whom I would like to bring here for 3 months, is not able to find local funds to bring him here even though I could provide for him on a grant while he is here. He *might* (and that is questionable) be able to get a Fulbright, which would take him to the States. He is in charge of the EM at the SSI in Copenhagen and cannot manage to get away for a whole year, which is in the terms of most fellowships. ... Most *good* men can't afford and do not want to be away from their own lab for a year. ... It is no damn good if you have to spend hours, weeks and months trying to find the man's fare!

All our love
Your son,
Bob

McGill University, Montreal
7 December 1954

Dear old Son,

We are glad to have the news that Doris is home. I suppose the nausea is aureomycin & will soon pass off.

If it will not upset your affairs we have planned to arrive in London by CPR on 23rd December & have booked a room at the Hotel London. Mum will be able to help Doris at that busy & important time.

185 LAC, E.G.D. Murray Papers, vol. 19.
186 G. Harold Ettinger was a professor in the Department of Physiology at Queen's University and at that time he was the secretary of the Medical Committee of the National Research Council. J.B. Collip was the chairman of that committee. RGEM.
187 The Danish microbiologist Aksel Birch-Andersen from the Statens Seruminstitut (S.S.I.), Copenhagen, Denmark, was an expert in the field of electron microscopy. He spent several months in Robert Murray's lab assisting in the preparation for the arrival of Western's first electron microscope. Robert Murray also visited Andersen's labs in Copenhagen and remembers him as "a great friend and colleague." RGEM.

We could not come earlier because I have lectures & other engagements to the last minute. We shall stay about a week. There will be things to settle with Bert Collip.

I received my goodbye letter from Principal (Cyril) James today.[188] It is in rather different form to those I have heard of received by others. It has a personal touch and says things won't be the same without me & praises what I have done; it has a wish for future happiness.

I also received a letter from J.H. Bailey, Secretary of SAB, giving the voting for Vice-President: CA Stuart 1111 & E.G.D. Murray 964. I am surprised to have got so many votes & be only 147 behind. It is good for a Canadian against an American ... In that light the vote is a real compliment. I am not sorry over the result as I have enough to do without the big job the SAB would have involved. Bailey wrote a nice letter about it.

I have just received "Review of Medical Microbiology" by Jawetz, Melnick & Adelberg price $4.50. It has photographs by you, Robinow, van Iterson[189] & others, all very well reproduced. I have not read it yet, but glancing through it is up to date & concise though brief. In a few spots the brevity is carried too far to be informative. My impression is that it will be excellent for medical students if combined with Burnet's Natural History of Infectious Disease.[190] I must read it through.

We are looking forward to seeing you; we hope at the end of this week.

Give Doris our special love & good wishes.

Love, Dad

London, Ont.
3rd January 1955

My dear Mum and Dad:[191]

It was so nice to have you here for Xmas and we hope that you really did enjoy yourselves. Of course, the problems of housing must have been on your minds and we were delighted that you were able to find something to your liking so quickly. ... It looks like a very neat little house and it is certainly convenient for you and near to the university. ...

Xmas was a grand celebration for the kids with both of you here and you certainly did everything to see that it was a Merry Xmas. You were much too generous and, although every bit was both welcomed and appreciated, we feel somewhat overwhelmed. But I can assure you that everything is enjoyed. ...

---

188 See chapter 2, note 6.
189 The Dutch microbiologist Woutera van Iterson. See chapter 6, note 53.
190 Sir Frank Macfarlane Burnet, *The Natural History of Infectious Disease* (New York: Cambridge University Press, 1953). See chapter 4, note 10 for information on F.M. Burnet.
191 LAC, E.G.D. Murray Papers, vol. 19.

We have to get Doris' hip X-rayed again tomorrow and see if there is any visible damage. For a few days she was quite pain free – but she has done a bit too much the past two days (or it is the wet weather?) and has had to adjust to pain again.

It is nice to get the lab working along again. Seemed very quiet for two weeks and I can now get back to some active things.

Let us know how your plans and plots are coming and, particularly, if there is anything that we can do to help you.

All our love to you both.

Your son,

Bob

McGill University, Montreal
6 January 1955

Dear old Son,

Thank you for giving us a happy time over Xmas. It was splendid being with you & we enjoyed the family holiday.

I was glad to see the improved conditions in the lab; the change is important to both you & [Robinow]. However, the clinical lab staff seems to me an urgent matter. The other sides of your work are more important, though reliability and progressiveness in the clinical lab must be secured & watched. ...

Our lease here ends 31 March & we shall have to move our effects (household) by then. The lab things can wait a while longer as Duff tells me my time ends 31st August & it would be better for Mum to get her house straight first. The packing up will be wearisome & the sorting difficult.

If you found us a bit strained it is because we find the impending change of circumstances difficult to adjust ourselves to. The end of the road with an uncertain path beyond it is not very encouraging. We are sure we can make something of it in time, but the queasy shadows have to be looked at and we have to test the strength of our own resourcefulness. It will be rather like packing an unfamiliar trail in the dark, feeling for the hard ground with your feet, but we will find the lake.

Gertrude broke her leg last Sunday at their camp in the Laurentians. They were walking in the snow & she fell. I gather the fibula is broken & the tibia split & she is now in a walking cast. Very hard luck.

I went over the paper with Morganti & I hope improved it sufficiently. She will send it to you by the end of this week.

Love, Dad

London, Ont.
January 12, 1955

My dear Mum and Dad:[192]

It was grand to get your letters and to hear that all is well. You will be so relieved to have a house all set here in case your lease is not extended. ... At any rate, there is a lot of future to be happy in and we look forward to enjoying your enjoyment of it. ...

Peter was thrilled to get the envelope of stamps from his Opa, and I believe that he has written a letter to you already. That is some stimulus! We shall have to be prepared for a spate of collections. My old stamp album will come in handy and if you have a chance to pack it up I would be grateful. For the time being we will have to find him some simpler way of organizing his stamps. ...

The report season is coming up for its finish and we are just getting down to some real work again in the lab. I have been looking at Seeliger's *Diplococcus mucosus* strain and they are quite interesting.

    All the best to you both.
        Your loving son,
            Bob

McGill University, Montreal
14 January 1955

Dear old Son,

Tomorrow is your Wedding Day & we wish you & Doris every continued happiness.

I had a letter from Geoffrey Edsall[193] & he speaks very highly of what you did at French Lick & of yourself. He says "he was one of the liveliest and most effective contributors to the gathering".

I am trying to get the address written that I have to give to the Academy of Medicine at Ottawa. It is not coming easily yet.

My brother, Rory, sent me some S.A. stamps which I have sent to Peter. I have not seen some of them before. I am sending your album.

When I was quickly getting out after my lecture yesterday, one of the students stood up & called to me. He then made a very nice little speech on the theme that

---

192 LAC, E.G.D. Murray Papers, vol. 19.
193 A Harvard Medical School graduate, Geoffrey Edsall was assistant director of the Division of Biologic Laboratories of the Massachusetts Department of Public Health (1940-1942) and its director from 1942 to 1949. He was also for many years professor and chairman of the Department of Microbiology at Boston University School of Medicine. In 1951 he was appointed director of the Division of Immunology at Walter Reed Army Institute of Research.

this was my last lecture to the medical students at McGill. He put it very nicely on behalf of the class, and they were all standing. They applauded for a long time before I could thank them. ...

I'm sorry Doris is still troubled. Why not send some serum & let us see if she has agglutinins for PPLO. ...[194]

Love, Dad

McGill University, Montreal
14 January 1955

Dear old Son,

In my letter yesterday I omitted to tell you that Joan & Douglas have herpangina. It started with Joan & the little girl had some 40 spots in her mouth ranging from small ones to about eight close to ¼ inch in diameter. Her mouth was terribly sore & her temperature went to 105. She was quite ill & could not eat or drink. Doug became ill 3 days later & had many fewer lesions, was less ill than Joan, though he had a high temperature & headache & great difficulty in eating & drinking. They are convalescent now & the lesions are healing. ...

They have had quite a bad time with it. I sent specimens to Andy Rhodes in case he was interested & will hear results in time. I thought it worthwhile as Joan's case was unusually severe.

I also forgot to enclose what we owed you & did not pay when we were in London. Don't argue!

I happened to hear a part of Rawhide's vagaries last night. He was making fun of psychiatry & invented a new drug – Pentothal with Solium, & the truth comes out whiter.

Love, Dad

London, Ont.
January 20, 1955

My dear Dad:[195]

The album arrived today and Peter and I had a very happy time looking through it this evening after supper. ...

Your greetings on our Wedding Anniversary were much appreciated. We did not do much special but we did go out in the evening and had a pleasant private dinner party with friends. ... Last night I went to Thames Hall to see the Swedish National

---

194 Abbreviation for pleuropneumonia-like organism.
195 LAC, E.G.D. Murray Papers, vol. 19.

Gym Team give an exhibition to a packed house. They were marvelous and put on a grand show. ...

It was very kind and considerate of the medical students to make a pleasant ceremony of your last lecture to them. They are a good lot, in general, and really do appreciate good people and what is done on their behalf.

The clinical lab has taken quite a surge of business the last few days and we have been dealing with 50-60 spec. a day this week – not all routine stuff either.

    Love from us all.

       Yours,

          Bob

Faculty of Medicine
Medical Building
McGill University
January 27, 1955

Professor R.G.E. Murray[196]
Department of Bacteriology & Immunology
The Hamilton King Meek Medical Laboratory
University of Western Ontario
375 South Street,
London, Ontario.

Dear Bob,

As I thought might be the case, the Selection Committee to choose a successor to your Father in the Chair of Bacteriology and Immunology here has decided unanimously that they would like to recommend your appointment to this post. Before placing this recommendation formally before the Board of Governors there is a natural desire to know whether you would be willing to accept it. The appointment would be as Professor of Bacteriology and Immunology on permanent tenure subject to the age retirement regulations of the University, and at an initial salary of $9,000.00 per annum. ... The Professorship does not automatically carry with it Chairmanship of the Department, which is an administrative appointment made annually, but you could rest assured that you would certainly be appointed and continued as chairman of the Department each year save in the most exceptional circumstances. ... At the moment $10,000.00 per annum is about the top for heads of departments in the faculty of Medicine, but there are a few instances in which the figure is now as high

---

196 UWO Archives, Robert Murray Collection.

as $12,000.00 per annum. This is merely to indicate the present range within which increases of salary might be legitimately expected. The University would contribute to the extent of $750.00 (tax free) toward the expense of moving your household from London to Montreal. ...

I am sure that you are familiar with the Department in a general way, but I am equally sure there will be a good many details and perhaps important points on which you will wish to have further information, especially in relation to your own interests as a prospective successor. Accordingly, I very much hope you will come to Montreal as soon as may be convenient in order to collect any information that you may wish to have and to discuss with us any points that you may wish to bring up. Naturally we would cover the expenses of such a trip.

I shall look forward to seeing you in the near future and meanwhile may I say how much I hope that you will decide to join us.

With kindest regards.
Yours sincerely,
G. Lyman Duff, M.D.
Dean, Faculty of Medicine

McGill University
February 14, 1955

Dear Bob,[197]

Thank you very much for your letter of February 9th, and for your promptness in replying. That is an advantage in the circumstances, but I probably do not need to tell you how disappointed I am that you have decided not to come. I would greatly have enjoyed the association which might have materialized, both in our related work in Pathology and Bacteriology in the Pathological Institute and as Dean. However, I can well understand how difficult the decision was and equally I can understand the reasons for your decision to remain at Western.

I hope that we shall continue to see you from time to time at McGill, even after your father's departure removes one of the reasons which might bring you here occasionally. ...

With kindest regards,
Sincerely yours,
Lyman
G. Lyman Duff, M.D.

---

197 UWO Archives, Robert Murray Collection.

> Principal and Vice-Chancellor
> McGill University
> February 21st, 1955

Dear Professor Murray,[198]

Dean Duff has forwarded me your letter in which you convey your decision to remain at the University of Western Ontario, and although we are very sorry indeed that you cannot accept our invitation I can realize that there [are] many considerations to be weighed in the balance.

I am attaching the University's cheque for $48.50 in reimbursement of your expenditure on the visit to Montreal, and with best personal wishes, remain,

> Cordially yours,
> F. Cyril James

Robert Murray reflects on why he did not succeed Joburg at McGill[199]

You ask me to explain why I did not accept the McGill posting and I will do so – but, I am sure I had to couch the missing letter in gentler terms than I may use in this statement. I was deeply honoured to be invited to replace my father and realise that others believed I was qualified to deal with the special complexities of the McGill situation. I had no doubts I could do the job. The assurances of Dean Duff and Principal James were clear that I would be supported more than adequately, would obtain the special equipment that I identified as essential, and develop some funding prospects. However, I had built up a thriving Department at UWO that was by 1955 showing its possibilities. It had been accomplished from scratch and the vista for my research group was as good as I could hope for in any place. The University had nowhere to go but up and my superiors in the Faculty, the Victoria Hospital and the University were on my side almost all the time and I had very encouraging friends. For my part, I was a touch concerned about how Montreal would view a "family succession" and how the paternal interaction would play (even though I knew that my father would encourage and not interfere with any decision of mine and would respond thoughtfully if ever asked for advice). He had built his Department from scratch and in a more complex working environment than I had at Western; it was clearly HIS Department. I looked forward to the UWO Department developing further to be as great a centre of effort in microbiology in terms of teaching, research, and service as the distinguished McGill Department and be in essence my accomplishment. Furthermore the move would have taken at least two years away from full involvement in research while having to initiate and modify in places a complex operation as well as see that

---

198 Ibid.
199 Robert Murray [or R.G.E. Murray] to Donald Avery, 8 December 2007.

a new set of people were happy with it. So, I made a selfish but sensible decision then and maintained it to retirement – it defines one's position and intent as does the old nautical maxim: "Hold your course and speed". I have no regrets.

McGill University, Montreal
14 February 1955

Dear old Son,

Your loving letter gives us every happiness because of yourself and pride in your reasoned judgments. Your decision is not unexpected & I am convinced it is correct. Your estimate is just and true and is not fogged by false images. I am sorry it had to be. The department deserves better but its setting is wrong. McGill was given the materials for a good fly but they tied it wrongly and fished it badly.

It is best to be where you are appreciated and get sympathetic understanding. In time you will get what you need & there is no reason why your lab should not take the premier place in the Dominion. In the long run it is the man that counts and a department depends completely on the leadership of its Head.

Your decision based on true values, with due right given to personal inclinations, and judicious comparisons, makes us more than ever proud of you.

It was splendid having a few hours with you & it made all of us very happy. That was the nicest lobster we have had for a long time & is to be remembered.

We are looking forward to a happy time with nice people. Sue's lot and the lake are our only regrets. The move will be a trouble & there will be an interval of "rooming" here that won't be pleasant.

I must slow down on the work and give time to my own affairs. There are quite a lot of details to arrange. ...

Guilford Reed had a stroke on Saturday night on his way through Montreal from Washington to Kingston. He was put in the R.V.H. with left sided hemiplegia and unconscious. Fortunately his secretary was at hand, she has been with him about 30 years, & she kept things on a right footing. Mrs. Reed is on a trip in Mexico & no one knew how to reach her; but John Orr[200] has found how to send her a message. No reply yet. Gwen Dawson phoned us midnight Saturday. Sunday he was barely conscious but had a little movement in his left foot & could be roused to answer a short question or two. Today he is alert, smiles crookedly but can move his left leg perfectly well. His left arm is paralyzed but is flaccid & not spastic. He clapped me & was more cheerful than

---

200 John Harland Orr (1899-1965) was an esteemed member of the medical faculty at Queen's University. He was a professor of biochemistry and bacteriology at the university and was eventually appointed head of the department.

one could expect. His brother, from Lindsay, is here today. Ray Brow[201] is looking after him & everything is under control. I have spent much time on long distance phone.

There was a caucus of the City Council this afternoon from 3:30 to 6:45 on the threat of the Mayor & the Chairman of the Executive Committee to cut off all contributions to charity. Many speeches, nearly all in the same direction (me too) & eventually the Executive was requested to restore the donations, with only two dissenting votes. The Mayor made a stupid legalistic speech; he has a lot to learn.

I have a pile of correspondence I have not touched, which will add to that of tomorrow.

I have polished up my address to the Ottawa Academy of Medicine & I think it will serve its purpose.

    Our love to all of you,
        Dad

                McGill University, Montreal
                       21 February 1955

Dear Old Son,

Guilford Reed died this morning.[202] He did not regain consciousness after the new episode on the afternoon of Wednesday 16th; which left him with involvement of his right side as well. He faded out gradually.

His brother came back here this evening (from Lindsay, Ont.), Roger Reed arrives tonight & John Orr came this evening. Arrangements have not been settled yet; they will decide tomorrow when they all get together with Elsie.

Mum has had rather a difficult time & has done a brave and well adjusted job. She is a wonderful person at all times & rises to an emergency marvelously.

It has been a worrying week.
    Love, Dad

                McGill University, Montreal
                       27 February 1955

Dear old Son,

Mum is pleased to have the snaps of the house & Sue will be interested to see them. I shall be glad to be settled there as the present is very disturbing and the near future will be difficult. ...

---

201 Ray Brow was a senior physician at the Royal Victoria Hospital and a practitioner of internal medicine in Montreal.
202 See *Globe and Mail*, 22 February 1955, p. 10.

The lawyer said the titles are in order & they will be sending me notice of the payments when they become due. They also got in touch with the Bell Telephone Co & the office here called up to arrange that all would be well. Mum will move the household effects about April 5th & will go to London to settle them in.

Doug hurt his knee yesterday skiing up at Ste. Agathe, in a fall. He has been doing very well & is becoming quite proficient. His knee was swollen, but there were no broken bones. It is still painful with movement & blown up today.

There is only one cutting about Guilford in Montreal papers, Sue has it & will send it to you. John Orr could get you those in the Kingston papers which will have more to them. I went to the funeral & some important people turned up. Phil Greey & van Rooyen from Toronto. ... Also a number from Ottawa (DRB, Fisheries) & friends like Charles Mitchell. Fredette & myself from here. He will be sorely missed.

I don't feel that things are moving either fast or with much judgment to provide my successor. They can't expect to get even as good as I have given for the price they have paid or are paying now, but they seem to want to get it cheaper. What do you know of Cheever, who wrote part of the Enterobacteriaceae in Dubos' book?[203] I'm told he is visiting the lab while I am in Ottawa. I feel strongly that any outsider (UK, US etc) must be vastly better than available Canadians to be satisfactory. I fear the selection of an American would end up like the Hoff situation.[204] Anyway, it is useless to worry as it is not any affair of mine in fact.

I hope the $2000.00 won't hamper you & that you are not having to make some arrangements to provide it.

 Love, Dad

Godwan River Estate
Ngodwana, E. Transvaal
7th March 1955

Dear Biff,[205]

This is to introduce Biddy Monckton, a very dear friend and the widow of a great friend of ours Claude Monckton who passed away in March 1954. Claude was very badly wounded in the 14-18 Show; but showed remarkable fortitude in spite of never being completely out of pain. He suddenly passed on from Coronary Thrombosis much to every ones grief and surprise.

---

203 F.S. Cheever of the Graduate School of Public Health and the Department of Microbiology, University of Pittsburgh School of Medicine. See René J. Dubos, ed., *Bacterial and Mycotic Infections of Man* (Philadelphia: J.B. Lippincott, 1948).
204 Robert Murray does not recall this specific controversy.
205 LAC, E.G.D. Murray Papers, vol. 19.

Biddy & Claude were both Canadians and as Biddy is proceeding to Canada, to see her folks, I have asked her to look you up. She will be able to give you first hand news of us and will take some snaps with her and a plan of the cutting up of the farm I told you about, and will be able to explain it all to you.

Biddy has a keen sense of humour and I am sure you will like her. She will also be able to put you wise on the Political situation in this country. She will also be able to bring us back news of you all.

Love from us all,
Your loving brother,
Rory

Chateau Laurier, Ottawa Ontario
9 March 1955

Dear old Son,

Mum tells me your cheque for $2000.00 has arrived & she is keeping it until I return. I hope this payment does not put you in any difficulties. We can meet the payment on the house & do the moving we have to do without being on the rocks.

The paper to the Academy of Medicine went quite well, though I think few of them could properly appreciate what it meant. I gave the M.S. to Ray Farquharson to read. He is impressed with it and is trying to persuade me to expand it into a small book. He thinks it would be good for clinicians & pathologists as well as bacteriologists. He has brought it up three separate times. At any rate I shall send it to the Canadian Med J & perhaps Ernest MacDermot[206] will accept it.

The talk with the "Bug Club" was good fun & they seemed to like it. There was a good attendance & a lovely discussion. They welcomed me in a very friendly way.

I enclose a cutting from the Montreal Star of last night. I had no warning of this & it is very nice of the boys to think of it. I did not intend to go to the Medical Ball as it clashes with the Cambridge Boat Race dinner but now I must change my plans.[207]

These are strenuous meetings over long hours. I hope to catch the 4:20pm train home on Friday & will have been here since 28 Feb. There will then be a couple of days to catch up on accumulations & then the budget meeting of the City Council

---

[206] Hugh Ernest MacDermot (1888-1983) was a McGill-educated doctor who returned after service in the First World War in the Canadian Army Medical Corps to teach in the Department of Anatomy (1921-1924) and as a demonstrator in medicine (1925-1949). He was also on the staff of the Montreal General Hospital. He became editor of the *Canadian Medical Association Journal* in 1942.

[207] An announcement of a tribute planned for E.G.D. Murray's retirement at the annual McGill Medical Ball appears in the *Montreal Star*, 8 March 1955.

starts on 15 March to go on daily until midnight 31st March. It is a bad time as we shall move our things to London 5th April & time must be found to pack some things & to select a few things to be kept back. From then on it will not be too easy until we go to London. ...

I hope you are all well & that your staff troubles have improved.
 Love, Dad

Godwan River Estate
Ngodwana
E. Transvaal
11th March 1955

Dear Biff,[208]

I received a wire from Graaff-Reinet that Old Aunt Nan had passed away yesterday. She had not been too well for a little time and from what I understand it was a merciful release. I arranged for a wreath to be sent from us all (That is yourself and myself). The Old Lady must have been over 84 and was crippled with Rheumatism. ...

Brother, look after yourself.
 Your loving Brother,
  Rory

London, Ont.
16 March 1955

My dear Dad:[209]

Both Hall and Collip have told me that you looked very well in the time that they saw you in Ottawa. I hope that you will try to take it a little bit easier and let go some of the burden. With the City Council meetings on you will have more than enough to do.

Glad to hear that the moving and other plans are settling a bit. We will look forward to seeing both Mum and Sue. It will be grand to have Sue here for a change – she has really hardly had a chance to see us or the kids and it will do her some good to get around a little. ...

Doris was away today to Toronto to see the magnificent collection of Dutch paintings being shown. ...

The problems of staff are very much on my mind and I am not in sight of any solution. Jack Rublee[210] seems not to be interested in moving. Another of the DRB

---

208 LAC, E.G.D. Murray Papers, vol. 19.
209 Ibid.
210 The Manitoba-born Jack Driscoll Rublee (1920-2001) received his medical education at McGill.

boys has written asking if he might be considered. A fellow called E.W.R. Campbell who worked with Steve (J.W. Stevenson) during the war. Took his MD at McGill '50 and has been at DRB since. Guilford apparently thought well of him. He seems to be interested in research but has little experience if any with clinical bacteriology, though he must be reasonably well founded in the subject. ...

Now that I have the Porter-Blum Microtome back in operation I am getting into high gear on the fixation problem. It was a nuisance having it away so long but, anyway, part of the time there would have been no opportunity to use it and in the remainder we got some useful spade work done. There are too many variable(s) for comfort in these fixation problems and we have to hope for some good fortune in making the right sorts of educated guesses! The role of ions in the changes during unraveling of normal structure is even more difficult. ...

   Lots of love.
      Your son,
         Bob

Dr. Phil Demers[211]
Continental Building,
Sherbrooke,
Quebec.

                                       May 1955

Dear Phil:

As you may know, Professor Murray retires from his post as Professor of Bacteriology & Immunology at McGill in July 1955.

We feel that it would be only fitting if his post-graduate students and colleagues, past and present, were to pay honour to a man who has served as so great a source of inspiration to those who have had the privilege of working with him.

We would suggest a subscription Banquet (preceded, of course, by a cocktail party) to be held in Montreal some time between May 15 and 30. It is our purpose, at present, to contact all of those people who have worked, at one time or another, with Professor Murray and those who have had close contact with him in his professional life so that we may determine the number who might attend.

     Yours very truly,
       THE STAFF
       Dept. of Bacteriology

---

211 LAC, E.G.D. Murray Papers, vol. 19.

Please address communications to:
Dr. J.W. Stevenson,
Dept. of Bacteriology & Immunology
3775 University Street,
Montreal, Quebec, Canada.

>The University of Western Ontario
>Faculty of Medicine
>Department of Bacteriology and Immunology,
>University of Western Ontario
>May 12, 1955

Dear Steve:[212]

I expect to be at the banquet on May 27th and enclose a cheque for $12.00 to cover the various expenses. The dinner sounds very nice and I am having trouble with decision (Friday or not!) – anyway, you had better put me down for *the meat*.

Let me know if there is anything that I can do.

>Yours sincerely,
>Bob
>R. G. E. Murray, M.D.
>Professor

Societe de Microbiologie de la Province du Quebec[213]

>July 1955

The annual meeting of the Quebec Society of Microbiology was held in the Dept. of Bacteriology McGill University, on April 15, 1955, at, 8pm. The following officers were elected: Retiring President – Dr. E.G.D. Murray; President – Dr. M. Saint-Martin; Vice-Presidents – Dr. P. Genest, Dr. F. Blank; Secretary – Dr. V. Portelance; Treasurer – M. J.M. Desranleau.

A symposium on New Techniques in Microbiology was presented ...

News of Members ...

Prof. E.G.D. Murray will retire as Chairman of the Dept. of Bacteriology and Immunology, McGill University, in August. ... On retirement Prof. Murray will move to London, Ont., and "set up shop" in Dr. Collip's department. Freed of teaching and administrative duties he will devote full time to research. ...

212 Ibid., vol. 20.
213 Ibid.

Dr. R.W. Reed, formerly Chairman of the Dept. of Bacteriology, Dalhousie University, will succeed Prof. Murray at McGill.

Prof. E.G.D. Murray was awarded the Honorary degree of Doctor of Medical Sciences, on June 4, by the University of Montreal.

Godwan River
E. Transvaal
16 August 1955

Dear Old Biff,[214]

Many thanks for the cuttings from the Press of the honour conferred upon you by the University of Montreal. Congratulations Old Brother, you deserve every bit of it and more. You should have had a knighthood for all the work you have done. ...

Things are much the same in this country. The present government still playing up like Idiots. I see trouble ahead; and as long as they are in power I see no way out. ...

Love from us all to you all.
> Your loving brother,
> Rory

---

214 Ibid., vol. 19.

CHAPTER NINE

FAMILY CORRESPONDENCE: 1956-1964

McGill University
Montreal
Jan. 28th, 1956

Dear Dr. Murray,[1]

Your visit last time was far too short but naturally I understand the reasons.[2] We do however, hope that you will soon come again and spend a little time with us, perhaps when Sue's baby is born. Try and make it for a seminar – we are missing you very much at these sessions.

Best wishes and regard to both of you.

Yours ever,
Gertrude [Kalz]

Godwan River Estate, Ngodwana
E. Transvaal
16th December, 1956

My dear Old Biff,[3]

Received your Air Mail letter yesterday – we were most pleased to see it. We sent you a card and we also send you all our love and best wishes to you all from all of us here.

I am sorry to hear that you have had such a bad go of neuritis. I hope that by the time this reaches you will have completely recovered.

I am glad to hear Freda is fit and has done so well with the garden. She must have a 'green thumb'. It surprises me to think of Bob and Sue each having 3 children. How time flies.

I wrote you a long letter thanking you for the books on de Gaulle and the Journal of Microbiology, which was such a tribute to your outstanding work, for which in my opinion you have not received nearly the honour due.

All our love to you all,
Your loving brother,
Rory

---

1 Unless otherwise specified, the correspondence in this chapter comprises part of Robert Murray's private collection.
2 LAC, E.G.D. Murray Papers, vol. 20.
3 Ibid.

Godwan River Estate
Ngodwana, E. Transvaal
24th June 1957

Dear Old Biff,[4]

Received your letter of 22/5/57 today with cutting about Bob. You have reason to be proud of him – give him our congratulations. I can see the likeness to Freda in him. I think it is the eyes and nose, and your forehead and chin. I like his looks. They show determination and intelligence. We are very pleased to have the cutting.

There is a good book from Lund University – "South African Animal Life" (Results of a Lund University Expedition of 1950 & 1951): 3 volumes: ... Published by Almqvist and Wiksell 1955, Stockholm. Highly scientific, too much so for the ordinary layman, might be of interest to you or some of your friends.

I am sending you four pamphlets (published monthly) by the United Party, which will interest you.

With love from us all to you all.
Your loving brother,
Rory

156 Sunnyside Ave.
Lakeside, Quebec
Oct 27, 1957

My dearest Opa,[5]

Mum phoned me today to tell me that you are in hospital. I am truly sorry, for I know how much you will miss Mum & your home. But it will only be for a little while ...

It is a great comfort to me, to know that Bob is so close, & able to get in to see you often. I do wish though, that I could do the same. I feel so far away from you, but I shall be with you in spirit. ...

... This is but a short note, to let you know that I am thinking of you, and that I love you.

Yours loving,
Sue

---

4 Ibid.
5 Ibid.

Dear Agie & Opa,

Needless to say we are considerable flattered to hear that anyone as tough as Opa would develop restricted pipe lines. It wasn't long ago we were swinging axes at the lake. I've been having pipe line trouble too – the unrestricted type. Perhaps we could arrange a swap. After all, you can't drive a motor for 67,000 miles without an overhaul ...

The rest of the family are all well. Joan and Doug seem to like school and are getting along without too many problems. Doug took the microscope to school – where he apparently lectured the school principal and his teachers on the fine points of using a microscope.

Chris is now completely mobile. His conversation is considerable – some of which is intelligent.

Sue is beginning to look around for more jobs for me to do. I may have to put my foot down – I refuse to paint these rooms more than three times a year.

I have no doubt – with Agie enforcing the rules of the road – some bottled good cheer and some relatively good looking nurses, you should be in axe swinging condition again. We will be keeping you posted on up to date family affairs, dirty politics etc.

    Blake (Robinson)

                              London, Ontario
                              18 August 1959

Dear old Son,

I saw Bob Noble yesterday & he said you had arrived safely & all were well, also that you were catching fish. We are glad of the news. The weather reports indicate that you were having good weather. ...

We went to Stratford & saw "Othello" with the Kinchs[6] & Robinows on Wednesday last. It was very well done & an interesting production of a gloomy tragedy; the outstanding performance was the part of Iago. We took a picnick supper & had a pleasant time. I would like to spend a day at the Book Exhibition. I got absorbed for the time we were there in the exhibition of originals of Shakespeare's source books & the 1, 2, 3, & 4 Folios. You could only read the pages open in the cases, of course, but I found some interesting passages. ...

Your letter has just arrived. We are glad to have it and to know you had a good journey, in spite of the Orangemen's Parade & King Billy. I understand that the Toronto

---

6 Reference to Robert and Pat Kinch. Bob Kinch was professor of obstetrics and gynecology at UWO and later moved to McGill. In 2005, the McGill Faculty of Medicine established the Robert Kinch Chair in Women's Health for visiting professors. RGEM.

Faculty try to keep an exclusive right to Go Home Bay. Thanks for the invitation, but Mum is itching to get to Bark Lake to put through plots & plans. She worries because the place may be suffering from neglect & wants all the time she can get to put things to rights. ...

    All our love,
        Dad

Godwan River Estate
Ngodwana
E. Transvaal
15th March 1960

Dear Old Biff,[7]

I was most grateful for the pamphlet setting out Honours, Degrees, and Scientific Achievements. I am very pleased to know and have all this in print and also most honoured by your formidable achievements. All I have done is to kill a few people and be able to lead men into battle. But I get annoyed as I feel that you have not received one iota of the honour due to you for your wonderful work.

Our Nationalist government is making an unholy mess of things here and has all the world against them. ...

I fear a war between Russia and the West is inevitable. That show down should have taken place at the end of World War II. We should never have allowed Russia to get into Nazi Germany. But that was America's fault in siding with Stalin instead of Churchill ...

We are fit. This year has been a funny one – droughts in places, floods in other places – in fact difficult.

All our love to you all.

I hope you get some good fishing.

    Your loving brother,
        Rory

---

7 LAC, E.G.D. Murray Papers, vol. 20.

January 24, 1961
Blue Boar,
Cambridge, England

Professor & Mrs. E.G.D. Murray
London, Ont.

Dearest Family,

This second Cambridge letter should retrace a bit ...

All or most colleges seem to have new or extended buildings. John's has a very handsome new & renovated court behind the chapel. Queens has a quite modern building in the old gap between them and Kings. And so on. There is a sort of Mall & shopping centre between the Drummond St. bus station & St. Andrews St. Blocks of modern flats in various places. But withal they do *not* seem out of place & some are even handsome. There are renovations – eg. Churchill is building a set of flats for their young research students *on* the site. Clare has built a series of modern rooming houses that they rent out & allow to act as "hotels" in vacation period. I am quite impressed with some of the lively things. ...

Was in the Path lab & saw Greaves, who has to run the place. He tells me that H.R.D. is in the Evelyn & *very* difficult.[8] He has been there 2 wks without signing himself out & this apparently is a record and looked on as a bad sign. He is decompensating & very oedematous which he does not help by drinking too much fluids & not taking his medicines. Sounds bad all round but sounds as if it would be a relief to him & to almost all concerned if the (situation) would become decisive. Department morale low, I think, apart from Coombs' immunological group.[9]

Saw Dowson this p.m. He is *very* well & sends greetings.

Yours,
Bob

---

8 R.I.N. Greaves was a lecturer in the Department of Pathology, Cambridge, while Bob Murray was a student. He acted as head of department when H.R. Dean was terminally ill and later became professor of pathology. He was well known for his important wartime work on the preservation of human plasma for field transfusions as well as for developing methods of freeze-drying plasma. RGEM.

9 Robert Royston Amos Coombs (1921-2006), known as "Robin" to most, began his PhD at Cambridge in 1944 and remained there for the remainder of his career. He is considered one of the founders of the discipline of clinical immunology and helped establish the British Society of Immunology in 1956. RGEM.

59 rue du Grand-Pré
(5eme Etage)
Geneva
11 February 1961

Dear Mum & Dad,

I have not written for an age but it really has seemed that there is too much to do, see or think about and not much energy left. Now, however, we are beginning to get used to our new surroundings and can breathe a little. Everything has been a bit strange and all to be learned – geography, bus routes, quickest walking routes, shops, people etc. And our French is not really up to all the circumstances we find ourselves in – eg. arguing with the concierge ... This would be hard enough but she is dumb, careless & lazy to boot which seems to add to the adventure. My own French has proved just a little more serviceable than I had expected; Doris, I fear, has found it a bit difficult but Alice has risen nobly to the needs and has helped her a lot.

Well, I don't know where I was in the narrative but I think I must still have been at Cambridge & well before we left at that – good heavens, how time passes. Greaves was pessimistic. ... After our lunch Peter and I went to see McQuillen again & he took us out to see the site & the layout for the new Churchill College – on the Madingly road just before the Observatory.[10] They have a big site & if they don't build a bypass through then they should have plenty of room for all buildings *and* their playing fields right there. They have decided on a modern form & I saw the scale model of the whole thing. It includes a small apartment block (now the only bldg with temporary JCR & SCR etc.) for married graduate students. It is to be 30% non-scientific students so it is not the "pure science" college that rumour puts forth. ...

Doris has written in fair detail of some of our doings on the way back to London, & onto Paris and here. Mum's letters to both G.Ho. & Paris arrived safely & in the nick of time. *Very* welcome. Just got one from Dad the other day, too.

Paris was a most successful visit – in part thanks to Bertha Delaporte & her sister Madeleine, but also because we had so little time we had to be selective.[11] The toughest part was a wet tramp to & through the damp cold (incredible!) of Notre Dame, Sainte Chapelle, Ste. Sulpice, St. Germaine etc etc. Impressive in all sorts of ways! We ended up in the Alsatian Restaurant on the Boul. Mich. that Dad and I had an excellent meal in in 1953. The afternoon cleared miraculously for a visit to l'Etoile, which impressed the children no end, via the Champs Élyseés. It was fun! ...

The very excellent train journey here was rather spoiled for me by the shock of

---

10 R.A. McQuillen was a Cambridge biochemist involved with bacterial systems and a Fellow of Churchill College. RGEM.
11 Berthe Delaporte was a senior bacteriologist at the Pasteur Institute in Paris and did pioneering cytological work on bacteria in the 1930s and 1940s. RGEM

almost mislaying the tickets for a few minutes before we left. However, all became well except for Doris who sickened and had to spend the day trying not to be sick (in fact the next two days, too!). We enjoyed a fine cross section of the French country – through hills & one flat river valley after the other. ...

We got through customs & all such matters without a murmur (& indeed nothing opened atall) & we went right to the hotel by the station. Clean, not too pretentious, and quite reasonable. So we overlooked the Place Cornavin which is one of the busy places of the city for traffic & quite exciting. Doris could then go to bed & get over the misery that afflicted her.

The people at the Lab had put their necks out & opted on a 3 bedroom apt just coming vacant. Despite Mrs. Lourie's kind offer I could not really let them down.[12] Since it really fits our needs at the relatively reasonable Geneva price of about $200.00 a month; furnished etc, I was well advised to take it – so we have a lease. ... We should be very comfortable in the end as well as a bit better off than in one of the v. expensive new places that are going up all over town. This is a really residential district & mostly apartment houses. In between, in places, there are some of the old houses left; often with fine old trees in their gardens. I can see one beauty from here; it is symmetrical & almost like a cedar, but is at least 25 feet higher than the 6 floor apartment house that is next to it. ... The predominant trees in the city proper are big plane trees that have been trimmed and pollarded – but you even see some pollarded oaks.

12 Feb.

We went up to a look-out just by the little village and the low sun came out in spots on the prospect of mountains between here & Mont Blanc (which wasn't quite visible although you could just see the Dents du Midi). Impressively beautiful & makes one wish to live on one of the high places (we at least see the tops of a few of the hills).

We have a trolley bus route right by us and it costs 0.40 (about 10 cents) to ride downtown. Actually the distances are not as great as you might think here & it is not too much of a walk to almost anywhere except the mountains. Tom has even walked to the Airport.

The town is split by the river Rhône which debuts from the lake and forms a very decorative centre in the city – complete with islands, many bridges and a camouflaged power house. It is joined about 1 mile downstream by the Arve which comes from France ... It is not small & nor is the Rhône. The lab is right on the banks of the

---

12 Bob Murray met with Mrs. Lourie to enquire about the possible rental of her apartment in Geneva. Her recently deceased husband was a top man in the World Health Organization and an expert on tuberculosis. In the end the Murrays opted for an apartment found for them by the department in Geneva. RGEM.

Arve and almost equal distance on the other side of the centre from us. I can get there in ½ hour on foot or 20 mins by bus & foot. On wet days the umbrella is an invaluable accessory.

Just across the river from the lab is the big hockey rink – an impressive brand new affair – in which some of the "world championships" will be held between March 2-12. I have got some tickets for us for the Canadian games with Sweden & with U.S.A., but those for the Canada-USSR game were sold out except for standing room by the time I got there this week. I have a few tickets for the Czech-USSR game. There are a lot of games of course which might be reasonable to go to but the Canadian, Russian & US games are nearly all in the evening and more expensive. We will be fairly high-up & I doubt if T.V. will show us!!!

There is a lot to tell you about the town, the University & other things but this letter is becoming of Pasternakian proportions & had better stop soon. I promise to be a bit more regular from now on and this should be possible now that we have some idea of where we are. ...

We miss you both & hope that you are both well & content. The Toronto trip sounded like fun even if the "Heart" is an odd lot.

All the best,
Your son, Bob

Rue du Grand-Pré, 59
Genéve
n.d. February 1961

Dear Mum & Dad,

Please tell all our friends that we *will* write, but time went so quickly we couldn't. As you can see we seem to have an address. ...

Today was early closing for the food shops. There is a self serve near by so I had to do my butcher shopping in French and managed to get the right amount of lamb chops and pay for it. However, one cannot get by in English. We will all have to use some French. Mine is purely theoretical & 25 or 30 yrs old, however, it is helping me more than I thought it would. It takes real courage to use it! The Swiss speak faster than the Parisian French people. I could *understand* better in France.

The weather is *spectacular*. Sun, rain, wind, clouds high, clouds low, mnt's in sight then none. Roaring wind then calm and the night before last quite a thunderstorm. ...

To bring you up to date since my last, I expect Robin has told you of our Cambridge doings. We had a wonderful time there & the children all loved it. Peter liked it better than London & thought that was what London would be like. ... (at the moment the buildings are white against a black sky)

Alice and I spent hours at the Tate on our one day in London. The Whitney Collection was there and had the most *wonderful* collection of impressionists. It was crowded and some of the remarks were fantastic. We saw many other paintings there and enjoyed it very much indeed.

Our trip to Paris was *super* comfortable and I did ... sketches on the train of the French countryside. The Channel was roughish with gales predicted. By the time we'd crossed I had decided that after this I'd fly! With *great* control I wasn't sick and a kind lady gave me a mint which helped. ...

This is all as I must cook some! Robin will tell you about Geneva. Your letters are most welcome and clippings. I don't think Eldon house would qualify as a drawing. I like the art news.

 Love to all & Sue & family too. More soon.
  Love, Doris

             Rue du Grand-Pré, 59
                Genéve
                19 Feb 1961

Dearest family:

... I am now about set to do some work at the Lab with most needs now satisfied & some cultures going. There are some useful cooperative projects to get going and I am keeping some side issues to fill in the blank spots. Some time will have to be given to trying to clear up the differences of interpretation that have arisen between our labs – and a helpful start may be made on conditions of fixation. de Haller has a fascinating problem with chromosomes in a dinoflagellate protozoon (Amphidinium) which show real linear structure and periodicity.[13] The development of this regularity can be followed from interphase into metaphase & provides an unrivalled chance to study arrangement – we are going to see whether the protein (histone) can be followed. This may give us a clue to the nature of the fine structure. A good many other things are in discussion but will have to be ... set aside for later. ...

 All my love,
  Your son, Bob

P.S. Peter has been studying up on Chess as well as history. He is now a cutthroat player & he & Doris play nightly.

---

13 Gerard de Haller was a member of Edouard Kellenberger's group in the Institute of Biophysics, University of Geneva, working on bacteriophage and coliform and other genetic problems. RGEM.

"London Doctor Given Award"[14]

OTTAWA – The Royal Society of Canada award, a senior research fellowship with a value of $6,000, has been awarded to Dr. R.G.E. Murray, professor of bacteriology at the University of Western Ontario.

Dr. Murray, who is already in Europe, will spend a year abroad studying latest developments in research relating to cellular organization and fine structure of bacteria. He will go first to the Institut de Biophysique, at the University of Geneva, and later will visit other noted centres of study in Denmark, The Netherlands and Great Britain.

Rue du Grand-Pré, 59
Genéve
20 March 1961

My dear Mum & Dad:

... We are trying to get all things straight & ready for the kids to go to school after the Pâques holidays, & this is quite soon. It is, for some reasons, extraordinarily hard to get this straight; there is a fairly rigid system (& quite good on the classical side I would judge) but even then it requires innumerable consultations to find out what it is! Hope is to get Peter settled tomorrow. I have written our London schools for immunization record & Bob Campsall for B.C.G. certifs. in case they are needed, and they might be – quickly![15] The Swiss have a remarkable democracy with a fantastic bureaucracy – what makes it work is the citizens' right to arrange a vote (by petition) & vote the offending bit out of existence. They have votes on everything from the taxes on gasoline, up & down. This & two other things ... [were] voted on two weeks ago! ...

The Chateau (de Chillon) is really the most impressive castle I have ever seen & as a whole gives an impression of what both a fortified & habitable castle was like, from dungeons to living rooms & from keep to latrines. ... It is a complicated place as you could guess from the plan & there are really *three* courtyards mostly separated by the keep. The restorations have been very thoughtfully & carefully done ...

The appointment of Harold Warwick as Dean is a pleasant surprise.[16] I was afraid that we might have had an internal appointment. I don't know him but several of

---

14 *London Free Press*, Friday, 17 March 1961.
15 The physician Bob Campsall joined Bob Murray's department to be in charge of the clinical laboratory since with the introduction of Ontario's hospital insurance program there was a very considerable increase in the amount of diagnostic work to be done. Earlier he had worked in Kingston under G.B. Reed at Queen's University. RGEM.
16 O. Harold Warwick was dean of medicine at the University of Western Ontario from 1961 to 1965. He did pioneering research in the area of cancer control and treatment, and was the first director of the National Cancer Institute of Canada. RGEM.

our people know him well enough. Rossiter thinks a lot of him. I am sure that Collip would be quite happy with this choice. Let's hope he is not too wedded to cancer!

The various bits of news about the University are very interesting. I think that Uffen will do a lot of good as Principal of Univ. College.[17] He is a good scientist and has some imagination and guts. However, I note that Stiling continues to be Dean: I was under the impression that he was seriously ill but maybe he has made a recovery.[18] I must say I have found him obstructive rather than helpful, & I have been impressed that general opinion in that Faculty seemed to put him in the Gaullist class. However, I do hope he is in better health.

The "radiation research" support is bound to come. I could wish equally for "freshwater biology & hydrography" but I suppose that it is a case of the squeakiest wheel. Incidentally, one of the things [Edouard] Kellenberger[19] has been doing in Calif. (with, among them, Harold Johns of Toronto) involves deciding about "non-genetic" damages of radiation ... So they took U.V. produced by a monochromator – according to wavelength (& specific absorbtion) they found quite specific series of damages to different parts of bacteriophage and these are expressed both in physical and/or biological effects. Most important & I think first real demonstration.

The Victoria Hospital Trust is extraordinary – it is hard to believe some of their shenanigans. One can only suppose that it is, in large part, sniping from Fisher who is,

17 R.A. Uffen was a UWO geophysicist in the 1950s who became a friend of Bob Murray. He later went to Queen's University where he served as dean of science. RGEM.
18 F. Stiling was dean of arts and principal of University College at the University of Western Ontario at this time. RGEM.
19 Born in Geneva, Switzerland, in 1901, Edouard Kellenberger studied physics at the University of Geneva, receiving his PhD in 1923. He remained in the Institute of Physics at Geneva into the mid-1960s, eventually becoming professor and director, and then moved to the Biozentrum at Basel until his retirement in 1980. He subsequently accepted a retirement post as an adjunct professor in Lausanne. Kellenberger was well known for his groundbreaking work in electron microscopy, and he was an important figure in the fields of bacterial cytology and genetics. Murray's department knew about him by 1952 from his electron microscopy work, but also because he had started work on trying to resolve cell structure problems, including those of bacterial cells, and he had been in touch with associates of Carl Robinow. Kellenberger and his associates did crucial work in phage genetics and phage structure with distinguished colleagues like W. Arber (Nobel laureate) and on the embedding, sectioning, and staining of bacterial cells from 1952 to 1964. Robert Murray and Kellenberger first met at the International Congress for Microbiology (Rome, 1953) where they took part in an extended symposium on bacterial structure; see Edouard Kellenberger, "Early Times of Electron Microscopy in Geneva," in *History of Electron Microscopy*, ed. J.R. Gunther (Basel: Birkhauser, 1990). The work carried out in Kellenberger's lab was important for Murray and others because, according to Bob Murray, he "developed crucial techniques of the day allowing for high resolution microscopy of cell structure in bacteria." On his first visit to Geneva during his sabbatical in 1960 Murray recalls learning a great deal about electron microscopy and about phage work. The two scientists became good friends and shared a mutual personal and professional respect for each other. RGEM.

alas, "representing" the Medical Executive for the year. This sort of short sight is next to inexcusable & then they want to be political appointees rather than "elected"!

The appointment of Robertson Davies to Massey College at Toronto is most interesting. He is most lively and articulate. I hope Ed Hall notes that they are giving him living space for himself *and* his family.[20] Hall opposed this sort of thing for our residences but I may have got him to agree to a proper house, when they build the other two.

Much as I like Gregg, I think that Cam Henry does some aspects of the job a lot more good as a practising academic man. I hope that we can get Hall to note this sort of thing and not hide behind Governors or the like when this sort of appointment opens up again.[21]

Well it is a beautiful cool morning – very clear and bright. I must away to the lab & I look forward to the walk.

Don't take my suggestion about a Swiss visit as having more weight than it has. Merely that you could have a less expensive headquarters and a change of scene (The lake [Bark] would look very attractive too!!!). This place is a tourist mecca I gather & they expect 85-100,000 of them during the summer.

Love to you all & good springing. The willow leaves are out here, so are daffs & some other bulbs & fruit blossoms. Will be quite gay if we get more warm weather.

    All my love.

        Your son, Bob

                          Rue du Grand-Pré, 59
                                      Genéve
                                  March 20, 1961

Dear Mum & Dad,

Yesterday was a half sunny, half rainy day so we chose an indoor spot and visited the castle of Chillon. This turned out to be a *huge* success. We happened to have the poem with us and a bit on its background, so even Tom was suitably prepared. …

The castle is much bigger and more complicated than one would expect and perfect for the children. They could go wherever they wished and they enjoyed the double walls, secret passages, dungeons, towers and there was a place where they could go down a ladder under the dungeons into the bedrock caverns. However, it was pitch

---

20 See chapter 3, note 73 for information on G. Edward Hall.
21 When the first residence was built President Hall persuaded Milton Gregg V.C. (a distinguished soldier) to be the first Master of Medway Hall; the appointment included a small residence for him. He stayed for two years and his place, as Bob Murray had hoped, was taken over by Cam Henry (1916-2004), a professor in the philosophy department at Western. RGEM.

black so we have to go back again some time with flash lights! Some friends of ours have seven children. They went down with flashlights and found rooms of another building there.

Not only was it exciting for the children, but it was charming for the adults as it is very beautiful and surprising sights were around every corner. We are lucky to see all the sights when the places aren't crowded with tourists.

The trains are heavenly. They seem lightly built and with taste. They are clean with huge *clean* windows and are very smooth and quiet. They are so fast that we have found it wisest to take the locals when we want to see the countryside. The trains are used by everyone and are kept up well, with new equipment, when needed. ...

Spring is *really* here now and I can see that things get lovelier & lovelier until November.

I have a present for Joan to come to her soon & more will be on the way for Doug & Christopher.

    Lots of love to all,
        Doris

R.G.E. Murray
c/o Asheshov
259 Swakeley Road
Lakenham, Middlesex
13 April 1961

Mrs. R.G.E. Murray,[22]
Rue du Grand-Pré 59
Genève
SWITZERLAND

Dearest family:

Well, I hear that Geneva is having weather that would do well on the Riviera, let alone the Rhône. As far as I am concerned it was a good thing that I was resigned to being not in Geneva and not *quite* cold here! This weather isn't bad but it is a good deal colder than we had been having. ...

Landed here and phoned Ash and Liz. They had tried very unsuccessfully to have a message for me at the Garden House! Insisted on my having the spare room and I am really very comfortable with them. Most kind. The journey in and out is not too

---

22 Correspondence between Robert Murray and Doris Murray, private collection, Robert Murray.

bad. Ash is about the same but he is finding it too hard to get out to his greenhouse. They both send their love to all of you and inquired in detail about your welfare.

I have thought of you all week and wondered how the first days of school have gone. Don't be too discouraged by the language problem. Make an effort to increase vocabulary and listen for the words and their associations at every opportunity. With that sort of approach it won't be long.

The meeting is very interesting and I have met lots of old and new friends. The lecture room is a historic place – Faraday's experiments – and there are collections of old apparatus. ...

I look forward to hearing from you when I get to Cambridge tomorrow.

All my love,
Dad

Christ's College,
Cambridge
14 April 1961

Mrs. R.G.E. Murray,[23]
Rue du Grand-Pré 59
Genève
SWITZERLAND

Dearest:

It seems odd to write on this but I have it on my pocket and I can write while I have my B'fast! Your #3 reached me this a.m. as I left the College. *Very* welcome to hear the news and to know how things are going. ...

It has been positively hot, and I even wished on two occasions that I didn't have my sweater on! The meeting was very interesting and other than a quick trip to get paints for you from Windsor and Newton, I stayed with it to the end. I had a hurried night on Wednesday because I left the book of the Symposium papers and the notes that I made (my brains!) on the underground. With characteristic thoroughness they saved them all and had them at Ruislip Sta. for me. (That was *my* bad day of losing things). ...

The news about the schools is a comfort to me. I wondered how the first days might go. I know they will be *very* tired (it is tiring to have to listen in another language) but I can see that this way, they will really learn. Peter may have the worst of it – but it may be that it only seems so – he will keep quiet until he is sure of things. ...
I am glad he has discovered the history books for they will help him very much with

---

23 Ibid.

vocabulary. If each of them can have French speaking friends the learning will go all the more quickly. ...

The College and Cambridge look lovely. Very quiet – term doesn't start until next week.

Will write immediately on the weekend.

    Love,

        Robin

19 April 1961

Dear Mum & Dad:

So nice to hear of your doings. Doris sent on your long letter, Dad, and it reached me at Cambridge where I have spent the last 5 days. The time just flew by and I really feel that I was there no time at all. For three days before that I was in London at the S.G.M. meeting and I stayed with Asheshovs while I was in London.[24] Now I am on the way to Manchester to visit with Milton Salton and to give a lecture. Then a couple of days with the Maitlands at Wolverhampton and then home on the 24th. I hope I will have time to drop in to ... see the Aunt, but time compresses & I have to be home a day earlier than I had planned.[25]

The flight to London was quite dramatic. A *Comet*. It climbed up, while we were still over the Rhône Valley, to 30 000 ft and I had a very good view of the folds ... of the Jura in the low sunlight. We got to London airport in 1 hour and 10 minutes, which seemed quite absurd. I phoned Liz & she came over to collect me in the car, for it is not much of drive to Ickenham. Ash was very glad to see me & I think he appreciated my staying with them. (It meant an hour in the tube to get in each day; but it wasn't bad). The spring is more than advanced here and we could see that estimates of 2-3 weeks ahead of usual must be correct. The gardens look lovely.

The SGM meetings were very interesting and well conducted. The Symposium was not quite as well discussed as I had hoped but not too bad. I will write separately about things in it. At the end there was a cocktail party ... for officers of the Soc. and others; I saw Nan Cowan & Sam (occasionally, because there was another meeting on sterilization problems at the same time, to be published by the Soc. of Applied Bacteriology).[26] They send their best love & regards. ...

I spent most of the evenings other than the party one quietly with Ash & Liz &

---

24 For information on Igor N. Asheshov and his wife Elizabeth Hall see Introduction, note 377; and chapter 4, note 122.
25 Aunt Betty Knight was Freda Murray's sister, who lived in Plymouth, England. RGEM.
26 The British bacteriologist Samuel Tertius Cowan (1905-1976). See description in chapter 6, note 99.

many discussions. He is much the same. Cheerful but almost incapacitated. Television, radio & planning his garden seems to keep him occupied.

I dropped into the Lister to see Miles & had a grand & amusing chat.[27] Then to Cambridge by the usual route.

Cambridge looked more than lovely for Spring suits it well & weather very fine. Stayed in College. Dined in Hall twice, Kings once & Churchill College once. Saw many friends including Hudson.[28] Will send the next instalment from Manchester!!

Your loving son, Bob

Rue du Grand-Pré, 59
Genéve
Mayday

Dear Mum & Dad:

... Robin's trip back happened at the same time as the Algerian Affair and he had to fly *around* France. So it took 3 hours in a Convair. He arrived at midnight instead of 9:15.

The world news isn't the most joyful as usual. I do wish the Kennedy administration would employ more strategy and fewer seemingly impulsive gestures. But who am I to criticize. It just looks like a familiar pattern to me. Alas!

The children are due home now so I must close.

This brings lots of love from us all.

Doris

Rue du Grand-Pré, 59
Genéve
May 12, 1961

Dear Mum & Dad:

... We have had some delightful trips ... We were in Chamonix between seasons and were very nearly the *only* tourists there! There are many téléphériques including the very famous one over ... Mt. Blanc, but we decided to go up across the valley so that we could get a better view of the mountain. As it turned out the long one is very expensive. The one we took was quite hair-raising enough for us all and we enjoyed it tremendously! We went to the first stop & sat out on a terrace with refreshments

---

27 See chapter 8, note 4 for information on Sir Ashley Miles (1904-1988).
28 Robert Murray remembers Hudson as the "Chief Labman" in the Department of Pathology at Cambridge. "To me he never had a first name (a strong British characteristic) [and] from childhood on I knew him and many others that way." RGEM.

& soaked up the beauty of the enormous mountain and a high glacier with other smaller ones. It was quite a sight. We were at 6,000 ft. Our ride was [illegible] & Chamonix looked very tiny & distant in the valley below. We walked in the snow, watched a few skiers come down from the top station and also admired the flowers & snow combined. This is another place to go back to …

 Lots of love, Doris

[Enclosed letter from Peter]

Dear Opa and Agie,

 I'm sorry I haven't written sooner but I've been quite occupied with school and trips. We've gone to Chamonix, Sion, Bern, and up Mt. Salève and other places. Chamonix is spectacular. We went on a perfectly clear day although some clouds came in the late afternoon. Mt. Blanc seems twice as high as it seems safe. A very high jagged mountain to the left of it seemed to be the most spectacular to me though. We went up 2000 meters on the mountain opposite Mt. Blanc and got an excellent view, Sion has two castles on the tops of hills looking very fairy-tale-like. We went up to the ruins of the castle of Tourbillon – a strenuous climb.

 I can't speak French fluently yet but I can get along with the other boys who are anxious to know all about the Indians and Eskimos. There is a tremendous demand here for Canadian stamps and I would like it very much if you could send as many as you can as what I get through the mail runs out very quickly. If you can get a used copy of the orange 6 cent stamp in the Queen Elizabeth set for me. The post office should have it.

 I was surprised at the size you said the Thurtell and Everitt pedigrees were. They must be awfully detailed. How far back do they go?

 Write soon, and send news from home.

 Peter

 P.S. – Save the Swiss stamps we send you as I have started a Swiss collection.

Genève
20 May 1961

Dearest Mum & Dad:

 I have had a very nice birthday and thank you so much for all your kind thoughts. Dad's letter with the very generous cheque from you both came several days ago and Mum's letter arrived on the morning of the day – as calculated. So nice to have your letter and we could only wish that we could all [have] been together (I nearly said "celebrate" but since, with all these Swiss calories, I am trying to assume a somewhat "spreading shape", it is a bit of a prick to the conscience!). Everyone seemed to enjoy it as much as I did and it certainly ended with a grand feast. ...

We are now making plans for the Danish part of this expedition and we shall be leaving in only 6-7 weeks. Aksel & Inga [Birch-Andersen] seem to have got the ... house next to them for a couple of weeks.[29] This is up in Jutland near Aarhus, which is an ancient university that I want to see. Now we have to decide how we are going to do things either for a short time or go there for nearly the whole balance of the year. Needless to say Aksel is very keen for us to stay & says we will be the more welcome the longer we stay. I had a long talk with Edouard about this & the various problems he has here (with the histological and pathological groups, among others) with demands on his facilities & his people. We are nearly agreed that it would be best to up stakes and move up to Copenhagen. This is too bad from the point of view of schools & the children's French but at least they have a good basis. I think it would be too expensive to maintain two bases and this place at $200 a month is quite a drain. Will let you know the exact decision in the next letter or two. ...

Directly I have the time I will do something interesting with the cheque you sent!

All my love & thanks.

Your son, Bob

28 May 1961

My dear Mum:

... I must say it was a busy week for I had to give two university lectures on Tuesday & Wednesday. In the fashion here they were at 6 pm! after a longish day. This went well but in rather characteristic fashion the biologists were thin on the ground & the histological group at the Med. Sch. just didn't turn up. ... Edouard just holds his hands up in despair & I rather agree that it *is* impossible. However, the group that was there was attentive, thoughtful & gave me a good discussion.

I am up to my ears trying to complete one phase of the cell wall structure work and in trying to get things written down (& this *is* slow, damn it). ...

A lot of writing to do before I go to Munich for 2 days on Tuesday a.m. Must also get our arrangements straight for Copenhagen. Aksel has got us a place next to their summer place for 2 wks from July 15th. Then we must settle in C'hagen & get all going again. Too bad after a rather nice settlement here but the fact is that the work really demands a move & we can't keep two establishments!

Aksel is very glad about the decision & we do look forward to seeing gay Copenhagen again. The kids are resigned to it & I hope we can keep up their French there.

---

29 See chapter 8, note 187.

We went to a grand concert on Wed. night after my lecture. Fricsay[30] conducting an all Beethoven programme. Backhaus[31] playing 4th Concerto & very good it was too.

Göran Hedén[32] here for a few hours & I managed to see him for a short time. He sends his best.

All my love.

Your son, Bob

Laboratoire de Biophysique Université de Genève
Quai de l'Ecole-de-Médecine 24
3 June 1961

Prof & Mrs. E.G.D. Murray
c/o C.W.B. Robinson
156 Sunnyside Lakeside
Montreal 33, P.Q., Canada

Dear Mum & Dad and Sue & Blake:

I hope this letter gets to you all at once (!) for I gather from Opa's last letter that you will probably be together in Montreal – hence the address. ...

Well it has been a busy week and I don't feel I have quite got on top of the work that I should have done. However, it was quite exciting – particularly the trip to Munich. It was only for 2 days but I doubt if I could have taken much more – straining at and around a language you don't speak or understand is *very* tiring – and it is *not* true that they all speak English or French!

Edouard & I went together & this was nice for me (He speaks German as well or better than he speaks French & English) although no doubt confusing to him since he started speaking English to the people & German to me! ... Kurt Liebermeister & his assistant Schmeidel (Bacteriologists for the Municipal Hospital – *not* the Univ. Hosp. which is served by the *Pettenkofer Inst.* & I didn't have time to see over it) took me out for a tour of Munich in the sunlight, for the skies were clear, tea & then a couple of hours in their lab. There are big contrasts of old & new, of bomb-damage (lots, still) & rebuilding, memorials of all sorts to the Bavarian monarchy, esp. Maxmilian I. But only a few reminders of Hitler (It gave me quite a turn to hear a train announcement to Berchtesgarten in the RR Sta) (also the film "Eichmann, Man of the Riech"

---

30 The Hungarian conductor Ferenc Fricsay (1914-1963) was known for his interpretations of Beethoven.
31 The German pianist Wilhelm Backhaus (1884-1969) made numerous recordings of Beethoven's sonatas and concertos, including the Fourth.
32 See chapter 7, note 5.

playing there). We looked at city halls (2), city walls, Residenzes, Nazi Party headquarters with famous balcony, innumerable Platzes, & some very handsome modern buildings (tremendous activity). Looked at Hofbrauhaus & had a view of Munich, the DOM kirke etc. from the front of the Bav. govt Bld.

Liebermeister's lab [is] most interesting and I think doing a very good job on the Clinical work. He has a new building of basement & 1st floor, beautifully fitted & he says they are going to add another floor next year. He has a virus lab and diagnostic serology. Very simple approach & well worked out. Pictures on the wall apparently provided all municipal & state buildings by the State Museum! They are very concerned with "Hospitalism" – both Staphs and Enteropathogenic *E. coli*. The former is neither better nor worse than anywhere else – but they do have a problem & are doing things about it. They have a more serious difficulty with the *E. coli* in children & infants. ... They have had a number of cases, some have appeared in 3 other hospitals locally & 1 some distance away, & cases that come in often mean more in the hospital. They can't find a good reason for the degree of spread in their particular environment. By the time we had done the building and hashed this out I was tired but then Hofschneider phoned to say he *might* not be able to be in the next day & so they would discuss some more of his phage & protoplast work (it was then 7 and I thanked my stars for a Roma Special – a huge cake built up of trifle – that I had at tea). By close to 9 the pangs caught up with the rest & we all adjourned to the Hofbrauhaus for Löwenbrau beer (one of many special brews, every month or so), two different sorts of sausages & a good bit of beef. Without Edouard as an occ. interpreter & a nice American & wife, who had learned German well, I would have been very lost. ...

The work is going poorly, largely because writing goes slowly & not too well. Not in the mood! Have some nice pictures of cell wall & about ready to write it up – however the interpretation of the edge is still subject to great uncertainty. At least the good resolution I have on the Siemens has helped by forcing an alternative explanation. I will be glad when the writing task is done. ...

 Love to you all.
  Your son, Bob

Bruinrug, C.P. (Cape Province – mine)
August 16th, 1961

My dear Biff:[33]

I am ashamed that I have been so long in answering your letter. I have no excuse to offer, except that like most people who do little of importance in a lifetime, am

---

33 LAC, E.G.D. Murray Papers, vol. 21.

perpetually in a state of "having no time". I was very pleased to hear from you, and to receive Bob's Curriculum Vitae – you and Freda must be very proud parents. Thank you for keeping me posted. I have been able to fill in a lot on the Family Tree and Scrap Book about your family, which is as it should be, as you are head of our branch. I shall always be delighted to receive any hints of news from you. What a great honour to Bob to be given a Royal Society of Canada Traveling Award and as you say a great experience for his family. ...Our years of drought ended in March with torrential rains – and great flood damage – which will take years to repair. ...Mother keeps very well, and is in her 94th year. She is deaf and has sight in only one eye, otherwise has not an ache or a pain anywhere. ...

I hope Rory has sent your Dad's letters on to you.

I am enclosing one or two extracts that may interest you. They come from Dr. Scott Miller M.O.H. Johannesburg – his wife Hilda is a member of the East London Murray family.

Love to you and Freda from all of us.

Marge[34]

Extract from A History of Medicine in S.A. by Edmund Burrows[35]
George Alfred Everitt Murray LRCP (London.) M.B. C.M. (Durham 1887) F.R.C.S. (England 1887) – d. 1941. was one whose name will always stand in the Medical History of Johannesburg. Born at Graaff Reinet, the son of the Commanding Officer of the Volunteers in the Galeka War. Dr. Murray returned to S. Africa in 1888 and settled in Johannesburg, where his surgical experience soon placed him in the foreground. He had been a House Surgeon at St. Bartholemew Hospital, London, and was soon recognised as a leading surgeon. ... In 1889 he was called to Pretoria to give professional evidence in the celebrated hospital scandal. In 1890, the year in which he was licensed for practice in the Republic he was appointed Medical Attendant to Nazarath House. In the S.A. War he was a consulting surgeon to the British Forces. In addition he was twice elected President of the Transvaal Medical Society and was actually the 1st President of the S.A. Medical Association.

Dr. Murray's son Everitt George Dunne Murray O.B.E, M.A., L.M.S.S.A., M.D., and D.S.C. retired from the chair of Bacteriology and Immunology in McGill University Canada in 1955.

---

34 Marge Shearing was a distant cousin of the Murray family and a genealogical historian. RGEM.
35 Edmund Burrows, *A History of Medicine in South Africa, Up to the End of the Nineteenth Century* (Cape Town: A.A. Balkema, 1958).

c/o Aksel Birch-Andersen
Statens Serum institute
Copenhagen, Denmark
6 Sept 1961

Dear Mum & Dad:

It is a terribly long time since I wrote last and then I think it must have gone to London well ahead of your travelling home. But times have been very full of business and very interesting. Doris said she would write about general matters. At the moment I am in Berne for 4 days of an international symposium on cellular fine structure and it is a very interesting small meeting. About 50 people all told and a very select group. Good discussions but the sessions, although interesting, are very long and concentrated. We do have a bit of relaxation with good food & table conversation. I have learned a lot.

At Copenhagen the lab is most congenial & I have got several experiments under way. The electron microscopy is very good and, at last, I cannot complain about insufficient access to instruments of the highest quality.

The family is very well. Doris is much better in the climate of Denmark, which is odd but true. However, she does have some troubles still and occasional devastating headaches with nausea. I have arranged for her to see an internist. ... She has done some more painting but not as much as one would expect if she were quite well.

Alice is going to the Rygaards French School, but the boys don't have anything but their Canadian schoolwork. This is too bad, because it is the plainest, most dully repetitive stuff (and I'm quite embarrassed to think what this means in terms of Canada's educational future) and they don't have companions. Aksel's boys can meet them, part of weekends, but even that is difficult because of their games & normal companions.

We have to move at the end of the month. We are dickering for a lovely apartment downtown & near work. High up & behind Tivoli. There are some difficulties but I hope they will clear. The owner is one of the profs of Physiology.

Well I must rush off to another session!
    Love to you all.
        Your son, Bob

c/o Aksel Birch-Andersen
Statens Serum institute
Copenhagen, Denmark
13 & 15 Sept. 1961

My dear Mum & Dad:

... The trip to Switzerland and Belgium was very successful and was most worthwhile. (You should have had a letter from me in Berne). The social & meeting aspects were very nicely handled and with only about 50 participants there was a great discussion all the time. I met a number of people that I had wanted to know for years – such as Frey-Wyssling[36] ... and many others. I managed to contribute a little and made a few useful points. ...

... The Belgians are not so easily multilingual as the Swiss and my French was really worked to the limits. The talk (on structural and biochemical organisation exterior to the cytoplasmic membrane) went quite well and there was some lively discussion. They are remarkably well equipped when you consider the difficulties they are up against, and they can do very refined work. It is not a very gay place but they have tried to do a bit of landscaping. The same sort of comment applies to Brussels itself which is more like an industrial city in the English midlands – a bit grimy and dirty; full of depressing row houses, and more than enough rain. ...

Just had a letter from Liz Asheshov to say that Ash died on Sept 3rd after a day or two of complete paralysis and uncertain consciousness. This has, of course, been hanging over them for some time & they were prepared. I'm glad I had seen him twice this year and apparently he enjoyed the visits. It is good that Liz has such a good post at Colindale & is appreciated by one & all. (the address is Northrepps, 259 Swakeleys road, Ickenham)

Aksel & Inga send their love – we are all (11 of us) going to the circus tonight with Tom Anderson. Promises well. Copenhagen is lovely and the more I see of it the more I like its spirit.

  Love to you all.
   Your son, Bob

---

36 The Swiss botanist Albert F. Frey-Wyssling (1900-1988) was professor in the Department of Botany and Plant Physiology at the Federal Institute of Technology Zurich (ETH Zurich) from 1938 to 1970 and also served as head of the institution (1957-1961). He is regarded as a pioneer of submicroscopic morphology who helped initiate the field of molecular biology.

Statens Seruminstitut
20-21 Sept 1961

Dear Mum & Dad:

We are all very sad at the death of Dag Hammarskjold – a most untimely and unfortunate accident – and we are all rather concerned at the various turns of events in the world. Berlin is very much before us and the "Aktuelt" on the Danish T.V. provides a considerable coverage each night. We seem to have developed a situation where rights, the law and practicality each provide different solutions and none really uphold *anyone's* stand in the matter. Damned odd & uncomfortable. ...

As I told you we had a very interesting week with T.F. Anderson of Philadelphia.[37] He is good company and he was here one evening; we all had dinner together last Friday & he took our 2 families to the Schuman Circus. A "permanent" one ring circus in the city. The Schuman family are equestrians and they have lovely horses, beautifully trained. This year they also had a trained Giraffe – most odd & the first I have seen. They had a marvelous clown act, rather like "A drop of water", but concerning painters – messy and very funny. ...

Your two letters came this morning and we are very glad to have them with their news. I must have omitted to explain, when we came here, that Aksel was only able to get this nice suburban house for 3 months – but it was so nice (& we really agree) that it was better than taking a poor substitute for the whole time. ...

I hope Robin [Robinow] will see the paper before you send it back. I suppose he is immersed in teaching so I have not heard from him (I have been hoping ... [to receive] his *B. mycoides* strain that he was going to send), but I had a grand letter from him in August.

Tivoli is no temptation anymore & it is closed until May. Too bad, it is much fun. There are lots of other things of all sorts. Theatre, ballet and all; so I hope we will have energies for some of them. ...

I am in the midst of experiments & must away.

    All the best love.

        Your son, Bob

---

[37] Tom Anderson, from the Institute for Cancer Research, the Fox Chase Cancer Center, was an important contributor to the biological revolution occurring in the 1940s and 1950s, both as a pioneer electron microscopist and as a researcher on bacteriophages (both structurally and biologically). Robert Murray recalls Anderson's Copenhagen visit very well because "he showed me that it was possible to see the general conformation of antibody molecules in negatively stained preparations with the electron microscope. We had some good times together." RGEM.

c/o Crone
Niels Brocksgade 4
Copenhagen, Denmark
2 Oct. 1961

My dear Mum & Dad:

... Work is going quite well at the lab & quite a lot of work is well along the way with some results rolling in as we go. There are some nice pictures of the "*B. anitratum*" strain B5W and it has an interesting fine structure – very much like *Vitreoscilla* in cell wall. Still trying to define conditions for an experiment on the plasma membrane intrusions in *Spirillum*. Want to abolish them, then mark the surface with something, and watch them grow. We will have a chance of finding out if the existing membrane "flows" into these structures and carries things with it. This may be the basis of some of the seemingly specific but useless (or mysterious) concentrations of rare elements in the environment: chelation at a surface and then dragged in willy-nilly. The needed trials are done but not all looked at yet. Takes time. ...

It was fun & very interesting to work with Tom Anderson for a week with phage antibodies. It was really seeing that he got good instrument conditions for the best resolution – seems odd to be doing that since he is both a master & pioneer at electron microscopy. The experiment was to see if it could be done – now he will take the horse antibody (slightly larger molecule?) that Jerne made years ago, fractionate it, purify elements by differential absorption and see what sort of patterned distribution of substances (presumably antigens most of them) both in plane and in depth. Not surprising that things can be covered up especially when the "holes" (for nutrients etc to approach the physiological surface of the plasma membrane & for antibody to approach deeper layers) cannot be very much bigger than one or a few antibody molecules. ...

The family keeps well and there are still things to interest them. Glad to hear news of Sue & Co. Wish you could see Copenhagen with us – it is a fascinating place and most cheerful.

Keep well.

All our love.

Your son, Bob

P.S. Kauffman[38] was in yesterday holding a letter from you & full of vague stories of not being sure that he would go to Montreal. He is an odd fellow. Told him he should be there & people would expect him since they were likely to discuss his sort of contention in principle.

---

38 Bob Murray describes Fritz Kauffman as a very serious minded worker on the enteric pathogens, notably the Salmonella species. RGEM.

Niels Brocksgade 4
Copenhagen, Denmark
October 11, 1961

Dear Agie and Opa,

Thank you very much for the fifteen dollars. I received the check yesterday night and this afternoon I spent three dollars of it on a small electric motor with changing gears. The rest will go towards a book of ships (called Skibet in Danish) which is coming out in English this month and is an inch and a half thick!

Next week we are going to take a trip to Oslo and from there we will fly to Sweden (two or three days in each). This takes place over my birthday so I'm sure we'll have plenty of fun.

It is rather tiring not going to school and the correspondence courses are getting on my nerves but I still enjoy it here.

I'm looking forward to getting back to London and I hope things are fine with you.
With love from,
Tom

c/o Crone
Niels Brocksgade 4
Copenhagen, Denmark
10-11 October 1961

My dear Mum and Dad:

You must be getting off for the short trek to. It must be rather hard getting the script into order but I am sure it is worth it. It is really a great honour to you to be asked to give the Wadsworth Lecture. I suppose it will be printed. Someone was through here a while ago from the Albany lab but I'm not sure of his name – Bourdillon, I think it was.

You will be off at about the same time that we will be in Oslo & Stockholm. We will spend the 17th & 18th in Oslo – I hope to see Henricksen[39] & the family want to see the Viking ships & other treasures. (Peter has been doing more searching into Scandinavian history as well as reading Sagas). Then we fly to Stockholm on the a.m. of Tom's birthday – hope to visit about a bit as well as see the sights. The complete ship raised up out of the depths of the harbour will be something.

Tom was thrilled to get your present early and it is appropriate because he will

---

39 S.D. Henriksen (Oslo) was an authority on the Neisseria and the Moraxella, and Robert Murray visited him to discuss these and taxonomic matters. There was an interesting Canadian connection, for during the war, Henriksen was a medical officer with a Norwegian wing of the Air Force and was stationed at Ajax, Ontario. RGEM.

celebrate here (as well as abroad) on Sunday. We can have a little bit of a party with the Birch-Andersen family. ...

The whole institute was very upset this morning because von Magnus lost his 18 year old son, who took his own life for quite unknown reasons.[40] von Magnus is a delightful person & much liked, as all his family. There is no known reason for this tragedy & can only guess at some fit of depression that the family did not spot.

I have just come back from an evening lecture at the University given by another visitor – a biochemical type named Rupert from Johns Hopkins. Interesting enough as a problem – ultraviolet damage & the photoreactive repair process – but he was a poorish speaker. After we adjourned to the Anatomy Museum (much comparative anatomy) & sat down to tables among the skeletons (I had a series of intestinal models behind me & a giant turtle carapace in front of me) to a *feast* of Smørrebrod, beer & snaps. Some gay talk. My neighbour was the delightful H.L. Jensen from Lyngby whom I should have visited long ago.[41] He is now Director (newly appointed) of the lab and will have the "Apartment" in the beautiful grounds next the horticultural research station and close to the Frilandsmuseet.

The work goes well & in between the more formal experiments ... am doing some checking work on the cell wall observations. I have obtained some lovely micrographs in doing this with the new EM 200 and it is quite revealing. We need badly a new generic name for this tetrad forming, sheet forming unique sort of "micrococcus" – the simple things such as Tetracoccus ... seem to be taken up and it is a pity to mix roots ... Any ideas? ...

How are you & Ken Carroll coming along with the Listeria lipid problem? – apart from needing new stuff for later.[42] How lucky to have "forgotten" a batch of organisms – 45 g. is quite a lot! ...

The days are shortening – sunrise at 7:00 & sunset at 3:00 here at the shortest day. The flowers & the gardens are still lovely; leaves begin to fall but not badly yet.

Love to you both.

Your son, Bob

---

40 Preben C.A. von Magnus (1912-1973) worked in the Serum Institute and by the time of Bob Murray's second visit to Copenhagen in 1968 he had become its director. RGEM.

41 H.L. Jensen was head of the State Plant Pathology Institute at Lyngby in Denmark. Bob Murray remembers he "was a remarkably interesting man" and "a mine of information on many bacteria in nature as well as an authority on plant diseases." RGEM.

42 Kenneth K. Carroll (1923-1998) moved to Western as a graduate student of Collip's when the latter arrived as dean of medicine. Aside from two years as a post-doctoral fellow at Cambridge, Carroll spent his entire scientific career at the University of Western Ontario, primarily as professor of biochemistry. He was an expert on lipids and lipid metabolism. He worked with Everitt Murray to determine if there were significant lipids contributing to features of Listeria monocytogenes.

EGD Murray
126 Regent St.
London, Ontario, Canada
22 October 1961

Professor R.G.E. Murray
c/o Crone Niels Brocksgade 4
Copenhagen V, Denmark

Dear Son & Doris,

We have your letters (Bob 10th Oct & Doris card 18 Oct from Oslo) & and we are glad of all the news. You are seeing & doing interesting things. We are thinking of you on these excursions.

We had a nice letter from Tom & about his special purchase & we are pleased he is happy about it.

Geoffrey Edsall gets about a lot.[43] He & his wife dined with us not long ago & I think I told you about it. The partaking among the skeletons must have been amusing.

I have just returned from Albany & [am] relieved that everyone seemed pleased with my Wadsworth Lecture. Many seem to have asked how they could get a copy so Dr. Tompkins (Director of Division of Labs and Res) asked me for the MS to publish it. I had not thought to do so. Many asked me questions after it. They had a table full of journals & a to do with me & some asked for reprints! Doris (Nunes) Collins sends her kind regards to you, she is very highly thought of here. I saw some old friends & made some new ones. Everyone was very kind to me & I had some very good talks with several people, especially those interested in Listeria.

A Dr. M. Edwards (their Electron-microscopist) has cut some very good sections of Listeria & I think she felt the wrong Murray had come there. She is most anxious to talk to you about them when you return. There seems to be quite a heavy cell-wall membrane & the cytoplasm seems full of things, but there is one or more very large *reticulate* organelles (almost like a fish-net) very distinct and with an attachment "pedicle" to the cell membrane or cell wall. I don't know enough about the subject to say what there is to see but I am sure you will be very interested in what she has. She & others are quite excited about Listeria with structure and about its antigens & its behaviour.

Altogether I enjoyed the venture; & Tomkins had a dinner with a number of nice people the night I arrived, which was very good fun. The virologist (Englishman) who went with Dolman a little while & left to go to Albany was there & is a very good chap.

---

43 Geoffrey Edsall of the Division of Immunology, Walter Reed Army Institute of Research, Washington, DC. See chapter 8, note 193.

I'm glad indeed your work is going so well & interestingly. I don't know what character you wish to emphasize; plagula = a curtain, Codix = rug, sheets, counterpane, corium = a layer of some kind (eg. of lime) & the leather body-armour made of overlapping flaps [perhaps "Coriococcus" would not sound so bad, though a mixture of Latin and Greek to annoy purists, but coccus has taken on a special meaning in bacteriology]. You could have "Coriofex". But whatever you choose as a *genus* name you have to accompany it with a *species* name to make it valid by the rules.[44]

I am glad to hear Beetles Dowson is so cheerful & active.[45] I don't know what to say about A.F.L. Maclean, he could be a very good man but I don't know where Collip could get the money or if it would interest him.[46] We would want to know more about Maclean & his particular interests before broaching the subject. It would have to wait your return in any case, as I am not in a position to make suggestions when asked. ...

We are well & enjoying the hockey on Saturdays. The weather is still good, but a little cooler.

Love to everyone,
Dad

c/o Crone
Niels Brocksgade 4
Copenhagen, Denmark
25 October 1961

Dear family:
... It was really quite exciting for everyone and we were quite charged up by the time we left [for Oslo] after much packing activity the night before. We had to be up before 6 for the early plane; phone & get a taxi (an adventure in Danish) and have all the excitement of the big airport despite the early hour. A lovely day it was too & the sun came up only just before we took off. Breakfast ... was lightish & the only patch of rough weather came at the moment that we had our coffee cups full. Doris & one other lady had nearly baths but no harm done. The skies were cloudy but there was enough clear for us to see much ground of the north of Zealand. ... The airport is just before Oslo (on the west side) so we could do no more than see the wide waterfront from the distance & the backing of hills. ...

The next day I spent entirely in his lab up to 3 o'clock & then a short walk about

---

44 This bacterium was eventually named Deinococcus in 1978.
45 Dowson was a member of the Department of Botany at the University of Cambridge with a particular interest in plant pathology and the diseases caused by bacteria. He visited Murray's UWO lab in the 1950s. E.G.D.Murray always referred to him by his nickname, Beetles. RGEM.
46 Maclean was probably a job applicant at UWO. RGEM.

town with Doris. ... Henricksen's lab is really very nice. It is quite big – 4 floors & a basement, a big teaching lab (about 50 students) and a *large* lecture room. Very little of it has to be shared. He has the bacteriology & serology for the Rikshospitalet next door. ... There is a separate virus laboratory & he does not worry about it. He seems to do a more than ordinarily sound job of bacteriology & takes a very direct interest in urinary infections and the detection of urinary pathogens. Most interesting is an odd "paracolon" ... that is mucoid & has a most odd butyrous smell. He has an immunologist & a couple of biochemists all working on practical problems; the nature of some capsules and isolating antigens. They told me they just had their first undoubted *Listeria* ... a few months ago and [H.P.R.] Seelinger typed it for them (I think). He seems to have nice people, quiet & effective. ...

The flight to Stockholm did not have the beauty of the trip to Oslo – it was all cloud & when we landed there was a drizzle. Göran Hedén met us with a Volkswagen bus owned by one Miller from Hamilton, Montana who exchanged houses with [B.] Malmgren who is away for the year... So we got to the Malmen Hotel with a minimum of trouble. It proved both comfortable and convenient since a bus takes one to the door of the Karolinska in 20 minutes and the T-train has a station in the basement! ...

The lab is most interesting and a gold mine for apparatus to do nearly everything possible for bacterial comfort! As usual it is very active & I spent all of three days in it & still there was stuff left over that I had not a chance to see. So I didn't see any other labs except for Joe Bertani's new one in building.[47] I learned a lot from Holme & Zacharias & others about small scale continuous cultivation and I think they have *quite* simple apparatus developed for producing modest quantity in a small space. Most ingenious aeration and stirring devices. But I can now see some of the experiments that are possible with these devices and I think they are most interesting. They can be run in cascade so that an output in one physiological state can be translated rapidly to a new situation. Even small outfits can run for 30 days with output of up to 1 litre an hour!

Great pleasure to see Joe Bertani and his wife Betty again. He went there from Univ. of S. California to give them a course & a bit of a start on Bacterial genetics. They liked him & his work so much & his teaching was so much appreciated that they have offered him a personal research Professorship & have built him a lab. ... This is a great honour to him & it involves a lot of work for the Swedes for they have to persuade their colleagues & the government. Then the professorship is officially granted by the King. They seem very happy there and Joe says the worst is to have accounting & have to do it in Swedish. He is noted now as a first example of reverse flow – a European who went to the US & now has returned. It is quite a triumph.

On the Friday night they arranged for me to give a lecture to members of the

---

[47] See chapter 8, note 90 for information on the microbiologist Giuseppe Bertani.

Swedish Microbiological Society etc. Quite a small group were down from Uppsala and there was a modest crowd. Doris & Alice came along to listen. It went quite well and was seemingly successful; quite a few questions after and some useful discussion points. I talked about the interpretation of cell wall structure and was very glad to have Weibull there.[48]

On Saturday morning I watched Hedén and one of his people start up the continuous culture of your *listeria*. It was in a vessel of intermediate complexity. A great big jug of medium strands over all and there are many gadgets for continuous control of pH, gas inflow & the like and a water jacket temperature maintenance system. It looks like the devil to get properly balanced on starting but once going it almost looks after itself. ...

The Sunday morning was overcast (as usual) when we packed up & took our bags down to the air terminal. We could then have a look around at some special things. ... We were all *very* disappointed that the "Vasa", the ship of the line from the 17th century just raised this year, could not be seen. They were putting the roof on the house being built around it as a special museum and would not even allow a peek. ...

The Nordic Museum had a marvellous collection of historical pieces of the stone & bronze age. Very carefully selected and arranged. The much larger collection here is much more confusing; Peter, who is becoming very knowledgeable in Scandinavian stuff, was quite entranced with the whole place.

Anyway all agreed that the northern capitals tour was most worthwhile and a real success. ...

I must stop – too much already.
 Love to you both.
  Your son, Bob

c/o Crone
Niels Brocksgade 4
Copenhagen, Denmark
Nov. 8, 1961

Dear Dad:
 ... Robin is working very hard and is finding success esp. with the flagella work and our apartment is wonderful! The boys find no school very difficult, more for lack of friends, incentives and deadlines etc., but they are working on the correspondence anyway, of course.

---

48 Claes Weibull was in the Department of Microbiology, University of Lund, in Sweden and performed many important studies of structure and function in bacterial cytology. RGEM.

Alice & I made a trip to Germany last weekend to see my uncle who is losing his sight which is very difficult as reading & writing were his chief occupation. The 4 children (his grandchildren) are adorable so he is surrounded by a thriving family. His son Roland (my cousin) is a surgeon and my uncle's hospital is just next door so his friends can drop in. He was head of it and everyone must have liked him from the way they greet him!

Our weather is much like home in Nov. but not quite as cold. This fall has been beautiful. I haven't done so much sketching as I should. The days go by so fast it is unbelievable. Tom particularly wishes he was home & in school there. Perhaps even a Danish school would have been better in hind thought!!

Much love to you & Mum. The present is a pipe holder! or ornament. It is an example of Sweden's finest craftsmanship.

  Love, Doris

c/o Crone
Niels Brocksgade 4
Copenhagen, Denmark
11-13 November 1961

Dear Mum & Dad:

Delighted to have your long letter & to know all the things that you are up to. A real pair of travelers – all over the map and it sounds like fun. Too bad you could not have managed a day or two here as well. Doris says she wrote immediately to say something of plans for our return (I must say that it is hard to realize that we have only 6 weeks more here & so much undone!). We intend returning via the Mediterranean & New York on the Leonardo da Vinci. At least the route is in style even if we will be Tourists. The sailing is from Genoa on the 29th & she is due in N.Y. on Jan. 9th at 9:00 a.m., so we should be back on the 10th all being well. I wrote to Hal Warwick about this plot quite a while ago & he raised no objection so we will go through with it. It means quite a journey for us but after the darkling afternoons of Denmark it may be nice to have a while in comparative warmth & light. ...

The work is going quite well and I shall be sorry to have to choke it off. There are lots of rewarding pictures and little details. Some of the bigger ideas have their troubles of course, & I have not found a way of marking the surface of parts of the plasma membrane yet – *encouraging* attempts have been followed by equally *discouraging* attempts to repeat them! The most worthwhile thing is a real try to find out what happens at the cell end of flagella. This has always been discouraging & the tracing ends at the cell wall. I decided to try again in Spirillum where a great fascicle of flagella go in at one spot. ...

We have just started a comparison (Alice Reyn[49] is providing the cultures) of Meningo, Gono, and two of the other *Neisseria* – prob. *catarrhalis* and *flava*. Have good embedding of the first two which, like some I did years ago, are very interesting & quite unlike most cocci.

12 Nov.

... There have been a lot of lectures & interesting ones. I was involved in giving the monthly lecture to the Polymorphien (the sort of club of the Serum institute) and this was one reason for needing to be here rather than spending the weekend in Germany. It went very well indeed with good discussion from Ole Maaløe.[50]

Sakin, the immunologist of the Serum Inst., talked about ideas on immunization & the nature of immune processes. A very good review with some nice experiments.

We went (Doris & I) to a lecture by Preben Hansen (a famous architect) on Danish art & architecture at the International House. It was a very charming effort illustrated by the most beautiful colour photos that he took himself. He illustrated the association of the land and the work of the people & the things they depend on & the design of the houses they live & work in. It really *was* art. Most amusing that the first *railway stations* were designed & brought from England.

G. Pontecorvo[51] was here from Glasgow to get the Ensil Christian Hansen Prize (for research bearing in Applied & Industrial Microbiology). He gave a fairly pretentious but nevertheless interesting talk on fungal genetics & gave some of his ideas of what goes on in human genetics. ...

H.G. Callan gave two lectures on the "Lampbrush" chromosomes from Oocyte nuclei & they were very interesting (even if too speculative at the end). Nice to hear of this work from himself. He is prof. of Natural History at St. Andrews.

Robert Cruickshank[52] was here & gave a talk on Clin. epidemiological research to

---

49 Alice Reyn was a good friend and colleague of Bob Murray in the Neisseria Department, Statens Seruminstitut, Copenhagen. She was an authority on the genus Neisseria, which causes gonorrhea and meningococcal meningitis. In his subsequent visit in 1968, Robert Murray and Reyn worked on several joint projects. RGEM. See A. Reyn, A. Birch-Andersen, and R.G.E. Murray, "The Fine Structure of Cardiobacterium hominis," *Acta Pathologica, Microbiologica et Immunologica Scandinavica* (APMIS), Section B, 79 (1971): 51-60.
50 Ole Maaløe was head of the University Institute of Microbiology in Copenhagen, which was a major contributor to the development of molecular biology in the 1950s and 1960s. Bob Murray remembers that he often visited the Serum Institute as he was a former member and had many friends there. RGEM.
51 Guido Pontecorvo (1907-1999) was a pioneer in fungal genetics. In 1945, he was appointed as the first Lecturer in Genetics at the University of Glasgow, and he headed the subsequently formed Department of Genetics until 1968. RGEM.
52 See chapter 8, note 20.

the Path. Soc. He sent his regards and asked after you. He seemed very happy with his move from too many troubles at St. Mary's to the old stomping ground at Edinburgh in T.J. Mackie's chair. He says he can even get some work done now & then; which was more than he could say for London. He is still very much interested in "illness", in measuring the complicated interactions of environment & illness, and in chronic infections, esp. bronchitis. ...

Today we had a seminar from a nice small Japanese (? Akama, I think his name is) who has done some very solid work on phage mediated transductions of the 10, 15 & 34 antigens in the E group *Salmonella*. Kauffmann seemed quite content to explain that he was the only reason 34 is known & he had to persuade P.R. Edwards of its existence. He is an odd bod & I think it is just as well he has decided not to go to Mtl. ... the locals say that he won't go because he knows he would be on the losing end of a battle. Anyway he would probably use up time better given to more significant discussion and better speakers. I wrote Cowan asking him to try *not* to replace him!!

I would help you with the Cttee of Honour but I don't know what is to guide the selection. There could be several sorts:- the giants *or* the "new men". If the latter is it to be a special honour for service to microbiology research or what? There may be reason for IAMS to consider if there should be a mechanism for doing special honour to microbiologists who contribute most importantly to development.

Those from Denmark certainly should consider J. Ørskov[53] (who is still a lively devil, but may not go to Mtl), H.C. Jensen who is distinguished in many ways, & Ole Maaløe who contributes tremendously.

The trouble with these things is avoiding unnecessary hurts and yet not making a fatuously big group. We have a few Nobel prize winners: Enders, Lederberg, Burnet ... & they should be there if not nearby.[54] I think these & others should be chosen whether or not they are likely to go to Montreal. Those that are there can form a group, if needed. ...

Well, this is long enough!

    Regards to all old friends along the way and love to you both.

        Your son, Bob

17 Nov. 1961

Dear Agie and Opa:

Many, many thanks for the lovely birthday note and $15 that you sent. I received them, with a great deal of pleasure, about three or four days before the big day. $15

---

53 Jeppe Ørskov was the director of the Serum Institut, Copenhagen.
54 John F. Enders, Joshua Lederberg, and Sir Frank MacFarlane Burnet won the Nobel Prize in Physiology or Medicine in 1954, 1958, and 1960 respectively.

goes a long way here, and Denmark is full of beautiful things. I shall have a hard time choosing what to buy!

I had a most delightful and quiet birthday (they get quieter every year!) with not too many, but wonderful, gifts. The major one was a very gay Norwegian sweater which I chose when we were there but was not allowed to wear until the twelfth. I now appreciate it very much because it's been bitterly cold for the last week or so.

I can see now that what of the Danish winter we do experience will be *quite* an experience if it keeps up like this. We have been having very clear weather but the sun never gets much above the trees. Already it rises in the morning at 8:00 when I go to school, and sets at 4:30 in the afternoon when I get home, which doesn't leave me much daylight!

My French school has worked out beautifully; I am learning *more* French here than I did in Geneva and I do a fair amount of reading and writing as well as speaking. It's a wonderful language and I shall certainly miss it at home – but I don't think I shall ever forget it now.

We're all beginning to get excited about going home now. It will be welcome to have our own things after a year of living in other people's homes – and it'll certainly be *lovely* to see you again – it's not long now!

Thank you very, very much
  With all my love, Alice

London, Ontario
27 November 1961

Dear old Son,

We are anticipating Christmas because it is near time for you to come home and there may be something you wish to bring back under the duty-free allowance. In any case, it goes with our love & every wish for your happiness.

We had an excellent & interesting letter from Alice which gave us great pleasure. It was happy & affectionate & perfectly composed.

We had a very busy time in Vancouver (after a vicissitudinous journey dictated by fog) with business meetings of Canadian Heart Directors concerned with funds & scientific (Chemical mostly) meetings out of my understanding for the most part. These were relieved by a very pleasant free discussion arranged by Bowmer on bacteriological problems with a number of old friends & some I had not met before.[55] It passed a very pleasant afternoon. ... The new medical buildings are fairly large (? if

---

55 Ernest J. Bowmer (1915-2001) was an English-born and trained pathologist with a particular interest in parasitology, who moved to Canada in 1956 to take up the position of director of the Division of Laboratories of British Columbia.

enough) & ugly outside but said to be satisfactory inside. The weather was good to us too. I saw a number of old friends & old students at the meetings & enjoyed it on the whole. Dolman was ill.[56] Returning we stayed 4 days at Winnipeg & Jack Wilt[57] & everyone else saw to it that we had a royal time though very full & much discussion; also dinners & parties. We enjoyed it very much. Jack is still keeping his department moving & growing & the addition to their building is nearly finished, & about time too as they are frightfully cramped for space in the main department. Peter Warner has now got adequate space in the hospital and Lansdown has what he needs in the Public Health labs. Lees (who succeeded Payne) seems a good fellow and is much liked by his staff. He has some good people & a number of graduate students, very large undergraduate classes (up to 300) and they have moved out of the old building into adequate newly re-constructed space in the Biology Building. They have two floors there & happy with it. ...

I had a letter yesterday from V.B.D. Skerman; he has been appointed to the newly formed Chair of Microbiology in the University of Queensland & is very happy. He has been there quite a while & deserves the job as he has done a lot for them over the years.

They told me Larry Mackenzie is retiring at UBC and they are looking round for a new President, but I did not hear any rumours about who it might be.[58] Gordon Schrum has resigned as Professor of Physics & also from B.C. Power at $40,000 a year (less than half the salary of his predecessor!) & a big business shot.[59] He did a lot for U.B.C. in his time, especially about the student residences, one of which is called after him.

A package of your photographs arrived safely & I have them with others. They are very good. We look forward to you telling us about some of them.

A. St. G. Huggett, Professor of Physiology at St. Mary's, was here to give the Merck Lecture. We met him at dinner with the Kinchs & he is very good fun.

The Med. Faculty Council Dinner is on tomorrow night and I have been invited. There is to be a presentation to Bert Collip of a U.W.O. Shield done by Mrs. Warwick, who has a real talent in raised leather work.[60] [The Hunt Club is said to be "dry" but

---

56 Claude Ernest Dolman (1883-1976) was head of the Department of Bacteriology and Immunology at the University of British Columbia. See chapter 7, note 88.
57 Jack Wilt was head of the Department of Medical Microbiology in the Medical School of the University of Manitoba. RGEM.
58 The international lawyer Norman A.M. "Larry" Mackenzie (1894-1986) served as president of the University of British Columbia from 1944 to 1962. See P.B. Waite, *Lord of Point Grey: Larry Mackenzie of U.B.C.* (Vancouver: University of British Columbia Press, 1987).
59 Gordon Schrum was a physicist, teacher, and university administrator. He served as the first chancellor of Simon Fraser University. RGEM.
60 See Introduction, note 317 and chapter 1, note 56 for information on J.B. Collip.

we had wine & liquors at Kinch's dinner for Huggett on Saturday & we were the only people in that big place]. Anyway, there were cocktails at Warwick's beforehand.

Cold weather seems to have caught up with us now & the garden is frosted. We have had only one snow flurry, but it is down to 25-30 & cold in the North & there is a ... [bite] to the wind.

Coming back from Winnipeg fog prevented our landing at Toronto so they took us to Montreal & we spent a night with Sue & Family. A happy contretemps! They are well & happy.

All out love & best wishes,
    Mum & Dad

c/o Crone
Niels Brocksgade 4
Copenhagen, Denmark
1 December 1961

My dear Mum & Dad:

We were so glad to get your letters today and quite floored by your magnificent gifts for Xmas. You are right that there are lots of lovely things to buy in Copenhagen, so many in fact that the mind boggles & it is very hard to make a choice. The general level of taste in "things" is very high. ...

It is an interesting thing to visit the labs around here for they are a most curious blend of the most ancient & the very modern ...

The Carlsberg lab. with its fantastically able past & present staff retains its old fashioned facade and many rooms. The room they have seminars in is characteristic with busts, makeshift blackboards, 8 huge portraits of former worthies in their labs, and it is only big enough to crowd in about 30. The labs are full of fine equipment & people but they are very small. ... they keep it like this largely so that they can keep the number of visitors down to a small enough number. They can honestly say "no room"! But it is a lively, intellectual haven & well worth its reputation. ...

It looks as if we will have some nice & useful general description of structure from the work with Lautrup on "*B. anitratum*". We only had time to work with the B5W strain but it looks very lovely. In the physiological state ... it does definitely glide – it has a beautifully wavy outer cell layer over a taut inner layer. This is very similar to *Vitreoscilla*. Also there are many tube like short protrusions from this outer layer – might these be "cirri" of a special sort? Of course this odd structure does turn up in other cells, and the best current example is *Spirillum* that has never been said to glide although helical, gliding blue-greens do occur. The other day I did a small experiment on this & bashed off the flagella of a lively culture so that 90% were not motile (but alive). These were watched in a wet mount where a high proportion settled onto the glass or on a thin agar surface. I did not see anything definite as glides but Aksel swears he thinks he saw one in the act. It is too easy to misinterpret. I will try again some day. I think there is still a possibility that there is some unity of action in gliding & flagella motility. ...

We leave on the 26th. We will stop over a day in Switzerland before we go to Genoa on the 29th. Sail on a.m. of 30th for Cannes, then to Naples (4 hrs only!) on 31st, Gibraltar Jan 2nd & N.Y. on 8th. Quite a journey.

Had a nice cheerful letter from Robin [Robinow] who seems to have done some good things despite the fact that he doesn't love administering. We are certainly lucky to have such a friend & colleague.

Well, enough of this! Love to you both.

    Your son, Bob

> c/o Crone
> Niels Brocksgade 4
> Copenhagen, Denmark
> Dec. 3, 1961

Dear Mum & Dad:

Your checks and fascinating letters have arrived in good time and are much appreciated by us all. Your flying sounded most adventurous! What a series of events! ...[61]

Peter told me the other day that I was not only getting old, but that I was losing my memory too! So I told him I would be getting older & older until he was around 20 or 22, then I might get younger again and eventually of course reach my second childhood! ...

Tom will be *very* glad to get home. He has missed it most of all. His school, friends, room & things. ...

This is all I can write as I have to get lunch. So many thanks for your letters & checks and take good care of yourselves & we'll hope the next 6 weeks zip by happily. We'll write again in more detail about our actual arrival plans of course. ...

    Love to you both,

        Doris

---

61 Everitt and Freda Murray had encountered some very heavy fog on a flight during their 1961 trip to western Canada.

London Ont.
9 December 1961

Dear Bob & Doris:

Doris' letter of Dec. 3rd & Bob's of Nov. 11, 12, 13 via Vancouver both arrived on Dec. 7th & yesterday (Dec. 8th) Bob's letter of Dec. 1 with 3 packets of negatives. All of them most interesting & gave us great pleasure. We were not at Hotel Vancouver so did not get your letter there. We thought the Xmas cheques had better get to you early to be useful, and we are glad to know they arrived & you are having fun choosing which to buy; I'm sure there must be quite a choice.

The trips through the Mediterranean will be good fun, with added places of interest if you have time. The ship should be interesting too & I expect that field glasses will be in full use all day. It makes a long but worthwhile journey. I gather that the Duncans expect things to work out in good time.[62] The Vaughans, as you say, are building a new house too & seem happy with it.[63]

We shall be back from Montreal in good time to welcome you home. The Danish ... [jul] & and birthday festivities seem to be gay & happy affairs. Also the various expeditions have been fun & interesting. It has been a very good year for all of you that will carry through for years to come.

It is splendid that work has gone so well and the many conferences & special lectures seem to have brought good people & stirred interesting discussions.

The quick trip to Scotland should be interesting & I hope the weather does not make difficulties. I am sure you will make it a success.

We are back 4 days now from Toronto where we spent 3rd, 4th, & 5th Dec. at CPHA Lab Section, Congress Executive Cttee & C. Ass. Med. Bact. meetings. On Wed 13th we go to Ottawa & return on 16th for a meeting of the Hospital Infections Cttee which I am not looking forward to as it is so hard to get results. The Lab. Sect. meeting was quite good, with a number of worthy papers & many friends. Discussion is not as brisk as it should be. Dolman gave a splendid luncheon address on "Tidbits of Bacteriological History", very well balanced evaluations & good information, especially of personalities & conditions, nice touches of humour, good slides and perfect eloquence. Very many people asked after you & were glad of good news. The Med. Bacteriologists were mainly concerned with the construction & presentation of a brief to the Hall Royal Commission on Health Services in order to secure proper standing & provision for bacteriology. Many sensible suggestions were made covering points to be made in the brief. The draft will be circulated to members. It seems necessary.

---

62 This refers to Ian B.R. Duncan and Norma Duncan. Trained in Scotland, Ian Duncan moved to London and became the leading microbiologist at St. Joseph's Hospital. RGEM.

63 Gertrude Vaughan was Robert Murray's personal research technician for twenty-five years and a good family friend. RGEM.

The Congress Executive meeting went quite well & reports from the Committees indicated their work was going well. There is need to stir up those who are to help to raise money; Provincial Gov'ts are hanging fire & only P.Q. has given & that too small (2000), the others are waiting to see what Que. & Ont. do & Ont. is likely to be sticky. Industry is slow & rather stingy. I.U.B.S. treated us better than it has any other Congress & Héden must have worked on them most helpfully.

... Nothing new from Programme Cttee except the "Film Session" seems messy. Publications seem in good order. Publicity Cttee a bit shocked by the job before it & not anxious to take on the daily news-sheet.

Honorary Officers will be listed as the Past Presidents of Congresses & of E.C.I.A.M.S. & leave it at that. There will *not be* an "Honorary Committee" because of the danger of invidious distinctions & the difficulty of composing even a large list. It won't be missed & troubles are avoided. Something must be done about "*The International Biological Programme*", but we lack information about it & its organization. Quite evidently microbiology must be represented and I.A.M.S. must see to it that it is represented effectively. The Congress is the right place to initiate suitable & proper action. I am glad to see some order to the "Opening Ceremonies" on to the "Banquet" so that I can know what my duties are. There is difference of opinion on "entertainment" at the Banquet & it hinges on taste & appreciation of individuals (Some liked the Stockholm production & some thought it uninteresting & out of place. Personally I thought Stockholm was better than any other I had seen). "Simultaneous Translation" has quieted down for the present. Norm will send you the Minutes.

We shall go to Montreal on 22 Dec. and stay with Sue until 29th. You will be much in our thoughts over Xmas & we shall drink to your health & happiness & to your pleasant journey. We shall be back in good time to welcome you home.

Dad

London, Ontario
10 December 1961

We have seen excellent hockey on your TV (that is about all we have used it for & we have been glad of it). The reshuffle of players between the teams has evened up the top ones more than they were. The order is Montreal 36, Toronto 33, New York 30, Chicago 22, Detroit 22 & Boston 13. Last night Leafs walloped Boston 9 to 2, and at the beginning of the 2nd period Leafs scored 3 goals in 43 seconds (official timing) in a flurry of excitement. Good clean very fast game & Boston played better than the score shows. ...

Mum is busy making Christmas Pudding & Mincemeat & they smell very good. She is a bit late with it this year because of our unusual travels, but I expect she will get it done as well as all the other seasons shopping & arrangements. She does an amazing amount without any fuss. ...

Penfield[64] & Mrs. P. were here for a couple of days, brought by the Undergraduate Lecture Series; he had a good audience (Thames Hall) but I did not find anything outstanding in what he said. We met them at a tea reception at Medway Hall & had a brief few friendly words. We have not been to others in the series.

 Love to all of you from both of us & happy days.
  Dad

Godwan River Estate
Ngodwana
E. Transvaal
15th December 1961

Dear Biff,[65]

I am more than glad that Bob is doing so well. He richly deserves the cudos (sic) he is getting and more. Give him my Salaams ...

The children are well and happy. But I fear for the future for both yours and mine and their offspring. The World is in chaos. It appears to me that not even the Military Commanders know what they are doing (they are befogged by political influence ...). God knows we soldiers are accused of death and destruction but even with our worst faults, which are many, we have not made the faults of the politicians. ...

This country is in a mess. With bad leaders. If I was ten years younger there would be trouble but I realise I could not stay the course of hard fighting. ... The Romans said 'Ex Affrica aliquid semper novi'[66] and by God they were right.

Will write again in a couple of days.
 With love to you all and best wishes for the festive season.
  Your loving brother,
   Rory

---

64 Wilder Penfield, director of the Montreal Neurological Institute, McGill University. See chapter 3, note 130.
65 LAC, E.G.D. Murray Papers, vol. 21.
66 "Africa will always bring something new," attributed to the Roman writer, philosopher, naturalist, and military commander Pliny the Elder (AD 23-79).

London, Ontario
17 December 1961

Dear Bob & Doris,

We are just back last night, from Ottawa, & a hard working Infectious Committee & evenings with friends. Meeting for two days from 9 am to 5 pm with one hour for lunch. This is the last of a series of meetings wide distances apart & I am glad they are over for a time. I am tired by the end of the day & not very bright at the kindly dinner we are invited to in the evening (Wed. Gibbons, Thurs. Cooks ...).[67] We saw a number of old friends here & there which was pleasant. It was quite cold there but no snow, two nights it dropped to 10° below & only a high of 10 above during the day, but it was nice & bright & dry.

We went both days by air as you recommended & it is a much easier journey than by train & little more costly. A couple of hours waiting planes at Malton is tiresome.

When we got back we quickly found your letter from ... [Robert] and a nice letter from Alice. We are glad Alice has her eye on something & we were able to help her to get it. Also that Xmas ... [jul] in Copenhagen is so gay.

Bob's (trip) to Edinburgh was most satisfactory (Prestwick seems as inadequate as Malton; there were at least as many people looking for a seat as there were seats for at Toronto & the place was stuffy & dirty and noisy. When we were there for the Lab Section & the Exec. it was not an attractive drive, with only 3 passengers too. It seemed to be a scheme to exasperate people into taking taxis of the same company at $7.50 or more. Some competition would be a good thing.)

The Edinburgh Univ Staff Club seems to be a good place & it is well for you must have needed all the comfort you could get after so long in bus & train.

I suppose it is R. Cruickshank, & I remember Duguid when he was with Kettle in Cardiff.[68] Their arrangements for you were pleasant to see old friends & make new ones, discussion of mutual interests ... It is good news that you had good & interesting audiences & that there was good discussion. The lively group of Honours students must have been good fun & stimulating. Edinburgh seems to have thought up a worthwhile scheme & doing good with it.

Schoenfeld will be interested that you met their old friends in Glasgow.[69] I shall watch for Lominsky's paper in the J Path & Bact on Staph pathogenicity.[70] His immunological

---

67 Norman E. Gibbons was a microbiologist with the National Research Council, while Bill Cook was head of the Applied Biology section. See chapter 5, note 80 for information on Gibbons.
68 J.P. Duguid worked in Robert Cruickshank's laboratory. RGEM.
69 Henrik Schoenfeld and his wife, a weaver, were good friends of Robert and Doris Murray. He was the winemaker to the London Winery (now defunct), and Bob Murray helped him by freeze-drying his strains of yeast to preserve the cultures. RGEM.
70 This refers to Iwo Lominsky, a bacteriologist.

chain formation with E. coli is excitingly odd. Others at Glasgow seem to have contributed too; & Stoker & Howie are nice people.[71] All told the trip was worthwhile, though, as usual, you are a bit rushed for time & that is tiring as well as annoying.

It is good to know that Nan & Jimmy are flourishing & I am sure they were glad to see you too. Pity the day had such bad weather to veil such a nice place & make it unpleasant to be out.

We shall be leaving here evening of 21st Dec for Montreal & will stay a week. We have been invited again by the Montreal Med-Chir[72] to spend a week there in February at the Society's expense. Very generous & pleasant.

Today is our Wedding Day, 44 years.

 Love to all of you,
  Dad

Montreal, Quebec
January 17, 1964

Dear Mum and Dad,[73]

There is very little news. Things have been quite as usual – lectures, work and Saturday night out! …

The marks for the dissection exam came out yesterday, and I got a B!! They almost gave up on the results, but in the end I presume they decided to mark it on a curve. The original results were such that only 4 in our Lab of 90 passed! I don't believe that they will ever give us such small frogs again.

Chris and I saw an extremely interesting cartoon last week, which was considerably better than the movie that went with it. It was an Italian cartoon depicting the Unification of Italy. The artwork that went into it was excellent and certainly original. There was no talking and the music was sufficient and I was tempted to suggest the schools use it to teach that particular section of history. They didn't leave out any details, all the factions were represented and the passions and sorrows of that time were beautifully, and effectively indicated.

I still haven't received my Genetics mark but I'm sure I did well. Right now we're discussing triple coding and the general chemical mechanism of heredity. That

---

71 Sir James W. Howie (1907-1995) was professor in the Department of Pathology and Bacteriology at the University of Glasgow and an important figure in the development of the National Health Service (UK). From 1963 to 1973 he served as director of the Public Health Laboratory Service in England. M.G.P. Stoker was a distinguished virologist at Cambridge and was knighted for his scientific achievements.
72 Reference to the Montreal Medico-Chirurgical Society.
73 Correspondence between Robert Murray and Doris Murray, private collection, Robert Murray.

article in Scientific American came in handy since I understood most of it before it came up in lectures. Carol Kemp had to read the article also, and we've had some lively discussions about it. We have a young lecturer who is teaching for the first time this year. He is quite good, except that he has an aversion to looking at the class!

Dr. Gibbs is an interesting and amusing lecturer, and I am still enjoying the course.

It was marvelous to talk to you on Wednesday, Dad. I would have come out to meet you – I knew you were going to be there but I was still under semi-house arrest, due to a rather persistent fever, and I really wasn't feeling too healthy.

I have to go off to a lecture now. I hope all is well. Give my love to the boys, and Agie and Opa.

    Love,

        Alice

SELECT BIBLIOGRAPHY

Adami, Marie, ed. *J. George Adami: A Memoir, Together with Contributions from Others.* London: Constable, 1930.

Allan, Ted, and Sydney Gordon. *The Scalpel, the Sword: The Story of Dr. Norman Bethune.* Toronto: McClelland & Stewart, 1952.

Appel, Toby. *Shaping Biology: The National Science Foundation and American Biological Research, 1945-1975.* Baltimore: Johns Hopkins University Press, 2000.

Austoker, Joan, and Linda Bryder, eds. *Historical Perspectives on the Role of the MRC: Essays in the History of the Medical Research Council of the United Kingdom and Its Predecessor the Medical Research Committee, 1913-1953.* New York: Oxford University Press, 1989.

Avery, Donald. *The Science of War: Canadian Scientists and Allied Military Technology during the Second World War.* Toronto: University of Toronto Press, 1998.

Axelrod, Paul. *Scholars and Dollars: Politics, Economics and the Universities of Ontario 1945-1980.* Toronto: University of Toronto Press, 1982.

Axelrod, Paul, and John Reid, eds. *Youth, University and Canadian Society: Essays in the Social History of Higher Education.* Kingston: McGill-Queen's University Press, 1989.

Baillargeon, Denyse. *Making Do: Women, Family and Home in Montreal during the Great Depression.* Waterloo: Wilfrid Laurier Press, 1999.

Barr, Murray L. *A Century of Medicine at Western.* London: University of Western Ontario, 1977.

Barry, John. *The Great Influenza: The Epic Story of the Deadliest Plague in History.* New York: Penguin Books, 2004.

Bator, Paul, and Andrew Rhodes. *Within Reach of Everyone: A History of the University of Toronto School of Hygiene and the Connaught Laboratories.* Vol. 1, *1927-1955.* Ottawa: Canadian Public Health Association, 1990.

Berliner, H.S. *A System of Scientific Medicine: Philanthropic Foundations in the Flexner Era.* New York: Tavistock, 1985.

Blair, John S.C. *In Arduis Fidelis: Centenary History of the Royal Army Medical Corps.* 2nd ed. Edinburgh: Lynx Pub, 2001.

Bliss, Michael. *Banting: A Biography.* Toronto: McClelland & Stewart, 1984.

———. *Plague: How Smallpox Devastated Montreal.* Toronto: Harper-Perennial Company, 2003.

———. *William Osler: A Life in Medicine.* Toronto: University of Toronto Press, 1999.

Bonner, T.N. *To the Ends of the Earth: Women's Search for Education in Medicine* Cambridge, MA: Harvard University Press, 1992.

Bourne, Geoffrey. *We Met at Bart's: The Autobiography of a Physician*. London: Muller, 1963.

Bradbury, Bettina, and Tamara Myers, eds. *Negotiating Identities in Nineteenth- and Twentieth-Century Montreal*. Vancouver: University of British Columbia Press, 2005.

Broadfoot, Barry. *The Veterans' Years: Coming Home from the War*. Toronto: Douglas & McIntyre, 1985.

Brooke, Christopher. *A History of the University of Cambridge*. Vol. 4. New York: Cambridge University Press, 2007.

———. *Oxford and Cambridge*. New York: Cambridge University Press, 1988.

Brown, Kevin. *Penicillin Man: Alexander Fleming and the Antibiotic Revolution*. Sparkford, UK: Sutton, 2004.

Bud, Robert. *The Uses of Life: A History of Biotechnology*. Cambridge: Cambridge University Press, 1993.

Bunge, Mario, and William Shea, eds. *Rutherford and Physics at the Turn of the Century*. New York: Cambridge University Press, 1979.

Cairns, John. *Phage and the Origins of Molecular Biology*. Centennial ed. Cold Spring Harbor, NY: Cold Spring Laboratory Press, 2007.

Chadarevian, Soraya de. *Designs for Life: Molecular Biology after World War II*. New York: Cambridge University Press, 2002.

Chick, Harriette, Margaret Hume, and Marjorie MacFarlane. *War on Disease: A History of the Lister Institute*. London: A. Deutsch, 1971.

Cirillo, Vincent. *Bullets and Bacilli: The Spanish-American War and Military Medicine*. New Brunswick, NJ: Rutgers University Press, 2004.

Copp, Terry. *Anatomy of Poverty: The Condition of the Working Class in Montreal, 1897-1929*. Toronto: McClelland & Stewart, 1974.

Copp, Terry, and Bill McAndrew. *Battle Exhaustion: Soldiers and Psychiatrists in the Canadian Army, 1939-1945*. Montreal: McGill-Queen's University Press, 1990.

Cowdrey, Albert. *Fighting for Life: American Military Medicine in World War II*. New York: Free Press, 1994.

Cunningham, A.R., and J.P. Williams. *The Laboratory Revolution in Medicine*. Cambridge: Cambridge University Press, 1992.

Davis, Mike. *The Monster at Our Door: The Global Threat of Avian Flu*. New York: Henry Holt & Co., 2005.

Digby, Anne. *Diversity and Division: Health Care in South Africa from the 1800s*. New York: Lang, 2006.

———. *Making a Medical Living: Doctors and Patients in the English Market for Medicine*. Cambridge: Cambridge University Press, 1994.

Dolev, E. *Allenby's Military Medicine: Life and Death in World War I Palestine.* London: I.B. Tauris, 2007.
Dubos, René, ed. *Bacterial and Mycotic Infections of Man.* Philadelphia: J.B. Lippincott, 1948.
———. *The Bacterial Cell in Its Relation to Problems of Virulence, Immunity and Chemotherapy.* With an addendum by C.F. Robinow. Cambridge, MA: Harvard University Press, 1945.
———. *The Professor, the Institute, and DNA.* New York: Rockefeller University Press, 1976.
Fahrni, Magdalena. *Household Politics: Montreal Families and Postwar Reconstruction.* Toronto: University of Toronto Press, 2005.
Fedunkiw, Marianne. *Rockefeller Foundation Funding and Medical Education in Toronto, Montreal and Halifax.* Montreal and Kingston: McGill-Queen's University Press, 2005.
Feldberg, Georgina. *Disease and Class: Tuberculosis and the Shaping of Modern North American Society.* New Brunswick, NJ: Rutgers University Press, 1995.
Fischer, Ernest Peter, and Carol Lipson. *Thinking about Science: Max Delbrück and the Origins of Molecular Biology.* New York: Norton, 1988.
Flexner, Abraham. *Medical Education: A Comparative Study.* New York: MacMillan, 1925.
Foster, W.D. *A History of Bacteriology and Immunology.* London: William Derck, 1970.
Fowler, Laurence, and Helen Fowler. *Cambridge Commemorated: An Anthology of University Life.* Cambridge: Cambridge University Press, 1984.
Frappier, Armand. *Un Rêve, une lutte: autobiographie.* Sillery: Presses de l'Université du Québec, 1992.
Fredrickson, Donald. *The Recombinant DNA Controversy, A Memoir: Science, Politics, and the Public Interest, 1974-1981.* Washington, DC: ASM Press, 2001.
Friedland, Martin. *The University of Toronto: A History.* Toronto: University of Toronto Press, 2002.
Frost, Stanley B. *McGill University for the Advancement of Learning.* Vol. 2, *1985-1971.* Kingston and Montreal: McGill-Queen's University Press, 1980.
Gabriel, Richard. *A History of Military Medicine.* Boulder, CO: Greenwood Press, 1992.
Gaudillière, Jean-Paul, and Ilana Löwy, eds. *Heredity and Infection: The History of Disease Transmission.* London: Routledge, 2001.
Geison, G.A. *Michael Foster and the Cambridge School of Physiology: The Scientific Enterprise in Late Victorian Britain.* Princeton: Princeton University Press, 1978.
Geissler, Erhard, and John Ellis van Courtland Moon, eds. *Biological and Toxin Weapons: Research, Development and Use from the Middle Ages to 1945.* Oxford: Oxford University Press, 1999.

Germaine, Anne. *Montreal: The Quest for a Metropolis.* New York: Wiley, 2000.

Gordon, Alan. *Making Public Pasts: The Contested Terrain of Montreal's Public Memories, 1891-1930.* Montreal: McGill-Queen's University Press, 2001.

Goulet, Denis. *Histoire de la Faculté de médecine de l'Université de Montréal, 1843-1993.* Montreal: VLB, 1993.

Gridgeman, N.T. *Biological Sciences at the National Research Council of Canada: The Early Years to 1952.* Waterloo: Wilfrid Laurier University Press, 1979.

Gwynne-Timothy, J.R.W. *Western's First Century.* London, ON: University of Western Ontario, 1978.

Hager, Thomas. *Force of Nature: The Life of Linus Pauling.* New York: Simon & Schuster, 1995.

Hanaway, Joseph. *McGill Medicine.* Vol. 2, *1885-1936.* Montreal and Kingston: McGill-Queen's University Press, 2006.

Hare, Ronald. *The Birth of Penicillin and the Disarming of Microbes.* London: Allen & Unwin, 1970.

Harris, Sheldon. *Factories of Death: Japanese Biological Warfare 1932-45 and the American Cover-up.* London: Routledge, 1994.

Harrison, Mark. *Medicine and Victory: British Military Medicine in the Second World War.* Oxford: Oxford University Press, 2004.

———. *Public Health in British India: Anglo-Indian Preventive Medicine 1859-1914.* Cambridge: Cambridge University Press, 1994.

Hartcup, Guy. *The Challenge of War: Britain's Scientific and Engineering Contributions to World War Two.* New York: Palgrave, 1970.

Hershey, Alfred Day. *We Can Sleep Later: Alfred D. Hershey and the Origins of Molecular Biology.* Cold Spring Harbor, NY: Cold Spring Harbor Laboratory Press, 2000.

Howarth, T.E.B. *Cambridge between Two Wars.* London: Collins, 1978.

Jacobson, Jeffrey, ed. *Immunotherapy for Infectious Diseases.* Totowa, NJ: Humana Press, 2002.

Johnson, G. *University Politics: F.M. Cornford's Cambridge and His Advice to the Young Academic Politician.* Cambridge: Cambridge University Press, 1994.

Jonas, G. *The Circuit Riders: Rockefeller Money and the Rise of Modern Science.* New York: Norton, 1989.

Judson, H.F. *The Eighth Day of Creation.* New York: Simon & Schuster, 1979.

Lappé, Marc. *Evolutionary Medicine: Rethinking the Origins of Disease.* San Francisco: Sierra Club Books, 1994.

Latour, Bruno. *The Pasteurization of France.* Cambridge, MA: Harvard University Press, 1988.

Lechevalier, Hubert. *Three Centuries of Microbiology.* New York: Dover Publications, 1974.

Leedham-Green, E.S. *A Concise History of the University of Cambridge.* New York: Cambridge University Press, 1996.
Leslie, Stuart. *The Cold War and American Science.* New York: Columbia University Press, 1993.
Lewis, Sclater. *Royal Victoria Hospital, 1887-1947.* Montreal: McGill University Press, 1969.
Li, Alison. *J.B. Collip and the Development of Medical Research in Canada.* Montreal and Kingston: McGill-Queen's University Press, 2003.
Linteau, Paul-André. *Histoire de Montréal depuis la Confédération.* Montreal: Boreal, 2000.
Luria, S.E. *A Slot Machine, a Broken Test Tube: An Autobiography.* New York: Harper & Row, 1984.
Macfarlane, Gwyn. *Howard Florey: The Making of a Great Scientist.* Oxford: Oxford University Press, 1979.
Maclennan, Hugh, ed. *McGill: The Story of a University.* London: Allen & Unwin, 1960.
McKillop, Brian. *Matters of Mind: The University in Ontario 1791-1951.* Toronto: University of Toronto Press, 1994.
McPhedran, N. Tait. *Canadian Medical Schools: Two Centuries of Medical History, 1822 to 1992.* Montreal: Harvest House, 1993.
Mendelsohn, Everett, et al., eds. *Science, Technology and the Military.* 2 vols. Boston: Kluwer Academic Publishers, 1988.
Moberg, Carol. *René Dubos, Friend of the Good Earth.* Washington, DC: ASM Press, 2005.
Moberg, Carol, and Zanvil Cohn. *Launching the Antibiotic Era: Personal Accounts of the Discovery and Use of the First Antibiotics.* New York: Rockefeller University Press, 1990.
Modin, Yuri. *My Five Cambridge Friends.* Paris: Headline, 1994.
Naylor, C. David. *Private Practice, Public Payment. Canadian Medicine and the Politics of Health Insurance 1911-1966.* Kingston: McGill-Queen's University Press, 1986.
Nicholson, G.W.L. *Seventy Years of Service: A History of the Royal Canadian Army Medical Corps.* Ottawa: Borealis Press, 1977.
Owram, Doug. *Born at the Right Time: A History of the Baby Boom Generation.* Toronto: University of Toronto Press, 1996.
Penfield, Wilder. *No Man Alone: A Neurosurgeon's Life.* Boston: Little Brown, 1977.
Rawling, Bill. *Death Their Enemy: Canadian Medical Practitioners and War.* Ottawa: AGMU Marquis, 2001.
———. *The Myriad Challenges of Peace: Canadian Forces Medical Practitioners since the Second World War.* Ottawa: Canadian Government Publishing, 2004.

Rosenberg, Charles. *Explaining Epidemics and Other Studies in the History of Medicine*. New York: Cambridge University Press, 1992.

Rothstein, William. *American Medical Schools and the Practice of Medicine: A History*. New York: Oxford University Press, 1987.

Shortt, S.E.D., ed. *Medicine in Canadian Society: Historical Perspectives*. Montreal: McGill-Queen's University Press, 1981.

Siddiqi, Javed. *World Health and World Politics: The World Health Organization and the U.N. System*. London: Hurst & Company, 1995.

Silverstein, A.M. *A History of Immunology*. San Diego, CA: Academic Press, 1989.

Sinclair, Andrew. *The Red and the Blue: Intelligence, Treason and the Universities*. London: Weidenfeld and Nicolson, 1986.

Smith, Jonathan. *Teaching and Learning in Nineteenth Century Cambridge*. Rochester, NY: Boydell Press, 2002.

Sorkes, Theodore, and Gilbert Pinard, eds. *Building on a Proud Past: 50 Years of Psychiatry at McGill*. Outremont, Quebec: Productions Immeda, 1995.

Stanier, Roger, Mike Doudoroff, and Ed Adelberg. *The Microbial World*. Englewood Cliffs, NJ: Prentice-Hall, 1959.

Starr, Paul. *The Transformation of American Medicine: The Rise of a Sovereign Profession and the Making of a Vast Industry*. New York: Basic Books, 1982.

Steiner, Paul. *Disease in the Civil War, the Natural Biological Warfare in 1861-1865*. Springfield, IL: C.C. Thomas, 1968.

Sullivan, John, and Norman Ball. *Growing to Serve: A History of Victoria Hospital, London, Ontario*. London, ON: Victoria Hospital Corporation, 1985.

Summers, William. *Felix d'Herelle and the Origins of Molecular Biology*. New Haven: Yale University Press, 1999.

Swettenham, John. *McNaughton*. 2 vols. Toronto: Ryerson University Press, 1968.

Taper, Ted. *Oxford, Cambridge, and the Changing Idea of the University: The Challenge to Donnish Domination*. Philadelphia: Society for Research into Higher Education, 1992.

Taylor, Kevin. *Central Cambridge: A Guide to the University and Colleges*. New York: Cambridge University Press, 1994.

Tom Rivers: Reflections on a Life in Medicine and Science; an Oral History Memoir, prepared by Saul Benison. Cambridge, MA: MIT University Press, 1967.

Travill, A.A. *Just a Few: Queen's Medical Profiles*. Kingston: Faculty of Medicine, Queen's University, 1992.

Tucker, Jonathan. *Scourge: The Once and Future Threat of Smallpox*. New York: Atlantic Monthly Press, 2001.

Twohig, Peter. *Labour in the Laboratory: Medical Laboratory Workers in the Maritimes*. Montreal: McGill-Queen's University Press, 2005.

Ullman, Agnes. *Origins of Molecular Biology: A Tribute to Jacques Monod*. Rev. ed. Washington, DC: ASM Press, 2003.

Waddington, Keir. *Medical Education at St. Bartholomew's Hospital 1123-1995*. Suffolk: Bury St. Edmonds, 2003.

Wang, Jessica. *American Science in an Age of Anxiety: Scientists, Anticommunism and the Cold War*. Chapel Hill: University of North Carolina Press, 1999.

Warner, J.H. *The Therapeutic Perspective: Medical Practice, Knowledge and Identity in America, 1820-1885*. Princeton, NJ: Princeton University Press, 1986.

Watson, James. *The Double Helix*. New York: Atheneum, 1968.

Weatherall, Mark. *Gentlemen, Scientists and Doctors: Medicine at Cambridge 1800-1940*. Rochester, NY: Boydell Press, 2000.

Weintraub, William. *City Unique: Montreal Days and Nights in the 1940s and '50s*. Toronto: McClelland & Stewart, 1996.

Werskey, Gary. *The Visible College: A Collective Biography of British Scientists and Socialists of the 1930's*. London: A. Lane, 1988.

Westley, Margaret. *Remembrance of Grandeur: The Anglo-Protestant Elite of Montreal, 1900-1950*. Montreal: Libre expression, 1990.

Whitaker, Reginald, and Gary Marcuse. *Cold War Canada: The Making of a National Insecurity State, 1945-1957*. Toronto: University of Toronto Press, 1994.

Wilkinson, Lise, and Anne Hardy. *Prevention and Cure: The London School of Hygiene & Tropical Medicine: A 20th Century Quest for Global Public Health*. London: Kegan Paul, 2001.

Willis, Ross. *Western 1939-1970: Odds and Ends*. London, ON: University of Western Ontario, 1980.

Wills, Christopher. *Plagues: Their Origins, History and Future*. London: Flamingo, 1997.

Woroboys, Michael. *Spreading Germs: Disease Theories and Medical Practices in Britain 1865-1900*. Cambridge: Cambridge University Press, 2000.

Young, Brian. *Respectable Burial: Montreal's Mount Royal Cemetery*. Montreal: McGill-Queen's University Press, 2003.

Zweiger, Gary. *Transducing the Genome: Information, Anarchy, and Revolution in the Biomedical Sciences*. New York: McGraw-Hill, 2001.

APPENDIX A

THE MURRAY FAMILY PROFILE[1]

THE CHILDREN OF ALFRED THURTELL MURRAY (1798-1875)
AND MARY EVERITT (1799-1863)
(BOTH FROM THE NORFOLK REGION)
MARRIED 1829 IN LONDON, ENGLAND

Born with the surname Thurtell, Alfred changed his name to Murray in 1829 at the time of his wedding, because of a murder scandal involving his cousin John Thurtell. In 1835, Alfred and Mary emigrated to the Salem district, Cape of Good Hope, South Africa, and later to the Graaff Reinet region, in the interior of the Cape. Their home, Roode Bloem, built in 1875, later became the primary focus of Murray family life in South Africa.

1. George Everitt   born: 1830   died: 1901
2. Alfred Everitt   born: 1832   died: 1909
3. **Walter Everitt**   **born: 1837**   **died: 1924**
4. Jane H.   born: 1842   died: 1926

---

[1] This information has been obtained from the thorough genealogical research of Marge E. Shearing in South Africa and Peter Murray in Markham, Ontario, Canada. Shearing's work, called *My Ain Folk (Family Book)*, was privately published after she died in 1992. Peter Murray has built on this previous research, having corresponded extensively with Marge Shearing, along with other Murray and Thurtell relatives in South Africa. He has also traced some of the experiences of members of the Thurtell Murray family, who dispersed widely, including a branch who emigrated to San Luis Obispo, California, during the 1840s. All of this information remains in the possession of Peter Murray.

THE CHILDREN OF WALTER MURRAY (1837-1924)
AND ANNA ELIZABETH SOUTHEY (1836-1914)
MARRIED 1859 IN GRAAFF REINET, CAPE TOWN, SOUTH AFRICA

1. William Everitt   died: 1861   infant
2. Ellie Harriett   died: 1861   infant
3. **George Alfred Everitt**   **born: 1862**   **died: 1941**
4. Edith Mary   born: 1863   died: 1923
5. Walter Everitt   born: 1865   died: 1936
6. Harriett Eleanor   born: 1866   died: 1942
7. Arthur Everitt   born: 1869   died: 1939
8. Thomas Everitt   born: 1870   died: 1941
9. Frank Everitt   born: 1873   died: 1907
10. Annie Elizabeth   born: 1876   died: 1955
11. Herbert Everitt   born: 1879   died: 1880   infant

George and his brother Frank both studied at St. Bartholomew's Hospital, London, England.

THE CHILDREN OF GEORGE ALFRED MURRAY (1862-1941)
AND KATHLEEN DUNNE (1862-1936)
MARRIED IN 1889 IN SOUTH AFRICA

                                                                                                                             Nicknames

1. **Everitt George Dunne**   **born: 1890**   **died: 1964**   **Biff, Joburg**
2. Thorkill Howard Everitt   born: 1892   died: 1968   Dink
3. Roger Hi-Regan Everitt   born: 1895   died: 1968   Rory
4. Allan Cameron Everitt   born: 1901   died: 1982   Lal

THE CHILDREN OF EVERITT GEORGE DUNNE MURRAY (1890-1964)
AND HARRIET WINIFRED HARDWICK (FREDA) WOODS, (1895-1990)
MARRIED IN 1917 IN LONDON, ENGLAND

1. Robert George Everitt   born: 1919   Bob
2. Susan Ann   born: 1926   Sue

THE CHILDREN OF ROBERT GEORGE EVERITT MURRAY (B. 1919) AND DORIS MARCHAND (1916-84) MARRIED IN 1944 IN MONTREAL, CANADA

1. Alice Blair   born: 1944   Married Norman Francis Rae
2. Peter Everitt   born: 1946   Married Dianne Grace Hart
3. Thomas Everitt   born: 1948   Married Daisy Romero-Fernandez

After the death of Doris in 1984, Robert Murray married Marion Luney.

THE CHILDREN OF SUSAN MURRAY (B. 1926) AND C.W.B. (BLAKE) ROBINSON (DIED 1968) MARRIED IN 1945 IN MONTREAL, CANADA

1. Murray Douglas   born: 1947   Married Susan Kovacs & Dee McMullen
2. Joan Diana   born: 1951   Married Philip Smith
3. Christopher Blake   born: 1956   Married Diane Russell
4. Sarah Winifred   born: 1964   Married Thomas Schmidt

After the death of Blake in 1968, Susan married Don King.

APPENDIX B

PROFESSIONAL ACHIEVEMENTS OF EVERITT AND ROBERT MURRAY

E.G.D. MURRAY: Born 21 July 1890, Johannesburg, South Africa

Degrees
Bachelor of Arts   Cambridge University (Honours 1912)
Master of Arts   Cambridge University (1918)

University and Medical School
University of Cambridge (Christ's College), England
St. Bartholomew's Hospital, London

Professional
Licentiate in Medicine and Surgery of the Society of Apothecaries (L.M.S.S.A): London, 1916
Specialist Certification as a Pathologist and Bacteriologist, Royal College of Physicians and Surgeons of Canada, 1946

Honours
Officer of the British Empire (O.B.E.), Military Division, 1918
Fellow of the Royal Society of Canada (F.R.S.C.), 1938
Medal of Freedom: awarded by the United States War Department, 1947
Flavelle Medal, Royal Society of Canada, 1953
Coronation Medal, Queen Elizabeth II, 1953
Hon. Fellowship, Montreal Medico-Chirurgical Society, 1955
D.Sc. (Hon.), University of Montreal, 1955
D.Sc. (Hon.), McGill University, 1957
Hon. Membership, Canadian Society of Microbiologists, 1957
Hon. Life Member, Canadian Public Health Association, 1957
Hon. Membership, Society for General Microbiology (United Kingdom), 1960
Hon. Life Member, American Society for Microbiology (formerly SAB), 1961
Hon. Member, Canadian Association of Medical Bacteriologists, 1962

War Research and Service (1914-1918)
Captain, Royal Army Medical Corps
Research Staff of the War Office Central Cerebro-Spinal Fever Laboratory, 1915-16
Staff of the War Office Vaccine Department, R.A.M. College, 1917-19
Member of the War Office Committee on Dysentery

Medical Research Council, Great Britain
Research Bacteriologist 1920-26 (then given a Research Grant)
Committee on Meningococcus and Pneumococcus Serum

Academic and Teaching Appointments in the United Kingdom
St. Bartholomew's Hospital, London: Senior Demonstrator in Pathology, 1919
University of Cambridge:
    Fellow of Christ's College, 1923-31
    Lecturer in Pathology, 1925-30
    Director of Medical Studies, Christ's College, 1925-30
    Represented Cambridge at the installation of the new Chancellor of the University of Toronto, 21 November 1947

Academic and Teaching Positions in Canada
McGill University:
    Professor and head of Department of Bacteriology and Immunology, 1930-55
    Bacteriologist-in-chief, Medical Board of Royal Victoria Hospital, 1931-46
    Bacteriologist-in-chief, Medical Board of Children's Memorial Hospital, 1938-48
    Member of the Medical Board of Royal Edward Laurentian Hospital, 1940-46
    Member of the Medical Board of the Alexandra Hospital, 1932-55
    Member of the McGill Senate, 1943-53 (representing Faculty of Medicine)
University of Western Ontario: Visiting Professor in Medical Research, 1955-1964

War Research and Service: Second World War and Post-War
Member of the National Research Council (NRC) Subcommittee on Infections
Member of the NRC Subcommittee on Shock and Blood Substitutes
Member of the Special Canadian-U.S. Committee on Gas Gangrene
Canadian Chairman of the Joint U.S.-Canada Commission (War Disease Station)
Chairman Biological Warfare Committee (C-1), Directorate of Chemical Warfare and Smoke, National Defence Headquarters, Ottawa
Chairman of the BW Advisory Board, Defence Research Board, 1946-48

Scientific Societies and Public Service (select citations)

American Journal of the Medical Sciences
    One of the Associate Editors, 1940-55

Arctic Institute of North America
    Charter Associate, 1948-64

Bergey's Manual of Determinative Bacteriology
    Member of the Trustee Board of Editors, 1934-1964

Canadian Public Health Association
    Chairman of the Laboratory Section, 1937-38
    Member of Council, 1952-53

Canadian Society of Microbiologists
    President, 1955-57

City of Montreal
    Councillor (Class C), representing McGill University, 1947-50, 1950-54, 1954-55
    Appointed member of the Board of Health of the City of Montreal, 1948

International Federation of Culture Collections of Microorganisms
    Member of the Permanent Commission

International Botanical Congress
    Recorder Bacteriology Section, 1930

International Association of Microbiological Societies
    Representative for Canada on Permanent International Commission, 1938-53
    Chairman of the Committee on Bacteriological Nomenclature at the 3rd International Congress, September 1939, New York City
    Member of the Judicial Commission of the Permanent Committee of Nomenclature, 1943-53, and 1953-63
    Vice-President, Section 1 (General Microbiology) Fifth International Congress (Rio de Janeiro), 1950
    Vice-President, Section on Taxonomy and Nomenclature, Sixth International Congress, Rome, 1953
    Elected Chairman of the International Committee on Bacteriological Nomenclature, 1953-1962

Appointed to the Committee of Honour at the Seventh International Congress, Stockholm, 1958
President, Eighth International Congress of Microbiology, Montreal, 1962

Montreal Medico-Chirurgical Society
President, 1942-44; Member of Council, 1946; Hon. Fellow, 1955

National Research Council of Canada
Member, Scientific Advisory Sub-Committee on the Institute of Parasitology, 1947-53
Member of the Medical Advisory Committee and its Executive Committee, 1952-56
Member of the National Research Council, 1953-56
Chairman, Associate Committee on Control of Hospital Infections, 1957-62

Osler Society, McGill University
Honorary President, 1950

Royal Society of Canada
Member, 1938; President of Section V, 1954; Flavelle Medalist, 1953

Societe de Microbiologie de la Province de Quebec
President, 1955; Hon. Life Member, 1957

Society of American Bacteriologists (now American Society of Microbiology)
Committee of Bergey's Manual of Determinative Bacteriology, 1932-36
Member of Council, 1940-42
Member of the Committee on Teaching Bacteriology, 1944
Member of the Nominating Committee, 1947
Nominated for President, 1954

ROBERT MURRAY:[2] Born 19 May 1919, London, England

Degrees:

Bachelor of Arts	University of Cambridge (1941)
Master of Arts	University of Cambridge (1945)
M.D.C.M. Medical Degree (Medicinea Doctorem et Chirurgiae Magistrum)	McGill University (1943)

Honours and Scholarly Awards

Harrison Prize, Royal Society of Canada (shared with C.F. Robinow), 1957
F.R.S.C., Royal Society of Canada, 1958
Royal Society of Canada, Traveling Award, 1961
C.S.M. Award, Canadian Society of Microbiologists, 1963
F.A.A.M., American Academy of Microbiology, 1973
Flavelle Medal, Royal Society of Canada, 1984
D.Sc. (Hon.), University of Western Ontario, 1985
J. Roger Porter Award, American Society for Microbiology (U.S. Federation of Culture Collections), 1987
D.Sc. (Hon.), University of Guelph, 1988
Bergey Medal of the Bergey's Manual Trust, 1993
D.Sc. (Hon.), University of Victoria, 2002
D.Sc. (Hon.), McGill University, 2007

Decorations

Coronation Medal, 1953
Centennial Medal, 1967
Queen's Jubilee Medal, 1978
Officer, Order of Canada, 1998
Queen's Jubilee Medal, 2002

Scientific Organizations and Committees (select citations)
American Academy of Microbiology, Member, Board of Governors, 1980-1983

American Society for Microbiology, Vice-President, 1971-1972 and President, 1972-1973. Editorial Board, Journal of Bacteriology, 1951-1954 and 1980-1986. Co-Editor, Methods for General and Molecular Bacteriology, ASM Press, Washington, DC, 1994.

---

2 Robert Murray, "A Structured Life," American Review of Microbiology 42 (1988): 1-34

Bergey's Manual Board of Trustees, 1964-1990; Vice-Chairman, 1966-1967; Chairman, 1976-1990; Trustee Emeritus, 1990-date

Biological Council of Canada, Governing Board, 1966-1973

Canadian Bacterial Diseases Network of Centres of Excellence, Member, Board of Directors, 1990-1998

Canadian Society of Microbiologists Council, Chairman, Founding Committee, 1950-1951; Founding President, 1951-1952; Council Member, 1951-1960.Editor, *Canadian Journal of Microbiology*, 1954-1960.
Editorial Board of *Bacteriological Reviews*, 1967-1969, and Editor-in-Chief of *Bacteriological Reviews* (later *Microbiological Reviews*), 1969-1979.

International Association of Microbiological Societies:
    International Committee on Systematic Bacteriology, 1962-date; Executive Committee Member, 1966-1973 and 1978.
    Member of Judicial Committee, 1966-1973; Chairman, ICSB, 1982-1990 and Ex Officio member of the Judicial Commission, Life Member 1990.
    International Journal of Systematic Bacteriology, Associate Editor, 1982-1990; Editor-in-Chief, 1990-1994.
    Canadian Representative Plenary Session, International Union of Microbiological Societies, 1953, 1958, 1966, 1978, and 1991

International Review of Cytology, Advisory Editor, 1975-1985
Editorial Board, *Manual of Methods for General Bacteriology*, 1977-1979

National Research Council of Canada, Member, National Advisory Board on Scientific Publications, 1987-1990

Royal Society of Canada, Member, Hannah Medal Award Committee, 1997-2000; Chair, 1999

University of Western Ontario, Faculty of Medicine, Department of Microbiology (Bacteriology) and Immunology: Lecturer, 1945-48; Head of Department, 1949-74; Acting Dean of Science, 1973-74; Professor Emeritus, 1984 to present

APPENDIX C

SELECTION OF FREDA MURRAY'S POETRY

DAYBREAK[1]

A blue-grey mist veils all the lake
In chill embrace:
Slowly it swirls outside the trees
Standing tall along the coast,
Dark hemlock, ghostly birch.

The dawn-wind rustles through the leaves
Scentless and pure
And stars fade out across the sky;
A rosy feather in the east
Foretells the day to come.

Long before the golden sun
Touches the hills
The birds' full-throated morning song,
Ringing over the lake,
Echoes in shadowed bays.

The loon's long lingering cry:
Enchanted call
Of quiet lonely places,
The whisper of a foaming fall
One hears, so far away.

The coiling mist lifts up
And gathers fast
Into a drifting silver scarf,
Over the purple hills
New-flushed with rose.

---

[1] Published in *The Bulletin* of the Federation of Ontario Naturalists (Toronto), no. 85, Sept. 1959, p. 14.

Westward the penumbra flees:
Arch within arch,
Violet, rose and deepest blue,
In earth's own curve
It moves away.

Now hilltops flash with sudden gold
And trees are fringed
With lambent many-jewelled light;

The mist-men form and march and fade;
Dawn becomes day.

FULL CYCLE[2]

The ice-bound lakes lie still,
Snow covers all the land;
Now is the time of rest,
Surely and unhurried
Life's changes follow on.

Rooted in dark leaf-mold
The trees stand, starkly grey,
Asleep, biding their time;
Their life-sap lying still
Throughout the long dark days.

When Winter Solstice is passed
Each day the sun swings wider arcs
And shadows shorter grow;
The peepers shout from swamp and stream,
Their tiny voices high and shrill,
Freed from their icy bonds.

---

2 Unpublished poem, included in the desktop published book by Sara Schmidt, "Reflections-HWH Murray (n.d). There are twenty-four of Freda's poems in this collection.

Swiftly the warm sun melts the snow;
The ice is gone, the lakes are free;
Day follows day with gentle airs,
Gold sun, blue sky, from dawn to dusk;
The trees awake, their twigs alive
With purple-misted brown
Until such time the warming sun
Prompts them to don, all shining gold,
The marriage veil of Spring.

It passes, that brief glorious time
And Summer comes anon;
The hills look softer, rounded now
With trees all green with lusty life,
Their ripe seed falling slow
On leaves of yesteryear.

Then, gently, Autumn changes all
To flame of red and gold;
But wind and rain tear from their twigs
The fluttering, dying leaves,
And down to earth they fall, to build
Anew the Rites of Spring.

REFLECTIONS[3]

Still is the night
No breeze is stirring on this shore
Thought on the island in the bay
A ripple laps against a rock;
But for that little sound out there,
Still is the night.

---

3 Published in "The Bulletin" of the Federation of Ontario Naturalists (Toronto), no. 89, Sept. 1960, p. 19.

Bright is the sky!
All patterned with the brilliant stars:
Far-off, shinning points of light
In constant forms well loved by man;
Unchanging guides throughout the years,
Bright is the sky.

Dark are the hills!
No tree, no valley can be seen;
But in their gentle well-known shapes
Smooth as black velvet where they lie.
Softly curving against the sky,
Dark are the hills.

Is man so great?
Compared with all these timeless things
His life is brief, its impact slight;
Without immeasurable force
Such as moves mountains, makes new stars
And slowly changes man himself.

APPENDIX D

OVERVIEW OF DORIS MURRAY'S PAINTINGS

TRIBUTE TO DORIS MURRAY

May 8 - June 2, 1985

Cover of the 1985 Exhibit

DORIS MARCHAND MURRAY, 1916–1984

Born Princeton, New Jersey, 1916
Died London, Ontario, 1984

Education
Smith College, Northhampton, MA, BA, 1938
Private study, Denmark and Switzerland, 1968–69

Selected Exhibitions
Individual
1963 London Public Library and Museum, Crouch Branch, London, Ontario
1970 Glen Gallery, London
1971 McIntosh Gallery, University of Western Ontario
1972 Scarborough College, Toronto
1973 Gallery 93, Ottawa
1974 Glen Gallery, London
1978 Glen Gallery, London

Group
1958 Waverly Gallery, London, Ontario
1966 London Public Library and Art Museum, Landon Branch, London
1971 Woodstock Art Gallery, Woodstock, Ontario

Collections
University of Western Ontario
Private collections in Canada, U.S.A, and Europe

Doris Murray and her paintings

INDEX

Adami, John George, xxxix, xxxixn118, lv, lv-lvin209-10, lxin241
Adams, Mark H., xciiin406, 614, 614n105, 620
Addenbrooke's Hospital, xxxviii, xxxix, lxxvi, 157-58, 387
Adrian, E.D. (Lord), xlixn176, lxxv, 23, 163, 469n30
Alexander, Hattie Elizabeth, 436n118
Allbutt, Sir Clifford, xxxviiin105, xxxviiin110
Allin, A.E., 452, 452n4, 498
American Association for the Advancement of Science, 50, 452
American science: and Vannevar Bush, 97n17; and Camp Detrick, 128-29, 132, 139; and Cold War hysteria, lxxxii, lxxxiin341, ciii, 591, 658; and Office of Scientific Research and Development, 99n17, 113n35, 247. See also biological warfare and Society of American Bacteriologists.
Anderson, Thomas, 743-744, 744n37, 745
Andrewes, Christopher H., xlin130, 535, 535n98, 535n100
Andrewes, Frederick W., xl, xlin130, xlii-xliii, xlvin159, 1, 4, 4n6, 7, 7n12
Armour, John, 345, 345n140
arctic bacteriology, xvi, lxii, lxvi, 48-49, 502n73, 663
Arctic Institute of North America, lxvi, lxxxv, lxxxvn359, 153n104, 500, 503, 517, 573, 626, 694, 778
Asboe-Hansen, G., 689, 689n158
Ashdown, David, 185, 185n46
Asheshov, Elizabeth Hall, lxxxixn378, cx, 326, 326n122, 479, 655, 735
Asheshov, Igor N. (Ash), xviii-xixn19, lxxvii, lxxviin315, lxxviii, lxxxviii, lxxxviiin377, lxxxix, lxxxixn378, xci, xcvi, cix, cx, cxn508, cxxiv, 206, 208, 210-17, 217n90, 219, 222, 222n100, 223, 223n103, 224, 226, 228-33, 235, 235n122, 249-51, 253, 263, 269, 275, 278, 283, 285, 287, 296-98, 310, 321, 326, 326n122, 328, 328n124, 333-35, 340, 343, 352, 364, 379, 388, 395, 399, 399n79, 403-07, 410, 413, 421, 428-429, 437, 440-41, 447, 452, 457, 466, 476, 479-80, 517, 614, 620, 655, 733-35, 743
Athlone, Earl of (Governor-General of Canada), xxx, 95n18
atomic research and bomb, xlixn175, lxviii, lxxvn302, 67n109, 89n4, 100n18, 130n63, 149, 219, 220n18, 359, 590n76, 596n84
Attlee, Clement, cvii
Avery, Oswald, xviiin17, lxin243, xciv, xcivn409-10, 24, 24n41, 44, 44n71, 98n15
Ayerst, McKenna and Harrison: and Everitt Murray, consulting, lxn237, lxi, cxxi, 43-46, 43n69

bacterial taxonomy, xv, xvn9, xxiii, xlvii, lxxx, c, cx, 2
bacteriophage: general, xix, xcii-xciiin402, 51, 51n81, 261; and DNA discovery, xcii-xcvi; and Robert Murray, xiv, xci-xcii, 233, 399, 404, 428, 452, 454, 475-76, 511-12, 516, 521, 525,

533-34, 544-46, 546n9-10, 548, 550, 554, 557, 568-69, 575, 586, 613-14, 627, 633n27, 634-36, 635n31, 639, 642, 657, 659, 690, 731, 731n19, 740, 744n37, 745, 754
Bailey, Hamilton, 321, 321n117, 667, 706
Baldwin, Ira, lxxi, 128, 702
Ball, Sir William Girling, 156, 156n3
Banting, Frederick G., xviii, xlii, lxii, lxivn248, lxviii-lxix, lxxi-lxxii, lxxviiin317, 66, 66n106-07, 93n9, 100-01, 101n19, 108-09, 628n15
Bark Lake: and Bark Lake Protective Association, xxin27, 274, 278, 324n119; and Columbus Club, lxvii, 155, 155n2, 262, 274, 274n72, 303, 307, 310, 313-14, 316, 366, 386, 392, 400, 424, 426, 470, 476, 477, 483, 488-91, 492n57, 499, 503, 522, 535-36, 568, 574, 576, 609-11, 651-52, 701; and family cottage life, xx-xxi, lxvi-lxvii, lxxii, 78
Barcroft, Joseph, lxxxiiin344, 23, 391, 391n70, 393, 691
Barr, Murray, lxxviii, 314, 314n107, 329, 329n126, 338-39
Bausch and Lomb, 362
Beal, Bob, 301, 301n98
Beattie, W.W., 19n31, 24
Beatty, Edward, lx
Beaudette, Fred Robert, 363, 363n32
Behring, Emil, 470, 470n31
Benson, Edgar J., 330n128
Bergey, David H., xvn9, 244n138
Bergey's Manual of Determinative Bacteriology: and Everitt Murray, xv, lxxxi, lxxxv, lxxxvn363, lxxxvi-lxxxvii, 48, 57, 244n138, 260, 297, 331, 349, 433, 458, 535, 535n100, 778-79;

general, xv, xvn9, lxxxviin370; and Robert Murray, xv, lxxxvii, lxxxviin371, 690, 780
Bernard, Claude, 255, 255n30
Bertani, Giuseppe, xcii, 659, 659n90, 664, 750
Best, Charles, lxxviiin317
Bickell Foundation, xci
Billingham, Rupert Everitt, 685, 685n143
biological warfare (Weapons): and anthrax research and production, lxx, lxxin284, 113-32; and botulinium toxoid, lxx; and Canada policy, lxviii-lxxii; and Everitt Murray, 87-153; and Experimental Station Suffield, lxix, lxxin284, 115, 119-21, 124-27, 131, 139, 217n92; and Grosse Île (War Disease Control Station), lxix-lxx, 108, 113-132; and Japanese balloons, lxx, 130n62, 132-37; and Porton Down Experiment Station, lxxi, 116n39, 391n70; and post-war developments, lxxi-lxxii, lxxin284, lxxxiii; and Rinderpest vaccine, lxx, 73n115, 105, 107-122, 133, 148, 151; and trilateral (Can., U.K., U.S.) co-operation, lxviii-lxxi, 87-153; and wartime mobilization, lxvii-lxxii
Birch-Andersen, Aksel, xci, cixn505, cxiiin524, 705, 705n187, 738, 747. See also electron microscopy.
Bisset, K.A., 559, 586, 589n69
Bittner, Josephine, lxxxix, 217n90, 249, 249n8, 250, 254, 305, 344, 376, 421, 517, 520, 532, 647, 696
Blackett, Patrick, xlixn174, lxxivn300
Blaisdell, J.L., 137, 137n74, 468, 468n28, 497

Blay, Barbara, 377n57
Bocking, Douglas, 317n112
Bohr, Niels, xciiin403
Boivin, André, 540, 540n105
Boltjes, Kingma, 509, 513, 513n82, 518, 520
Bonner, David M., ciii, 636-37, 636n37
Born, Max, xciiin403
Bowmer, Ernest J., 755, 755n55
Boyd, Sir John Smith Knox, 534, 534n96, 636
Boyd, William, 142, 242, 242n135, 243, 365, 367, 372, 500
Bragg, Sir Lawrence, xcvn420
Branham, Sarah, cii, ciin464-466, 693, 693n167
Breed, Robert S., xvn9, lxii, 61n100, 244, 244n138, 298n95, 366, 405, 407, 573, 595, 597, 607, 613, 634, 636, 640-41, 649, 651, 653, 667, 685, 690
Brenner, Sydney, xxxn66
British medicine: and British Institute of Preventative Medicine, xxxvin100; and British Medical Association, xxvi; and developments in First World War, xli-xlviii; and early 1900's trends, xxxii-xxxiii, xxxvi-xxxix; and Everitt Murray's military service, xli-xlv; and inter-war developments, xlviii-lix; and Medical Research Council (U.K.), xiv, xxxii, xxxvin99, xlvi, xlviii-xlix, lii, lxxxviiin377; and National Institute for Medical Research, xlviii, xlviiin172
Brow, Ray, 295, 295n92, 714, 714n201
Brown, M.H., 54-55, 54n85
Brown, N.N., 320, 320n116, 360, 367-68, 397, 411
Brown, Tony, 468, 468n29, 580, 629, 629n19, 657
Browne, J.S.L., 85, 189, 335, 335n134, 435
Bruce, Sir David, xliii, xliv-xlv, xliv-xlvn151-52, 2, 4, 4n8, 5, 22n38, 38
brucellosis, lxi, 37, 149
Bruere, Andrew A., 18, 18n29
Buchanan, R.E., lxii, 48, 61n100, 245, 245n140, 246, 640, 640n53, 641
Bulgakov, Nikolai, 128, 128n124
Bunche, Ralph, 578, 578n58
Burkholder, W.H., 630, 630n22
Burnet, Frank Macfarlane, 249, 249n10, 573, 575-76, 579, 754, 754n54
Burton, Alan C., lxxviii, 329n125, 402, 402n86, 668

Cameron, G.D.W., lxxxiii, lxxxiiin350, 56-57, 56n91, 96, 109, 327, 491
Cameron, T.W.M., 49, 552, 573, 596, 687
Campbell, Archie, 650, 650n76
Campsall, Robert, 730, 730n15
Canadian Medical Association, lxii, lxxxvi, cxxi, 50-53, 239, 240n131, 241
Canadian Medical Hall of Fame, xiv, cxxi, 314
Canadian medicine: and development of, lvn205; and Everitt Murray, Health Sciences Inquiry of Quebec, lxxxiv, lxxxivn352-53; and Medical Research Council, lxxx, xci; and Quebec medicine, lxxxiii-lxxxiv; and Rockefeller Foundation grants, lv; and Second World War and post-war trends, lxxix-lxxx; and universal health care, xix, lxxx, lxxxn330-31, lxxxiii, 432-33, 513-14
Canadian Public Health Association, xv, xxii, lxxxiii, lxxxvi, 53-58
Canadian Public Health Association

(Laboratory Section), xviin11, lxii, lxxxiiin350, 53-58, 235, 235n121, 310, 313, 318-19, 331, 425, 428-29, 431, 433, 435-36, 489-91, 498, 502, 537, 581, 584, 612, 653, 759, 762
Canadian Physiological Society, 76-7, 435
Canadian Society of Microbiologists: and *Canadian Journal of Microbiology*, xcviiin436, xcixn445, cxvi, cxvin532, cxxi, 657-58, 681-83, 686-91, 696; and Everitt Murray, lxxxvi, xcix, 70-75, 440n122; general, xxii; and Robert Murray, establishment of, xvi, lxii, xcix, 67n108, 356, 356n22, 552, 570, 584-98; and Robert Murray, CSM Award (1963), xcvii-xcviii
Cannon, Walter B., 266, 266n57, 268
Carmichael, Michael, 262, 262n53, 412, 424, 426
Carroll, Kenneth K., 747, 747n42
Carthew, Rex, 400, 400n82
Chadwick, James, xlixn174
Chain, Ernest Boris, ln182, 561, 561n42, 653, 653n81
Chambers, Robert, 547, 547n12, 549, 555
Chaplin, Charles, 691, 694, 694n169
Chapman, Jim, 217, 217n93
Chase, Martha, xcv
Cheever, F.S., 715, 715n203
Churchill, Winston, cvii, 629, 631, 724-25
Clark, Willy, lxn237, 85
Cockcroft, John, xlixn174
Coffey, Theodore, lxxviiin316
Cohen, Seymour S., 620, 620n116
Collip, James B., xvi, xviiin17, xxiin32, lvi, lvin212, lxi, lxxin281, lxxviii, lxxviiin316-18, xcii, cxxiv, 32, 32n56, 43n69, 85, 90, 93, 93n9, 94, 102-04, 106, 109-13, 123, 387, 391, 395, 401-02, 404, 417, 420-22, 430-31, 441, 443, 455, 460, 469-74, 476, 478-79, 481, 484, 499, 503, 505, 507, 510, 514, 516, 518, 523, 526-27, 538, 547, 551, 558, 562, 581, 604, 608, 621, 642, 658, 673, 681, 685, 691, 702, 704, 705n186, 706, 717, 719, 731, 747n42, 749, 756
Communist Party of Canada, lix
Cone, William, lvin213, 85, 343, 343n138, 353
Conn, H.J., 640-41, 640n52
Cook, William Harrison, 558, 558n37, 762n67
Coombs, Robert Royston Amos, 725, 725n9
Co-operative Commonwealth Federation (C.C.F.), lix, lixn229, lxxxiii
Cowan, Samuel T., cn449, 535, 535n99-100, 595, 662, 662n97, 693, 735, 754
Craigie, James Hubert, xxiin32, lxxxiiin347, 22, 22n37, 64, 64n103, 106, 109, 123, 150, 256, 313, 316, 328-29, 331, 340, 348, 405, 409, 429
Creech, H.J., 65, 65n105
Creite, Joachim, 255, 255n32
Crick, Francis, lxxvn304, xcv, xcvn420, xcvn422, 163. See also DNA.
Cruickshank, Robert, 629-30, 630n20, 753, 762, 762n68
Currie, Sir Arthur, liv, lviin215, lix, lx, cxxii, 10n18, 12, 16-9, 33, 37-40
Cushing, Harvey B., xviii, 405, 405n90, 440

Dale, Sir Henry Hallett, xln128, xlviiin172, lxxin282, cvii, 10, 428, 428n110, 430-431, 434-37, 439, 551
Davidson, James Norman, 578, 578n59, 587

INDEX

de Haller, Gerard, 729, 729n13
Dean, Henry Roy, xlix, xlix-ln178-180, l-lii, ln183, liii, lvi-lvii, lxn237, lxii, lxxv, lxxxv, 12-13, 13n25, 14, 17, 23-24, 24n42, 162, 323, 354, 363, 534, 625n4, 725n8
Defries, R.D., lxii, lxxxiii, xcviin431, 65, 65n104, 74-75, 495, 648, 648n68, 651
Delaporte, Berthe, 726, 726n11
Delbrück, Max, xviiin17, xciii, xciiin403-06, xciv, xcivn414, xcvn418, 639n46
den Dooren de Jong, L.E., 545, 546n7
Deoxyribonucleic acid (DNA), xiii, xiiin1-3, xiv, xcii-xcvi, 525. See also Francis Crick and James Watson.
Diefenbaker, John G., cxi, cxin516
d'Herelle, Félix, xiv, xviii, xviii-xixn19, xciiin404, 51, 51n81
Dolman, Claude Ernest, cxii-cxiii, 601, 601n88, 613, 699, 748, 756, 756n56, 759
Douglas, Howard C., 528, 528n91, 531, 533, 536
Douglas, Tommy, lxxxiii, cxin516
Downes, Prentice G., 234, 234n120
Downshide Jesuit boarding school, xxviii, xxxi
Dowson, Beetles, 609, 725, 749, 749n45
Drake, Charles G., 317, 317n113
Dubos, René, xviiin17, lxxxii, lxxxiin338, lxxxivn356, xc, xcn388, xcivn410, xcvn417, 269, 269-270n66, 275, 313, 318, 430-31, 453, 456, 640, 715
Duff, George Lyman, xxiin32, xcviin433, 143, 143n84, 189, 499, 499n67, 549, 608-09, 626, 650, 664, 695, 697, 702, 707, 711-12
Duguid, J.P., 762, 762n68

Duncan, Ian B.R., 759, 759n62
Duncan, Norma, 759, 759n62
Dunn, Sir William, xlix, 561n42
Dunne, Captain John J., xxvi, xxvin53, xxvii, 167
Dunne, Mary Chavelita, xxvi, xxvi-xxviin54
Duplessis, Maurice, xvii, lxxix, lxxixn325, cvii, 263, 660, 664
Dyer, Rolla E., 150, 340n137
Dysentery, xxii, xli, xlii, xliii, xliv-xlv, xlvii, lii, lxxxvn359, 1-2, 5, 7, 7n12, 90, 126, 292, 294, 534n96, 557, 659, 663, 777

Eagles, Bylthe Alfred, 649, 649n71
Ebisuzaki, Kaney, 546, 546n10, 549, 555, 569
Eccles, W. McAdam, 2, 3n2
Edholm, Otto G., lxxviii, lxxxiin343, 375, 375n54, 393, 409
Edsall, Geoffrey, 389, 389n67, 708, 708n193, 748, 748n43
Ehrlich, Paul, xxxiiin84, xxxvi
Elder, Robert H., xcin396, 489, 489n55, 490, 492-493, 495, 499-501, 540, 546, 551, 555, 582, 600, 603-604, 608, 689, 693, 693n168, 694, 704
Elder, Ruth, 572, 587, 603-04, 608-09
electron microscopy: and Robert Murray, xiv, xci, xcin392, xcvii, xcviin429, ci, cx, cxv, 314n107, 475, 475n39, 486, 489, 509, 544, 579, 579n60, 587, 633, 642, 656, 684, 705n187, 731n19, 742, 744n37, 745
Eli Lilly, 640, 640n50
Elizabeth II, Queen, cvii, 615-619, 615n106, 629, 631
Enders, John F., 754, 754n54

Entin, Martin, 325, 325n121, 352, 370-71, 399
Ettinger, G. Harold, 76, 76n119, 77, 705, 705n186
Evans, Charles A., xcn384, 507-08, 507n76, 510, 512, 528
Eve, A.S., 37, 37n62, 39-42
Evelyn, Kenneth E., 289-90, 289n85, 635

Farquharson, Ray, 674, 674n 116, 716
Farren, Sir William Scott, 625, 625n3
Fayette, Jonathan, 14n26
Fell, Dame Honor Bridget, 621, 621n118, 661
Ferguson, R.G., 369, 369n46
Fildes, Sir Paul G., lxxi, lxxiin286, 107-08, 115-16, 116n39, 119-120, 120n46, 124-25, 125n55, 131, 132n66, 259, 454, 516
First World War: and British medical research, xli-xlviii; and Everitt Murray, xix, xlii-xlviii
Fisher, John Heber, 217, 217n91, 344, 731-32
Fitz-James, Philip C., xcin396, 575, 575n55, 593, 639, 648, 650, 667, 670, 673-74, 677, 684
Fitzgerald, John G., 58-59, 58n94
Flaherty, Robert J., 186, 186n48
Fleming, Alexander, xviiin17, xliiin142, xlviin166, ln182, 175n39, 176, 290n87, 561n42. See also penicillin.
Fleming, A. Grant, lxviii, 87-89, 239
Fleming, Donald S., 235n123, 271, 298, 312-313, 316, 413, 419, 424, 555, 557, 670
Fletcher, Walter Morley, xlviiin170, xlixn173
Flexner, Alexander, xlv, lvn205

Florey, Howard W., xviiin17, l, ln182, lvii, lviin217, 116, 159, 561n42
Flower, Sir William, 21n35
Flower, Stanley Smyth, 21n35
Forsey, Eugene, lixn229
Foster, J.W., 639, 639n45
Fox, W. Sherwood, 215-216, 257, 257n38, 417-18, 421, 432
Franks, W.R., 65, 65n105
Frappier, Armand, lxvi, lxxxiv, lxxxivn353, ic, icn443, cxin515, 373, 373n48, 495, 570, 584, 595, 629-30, 632, 647, 650, 657, 690
Fraser, Donald T., 109, 351, 351n9, 355, 551, 581, 613
Fraser, Freda Helen, 625, 625n2
Fraser, J.R., 94-95, 95n12
Fred, E.B., lxxi, lxxiin286, 98, 113-14, 113n35, 149-50
Frey-Wyssling, Albert F., 743, 743n36
Fricsay, Ferenc, 739, 739n30
Fritz Blank, C.S.M., 588-89, 591, 638, 638n39, 700, 703

Garrod, Sir Archibald, xliii, l
Gask, George Ernest, 157-59, 157n5
George VI, King, cvii, 631
Geneva Protocol (1925), lxviii
gas gangrene, xliii, xliiin142, lxviiin269, 73n115, 90, 96-100, 116n39, 156, 466, 777
Gibbard, James, 472, 472n36, 474, 4476, 498, 501, 552, 667, 693
Gibbons, Norman E., cn449, cxiiin518, 67, 67n108, 399, 399n80, 552n26, 593, 608-09, 630, 632, 636, 657, 692, 699, 762
Gibson, John, 390, 390n69
Gillen, D.H., 546, 546n9, 549-51, 555,

593-94, 600, 604, 644, 650
Goford, Colonel Wally, 131, 131n65
Goldbloom, Alton, 674, 674n117
Gordon, Mervyn H., xl-xli, xln127, xliii-xliv, lii, lxxxiin344, 1, 4, 4n7, 7, 7n12, 274, 344, 369
Gordon, Ruth E., 683, 683n139
Gouzenko, Igor, 257n37
Gower-Rees, Arch Deacon, 662, 662n98
Graham, Angus, 96, 142, 463, 463n22, 541, 551, 551n20, 613
Greaves, R.I.N., 540, 725n8, 726
Greey, Philip, lxxxiiin347, ciiin471, 96, 109, 715
Gregg, Milton, 732, 732n21
Gregoire, Adelard, 274, 274n72
Grose, S.W., 80, 80n126, 81-81, 162, 236, 289, 309
Gulline, Brian, 262, 262n50, 264, 604
Gurd, Fraser, 393, 393n76, 394, 400, 462-63

Hagan, W.A., 363, 363n31
Hahn, Otto, xlixn174, lin174
Haig-Brown, Roderick, 342, 345, 376, 376n55, 508
Haldane Commission (on University Reform), xxxiii
Haldane, J.B.S., lxxivn300
Hall, G. Edward, lxxvii-lxxviii, lxxviin313, lixivia-lxxxix, xcii, vci, vcin427, civ, civn473, cixn502, cxxiv, 205, 205n73, 209-16, 222-24, 233, 251, 257-58, 261, 263, 265, 268, 275, 280, 309, 314-15, 321, 324-35, 325n121, 326, 332, 344, 352, 357, 365, 370-71, 374, 376, 378-86, 388-89, 392-95, 408-09, 437, 441, 443, 455, 467-68, 470, 473, 484, 497-98, 507, 514, 583, 540, 660, 664-65, 673, 684, 717, 732. See also University of Western Ontario.
Hale, C.W., 422, 422n105
Hamilton, Alvin, cxin516
Hamilton, Jim, 329, 329n125-26, 338, 418
Hanes, Charles Samuel, 665, 665n86
Hannay, Christopher, vciin429, 623, 623n120, 638, 656, 676, 678, 682, 684-85, 696, 723
Hannay, Hilda, 623n120, 678, 741
Hare, Ronald, 57, 57n92
Harrison, F.C., 10, 10n17, 11-13, 18-19
Hartley, Percival, 357, 357n25
Harvey, Lieutenant Colonel David, xliv, 4-5, 7, 7n12. See also Royal Army Medical College.
Heagy, Fred C., xcin396, 479, 479n46, 512, 516, 521, 525, 544, 546-47, 560, 578, 578n59, 580, 587
Heard, Nancy E., 171n29, 281, 281n79, 293n90, 668, 668n107, 678, 678n128, 679, 763
Heard, "Roddy", 293n90, 338, 423, 435
Heard, W.S.N., 171n29, 354n16
Hedén, Carl-Gören, c, cn452, 545, 545n5, 608, 608n95, 679-81, 739, 750-51, 760
Henriksen, S.D., 746, 746n39
Henry, Stewart, 386, 386n65
Hercus, Sir Charles Ernest, 643, 643n57
Hershey, Alfred, xciiin405, xcv, 639n46
Hillier, James, 475, 475n39
Himmelweit, Fred, 176, 176n35
Hobbs, Edgar, lixivia, 463, 463n20, 506
Hooker, Sanford B., 411, 411n95
Hopkins, Sir Frederick Gowland, xlix, xlixn176, 162
Horsfall, Frank L., 44, 324, 324n120, 498, 620
Houde, Camillien, xvii, lviii, lviiin225,

civ, civ-cvn479-480, cvii, cviin493, 351, 351n10, 352, 355, 446, 459, 459n16, 461, 617, 667
Howie, Sir James W., 763, 763n71
Huskins, Leonard, 424, 424n107, 676
hyaluronidase: and Robert Murray, research, xci, 259, 259n40, 263, 265, 268-69, 291, 297, 297n94, 299, 310, 333, 348, 358, 390, 421, 436, 438, 440, 457, 462-64, 467, 477-78, 495, 499, 514, 538

International Association of Microbiological Societies (IAMS): and Everitt Murray, lxxxi, lxxxv- lxxxvi, 58-70; general, lxii, 58-70; and Montreal Congress, 1962, xvi, cxi-cxii; and New York City Congress, 1939, lxxvin308, 61-70; and Nomenclature Committee, xv, lxii, cn449, 61n100, 244-46; and Rome Congress, 1953, c

James, Cyril J., lxx, lxxii, lxxiin288, lxxxin334, cviiin497, cxxii, 11n21, 91-92, 91n6, 93-95, 143-48, 260, 262, 402, 442, 706, 712
Jensen, H.A., 747, 747n41, 754
Johannesburg (South Africa), xxiv, xxivn41, xxv, xxvii-xxx, 2-3, 28-29, 165-68
Johannesburg Turf Club, 28, 168n26,
Johns, C.K., lxiin247, 71-72, 74-75
Johnson, Frank H., 688, 688n156
Johnson, F.M.G., 18-19
Johnson, Very Reverend Hewlett, ciii-civ, 660, 660n92
Johnson, Walter, 261-62, 261n29, 463, 467
Johnston, Wyatt G., lxxxvi, 664

Kalz, Fred, 222n101, 536
Kalz, Gertrude (*Appears throughout volume, limited references listed here*): lxxii, lxxvii, lxxxin335, lxxxii, cii, ciin463, 190, 222, 222n101
Kantrack, Alfredo, xxxix, xxxixn121
Kauffman, Fritz, 745, 745n38
Keefer, Chester Scott, 392, 392n74
Keeping, Eleanor Silver (Dowding), 687, 687n151
Kellenberger, Edouard, 729n13, 731, 731n19
Kelly, Cliff, 298, 298n95, 391-92, 591, 670
Kelser, General Raymond A., lxxi, 117-119, 118n43, 121-22, 138, 148, 150-51
Kendrew, John, xcvn420, xcvn422
Kennedy, Frank, 251, 251n16, 254-55, 514, 547, 632
Keys, David, lxvii-lxviii, 89-91, 89n4, 359, 361, 694, 694n170-71
Kinch, Pat, 723, 723n6, 756-57
Kinch, Robert, 723, 723n6, 756-57
King, Don, cxiv
King, William Lyon Mackenzie, lvn206, cviiin493, 247, 247n2, 257, 352, 355
Kipling, Rudyard, 404n88
Kluyver, Albert Jan, 546n7, 696, 696n173
Knaysi, Georges A., 586-587, 586n69, 639
Knight, Betty, 735n25
Koch, Robert, xxxiii, xxxv, xxxvn93

Lac Nomininigue (Quebec), lxvii, 155, 611
Landsteiner, Karl, 294-95, 295n91
Larmonth, Jonathan, 14n26
Lederberg, Joshua, xcn388, xciv-xcv, xcvn416, xcvin426, ci, 583, 583n66, 584, 754, 754n54
Leopold, Mitzi, 301, 301n89

Lesage, Jean, cxi, cxin515
Lewis, D. Sclater, 239, 239n131
Lewis, John, 403, 403n87
Lichte, Delia, 271, 279, 324, 324n119
Ling, George, civn477, 326, 330, 333
Lister, Joseph, xxxiii, 34, 60n97
Lister, Sir Spencer, xxxn66, lin189
Lister Institute, ln178, lxii, lxxxviiin377
listeria, and Everitt Murray, research, xxii, xxxn66, li, lin189, liii, lxii, lxxxi, lxxxv, 46-48, 537, 539, 543-45, 550, 555, 586, 596-97, 653, 661, 661n96, 667, 681, 747-48, 747n42, 750-51
Lochhead, Allan Grant, 109, 610, 610n97, 641, 669, 687, 691, 694, 704
Loeb, Lazarus J., 223n103, 481-82, 481n48, 486-88, 490, 511, 521, 523, 525-26, 549, 555
Lominsky, Iwo, 762-63, 762n70
London Hospitals Act, 1893 (Britain), xxxvi
London School of Hygiene and Tropical Medicine, xxxv, liiin196, lxxxviiin377
Lower Canada College: and Robert Murray, education, lxiv, lxvi
Luria, Salvador E., xviiin17, xcii, xciin401, xciii, xciiin405, xciv, xcivn411, xcvi, 548, 639, 639n46, 659, 659n90
Lwoff, André Michel, 688, 688n155
Lyman, Warren S., 241-42

Maaløe, Ole, 753-54, 753n50
Maass, Carol Edna Robertson, 311n102
Maass, Otto, lxi, lxviiin268, lxxn277, lxxin289, ciiin471, 107, 107n30, 108, 110, 112-16, 119-125, 133-34, 136, 138-40, 147, 311n102, 395, 631
Macallum, Bruce A., 314, 314n108

MacDermot, Hugh Ernest, 716, 716n206
Mackenzie, C.J., lxviiin268, 100-01, 100n18, 103-04, 112-13
Mackenzie, David, 378, 378n58, 386
Mackenzie, Norman A.M. (Larry), 756, 756n58
Mackie, Thomas Jones, 648, 648n70, 654
MacKinnon, Jean, 376-77, 377n57
MacLennan, Hugh, lxivn251, 79
MacLeod, J.J.R., lxxviiin317, 66n106
MacLeod, J. Wendell, lxxxiiin346, 700-01, 700n181
MacPherson, Kay, 85, 340, 436, 436n118
Maitland, K.B., 10-12, 11n21
Mankiewicz, Edith, 692, 692n164
Marchand, Erich, xxin29, 218n94, 264n55
Marchand, Grace Blair Wilkinson (Mother Grace), xxi, 218, 218n94, 234, 248, 250-51, 251n16, 254, 255, 257-258, 260, 265, 404, 406, 418, 450, 483, 485-486, 506, 527, 531, 565
Marchand, John F., xxin29, 218n94, 404, 406, 411, 620n111
Marchand, Richard Werner, xxi, xxin29, 187-88, 218n94
Marion, Leo, 681, 681n133, 682-83, 689-91
Marsh, Leonard, lixn229
Marshall, Francis (Tibby) H.A., xlvi, lxxv, 162, 517, 517n89, 518
Martin, Charles, liv-lvi, lviin215, lix-lx, lxvii, 10, 10-11n20, 12-13, 14n26, 18-19, 24, 32-33, 41
Martin, Médéric, lviii, lviiin225
Massey Commission (Royal Commission on National Development in

the Arts, Letters and Sciences, lxxix, xcix, cvii, 664
Massey, Vincent, lvn206
Masson, Pierre, 471-72, 471n35
McDonald, Ian J., 594, 594n82
McGill University: and Department of Bacteriology and Immunology, development of, xvi-xvii, lix-lxi, 31-36; and Donner Research Institute, 289n85; and Everitt Murray, appointment, xxxvii, liv, lvii-liv, 10-30; and Everitt Murray, "magic serum" airlift, lxi-lxiii, 42-43; and Everitt Murray, retirement, cviii-cix, 695, 697-700; and "Greenhouse Follies", lxvi; and University Senate, xvii; and Robert Murray, education, lxiv, lxxvi-lxxvii, 179-190; and Robert Murray, offer to succeed father, cviii, 710-713; and Rockefeller Foundation grants, lv, lvn208, lvi; and Second World War, lxvii-lxviii; and University Club, lxiv, lxvi; and War Advisory Board, lxvii
McFarlane, D. Cecil, lxxviiin316
McIntosh, J.F., 161, 161n14, 238-39
McLachlin, A.D., 299-300, 299n96, 303-04, 383, 392
McLean, Norman, 20, 20n33, 312, 312-13n105
McNaughton, General A.G., lxvii, 66-68, 67n109, 100-01, 100-01n18-19, 132, 132n68, 134n70. See also National Research Council (Canada)
McQuillen, R.A., 726, 726n10
Meakins, Jonathan, lvi, lxi, lxin240, lxxiin288, lxxxin334, 11n20, 14, 14n26, 95n12, 142-43, 160, 189, 262, 275, 354, 357, 378, 383, 405, 412, 470, 611, 650, 650n74, 661n94

Means, James Howard, 274, 274n71
Medawar, Peter B., 249n10, 685n143
Mellanby, Lady, 621, 621n117
Mellanby, Sir Edward, 621, 621n117
Mendel, Bruno, 65, 65n105
Meningitis, xxii, xlii-xliv, xlivn146
Menshikov, I.I., cxii
Merck, George, lxxi, cxn510, 123, 123n51, 151, 151-52n101, 287n83, 290
Mintun, Herbert Delwyn (Del), 189, 227, 227n108
Miles, A. Ashley, li, lin186, xcviii, 389n68, 625, 625n4, 638, 640-41, 643, 688, 699, 736
Miller, Rosario (Rosie), 279, 279n77, 308, 652
Mitchell, Charles A., lxix, lxxxiiin347, lxxxv, lxxxvn361, xcvii, ciiin471, 109, 114, 114n36, 115, 118, 123, 132-35, 153, 392, 425, 429, 447, 453, 489, 573, 597, 688, 688n153, 715
Munroe, Eugene, 249-50, 249-50n11, 584, 629
Montreal: cultural organizations, lxiv-lxv; recreational clubs, lxv; society, lvii-lix, lxiv-lxvii
Montreal Canadiens, cv, 26n48
Montreal City Council: and Everitt Murray, xvii, civ-cv, 441-42, 446, 458-59-461-62, 465-67, 493, 502, 519, 550 558, 568, 572, 580, 603, 651, 661, 665, 692-93, 714, 716-17, 778. See also Camillien Houde.
Montreal Medico-Chirurgical Society, lxxxvi, 161n14, 238-241, 428-29, 703, 763, 776, 779
Montreal Neurological Institute, lvi, lvin213, lx, 190, 343n138. See also Wilder Penfield.

INDEX

Morgan, Councilman, 689, 689n159
Morganti, Odesca, 692, 692n163, 702, 707
Morley, Thomas, xviiin18
Mount Desert Biological Laboratory, xxi, lxxvi, 85, 185, 185n45
Mudd, Emily Hartshorne, 620n114
Mudd, Stuart, 475, 475n38, 620, 620n114, 656
Murray, George Alfred Everitt, xix, 741, 774
Murray, Alfred Thurtell, liiin198, 773
Murray, Alice Blair (*Appears throughout volume, limited references listed here*): xxn23, xxi, lxxvi, cv, cvin487, cxvn529
Murray, Allan (Lal) Cameron Everitt, xx, xxvii-xxviii, xxviin56, xxviiin59-60, 28, 28n51, 774
Murray, Doris Marchand (*Appears throughout volume, limited references listed here*): and artistic and academic interests, xxi, cxiv-cxv, 319, 334, 355, 358, 360, 410, 682, 688, 694, 717, 752, 786-88; and marriage to Robert Murray, xxi, 185, 187-188
Murray, Frank Everitt, xxvi, 774
Murray, George Alfred Everitt., xix, xxiv-xxx, xxxii, xxxvin101, xlii, xlv, lin189, cxiii-cxiv, 28, 28n50, 28n52, 29n53, 95n13, 164-68, 741, 774
Murray, Gordon, 240, 240n132
Murray, Harriet Winifred (Freda) Harwick Woods (*Appears throughout volume, limited references listed here*): and Bark Lake, 78, 271, 274, 276-77, 281; and Everitt Murray, xlv-xlvi, liii, 2-3, 190-195; and First World War experiences, xlvn154; general, xx, liiin197; and Imperial Order of the Daughters of the Empire, Montreal Chapter (I.O.D.E.), lxiii, 197-204; and organization memberships, cxxi-cxxii; and poetry, xx-xxi, 782-85; and Second World War activities, lxxii-lxxiii, lxxiiin293, 197-204; and South Africa, xxxi, cxiii-cxiv; and travel, cxiii, cxiiin524
Murray, Kathleen (Kate) Dunne, xx, xxvi-xxviii, xxixn65, xlvi, xlvin158, 28n52, 774
Murray, Marion Luney, cxv, cxvn529, cxxv, 775
Murray, Mary Everitt, liiin198, 773
Murray, Peter Everitt (*Appears throughout volume, limited references listed here*): xxi, liiin198, cv, cvin487, cxvn529, cxviii, cxxv, 289-90
Murray, Roger (Rory) Hi-Regan Everitt (*Appears throughout volume, limited references listed here*): xx, xxvii, xxviin56, xxviiin58-60, xxx-xxxi, xxxin75, xlv, lviin221, cvn483, cvi, cvin485-86, cxiii, cxiiin523, cxiv
Murray, Susan Ann (*Appears throughout volume, limited references listed here*): xxin27, lxvin258, lxxiii, lxxiiin291, cix, cxiii-cxiv
Murray, Thorkill (Thor) Howard Everitt, xx, xxviii, xxviiin59-60, xlv
Murray, Thomas Everitt (*Appears throughout volume, limited references listed here*): xxi, cv, cvin487, 470, 470n33, 496
Murray, Unita (Nita) Thorburn, xxviiin59-60, cvn485, cxiii, 164n21, 529, 529n94, 690
Murray, Walter Everitt, xxiii, xxvi, xxviii, 29, 167, 773-74

Murray family: and Bark Lake cottage, cvi, 78; and Cambridge life, liii-liv,8-10; and Everitt Murray, death, cxii-cxiii; and Everitt Murray, retirement, cviii-cix; and Everitt Murray, professional achievements and honours, 776-79; and Everitt and Robert Murray, professional advice and shared research interests, lxxxix-xc, c-ci; and fishing and other outdoor activities, lxv, lxvn256, lxvi-lxvii, lxviin, 264, 8, 27, 31, 155, 183, 276-77, 279-80, 283-84, 287-89, 299-301, 313, 351, 376, 426, 476-77, 522-23, 529-31, 535, 605-606, 611-12, 701-02; and meningitis, Robert Murray attack, lxxxiiin345, 598-606; and moving and adjusting to Canada, liv-lviii, lxiv-lxvii, 10-31; and Robert Murray, birth of, xlv, 26n45; and Robert Murray, early education, liv, lxvi, 8-10; and Robert and Doris Murray, move to London, 204-207, 216-237; and Robert Murray and Doris Marchand, courtship and marriage, lxxvi, lxxvin308, 168-90; and 1951 Royal Visit, cvii, 615-19; and Second World War, lxxii-lxxix, 155-204; and social and political issues, cii-civ; and post-Second World War planning, cv-cviii, 204-237

National Collection of Type Cultures (Lister Institute), lxii, cn449, 61n100, 346, 403, 535n99, 699
National Foundation for Infantile Paralysis (U.S.), lxxxviiin377, cix, cx
National Institutes of Health (U.S.), lxxx, lxxxin336, xciin401, cviii-cix, 340n137, 693n167
National Insurance Act, 1911 (Britain), xxxvi, xxxvin99, xxxix
National Research Council of Canada: and Everitt Murray, NRC Associate Committee on Control of Hospital Infections, lxxxiv759, 779; and Associate Committee of Medical Research, 92-94; general, lxiv, lxvii-lxix; and Robert Murray, early applications, xci
Needham, Joseph, lxxiv-lxxvn300-301, 162, 266
Nègre, Léopold, 579, 579n61
neisseria (meningitis): general, 588, 746; and Everitt Murray, xlvii, lxxxv, cii, ciin466, cxiin517, 2, 693n167; and Robert Murray, 753, 753n49
Newcombe, Howard B., 596, 596n84
Newman, Bill, 155, 310, 313, 373-74, 392, 400, 488-91, 576, 627-28
New York Botanical Garden, lxxxviiin377, 326n122
Nickerson, Walter John, 683, 683n138
Nicolle, Charles, 5n9
Nicolle, Maurice, 5, 5n9
Nobbs, Percy Erskine, 381, 381n59
Nobel Prize, xxxn66, xxxiii, xlix, xlixn174, xlixn176, ln182, lxviii, lxxviiin317, xcn388, xcicn409, xcvn416, xcvi, xcvin428, 5n9, 37n62, 66n106, 163, 249n10, 287n83, 295n91, 385n62, 469n30, 561n42, 578n58, 583n66, 620n113, 639n46, 685n143, 731n19, 754, 754n54
Noble, Robert L., lxxviii, lxxviiin316, 85, 402, 402n85, 417, 438n119, 540, 564, 605-06, 646, 673, 723
Notification of Diseases Act, 1889 (Britain), xxxvi

Nunes, Doris, 471, 471n34, 478, 478n43, 748
Nungester, Walter J., 589, 589n74
Nuttall, George, xlvi, lin185, 7, 7n12

Oertel, Horst, lv, lvin209, lxi, 10, 10n19, 12-13, 17-19, 24, 558
Ordal, Erling J., 528, 528n92
Order of the British Empire: and G. Edward Hall, 205n73; and Everitt Murray, xlvi, 2, 776
Orr, John Harland, 713, 713n200
Orr, Mort, 648, 648n69
Ørskov, Jeppe, 754, 754n53
Osler, Sir William, xvn8, xviii, xxxii-xxxiii, xxxiiin82-84, lvn206, 11n20, 189
Osler Society, xlvn152, 4, 274
Osnos, Max, lxii, 43
Owens, O.N., 304, 304n99, 314

Pamplin, Rosemary, 260, 260n47, 264, 269
Panisset, Maurice, 650, 650n73, 658
Pappenheimer, Alwin M., 97-99, 98n15
Parker, Raymond, 325n121, 329, 329n126, 331, 339-40
Parry, Sir David Hughes, 468, 468n27
Pascoe, Charles, 155, 155n2, 262, 424, 476, 489, 501-02, 504, 622, 701-02
Pasteur Institute, xviiin19, xxxix, xliii, xlv, c, 2, 5, 5n9, 58n94, 60n97, 346n141, 576, 579, 583, 656, 676, 688n155, 726n11
Pasteur, Louis, xxxiii, 5, 34
Patterson, James C., 217, 217n92, 228, 290, 326, 333-34, 383-84, 389, 406, 443, 460
Pauling, Linus, xviiin17, xcv, xcvn419, 290

Pearce, R.H., lxxxix, 297, 297n94, 299, 305, 312, 358, 438, 538n103
Pearson, James (Jimmie) Stuart Abercrombie, 281, 281n79, 668n107
Pearson, Lester B., lxxx
Penfield, Wilder, xviiin17, lvi, lvin213, lxi, lxin240, 11n20, 239, 239n130, 343n138, 761, 761n64
penicillin, xliiin142, ln182, lxxxivn356, lxxxv, cxn510, 57n92, 116n39, 152n101, 206, 228, 235n122, 253, 257, 263, 290n87, 291, 333, 370, 372, 393-94, 396, 464, 486-88, 513.15, 561n42, 564, 568, 674
Pennington, Derrol, 613, 613n103
Penso, Guiseppe, 561, 561n41, 692
Perutz, Max, xcvn420, 422
Philpott, N.W., 458n14
Pijper, Adrianus, 513, 513n82
Pirie, N.W., 46-47, 389, 389n68
Pontecorvo, Guido, 753, 753n51
Porter, Keith R., 683-84, 683-84n141
Potel, J., 653, 653n78, 661
pneumonia, xxxv, xlivn146, xlviii, lxi, lxin243-244, lxxxiv, xciv, 1, 18, 44-45, 44n71, 90, 98n15, 173, 218n94, 230, 235, 251n16, 257, 257n33, 259-60, 265, 271, 280, 297, 320, 324, 349, 358, 361, 515, 533, 541, 588, 632, 709n194
Prévot, André-Romain, 583, 583n65
Public Health Act, 1875 (Britain), xxxvi

Queen's University, lvn205, lix-lxn233, lxii, lxviiin269, lxix, lxixn275, lxxixn326, lxxxiiin347, 56n91, 73n115, 76n119, 104, 107, 140, 326n122, 435, 591, 611, 669n108, 705n186, 713n200, 730n15, 731n17

Ralston, J.L., 132n68
Ramsay, G.H., 289, 299-300, 299n96, 321
Reed, Guilford, xviiin17, xxiin32, lix-lxn233, lxii, lxviiin269, lxix, lxxn277, lxxxiii, lxxxiiin347-48, xcn382, ciiin471, cviii, cviiin499, 73, 73n115, 96-97, 98n15, 104, 104n25, 106, 109, 115-19, 123, 128, 138, 140, 193, 275, 319, 326n122, 396, 410, 429, 452, 474, 476, 503, 517, 539n104, 669, 673-74, 713-14, 730n15
Reed, Roger, 539, 539n104, 692, 698, 714, 720
R.G.E. Murray Lectureship, xiii, xiiin36
Renouf, Louis, 20, 20n34
Renwick, Major General Charles, 213
Reyn, Alice, 753, 753n49
Rhoads, C.P., 620, 620n112
Rhodes, Cecil, xxv, xxvn45
Rhodes, A.J., cin457, 58n94, 428, 428n109, 429, 458, 491, 497, 551, 668, 698, 709
Ribonucleic acid (RNA), xiv, xivn6, 556, 660
Richards, Guy, 485, 485n51, 487, 560
Rivers, Thomas, xviiin17, lxii, lxxxiiin338, 61, 63, 63n102, 64, 68-70, 313, 318, 497
Robertson, Hilary, 155, 155n2, 194, 202-03, 313, 366, 476, 571, 576
Robinow, Carl Franz (*Appears throughout volume, limited references listed here*): xc-xci, xcn386-88, xci-xcii, xcvi-xcvii, xcviin429, cn452, cxvi, cxvin531, 475-76, 475-76n40, 509-10, 516-18, 520, 523, 527, 532-34, 536-37, 539, 541, 544-49, 546n10, 556, 559, 564, 585-86, 594, 597-99, 601-04, 608, 613-14, 623, 629, 631-33, 636, 639-40, 656, 662-63, 677, 682-84, 686, 689, 696, 698, 706-07, 723, 731n19, 744, 758
Robinow, Rosy, 569, 569n46, 571
Robinson, Blake (*Appears throughout volume, limited references listed here*): cv, cvn481-82, cvin487, cxiv
Rockefeller Foundation, xvn8, xlix, xlixn177, lvn206, lvn208, lviin213, lx, lxn238, 158, 187-88, 328n124
Rockefeller Institute of Medical Research, xxi, xxxv, xlviii, lvn205, lxii, lxxxii, 19n31, 44-45, 44n71, 54n87, 60n97, 63n102, 64, 98n15, 187-88, 218n94, 270n66, 324n120, 498n64, 535n98, 579n60, 614n105, 620, 684n141
Rogers, Leonard, 7, 7n12
Romeyn, A.J., civ
Roodebloem Estate (Graaff Reinet region, South Africa): and xxiii, xxvi, xxvii, xxviii, xxxi
Roosevelt, Franklin D., 99n17
Rose, Sylvia, 187n50, 196, 230n114, 231, 342, 373, 376, 433, 462, 646
Rose, Ted, 187n, 187n50, 196, 331, 342, 373, 376, 433, 462, 464, 646
Rossiter, Helen, 403, 408, 565, 565n43, 635-36
Rossiter, Roger, lxxviii, lxxxiin343, cvi-cvii, 314-15, 315n109, 375, 390, 392, 403, 408, 424-25, 438n119, 505-07, 510, 515, 525, 543, 551, 554, 560, 562, 565n43, 580, 592, 605-07, 635-36, 638, 642-43, 673, 676, 686, 688, 695, 731
Rous, Peyton, 620, 620n113
Roy, Charles, xxxix
Roy, Theodore (Ted) E., lxxxiiin348, 55n89, 85, 191, 268, 271, 275, 291, 339-40, 445, 449-50, 452, 551, 627, 681,

687, 697-98
Royal Army Medical College (U.K.), cxxiii, 1, 4-5
Royal Army Medical Corps (R.A.M.C.), xix, xlii-xlviii, xliin135, xliiin142, cxxiii, 1-2, 7, 22n38, 60n96, 77
Royal College of Physicians and Surgeons (Canada), cxxiii, 240n131, 241-244, 325n121, 371, 653, 776
Royal College of Physicians and Surgeons (U.K.), xxiv, xxixn61, xxxii, xxxviiin111, xxxixn121, xlviii, 3n2, 164n23, 165-68
Royal Commission on Health Care (Emmett Hall Commission), lxxx, 759
Royal Commission on the South African War, xlii
Royal Commission on Tuberculosis, 1901, xxxix
Royal Society of Canada: and Flavelle Medal, xxii, lxxxvi, cxv, 143n84, 610n97, 655n86, 665-67, 777; general, lxii, lxxviiin317, 59, 59n95, 67n108, 68, 73n115, 107n30, 114n36, 143n84, 249n11, 260, 272, 405, 457, 465, 476, 580, 610n97, 625n3, 638, 655n86, 626-63, 665-66, 669; and Harrison Prize, xcvii-xcviii; and Joburg and Robert Murray, cxxiii, 666, 672, 693, 730, 741, 776-77, 780
Royal Victoria Hospital, xv, liv, lvi, lviin213, lix, lixn231-232, lxi, lxin239-240, lxin243, lxxvi-lxxvii, cxxiii, 14n26, 15, 18n29, 31, 43, 55n89, 146, 189-90, 211, 216, 222n101, 239n131, 295n92, 320n116, 324n120, 325n121, 345n140, 458n14, 650n76, 714n201, 777

Rublee, Jack Driscoll, 717, 171n210
Rutherford, Ernest, xviiin17, xlix, xlixn175, lviin215, 37n62
Ryle, John A., xl, lxxiv, lxxivn298, 385, 385n62, 387
Ryle, Martin, 385n62

Salton, M.R.J., cxv-cxvi, 735
Sauer, Max, 262, 262n49, 554, 554n30
Saunders, J.T., 7, 7n13, 9, 162
scarlet Fever, xxxv, 351n9, 487, 490, 674
Scrimger, Lt. Colonel Francis A. C., 197-98, 202, 202n68
Scriver, Walter, 155, 476
Schatz, Albert, 287, 287n83
Scherp, Henry W., 640, 640n49
Schlesinger, R. Walter, 639, 639n47
Schoenfeld, Henrik, 552n28, 615, 762, 762n69
Schrödinger, Erwin, xcii-xciiin408
Schrum, Gordon, 756n59
Scott, Frank, lixn229
Seale, Elizabeth, 549, 549n18, 553
Second World War: and Canadian biological weapon research, lxviii-lxxii, 87-153; and Canadian universities, contributions of, lxvii-lxix; and Defence Research Board (Canada), ciii, ciiin471, 153, 153n104, 355n21, 377, 629, 631, 669, 694, 715, 717-718, 777; and Directorate of Chemical Warfare and Smoke, lxix, lxx, cxxii, 106-07, 107n30, 131, 141, 357; and Everitt Murray and biological weapons program, xix-xx, lxvii-lxxii, lxxiin286; and Everitt Murray, United States Medal of Freedom, lxx-lxxi, 152, 423, 434-36, 442; and Robert Murray, Royal Canadian

Army Medical Corps, xliin135, xliiin144, lxxvi, cxxiii, 217n92, 204-224; and veteran education assistance, lxxxviiin375, 311-12. See also, Murray family.
Selye, Hans, lvin212, 635, 635n33
Sertic, V., 328, 328n124
Shaw, Norman, 359, 359n28, 694
Shaw, Robert McLeod, 498-99, 499n66
Shearing, Marge, 741, 741n34, 773
Sherman, J.M., 267, 267n61, 299
Sherrington, Sir Charles Scott, xxxiiin83, xlixn176, ln182, lviin217, 469-70, 469n30
Shipley, A.E., xlvi, 312-13n105
Shope, Richard, lxxi
Simpson, J.C., 91-92, 91n5
Sketch, Mary, 449
Sketch, Ralph, 353-54, 353-54n16, 449
Skey, A.J., 130-32, 130n60
Sloan-Kettering Cancer Research Center, 324n120, 620n112
Smith, Frederick E., lin186, lx, lxn238, lxxii, lxxvi-lxxvii, lxxviiin318, lxxxiin334, cviiin498, 54, 54n87, 58, 85, 109, 145-46, 231, 235, 259, 259n43, 260, 291, 298, 310, 324, 346, 363, 372, 374, 385-86, 386n64, 387, 390-91, 413, 419, 424, 435, 445, 533, 533n95, 549, 554, 557, 611, 650,
Smith, Ivan, 283n81, 301-02
Smith, Nathan R., 411, 411n97, 464, 573, 607, 640-41, 651
Smuts, Jan, xxxn69, 484, 671
Snow, Charles Percy, li, lxxv, lxxvn302, 162
socialized medicine, xix, 432-33, 513-14
Society of American Bacteriologists (SAB): general, xvn9, xvi, xvin12, lxxi, lxxxi, lxxxi-lxxxiin337, lxxxvi, 22n37, 244n138, 267n60, 589n74, 640n49-50, 683n140; and *Journal of Bacteriology*, xcvii, xcviin434-35, 71, 75, 589-90, 780; and Everitt Murray, 46, 56, 70-75, 92-93, 256, 260, 271-72, 405, 407, 465-66, 523, 555, 597, 702, 706, 776, 779; and Robert Murray, xcvii, 253, 374, 394, 409-11, 464, 468, 495, 520, 547-48, 551, 561, 589, 590n75, 595-96, 598, 614, 634, 638, 641, 780
Society for General Microbiology, 61n100, 776
Solandt, Omond McKillop, 153, 153n104, 357, 380
South Africa: and Anglo-Boer War, xix, xxvii-xxviii; and Everitt Murray, xiiin4, xxvii-xxxi, xxixn65, xliv, lvii, lxiv, 2-3, 663; and institutionalized racism, xxix, cvi, cvin485-86, 670-72; and medicine, xxiv-xxvi, xxvn46-51, xxxn66, 47, 252-53, 318; and Murray family background, xxiii-xxxi, xxxin74, 166-68; and politics, xxiiin37, xxxi, 151, 670-72, 720; and South African Medical Association, xxv-xxvi; and veterinary diseases, xxiii, 722
Sonneborn, Tracy Morton, 583, 583n68
Southey, Anna Elizabeth, xxiii, 167, 779
Southey, Richard, xxiii
Southey, Robert, xxiii
Spanish Civil War, lixn229
Spanish influenza, xlvii-xlviii, xlviin166-167
Spooner, E.T.C., li, lin186, lxxv, lxxxiin342, 162, 162n18, 374, 380
Stamp, Lord Edgar, lxxin282, lxxiin286, 120, 12n46, 124-25, 128, 139

INDEX

Stanier, Roger: general, lxxvii, lxxxvii, xcivn414, cx-cxin510-12, 367-68, 411, 545-46, 621; and Robert Murray, scientific co-operation, cx-cxi

Stanley, Carleton, 34n59

Stanley, Neville F., 537, 537n101, 539, 543, 545, 550, 555, 701

staphylococcal infections, xxii, xxii-xxiiin34-35, xxxvi, xlivn146, xlviii, lxi, lxxxiv, 45, 55n89, 61, 90, 250, 250n13, 251, 253, 259, 265, 272, 297, 323n122, 329, 311, 333-34, 372, 394, 400, 458, 461, 464, 477, 480-81, 499, 511, 523, 549, 588, 637, 692-93, 740, 762-63

Starkey, Hugh, lxxxin335, 31, 54-55, 54n88, 222n102, 438, 440, 626, 698

Statens Seruminstitut (Copenhagen, Denmark), 310, 705n187, 742-58

Strathcona, Lord, lvin210

Stavraky, George W., 248, 248n5, 314, 317, 421-22

Stern, Karl, 626, 626n8, 628

Strelitz, Frieda, lxxxix, lxxxixn378, 232, 235, 326, 399, 399n79, 415, 421, 479, 482

Stevenson, Ian P., 184, 184n44, 187, 196, 290, 297, 325, 435

Stevenson, J.W., lxxii, lxxviii, lxxviiin316, lxxxiiin348, 85, 275, 275n74, 278, 290, 297-98, 590, 590n77, 643, 657, 670, 698, 698n175, 718-19

Stevenson, Lloyd, 628, 628n15, 688, 691, 700

Stiling, F., 731, 731n18

Stockdale, David, 9, 155, 162

Stockdale, Joan, 9

Stoker, M.G.P., 763, 763n71

streptococcal infections, xxxvi, xli, xliiin142, xlviii, lxxxvn363, xci, 4, 19n31, 54, 57, 90, 96, 173n32, 206, 223n103, 252, 254-55, 257n36, 258-59, 265-67, 269, 275, 286-87, 291, 297, 299, 327, 358, 395, 398-99, 427, 452-53, 464, 467, 480-81, 486-88, 490, 497-99, 498n64, 513, 515, 521, 549, 566-68, 576, 588, 661, 674, 691

streptomycin, xcvin428, 287n83, 322, 324, 392-94, 396, 430n113, 683

St. Bartholomew's Hospital: and Abernethy Prize in Surgery, xxiv; and Brackenbury Surgical Scholarship, 164n23, 164; and British medical education, xxxi-xxxii, xxxiin78, 315, 322, 325; and Everitt Murray, xxix, xl-xli, xlvi, cxxiii, 1-4, 776-77; general, xxxixn121, xliii, xcn386, 3n2, 4n6-7, 156n3, 305, 535n98; and George Murray, xxiv, xxix, 164-68, 741, 774; and Robert Murray, 156-57

St. John-Brooks, Ralph Terence, lxii, 61, 61n100, 62-63, 142-43, 245-46, 346, 348, 445

St. Laurent, Louis, xvii, cviin493, 352

Summerfields boarding school, liv, 9, 79

Swann, Meredith, l-li, lin188, 661, 661n96

Swift, Homer, 498, 498n64

syphilis, xxxiiin84, xxxvin97, 18n29, 51-52, 116n39, 253n26, 498n64

Szilard, Leo, xviiin17, xcvin423

Talman, James J., xxxn67

Tatum, Edward, xciv, xcv, xcvn416

tetanus, xliii, xlvii, 22n38, 73n115, 90, 116n39, 317-19, 351, 351n9, 370, 372, 399, 458, 464-67, 498, 514

TeWalt, Louise, 411, 411n96, 419

Theiler, Max, xxxn66

Thomson, David, lvin212, 468, 635, 635n35, 637, 674, 674n115
Thompson, Charles, 483, 483n50
Thompson, Jim, 566-67, 566n45
Thompson, Peggy, 566-67, 566n45
Tulloch, W.J., xlvii, xlviin163, 13, 22, 22n37-38
Topley, W.W.C., xxxv, xxxvn95, xlvii, xlviin162, xlviin166, liiin196, lvi, lxxin282, lxxxviiin377, xcviii, 11, 11n22-24, 12, 25, 422, 422n104
Treffers, Henry P., 498, 498n65
Truant, J.P., 638, 638n42, 645, 668, 688
tuberculosis, xxxv, xxxix, lii, lvin210, lxxxv, xcvin428, 38, 73n115, 189, 223n103, 250, 259n39, 321, 346n141, 369, 369n46, 371-73, 373n48, 405, 468n28, 574, 579, 602n89, 727n12
Tymms, Sir Frederick, 667, 667n104
Typhoid, xli, xlii, xliii, xlvii, lxv, lxvn255, lxxxviiin377, 5, 5n9, 38, 126, 235, 322, 351, 409n93, 692, 701

Uffen, R.A., 731, 731n17
Uncle Fudge, lxvii-lxviiin262, 260, 260n45, 288, 304, 306, 343, 350, 387, 389, 412, 429, 441, 448-49, 504, 506-07, 542, 585
University of Cambridge: and British medical education, xxxvi-xxxix, lxxiv; and Cavendish Laboratory, xlix; and Christ's College, xxix, xxxvii, xlvin160, li, lxiv-lxvii, lxivn250; and "The Eagle", 390n69; and Everitt Murray, xxix, xxxi, xxxvi-xxxix, xl, xlivn147, xlvi, xlviii-xlix, l-liv, lxxxii, 1-3, 6-7; general, lvii; and Molteno Institute of Biology and Parasitology, 163n19; and Natural Sciences Tripos, xxxi, xxxvii, xli, xlix, ln179, ln184, li, lx, lxxiv-lxxv, 7, 14, 23, 81, 83, 160-62, 172, 211, 3232; and Pathology Department, xiv-xv, xxxix, xlix-liv, lvii; and Robert Murray, education, xx, xxxviin102, lxiv, lxxiv-lxxvi, 155-64, 168-79; and Second World War, impact, lxxv, 163; and Strangeways Research Laboratory, xcn386, 475, 475n40, 517, 621, 641
University of Montreal, lvn208, lxvi, lxxixn326, lxxxiv, lxxxivn353, 373, 373n48, 335n33, 650n73, 674, 720, 776
University of Oxford, xxxiii, xln127, ln178, ln182, lvii, lviin217, lxxxii, 9, 30, 116n39, 158, 315, 315n109, 375, 385, 385n62, 387, 437, 439, 561, 561n42, 565n43, 574, 574n54
University of Toronto: and biological weapons defence research, lxix; and Connaught Laboratories, lxii, lxxxiii, xcvii, cx, cxn508, 22n37, 42, 54n85, 56n91, 57n92, 58n94, 65n104, 96, 106, 290n87, 329n126, 350, 351n9, 428n109, 432, 432n114, 463n22, 495, 497, 497n61, 551, 551n20, 601n88, 625, 625n2, 648n68-69, 655
University of Washington (Seattle): and Robert Murray, research visit, xc, xcn384, 507-08, 507n76, 510-11, 514, 516, 526-28
University of Western Ontario (UWO): and Chiron Club, cvii, 438n119, 594, 691; and Department of Bacteriology and Immunology, Robert Murray appointment, xvii, lxxviii, 208-216; and Faculty of Medicine, xv, lxxvii-lxxviii, lxxviin314, lxxxviii-lxxxix; and Robert Murray, Department of

Bacteriology and Immunology, lxxix, lxxxvii-xcii, xcvi-xcvii, xcvin427; and Robert Murray, appointment as chair, xc, xcn385, 512-14, 538; and Robert Murray, assistance to Igor Asheshov, cix-cx; and Robert Murray, honours and scientific legacies, cxv-cxvi, 779-81; and Robert Murray scientific partnership with Carl Robinow, xc-xcii, xcvii; and Robert Murray, public lectures and radio talks, xcviii, xcviiin437, 332, 360, 374, 380, 394; and Robert Murray, editorial work, xcvi-xcvii, xcviin435; and Rockefeller Foundation grants, lvn208

Vanier, Georges P., cxi
van Iterson, Woutera, 486-87, 486n53, 489-90, 509, 513n82
van Niel, C.B., lxxxvii, xciv
van Rooyen, C.E., cin457, 497, 497n61-62, 541, 551, 612-13, 621, 627, 629, 631, 648, 687, 698, 715
veterinary medicine, xvi, lxx-lxxi, lxxxv, 7n12, 118n43, 151, 291, 346, 348, 354, 356, 359, 363, 366, 488, 572-73, 687
virology, xiii, lxxxi, cin457, 63n102, 307, 405-06, 497n61-62, 611
Vivian, R.F., lxxxi, lxxxin333, 354, 354n18, 405, 555, 557
Vollum, R.L., 574, 574n54
von Magnus, Preben C.A., 747, 747n40
von Wassermann, August, xxxiiin84, ln178

Waksman, Selman, xcvi, xcvin428, 287, 287n83, 452
Waksman Institute (Rutgers University), xcvi, 287n83, 682-84

Walter Reed Army Institute of Research, 389n67, 708n193, 748n43
Wansbrough, V.C., 79, 82
Warburton, Cecil, 4, 4n5
Warwick, O. Harold, cixn502, 730-31, 730n16, 752, 757
Watson, James, xviiin17, lxxvn304, xcivn411, xcv, xcvn20-21, ciin462, 163, 396, 689
Webb, Robert, l-li, 661n96
Webster, Leslie, xxxv
Weed, Lewis H., 434, 434n116
Wernicke, Erich, 470, 470n31
Whitfield, James F., xcin396, xcii, 590, 590n76, 614, 652, 690, 695
Williamson, Norman, 316, 316n100
Wilson, G.S., xlv, l, liii, liiin196, cxii, 3, 11, 11n22-24, 13, 422, 422n104, 485
Wilt, J.C., 502, 502n73
Wishart, F.O., 432, 432n114
Wolman, Lionel, 588, 588n72
Woodhead, German, xxxix, xxxixn121
Woods, Thomas Cecil Hardwick, xlv, liiin197, 84n132
Woolwich Military College, xxviii
World Health Organization, xix, 396n46, 727n12
Wright, Sir Almroth Edward, xliii, xliiin142, xlvii, lxxxiin344, 176n35, 409, 409n93
Wyckoff, Ralph W.G., xciin401, 509, 579, 579n60, 633-34, 636, 641
Wynne-Edwards, Vero C., 267, 267n58, 291, 573

zoology, xxi, xxviii, xxix, xxxi, xl, liv, 4, 7, 7n13, 79-80, 82-85, 210, 218n94, 267n59, 276, 291, 468n29, 501, 515, 540, 666

The Champlain Society

OFFICERS OF THE SOCIETY, JANUARY 2007

Honorary President
John Warkentin, OC, PhD, LLD, FRSC, York University, Toronto, Ontario

President
Michael Moir, MA, University Archivist & Head, Special Collections, York University, Toronto, Ontario

Honorary Vice-Presidents
Frederick H. Armstrong, PhD, FRHistS, London, Ontario
George W. Brigden, QC, Toronto, Ontario
The Hon. Alastair Gillespie, OC, PC, MComm, Toronto, Ontario
Conrad Heidenreich, PhD, Toronto, Ontario
Morris Zaslow, PhD, FRSC, London, Ontario

Vice-Presidents
Patrice A. Dutil, PhD, Toronto, Ontario
Allan Hoyle, MA, Newmarket, Ontario
Frits Pannekoek, PhD, Calgary, Alberta
Germaine Warkentin, PhD, Toronto, Ontario

Secretary of the Board
William Moreau, PhD, Toronto, Ontario

Treasurer
John T. Pepall, LLB, Toronto, Ontario

Members of Council
Jeanne R. Beck, PhD, Dundas, Ontario
Sarah Carson, MISt, Toronto, Ontario
Sarah Carter, PhD, Edmonton, Alberta
Bryan P. Davies, MPA, Toronto, Ontario
Olive P. Dickason, OC, PhD, Ottawa, Ontario
E. Jane Errington, PhD, Kingston, Ontario
Doug Gibson, Toronto, Ontario
A. Ian Gillespie, MBA, Ottawa, Ontario

Roger Hall, PhD, Toronto & London, Ontario
Robynne Healey, PhD, Langley, British Columbia
Bosko D. Loncarevic, PhD, Bedford, Nova Scotia
The Hon. Roy MacLaren, PC, MDiv, LLD, Toronto, Ontario
Donald W. McLeod, MLS, Toronto, Ontario
Carolyn Podruchny, PhD, Toronto, Ontario
Lutzen Riedstra, BA, Stratford, Ontario
William A. Waiser, PhD, Saskatoon, Saskatchewan
J. David Wood, PhD, Toronto, Ontario

General Editor

Roger Hall, PhD, University of Western Ontario, London, Ontario

Secretariat

Christina Becker, MBA

The Champlain Society Office

10 Morrow Avenue, Suite 202
Toronto, Ontario M6R 2J1
Tel. 416-482-9635
Fax 416-482-9341
info@champlainsociety.ca
www.champlainsociety.ca

The Champlain Society was founded in 1905 to publish original Canadian historical documents. Between 1907 and 2007 it has published more than a hundred casebound books and several softcover Occasional Papers. The 2007 publications was *The Letters of Adam Hope, 1834-1845*, edited by Adam Crerar, and in 2006 *Champlain and the Champlain Society: An Early Expedition into Documentary Publishing*, by Conrad Heidenreich of York University, commemorating the 100th anniversary of the formation of the Champlain Society. A limited number of copies are available for purchase by members of the public.

Our website at www.champlainsociety.ca carries a list of all publications. The website also carries information on the availability of back issues for purchase and provides access to eighty-three Society publications which are displayed in our digital collection in a full text, searchable format. (Please note that volumes still available for purchase from the Society are not customarily included in this digital collection.)

DONORS 2007

ORDER OF GOOD CHEER

ASTROLAB
Pringle, Andrew M., Toronto ON

CAPITAINE
Bethell, Dr. Walter, Port Medway NS
Roy, Curtis L., Bloomington MN USA

DONORS

Brown, Rodney C., Hamilton ON
Delworth, Mrs. W. T., Ottawa ON
Fraser, John A., Toronto ON
Johnston, C. Fred, Kingston ON
Loncarevic, Bosko D., Bedford NS
MacDonald, Don, Edmonton AB
MacFarlane, Anthony L., Hamilton ON
M'Gonigle, Shelagh, Ottawa ON
Saunders, Robert E., Scarborough ON
Smiley, Trustee, J. J., Eureka MT
Smye, Mrs. A. M. L., Portimâo, Portugal
Wilson, Ian E., Ottawa ON

INDIVIDUAL MEMBERS 2007

LIFE MEMBERS
Crooke, Kenneth W., Toronto ON
Crowley, Terry, Guelph ON
Friedman, Frank A., Robesonia PA, USA
Gillespie, A. Ian, Ottawa ON
Gmoser, Margaret G., Banff AB
Gough, Prof. Barry M., Victoria BC
Pringle, Andrew M., Toronto ON

CANADA
Aagaard, A. Kim, Newmarket ON
Abel, Kerry, Ottawa ON
Abler, Mr. Thomas S., Waterloo ON
Adamchick, Tom, Eganville ON
Agnew, J. Gurd, Toronto ON
Agranove, Dr. Larry M., London ON
Aitken, Donald B., Toronto ON
Allenbick, Jeanette, Millbrook ON
Alston, Sandra, Toronto ON
Andrews, Tom, Yellowknife NT
Andrews, Robert James, Gananoque ON
Andrews Sayle, Timothy, Toronto ON
Anglin, F. M., Ottawa ON
Angus, John F., Beaconsfield QC
Angus, Stephen F., Erin ON
Arkin, Harold J., Toronto ON
Armour, Andrew, MD, Perth ON
Armstrong, Frederick H, London ON
Armstrong, Prof. C., Toronto ON
Arthurs, David, Winnipeg MB
Bacque, James, Toronto ON
Bain, James R., Toronto ON
Baird, Donald W., St. Catharines ON
Baker, Dr. Melvin, St. John's NF
Baker, Warren, Montreal QC
Baragar, Robert, Greely ON
Barker, Ruth B., Calgary AB
Barnett, H. J. M., MD, King City ON
Bartlett, Will, Ilderton ON
Barwick, Clifford H., St. Thomas ON

Baskerville, Peter, Victoria BC
Beasley, Tom, Vancouver BC
Beaulieu, Michel S., Thunder Bay ON
Beck, Jeanne R., Dundas ON
Beers, Donald W., Calgary AB
Benn, Dr. Carl, Toronto ON
Berry, Michael J., Vernon BC
Best, Janna Ramsay, Sudbury ON
Bethell, Dr. Walter, Port Medway NS
Binnema, Theodore, Prince George BC
Bird, Malcolm, Ottawa ON
Birks, G. Drummond, Montréal QC
Blackstock, Cicely, Toronto ON
Blain, Kerry, Sooke BC
Bliss, Prof. Michael, Toronto ON
Bloore, R. L., Toronto ON
Bohaker, Heidi, Toronto ON
Bourgeois, Donald J., Kitchener ON
Boxer, Richard, Toronto ON
Bradshaw, Graham, Milton ON
Brands, Andrew D., Toronto ON
Brennan, Terence, Montreal QC
Brierley, James D.M., Dunham QC
Brock, Daniel J., London ON
Brown, Rodney C., Hamilton ON
Brown, R. Craig, Toronto ON
Brown, Jennifer, Winnipeg MB
Brown, Ellen K., London ON
Bryan, Prof. Alan Lyle, Edmonton AB
Bryden, D. G., Guelph ON
Buchanan, Dr. Roberta, St. John's NF
Buckmaster, Dr. H. A., Victoria BC
Burnett, Frederick C., Upper Brighton NB
Burnett, Brandyn, Stouffville ON
Burns, Rev. L. D., Toronto ON
Caldwell, Robert C., Waterloo ON
Cameron, C. Jean, North Vancouver BC
Carson, Sarah, Toronto ON
Carter, Dr. Sarah A., Edmonton AB
Caya, Dr. Marcel, Roxboro QC

Charlebois, Eloise, Penetanguishene ON
Charlebois, John Paul, QC, Burlington ON
Chochla, Mark W., Thunder Bay ON
Clark, Michael J., DVM, Ottawa ON
Clarke, Grant, Yellowknife NT
Cleghorn, John E., Toronto ON
Clipperton, Robert, North Battleford SK
Coates, Colin, Toronto ON
Collier, Arthur J., East Preston NS
Collingwood, Norman C.M., Surrey BC
Collins, John, MD, Mahone Bay NS
Conaty, Gerald, Calgary AB
Connolly, Dr. John G., Toronto ON
Converse, John L., Kingston ON
Cooper, Lawrence R., Niagara on the Lake ON
Coutts, Jim, Toronto ON
Craig, Barbara Lazenby, Don Mills ON
Cramer, W. A., Rosedale BC
Craven, R. J. N., Mississagua ON
Cuthbertson, Brian, Halifax NS
Dalglish, Camilla, Toronto ON
Davenport, Mrs. Alan, London ON
Davies, Bryan P., Etobicoke ON
Davis, Prof. Richard C., Calgary AB
Davis, John N., Toronto ON
Davis, Peter, Mississauga ON
Dean, William G., Lunenburg NS
Deeks, Gordon D., Toronto ON
Delworth, Mrs. W. T., Ottawa ON
deMille, Dr. Evelyn, Calgary AB
Dibb, Mr. G., Peterborough ON
Dickason, Olive, Ottawa ON
Distad, Dr. Merrill, Edmonton AB
Douglas, W. A. B., Ottawa ON
Dove, Michael, St. Thomas ON
Duckworth, Harry W., Winnipeg MB
Duke, Scott, Yellowknife NT
Dunfield, R. W., Granville Ferry NS
Dunn, Jack, Calgary AB
Dutil, Patrice A., Toronto ON
Dyck, Ian, Ottawa ON

Eayrs, James, Toronto ON
Ellis, John, Toronto ON
England, Dr. & Mrs. R. E., Winnipeg MB
English, John, Waterloo ON
Ens, Prof. Gerhard, Edmonton AB
Errington, E. Jane, Kingston ON
Fair, Dr. Ross D., Toronto ON
Fairley, Bruce F., Golden BC
Fairlie, Thomas F., Toronto ON
Fecteau, Rodolphe D., Dundas ON
Ferguson, Kerry G., Toronto ON
Fiell, John, Invermere BC
Fisher, Robert, Ottawa ON
Foran, Thomas, Ottawa ON
Fortney, Thomas A.C., Sarnia ON
Franks, Dr. W. M., Duntroon ON
Fraser, John A., Toronto ON
French, Goldwin, Dundas ON
Fulford, James, Yellowknife NT
Gaba, Robert G., Victoria BC
Galloway, Ray H., St. Thomas ON
Gerhard, Howard, Toronto ON
Getty, Ian A. L., Calgary AB
Gibson, Doug, Toronto ON
Gibson, Douglas S., Toronto ON
Gibson, James R., North York ON
Gibson, John D., Toronto ON
Gilbert, Jeremy, Toronto ON
Gillespie, Alastair, P.C., Toronto ON
Gilmor, R. Paul, Moffat ON
Glover, Dr. William, Kingston ON
Godfrey, Paul, Port Hope ON
Godfrey, Sheldon J., Toronto ON
Goodwin, George, Toronto ON
Grace, A., Calgary AB
Grant, Shelagh, Peterborough ON
Gray, Charlotte, Ottawa ON
Green, Larry, Saskatoon SK
Greene, Marianne, Toronto ON
Greene, Ronald A., Victoria BC
Grenville, John H., Kingston ON
Griffiths, N. E. S., O.C., Ottawa ON

Grover, Malcolm, Alvinston ON
Hahn, Paul D., Toronto ON
Hale, Nancy R., Montreal QC
Hall, David, Picton ON
Hall, Alan William, Calgary AB
Hall, Roger, Toronto ON
Hamilton, William B., Sackville NB
Hardin, Dr. Harry T., Toronto ON
Hawken, Edwin F., Toronto ON
Haworth, Donald, Aurora ON
Healey, Robynne Rogers, Langley BC
Heather, Earl, Calgary AB
Heidenreich, Conrad J., Toronto ON
Heidenreich, Robert W., Toronto ON
Heidenreich, Conrad E., Lefroy ON
Hele, Karl, London ON
Hertzman, Prof. Lewis, Toronto ON
Hess, Dr. M. P., Calgary AB
Hewett, F. Robert, Aurora ON
Hillier, Bryan A., Etobicoke ON
Hobbs, Brenda, Toronto ON
Hodges, Dr. R.D., Barrie ON
Hodsoll, E. C., Etobicoke ON
Hogarth, Donald D., Ottawa ON
Hopcroft, Grant, London ON
Hordelski, Mrs. I., London ON
Horn, Prof. Michiel, Toronto ON
Horne, Arthur G., Gadshill Stn. ON
Houston, Dr. C. Stuart, Saskatoon SK
Houston, Prof. Susan E., Toronto ON
Howard, Ernest, Toronto ON
Hoyle, Allan L., Newmarket ON
Humphreys, D. J. R., Manotick ON
Hunter, Hope, Edmonton AB
Hunter, Jamie, Midland ON
Hutchins, Peter W., Montréal QC
Innis, John, Calgary AB
Jackman, The Hon. Henry N.R., Toronto ON
Jackman, Frederic L. R., Toronto ON
Jackson, C. Ian, PhD, Montreal QC
Jacobs, Dean, Wallaceburg ON

Jacques, Glen D., North Battleford SK
Jaenen, Cornelius J., Gloucester ON
Jephcott, Amy H., Toronto ON
Johnston, C. Fred, Kingston ON
Jones, Prof. Elwood, Peterborough ON
Kealey, G. S., Fredericton NB
Keane, David R., Hamilton ON
Kerr, Don, Toronto ON
Killan, Dr. Gerald, London ON
Koerner, S.T., Victoria BC
Kofman, Dr. Oscar, Toronto ON
Koundakjian, Vicken, Ottawa ON
Kowal, W. N. & M. A., Milton ON
Lamb, Rev. J. William, Etobicoke ON
Lampard, Dr. Robert, Red Deer AB
Langille, Lyndon R., River John NS
Lapierre, Paula, Montreal QC
Lash, A. B., Toronto ON
Lash, Timothy J. F., Ottawa ON
Latham, Brian, Yellowknife NT
Latimer, W. H., Islington ON
Laverdure, Paul, Yorkton SK
Lazier, The Hon. Colin S., Hamilton ON
Lazier, R. Douglas, Ottawa ON
LeBlanc, Dr. Raymond, Edmonton AB
LeBlanc, M. C., Calgary AB
LeClair, Laurie, Toronto ON
Leech, Robert, Toronto ON
Leighton, J. Douglas, London ON
Lennox, Prof. John, Newmarket ON
Lennox, Brian W., Ottawa ON
Litt, Paul, Ottawa ON
Little, George, Mansfield ON
Loncarevic, Bosko D., Bedford NS
Long, Benoit, Mississauga ON
Lovisek, Dr. Joan A., Surrey BC
Lucas, Mrs. Alec, Plaisance QC
Lund, K. A., Toronto ON
Lundell, Liz, Toronto ON
Luste, George J., Toronto ON
Lysecki, Burton, Winnipeg MB
MacCallum, James D., Toronto ON

MacDonald, Graham A., Victoria BC
MacDonald, Dr. Robert, Calgary AB
MacDonald, Don, Edmonton AB
MacFarlane, Dr. Anthony L., Hamilton ON
MacFeeters, Ronald L., Toronto ON
Machnacky, Christina, Etobicoke ON
MacKay, D. S. C., Ottawa ON
Mackenzie, Hector, Ottawa ON
MacKenzie, David, Toronto ON
Macklem, O. R., Westmount QC
MacLaren, Roy, P.C., Toronto ON
MacLean, Duart A., Quathiaski Cove BC
Mactaggart, Sandy A., Edmonton AB
Magee, Marion E., Toronto ON
Malaher, David G., Whistler BC
Marchildon, Dr. Gregory P., Regina SK
Martin, Joe, Toronto ON
Matavic, Ivona, Stouffville ON
Mathews, Lynn, MLS, Kitchener ON
Matthews, R.O., Roches Point ON
Matthews, F. Richard, Calgary AB
McCalla, Dr. Douglas, Guelph ON
McCormack, Dr. Patricia, Edmonton AB
McCreery, Christopher P., Kingston ON
McDonald, Patrick, Rocky Mountain House AB
McDougall, Douglas G., Montreal QC
McGarr, Richard G., Niagara-on-the-Lake ON
McIntyre, Sharyn, Stouffville ON
McKillop, Duncan C., St. Thomas ON
McLeod, Donald W., Toronto ON
Mclernon, John R., Vancouver BC
McMillan, John B., Burlington ON
McNab, David T., Toronto ON
McPhail, Ian D.C., Toronto ON
M'Gonigle, Shelagh, Ottawa ON
Millar, Laura, New Westminster BC
Miller, Jim, Port Lambton ON
Moir, Michael B., Toronto ON
Money, J. D., Agincourt ON
Moore, Heather-Jane, Ottawa ON
Moreau, Bill, Woodbridge ON
Morgan, W. Charles, Penticton BC
Morin, Ruth, Ottawa ON
Morris, Mary Eleanor, Toronto ON
Morrison, James, Winnipeg MB
Moysa, N., North Vancouver BC
Muldrew, Ken, Calgary AB
Muncaster, Ian, Halifax NS
Murray, Jeffrey, Merrickville ON
Narhi, Brian K., St. Catharines ON
Neeley, Alastair, London ON
Nelles, H. V., Toronto ON
Nicholls, William T. A., Port Hope ON
Nicholson, Dr. A. J., Fort McMurray AB
Nicol, James Allan, Moncton NB
Norton, David J., Belmont ON
Oleson, Robert V., Winnipeg MB
O'Neil, Daniel Murray, Denfield ON
O'Reilly, James M., Toronto ON
Osborne, Brian S., Kingston ON
Osborne, Dave, Orillia ON
Osler, Glyn W., Etobicoke ON
Page, Robert, Calgary AB
Pannekoek, Frits, Athabasca AB
Pathy, Barbara A., Toronto ON
Pearson, J. Bruce, Coldwater ON
Peeling, Prof. James, Winnipeg MB
Penney, Gerald, St. John's NF
Pepall, John, Toronto ON
Perron, Marie L., Ottawa ON
Pilon, Dr. Jean Luc, Gatineau QC
Plimer, B. H., Ottawa ON
Pollock, John H., Toronto ON
Pond, David, Mississauga ON
Radford, John, Toronto ON
Rafuse, Audrey J., London ON
Rath, N. S., Burlington ON
Reed, R. Keith, Victoria BC
Riedstra, Lutzen, Stratford ON
Ripmeester, Michael, St. Catharines ON
Risley, W. Cary, Halifax NS
Robertson, E. Suzanne, Burlington ON

Robinson, Sidney Ivor, La Ronge SK
Ronaghan, Allen, Edmonton AB
Rooke, Richard, Kitchener ON
Rotstein, Abraham, Toronto ON
Roy, Patricia E., Victoria BC
Roy, Jacques, Quebec QC
Rudd, Jeffrey M., North Vancouver BC
Ruddell, Rosemary A., Osgoode ON
Rudnicki, W., Ottawa ON
Russell, Dale, Saskatoon SK
Sanagan, Chris, Toronto ON
Saunders, Robert E., Scarborough ON
Saunders, Richard C., Renfrew ON
Scace, Dr. Robert C., Calgary AB
Scott, K.W., Q.C., Cobourg ON
Shank, Robert J., Kanata ON
Sloan, William A., Castlegar BC
Smith, Ralph G., Oakville ON
Smith, Donald B., Calgary AB
Smith, Mark J., Uxbridge ON
Smith, Charles H., Ottawa ON
Smith, David B., London ON
Smith, Stephen, Toronto ON
Smith, A. Britton, Q.C., Kingston ON
Snyder, R. B., Ottawa ON
Spears, Mrs. John C., Toronto ON
Spraakman, Gary, Toronto ON
St.John, Edward S., Cornwall ON
Standen, S. Dale, Peterborough ON
Staveley, Dr. Michael, St. John's NF
Stevens, John A., Toronto ON
Stewart, Dugald, Toronto ON
Strathy, John G. B., Toronto ON
Struthers, Prof. J.R., Guelph ON
Stuewe, Paul, Toronto ON
Sutherland, David A., Halifax NS
Swainger, Jon, Prince George BC
Symons, Prof. T. H. S., Peterborough ON
Tadman, Megan, Kitchener ON
ten Caté, Jill, Toronto ON
Thoburn, Weldon J., Toronto ON
Thom, Graeme, Orangeville ON

Thomas, Morley K., Toronto ON
Tingley, Ken, Edmonton AB
Tokerud, Bjarne, Victoria BC
Traviss, Dr. Brian, Waterloo ON
Turner, William I., Westmount QC
Vernon, John A.H., Innisfil ON
Waisberg, Lorie, Toronto ON
Waiser, Bill, Saskatoon SK
Walden, Keith, Peterborough ON
Walker, Caroline, Toronto ON
Wallot, Jean-Pierre, Ottawa ON
Warkentin, Germaine, Toronto ON
Warkentin, John, Toronto ON
Watkins, Robert E., Ottawa ON
Weekes, Justice Robert N., Gravenhurst ON
Westcott, Beverley B., Toronto ON
Wetherell, D., Calgary AB
Williams, Richard M., Kanata ON
Williamson, Dr. Ron, Toronto ON
Willson, William A., Windsor ON
Wilson, Ian E., Ottawa ON
Wilson, L.R., Toronto ON
Wilson, R. B., Bancroft ON
Winegard, Dr. W. C., Guelph ON
Wishart, Rev. Ian S., St. John's NF
Wishart, D. H., Toronto ON
Witham, John, Sylvan Place AB
Wood, J. David, Toronto ON
Wright, Sherwood, Toronto ON
Wright, Joyce M., South Mountain ON
Wright, Justice J. deP., Thunder Bay ON
Wright, Glenn T., Ottawa ON
Zimmerman, Adam H., Toronto ON

UNITED STATES

Alsip, Edward, Yucca Valley CA
Ashquabe, Torrence, Depew NY
Axtell, James L., Williamsburg VA
Banks, Robert, M.D., San Antonio TX
Beall, Thomas R., Middleton RI
Bigart, Robert, Missoula MT
Brandao, Jose A., Kalamazoo MI

INDIVIDUAL MEMBERS 2007

Breckenridge, Andrew, Erie PA
Coffman, Prof. Ralph, Marblehead MA
Cowger, Thomas, Thornton WA
DeMallie, Raymond J., Bloomington IN
Dunbar, Gary S., Cooperstown NY
Elk, Charles, Fenton MI
Erickson, David, Crestwood NY
Espley, Derek J., Riverview MI
Fenn, Elizabeth A., Hillsborough NC
Ferguson, Peter A., Middleburgh NY
Filemyr, Albert, Meadowbrook PA
Ford, Harold, Stone Mountain GA
Ginsberg, Michael, Sharon MA
Gregg, John R., York ME
Hanks, Christopher C., Cripple Creek CO
Hogan, James A., Moss Beach CA
Keith, H. Lloyd, Arlington WA
Kelly, Charles A., Chicago IL
Kurz, Dr. Joseph L., Lonedell MO
Lovis Jr., Dr. William, East Lansing MI
Malouf, Richard T., Twin Falls ID
Mandel, Larry, Ann Arbor MI
Marcolin, Lorenzo, MD, Rockville MD
McDougall, Dr. A. K., Oak Harbor WA
Muchka, Albert, West Allis WI
Paxton, James, Bethlehem PA
Pearson, Mark Landell, PhD, Scottsdale AZ

Priebe, Chandler, Pullman WA
Putnam, William L., Flagstaff AZ
Roy, Curtis L., Bloomington MN
Siple, Kenneth J., Watervliet NY
Smiley, Trustee, J. J., Eureka MT
Sollish, George, Baldwinsville NY
Tingwall, Douglas F., Everett WA
Turner, Peter M., Muskegon MI
Ugarenko, Leonard G., Silver Spring MD
Vannah, Alison, PhD, Arlington MA
Walsh, Thomas H., Quartz Hill CA
Weber, Edgar L., San Francisco CA
West, James D., St. Augustine FL
White, Bruce M., St. Paul MN
Willyard, Bruce, Carrington ND

OVERSEAS

Baltes, Dr. Henry, Zurich, Switzerland
Baylis, John, Guanajuato, Mexico
Black, Dr. John B., Habère-Lullin, France
Codignola, Luca, Genoa, Italy
Feest, Dr. Christian F., Vienna, Austria
Montague, Richard W., Ebenhausen, Germany
Smye, Mrs. A. M. L., Portimâo, Portugal
Warkentin, Juliet M., London, England